APPLIED AND COMPUTATIONAL COMPLEX ANALYSIS

VOLUME 2

Special Functions—Integral Transforms

—Asymptotics—Continued Fractions

PETER HENRICI

Professor of Mathematics
Eidgenössische Technische Hochschule, Zürich

Wiley Classics Library Edition Published 1991

A WILEY-INTERSCIENCE PUBLICATION

JOHN WILEY & SONS, New York • Chichester • Brisbane • Toronto • Singapore

To
MARIE-LOUISE

Library of Congress Cataloging in Publication Data:
Henrici, Peter, 1923-
   Applied and computational complex analysis / Peter Henrici.—
Wiley classics library ed.
      p.   cm. — (Wiley classics library)
   "A Wiley-Interscience publication."
   Includes bibliographical references (p.   ) and index.
   Contents: v. 2. Special functions—integral transforms—
asymptotics—continued fractions

   ISBN 0-471-54289-X (paper)
   1. Analytic functions. 2. Functions of complex variables.
3. Mathematical analysis. I. Title.
QA331.H453 1991
515'.98—dc20   90-25198
                                                       CIP

Printed in the United States of America
10 9 8 7 6 5 4 3 2 1

PREFACE

In the present Volume II of our three-volume work we continue to discuss algorithmic techniques that can be used to construct either exact or approximate solutions to problems in complex analysis. A focal point for these applications is the evaluation and manipulation of solutions of analytic differential equations. Successive chapters deal with the representation of solutions by (convergent or divergent) series expansions, with the method of integral transforms, with asymptotic analysis, and with the representation of special solutions by continued fractions. The gamma function is dealt with in the opening chapter in the context of product expansions of analytic functions.

Together with its companions, this volume provides a fair amount of information on some of the more important special functions of mathematical physics. However, our treatment of these functions is unconventional in its organization. Whereas the conventional treatment proceeds function by function, giving to each function its due share of series and integral representations, and of asymptotic analysis, our treatment proceeds by general methods and problems rather than by individual functions. Special results thus appear mainly as applications of general principles. The same methodology will be followed in Volume III; for instance, addition theorems will be considered in the context of partial differential equations.

Although I hope that my program has enabled me to illuminate the basic properties of special functions such as the gamma function, the hypergeometric function, the confluent hypergeometric function, and the Bessel functions, it must be pointed out that a full in-depth treatment of any class of special functions was neither intended nor possible. For more detailed information the reader should turn either to specialized treatises or to the monumental Bateman manuscript project (Erdélyi [1953], [1955]), which provides an essentially complete collection of results known up to the early 1950s.

To call this treatment of complex analysis computational is not meant to imply that I deal exhaustively with the problem of obtaining numerical values of a given special function for all possible values of the variable and of the parameters. This topic has grown into a far too specialized and refined science to be treated thoroughly in a book that also must deal with many other topics. The reader is referred to Gautschi [1975] for an excellent survey of the methods that are currently employed. Questions of computational efficiency, including the manipulation of power series, will be dealt with in Chapter 20 (Volume III) of the present work.

The contents of individual chapters are, briefly, as follows. Chapter 8, on infinite products, features, after the necessary preliminaries, some products of importance in number theory, including Jacobi's celebrated triple product identity. The striking combinatorial implications of this identity seem appropriate as an eye-opener to the joys of classical analysis. We then proceed to a standard treatment of the gamma function, proving the equivalence of the definitions by Weierstrass, Gauss, Euler, and Hankel. Stirling's formula is obtained via the Weierstrass definition; derivations from the other three definitions are contained in Chapter 11. The chapter concludes with a discussion of integrals of the Mellin–Barnes type and their application to hypergeometric functions.

The next chapter, on ordinary differential equations, begins with a standard presentation of the analytic theory from the matrix point of view. Here we can apply some of the material given in Chapter 2 on analytic functions with values in a Banach algebra. The treatment of the confluent and of the standard hypergeometric equations is more detailed than is customary in more theoretically oriented texts. In particular we present, on the basis of Riemann's epochal paper [1857], a complete theory of the linear and the quadratic transforms of the hypergeometric series. Because Legendre functions are merely hypergeometric functions permitting quadratic transforms, written in a different notation, we can dispose of these functions very quickly.

Chapter 10, on integral transforms, begins with a broad discussion of the Laplace transform from an elementary point of view, avoiding advanced real variable theory. To present a clean solution of the inversion problem, we provide a self-contained discussion of the Fourier integral theorem (for piece-wise continuous L_1 functions). We next apply the Laplace transform to Dirichlet series and use this opportunity to give a short account of the Riemann zeta function and its connection with the prime number theorem. A presentation of Polya's theory of Laplace transforms of entire functions of exponential type, with its fascinating link between the growth of the original function in a given direction and the location of the singularities of the image function, follows. The next section, on discrete Laplace transforms, contains some generalizations of Polya's theory suggested by the late H. Rutishauser. The chapter concludes with a discussion of the Mellin transform, and of some simple applications of the integral transform idea to problems in mathematical physics.

In Chapter 11, on asymptotics, we have tried, first of all, to give a clear definition of asymptotic series, a concept that is notoriously difficult to absorb for the beginning student. We then prove the important result that the (generally diverging) formal series solutions to differential equations with irregular singular points are asymptotic to appropriate actual solutions. In addition to standard topics, such as Watson's lemma, Laplace's method, the method of steepest descent, Darboux's method, and the Euler–Maclaurin sum formula, we then present some less orthodox subjects such as general asymptotic series (in particular, asymptotic factorial series, for which a useful analog of Watson's lemma is given), and the numerical evaluation of limits by the Romberg algorithm.

The last chapter of this volume, on continued fractions, presented a special challenge to the expositor because the analytical theory of continued fractions is seldom presented in a larger context in a textbook. A novel feature here is the prominence given to Moebius transformations, and with them to the geometric point of view. This not only enables us to deal efficiently with the formal aspects of continued fractions, but also permits us to treat questions of convergence in an intuitively appealing manner. Once again, the qd algorithm makes its appearance; here it is used to establish some classical continued fractions representing hypergeometric functions. We then discuss the division algorithm and use it to give an alternate solution of the stability problem for polynomials. The second half of the chapter is devoted to continued fractions of the Stieltjes type. Contrary to other presentations, in which such fractions are merely incidental to a discussion of the moment problem, continued fractions and the functions represented by them here are at the center of interest. Our approach enables us to encompass in a very natural way topics of general interest such as the Stieltjes integral, normal families, Vitali's theorem, and the representation formulas of Herglotz, Hamburger, and Nevanlinna for functions with values in a circular region. Some of these topics will be required again in Volume III. We then proceed to the Carleman convergence criterion and more generally to various estimates for the truncation error, valid also when the corresponding power series has radius of convergence zero. Numerous applications, some of them new, should demonstrate the usefulness of the theory.

As in Volume I, I have restrained myself from using an excess of specialized mathematical notation and terminology to make my subject matter accessible to readers with a variety of backgrounds. Although power series are still favored, this volume can be read without knowing in detail the formal power series approach to complex analysis presented in Volume I. Even within this volume, the chapters are reasonably self-contained to make our text useful also to the casual peruser. Courses of varying length on aspects of applied and computational analysis could be based on almost any combination of chapters; in fact, most of the material was presented in the form of such courses at the ETHZ. By exposing the student to a variety of techniques and applications, we are trying to educate applied mathematicians who are able to contribute to the progress of science by their general expertise as well as by specialized research.

Once more, it is my pleasure to express my thanks to the many individuals who have helped me along in my expository endeavors. In addition to the teachers and collegues mentioned in the preface to Volume I, I wish to record my indebtedness to J.-P. Berrut, P. Geiger, M. Gutknecht, E. Häne, M.-L. Henrici, and J. Waldvogel, who have read parts of the manuscript, corrected errors, and suggested numerous improvements. M. Gutknecht, in addition, wrote the programs for drawing the graphs of the gamma function that appear in Chapter 8. R. Askey and J. F. Kaiser supplied valuable information. R. P. Boas provided not only encouragement but also some important references. During my stay at the Bell Laboratories in 1975, D. D. Warner substantially deepened my understanding of continued fraction theory.

I also wish to express my appreciation to the staff of John Wiley & Sons, who once more handled all problems that arose in the production of a manuscript of mine in the most expert and professional manner.

I dedicate this volume to my wife, who by her optimism and good judgment has been of invaluable help in making the many decisions that were necessary to shape my manuscript into its final form.

PETER HENRICI

Zürich, Switzerland
September 1976

CONTENTS

8

INFINITE PRODUCTS

§8.1. DEFINITION AND ELEMENTARY PROPERTIES

Let $\{a_n\}$ be a sequence of complex numbers. It is intuitively clear what is to be understood by the infinite product:

$$\prod_{n=1}^{\infty} a_n = a_1 a_2 a_3 \ldots . \tag{8.1-1}$$

We are to form the sequence of partial products $\{p_n\}$, where $p_1 := a_1$, $p_2 := a_1 a_2$, $p_3 := a_1 a_2 a_3, \ldots$. This sequence is somehow to be identified with the infinite product $a_1 a_2 a_3 \ldots$. It is clear, however, that to write down the factors of the product conveys more information than to write down merely the sequence of partial products. If one factor, say a_n, is zero, then all partial products p_m are zero for $m \ge n$, and it is impossible to recover the values of the factors a_m from the sequence of partial products for $m > n$. (Contrary to this, the terms of an infinite series can always be recovered from the sequence of its partial sums.) For this reason we shall adopt the following formal definition (see Buck [1965], p. 158):

An **infinite product** is an ordered pair $[\{a_n\}_1^{\infty}, \{p_n\}_1^{\infty}]$ of sequences, where $a_1, a_2, \ldots$ are complex numbers, and where $p_n := a_1 a_2 \cdots a_n$, $n = 1, 2, \ldots$.

The numbers a_n and p_n are, respectively, called the nth factor and the nth partial product of the infinite product $[\{a_n\}, \{p_n\}]$. Once this definition is understood, it is completely acceptable to denote an infinite product by a symbol such as (8.1-1), which exhibits only the factors.

Some difficulties also arise if we try to define the concepts of convergence and of value for infinite products. Proceeding as in the case of infinite series, it would be tempting to call the product (8.1-1) convergent if the limit

$$\lim_{n \to \infty} p_n =: p \tag{8.1-2}$$

exists, and to define p as the value of the product. In the interest of formulating simple necessary and sufficient conditions for convergence, it

1

is advantageous, however, to call a product of *nonzero* factors convergent only if the limit (8.1-2) exists *and is different from zero.* If a product has zero factors, the limit of its partial products always exists and has the value zero. Convergence would thus not depend on the whole sequence of factors. To avoid this exceptional situation, we call a product with zero factors convergent *if the product of the nonzero factors converges* in the foregoing sense. Thus, in summary, we adopt the following

DEFINITION

The product (8.1-1) *is said to* **converge** *if and only if at most a finite number of its factors are zero and if the sequence of partial products formed with the nonzero factors has a limit which is different from zero.*

Let $\prod_{n=1}^{\infty} a_n$ be a convergent infinite product, and let $p_n := a_1 a_2 \cdots a_n$, possible zero factors excluded. Then we have, for n sufficiently large, $a_n = p_n / p_{n-1}$. Because $p_n \to p \neq 0$ there follows

$$\lim_{n \to \infty} a_n = 1. \tag{8.1-3}$$

Thus in a convergent infinite product the factors must tend to one. In view of this it is customary to write infinite products in the form

$$\prod_{n=1}^{\infty} (1 + a_n),$$

so that $a_n \to 0$ now is a *necessary* condition for convergence. It is easy to see that this condition is not sufficient by considering the example $a_n := 1/n$, $n = 1, 2, \ldots$. Here

$$p_n = \left(1 + \frac{1}{1}\right)\left(1 + \frac{1}{2}\right) \cdots \left(1 + \frac{1}{n}\right) > 1 + \left(1 + \frac{1}{2} + \cdots + \frac{1}{n}\right),$$

and the product is divergent, because the harmonic series is divergent.

The logarithm of a finite product equals the sum of the logarithms of the factors. We thus may expect to derive convergence criteria for products from convergence criteria for sums by taking logarithms. We are led to consider the infinite series

$$\sum_{n=1}^{\infty} \text{Log}(1 + a_n), \tag{8.1-4}$$

where, for any $z \neq 0$, $\text{Log } z$ denotes the principal value of the logarithm, here defined by the condition $-\pi < \text{Im Log } z \leq \pi$. Let s_n be the nth partial sum of (8.1-4). Then $p_n = e^{s_n}$, and if $s_n \to s$, it follows from the continuity of the exponential function that $p_n \to p := e^s \neq 0$. Thus the convergence of (8.1-4) is a *sufficient* condition for the convergence of the infinite product. We now shall show that this condition is also necessary. Suppose that

$p_n \to p \neq 0$. We let $\phi := \operatorname{Im} \operatorname{Log} p$ and define a single-valued branch $\log^* z$ of $\log z$ by the condition $\phi - \pi < \operatorname{Im} \log^* z \leq \phi + \pi$. Then $\log^* z$ is continuous in the vicinity of $z := p$, and there follows

$$\log^* p_n \to \operatorname{Log} p \ (n \to \infty). \tag{8.1-5}$$

We cannot be sure that $s_n = \log^* p_n$ [because the branches of the logarithms in (8.1-5) have already been chosen] but it is certainly true that

$$s_n = \log^* p_n + h_n 2\pi i, \tag{8.1-6}$$

where h_n is some well-determined integer. We wish to show that $\lim_{n \to \infty} s_n$ exists, and in view of (8.1-5) this amounts to showing that $h_n = h_{n-1}$ for all sufficiently large n. Taking the difference of two consecutive terms in (8.1-6), we find

$$\operatorname{Log}(1 + a_n) = \log^* p_n - \log^* p_{n-1} + (h_n - h_{n-1})2\pi i$$

which we write in the form

$$(h_n - h_{n-1})2\pi i = \operatorname{Log}(1 + a_n) + [\log^* p_{n-1} - \operatorname{Log} p]$$
$$- [\log^* p_n - \operatorname{Log} p].$$

For sufficiently large values of n, $|\operatorname{Im} \operatorname{Log}(1 + a_n)| < 2\pi/3$ in view of $a_n \to 0$, and $|\operatorname{Im}(\log^* p_{n-1} - \operatorname{Log} p)| < 2\pi/3$ in view of (8.1-5). Thus ultimately

$$|h_n - h_{n-1}|2\pi < 2\pi,$$

which implies that $h_n = \text{const}$ and $s_n \to s$, as desired. ∎

Altogether we have proved:

THEOREM 8.1a

An infinite product $\prod_{n=1}^{\infty} (1 + a_n)$ with nonzero factors converges if and only if the series $\sum_{n=1}^{\infty} b_n$ converges, where $b_n := \operatorname{Log}(1 + a_n)$ (principal value).

A necessary condition for the convergence of the product $\prod (1 + a_n)$ or of the series $\sum \operatorname{Log}(1 + a_n)$ is that $a_n \to 0$. Now if $a_n \to 0$, $\operatorname{Log}(1 + a_n)$ asymptotically behaves like a_n. In fact, from

$$\operatorname{Log}(1 + z) = \int_0^z \frac{1}{1+t} \, dt = \int_0^z \left(1 - \frac{t}{1+t}\right) dt$$

we have for $|z| \leq \frac{1}{2}$, integrating along the straight line segment,

$$\operatorname{Log}(1 + z) = z(1 + wz),$$

where $|w| \leq 1$. Thus if $|a_n| \leq \frac{1}{2}$, then

$$\tfrac{1}{2}|a_n| \leq |\operatorname{Log}(1 + a_n)| \leq \tfrac{3}{2}|a_n|. \tag{8.1-7}$$

Hence $\Sigma |\text{Log}(1+a_n)|$ converges and diverges simultaneously with $\Sigma |a_n|$. An infinite product for which the series $\Sigma \text{Log}(1+a_n)$ converges absolutely will be called **absolutely convergent**. In this terminology, we have obtained:

THEOREM 8.1b

A necessary and sufficient condition for the absolute convergence of the product $\prod_{n=1}^{\infty} (1+a_n)$ is the absolute convergence of the series $\sum_{n=1}^{\infty} a_n$.

The emphasis here is on absolute convergence. Simple examples (see problems 9 and 10) show that the theorem is not true if the words "absolute convergence" are replaced by "convergence."

These definitions and theorems also apply to the pointwise convergence of infinite products whose factors depend on a variable. A difficulty arises if we wish to define uniform convergence, because of the vanishing of factors. For definiteness, assume that the functions $a_n(z)$ are analytic on a region S, that none of the functions $1+a_n(z)$ vanishes identically on S, and that at most finitely many of these functions assume the value zero on S. The product

$$\prod_{n=1}^{\infty} (1+a_n(z))$$

is said to **converge uniformly** on S if the sequence of partial products formed with those factors that do not vanish on S converges to a limit $\neq 0$ uniformly for all $z \in S$. With this convention, the following analog of Theorem 8.1b holds and is proved similarly:

THEOREM 8.1c

A necessary and sufficient condition for the absolute and uniform convergence of the product $\prod (1+a_n(z))$ is the absolute and uniform convergence of the series $\Sigma a_n(z)$.

The fundamental theorem on uniformly convergent sequences of analytic functions (Theorem 3.4b) shows that the values of a product of analytic factors that is uniformly convergent on a set S define an analytic function on S, even if the factors with zeros are included.

EXAMPLE

Let $a_n(z) := -z^2/n^2$. We shall show that the product

$$\prod_{n=1}^{\infty} \left(1 - \frac{z^2}{n^2}\right) \tag{8.1-8}$$

converges uniformly on every bounded set S. Indeed, let k be an integer such that S is contained in the disk $|z| \leqslant k$. Omitting the first k factors, we obtain the product

$$\prod_{n=k+1}^{\infty} \left(1 - \frac{z^2}{n^2}\right),$$

whose factors do not vanish on S. The series

$$\sum_{n=k+1}^{\infty} \left(-\frac{z^2}{n^2}\right)$$

converges uniformly and absolutely on S, because it is majorized by the converging series $k^2 \Sigma (1/n^2)$. Hence the uniform convergence follows by Theorem 8.1c. We conclude that (8.1-8) represents an entire analytic function. Because the zeros of this function are located at $z = \pm 1, \pm 2, \ldots$, we may expect it to be closely related to $(\pi z)^{-1} \sin(\pi z)$, which has the same zeros and the same value at $z = 0$. It is shown in §8.3 that the two functions are, in fact, identical.

PROBLEMS

1. Show that

$$\prod_{n=2}^{\infty} \left(1 - \frac{(-1)^n}{n}\right) = \frac{1}{2}.$$

2. In calculus it is shown that

$$\frac{\pi}{2} = \lim_{n \to \infty} \frac{2 \cdot 2 \cdot 4 \cdot 4 \cdot 6 \cdot 6 \cdots 2n \cdot 2n}{1 \cdot 3 \cdot 3 \cdot 5 \cdot 5 \cdot 7 \cdots (2n-1)(2n+1)}$$

    (Wallis' formula). Show that this may be written

$$\frac{2}{\pi} = \prod_{n=1}^{\infty} \left(1 - \frac{1}{(2n)^2}\right).$$

3. Prove that

$$\prod_{n=0}^{\infty} (1 + z^{2^n}) = \frac{1}{1-z}$$

    uniformly on every compact set contained in $|z| < 1$.
4. Show that

$$\prod_{n=1}^{\infty} \left(1 + \frac{z}{n}\right) e^{-z/n}$$

    represents an entire analytic function with zeros at the negative integers.
5. Let the real number ξ be given in decimal representation,

$$\xi := a_{-m} a_{-m+1} \cdots a_0 \cdot a_1 a_2 a_3 \cdots .$$

    Show that

$$e^{\xi} = \prod_{k=-m}^{\infty} e^{10^{-k} a_k}.$$

6. Show that for any $z \neq 0$,

$$\prod_{k=0}^{\infty} \cos(2^{-k}z) = \frac{\sin 2z}{2z}.$$

[Using trigonometric identities, express the partial products as sums of cosines. Then use the definition of the Riemann integral.]

7. Find the value of Vieta's product,

$$\sqrt{\frac{1}{2}}\sqrt{\frac{1}{2}+\frac{1}{2}\sqrt{\frac{1}{2}}}\sqrt{\frac{1}{2}+\frac{1}{2}\sqrt{\frac{1}{2}+\frac{1}{2}\sqrt{\frac{1}{2}}}}\cdots.$$

8. Following D. H. Lehmer (*Amer. Math. Monthly*, 1935), show that the value of the infinite product

$$\left(1-\frac{i}{3}\right)^4\left(1-\frac{i}{17}\right)^4\left(1-\frac{i}{99}\right)^4\left(1-\frac{i}{577}\right)^4\cdots,$$

in which the successive denominators satisfy $d_n = 6d_{n-1} - d_{n-2}$, is purely imaginary.

[Solve the recurrence relation for the d_n. The resulting formula has a meaning for nonintegral n, and there follows

$$\lim_{n\to\infty} \frac{d_{n+1}}{d_n} = 3+\sqrt{8}.$$

Letting $t_n := \tan(\arg p_n^{1/4})$, where p_n is the nth partial product, show that-

$$t_n = \frac{d_{n/2+1} - d_{n/2}}{2d_{n/2}}.\bigg]$$

9. Show that the product

$$\prod_{n=2}^{\infty} \left(1+\frac{(-1)^n}{\sqrt{n}}\right)$$

is divergent, although the series $\Sigma (-1)^n/\sqrt{n}$ is convergent.

10. Let

$$a_{2n-1} := -\frac{1}{n^{1/3}}, \ a_{2n} := \frac{1}{n^{1/3}-1}, n = 2, 3, \ldots.$$

Show that the product

$$\prod_{n=3}^{\infty} (1+a_n)$$

is convergent (and, in fact, has the value 1), although the series Σa_n is divergent.

§8.2. SOME INFINITE PRODUCTS RELEVANT TO NUMBER THEORY

The main purpose here is the study of certain classical infinite products with variable factors. Although the functions defined by these products do not lie in the mainstream of general complex analysis, some of the identities that exist between them have striking combinatorial and numbertheoretical applications.

The products in question are

$$p(z) := \prod_{n=1}^{\infty} (1+z^n), \qquad q(z) := \prod_{n=1}^{\infty} (1-z^n). \tag{8.2-1}$$

By the criterion of Theorem 8.1c, both products are uniformly and absolutely convergent in any disk $|z| \le \rho$ where $\rho < 1$. Hence they represent analytic functions that can be expanded in Taylor series for $|z| < 1$:

$$p(z) =: \sum_{n=0}^{\infty} a_n z^n, \qquad q(z) =: \sum_{n=0}^{\infty} b_n z^n. \tag{8.2-2}$$

Because the products contain no zero factors, they are (by the definition of convergence!) different from zero for $|z| < 1$. Thus their reciprocals are likewise analytic for $|z| < 1$; we put, in particular,

$$\frac{1}{q(z)} =: \sum_{n=0}^{\infty} c_n z^n.$$

The coefficients a_n, b_n, c_n can be evaluated very easily. Consider, for example, the nth partial product of $p(z)$,

$$p_n(z) := \prod_{k=1}^{n} (1+z^k).$$

This is a polynomial of degree $\frac{1}{2}n(n+1)$, which we write as

$$p_n(z) =: \sum_{k=0}^{(\infty)} a_k^{(n)} z^k.$$

Because $p_n(z) \to p(z)$ locally uniformly in $|z| < 1$, the basic theorem on convergence of sequences of analytic functions (Theorem 3.4b) implies that for each $k = 0, 1, \ldots,$

$$a_k = \lim_{n \to \infty} a_k^{(n)}.$$

By comparing coefficients of $1, z, \ldots, z^n$ in the relation

$$p_{n+1}(z) = (1+z^{n+1})p_n(z)$$

we see, however, that

$$a_k^{(n+1)} = a_k^{(n)}, \qquad k = 0, 1, \ldots, n.$$

Thus for any fixed k the coefficients $a_k^{(n)}$ no longer change once n has reached the value k, and we find

$$a_k = a_k^{(k)}, \qquad k = 0, 1, \ldots. \tag{8.2-3}$$

Thus we have, for instance,

$$(1+z)(1+z^2)(1+z^3) = a_0 + a_1 z + a_2 z^2 + a_3 z^3 + a_4^{(3)} z^4 + \cdots.$$

As in the discussion of the "drawer problem" in §7.3 we see that a_k for $k > 0$ equals the number of ways (without regard for the order) in which the integer k can be written as a sum of distinct positive integers. For example,

$$1 = 1, \qquad \text{hence} \quad a_1 = 1$$
$$2 = 2, \qquad \text{hence} \quad a_2 = 1$$
$$3 = 3$$
$$ = 1 + 2, \qquad \text{hence} \quad a_3 = 2$$
$$6 = 6$$
$$ = 1 + 5$$
$$ = 2 + 4$$
$$ = 1 + 2 + 3, \quad \text{hence} \quad a_6 = 4.$$

Putting

$$q_n(z) = \prod_{k=1}^{n} (1 - z^k) =: \sum_{k=0}^{(\infty)} b_k^{(n)} z^k,$$

we find in an analogous manner that

$$b_k = b_k^{(k)}, \qquad k = 0, 1, 2, \ldots.$$

The combinatorial interpretation analogous to the above is as follows: For $k > 0$, b_k represents the excess of the number of ways in which k can be written as a sum of an even number of distinct positive integers over the number of ways it can be written as a sum of an odd number of distinct positive integers. For instance, from the data given we see that $b_6 = 0$.

A combinatorial interpretation also exists for the coefficients c_k. Formally, this can be seen by expanding each term of the product $1/q(z)$ in a geometric series:

$$\frac{1}{(1-z)(1-z^2) \cdots} = (1 + z + z^2 + \cdots)(1 + z^2 + z^4 + \cdots)$$
$$\times (1 + z^3 + z^6 + \cdots) \cdots.$$

Multiplying the series on the right, we see that only the first $k-1$ factors can contribute to the coefficient of z^k. This is confirmed by letting

$$\frac{1}{q_n(z)} = \prod_{k=1}^{n} \frac{1}{1-z^k} =: \sum_{k=0}^{\infty} c_k^{(n)} z^k$$

and noting the recurrence relation

$$\frac{1}{q_{n+1}(z)} = \frac{1}{q_n(z)} \frac{1}{1-z^{n+1}} = \frac{1}{q_n(z)}(1+z^{n+1}+\cdots)$$

which implies, as before, that

$$c_k^{(n+1)} = c_k^{(n)}, \qquad k = 0, 1, \ldots, n.$$

Each contribution to a coefficient c_k where $k>0$ stems from a product of the form

$$z^{n_1 \cdot 1} z^{n_2 \cdot 2} \cdots z^{n_k \cdot k} \tag{8.2-4}$$

where the n_i are nonnegative integers such that

$$n_1 \cdot 1 + n_2 \cdot 2 + \cdots + n_k \cdot k = k \qquad (n_i \geq 0), \tag{8.2-5}$$

and each such product adds 1 to c_k. A representation of a positive integer k of the form (8.2-5) is called a **partition** of k. Because there is a one-to-one correspondence between products (8.2-4) equalling z^k and partitions of k, the coefficient c_k is equal to the total number of partitions of k. In the number-theoretical literature, the function $\pi(k) := c_k$ is called the **partition function**. We have established $1/q(z)$ as the generating function of the partition function.

EXAMPLE **1**

The number 6 has the partitions

$$6 = 6$$
$$= 5+1 = 4+2 = 3+3$$
$$= 4+1+1 = 3+2+1 = 2+2+2$$
$$= 3+1+1+1 = 2+2+1+1$$
$$= 2+1+1+1+1$$
$$= 1+1+1+1+1+1.$$

There are 11 partitions; we conclude that $c_6 = 11$.

The partition function is of considerable interest in number theory. To calculate the function from its definition is extremely laborious. For very large k, asymptotic formulas discovered by Rademacher are sufficiently accurate to permit the numerical determination of $\pi(k)$ without error. Here we shall derive certain functional relations between the functions p and q that can be used to compute the partition function rapidly by recursion.

As a first result we prove

THEOREM 8.2a (Euler):

$$p(z) = \frac{q(z^2)}{q(z)}, \qquad |z| < 1. \qquad (8.2\text{-}6)$$

Written out explicitly, this formula reads, after canceling common factors,

$$(1+z)(1+z^2)(1+z^3)\cdots = \frac{1}{(1-z)(1-z^3)(1-z^5)\cdots}.$$

Two proofs of this formula, both classical, are presented.

First Proof. By problem 3 of §8.1 we have

$$(1+z)(1+z^2)(1+z^4)\cdots = \frac{1}{1-z},$$

$$(1+z^3)(1+z^{2\cdot3})(1+z^{4\cdot3})\cdots = \frac{1}{1-z^3},$$

$$(1+z^5)(1+z^{2\cdot5})(1+z^{4\cdot5})\cdots = \frac{1}{1-z^5},$$

Multiplying all these expansions, we get on the right-hand side just the product on the right of (8.2-6), which is known to be absolutely convergent for $|z| < 1$. The totality of the factors of all products on the left-hand side just comprises the factors of $p(z)$, because every positive integer can be represented as a power of two times an odd integer in just one way. Because the product of all these factors is absolutely convergent, it follows from the corresponding theorem on series (quoted in the proof of Theorem 3.1b) that the factors may be arranged in any order. Thus the product actually equals $p(z)$. ∎

Second Proof. We consider the function

$$k(z) := \frac{p(z)q(z)}{q(z^2)} = \prod_{n=1}^{\infty}(1+z^n)(1-z^{2n-1})$$

where $|z| < 1$. Now

$$k(z^2) = \prod_{n=1}^{\infty}(1+z^{2n})(1-z^{4n-2})$$

$$= \prod_{n=1}^{\infty}(1+z^{2n})(1+z^{2n-1})(1-z^{2n-1})$$

$$= \prod_{n=1}^{\infty} (1+z^{2n})(1+z^{2n-1}) \prod_{n=1}^{\infty} (1-z^{2n-1})$$

$$= k(z),$$

hence $k(z) = k(z^2) = k(z^4) = \cdots = k(\lim_{n\to\infty} z^{2n}) = k(0) = 1$, which is equivalent to the required result. ∎

By considerations entirely analogous to those above we see that the coefficient of z^k in the expansion of

$$\frac{1}{(1-z)(1-z^3)(1-z^5)\cdots}$$

in powers of z equals the number of partitions of k into odd integers. Considering the combinatorial interpretation of the expansion of $p(z)$, Theorem 8.2a thus states that the number of partitions of $k > 0$ into sums of *distinct* positive integers equals the number of partitions into *odd* (but not necessarily distinct) integers. (See also §7.3.)

EXAMPLE **2**

$k = 9$ has the following partitions into *unequal* integers:

$$9 = 1+2+6$$
$$= 1+3+5$$
$$= 2+3+4$$
$$= 1+8$$
$$= 2+7$$
$$= 3+6$$
$$= 4+5$$
$$= 9,$$

thus $a_9 = 8$. And indeed, 9 has the following partitions into odd parts,

$$9 = 1+1+1+1+1+1+1+1+1$$
$$= 1+1+1+1+1+1+3$$
$$= 1+1+1+3+3$$
$$= 3+3+3$$
$$= 1+1+1+1+5$$
$$= 1+3+5$$
$$= 1+1+7$$
$$= 9$$

and no others.

The next theorem introduces a function defined by an infinite product depending on two variables.

THEOREM 8.2b (Jacobi's triple product identity)

Let

$$g(z, t) := \prod_{n=1}^{\infty} (1+z^{2n-1}t)(1+z^{2n-1}t^{-1}). \qquad (8.2\text{-}7)$$

For each $t \neq 0$, $g(z, t)$ defines a function of z analytic for $|z| < 1$ and for each z such that $|z| < 1$, it defines a function of t analytic for $0 < |t| < \infty$. As a function of t it has the Laurent series

$$g(z, t) = \frac{1}{q(z^2)} \sum_{k=-\infty}^{\infty} z^{k^2} t^k. \qquad (8.2\text{-}8)$$

Proof. The statements concerning analyticity follow directly from Theorem 8.1c, because for fixed $t \neq 0$ the product (8.2-7) converges uniformly in $|z| \leq \rho$ for each $\rho < 1$, and for fixed z, $|z| < 1$, uniformly in $\rho \leq |t| \leq \rho^{-1}$ for each $\rho \in (0, 1)$. From the uniform convergence on $|t| = 1$ (for instance) it follows that the Laurent coefficients in the expansion

$$g(z, t) =: \sum_{k=-\infty}^{\infty} h_k t^k$$

equal the limits as $n \to \infty$ of the Laurent coefficients of the nth partial product of $g(z, t)$,

$$g_n(z, t) := \prod_{k=1}^{n} (1+z^{2k-1}t)(1+z^{2k-1}t^{-1})$$

$$=: \sum_{k=-n}^{n} h_k^{(n)} t^k.$$

We shall calculate the $h_k^{(n)}$ explicitly. In view of $g(z, t) = g(z, t^{-1})$ we have $h_k^{(n)} = h_{-k}^{(n)}$, and it suffices to determine these coefficients for $k = 0, 1, \ldots, n$. Evidently,

$$h_n^{(n)} = z^{1+3+5+\cdots+(2n-1)} = z^{n^2}. \qquad (8.2\text{-}9)$$

From the functional relation

$$g_n(z, z^2 t) = \prod_{k=1}^{n} (1+z^{2k+1}t)(1+z^{2k-3}t^{-1})$$

$$= \frac{1+z^{2n+1}t}{1+zt} \frac{1+z^{-1}t^{-1}}{1+z^{2n-1}t^{-1}} g_n(z, t)$$

$$= \frac{1+z^{2n+1}t}{zt+z^{2n}} g_n(z, t)$$

we find

$$(zt + z^{2n}) \sum_{k=-n}^{n} h_k^{(n)} z^{2k} t^k = (1 + z^{2n+1} t) \sum_{k=-n}^{n} h_k^{(n)} t^k,$$

and hence, comparing coefficients of t^k,

$$z^{2k+2n} h_k^{(n)} + z^{2k-1} h_{k-1}^{(n)} = h_k^{(n)} + z^{2n+1} h_{k-1}^{(n)},$$

$$h_{k-1}^{(n)} = \frac{1 - z^{2k+2n}}{z^{2k-1}(1 - z^{2n-2k+2})} h_k^{(n)},$$

$k = 1, 2, \ldots,$ and thus, using (8.2-9),

$$h_k^{(n)} = \frac{(1 - z^{2k+2n+2})(1 - z^{2k+2n+4}) \cdots (1 - z^{4n})}{(1 - z^2)(1 - z^4) \cdots (1 - z^{2n-2k})} z^{k^2}.$$

In terms of the partial products

$$q_n(z) := \prod_{k=1}^{n} (1 - z^k)$$

of $q(z)$, $h_k^{(n)}$ may be expressed as follows:

$$h_k^{(n)} = \frac{q_{2n}(z^2)}{q_{n+k}(z^2) q_{n-k}(z^2)} z^{k^2}.$$

If k is fixed and $n \to \infty$, each of the partial products tends to $q(z^2)$, and on cancellation we find

$$h_k = \lim_{n \to \infty} h_k^{(n)} = \frac{1}{q(z^2)} z^{k^2},$$

as asserted.   ∎

Let us now consider some consequences of Theorem 8.2b. Multiplying (8.2-8) by $q(z^2)$ yields

$$\prod_{n=1}^{\infty} (1 + z^{2n-1} t)(1 + z^{2n-1} t^{-1})(1 - z^{2n}) = \sum_{n=-\infty}^{\infty} z^{n^2} t^n.$$

Let $0 \le x < 1$, and set $z := x^{3/2}$, $t := -x^{1/2}$. On the left there results

$$\prod_{n=1}^{\infty} (1 - x^{3n-1})(1 - x^{3n-2})(1 - x^{3n}).$$

All positive integral powers of x occur here; hence this equals $q(x)$. We thus have proved

$$q(z) = \sum_{n=-\infty}^{\infty} (-1)^n z^{(3n^2+n)/2} \qquad (8.2\text{-}10)$$

for nonnegative real $z < 1$. Only nonnegative integral powers of z occur on the right; hence the series on the right represents an analytic function for $|z| < 1$. Because $q(z)$ is likewise analytic for $|z| < 1$, the series on the right is the Maclaurin series of q, and (8.2-10) holds for all z such that $|z| < 1$. Written out in full, the equation reads

$$(1 - z)(1 - z^2)(1 - z^3) \cdots$$
$$= 1 - z - z^2 + z^5 + z^7 - z^{12} - z^{15} + \cdots.$$

In the notation used earlier, we thus have

$$b_k = \begin{cases} (-1)^n, \text{ if } k = \frac{1}{2}(3n^2 + n) \text{ for some integer } n; \\ 0, \text{ if this is not the case.} \end{cases}$$

It is remarkable that only relatively few of the coefficients b_k are different from zero, and that the nonzero coefficients have one of the values $+1$ and -1. In view of the combinatorial interpretation of the b_k given earlier, we have:

THEOREM 8.2c

For any integer $k > 0$, let e_k and o_k denote the number of its partitions into an even and into an odd number of distinct positive integers. If k is not of the form $\frac{1}{2}(3n^2 + n)$, then $e_k = o_k$. If $k = \frac{1}{2}(3n^2 + n)$ for some integer n, then $e_k - o_k = (-1)^n$.

EXAMPLE **3**

The partitions of $k = 7$ into unequal terms are

$$7 = 7 \qquad\qquad = 1 + 6$$
$$= 1 + 2 + 4 \qquad = 2 + 5$$
$$= 3 + 4$$

There are three partitions into an even number of terms and one less partition into an odd number of terms, which must be so because $7 = \frac{1}{2}(3 \cdot 2^2 + 2)$.

We next establish a recurrence relation for the partition function c_k. Because

$$(c_0 + c_1 z + c_2 z^2 + \cdots)(b_0 + b_1 z + b_2 z^2 + \cdots) = 1,$$

we find in view of $b_0 = 1$

$$c_n = -\sum_{k=1}^{n} c_{n-k} b_k, \qquad n = 1, 2, \ldots,$$

where $c_0 = 1$, and thus, using the values of b_k found above and writing $c_k =: c(k)$ for clarity,

$$c(n) = - \sum_{\frac{1}{2}(3m^2+m)<n} (-1)^m \left\{ c\left(n - \frac{3m^2-m}{2}\right) + c\left(n - \frac{3m^2+m}{2}\right) \right\},$$

$$n = 1, 2, \ldots .$$

Using this relation and working by hand, Mac Mahon, the British number theoretician, computed $c(n)$ up to $n = 200$, and found that

$$c(200) = 3{,}972{,}999{,}029{,}388.$$

We conclude by discussing briefly another infinite product of even greater importance in number theory. The symbol p now stands for a prime number. For Re $z > 1$, we define

$$f(z) := \prod_p \frac{1}{1-p^{-z}}. \tag{8.2-11}$$

The product here comprises the prime numbers only; $p^{-z} := e^{-z \, \mathrm{Log}\, p}$. To prove the convergence of the product, it suffices (by our definition) to establish the convergence of the reciprocal product

$$\prod_p (1 - p^{-z}).$$

If $z = x + iy$, we shall show that this product converges, in fact, uniformly for $x \geq x_0$, where $x_0 > 1$. This follows from Theorem 8.1c, because

$$\sum_p |p^{-z}| = \sum_p p^{-x} < \sum_{n=1}^{\infty} n^{-x} \leq \sum_{n=1}^{\infty} n^{-x_0}.$$

We now shall derive an alternate representation for f. Taking the partial product up to the greatest prime $\leq k$, we get

$$\prod_{p \leq k} \frac{1}{1-p^{-z}} = (1 + 2^{-z} + 2^{-2z} + \cdots)$$

$$\cdot (1 + 3^{-z} + 3^{-2z} + \cdots)$$

$$\cdot \cdot \cdot$$

$$\cdot (1 + p_k^{-z} + p_k^{-2z} + \cdots),$$

where p_k is the greatest prime $\leq k$. Multiplying out, we obtain products like

$$2^{-m_2 z} \cdot 3^{-m_3 z} \cdot 5^{-m_5 z} \cdots p_k^{-m_{p_k} z},$$

where $m_2, m_3, m_5, \ldots, m_{p_k}$ are arbitrary nonnegative integers. Thus the result is a number of the form n^{-z}, where n is an integer having prime factors $\leq k$ only, and every such number is obtained exactly once in view of the uniqueness of the prime number decomposition. The integers with prime factors $\leq k$ certainly comprise all integers up to k, and some that are greater than k. We thus have

$$\prod_{p \leq k} \frac{1}{1 - p^{-z}} = \sum_{n=1}^{k} n^{-z} + r_k(z),$$

where, for $x > 1$,

$$|r_k(z)| \leq \sum_{n > k} |n^{-z}| \leq \int_k^\infty v^{-x} \, dv = \frac{k^{-x+1}}{x - 1}.$$

We see that $|r_k(z)| \to 0$ as $k \to \infty$, and thus we find

$$f(z) = \sum_{n=1}^{\infty} n^{-z} \quad (\operatorname{Re} z > 1). \qquad (8.2\text{-}12)$$

The function f defined by either (8.2-11) or (8.2-12) is known as the **zeta function of Riemann**. (Another zeta function was introduced by Jacobi in the theory of elliptic functions.) Further properties of the Riemann zeta function are established in §10.8 and are applied to obtain a deep result on the distribution of prime numbers.

PROBLEMS

1. Establish the following identities, all valid for $|z| < 1$, as special cases of Theorem 8.2b:

   (a) $\dfrac{1 - z^2}{1 - z} \dfrac{1 - z^4}{1 - z^3} \dfrac{1 - z^6}{1 - z^5} \cdots = \displaystyle\sum_{n=0}^{\infty} z^{(n^2+n)/2}$

   (b) $\dfrac{1 - z}{1 + z} \dfrac{1 - z^2}{1 + z^2} \dfrac{1 - z^3}{1 + z^3} \cdots = 1 - 2z + 2z^4 - 2z^9 + 2z^{16} - \cdots$

   (c) $\dfrac{q(z)q(z^4)}{q(z^2)} = \displaystyle\sum_{m=-\infty}^{\infty} (-1)^m z^{2m^2+m}$

   (d) $\dfrac{q(z^2)^5}{q(z)^2 q(z^4)^2} = 1 + 2 \displaystyle\sum_{m=1}^{\infty} z^{m^2}.$

2. Show that the Maclaurin series of $\{q(z)\}^3$ is

$$\{q(z)\}^3 = \sum_{n=0}^{\infty} (-1)^n (2n + 1) z^{(n^2+n)/2}.$$

   [Consider $g(z, -t)$.]

3. Use reasoning analogous to the proof of (8.2-12) to show that the number of primes is infinite.

§8.3.   PRODUCT REPRESENTATIONS OF ENTIRE FUNCTIONS

Let $n \geqslant 1$, and let

$$p(z) = a_0 + a_1 z + \cdots + a_n z^n$$

be a polynomial of degree n, $a_n \neq 0$. If the zeros of p are denoted by $z_1, z_2, \ldots, z_n$ (a zero of multiplicity m being counted m times), it is well known that p can be represented in the form

$$p(z) = a_n (z - z_1)(z - z_2) \cdots (z - z_n). \qquad (8.3\text{-}1)$$

Our aim is to obtain similar expansions as products of linear factors also for other, more general classes of analytic functions, in particular for entire functions.

As is shown, for example, by e^z, an entire function need not have any zeros. Furthermore, it would be quite wrong to think that the exponential function is exceptional in this respect, for if $g(z)$ is any entire function whatsoever, then $e^{g(z)}$ is an entire function without zeros. This fact permits the following converse:

THEOREM 8.3a

Let f be an entire function without zeros. Then there exists an entire function g such that $f(z) = e^{g(z)}$ for all z.

Proof.   Because f has no zeros, the function $1/f$ is likewise entire, and so is the function f'/f. We define g by

$$g(z) := \int_0^z \frac{f'(t)}{f(t)} \, dt,$$

where the path of integration is irrelevant, because the integrand is analytic everywhere. Evidently,

$$g'(z) = \frac{f'(z)}{f(z)}.$$

We compute the derivative of $f e^{-g}$ and find

$$(f e^{-g})' = f' e^{-g} - f g' e^{-g} = e^{-g}\left(f' - f\frac{f'}{f}\right) = 0.$$

Hence

$$f(z) e^{-g(z)} = f(0) e^{-g(0)} = f(0),$$

and there follows

$$f(z) = f(0) e^{g(z)}.$$

Because the constant factor $f(0)$ can be absorbed into the exponent, this is a representation of the required type. ∎

Now let f be an entire function with finitely many zeros $z_1, z_2, \ldots, z_n$ (a zero of multiplicity m being counted m times). Then the function

$$f_1(z) := \frac{f(z)}{(z - z_1) \cdots (z - z_n)}$$

is still entire, but without zeros; hence it is of the form $e^{g(z)}$, where g is a suitable entire function. There follows

$$f(z) = e^{g(z)} \prod_{k=1}^{n} (z - z_k). \tag{8.3-2}$$

We now turn to the general case in which the entire function has infinitely many zeros z_i, which we number according to increasing moduli. We also assume that 0 is not a zero, so that

$$0 < |z_1| \leq |z_2| \leq \cdots.$$

Because the zeros cannot have a finite point of accumulation, $z_n \to \infty$ for $n \to \infty$. This shows that the form (8.3-2) of the product representation is not suitable for generalization to the case of infinitely many zeros, because the factors in the product

$$\prod_{m=1}^{\infty} (z - z_m)$$

do not approach 1 for any value of z. It is easy, however, to find the necessary modification by considering the product (8.3-1) where we assume that all $z_k \neq 0$. Dividing the kth factor by $z_k (k = 1, 2, \ldots, n)$ and observing that $a_n z_1 z_2 \cdots z_n = (-1)^n a_0$, we obtain for the polynomial p the alternate representation

$$p(z) = a_0 \prod_{m=1}^{n} \left(1 - \frac{z}{z_m}\right).$$

The formal generalization of this product,

$$\prod_{n=1}^{\infty} \left(1 - \frac{z}{z_n}\right), \tag{8.3-3}$$

now at least satisfies the necessary condition for convergence that the general factor tends to 1. According to Theorem 8.1c, it will converge (absolutely) for all z if and only if

$$\sum_{n=1}^{\infty} \frac{1}{|z_n|} < \infty. \tag{8.3-4}$$

If this condition is satisfied, (8.3-3) evidently represents an entire function with exactly the same zeros as f; hence the ratio of the two is an entire function without zeros; hence f may be represented in the form

$$f(z) = e^{g(z)} \prod_{n=1}^{\infty} \left(1 - \frac{z}{z_n}\right).$$

In general, condition (8.3-4) is not satisfied. (It is not satisfied, for instance, for such a simple entire function as sin z.) As in the Mittag–Leffler expansion of a meromorphic function in partial fractions (Theorem 7.10b), the situation will be saved by introducing convergence-producing terms.

In the case of meromorphic functions, the convergence-producing terms must be such that they introduce no additional poles. Thus they must be chosen as entire functions. Also, we want them to be as simple as possible. Thus we chose them as polynomials. In the present situation, the convergence-producing terms should introduce no new zeros. Thus we should choose them in the form $e^{g_n(z)}$, where the g_n are entire functions. Also, we again want them to be as simple as possible. Thus we are led to seek them in the form $e^{p_n(z)}$, where the p_n are polynomials. We are thus led to investigate the following question: Given a sequence $\{z_n\}$ of zeros such that $\{|z_n|\}$ is increasing, $|z_1| > 0$, $z_n \to \infty$, is it possible to find polynomials $p_n(z)$ such that the product

$$\prod_{n=1}^{\infty} \left(1 - \frac{z}{z_n}\right) e^{p_n(z)} \tag{8.3-5}$$

converges uniformly in every compact set of the plane?

The product converges uniformly if and only if this is the case for the series $\Sigma r_n(z)$, where

$$r_n(z) := \text{Log}\left\{\left(1 - \frac{z}{z_n}\right) e^{p_n(z)}\right\} = \log\left(1 - \frac{z}{z_n}\right) + p_n(z), \tag{8.3-6}$$

where the branch of log is chosen such that the imaginary part of $r_n(z)$ lies between $-\pi$ and $+\pi$. We wish to choose the p_n such that the series (8.3-6) converges uniformly on every set $|z| \leq \rho$.

For $|z| < |z_n|$ we have

$$\text{Log}\left(1 - \frac{z}{z_n}\right) = -\frac{z}{z_n} - \frac{1}{2}\left(\frac{z}{z_n}\right)^2 - \frac{1}{3}\left(\frac{z}{z_n}\right)^3 - \cdots.$$

We choose for p_n the negative of a partial sum of this series,

$$p_n(z) := \frac{z}{z_n} + \frac{1}{2}\left(\frac{z}{z_n}\right)^2 + \cdots + \frac{1}{m_n}\left(\frac{z}{z_n}\right)^{m_n},$$

whose degree m_n is yet to be determined. The function

$$r_n^*(z) := \mathrm{Log}\left(1 - \frac{z}{z_n}\right) + p_n(z)$$

[equal to $r_n(z)$ up to a multiple of $2\pi i$] then has the representation

$$r_n^*(z) = -\frac{1}{m_n + 1}\left(\frac{z}{z_n}\right)^{m_n+1}\left\{1 + \frac{m_n + 1}{m_n + 2}\frac{z}{z_n} + \cdots\right\}.$$

Replacing $\{\ \}$ by a geometric series, we easily obtain the following estimate, valid for $|z| < |z_n|$:

$$|r_n^*(z)| \leq \frac{1}{m_n + 1}\left(\frac{|z|}{|z_n|}\right)^{m_n+1}\frac{1}{1 - |z|/|z_n|}. \tag{8.3-7}$$

Let now the m_n be chosen *such that the series*

$$\sum_{n=1}^{\infty}\frac{1}{m_n + 1}\left(\frac{z}{|z_n|}\right)^{m_n+1} \tag{8.3-8}$$

converges for all z. Such a choice is possible, for if $m_n = n$, then the series is the integral of the power series $\sum_{n=1}^{\infty} z^n|z_n|^{-n-1}$ whose radius of convergence is infinite because $z_n \to \infty$. In special cases (see below) it may not be necessary to choose m_n as large as n. We assert that with any choice of the m_n such that (8.3-8) converges for all z, the product (8.3-5) converges uniformly on every set $|z_n| \leq \rho$.

Let $\rho > 0$ be arbitrary, and let $|z| \leq \rho$. By the definition of uniform convergence of an infinite product, only the terms in (8.3-5) where $|z_n| > \rho$ are to be taken into account. For these n we have by (8.3-7)

$$|r_n^*(z)| \leq \frac{1}{m_n + 1}\left(\frac{\rho}{|z_n|}\right)^{m_n+1}\left(1 - \frac{\rho}{|z_n|}\right)^{-1},$$

and because

$$\sum_{n=1}^{\infty}\frac{1}{m_n + 1}\left(\frac{\rho}{|z_n|}\right)^{m_n+1}$$

converges, it follows that $r_n^*(z) \to 0$ uniformly. In particular, then, $|\mathrm{Im}\, r_n^*(z)| < \pi$, and thus $r_n^*(z) = r(z)$, for n sufficiently large. Again the estimate (8.3-7) then shows that $\Sigma\, r_n(z)$ converges uniformly for $|z| \leq \rho$, implying the uniform convergence of the product.

In the foregoing construction, no assumption concerning the zeros z_n was made except that they have no finite point of accumulation, and that all $z_n \neq 0$. A possible zero at the origin can be handled by inserting a suitable factor z^m. We thus have proved:

THEOREM 8.3b

There exists an entire function with an arbitrarily prescribed sequence of zeros $\{z_n\}$ provided only that, in the case of infinitely many zeros, $z_n \to \infty$. Every such function can be represented in the form

$$f(z) = z^m \, e^{g(z)} \prod \left(1 - \frac{z}{z_n}\right) e^{p_n(z)}, \qquad (8.3\text{-}9)$$

where

$$p_n(z) := \frac{z}{z_n} + \frac{1}{2}\left(\frac{z}{z_n}\right)^2 + \cdots + \frac{1}{m_n}\left(\frac{z}{z_n}\right)^{m_n},$$

the product is with respect to all zeros $z_n \neq 0$, m and the m_n are suitable integers, and g is an entire function.

From this we easily obtain a new representation for meromorphic functions.

COROLLARY 8.3c

Every function that is meromorphic in the whole plane can be represented as a quotient of two entire functions.

Proof. Let h be meromorphic in the plane, and let f be an entire function that has the poles of h as its zeros. Then the function hf has removable singularities at all poles and hence can be extended to an entire function g. We thus find for every z that is not a pole of h

$$h(z) = \frac{g(z)}{f(z)},$$

as desired. ■

Corollary 8.3c shows that every meromorphic function can be written as a product of a certain integral power of z, an entire function of the form $e^{g(z)}$, and the ratio of two products. Conversely, it is very easy to find a partial fraction expansion of the logarithmic derivative f'/f of an entire function f if we know a product representation of f.

COROLLARY 8.3d

Let the entire function f be represented in the form (8.3-9). Then the following expansion holds uniformly on every compact set avoiding the zeros of f:

$$\frac{f'(z)}{f(z)} = \frac{m}{z} + g'(z)$$

$$+ \sum_{n=1}^{\infty} \left[\frac{1}{z - z_n} + \frac{1}{z_n}\left(1 + \frac{z}{z_n} + \cdots + \left(\frac{z}{z_n}\right)^{m_n - 1}\right)\right].$$

Proof. We denote by $p_k(z)$ the expansion (8.3-9) with the product replaced by its first k factors. Then by the ordinary rules of logarithmic differentiation,

$$p_k'(z) = p_k(z)\left\{\frac{m}{z} + g'(z) + \sum_{n=1}^{k} [\cdots]\right\}.$$

If S is any compact set, then $p_k(z) \to f(z)$ and $p_k'(z) \to f'(z)$ uniformly on S. If S is any compact set avoiding the zeros of f, then the greatest lower bound of $|f(z)|$ on S is positive, and it follows that also $p_k'(z)/p_k(z) \to f'(z)/f(z)$ uniformly on S, as asserted. ∎

Before giving examples for the product representation (8.3-9), we discuss a simplification and standardization which is frequently applicable.

In the representation (8.3-9) the integers m_n are to some extent arbitrary. If $\{m_n\}$ is a sequence of integers for which the product converges, then it will converge, according to our construction, for any sequence of larger integers. On the other hand, we can always set finitely many of these integers equal to zero without disturbing the convergence. [Naturally, all these operations will change the function $g(z)$.] Since Weierstrass it has become customary to standardize the representation (8.3-9), if possible, by the following two postulates: (1) The m_n are all set equal to each other, $m_n = h$; (2) the common value h is taken as small as possible. If $m_n = h$, (8.3-8) is tantamount to the condition

$$\sum_{n=1}^{\infty} \frac{1}{|z_n|^{h+1}} < \infty. \tag{8.3-10}$$

If there is an integer h such that (8.3-10) holds, and if h is the smallest such integer, then the product in (8.3-9) where $m_n = h$,

$$\prod\left(1 - \frac{z}{z_n}\right)\exp\left(p_h\left(\frac{z}{z_n}\right)\right),$$

$(p_h(z) := z + \frac{1}{2}z^2 + \cdots + z^h/h)$ is called a **canonical product**, and h is called the **genus** of that product. If an entire function f has a representation of the form (8.3-9) where g is a polynomial and the product is canonical, it is said to be of finite genus, and the genus of f is defined to be the degree of g or the genus of the product, whichever is larger. This definition, which at first sight may seem arbitrary, is motivated by a close connection between the genus of an entire function and its growth as $|z| \to \infty$. For instance, it can be shown that for every $\varepsilon > 0$ there exists a constant μ such that

$$|f(z)| \leq \mu \, e^{|z|^{h+\varepsilon}},$$

for all z, but that such an inequality does not hold if the exponent $h + \varepsilon$ is replaced by a number $< h$. For further details see Ahlfors [1966], Chapter 5, Section 2.3, or Titchmarsh [1939], Chapter 8.

For an example of a canonical product representation we consider the function $\sin \pi z$. The zeros are the integers $z = \pm n$. Because $\Sigma \, 1/n^{h+1}$ diverges for $h = 0$ and converges for $h = 1$, we must take $h = 1$ in the canonical product. There follows

$$\sin z = z \, e^{g(z)} \prod_{n \neq 0} \left(1 - \frac{z}{n}\right) e^{z/n}.$$

As always, there remains the problem of determining the entire function g, and as always, we hope that g will turn out to be constant. The simplest way to establish this fact is by appealing to a known result. By logarithmic differentiation (Corollary 8.3d) we have

$$\pi \cot \pi z = \frac{1}{z} + g'(z) + \sum_{\substack{n = -\infty \\ n \neq 0}}^{\infty} \left(\frac{1}{z - n} + \frac{1}{n}\right).$$

A similar formula for $\cot \pi z$ has been established in §7.10. A comparison with (7.10-11) yields $g'(z) = 0$. It follows that g is a constant function. Because $\sin \pi z / z \to \pi$ as $z \to 0$ we must have $e^{g(z)} = \pi$. Hence we have found the representation

$$\sin \pi z = \pi z \prod_{n \neq 0} \left(1 - \frac{z}{n}\right) e^{z/n}.$$

This can be simplified by multiplying together the terms corresponding to n and $-n$. The convergence-generating factors then disappear and we find

$$\sin \pi z = \pi z \prod_{n=1}^{\infty} \left(1 - \frac{z^2}{n^2}\right). \qquad (8.3\text{-}11)$$

Although convergent for all z, this product is never used for the numerical evaluation of the sine function, because the convergence is uniformly slow in any closed set avoiding the zeros.

PROBLEMS

1. Show that $\sin \pi z$ has genus 1.
2. Show that every entire function of finite genus that is periodic with period 1 and has its zeros at the integers is a constant multiple of $\sin \pi z$.
3. What is the genus of $\cos \sqrt{z}$?
4. Show that

$$\cos \pi z = \prod_{n=-\infty}^{\infty} \left(1 - \frac{z}{n + \frac{1}{2}}\right) \exp\left(\frac{z}{n + \frac{1}{2}}\right).$$

5. Determine the product expansion of $\cos \sqrt{z}$.

§8.4.　THE GAMMA FUNCTION

After the so-called elementary functions, the gamma function, since Euler denoted by Γ, is the most important function of classical analysis. This function can be introduced in a number of ways, each of which has its own advantages. One popular approach is to start with the problem of interpolating the function f defined at the positive integers by $f(n) := n!$ by an analytic function which, in a certain sense, is as smooth as possible. This approach can be carried through entirely within the realm of real analysis; see Theorem 8.4a below. To be more in line with the preceding material, however, we first characterize this function by its zeros and its singularities. The function Γ has no zeros, has simple poles at 0 and the negative integers, and is analytic everywhere else. It follows from Definition (I) that $1/\Gamma$ is an entire function with zeros at 0 and the negative integers.

The simplest function with zeros at the negative integers is given by the canonical product

$$g(z) := \prod_{n=1}^{\infty} \left(1 + \frac{z}{n}\right) e^{-z/n}. \tag{8.4-1}$$

It is evident that $g(-z)$ has zeros at the positive integers, and by comparison with the product representation of $\sin \pi z$ we see that

$$g(z)g(-z) = \frac{\sin \pi z}{\pi z}. \tag{8.4-2}$$

The function $g(z-1)$ has the same zeros as $g(z)$; in addition, one simple zero at $z = 0$. Thus $g(z-1)/zg(z)$ is an entire function without zeros. By Theorem 8.3a we have

$$g(z-1) = z e^{h(z)} g(z), \tag{8.4-3}$$

where h is an entire function. To determine h we take logarithmic derivatives. By Corollary 8.3d we get

$$\sum_{n=1}^{\infty} \left(\frac{1}{z-1+n} - \frac{1}{n}\right) = \frac{1}{z} + h'(z) + \sum_{n=1}^{\infty} \left(\frac{1}{z+n} - \frac{1}{n}\right).$$

In the sum on the left we replace $n-1$ by n:

$$\sum_{n=0}^{\infty} \left(\frac{1}{z+n} - \frac{1}{n+1}\right) = \frac{1}{z} + h'(z) + \sum_{n=1}^{\infty} \left(\frac{1}{z+n} - \frac{1}{n+1}\right).$$

Subtracting the two series, we find

$$h'(z) = \sum_{n=1}^{\infty} \left(\frac{1}{n} - \frac{1}{n+1}\right) - 1 = 0.$$

It follows that $h(z)$ is a constant, which we denote by γ. The recurrence relation (8.4-3) thus becomes

$$g(z-1) = z \, e^{\gamma} g(z). \tag{8.4-4}$$

To determine γ, we set $z = 1$ and obtain

$$1 = g(0) = e^{\gamma} g(1); \tag{8.4-5}$$

hence

$$e^{-\gamma} = g(1) = \prod_{n=1}^{\infty} \left(1 + \frac{1}{n}\right) e^{-1/n}.$$

Taking logarithms yields

$$\gamma = \sum_{n=1}^{\infty} \left(\frac{1}{n} - \text{Log}\left(1 + \frac{1}{n}\right)\right)$$

or, because $\text{Log}(1 + 1/n) = \text{Log}(1+n) - \text{Log } n$,

$$\gamma = \lim_{n \to \infty} \left(1 + \frac{1}{2} + \cdots + \frac{1}{n} - \text{Log } n\right). \tag{8.4-6}$$

The real number γ is called **Euler's constant**. Numerical computations (see §11.12) yield $\gamma = 0.5772156649 \ldots$. It is one of the outstanding problems of number theory to prove the (highly likely) conjecture that γ is irrational. See example 6 of §4.9 for another representation of γ.

Following Weierstrass, we now define the function Γ by

$$\Gamma(z) := \frac{1}{z \, e^{\gamma z} g(z)}, \tag{8.4-7}$$

which is to say:

$$\Gamma(z) := \frac{e^{-\gamma z}}{z} \prod_{n=1}^{\infty} \left(1 + \frac{z}{n}\right)^{-1} e^{z/n} \tag{I}$$

The reason for introducing the factor $e^{-\gamma z}$ is simplicity. In view of (8.4-4) we have

$$\Gamma(z+1) = \frac{1}{(z+1) \, e^{\gamma(z+1)} g(z+1)} = \frac{1}{e^{\gamma} g(z)};$$

thus Γ satisfies the fundamental recurrence relation

$$\Gamma(z+1) = z\Gamma(z). \tag{8.4-8}$$

Iterating this relation, we get for $n = 0, 1, 2, \ldots$

$$\Gamma(z+n) = (z)_n \Gamma(z) \tag{8.4-9}$$

or

$$(z)_n = \frac{\Gamma(z+n)}{\Gamma(z)}, \tag{8.4-10}$$

where $(z)_n := z(z+1) \cdots (z+n-1)$ denotes the Pochhammer symbol. For $z = 1$ this yields, in view of $\Gamma(1) = 1$,

$$n! = \Gamma(n+1). \tag{8.4-11}$$

Thus the Γ function does enjoy, apart from a trivial change of variable, the desired interpolating property for the factorial.[1]

Rewriting (8.4-2) in terms of Γ, we find

$$\Gamma(z)\Gamma(1-z) = \frac{\pi}{\sin \pi z}. \tag{8.4-12}$$

Setting $z = \frac{1}{2}$, we get

$$\Gamma(\tfrac{1}{2})^2 = \pi.$$

The product representation shows that $\Gamma(z) > 0$ for $z > 0$, which yields another important value of Γ:

$$\Gamma(\tfrac{1}{2}) = \sqrt{\pi} = 1.77245\ 38509 \cdots \tag{8.4-13}$$

From (8.4-9) there follows

$$\Gamma(n+\tfrac{1}{2}) = \sqrt{\pi}(\tfrac{1}{2})_n, \qquad n = 0, 1, 2, \ldots. \tag{8.4-14}$$

By its very construction, the Γ function has simple poles at the points $z = -n$ for $n = 0, 1, 2, \ldots$. We compute the residue r_n at the pole $z = -n$.

---

[1] If the interpolation of $n!$ were the only reason for introducing the Γ function, it obviously would have been more practical to consider $f(z) := \Gamma(z+1)$ as the basic function in place of Γ. Many authors, including C. F. Gauss and some of the older British analysts, indeed preferred the function $\Gamma(z+1)$, which they denoted by Π. There are many relations, however, where the use of the Γ function as defined yields simpler results. Today, the Γ standardization is almost universally adopted. In the interest of economizing mathematical symbolism, it would be of dubious value to introduce two functions that differ only trivially.

From (8.4-12) we have

$$\Gamma(z) = \frac{1}{\Gamma(1-z)} \frac{\pi}{\sin \pi z}.$$

Because $1/\Gamma(1-z)$ is analytic at $z = -n$ with value $1/n!$ and because $\pi/\sin \pi z$ has the residue $(-1)^n$, we find

$$r_n = \frac{(-1)^n}{n!}, \qquad n = 0, 1, 2, \ldots. \qquad (8.4\text{-}15)$$

Taking the logarithmic derivative of (I) (see Corollary 8.3d) we find

$$\frac{\Gamma'(z)}{\Gamma(z)} = -\frac{1}{z} - \gamma + \sum_{n=1}^{\infty} \left(\frac{1}{n} - \frac{1}{z+n} \right). \qquad (8.4\text{-}16)$$

A particular value is

$$\Gamma'(1) = -\gamma,$$

and from (8.4-9) or directly from (8.4-16) there follows

$$\Gamma'(n+1) = n! \left(1 + \frac{1}{2} + \cdots + \frac{1}{n} - \gamma \right),$$

where $n = 1, 2, \ldots$. Differentiating (8.4-16) once more, we obtain

$$\left[\frac{\Gamma'(z)}{\Gamma(z)} \right]' = \sum_{n=0}^{\infty} \frac{1}{(z+n)^2}. \qquad (8.4\text{-}17)$$

We conclude that for z real, $z \neq 0, -1, -2, \ldots$, $\Gamma''(z)$ always has the same sign as $\Gamma(z)$. The basic descriptive properties of $\Gamma(x)$ for x real are now apparent; see Fig. 8.4a. Because of the interpolating property, $\Gamma(x)$ increases very rapidly for $x \to \infty$, more rapidly than e^x. The precise rate of increase will be determined in §8.5.

We now shall construct two further representations of the Γ function, which historically predate that given by Weierstrass. From the definition of an infinite product it follows that $\Gamma(z)$ is the limit as $n \to \infty$ of the functions

$$p_n(z) := \frac{e^{-\gamma z}}{z} \prod_{k=1}^{n} \left(1 + \frac{z}{k} \right)^{-1} e^{z/k}.$$

This is identical with

$$\exp\left[\left(1 + \frac{1}{2} + \cdots + \frac{1}{n} - \gamma \right) z \right] \frac{n!}{z(z+1)\cdots(z+n)}.$$

In view of (8.4-6) the exponential factor may be replaced by

$$e^{z \operatorname{Log} n} = n^z.$$

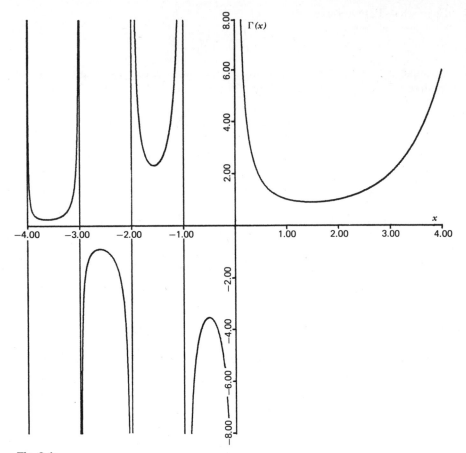

Fig. 8.4a.

without changing the limit. We thus have, for all $z \neq 0, -1, -2, \ldots$,

$$\Gamma(z) = \lim_{n \to \infty} \frac{n^z n!}{(z)_{n+1}} \qquad (II)$$

This relation was Gauss' starting point of his investigations of the Γ function. It was the preferred definition of the late Hermann Weyl, who once stated that this definition was "most centrally located" and that it led most easily to the important properties of the function.

As an example of Weyl's thesis, we use (II) to prove the so-called duplication formula of Legendre. This expresses the function

$$f(z) := \Gamma(z)\Gamma(z + \tfrac{1}{2})$$

in terms of a single Γ function. It is clear that f has simple poles at the points $z = 0, -\tfrac{1}{2}, -1, -\tfrac{3}{2}, \dots$. The function $\Gamma(2z)$ has simple poles at exactly the same points. Hence

$$g(z) := \frac{\Gamma(z)\Gamma(z + \tfrac{1}{2})}{\Gamma(2z)}$$

is an entire function. We evaluate g by expressing the various Γ functions by (II). To get a manageable expression, we replace n by $2n$ in the limit for $\Gamma(2z)$. The quotient whose limit as $n \to \infty$ must be computed then is the following:

$$\frac{n^z}{(z)_{n+1}} \cdot \frac{n^{z+1/2}}{(z + \tfrac{1}{2})_{n+1}} \cdot \frac{(2z)_{2n+1}}{(2n)^{2z}} \cdot \frac{(n!)^2}{(2n)!}.$$

Using the identities

$$(z)_{n+1}(z + \tfrac{1}{2})_{n+1} = 2^{-2n-2}(2z)_{2n+2},$$

$$(2n)! = 2^{2n}(\tfrac{1}{2})_n n!,$$

this simplifies to

$$\frac{1}{2^{2z-2}} \cdot \frac{1}{2z + 2n + 1} \cdot \frac{n^{1/2}n!}{(\tfrac{1}{2})_n},$$

which may be written as

$$2^{1-2z} \cdot \frac{n + \tfrac{1}{2}}{z + n + \tfrac{1}{2}} \cdot \frac{n^{1/2}n!}{(\tfrac{1}{2})_{n+1}}.$$

Here the first factor is independent of n, the second approaches 1, and the third by (II) approaches $\Gamma(\tfrac{1}{2}) = \sqrt{\pi}$. We thus have established the formula

$$\frac{\Gamma(z)\Gamma(z + \tfrac{1}{2})}{\Gamma(2z)} = 2^{1-2z}\sqrt{\pi}, \tag{8.4-18}$$

known as **Legendre's duplication formula**.

A third representation of $\Gamma(z)$, due to Euler, is, however, equally important. As an example of Weyl's principle we derive it from (II) and begin by obtaining an integral representation for the expression

$$\Gamma_n(z) := \frac{n^z n!}{(z)_{n+1}},$$

whose limit for $n \to \infty$ is sought. We suppose that $z = x > 0$. To begin with, we note that the factor $1/(x)_{n+1}$ is obtained by integrating the function τ^{x-1} $n + 1$ times, the last time between the limits 0 and 1:

$$\frac{1}{(x)_{n+1}} = \int_0^1 \int_0^{\tau_n} \cdots \int_0^{\tau_1} \tau^{x-1} \, d\tau \, d\tau_1 \cdots d\tau_n.$$

Now by a well-known formula of calculus (see also problem 2, §10.4) the $(n + 1)$-fold integral of a function $f(\tau)$ can be expressed by a single integral as follows:

$$\int_0^\xi \int_0^{\tau_n} \cdots \int_0^{\tau_1} f(\tau) \, d\tau \, d\tau_1 \cdots d\tau_n = \frac{1}{n!} \int_0^\xi (\xi - \tau)^n f(\tau) \, d\tau.$$

(The proof is by differentiation with respect to ξ or by operational calculus.) This yields

$$\frac{n!}{(x)_{n+1}} = \int_0^1 (1 - \tau)^n \tau^{x-1} \, d\tau.$$

Letting $\sigma := n\tau$, we find

$$\Gamma_n(x) = \frac{n! \, n^x}{(x)_{n+1}} = \int_0^n \left(1 - \frac{\sigma}{n}\right)^n \sigma^{x-1} \, d\sigma. \qquad (8.4\text{-}19)$$

We now recall the well-known limit, valid for all real or complex s:

$$\lim_{n \to \infty} \left(1 - \frac{s}{n}\right)^n = e^{-s}.$$

If it is possible to pass to the limit under the integral sign and simultaneously to let the upper limit of integration tend to ∞, then in view of $\Gamma_n \to \Gamma$ we have obtained **Euler's integral** for $\Gamma(x)$:

$$\Gamma(x) = \int_0^\infty e^{-\sigma} \sigma^{x-1} \, d\sigma. \qquad (\text{III})$$

The result is correct, but the dual passage to the limit leading to it calls for justification. It turns out that this justification is surprisingly difficult if we confine ourselves to the tools of classical ε-and-δ-analysis (see Whittaker and Watson [1927], p. 242). In view of the importance of (III), we now present three other proofs. The first uses modern real variable theory; the second, due to Erhard Schmidt, except for an ingenious idea uses nothing but calculus. The third proof, to be given at the end of §8.5, uses complex variable theory plus some knowledge of the asymptotic behavior of the Γ function.

(i) *Real Variable Proof.* Here we make use of the so-called **theorem of Beppo Levi** (see, e.g., Natanson [1961], p. 156): Let $\{f_n\}$ be a monotonically increasing sequence of functions defined and Lebesgue-integrable on some interval I (which may be infinite). If $f := \lim_{n \to \infty} f_n$, then

$$\lim \int_I f_n = \int_I f.$$

We apply this result to the interval $I := (0, \infty)$ and define the functions f_n by

$$f_n(\sigma) := \begin{cases} \left(1 - \dfrac{\sigma}{n}\right)^n \sigma^{x-1}, & 0 < \sigma \leqslant n, \\ 0, & \sigma \geqslant n. \end{cases}$$

It remains to be shown that this sequence is increasing, or that $f_{n+1}(\sigma) \geqslant f_n(\sigma), 0 < \sigma < \infty$. This is clear if $\sigma \geqslant n$, and is proved in calculus if $\sigma < n$. Thus the conditions of Beppo Levi's theorem are met, and the result follows.

(ii) *Calculus Proof of E. Schmidt.* It was shown in example 3, §4.1, that the function g defined for Re $z > 0$ by

$$g(z) := \int_0^\infty e^{-\tau} \tau^{z-1} \, d\tau \tag{8.4-20}$$

is analytic where defined, and that its derivatives are

$$g^{(k)}(z) = \int_0^\infty e^{-\tau} (\mathrm{Log}\ \tau)^k \tau^{z-1} \, d\tau, \qquad k = 1, 2, \ldots . \tag{8.4-21}$$

To show that $g = \Gamma$ it suffices, by analytic continuation, that $g(x) = \Gamma(x)$ for real $x > 0$. For such x, the function g has the following properties in common with Γ:

(A) $\qquad\qquad\qquad g(x) > 0,$

(B) $\qquad\qquad\qquad g(1) = 1,$

(C) $\qquad\qquad\qquad g(x+1) = xg(x),$

(D) $\qquad\qquad\qquad g''(x)g(x) - [g'(x)]^2 \geqslant 0.$

The functional relation (C) is obtained by integrating by parts in (8.4-20). The inequality (D) is a consequence of the Schwarz inequality for integrals,

$$\left(\int uv\right)^2 \leqslant \int u^2 \int v^2,$$

if the limits of integration are 0 and ∞, and if we set

$$u(\tau) := (e^{-\tau} t^{x-1})^{1/2}, \qquad v(\tau) := (e^{-\tau} t^{x-1})^{1/2} \mathrm{Log}\ \tau.$$

That g is identical with Γ now follows from

THEOREM 8.4a

Let g be any real-valued function that is twice continuously differentiable on $(0, \infty)$ *and satisfies the conditions* (A), (B), (C), (D). *Then* $g(x) = \Gamma(x)$ *for all* $x > 0$.

Proof. Let $d := \mathrm{Log}\, g - \mathrm{Log}\, \Gamma$. Then, by (B),

$$d(x + 1) = \mathrm{Log}[xg(x)] - \mathrm{Log}[x\Gamma(x)] = d(x), \qquad x > 0.$$

Thus d is periodic with period 1, and so are $d'(x)$ and $d''(x)$. From (8.4-17) we have (integral test) for $x > 1$

$$(\mathrm{Log}\,\Gamma(x))'' = \sum_{n=0}^{\infty} \frac{1}{(x+n)^2} < \frac{1}{x^2} + \int_x^{\infty} \frac{1}{\nu^2}\, d\nu \leq \frac{2}{x},$$

and hence

$$\lim_{x \to \infty} (\mathrm{Log}\,\Gamma(x))'' = 0.$$

By virtue of (D), $(\mathrm{Log}\, g(x))'' \geq 0$. It follows that $d''(x) \geq -2/x$, and consequently, by virtue of the periodicity,

$$d''(x) \geq 0, \qquad x > 0.$$

But, again using periodicity,

$$\int_x^{x+1} d''(\tau)\, d\tau = d'(x+1) - d'(x) = 0;$$

thus $d''(x) \equiv 0$ and $d(x) = ax + b$. Using periodicity once more, we have $a = 0$, and in view of (A), $b = d(1) = 0$. It follows that $d(x) \equiv 0$, proving Theorem 8.4a. ∎

It now has been shown that Euler's representation

$$\Gamma(z) = \int_0^{\infty} e^{-s} s^{z-1}\, ds \tag{III}$$

where $s^z := e^{z\, \mathrm{Log}\, s}$, holds for all z such that $\mathrm{Re}\, z > 0$.

In (III), s may be regarded a complex variable. For $\rho > 0$ and $\alpha \in [-(\pi/2), \pi/2]$, let $\Lambda_{\rho,\alpha}$ denote the closed curve in the s plane consisting of the straight line segments $s = \tau$, $s = e^{i\alpha}\tau$ ($0 \leq \tau \leq \rho$) and of the circular arc ($0 \leq \phi \leq \alpha$ or $\alpha \leq \phi \leq 0$) (see Fig. 8.4b).

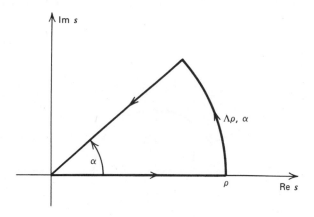

Fig. 8.4b.

For every fixed z such that Re $z > 0$, the integral of $f(s) := e^{-s} s^{z-1}$ along $\Lambda_{\rho,\alpha}$ exists and is zero, by Cauchy's theorem. For $\rho \to \infty$ the integral along the real axis tends to $\Gamma(z)$, and the integral along the circular arc is easily shown to tend to 0. It follows that in (III) the integral may also be extended along the ray arg $s = \alpha$, which we indicate symbolically by writing

$$\Gamma(z) = \int_0^{e^{i\alpha}\infty} e^{-s}s^{z-1}\, ds. \tag{8.4-22}$$

We now establish yet another integral representation for $\Gamma(z)$, due to Hankel, which is similar in appearance to (III) but which is free from the restriction Re $z > 0$. For $\delta > 0$ and $\pi/2 < \phi < \pi$, now let $\Lambda_{\delta,\phi}$ denote the improper curve sketched in Fig. 8.4c consisting of pieces of the two rays arg $s = \pm \phi$ extending to infinity, and of the circular arc $s = \delta e^{i\tau}, -\phi \leq \tau \leq \phi$.

Now let z be an arbitrary complex number, and for $|\arg s| < \pi$, let s^{-z} be defined by its principal value. The integral

$$I(z) := \int_{\Lambda_{\delta,\phi}} e^s s^{-z}\, ds$$

then clearly exists for every choice of $\delta > 0$ and $\phi \in (\pi/2, \pi)$. In fact, the integral converges uniformly with respect to z in every bounded set of the z plane and thus, because the integrand depends analytically on z, is itself an entire analytic function of z. By Cauchy's theorem the integral clearly does not depend on δ, and by estimating the integrand on large circles we see that the integral is also independent of ϕ in the permitted range of ϕ. To evaluate the integral, we suppose that Re $z < 1$. Then if $\delta \to 0$, the integral along the circular part of $\Lambda_{\delta,\phi}$ tends to zero. Substituting $t := -s$ we see that the

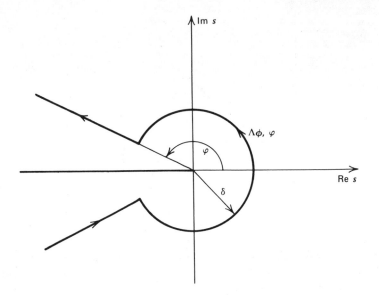

Fig. 8.4c.

integral along the lower leg of $\Lambda_{0,\phi}$ by (8.4-22) equals

$$\int_0^{-e^{-i\phi}\infty} e^{-t}(e^{-i\pi}t)^{-z}\,dt = e^{i\pi z}\Gamma(1-z),$$

and the integral along the upper leg

$$-\int_0^{-e^{i\phi}\infty} e^{-t}(e^{i\pi}t)^{-z}\,dt = -e^{-i\pi z}\Gamma(1-z).$$

We thus find

$$I(z) = (e^{i\pi z} - e^{-i\pi z})\Gamma(1-z) = 2i\,\sin\,\pi z\,\Gamma(1-z).$$

By (8.4-12) there follows

$$I(z) = \frac{2\pi i}{\Gamma(z)}.$$

Thus we have obtained the following result due to Hankel:

THEOREM 8.4b

If $\Lambda_{\delta,\phi}$ denotes the path depicted in Fig. 8.4c, and s^{-z} has its principal value,

$$\frac{1}{2\pi i}\int_{\Lambda_{\delta,\phi}} e^s s^{-z}\, ds = \frac{1}{\Gamma(z)} \qquad\qquad \text{(III')}$$

for all complex z, independently of the choice of $\delta>0$ and $\phi\in(\pi/2,\pi)$.

One may even choose $\phi=\pi$, provided that arg $s=\pi$ on the upper edge of the cut and arg $s=-\pi$ on the lower edge. If z is an integer, the integrals along the upper and lower edges cancel each other, and we have for $n=0,\pm1,\pm2,\ldots$

$$\frac{1}{\Gamma(n)}=\frac{1}{2\pi i}\int_{|s|=\delta} e^s s^{-n}\, ds.$$

This makes evident that $1/\Gamma(n)=0$ for $n=0,-1,-2,\ldots$ and, by virtue of the residue theorem, that $\Gamma(n)=(n-1)!$ for $n=1,2,\ldots$.

PROBLEMS

1. Deduce the basic properties of $\Gamma(z)$ from the Gauss definition (II).
2. Establish the Legendre duplication formula by means of (I).
3. Show that for real y

$$|\Gamma(iy)|^2=\frac{\pi}{y\sinh\pi y},\quad |\Gamma(\tfrac{1}{2}+iy)|^2=\frac{\pi}{\cosh\pi y}.$$

4. Using the product representation, show that for real x and y, $x>0$, $y\neq0$

$$\left|\frac{\Gamma(x+iy)}{\Gamma(x)}\right|<1.$$

5. Prove the following generalization of (8.4-18), valid for $n=1,2,\ldots$:

$$\sum_{k=0}^{n-1}\Gamma\!\left(z+\frac{k}{n}\right)=(2\pi)^{(n-1)/2}n^{1/2-nz}\Gamma(nz).$$

6. Establish

$$\prod_{k=1}^{8}\Gamma\!\left(\frac{k}{3}\right)=\frac{640}{3^6}\left(\frac{\pi}{\sqrt{3}}\right)^3.$$

7. Prove that for $m = 2, 3, \ldots$

$$\prod_{n=1}^{\infty} \left(1 - \left(\frac{z}{n}\right)^m\right)^{-1} = -z^m \prod_{k=1}^{m} \Gamma(-ze^{(2\pi i k)/m}).$$

8. Show that the conditionally convergent product

$$(1-z)\left(1+\frac{z}{2}\right)\left(1-\frac{z}{3}\right)\left(1+\frac{z}{4}\right)\cdots$$

has the value

$$\frac{2\sqrt{\pi}}{z\Gamma\left(\frac{z}{2}\right)\Gamma\left(\frac{1-z}{2}\right)}.$$

9. Establish the following representations of Euler's constant:

$$\gamma = -\int_{0}^{\infty} e^{-\tau} \operatorname{Log} \tau \, d\tau$$

$$= 1 - \sum_{p=2}^{\infty} \left(\frac{1}{p} \sum_{q=2}^{\infty} \frac{1}{q^p}\right)$$

$$= \sum_{p=2}^{\infty} \frac{(-1)^p}{p} \sum_{q=1}^{\infty} \frac{1}{q^p}.$$

10. The algorithm implied in the proof of Theorem 7.10b furnishes for $\Gamma(z)$ the partial fraction decomposition

$$\Gamma(z) = \sum_{m=0}^{\infty} \frac{(-1)^m}{m!(z+m)} + g(z),$$

where $g(z)$ is a suitable entire function. Using (III), show that

$$g(z) = \int_{1}^{\infty} e^{-\tau} t^{z-1} \, d\tau.$$

11. Deduce from (III):

$$\int_{1}^{\infty} \xi^{-2} (\operatorname{Log} \xi)^{\alpha} \, d\xi = \Gamma(\alpha + 1), \, \alpha > -1.$$

12. Turning the path of integration in Euler's integral by 90°, obtain the integral, valid for $0 < \operatorname{Re} z < 1$,

$$\int_{0}^{\infty} \tau^{z-1} e^{i\tau} \, d\tau = e^{-i\pi z/2} \Gamma(z).$$

Deduce that if $\nu > 1$,

$$\int_0^\infty \cos(\tau^\nu)\, d\tau = \Gamma\left(1+\frac{1}{\nu}\right) \cos\frac{\pi}{2\nu},$$

$$\int_0^\infty \sin(\tau^\nu)\, d\tau = \Gamma\left(1+\frac{1}{\nu}\right) \sin\frac{\pi}{2\nu},$$

and obtain Fresnel's integrals (problem 1, § 4.3) as a special case.

13. Let $\alpha, \beta > 0$. By evaluating the double integral

$$\frac{1}{\Gamma(\alpha)\Gamma(\beta)} \int_0^\infty \int_0^\infty e^{-u-v-xuv} u^{\beta-1} v^{\alpha-1}\, du\, dv$$

in two different ways, prove

$$\frac{1}{\Gamma(\alpha)} \int_0^\infty \frac{e^{-u}u^{\beta-1}}{(1+xu)^\alpha}\, du = \frac{1}{\Gamma(\beta)} \int_0^\infty \frac{e^{-v}v^{\alpha-1}}{(1+xv)^\beta}\, dv.$$

14. From (II) there follows

$$\operatorname{Log}\Gamma(z+1) = \lim_{n\to\infty}\left\{z \operatorname{Log} n - \sum_{k=1}^n \operatorname{Log}\left(1+\frac{z}{k}\right)\right\}.$$

By expanding the logarithm in powers of z, show that for $|z| < 1$

$$\operatorname{Log}\Gamma(1+z) = -\gamma z + \sum_{m=2}^\infty (-1)^m \frac{z^m}{m}\sigma_m, \qquad (8.4\text{-}23)$$

where

$$\sigma_m := \sum_{k=1}^\infty \frac{1}{k^m}, \quad m = 2, 3, \ldots.$$

Also obtain (8.4-23) from (I) via (8.4-17).

15. Let $\alpha > 0$, $0 \leqslant x < 1$. Show that for $x \to 1-$,

$$\sum_{n=0}^\infty x^{n^\alpha} \sim (1-x)^{-1/\alpha}\Gamma\left(\frac{1}{\alpha}+1\right).$$

Verify the cases $\alpha = 1, 2$! (Approximate the sum by an integral.)

§8.5. STIRLING'S FORMULA

In the preceding section we have obtained four classical representations, designated by (I), (II), (III), (III'), of the Γ function. These are now supplemented by a fifth, called *Stirling's formula*. This formula has the special advantage that it puts in evidence the behavior of $\Gamma(z)$ for large values of $|z|$. This behavior is often of interest when the Γ function is applied.

Stirling's formula, or the approximation to $\Gamma(z)$ following from it, can be derived from any of the four representations of $\Gamma(z)$ given in §8.4, each derivation constituting a classical piece of analysis. Here we obtain it from (I); see §§11.6, 11.7, and 11.11 on how to obtain it from (III), (III'), and (II).

Our starting point is (8.4-17), which we now write

$$\left(\frac{\Gamma'(z)}{\Gamma(z)}\right)' = \sum_{n=0}^{\infty} f(z+n), \qquad (8.5\text{-}1)$$

where

$$f(z+t) := \frac{1}{(z+t)^2}. \qquad (8.5\text{-}2)$$

For Re $z > 0$, $f(z+t)$ as a function of t satisfies the hypotheses of the Plana summation formula (Theorem 4.9c). Summing the series on the right of (8.5-1) by the summation formula we thus get

$$\left(\frac{\Gamma'(z)}{\Gamma(z)}\right)' = \tfrac{1}{2}f(z) + \int_0^\infty f(z+\xi)\,d\xi + i\int_0^\infty [f(z+i\eta)-f(z-i\eta)]\frac{1}{e^{2\pi\eta}-1}\,d\eta$$

or

$$\left(\frac{\Gamma'(z)}{\Gamma(z)}\right)' = \frac{1}{2z^2} + \frac{1}{z} + \int_0^\infty \frac{4\eta z}{(z^2+\eta^2)^2}\frac{1}{e^{2\pi\eta}-1}\,d\eta. \qquad (8.5\text{-}3)$$

We assert that the integral may be integrated under the integral sign with respect to the parameter z, that is, that for Re $z > 0$

$$\frac{\Gamma'(z)}{\Gamma(z)} = a - \frac{1}{2z} + \mathrm{Log}\,z - \int_0^\infty \frac{2\eta}{\eta^2+z^2}\frac{1}{e^{2\pi\eta}-1}\,d\eta, \qquad (8.5\text{-}4)$$

where a is a constant of integration. All we have to show to justify this step is that the derivative of (8.5-4) once again yields (8.5-3). Evidently, the improper integral on the right of (8.5-4) converges uniformly with respect to z in every compact subset of Re $z > 0$. According to Theorem 4.1a, it thus represents an analytic function whose derivative can be calculated by differentiating under the integral sign. This process, in fact, yields (8.5-3).

Our goal is an integral representation for log $\Gamma(z)$, and to this end we wish to integrate (8.5-4) once more. To avoid the multivalued function arctan under the integral, we transform the integral with respect to η by integration by parts. In view of

$$\int \frac{1}{e^{2\pi\eta}-1}\,d\eta = \int \frac{e^{-2\pi\eta}}{1-e^{-2\pi\eta}}\,d\eta = \frac{1}{2\pi}\,\mathrm{Log}(1-e^{-2\pi\eta}),$$

this process, first applied to the proper integral between the limits $\delta > 0$ and

μ, yields

$$\int_\delta^\mu \frac{1}{e^{2\pi\eta}-1}\frac{2\eta}{\eta^2+z^2}\,d\eta = \left[\frac{1}{2\pi}\operatorname{Log}(1-e^{-2\pi\eta})\frac{2\eta}{\eta^2+z^2}\right]_\delta^\mu$$

$$+\frac{1}{\pi}\int_\delta^\mu \operatorname{Log}(1-e^{-2\pi\eta})\frac{z^2-\eta^2}{(\eta^2+z^2)^2}\,d\eta$$

The integrated part vanishes as $\delta \to 0$ and $\mu \to \infty$, and the last integral exists as an improper integral. Hence we have

$$\frac{\Gamma'(z)}{\Gamma(z)} = a + \operatorname{Log} z - \frac{1}{2z} - \frac{1}{\pi}\int_0^\infty \frac{z^2-\eta^2}{(\eta^2+z^2)^2}\operatorname{Log}(1-e^{-2\pi\eta})\,d\eta. \quad (8.5\text{-}5)$$

Now we can integrate and obtain, first operating formally,

$$\log\Gamma(z) = az + b + (z-\tfrac{1}{2})\operatorname{Log} z - z + J(z), \quad (8.5\text{-}6)$$

where b is another constant of integration and J denotes the **Binet function**,[2]

$$J(z) := \frac{1}{\pi}\int_0^\infty \frac{z}{\eta^2+z^2}\operatorname{Log}\frac{1}{1-e^{-2\pi\eta}}\,d\eta. \quad (8.5\text{-}7)$$

The justification is again by noting that the integral (8.5-7) converges uniformly for $\operatorname{Re} z \geq \delta > 0$ and hence may be differentiated under the integral sign. Thus the expression on the right of (8.5-6) indeed represents an analytic branch of $\log\Gamma(z)$. For a proper choice of b we obtain the branch that is real for $z > 0$.

Before determining the constants a and b, we collect some facts concerning the Binet function.

THEOREM 8.5a

The Binet function $J(z)$ (defined by (8.5-7) for $\operatorname{Re} z > 0$) can be continued to a function that is analytic in the cut plane $|\arg z| < \pi$. For every θ such that $0 \leq \theta < \pi$ there exists $\kappa(\theta)$ such that

$$|J(z)| \leq \frac{\kappa(\theta)}{|z|} \quad (8.5\text{-}8)$$

for all z in the sector $S_\theta : |\arg z| \leq \theta$. As a corollary, if $z \in S_\theta$,

$$\lim_{z\to\infty} J(z) = 0. \quad (8.5\text{-}9)$$

[2] It seems desirable to name this function, although no name appears to be generally used. J. P. M. Binet, a contemporary of Cauchy, obtained various representations of J.

If $\theta \leqslant \pi/4$, (8.5-8) *holds with* $\kappa(\theta) = 1/12$; *in particular, if* $z = x > 0$,

$$0 < J(x) < \frac{1}{12x}.\tag{8.5-10}$$

Proof. By virtue of Theorem 4.3f, $\log \Gamma(z)$ can be defined as an analytic function in $|\arg z| < \pi$, and the elementary terms on the right of (8.5-6) are likewise analytic there. Thus although the integral (8.5-7) defines an analytic function only for Re $z > 0$, it follows that $J(z)$ is continued analytically by

$$J(z) := \log \Gamma(z) - \{az + b + (z - \tfrac{1}{2}) \operatorname{Log} z - z\}.$$

To obtain an explicit representation of the continued function similar to (8.5-7), we use the familiar device of rotating the path of integration. Let α be any number such that $-(\pi/2) < \alpha < \pi/2$. We shall obtain a representation of $J(z)$ that is valid for $|\arg z - \alpha| < \pi/2$. Consider the closed curve $\Lambda_{\alpha,\delta,\mu}$ depicted in Fig. 8.5a, where $\alpha > 0$.

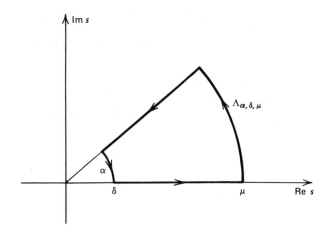

Fig. 8.5a.

For any fixed $z > 0$, the function

$$s \to \frac{z}{s^2 + z^2} \operatorname{Log} \frac{1}{1 - e^{-2\pi s}}$$

is analytic for Re $s > 0$; hence its integral along $\Lambda_{\alpha,\delta,\mu}$ is zero. Simple estimates show that the integrals along the circular arcs of radius δ and μ vanish as $\delta \to 0$ and $\mu \to \infty$. Hence it follows that for $z > 0$,

$$J(z) = \frac{1}{\pi} \int_0^{e^{i\alpha \infty}} \frac{z}{s^2 + z^2} \operatorname{Log} \frac{1}{1 - e^{-2\pi s}} \, ds,$$

where the integral is along the ray $\arg s = \alpha$. Parametrizing the ray by $s = w\sigma$, $0 \le \sigma < \infty$, where $w := e^{i\alpha}$, we obtain

$$J(z) = \frac{1}{\pi w} \int_0^\infty \frac{z}{\sigma^2 + \bar{w}^2 z^2} \operatorname{Log} \frac{1}{1 - e^{-2\pi w\sigma}} \, d\sigma. \qquad (8.5\text{-}11)$$

The integral on the right converges uniformly for $\operatorname{Re} \bar{w}z \ge \delta > 0$, and hence represents a function that is analytic for $\operatorname{Re} \bar{w}z > 0$, that is, for $|\arg z - \alpha| < \pi/2$. For $z > 0$ this function agrees with $J(z)$. It thus equals the analytic continuation of $J(z)$ into the set $|\arg z - \alpha| < \pi/2$.

To establish (8.5-8), let $w = e^{i\alpha}$ in (8.5-11) be chosen such that $\alpha = \frac{1}{2}\theta$. If $0 \le \arg z \le \theta$, then

$$0 \le \arg z^2 \le 2\theta, \qquad -\theta \le \arg \bar{w}^2 z^2 \le \theta,$$

hence if $\sigma \ge 0$,

$$|\sigma^2 + \bar{w}^2 z^2| \ge \sin\left(\max\left(\frac{\pi}{2}, \theta\right)\right)|z|^2.$$

Straightforward estimation of the integral now yields (8.5-8) where

$$\kappa(\theta) = \frac{1}{\pi \sin\left(\max\left(\frac{\pi}{2}, \theta\right)\right)} \int_0^\infty \left| \operatorname{Log} \frac{1}{1 - e^{-2\pi w\sigma}} \right| d\sigma < \infty.$$

Since $J(\bar{z}) = \overline{J(z)}$ by the reflection principle, this clearly also holds for $-\theta \le \arg z \le 0$.

If $|\arg z| < \pi/2$, we can use (8.5-7). If $|\arg z| \le \pi/4$, then $|\eta^2 + z^2| \ge |z|^2$, therefore

$$|J(z)| \le \frac{1}{\pi|z|} \int_0^\infty \operatorname{Log} \frac{1}{1 - e^{-2\pi\eta}} \, d\eta.$$

The integral may be evaluated by expanding in series, interchange of summation and integration being justified by Beppo Levi's theorem:

$$\int_0^\infty \operatorname{Log} \frac{1}{1 - e^{-2\pi\eta}} \, d\eta = \int_0^\infty \left(e^{-2\pi\eta} + \tfrac{1}{2} e^{-4\pi\eta} + \tfrac{1}{3} e^{-6\pi\eta} + \cdots \right) d\eta$$

$$= \frac{1}{2\pi} + \frac{1}{2} \cdot \frac{1}{4\pi} + \frac{1}{3} \cdot \frac{1}{6\pi} + \cdots$$

$$= \frac{1}{2\pi} \left(1 + \frac{1}{2^2} + \frac{1}{3^2} + \cdots \right)$$

$$= \frac{1}{2\pi} \frac{\pi^2}{6} = \frac{\pi}{12}$$

[see (4.9-5)], establishing (8.5-8) with $\kappa(\theta) = 1/12$. Equation (8.5-10) follows because $J(x)$ is clearly positive for $x > 0$. ∎

To obtain a representation of $\Gamma(z)$, it remains to identify the constants a and b in (8.5-6). This we do by appealing to some functional relations established in §8.4. Letting $x > 0$, and substituting (8.5-6) into the relation $\operatorname{Log} \Gamma(x+1) = \operatorname{Log} x + \operatorname{Log} \Gamma(x)$ derived from $\Gamma(x+1) = x\Gamma(x)$ yields

$$(x+1)a + b + (x+\tfrac{1}{2})\operatorname{Log}(x+1) - x - 1 + J(x+1)$$

$$= xa + b + (x+\tfrac{1}{2})\operatorname{Log} x - x + J(x),$$

which reduces to

$$a = 1 - (x+\tfrac{1}{2})\operatorname{Log}\left(1+\frac{1}{x}\right) + J(x) - J(x+1).$$

Letting $x \to \infty$, the expression on the right tends to zero by virtue of (8.5-10), and we conclude that $a = 0$.

We next apply (8.5-6) to the relation $\operatorname{Log} \Gamma(x) + \operatorname{Log} \Gamma(x+\tfrac{1}{2}) = \operatorname{Log} \Gamma(2x) + \operatorname{Log} \sqrt{\pi} - (2x-1)\operatorname{Log} 2$, which is the logarithmic form of the duplication formula (8.4-18). Using $a = 0$ there results, after simplification,

$$b = \tfrac{1}{2}\operatorname{Log} 2 + \operatorname{Log} \sqrt{\pi} + \varepsilon(x),$$

where $\varepsilon(x) \to 0$ for $x \to \infty$ by virtue of (8.5-9) and some calculus properties of the logarithm. We conclude that $b = \operatorname{Log} \sqrt{2\pi}$.

We thus have proved:

THEOREM 8.5b (Stirling's formula for $\Gamma(z)$).

For all z such that $|\arg z| < \pi$, the analytic branch of $\log \Gamma(z)$ that is real for $z > 0$ is

$$\log \Gamma(z) = \operatorname{Log} \sqrt{2\pi} + (z - \tfrac{1}{2})\operatorname{Log} z - z + J(z), \qquad (8.5\text{-}12)$$

where $J(z)$ denotes the Binet function defined by (8.5-7) and enjoying the properties described in Theorem 8.5a. Equivalently,

$$\boxed{\Gamma(z) = \sqrt{2\pi}\, z^{z-1/2}\, e^{-z}\, e^{J(z)}} \qquad\qquad \text{(IV)}$$

where the power of z has its principal value.

In (IV), $e^{J(z)}$ should be regarded as a correction factor by which the elementary function $\sqrt{2\pi}z^{z-1/2}e^{-z}$, called **Stirling's approximation** to $\Gamma(z)$, should be multiplied to furnish the exact value of $\Gamma(z)$. We know from (8.5-9) that the correction factor tends to 1 as $z \to \infty$ in any sector $|\arg z| \le \theta$ where $\theta < \pi$. Thus the relative error committed by omitting the correction factor tends to zero as $z \to \infty$. In fact, if $|\arg z| \le \pi/4$, the precise estimate

$$|(2\pi)^{-1/2}z^{-z+1/2}e^{z}\Gamma(z)-1| = |e^{J(z)}-1|$$

$$\le \frac{1}{12|z|}e^{1/(12|z|)}$$

is available for the relative error of Stirling's approximation. This makes it possible to compute $\Gamma(z)$ to arbitrary accuracy for any complex z not 0 or a negative integer by using the formula

$$\Gamma(z) = \frac{\Gamma(z+n)}{(z)_n}$$

[following from (8.4-9)] and choosing n such that (i) $|\arg(z+n)| \le \pi/4$ and (ii) Stirling's approximation to $\Gamma(z+n)$ yields the relative accuracy desired. Figure 8.5b shows a graph of $|\Gamma(z)|$ for complex z which was calculated by the foregoing method.

More precise asymptotic information on Binet's function $J(z)$ will be obtained in §11.1, §11.11, and §12.12.

Several computational applications of Stirling's formula are discussed in subsequent paragraphs. Here we use it to give a third proof of Euler's representation (III). Once more, let g be defined by

$$g(z) := \int_0^\infty e^{-\tau}\tau^{z-1}\,d\tau, \qquad (8.5\text{-}13)$$

where τ^{z-1} has its principal value. We have seen in example 3 of §4.1 that g is analytic in Re $z > 0$, and an integration by parts establishes the functional relation $g(z+1) = zg(z)$. As was shown in example 15 of §3.2, this relation enables us to continue g as a meromorphic function into the whole plane, with simple poles at $z = 0, -1, -2, \ldots$. Let now Γ be defined by (I). From the above it follows, as before, that $p(z) := g(z)/\Gamma(z)$ is an entire function, and that it is periodic with period 1. We wish to show that p is constant. Using Liouville's theorem and the periodicity, it suffices to show that p is bounded in a period strip, for instance in the strip $1 \le \text{Re } z \le 2$. By (8.5-13), if $z = x + iy$,

$$|g(z)| \le \int_0^\infty e^{-\tau}\tau^{x-1}\,d\tau = g(x);$$

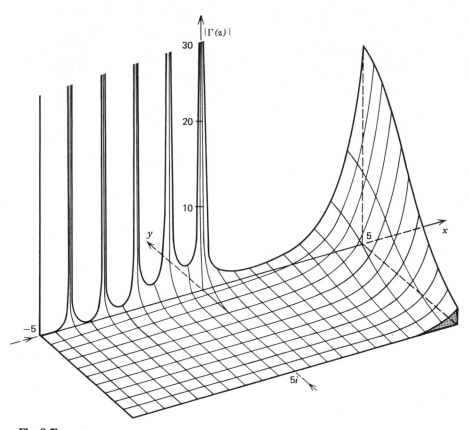

Fig. 8.5b.

thus $g(z)$ is bounded in the strip. The required lower bound for $|\Gamma(z)|$ is furnished by Stirling's formula. Taking real parts in (8.5-12), we find

$$\text{Log}\,|\Gamma(z)| = \text{Log}\,\sqrt{2\pi} - x + (x - \tfrac{1}{2})\,\text{Log}\,|z| - y\,\text{Arg}\,z + \text{Re}\,J(z).$$

For $1 \le x \le 2$, only the term $-y\,\text{Arg}\,z$ can become negatively infinite. In view of $|\text{Arg}\,z| < \pi/2$, $-y\,\text{Arg}\,z > -\pi|y|/2$. Thus $|p|$ is bounded by $\mu\,\exp(\pi|y|/2)$, where μ is a constant. For an arbitrary entire function this would not suffice to conclude that it is constant. Here, however, we can use the periodicity of p to define, for $w \ne 0$, $f(w) := p((1/2\pi i)\log w)$. Because p is periodic with period 1, it does not matter which branch of the logarithm is chosen, and because $\log w$ can be defined as an analytic function in the neighborhood of any $w_0 \ne 0$, it follows that f as a composition of analytic functions is analytic, with the sole possible exception of $w = 0$. Thus f may be expanded in a Laurent series,

$$f(w) = \sum_{n=-\infty}^{\infty} a_n w^n, \qquad (8.5\text{-}14)$$

where

$$a_n = \frac{1}{2\pi i} \int_{|w|=\rho} w^{-n-1} f(w)\, dw, \qquad (8.5\text{-}15)$$

ρ being arbitrary. To obtain an estimate for a_n, we estimate f on the circle $|w| = \rho$. The values of f on that circle are given by the values of p for $z = (1/2\pi i)\log w$ where $|w| = \rho$; that is, on the line $\operatorname{Im} z = (1/2\pi)\operatorname{Log}\rho$. The estimate obtained from Stirling's formula shows that for $\rho \to \infty$

$$|f(w)| \leqslant \mu\, \exp\!\left(\frac{\pi}{2}\frac{1}{2\pi}\operatorname{Log}\rho\right) = \mu\rho^{1/4},$$

and for $\rho \to 0$

$$|f(w)| \leqslant \mu\, \exp\!\left(-\frac{\pi}{2}\frac{1}{2\pi}\operatorname{Log}\rho\right) = \mu\rho^{-1/4}.$$

Using these estimates in (8.4-23) we immediately find that $a_n = 0$ for $n \neq 0$, and hence that f, and thus p, is constant.

PROBLEMS

1. Use Stirling's formula to compute $\Gamma(0.8)$ to six decimal places.
2. If a and b are any two complex numbers, show that

$$\frac{\Gamma(z+a)}{\Gamma(z+b)} \sim z^{a-b} \text{ (principal value)}$$

   as $z \to \infty$ uniformly in every sector S_θ where $\theta < \pi$.
3. Under what conditions on a does the binomial series

$$\sum_{n=0}^{\infty} \binom{a}{n} x^n$$

   converge (a) for $x = 1$, (b) for $x = -1$?
4. Show that the series

$$\sum_{n=0}^{\infty} \binom{\frac{1}{2}}{n}(1-x^2)^n$$

   converges to $|x|$ uniformly in the interval $-1 \leqslant x \leqslant 1$. (This result is used in Lebesgue's proof of the Weierstrass approximation theorem.)

5. Prove that if a is not an integer, and if Re $a < 1$, then

$$\sum_{n=0}^{\infty} (-1)^n \frac{\Gamma(a+n)}{\Gamma(1+n)} = 2^{-a}\Gamma(a).$$

6. For Im $z > 0$ the Binet function satisfies the functional relation

$$J(1-z) = J(z) + \text{Log}(1 - e^{2\pi i z}) - 1 - (z - \tfrac{1}{2})\, \text{Log}\left(1 - \frac{1}{z}\right).$$

How does this relation change for Im $z < 0$, and why?

§8.6. SOME SPECIAL SERIES AND PRODUCTS

In this section we evaluate a number of special series and products in terms of the Γ function.

To begin with, we consider the product

$$\prod_{n=0}^{\infty} \frac{(a_1+n)(a_2+n)\cdots(a_k+n)}{(b_1+n)(b_2+n)\cdots(b_k+n)} = \lim_{n\to\infty} \frac{(a_1)_n(a_2)_n\cdots(a_k)_n}{(b_1)_n(b_2)_n\cdots(b_k)_n}, \qquad (8.6\text{-}1)$$

where $a_1, a_2, \ldots, a_k$ and $b_1, b_2, \ldots, b_k$ are arbitrary complex numbers, $b_i \neq 0, -1, -2, \ldots, i = 1, 2, \ldots, k$. By Theorem 8.1a this product is convergent if and only if $\sum l_n$ converges, where

$$l_n := \text{Log}\frac{(n+a_1)\cdots(n+a_k)}{(n+b_1)\cdots(n+b_k)}.$$

As $n \to \infty$,

$$l_n = \text{Log}\left(1 + \frac{a_1}{n}\right) + \cdots + \text{Log}\left(1 + \frac{a_k}{n}\right)$$

$$-\text{Log}\left(1 + \frac{b_1}{n}\right) - \cdots - \text{Log}\left(1 + \frac{b_k}{n}\right)$$

$$= \frac{1}{n}(a_1 + \cdots + a_k - b_1 - \cdots - b_k) + O\left(\frac{1}{n^2}\right).$$

Because $\sum 1/n$ diverges and $\sum 1/n^2$ converges, a necessary and sufficient condition for the convergence of the product is

$$a_1 + a_2 + \cdots + a_k = b_1 + b_2 + \cdots + b_k. \qquad (8.6\text{-}2)$$

In case of convergence, the product is easily evaluated as follows. The mth partial product may be written

$$p_m = \prod_{n=0}^{m} \prod_{i=1}^{k} \frac{n+a_i}{n+b_i} = \frac{a_1 \cdots a_k}{b_1 \cdots b_k} \prod_{n=1}^{m} \prod_{i=1}^{k} \frac{1+a_i/n}{1+b_i/n}$$

or, on account of (8.6-2),

$$p_m = \frac{a_1 \cdots a_k}{b_1 \cdots b_k} \prod_{n=1}^{m} \prod_{i=1}^{k} \frac{(1+a_i/n)\exp(-a_i/n)}{(1+b_i/n)\exp(-b_i/n)}$$

$$= \prod_{i=1}^{k} \frac{a_i \prod_{n=1}^{m} (1+b_i/n)^{-1} \exp(b_i/n)}{b_i \prod_{n=1}^{m} (1+a_i/n)^{-1} \exp(a_i/n)}.$$

As $m \to \infty$, using (I), and observing (8.2-6) once more, we get

THEOREM 8.6a

The product (8.6-1) converges if and only if the parameters satisfy (8.6-2) and if none of the b_i is zero or a negative integer. In case of convergence,

$$\prod_{n=0}^{\infty} \frac{(n+a_1)\cdots(n+a_k)}{(n+b_1)\cdots(n+b_k)} = \prod_{i=1}^{k} \frac{\Gamma(b_i)}{\Gamma(a_i)}. \tag{8.6-3}$$

For $k=2$, $a_1=\frac{1}{2}$, $a_2=\frac{3}{2}$, $b_1=b_2=1$ we recover the *product of Wallis*,

$$\frac{1\cdot 3}{2\cdot 2}\frac{3\cdot 5}{4\cdot 4}\frac{5\cdot 7}{6\cdot 6}\cdots = \frac{1}{\Gamma(\frac{1}{2})\Gamma(\frac{3}{2})} = \frac{2}{\pi}.$$

Our next application concerns the generalized hypergeometric series,

$$_pF_q\left[\begin{matrix} a_1, a_2, \ldots, a_p; z \\ b_1, b_2, \ldots, b_q \end{matrix}\right] := \sum_{n=0}^{\infty} \frac{(a_1)_n \cdots (a_p)_n}{(b_1)_n \cdots (b_q)_n} \frac{z^n}{n!}, \tag{8.6-4}$$

where the $a_1, a_2, \ldots, a_p; b_1, b_2, \ldots, b_q$ are arbitrary complex parameters, $b_i \neq 0, -1, -2, \ldots, i=1, \ldots, q$. Some formal aspects of this series have already been dealt with in Chapter 1. We see in Chapter 9 that the solutions of some important differential equations can be expressed in terms of the series (8.6-4) or its analytic continuations. Here we merely evaluate the series in some selected special cases.

The series is convergent for all z if $p \leq q$ and divergent for all $z \neq 0$ if $p > q+1$. If $p = q+1$, the ratio tests shows that the radius of convergence of the series is 1. To study the convergence for $|z| = 1$ if $p = q+1$, we note that the nth term of the series equals

$$\frac{\Gamma(a_1+n)\cdots\Gamma(a_{q+1}+n)}{\Gamma(b_1+n)\cdots\Gamma(b_q+n)\Gamma(1+n)} z^n,$$

up to a factor independent of n. By Stirling's formula (see, in particular, problem 2 of §8.5) the quotient of Γ functions is asymptotic to

$$n^{a_1+\cdots+a_{q+1}-b_1-\cdots-b_q-1}.$$

Thus for the absolute convergence of the series if $|z| = 1$ it is necessary and sufficient that

$$\text{Re}(a_1 + \cdots + a_{q+1}) < \text{Re}(b_1 + \cdots + b_q). \tag{8.6-5}$$

We now evaluate several instances of the series where $p = q + 1$ and $z = 1$. If $p = 1$ and $q = 0$, then by the binomial theorem, if $|z| < 1$,

$$_1F_0[a;z] = \sum_{n=0}^{\infty} \frac{(a)_n}{n!} z^n = \sum_{n=0}^{\infty} \binom{-a}{n}(-z)^n = (1-z)^{-a}$$

(principal value), and hence if $\text{Re } a < 0$, using Abel's theorem (Theorem 2.2e)

$$_1F_0[a;1] = \lim_{z \to 1-} (1-z)^{-a} = 0.$$

We next evaluate the hypergeometric series of Gauss,

$$F(a, b; c; z) := {}_2F_1\begin{bmatrix} a, b; z \\ c \end{bmatrix}$$

for $z = 1$. The result is

THEOREM 8.6b (Gauss formula)

For $\text{Re}(a + b - c) < 0$,

$$F(a, b; c; 1) = \frac{\Gamma(c)\Gamma(c - a - b)}{\Gamma(c - a)\Gamma(c - b)}. \tag{8.6-6}$$

Proof. By comparing the coefficients of z^n it is easily verified[3] that for $|z| < 1$

$$c[c - 1 - (2c - a - b - 1)z]F(a, b; c; z) + (c - a)(c - b)zF(a, b; c + 1; z)$$

$$= c(c - 1)(1 - z)F(a, b; c - 1; z)$$

$$= c(c - 1)\left[1 + \sum_{n=1}^{\infty} (u_n - u_{n-1})z^n\right],$$

where u_n denotes the coefficient of z^n in $F(a, b; c - 1; z)$. We now let $z \to 1-$ through real values. The nth partial sum of the series obtained by letting $z = 1$ in the last expression is simply u_n. Stirling's formula shows that the hypothesis on $\text{Re}(a + b - c)$ implies that $u_n \to 0$ for $n \to \infty$. Thus by Abel's

---

[3] This is a case of Gauss' identities between "contiguous" hypergeometric functions.

theorem the limit of the series on the right as $z \to 1-$ is zero. Thus we find

$$F(a, b; c; 1) = \frac{(c-a)(c-b)}{c(c-a-b)} F(a, b; c+1; 1),$$

and by induction

$$F(a, b; c; 1) = \frac{(c-a)_m (c-b)_m}{(c)_m (c-a-b)_m} F(a, b; c+m; 1), m = 1, 2, \ldots.$$

Now if $m \to \infty$, then by Theorem 8.6a

$$\frac{(c-a)_m (c-b)_m}{(c)_m (c-a-b)_m} \to \frac{\Gamma(c)\Gamma(c-a-b)}{\Gamma(c-a)\Gamma(c-b)},$$

and it remains to show that $F(a, b; c+m; 1) \to 1$.

Let the coefficient of z^n in $F(a, b; c; z)$ be denoted by $v_n(a, b, c)$. Then

$$|F(a, b; c+m; 1) - 1| \leq \sum_{n=1}^{\infty} |v_n(a, b, c+m)|.$$

Now if $m > |c|$,

$$|v_n(a, b, c+m)| \leq v_n(|a|, |b|, m-|c|).$$

Hence

$$|F(a, b; c+m; 1) - 1| \leq \sum_{n=1}^{\infty} v_n(|a|, |b|, m-|c|)$$

$$= \frac{|a||b|}{m-|c|} \sum_{n=0}^{\infty} v_n(|a|+1, |b|+1, m-|c|+1)$$

$$= \frac{|a||b|}{m-|c|} {}_2F_1(|a|+1, |b|+1; m-|c|+1; 1).$$

The last series converges when $m > |c| + |a| + |b| + 1$ and, because its terms are getting smaller when m increases, its sum is a decreasing function of m. Thus the required relation

$$\lim_{m \to \infty} F(a, b; c+m; 1) = 1$$

follows, and Theorem 8.6b is proved. ■

In the special case $b = -n$ (n a nonnegative integer) the hypergeometric series in (8.6-6) terminates, and in view of $\Gamma(c+n)/\Gamma(c) = (c)_n$ we obtain the so-called **Vandermonde formula**

$$F(a, -n; c; 1) = \frac{(c-a)_n}{(c)_n}$$

already proved in §1.4.

Gauss' formula can be used to sum the series $F(a, b; c; z)$ for certain $z \neq 1$ if the parameters satisfy additional conditions. In problem 2, §1.6, we have obtained the formal result

$$F(a, b; 1+a-b; z) = (1-z)^{-a} F\left(\frac{a}{2}, \frac{1+a}{2} - b; 1+a-b; -\frac{4z}{(1-z)^2}\right)$$

$$(8.6\text{-}7)$$

Analytically, this is valid whenever the arguments of the two series have modulus <1, that is, in the interior of the inner loop of the curve defined by $|4z| = |1-z|^2$. Letting $z \to -1+0$, the argument of the series on the right tends to 1. The series still converges and can be summed by Theorem 8.6b. The limit thus equals

$$2^{-a} \frac{\Gamma(1+a-b)\Gamma(\tfrac{1}{2})}{\Gamma(1+a/2-b)\Gamma(a/2+1/2)} = \frac{\Gamma(1+a-b)\Gamma(1+a/2)}{\Gamma(1+a/2-b)\Gamma(1+a)},$$

by an application of the Legendre duplication formula. The series on the left of (8.6-7) converges for $z = -1$ if Re $b < 1$ (conditionally, unless Re $b < \tfrac{1}{2}$). We thus get

THEOREM 8.6c (Kummer's formula)

If Re $b < 1$, *and if* $a - b$ *is not a negative integer, then*

$$F(a, b; 1+a-b; -1) = \frac{\Gamma(1+a-b)\Gamma(1+a/2)}{\Gamma(1+a/2-b)\Gamma(1+a)}. \qquad (8.6\text{-}8)$$

If either a or b is a negative integer, the result is given in problem 15, §1.6. We prove one further result of this type.

THEOREM 8.6d (Dixon's formula)

If Re$(b+c) < 1 + $Re $a/2$, *and if neither* $b - a$ *nor* $c - a$ *is a positive integer, then*

$$_3F_2\left[\begin{array}{c} a, b, c; 1 \\ 1+a-b, 1+a-c \end{array}\right]$$

$$(8.6\text{-}9)$$

$$= \frac{\Gamma(1+a/2)}{\Gamma(1+a)} \frac{\Gamma(1+a-b)}{\Gamma(1+a/2-b)} \frac{\Gamma(1+a-c)}{\Gamma(1+a/2-c)} \frac{\Gamma(1+a/2-b-c)}{\Gamma(1+a-b-c)}.$$

Proof. The proof given below, due to G. N. Watson, is based on an ingenious rearrangement of a double series. The conditions given in the theorem are just sufficient for the absolute convergence of the series. For the proof we temporarily suppose that, in addition, the parameters a, b, c are real and that

$$a > 0, \qquad 0 < b < \tfrac{1}{2}, \qquad 0 < c < \tfrac{1}{2}. \qquad (8.6\text{-}10)$$

By the definition of a hypergeometric series,

$$s := \frac{\Gamma(a)\Gamma(b)\Gamma(c)}{\Gamma(1+a-b)\Gamma(1+a-c)} {}_3F_2\left[\begin{matrix} a, b, c; 1 \\ 1+a-b, 1+a-c \end{matrix}\right]$$

$$= \sum_{n=0}^{\infty} \frac{\Gamma(a+n)\Gamma(b+n)\Gamma(c+n)}{n!\,\Gamma(1+a-b+n)\Gamma(1+a-c+n)}.$$

Because $b + c < 1 + a$ we have, by Gauss' theorem,

$$\frac{\Gamma(1+a+2n)\Gamma(1+a-b-c)}{\Gamma(1+a-b+n)\Gamma(1+a-c+n)} = {}_2F_1\left[\begin{matrix} b+n, c+n; 1 \\ 1+a+2n \end{matrix}\right].$$

Hence

$$s = \sum_{n=0}^{\infty} \frac{\Gamma(a+n)\Gamma(b+n)\Gamma(c+n)}{n!\,\Gamma(1+a+2n)\Gamma(1+a-b-c)} {}_2F_1\left[\begin{matrix} b+n, c+n; 1 \\ 1+a+2n \end{matrix}\right]$$

$$= \sum_{n=0}^{\infty} \sum_{m=0}^{\infty} \frac{\Gamma(a+n)\Gamma(b+n+m)\Gamma(c+n+m)}{n!\,m!\,\Gamma(1+a+2n+m)\Gamma(1+a-b-c)}.$$

This is an iterated infinite series with positive terms. Because the series is convergent in one mode of summation, we may sum its terms in any order we please, for instance diagonally (see §9.2 or Kamke [1947], p. 67). This yields, putting $n + m =: p$,

$$s = \sum_{p=0}^{\infty} \sum_{n=0}^{\infty} \frac{\Gamma(a+n)\Gamma(b+p)\Gamma(c+p)}{n!\,(p-n)!\,\Gamma(1+a+n+p)\Gamma(1+a-b-c)}.$$

Using $1/(p-n)! = (-1)^n (-p)_n/p!$, this becomes, using Kummer's Theorem 8.6c,

$$s = \sum_{p=0}^{\infty} \frac{\Gamma(b+p)\Gamma(c+p)}{p!\,\Gamma(1+a-b-c)} \frac{\Gamma(a)}{\Gamma(1+a+p)} {}_2F_1\left[\begin{matrix} a, -p; -1 \\ 1+a+p \end{matrix}\right]$$

$$= \sum_{p=0}^{\infty} \frac{\Gamma(b+p)\Gamma(c+p)}{p!\,\Gamma(1+a-b-c)} \frac{\Gamma(a)\Gamma(1+a/2)}{\Gamma(1+a/2+p)\Gamma(1+a)}.$$

The last series is a multiple of an ordinary hypergeometric series of argument 1 that can be summed by the theorem of Gauss, yielding

$$s = \frac{\Gamma(a)}{\Gamma(1+a)} \frac{\Gamma(b)\Gamma(c)}{\Gamma(1+a-b-c)} {}_2F_1\left[\begin{matrix} b, c; 1 \\ 1+a/2 \end{matrix}\right]$$

$$= \frac{\Gamma(a)}{\Gamma(1+a)} \frac{\Gamma(b)\Gamma(c)}{\Gamma(1+a-b-c)} \frac{\Gamma(1+a/2)\Gamma(1+a/2-b-c)}{\Gamma(1+a/2-b)\Gamma(1+a/2-c)},$$

which is equivalent to (8.6-9). The result is thus proved under the restrictions (8.6-10).

It remains to extend the result into the domain $D := \{(a, b, c): \mathrm{Re}(b+c)< 1+\mathrm{Re}\, a/2,\, b-a \neq k,\, c-a \neq k,\, k = 1, 2, \ldots\}$ of the complex (a, b, c) space. Every term of the series clearly is an analytic function of (a, b, c) in D. Moreover, Stirling's formula shows that the series converges uniformly on every closed subset $D_\epsilon := \{(a, b, c): \mathrm{Re}(b+c) \leqslant 1+\mathrm{Re}\, a/2-\epsilon,\, |b-a-k| \geqslant \epsilon,\, |c-a-k| \geqslant \epsilon,\, k = 1, 2, \ldots\}$ of D $(\epsilon > 0)$. Hence its sum is analytic in D. On the subset of D described by (8.6-10) this function agrees with the expression on the right of (8.6-9), which is likewise analytic in D. Hence[4] the agreement persists throughout D. ∎

Finally we shall prove **Dougall's formula:**

THEOREM 8.6e

If a and b are not integers, and if $1+\mathrm{Re}(a+b)<\mathrm{Re}(c+d)$, *then*

$$\sum_{n=-\infty}^{\infty} \frac{\Gamma(a+n)\Gamma(b+n)}{\Gamma(c+n)\Gamma(d+n)} = \frac{\pi^2}{\sin \pi a \sin \pi b}\frac{\Gamma(c+d-a-b-1)}{\Gamma(c-a)\Gamma(d-a)\Gamma(c-b)\Gamma(d-b)}.$$

$$(8.6\text{-}11)$$

Proof. Stirling's formula in combination with (8.4-12) shows that the doubly infinite series on the left converges properly (and not merely as a principal value). Because $\pi \cot \pi z$ is a summatory function with residues $+1$, the sum of the series equals the sum of the residues of

$$f(z) := \pi \cot \pi z \frac{\Gamma(a+z)\Gamma(b+z)}{\Gamma(c+z)\Gamma(d+z)}$$

at the poles $z = 0, \pm 1, \pm 2, \ldots$ of $\pi \cot \pi z$. For $|z|$ large, $\mathrm{Re}\, z \geqslant 0$,

$$\frac{\Gamma(a+z)\Gamma(b+z)}{\Gamma(c+z)\Gamma(d+z)} \sim z^{a+b-c-d};$$

if $\mathrm{Re}\, z \leqslant 0$,

$$\frac{\Gamma(z+a)\Gamma(z+b)}{\Gamma(c+z)\Gamma(d+z)} \sim (-z)^{a+b-c-d}\frac{\sin \pi(z+c) \sin \pi(z+d)}{\sin \pi(z+a) \sin \pi(z+b)}.$$

Let ξ be a real number not congruent mod 1 to any of the poles of the function f. We integrate f along the square with vertical sides $\mathrm{Re}\, z = \xi \pm n$ and horizontal sides $\mathrm{Im}\, z = \pm n$, where n is an integer destined to tend to ∞.

[4] Here we use the basic facts on analytic continuation of functions of several complex variables, which are analogous to the corresponding facts about functions of one variable given in Chapter 3. See also Chapter 17.

On this square the absolute value of

$$\pi \cot \pi z \frac{\sin \pi(z+c) \sin \pi(z+d)}{\sin \pi(z+a) \sin \pi(z+b)}$$

is bounded, and the bound is independent of n. The integral

$$\int |z^{a+b-c-d}| \, |dz|$$

taken along the square tends to zero because of the condition $\text{Re}(a+b-c-d) < -1$. It follows that the total sum of the residues of f is zero, and hence that the value of the sum (8.6-11) equals the negative of the sum of the residues of f arising from the poles of $\Gamma(z+a)$ and $\Gamma(z+b)$. The residue of f at the pole $z = -a - m$ $(m = 0, 1, 2, \ldots)$ is

$$-\pi \cot \pi a \, \frac{(-1)^m}{m!} \, \frac{\Gamma(b-a-m)}{\Gamma(c-a-m)\Gamma(d-a-m)}.$$

By (8.4-10) we have, for instance,

$$\Gamma(b-a-m) = \frac{\Gamma(b-a)}{(b-a-m)_m}$$

$$= \frac{(-1)^m \Gamma(b-a)}{(a-b+1)_m}$$

and consequently the sum of all residues by Gauss' theorem (applicable here because the conditions of convergence are satisfied) equals

$$-\pi \cot \pi a \frac{\Gamma(b-a)}{\Gamma(c-a)\Gamma(d-a)} {}_2F_1 \left[\begin{matrix} a-c+1, \, a-d+1; \, 1 \\ a-b+1 \end{matrix} \right]$$

$$= -\pi \cot \pi a \frac{\Gamma(b-a)}{\Gamma(c-a)\Gamma(d-a)} \frac{\Gamma(a-b+1)\Gamma(c+d-a-b-1)}{\Gamma(c-b)\Gamma(d-b)}$$

$$= \pi^2 \frac{\cot \pi a}{\sin \pi(a-b)} \frac{\Gamma(c+d-a-b-1)}{\Gamma(c-a)\Gamma(d-a)\Gamma(c-b)\Gamma(d-b)}.$$

The residues arising from the poles at $z = -b - m$ can be obtained from the foregoing formula by interchanging a and b. The quotient of Γ functions is unchanged, and in view of

$$\frac{\cot \pi a}{\sin \pi(a-b)} + \frac{\cot \pi b}{\sin \pi(b-a)} = -\frac{1}{\sin \pi a \sin \pi b}$$

the negative of the sum of residues is just

$$\frac{\pi^2}{\sin \pi a \sin \pi b} \frac{\Gamma(c+d-a-b-1)}{\Gamma(c-a)\Gamma(c-b)\Gamma(d-a)\Gamma(d-b)},$$

as stated. ∎

PROBLEMS

1.   Express the following series in closed form, subject to convergence conditions:

(a) $\displaystyle\sum_{n=0}^{\infty} \binom{\alpha}{n}^2,$ (b) $\displaystyle\sum_{n=0}^{\infty} (-1)^n \binom{\alpha}{n}^2,$

(c) $\displaystyle\sum_{n=0}^{\infty} \binom{\alpha}{n}\binom{\beta}{n},$ (d) $\displaystyle\sum_{n=0}^{\infty} (-1)^n \binom{\alpha}{n}^3.$

2.   Using the identity (1.6-11),

$$(1-z)^{-a}{}_2F_1\left[\begin{matrix} a, b; & -\dfrac{z}{1-z} \\ c \end{matrix}\right] = {}_2F_1\left[\begin{matrix} a, c-b; z \\ c \end{matrix}\right],$$

prove the formulas

(a) $\displaystyle {}_2F_1\left[\begin{matrix} a, b; \frac{1}{2} \\ \dfrac{a+b+1}{2} \end{matrix}\right] = \frac{\Gamma(\frac{1}{2})\Gamma((1+a+b)/2)}{\Gamma((1+a)/2)\Gamma((1+b)/2)},$

(b) $\displaystyle {}_2F_1\left[\begin{matrix} a, 1-a; \frac{1}{2} \\ c \end{matrix}\right] = \frac{\Gamma(c/2)\Gamma((c+1)/2)}{\Gamma((c+a)/2)\Gamma((1+c-a)/2)}$

3.   Prove that

$$\sum_{n=-\infty}^{\infty} \frac{\Gamma(a+n)}{\Gamma(b+n)} = 0 \ (\mathrm{Re}\ a < \mathrm{Re}\ b - 1)$$

(a) as a special case of Dougall's formula; (b) directly by the calculus of residues.

4.   Prove:

$$\left(\frac{3\cdot 3}{1\cdot 5}\right)^4 \left(\frac{15\cdot 15}{13\cdot 17}\right)^4 \left(\frac{27\cdot 27}{25\cdot 29}\right)^4 \left(\frac{39\cdot 39}{37\cdot 41}\right)^4 \cdots = 12.$$

5.   Find the value of the product

$$\frac{3^2}{3^2-1}\frac{5^2-1}{5^2}\frac{7^2}{7^2-1}\frac{9^2-1}{9^2}\cdots.$$

6. Prove the following identities, subject to convergence conditions:

(a) $\quad 1+\left(\dfrac{x-1}{x+1}\right)^2+\left(\dfrac{(x-1)(x-2)}{(x+1)(x+2)}\right)^2+\cdots=\dfrac{2x}{4x-1}\dfrac{[\Gamma(x+1)]^4\Gamma(4x+1)}{[\Gamma(2x+1)]^4}$,

(b) $\quad 1-\dfrac{1}{3}\dfrac{x-1}{x+1}+\dfrac{1}{5}\dfrac{(x-1)(x-2)}{(x+1)(x+2)}-\cdots=\dfrac{2^{4x}[\Gamma(x+1)]^4}{4x[\Gamma(2x+1)]^2}$,

(c) $\quad 1-3\dfrac{x-1}{x+1}+5\dfrac{(x-1)(x-2)}{(x+1)(x+2)}-\cdots=0.$

7. Prove Lerch's formula, valid for $\mathrm{Re}(u-v)>1$:

$$\sum_{n=0}^{\infty}\frac{\dbinom{u}{n}}{\dbinom{v}{n}}=\frac{v+1}{v-u+1}.$$

§8.7. THE BETA FUNCTION

The integral

$$B(p,q):=\int_0^1 \tau^{p-1}(1-\tau)^{q-1}\,d\tau,\qquad(8.7\text{-}1)$$

where $\mathrm{Re}\,p>0$, $\mathrm{Re}\,q>0$, and where τ^{p-1} and $(1-\tau)^{q-1}$ have their principal values, is known as the **beta function** of Euler and occurs frequently in applications. The beta integral can easily be evaluated in terms of Γ functions. We assume temporarily that $\mathrm{Re}\,p>1$, $\mathrm{Re}\,q>1$. The binomial series

$$(1-\tau)^{q-1}=\sum_{n=0}^{\infty}\frac{(1-q)_n}{n!}\tau^n$$

then converges uniformly for $0\le\tau\le1$. Substituting into the integral and interchanging summation and integration, we find

$$B(p,q)=\sum_{n=0}^{\infty}\frac{(1-q)_n}{n!}\int_0^1\tau^{p+n-1}\,d\tau=\sum_{n=0}^{\infty}\frac{(1-q)_n}{n!}\frac{1}{p+n}$$

$$=\frac{1}{p}\sum_{n=0}^{\infty}\frac{(1-q)_n}{n!}\frac{(p)_n}{(p+1)_n}=\frac{1}{p}{}_2F_1\!\left[\begin{array}{c}1-q,p;\,1\\p+1\end{array}\right]$$

$$=\frac{1}{p}\frac{\Gamma(p+1)\Gamma(q)}{\Gamma(p+q)\Gamma(1)}=\frac{\Gamma(p)\Gamma(q)}{\Gamma(p+q)},$$

by Gauss' theorem 8.6b. Like $B(p,q)$, the function on the right is, in each of the variables p and q, analytic in the half plane of positive real parts. We thus have

THEOREM 8.7a.

For $\operatorname{Re} p > 0$, $\operatorname{Re} q > 0$,

$$B(p, q) := \int_0^1 \tau^{p-1}(1-\tau)^{q-1} \, d\tau = \frac{\Gamma(p)\Gamma(q)}{\Gamma(p+q)}. \qquad (8.7\text{-}2)$$

Some other forms of the beta function can be obtained by simple substitutions.

$$B(p, q) = 2 \int_0^{\pi/2} (\sin \phi)^{2p-1}(\cos \phi)^{2q-1} \, d\phi \quad (\tau = \sin^2 \phi) \qquad (8.7\text{-}3)$$

$$B(p, q) = \int_0^\infty \xi^{p-1}(1+\xi)^{-p-q} \, d\xi \left(\tau = \frac{\xi}{1+\xi} \right) \qquad (8.7\text{-}4)$$

$$B(p, q) = \int_0^1 \frac{\xi^{p-1} + \xi^{q-1}}{(1+\xi)^{p+q}} \, d\xi \left(\xi \to \frac{1}{\xi} \right). \qquad (8.7\text{-}5)$$

As an application of the beta integral we now derive an integral representation for the hypergeometric function. If $c \neq 0, -1, -2, \ldots$, and if $|z| < 1$,

$$\begin{aligned}
{}_2F_1(a, b; c; z) &= \sum_{n=0}^\infty \frac{(a)_n(b)_n}{(c)_n n!} z^n \\
&= \frac{\Gamma(c)}{\Gamma(a)\Gamma(b)} \sum_{n=0}^\infty \frac{\Gamma(a+n)\Gamma(b+n)}{\Gamma(c+n)n!} z^n.
\end{aligned}$$

If $\operatorname{Re} c > \operatorname{Re} b > 0$, then

$$\frac{\Gamma(b+n)\Gamma(c-b)}{\Gamma(c+n)} = B(b+n, c-b)$$

and thus

$${}_2F_1(a, b; c; z) = \frac{\Gamma(c)}{\Gamma(a)\Gamma(b)\Gamma(c-b)} \sum_{n=0}^\infty \frac{\Gamma(a+n)}{n!} z^n \int_0^1 \tau^{b+n-1}(1-\tau)^{c-b-1} \, d\tau.$$

If $|z| < 1$, the series

$$\sum_{n=0}^\infty \frac{\Gamma(a+n)}{n!} z^n \tau^n$$

converges uniformly to $\Gamma(a)(1 - z\tau)^{-a}$ on the range of integration. We thus may interchange integration and summation to obtain

$${}_2F_1(a, b; c; z) = \frac{\Gamma(c)}{\Gamma(b)\Gamma(c-b)} \int_0^1 \tau^{b-1}(1-\tau)^{c-b-1}(1-\tau z)^{-a} \, d\tau,$$

where $(1 - \tau z)^{-a}$ has its principal value. The result so far has been proved for $|z| < 1$. But the integral on the right has a meaning for any z in the complex plane cut along the real axis from $+1$ to $+\infty$, and it converges uniformly with respect to z in any compact subset of the plane thus cut. It therefore represents an analytic function in the cut plane. Because that function coincides with $_2F_1(a, b; c; z)$ for $|z| < 1$, it represents the analytic continuation of that function into the cut plane. We call that continuation the hypergeometric *function* (as opposed to the hypergeometric *series*, which is defined only for $|z| < 1$) and denote it by the symbol $F(a, b; c; z)$. We then have

THEOREM 8.7b.

For $\operatorname{Re} c > \operatorname{Re} b > 0$, *the hypergeometric function possesses the representation*

$$F(a, b; c; z) = \frac{\Gamma(c)}{\Gamma(b)\Gamma(c - b)} \int_0^1 \tau^{b-1}(1 - \tau)^{c-b-1}(1 - \tau z)^{-a} \, d\tau,$$

(8.7-6)

valid for all z *that are not real and* $\geqslant 1$.

It is a drawback of both representations given in the foregoing theorems that they are valid only under severe restrictions on the parameters. For the beta integral this drawback can be removed by a device due to Pochhammer (1890). In the complex t plane we consider a closed (but not simple) path Λ that originates, say, at $t = \frac{1}{2}$ and encircles each of the points $t = 1$ and $t = 0$ twice, first in the positive, then in the negative sense (see Fig. 8.7a).[5] It is customary to denote such a path symbolically by $(1+, 0+, 1-, 0-)$. The winding number (see §4.6) of Λ with respect to both $t = 0$ and $t = 1$ is zero. It follows that a continuous argument exists on Λ, and hence that $\log t$ can be defined as a locally analytic function on Λ, for instance by assigning it the principal value for $t > 1$. With this determination of $\log t$, $t^{p-1} := e^{(p-1)\log t}$ and $(1 - t)^{q-1} = e^{(q-1)\log(1-t)}$ can likewise be defined as locally analytic functions. Thus the integral

$$I(p, q) := \int_\Lambda t^{p-1}(1 - t)^{q-1} \, dt$$

has a meaning for arbitrary (complex) values of p and q, and by Cauchy's theorem is independent of Λ as long as Λ stays clear of the points 0 and 1. To evaluate $I(p, q)$, we temporarily assume $p > 1$, $q > 1$ and contract Λ to a path

---

[5] Such a path is given, for instance, by

$$t = \left\{ \cosh\left[\sin\left(\phi + \frac{\pi}{4}\right) + i\phi \right] \right\}^2, \quad 0 \leqslant \phi \leqslant 2\pi.$$

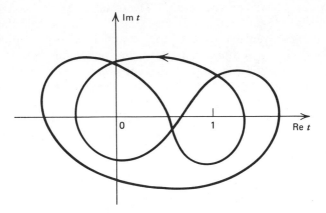

Fig. 8.7a.

consisting of four straight-line segments joining 0 and 1 and two pairs of small circles, one pair surrounding 0, the other surrounding 1. Schematically this path is indicated in Fig. 8.7*b*.

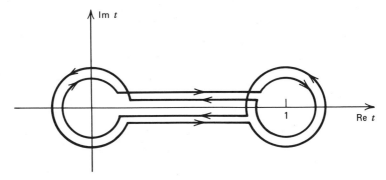

Fig. 8.7b.

As the radii of the circles tend to zero, the contributions of the circles vanish, and the integral reduces to the sum of the four integrals along the straight line segments. Each of these integrals equals $B(p, q)$ up to a factor of absolute value 1 because of the direction of integration and the momentary value of arg $\{t^{p-1}(1-t)^{q-1}\}$. Following the changes of the argument, we find

$$
\begin{aligned}
I(p, q) &= \{1 - e^{2\pi i q} + e^{2\pi i (p+q)} - e^{2\pi i p}\} B(p, q) \\
&= (1 - e^{2\pi i p})(1 - e^{2\pi i q}) B(p, q).
\end{aligned}
\tag{8.7-7}
$$

If neither p nor q is an integer, we can solve for $B(p, q)$. Because I clearly is an entire analytic function of p and of q, we obtain **Pochhammer's integral**,

THEOREM 8.7c.

For all complex p and q that are not integers,

$$B(p, q) = -\frac{e^{i\pi(p+q)}}{4 \sin \pi p \sin \pi q} \int_{(1+, 0+, 1-, 0-)} t^{p-1}(1-t)^{q-1} \, dt. \quad (8.7\text{-}8)$$

Pochhammer's integral may be cast into different forms by transformations of the variable. If, for instance, we set

$$t = \frac{s}{1+s}, \qquad dt = \frac{1}{(1+s)^2} \, ds,$$

we get

$$I(p, q) = \int_{\Lambda_s} s^{p-1}(1+s)^{-p-q} \, ds,$$

where the closed curve Λ_s winds once positively and once negatively around each of the points $s = 0$ and $s = -1$.

Setting further

$$s = e^{2iu}, \qquad ds = 2i \, e^{2iu} \, du,$$

we obtain

$$I(p, q) = 2^{1-p-q} \int_{\Lambda_u} e^{i(p-q)u}(\cos u)^{-p-q} \, du,$$

where Λ_u is a figure-eight-shaped loop having winding number $+1$ with respect to $\pi/2$ and -1 with respect to $-(\pi/2)$, see Fig. 8.7c. If $p + q < 1$, the

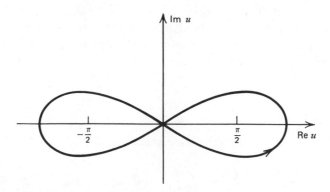

Fig. 8.7c.

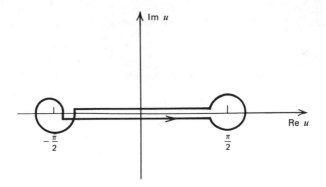

Fig. 8.7d.

loop may be contracted to a curve consisting of two straight-line segments joining $-(\pi/2)$ and $\pi/2$, and two small circles around $\pi/2$ and $-(\pi/2)$, as shown in Fig. 8.7d. The contributions of the circles tend to zero with the radii, and the integrals along the line segments together yield

$$2^{1-p-q}i(1-e^{-2\pi i(p+q)})\int_{-(\pi/2)}^{\pi/2} e^{i(p-q)u}(\cos u)^{-p-q}\,du.$$

The integral of the imaginary part is zero. Expressing I in terms of $B(p, q)$, we thus find

$$\int_0^{\pi/2}\cos[(p-q)u](\cos u)^{p+q}\,du = \frac{\pi}{2^{p+q+1}}\frac{\Gamma(1+p+q)}{\Gamma(1+p)\Gamma(1+q)}. \quad (8.7\text{-}9)$$

Loop integrals of the Pochhammer type, valid for less-restricted values of the parameters than (8.7-6), can also be obtained for the hypergeometric function; see Problem 6. Other representations of the hypergeometric function are obtained in §8.8.

PROBLEMS

1.   Compute the *volume of the n-dimensional unit ball,*

$$\omega_n := \int \cdots \int_{x_1^2+x_2^2+\cdots+x_n^2\leqslant 1} dx_1\,dx_2\cdots dx_n,$$

introducing n-dimensional polar coordinates $(\rho, \phi, \theta_1, \ldots, \theta_{n-2})$ where

$$x_n = \rho \cos \theta_{n-2},$$
$$x_{n-1} = \rho \sin \theta_{n-2} \cos \theta_{n-3},$$
$$x_{n-2} = \rho \sin \theta_{n-2} \sin \theta_{n-3} \cos \theta_{n-4},$$
$$\cdot \quad \cdot \quad \cdot \quad \cdot$$
$$x_3 = \rho \sin \theta_{n-2} \cdots \sin \theta_2 \cos \theta_1,$$
$$x_2 = \rho \sin \theta_{n-2} \cdots \sin \theta_1 \cos \phi,$$
$$x_1 = \rho \sin \theta_{n-2} \cdots \sin \theta_1 \sin \phi$$

$(0 \leqslant \phi < 2\pi, 0 \leqslant \theta_i < \pi, i = 1, \ldots, n-2)$ and using the transformation formula for definite integrals,

$$\omega_n = \int \cdots \int \frac{\partial(x_1, x_2, \ldots, x_n)}{\partial(\rho, \theta_1, \ldots, \theta_{n-2}, \phi)} \, d\rho \, d\theta_1 \cdots d\theta_{n-2} \, d\phi$$

involving the Jacobian determinant.
Result:

$$\omega_n = \frac{\pi^{n/2}}{\Gamma(1+n/2)}.$$

2. (Continuation) Let σ_n denote the *surface area* of the n-dimensional unit ball. Show that

$$\sigma_n = \frac{2\pi^{n/2}}{\Gamma(n/2)},$$

and conclude that

$$\lim_{n \to \infty} \frac{\omega_n}{\sigma_n} = 0.$$

[Volume and surface area of the n-dimensional ball of radius ρ clearly equal

$$\omega_n(\rho) = \rho^n \omega_n, \qquad \sigma_n(\rho) = \rho^{n-1}\sigma_n.$$

On the other hand, $\omega_n(\rho) = \int_0^\rho \sigma_n(\rho) \, d\rho$; hence

$$\sigma_n(\rho) = \frac{d}{d\rho} \omega_n(\rho) = n\rho^{n-1}\omega_n.] \ .$$

3. Prove Ramanujan's result, valid for $|z| < \pi/2$,

$$\int_0^\infty \frac{\cosh(2z\tau)}{\cosh(\pi\tau)} \, d\tau = \frac{1}{2 \cos z}.$$

4. Establish (8.7-9) directly by integrating $(z + z^{-1})^{p+q} z^{p-q-1}$ along the closed curve consisting of the right half of the unit circle and of the straight-line segment joining the points $\pm i$, indented at the origin.

5. Use Theorem 8.7b to establish (8.6-8) and the identities of Problem 2, §8.6.
6. Use Pochhammer's integral to obtain the representation

$$F(a, b; c; z) = -\frac{\Gamma(c)\, e^{-i\pi c}}{4\Gamma(b)\Gamma(c-b)\sin \pi b \sin \pi(c-b)}$$

$$\times \int_{(1+, 0+, 1-, 0-)} t^{b-1}(1-t)^{c-b-1}(1-tz)^{-a}\, dt,$$

$$b, c - b \neq 1, 2, \ldots.$$

7. Show that the **incomplete beta function,**

$$B_x(p, q) := \int_0^x \tau^{p-1}(1-\tau)^{q-1}\, d\tau$$

$(0 \leq x < 1)$ can be expressed in the form

$$B_x(p, q) = \frac{x^p}{p} F(p, 1-q; 1+p; x).$$

8. The numbers

$$\alpha := \int_0^1 (1-\tau^4)^{-1/2}\, d\tau, \qquad \beta := \int_0^1 \tau^2(1-\tau^4)^{-1/2}\, d\tau$$

are known as the **lemniscate constants.** Show that

$$\alpha = \frac{1}{4}\frac{1}{\sqrt{2\pi}}\left(\Gamma\!\left(\frac{1}{4}\right)\right)^2, \qquad \beta = \frac{1}{\sqrt{2\pi}}\left(\Gamma\!\left(\frac{3}{4}\right)\right)^2,$$

and deduce that

$$\alpha\beta = \frac{\pi}{4}.$$

§8.8. INTEGRALS OF THE MELLIN–BARNES TYPE

In this section we deal with a type of integral in which the integrand involves products of Γ functions, and in which the path of integration runs along the imaginary axis, with certain modifications to satisfy additional conditions. A typical **Mellin–Barnes integral** is as follows:

$$\frac{1}{2\pi i}\int_{-i\infty}^{i\infty} \frac{\Gamma(a+s)\Gamma(b+s)\Gamma(-s)}{\Gamma(c+s)}(-z)^s\, ds.$$

Up to a factor independent of z, this is later shown to be identical with the hypergeometric function $F(a, b; c; z)$. Integrals of this general form have proved to be extremely flexible tools for the study of the deeper properties of all kinds of hypergeometric functions, including their analytic continuation and their asymptotic behavior.

For a preliminary exploration we consider the function

$$f(s) := \Gamma(a+s)\Gamma(-s)z^s,$$

where $|\arg z| < \pi$, and where a is not a negative integer. The function f has two chains of poles: a *descending chain*, arising from the first Γ factor, at the points $s = -a - n$ ($n = 0, 1, 2, \ldots$), and an *ascending chain*, arising from the second Γ factor, at $s = n$ ($n = 0, 1, 2, \ldots$). We now integrate f along a closed curve Λ_m resembling a rectangle with the four corners $\pm im$, $m + \frac{1}{2} \pm im$, where m is a (large) integer. The horizontal sides and the right side of the rectangle are straight-line segments, but the left side is deformed in such a way that it separates the ascending and the descending chain of poles (see Fig. 8.8a). If a is not zero or a negative integer, it is evident that the curve Λ_m can always be chosen in this manner. We suppose that for sufficiently large values of $|\text{Im } s|$ the left part of Λ_m always coincides exactly with the imaginary axis. Thus the curvitious part of Λ_m shall not change with increasing m.

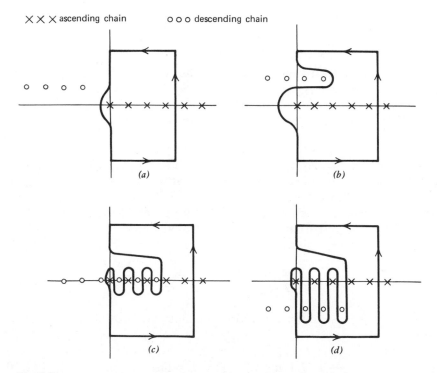

Fig. 8.8a.

By the residue theorem, $\int_{\Lambda_m} f$ equals $2\pi i$ times the sum of the residues at the ascending chain of poles. The residue at $s = n$ equals

$$(-1)^{n+1}\frac{\Gamma(a+n)}{n!}z^n,$$

by (8.4-15). We thus have

$$\frac{1}{2\pi i}\int_{\Lambda_m} f(s)\,ds = -\Gamma(a)\sum_{n=0}^{m}\frac{(a)_n}{n!}(-z)^n. \tag{8.8-1}$$

With a view toward letting $m \to \infty$, we now investigate the behavior of $f(s)$ for faraway points s on Λ_m. To this end we write

$$f(s) = \frac{z^s}{\sin \pi s}\frac{\Gamma(a+s)}{\Gamma(1+s)}.$$

By Stirling's formula, if $a - 1 =: \alpha + i\beta$,

$$\left|\frac{\Gamma(a+s)}{\Gamma(1+s)}\right| \sim |s|^\alpha\,e^{-\beta\,\arg s}.$$

Furthermore,

$$|z^s| = |z|^{\mathrm{Re}\,s}\,e^{-\arg z\,\mathrm{Im}\,s}$$

and, because Λ_m passes halfway between the positive integers, $|1/\sin \pi s| \leqslant e^{-\pi|\mathrm{Im}\,s|}$. On the faraway part of Λ_m, $|\arg s| \leqslant \pi/2$. We thus have, altogether,

$$|f(s)| \leqslant |z|^{\mathrm{Re}\,s}\,e^{-(\pi-|\arg z|)|\mathrm{Im}\,s|}|s|^\alpha \cdot \pi\,e^{|\beta|\pi/2}\,\mathrm{const}. \tag{8.8-2}$$

on the whole of Λ_m, with the exception of the wiggly part.

We now suppose that $|z| < 1$. On the right side of Λ_m, $|f(s)|$ is bounded by $\mathrm{const}\,|z|^m m^\alpha$; hence the integral along this part tends to zero as $m \to \infty$. On the top and bottom parts of the curve, $|f(s)| \leqslant \mathrm{const}\,m^\alpha\,e^{(|\arg z|-\pi)m}$, and again the integrals vanish for $m \to \infty$. The integral along the imaginary axis, including the wiggly part of the curve, exists as an improper integral (and not merely as a principal value integral). The sum of the residues in (8.8-1) tends to

$$-\Gamma(a)\sum_{n=0}^{\infty}\frac{(a)_n}{n!}(-z)^n = -\Gamma(a)(1+z)^{-a}.$$

Reversing the sense of the integration, we thus have proved that for $|z| < 1$, $|\arg z| < \pi$,

$$\frac{1}{2\pi i}\int_{-i\infty}^{i\infty}\Gamma(a+s)\Gamma(-s)z^s\,ds = \Gamma(a)(1+z)^{-a}, \tag{8.8-3}$$

where it is always understood that the path of integration separates the two chains of poles.

We now claim that (8.8-3) actually holds under much wider conditions on z than those stated previously, for it follows from (8.8-2) (where Re $s = 0$) that the integral exists for any z such that $|\arg z| < \pi$. Moreover, it converges uniformly for all z such that $|\arg z| \leq \pi - \epsilon$, where $\epsilon > 0$. The integral thus represents a function that is analytic for $|\arg z| < \pi$. Because this function agrees with $\Gamma(a)(1+z)^{-a}$ for $|z| < 1$, it must agree with that function throughout. If we did not know how to continue the binomial series beyond $|z| = 1$, we could use the integral to compute the continuation. This is unnecessary in the present situation, but the same method applies to more complicated hypergeometric functions.

To obtain a similar integral representation for the function defined by the hypergeometric series, we consider

$$f(s) := \frac{\Gamma(a+s)\Gamma(b+s)\Gamma(-s)}{\Gamma(c+s)}(-z)^s. \tag{8.8-4}$$

We suppose that $|\arg(-z)| < \pi$, that none of the numbers a, b, c is zero or a negative integer, and also that $a - b$ is not an integer. The function f, then, has two distinct descending chains of poles, namely at the points $s = -a - n$ and $s = -b - n$ ($n = 0, 1, 2, \ldots$), and one ascending chain, at the points $s = n$ ($n = 0, 1, 2, \ldots$). We integrate f along the rectangular curve Λ_m as defined, again deformed so as to separate the descending chains from the ascending chain. If m is sufficiently large, we have by the residue theorem,

$$\frac{1}{2\pi i} \int_{\Lambda_m} f(s)\, ds = -\frac{\Gamma(a)\Gamma(b)}{\Gamma(c)} \sum_{n=0}^{m} \frac{(a)_n (b)_n}{(c)_n n!} z^n. \tag{8.8-5}$$

To determine the behavior of the integrand when $m \to \infty$, we write

$$f(s) = \frac{(-z)^s}{\sin \pi s} \frac{\Gamma(a+s)}{\Gamma(c+s)} \frac{\Gamma(b+s)}{\Gamma(1+s)}.$$

From Stirling's formula we have as before

$$\left| \frac{\Gamma(a+s)\Gamma(b+s)}{\Gamma(c+s)\Gamma(1+s)} \right| \sim |s|^\alpha e^{-\beta \arg s},$$

where

$$a + b - c - 1 =: a + i\beta.$$

As in the derivation of (8.8-2), we thus find

$$|f(s)| \leq \text{const } |z|^{\text{Re } s} e^{(|\arg(-z)|-\pi)|\text{Im } s|} |s|^\alpha \tag{8.8-6}$$

on all of Λ_m, with the exception of the wiggly part.

If we suppose that $|z| < 1$, the integrals along the horizontal sides and along the right side of Λ_m again tend to zero as $m \to \infty$, and the sum of the right of (8.8-5) becomes a hypergeometric series. We thus have proved:

$$\frac{\Gamma(a)\Gamma(b)}{\Gamma(c)} F(a, b; c; z) = \frac{1}{2\pi i} \int_{-i\infty}^{i\infty} \frac{\Gamma(a+s)\Gamma(b+s)}{\Gamma(c+s)} (-z)^s \, ds. \quad (8.8\text{-}7)$$

As before, the range of validity of this formula may be extended. It is seen from (8.8-6) (where $\mathrm{Re}\, s = 0$) that the integral on the right of (8.8-7) converges uniformly in z on every compact set where $|\arg(-z)| < \pi - \epsilon$ $(\epsilon > 0)$. It thus represents an analytic function in the region $|\arg(-z)| < \pi$, that is, in the complex plane cut along the positive real axis. This function is the continuation of the function defined by the hypergeometric series in $|z| < 1$.

We now use (8.8-7) to obtain a more explicit representation of that continuation. We subject the path of integration in (8.8-7) to one more restriction: as always, the path should separate the ascending from the descending chains of poles, but if moved left by one unit, it should cross the first poles $s = -a$ and $s = -b$ of the two descending chains. A path satisfying these conditions is denoted by $\Lambda^{(0)}$ (see Fig. 8.8b). We denote by $\Lambda^{(k)}$ the

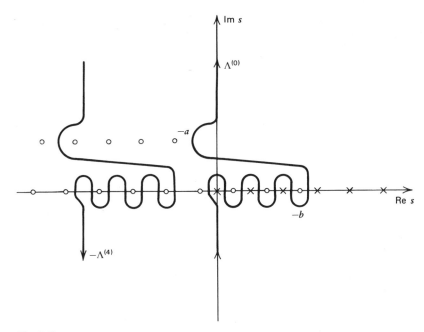

Fig. 8.8b.

path $\Lambda^{(0)}$ shifted by k units to the left. We now integrate f along the closed curve consisting of $\Lambda^{(0)}$, $-\Lambda^{(k)}$, and of two horizontal pieces Im $s = \pm\mu$, where μ is large. The contributions of the horizontal pieces are seen to vanish for $\mu \to \infty$ by (8.8-6). We thus find

$$\frac{1}{2\pi i}\int_{\Lambda^{(0)}} f - \frac{1}{2\pi i}\int_{\Lambda^{(k)}} f$$
$$= \sum_{n=0}^{k-1} \{(\operatorname{res} f)_{s=-a-n} + (\operatorname{res} f)_{s=-b-n}\}.$$

In the usual way we find

$$(\operatorname{res} f)_{s=-a-n} = (-1)^n \frac{\Gamma(a+n)\Gamma(b-a-n)}{n!\,\Gamma(c-a-n)}(-z)^{-a-n}$$

$$= (-z)^{-a}\frac{\Gamma(a)\Gamma(b-a)}{\Gamma(c-a)}\frac{(a)_n(1+a-c)_n}{n!\,(1+a-b)_n}z^{-n}.$$

A similar result, with a and b interchanged, holds for $(\operatorname{res} f)_{s=-b-n}$. Hence, by (8.8-7),

$$\frac{\Gamma(a)\Gamma(b)}{\Gamma(c)}F(a,b;c;z) = \frac{1}{2\pi i}\int_{\Lambda^{(k)}} f(s)\,ds$$

$$+ (-z)^{-a}\frac{\Gamma(a)\Gamma(b-a)}{\Gamma(c-a)}\sum_{n=0}^{k-1}\frac{(a)_n(1+a-c)_n}{n!\,(1+a-b)_n}z^{-n}$$

$$+ (-z)^{-b}\frac{\Gamma(b)\Gamma(a-b)}{\Gamma(c-b)}\sum_{n=0}^{k-1}\frac{(b)_n(1+b-c)_n}{n!\,(1+b-a)_n}z^{-n}.$$

The foregoing holds for every $k = 0, 1, 2, \ldots$. We now suppose that $|z| > 1$ and let $k \to \infty$. The sums of residues then tend to certain hypergeometric series of argument z^{-1}. To study the behavior of $\int_{\Lambda^{(k)}} f$, we note that

$$\int_{\Lambda^{(k)}} f(s)\,ds = \int_{\Lambda^{(0)}} f(s-k)\,ds.$$

In view of

$$\Gamma(z-k) = \frac{\Gamma(z)}{(z-k)_k} = (-1)^k\frac{\Gamma(z)}{(1-z)_k},$$

we have from the definition of f

$$f(s-k) = \frac{(1-c-s)_k(-s)_k}{(1-a-s)_k(1-b-s)_k}(-z)^{-k}f(s).$$

By Stirling's formula the ratio of Pochhammer symbols is for k large asymptotic to

$$\frac{\Gamma(1-a-s)\Gamma(1-b-s)}{\Gamma(1-c-s)\Gamma(-s)}k^{\alpha+i\beta} \quad (\alpha+i\beta := a+b-c-1),$$

uniformly for $s \in \Lambda^{(0)}$, and the ratio of Γ functions for $|s|$ large, $s \in \Lambda^{(0)}$ is asymptotic to const $|s|^{-\alpha}$. Using (8.8-6) we thus find that

$$\left| \int_{\Lambda^{(0)}} f(s-k)\, ds \right| \leq \mu |z|^{-k} k^{\alpha},$$

where μ is a constant independent of k. For $|z| > 1$ the expression on the right tends to zero if $k \to \infty$, and we thus have proved:

THEOREM 8.8a

If none of a, b, c is zero or a negative integer, and if $a-b$ is not an integer, then for $|\arg(-z)| < \pi$

$$F(a, b; c; z) = (-z)^{-a}\frac{\Gamma(b-a)\Gamma(c)}{\Gamma(c-a)\Gamma(b)}F\left(a, 1+a-c; 1+a-b; \frac{1}{z}\right)$$

$$+ (-z)^{-b}\frac{\Gamma(a-b)\Gamma(c)}{\Gamma(c-b)\Gamma(a)}F\left(b, 1+b-c; 1+b-a; \frac{1}{z}\right).$$

(8.8-8)

This was proved for $|z| > 1$, but holds without that restriction by virtue of analytic continuation.

The analytic continuation of any series of type $_pF_{p-1}$ beyond $|z| < 1$ can be obtained in exactly the same manner. Because there now are p descending chains of poles, p series of the form $_pF_{p-1}[\ldots; 1/z]$ will appear in the continuation formula.

We next apply the method of Mellin–Barnes integrals to the confluent hypergeometric function $_1F_1(a; c; z)$. All computations could be carried out in an analogous fashion for the function $_pF_p(a_1, \ldots, a_p; c_1, \ldots, c_p; z)$ for any integer p. By analogy it is easy to conclude that we now should integrate

$$f(s) := \frac{\Gamma(a+s)}{\Gamma(c+s)}\Gamma(-s)(-z)^s$$

along suitable closed curves. This function has one ascending and one descending chain of poles—at the points $-a-n$ and n $(n = 0, 1, 2, \ldots)$, respectively. We assume that neither a nor c are negative integers or zero, and begin by integrating along the rectangle Λ_m introduced previously, with the usual convention that the two chains should be separated. To determine the behavior of $f(s)$ for $|s|$ large, $\operatorname{Re} s \geqslant 0$, we again use the formula

$$\left|\frac{\Gamma(a+s)}{\Gamma(c+s)}\right| \sim |s|^{\alpha} e^{-\beta \arg s},$$

where, now, $a - c =: \alpha + i\beta$, and

$$|\Gamma(-s)| = \frac{\pi}{|s| \, |\sin \pi s|} \frac{1}{|\Gamma(s)|}$$

$$\sim \sqrt{\frac{\pi}{2}} \, e^{-\pi \operatorname{Im} s} |s|^{-\operatorname{Re} s - 1/2} e^{\operatorname{Re} s - \operatorname{Im} s(\pi - \arg s)}.$$

We thus find

$$|f(s)| \sim \sqrt{\frac{\pi}{2}} |s|^{\alpha - \operatorname{Re} s - 1/2} |z|^{\operatorname{Re} s} \cdot e^{[-\arg(-z) - \pi + \arg s] \operatorname{Im} s + \operatorname{Re} s}. \tag{8.8-9}$$

On the right side of Λ_m, $|s|^{-\operatorname{Re} s} \leqslant m^{-m}$. All other factors tend to ∞ at most like $e^{\operatorname{const} \cdot m}$, and it thus is clear that the integral tends to zero for $m \to \infty$. On the horizontal sides of Λ_m the integrand is bounded by

$$|s|^{\alpha - \sigma - 1/2} |z|^{\sigma} e^{\sigma} e^{[-\arg(-z) - \pi + \arg s]m},$$

where $\sigma := \operatorname{Re} s$, and the integral with respect to σ between the limits 0 and $m + \frac{1}{2}$ tends to zero for $m \to \infty$ only if $-\arg(-z) - \pi + \arg s = -\epsilon$, where $\epsilon > 0$. This can be guaranteed only if $|\arg(-z)| \leqslant \pi/2 - \epsilon$, and it thus becomes necessary to confine z to a more restricted region than in the case of the ordinary hypergeometric function.

If the foregoing conditions are met, then all integrals tend to zero except the integral along $-\Lambda^{(0)}$, in the above notation. The integral equals $2\pi i \sum (\operatorname{res} f)_{s=n}$. Now, evidently,

$$(\operatorname{res} f)_{s=n} = -\frac{\Gamma(a+n)}{\Gamma(c+n)} \frac{(-1)^n}{n!} (-z)^n.$$

Summing all these yields $_1F_1(a; b; z)$ times a Γ factor, and we have obtained:

$$\frac{\Gamma(a)}{\Gamma(c)} \, _1F_1(a; c; z) = \frac{1}{2\pi i} \int_{-i\infty}^{i\infty} \frac{\Gamma(a+s)\Gamma(-s)}{\Gamma(c+s)} (-z)^s \, ds, \tag{8.8-10}$$

with the usual separating convention about the path of integration.

We now stipulate, as before, that the path of integration is chosen such that by moving it to the left by one unit it crosses the first pole of the descending chain. Let $\Lambda^{(0)}$ be such a path, and let $\Lambda^{(k)}$ be the path obtained from it by shifting it to the left k units. If we integrate f along the closed curve defined by $\Lambda^{(0)}$, $\Lambda^{(k)}$, and two faraway horizontal links $\operatorname{Im} s = \pm \mu$, then (8.8-9) shows that the contributions of the links vanish as $\mu \to \infty$. (This is because on the faraway links $|\arg s|$ exceeds $\pi/2$ by arbitrarily little if μ is sufficiently large.) By the residue theorem we thus have

$$\int_{\Lambda^{(0)}} f - \int_{\Lambda^{(k)}} f = 2\pi i \sum_{n=0}^{k-1} (\operatorname{res} f)_{s=-a-n}$$

In view of

$$(\operatorname{res} f)_{s=-a-n} = (-1)^n \frac{\Gamma(a+n)}{n!\,\Gamma(c-a-n)}(-z)^{-a-n}$$

$$= (-z)^{-a} \frac{\sin(c-a)\pi}{\pi} \frac{\Gamma(a+n)\Gamma(1+a-c-n)}{n!} z^{-n},$$

we thus have

$$_1F_1(a;c;z) = (-z)^{-a} \frac{\Gamma(c)}{\Gamma(c-a)} \sum_{n=0}^{k-1} \frac{(a)_n(1+a-c)_n}{n!}(-z)^{-n}$$

$$+ \int_{\Lambda^{(k)}} f(s)\,ds. \tag{8.8-11}$$

It will be noted that the sum on the right is just the $(k-1)$st partial sum of the series

$$_2F_0[a, 1+a-c; -z^{-1}].$$

This series diverges for all z, and, of course, it would have been futile to hope for a convergent representation of $_1F_1$ in powers of z^{-1}, because the Laurent series at ∞ of any function defined by a power series P with radius of convergence ∞ is just P itself. However, we now show that the sum on the right of (8.8-11) serves as an *asymptotic expansion* (see §11.1) of the function on the left.

THEOREM 8.8b

Let none of $-a$, $-c$, $a-c$ be a nonnegative integer, and let $0 \leq \epsilon < \pi/2$. Then for every $k = 0, 1, 2, \ldots$, if $|\arg(-z)| \leq \pi/2 - \epsilon$,

$$\lim_{z \to \infty} (-z)^{a+k-1} \Big[_1F_1(a;c;z)$$

$$- (-z)^{-a} \frac{\Gamma(c)}{\Gamma(c-a)} \sum_{n=0}^{k-1} \frac{(a)_n(1+a-c)_n}{n!}(-z)^{-n} \Big] = 0. \tag{8.8-12}$$

In a technical notation the statement of Theorem 8.8b is written

$$_1F_1(a;c;z)$$

$$\approx (-z)^{-a}\frac{\Gamma(c)}{\Gamma(c-a)}{}_2F_0(a,1+a-c;-z^{-1}),\, z\to\infty,\, |\arg(-z)|\leqslant\frac{\pi}{2}-\epsilon;$$

$$(8.8\text{-}13)$$

it means that the error committed by replacing $_1F_1(a;c;z)$ by the $(k-1)$st partial sum of the series on the right tends to zero as $z\to\infty$, even if multiplied by $(-z)^{a+k-1}$.

Proof. By (8.8-11) the bracketed expression in (8.8-12) equals $\int_{\Lambda^{(k)}} f$. It thus is only necessary to show that

$$\left|(-z)^{a+k-1}\int_{\Lambda^{(k)}} f(s)\, ds\right|\to 0$$

as $|z|\to\infty$, $|\arg(-z)|\leqslant\pi/2-\epsilon$.

Let k be such that $\mathrm{Re}(-a-k+1)<0$. We then need not bother about the ascending chain of poles, and we may replace $\Lambda^{(k)}$ by a straight line, for instance by the line $\mathrm{Re}\, s := \sigma := -\mathrm{Re}\, a+k-\frac{1}{2}$. Then

$$\left|\int_{\sigma-i\infty}^{\sigma+i\infty} f(s)\, ds\right| = \left|\int_{\sigma-i\infty}^{\sigma+i\infty}\frac{\Gamma(a+s)\Gamma(-s)}{\Gamma(c+s)}(-z)^s\, ds\right|$$

$$\leqslant|z|^\sigma\int_{\sigma-i\infty}^{\sigma+i\infty}\left|\frac{\Gamma(a+s)\Gamma(-s)}{\Gamma(c+s)}\right| e^{(\pi/2-\epsilon)|\mathrm{Im}\, s|}\, ds.$$

The integral is easily seen to be convergent by Stirling's formula. Because it is independent of z, the assertion follows for $k\geqslant\max(0,-\mathrm{Re}\, a+1)$. For possibly remaining small values of k the assertion can now be deduced by writing terms on the other side. ∎

If a or c equals zero or a negative integer, the function $_1F_1(a;c;z)$ is either elementary or undefined. Thus Theorem 8.8b covers all interesting cases of that function except those in which $c-a$ is a positive integer. The excluded case can be dealt with similarly; for an example, see Problem 3. In the cases covered, Theorem 8.8b deals with the asymptotic behavior if z tends to ∞ in a sector of the left half-plane. From this the corresponding result for a sector in the right half-plane may be deduced by means of Kummer's identity (1.5-3), the result being

$$_1F_1(a;c;z)\approx z^{a-c}e^z\frac{\Gamma(c)}{\Gamma(a)}{}_2F_0(c-a;1-a;z^{-1}),\, |z|\to\infty,\, |\arg z|\leqslant\frac{\pi}{2}-\epsilon.$$

$$(8.8\text{-}14)$$

These results do not describe the asymptotic behavior of $_1F_1(a;c;z)$ in the neighborhood of the imaginary axis. This question is resolved in §11.5.

As another application of Theorem 8.8b we now discuss briefly the asymptotic behavior of the Bessel function of order ν, $J_\nu(z)$. For integer values of ν this function has been defined in §4.5. For ν arbitrary, $\nu \neq -1$, $-2, \ldots$ and for $|\arg z| < \pi$ one sets (see §9.7)

$$J_\nu(z) := \frac{(z/2)^\nu}{\Gamma(\nu+1)} {}_0F_1\left(\nu+1; -\frac{z^2}{4}\right). \qquad (8.8\text{-}15)$$

By Kummer's second transformation (9.9-17), this may be written in either of the forms

$$J_\nu(z) = e^{iz} \frac{(z/2)^\nu}{\Gamma(\nu+1)} {}_1F_1(\nu+\tfrac{1}{2}; 2\nu+1; -2iz), \qquad (8.8\text{-}16a)$$

$$J_\nu(z) = e^{-iz} \frac{(z/2)^\nu}{\Gamma(\nu+1)} {}_1F_1(\nu+\tfrac{1}{2}; 2\nu+1; 2iz). \qquad (8.8\text{-}16b)$$

From (8.8-16b) we may, for instance, deduce the behavior of $J_\nu(z)$ if $|z| \to \infty$, $\epsilon \leq \arg z \leq \pi - \epsilon$. Applying (8.8-13) we find

$$J_\nu(z) \approx e^{-iz} \frac{(z/2)^\nu}{\Gamma(\nu+1)} (-2iz)^{-\nu-1/2} \frac{\Gamma(2\nu+1)}{\Gamma(\nu+\tfrac{1}{2})} {}_2F_0\left(\nu+\tfrac{1}{2}, -\nu-\tfrac{1}{2}; -\frac{1}{2iz}\right)$$

or, using the duplication formula,

$$J_\nu(z) \approx \frac{e^{i(\nu+1/2)\pi} e^{-iz}}{\sqrt{2\pi z}} {}_2F_0\left(\nu+\tfrac{1}{2}, -\nu-\tfrac{1}{2}; -\frac{1}{2iz}\right) \qquad (8.8\text{-}17)$$

A similar expression is obtained for $-\pi + \epsilon \leq \arg z \leq -\epsilon$. Equation (8.8-17) shows, in particular, that for purely imaginary values of z, $z = iy$ $(y > 0)$

$$J_\nu(iy) \sim e^{i\nu\pi} \frac{e^y}{\sqrt{2\pi y}}, \qquad y \to \infty.$$

The behavior for real values of z (which is of great interest in many applications) is not covered by the present method. This is discussed in §11.5.

PROBLEMS

1. Show that for any $\sigma < 0$

$$\frac{1}{2\pi i} \int_{\sigma-i\infty}^{\sigma+i\infty} \Gamma(-s)(-z)^s \, ds = e^z.$$

2. Find a representation, valid for $|\arg(-z)| < \pi$ and for suitably restricted values of the parameters, of the generalized hypergeometric series,

$$_3F_2\left[\begin{matrix} a, b, c; z \\ d, e \end{matrix}\right]$$

and use it to determine the analytic continuation beyond $|z| < 1$ of the function defined by that series.

3. It is immediately verified by term-by-term integration of the Taylor series of e^{-t^2} that

$$\mathrm{erf}(z) := \frac{2}{\sqrt{\pi}} \int_0^z e^{-t^2}\, dt = \frac{2}{\sqrt{\pi}} z\ _1F_1(\tfrac{1}{2}; \tfrac{3}{2}; -z^2).$$

Find a Mellin–Barnes representation of the series on the right and use it to determine the asymptotic behavior of $\mathrm{erf}(z)$ for $|\arg z| \leq \pi/4 - \epsilon$.

4. If c is not an integer and $|\arg z| < \tfrac{3}{2}\pi$, show that for an appropriate path of integration

$$\frac{1}{2\pi i} \int_{-i\infty}^{i\infty} \Gamma(-s)\Gamma(1-c-s)\Gamma(a+s)\, ds$$

$$= \Gamma(1-c)\Gamma(a)\ _1F_1(a; c; z) + \Gamma(c-1)\Gamma(1+a-c)\ _1F_1(1+a-c; 2-c; z).$$

(The integral can be used to determine the asymptotic behavior of the function on the right for $|\arg z| < \tfrac{3}{2}\pi$; see Copson [1960], p. 262.)

5. Study the asymptotic behavior of the entire function

$$_2F_2\left[\begin{matrix} a, b; z \\ c, d \end{matrix}\right]$$

as $|\arg(-z)| \leq \pi/2 - \epsilon$.

SEMINAR ASSIGNMENTS

1. Using Stirling's formula, plus appropriate recurrence and functional relations, write a program to compute $\Gamma(z)$ for complex values of z, and use it to construct a graphical representation of the surface $z \longrightarrow |\Gamma(z)|$, using graphical display equipment.

2. The Lemniscate constants (see Problem 8, §8.7) can be computed numerically by a large variety of methods; see Todd [1975]. Who gets the most accurate values, using \$10 worth of computer time?

NOTES

Most of the material in this chapter is standard. A general reference on infinite products is Titchmarsh [1939].

§8.1. Definitions of convergence of infinite products in accordance with Ahlfors [1966].

§8.2. Proof of Jacobi's identity (Theorem 8.2b) taken from Polya & Szegö [1925]. Many further identities involving the functions p and q are given by Hardy and Wright [1954].

§8.4. H. Weyl made the remark concerning (II) on occasion of a talk on fractional differentiation by F. Moppert in the Mathematische Kolloquium Zürich. The proof of Euler's integral by the monotone convergence theorem is suggested, for instance, by Burkill [1951]. For E. Schmidt's proof see Schaefke [1963]. New properties of the Γ function are still being discovered; for instance, see Gautschi [1974].

§8.5. The proof of Euler's integral by Stirling's formula was suggested by Ahlfors [1966].

§8.6. For further series and products of this type see Bailey [1935] or Chapter 1 (by F. Oberhettinger) of Erdelyi [1953].

§8.7. The integral representation given in Theorem 8.7b is due to Riemann [1857]. On Lemniscate constants, see Todd [1975].

§8.8. For some striking applications of Mellin–Barnes integrals to problems in field theory see Buchholz [1957]. For applications to algebraic equations, see Hochstadt [1971], p. 81.

9

ORDINARY
DIFFERENTIAL
EQUATIONS

In this chapter we deal with some aspects of the theory of ordinary differential equations that can be treated by complex variable methods. With the exception of the first two sections, only linear differential equations are considered. The discussion is carried through for systems of differential equations of the first order, thus covering single equations of higher order as a special case. Vector and matrix notation is employed systematically. Although the discussion is logically complete, some familiarity with the elementary theory of differential equations is assumed.

§9.1. THE EXISTENCE THEOREM

The differential equation to be considered here is written in the form

$$\mathbf{w}' = \mathbf{f}(z, \mathbf{w}). \tag{9.1-1}$$

Here $\mathbf{w}$ denotes an n dimensional vector with complex components. Later it will be convenient to think of $\mathbf{w}$ as a column vector. To exhibit the components of the vector $\mathbf{w}$, we write

$$\mathbf{w} = \begin{pmatrix} w_1 \\ w_2 \\ \cdot \\ \cdot \\ w_n \end{pmatrix}$$

or, more practically, $\mathbf{w} = (w_1, w_2, \ldots, w_n)^T$. We define the norm of the vector $\mathbf{w}$ by

$$\|\mathbf{w}\| := \left(\sum_{k=1}^{n} |w_k|^2 \right)^{1/2}.$$

The symbol $\mathbf{f}$ in (9.1-1) denotes a function defined on some region (i.e., connected open set) D of the complex $(n+1)$ dimensional $(z, \mathbf{w})$ space. (The z space has one complex dimension, and the $\mathbf{w}$ space is n dimensional.) The values of $\mathbf{f}$ are complex n-vectors, $\mathbf{f} = (f_1, f_2, \ldots, f_n)^T$. For notational simplicity we occasionally write w_0 in place of z. The case $n = 1$ of (9.1-1) is called a **scalar differential equation.** If it is desirable to emphasize that $n > 1$ we call (9.1-1) a **system of** (first-order) **differential equations.**

In all that follows, only the case in which $\mathbf{f}$ is an *analytic function* of $(z, \mathbf{w})$ is considered. A single complex-valued function f defined on a subset D of $\mathbb{C}^{n+1}$ is called **analytic at a point** $\hat{\mathbf{w}} = (\hat{w}_0, \hat{w}_1, \ldots, \hat{w}_n) \in D$, if $\hat{\mathbf{w}}$ is an interior point of D, and if in some neighborhood $\|\mathbf{w} - \hat{\mathbf{w}}\| < \rho$, $\rho > 0$, of $\hat{\mathbf{w}}$ it is representable by a convergent power series,

$$f(w_0, w_1, \ldots, w_n) = \sum_{m_0=0}^{\infty} \sum_{m_1=0}^{\infty} \cdots \sum_{m_n=0}^{\infty} a_{m_0 m_1 \cdots m_n} k_0^{m_0} k_1^{m_1} \cdots k_n^{m_n},$$

where $k_i := w_i - \hat{w}_i$, $i = 0, 1, \ldots, n$, and the $a_{m_0 m_1 \cdots m_n}$ are complex constants. The function f is called **analytic in a region** $D \subset \mathbb{C}^{n+1}$ if it is analytic at each point of D. A vector-valued function $\mathbf{f}$ is called analytic if each component is analytic.

This is not the place to present a systematic theory of analytic functions of several complex variables. It may suffice at this point to mention that the analytic functions of several complex variables in a fixed region D form an integral domain, that (appropriately defined) compositions of such functions are analytic, and that the locally uniform limit of sequences of such functions is again analytic. See also §17.1.

A complex vector-valued function $\mathbf{w} = (w_1, \ldots, w_n)^T$ is called a **solution** of the differential equation (9.1-1) in some region D_0 of the complex plane, if each component w_i is analytic in D_0, if $(z, \mathbf{w}(z)) \in D$ for all $z \in D_0$, and if

$$\mathbf{w}'(z) = \mathbf{f}(z, \mathbf{w}(z)) \text{ for all } z \in D_0.$$

The **initial value problem** for the differential equation (9.1-1) is the problem of finding a solution $\mathbf{w}$ that at some preassigned point $\hat{z}$ takes a preassigned value $\hat{\mathbf{w}}$. The following theorem states conditions under which we can assert the existence of a unique solution of the initial value problem in a certain region D_0.

THEOREM 9.1

Let the function $\mathbf{f}$ be analytic and bounded in the region

$$R : |z - \hat{z}| < \alpha, \qquad \|\mathbf{w} - \hat{\mathbf{w}}\| < \beta,$$

where $\alpha > 0$, $\beta > 0$, and let

$$\mu := \sup_{(z,\mathbf{w}) \in R} \|\mathbf{f}(z, \mathbf{w})\|, \qquad \gamma := \min\left(\alpha, \frac{\beta}{\mu}\right).$$

Then there exists in the disk $D_0 : |z - \hat{z}| < \gamma$ a unique analytic function $\mathbf{w}$ which is a solution of (9.1-1) and satisfies the initial condition $\mathbf{w}(\hat{z}) = \hat{\mathbf{w}}$.

Proof. The method of successive approximations, familiar from the real theory of differential equations, can be applied also in the analytic case. As is well known, it consists in constructing a sequence of functions $\{\mathbf{w}_k\}$ by

$$\mathbf{w}_0(z) := \hat{\mathbf{w}}, \tag{9.1-2}$$

$$\mathbf{w}_{k+1}(z) := \hat{\mathbf{w}} + \int_{\hat{z}}^{z} \mathbf{f}(t, \mathbf{w}_k(t))\, dt,$$

$$z \in D_0, \qquad k = 0, 1, 2, \ldots, \tag{9.1-3}$$

where the integrals can be taken along the straight line segments joining $\hat{z}$ and z. The desired solution is then obtained as the limit function of the sequence thus obtained. To accomplish the proof along these lines, it is necessary to show (a) that the sequence $\{\mathbf{w}_k\}$ is well defined [i.e., that the construction (9.1-3) does not lead outside the domain of definition of $\mathbf{f}$], (b) that the sequence converges, and (c) that the limit function is a solution.

To prove (a), we shall show that for all $z \in D_0$ and for $k = 0, 1, 2, \ldots,$

$$\|\mathbf{w}_k(z) - \hat{\mathbf{w}}\| < \beta. \tag{9.1-4}$$

Since $\mathbf{w}_0(z) = \hat{\mathbf{w}}$, this is clearly true for $k = 0$. If it is true for some $k \geq 0$, then by (9.1-3) and by the definition of γ,

$$\|\mathbf{w}_{k+1}(z) - \hat{\mathbf{w}}\| = \left\| \int_{\hat{z}}^{z} \mathbf{f}(t, \mathbf{w}_k(t))\, dt \right\|$$

$$\leq \sup_{(t,\mathbf{w}) \in R} \|\mathbf{f}(t, \mathbf{w})\| \, |z - \hat{z}| \leq \mu\gamma \leq \beta.$$

Hence (9.1-4) is established for k replaced by $k + 1$, and hence by induction for all k.

It is clear that $\mathbf{w}_0$ is analytic and that if $\mathbf{w}_k$ is analytic, then so is $\mathbf{w}_{k+1}$, it being the integral of a composition of analytic functions. Thus all functions $\mathbf{w}_k$ are analytic for $|z - \hat{z}| < \gamma$.

To prove (b), and on some later occasions, we must estimate the norm of the difference

$$\mathbf{f}(z, \mathbf{v}) - \mathbf{f}(z, \mathbf{u}),$$

where $(z, \mathbf{v})$ and $(z, \mathbf{u})$ are arbitrary points in R. Its ith component may be written

$$f_i(z, \mathbf{v}) - f_i(z, \mathbf{u}) = \int_0^1 \sum_{j=1}^n \frac{\partial f_i}{\partial w_j}(z, \mathbf{u} + \tau \mathbf{d}) d_i \, d\tau,$$

where $\mathbf{d} = (d_1, \ldots, d_n) := \mathbf{v} - \mathbf{u}$. Hence

$$\mathbf{f}(z, \mathbf{v}) - \mathbf{f}(z, \mathbf{u}) = \int_0^1 \mathbf{J}(z, \mathbf{u} + \tau \mathbf{d}) \mathbf{d} \, d\tau,$$

where $\mathbf{J} := (\partial f_i / \partial w_j)$ is the **Jacobian matrix** of $\mathbf{f}$ with respect to the last n agruments.

We recall that for any matrix $\mathbf{A}$ the matrix norm induced by the vector norm $\|\cdots\|$ is defined by

$$\|\mathbf{A}\| := \sup_{\|\mathbf{w}\|=1} \|\mathbf{A}\mathbf{w}\|.$$

In the case of the euclidean vector norm employed here we have

$$\|\mathbf{A}\mathbf{w}\|^2 = \mathbf{w}^H \mathbf{A}^H \mathbf{A}\mathbf{w} \leqslant \lambda \|\mathbf{w}\|^2,$$

where λ is the largest eigenvalue of the hermitian matrix $\mathbf{A}^H \mathbf{A}$, with equality if $\mathbf{w}$ is an appropriate eigenvector. It follows that $\mathbf{A}$ equals the square root of the largest eigenvalue of $\mathbf{A}^H \mathbf{A}$. Because the sum of *all* eigenvalues of the positive semidefinite matrix $\mathbf{A}^H \mathbf{A}$ equals tr $\mathbf{A}^H \mathbf{A} = \sum_{i,j=1}^n |a_{ij}|^2$, we also have

$$\|\mathbf{A}\| \leqslant \left(\sum_{i,j=1}^n |a_{ij}|^2 \right)^{1/2}.$$

There follows

$$\|\mathbf{J}(z, u + \tau \mathbf{d}) \mathbf{d}\| \leqslant \|J(z, \mathbf{u} + \tau \mathbf{d})\| \|\mathbf{d}\|.$$

Hence we have

$$\|\mathbf{f}(z, \mathbf{v}) - \mathbf{f}(z, \mathbf{u})\| \leqslant \sup_{0 \leqslant \tau \leqslant 1} \|\mathbf{J}(z, \mathbf{u} + \tau \mathbf{d})\| \|\mathbf{v} - \mathbf{u}\|. \qquad (9.1\text{-}5)$$

Turning to the proof of assertion (b), we show that the sequence $\{\mathbf{w}_k\}$ converges uniformly on every disk $|z - \hat{z}| \leqslant \gamma_1$ where $\gamma_1 < \gamma$. By the proof of (a) we have on every such disk

$$\|\mathbf{w}_k(z) - \hat{\mathbf{w}}\| \leqslant \gamma_1 \mu < \mu \min\left(\alpha, \frac{\beta}{\mu}\right),$$

$k = 0, 1, 2, \ldots$. Thus for $|z - \hat{z}| \leqslant \gamma_1$ the points $(z, \mathbf{w}_k(z))$ all stay within a compact subset

$$R_1 : |z - \hat{z}| \leqslant \gamma_1, \qquad \|w - \hat{w}\| \leqslant \beta_1$$

of R, where $\beta_1 < \beta$. We define

$$\kappa := \sup_{(z,w) \in R_1} \|\mathbf{J}(z, \mathbf{w})\|.$$

By (9.1-3) the difference between consecutive iterates can be estimated as follows:

$$\|\mathbf{w}_{k+1}(z) - \mathbf{w}_k(z)\| = \left\| \int_{\hat{z}}^z [\mathbf{f}(t, \mathbf{w}_k(t)) - \mathbf{f}(t, \mathbf{w}_{k-1}(t))]\, dt \right\|$$

$$\leq \int_{\hat{z}}^z \|\mathbf{f}(t, \mathbf{w}_k(t)) - \mathbf{f}(t, \mathbf{w}_{k-1}(t))\| |dt|.$$

Using (9.1-5), we thus get for $|z - \hat{z}| \leq \gamma_1$

$$\|\mathbf{w}_{k+1}(z) - \mathbf{w}_k(z)\| \leq \kappa \int_{\hat{z}}^z \|\mathbf{w}_k(t) - \mathbf{w}_{k-1}(t)\| |dt|. \tag{9.1-6}$$

Clearly,

$$\|\mathbf{w}_1(z) - \mathbf{w}_0(z)\| \leq \mu |z - \hat{z}|,$$

(9.1-6) yields

$$\|\mathbf{w}_2(z) - \mathbf{w}_1(z)\| \leq \frac{\kappa \mu |z - \hat{z}|^2}{2!},$$

and an easy induction shows that generally

$$\|\mathbf{w}_k(z) - \mathbf{w}_{k-1}(z)\| \leq \frac{\kappa^{k-1} \mu |z - \hat{z}|^k}{k!},$$

$k = 1, 2, \ldots$. Hence the series of analytic functions

$$\sum_{k=1}^\infty (\mathbf{w}_k(z) - \mathbf{w}_{k-1}(z))$$

converges uniformly for $|z - \hat{z}| \leq \gamma_1$, which is to say that the limit

$$\mathbf{w}(z) := \lim_{k \to \infty} \mathbf{w}_k(z)$$

exists uniformly. Because all $\mathbf{w}_k$ are analytic, the limit function $\mathbf{w}$ by Theorem 3.4b is analytic in $|z - \hat{z}| < \gamma$.

We now prove (c). It is obvious that the function $\mathbf{w}$ satisfies the correct initial condition. To show that it satisfies the correct differential equation, we note that the existence of $\lim \mathbf{w}_k$ implies

$$\mathbf{w}(z) = \hat{\mathbf{w}} + \lim_{k \to \infty} \int_{\hat{z}}^z \mathbf{f}(t, \mathbf{w}_k(t))\, dt. \tag{9.1-7}$$

If we can deduce from this that

$$\mathbf{w}(z) = \hat{\mathbf{w}} + \int_{\hat{z}}^{z} \mathbf{f}(t, \mathbf{w}(t)) \, dt,$$

then we find by differentiation the desired identity

$$\mathbf{w}'(z) = \mathbf{f}(z, \mathbf{w}(z)), \qquad |z - \hat{z}| < \gamma.$$

To justify the interchange of integration and passage to the limit, we show that

$$\lim_{k \to \infty} \mathbf{f}(t, \mathbf{w}_k(t)) = \mathbf{f}(t, \mathbf{w}(t))$$

uniformly along the path of integration in (9.1-7). By an application of (9.1-5) we find, if $|t - \hat{z}| \leq \gamma_1$, taking into account that $\|\mathbf{w}(t) - \hat{\mathbf{w}}\| \leq \beta_1$,

$$\|\mathbf{f}(t, \mathbf{w}_k(t)) - \mathbf{f}(t, \mathbf{w}(t))\| \leq \kappa \|\mathbf{w}_k(t) - \mathbf{w}(t)\|.$$

The desired uniform convergence now follows from the uniform convergence of the sequence $\{\mathbf{w}_k\}$.

It remains to establish the uniqueness of the solution $\mathbf{w}$ constructed above. Let $\mathbf{v}$ be another analytic solution satisfying the same initial condition. By the identity principle for analytic functions, it suffices to show that $\mathbf{w}(z) = \mathbf{v}(z)$ for $|z - \hat{z}| < \epsilon$, where $\epsilon > 0$ may be arbitrarily small.

The solution $\mathbf{v}$ satisfies an identity similar to (9.1-7), and on subtracting the two identities, we find

$$\mathbf{v}(z) - \mathbf{w}(z) = \int_{\hat{z}}^{z} [\mathbf{f}(t, \mathbf{v}(t)) - \mathbf{f}(t, \mathbf{w}(t))] \, dt. \tag{9.1-8}$$

Applying (9.1-5), this implies

$$\|\mathbf{v}(z) - \mathbf{w}(z)\| \leq \kappa \int_{\hat{z}}^{z} \|\mathbf{v}(t) - \mathbf{w}(t)\| |dt|.$$

Now for $0 \leq \tau \leq \gamma_1$ let

$$\delta(\tau) := \sup_{|z - \hat{z}| \leq \tau} \|\mathbf{v}(z) - \mathbf{w}(z)\|.$$

Then $\delta(\tau) \geq 0$, and by the above

$$\delta(\tau) \leq \kappa \sup_{|z - \hat{z}| \leq \tau} \int_{\hat{z}}^{z} \|\mathbf{v}(t) - \mathbf{w}(t)\| |dt| \leq \kappa \tau \delta(\tau),$$

which is contradictory for any τ such that $\kappa \tau < 1$ and $\delta(\tau) \neq 0$. It follows that $\delta(\tau) = 0$ for $0 \leq \tau < \kappa^{-1}$, which is sufficient to establish uniqueness. ■

EXAMPLES

We show by means of examples that the assertion of the theorem concerning γ, the radius of the disk of analyticity of the solution, cannot be improved unless further assumptions on **f** are made.

For $\alpha \leq \beta/\mu$, this is illustrated by the scalar case $(n = 1)$ where f is independent of w and has a singularity on the circle $|z - \hat{z}| = \alpha$. Any solution

$$w(z) = \hat{w} + \int_{\hat{z}}^{z} f(t) \, dt$$

will then likewise have a singularity at the same point, or else $f(t) = w'(t)$ would have to be analytic.

For $\alpha > \beta/\mu$ consider the scalar problem

$$w' = f(w) := \mu \left[\frac{1}{2} \left(1 + \frac{w}{\beta} \right) \right]^{1/m}, \qquad w(0) = 0,$$

where $\beta > 0$, $\mu > 0$, and m is a positive integer > 1. Here $\alpha = \infty$. Clearly, f is analytic for $|w| < \beta$ and $|f(w)| \leq \mu$ for $|w| \leq \beta$. The solution can be determined by separating variables:

$$\frac{dw}{(1 + w/\beta)^{1/m}} = \mu 2^{-1/m} \, dz.$$

Taking into account the initial condition, we find

$$\left(1 + \frac{w}{\beta} \right)^{(m-1)/m} - 1 = \frac{z}{\gamma_m},$$

where

$$\gamma_m := \frac{\beta}{\mu} \frac{m}{m-1} 2^{1/m}.$$

Solving for w yields

$$w = w(z) = \beta \left[\left(1 + \frac{z}{\gamma_m} \right)^{m/(m-1)} - 1 \right].$$

The solution is analytic for $|z| < \gamma_m$ and has a singularity at $z := -\gamma_m$. Although $\gamma_m > \beta/\mu = \gamma$, it is true that

$$\lim_{m \to \infty} \gamma_m = \frac{\beta}{\mu}.$$

Thus the statement that the solution of a differential equation satisfying the conditions of Theorem 9.1 is analytic for $|z - \hat{z}| < \rho$ is false for all $\rho > \beta/\mu$.

The following example illustrates an even simpler, although equally important, point. Assume that the function f is entire, that is, that $\alpha = \beta = \infty$ in Theorem 9.1. Is it true that all solutions of $w' = f(z, w)$ are entire? Already the example

$$w' = w^2, \qquad w(0) = \hat{w}$$

($\hat{w}$ again one dimensional) shows that this is far from being the case. The explicit solution

$$w = w(z) = \frac{\hat{w}}{1 - z\hat{w}}$$

shows that no solution except $w \equiv 0$ is entire, and that the solutions can have singularities arbitrarily close to the initial point.

PROBLEMS

1. For the initial value problem

$$w' = w^2, \qquad w(0) = 1,$$

   determine the maximal disk of analyticity given by Theorem 9.1, and compare it with the actual disk of analyticity of the solution $w(z) = (1 - z)^{-1}$.
2. Let the function $\mathbf{f}$ in Theorem 9.1 depend analytically on a parameter vector $\mathbf{p}$ and assume that

$$\|\mathbf{f}(z, \mathbf{w}, \mathbf{p})\| \leqslant \mu$$

   for all $(z, \mathbf{w}) \in R$ and all $\mathbf{p}$ such that $\|\mathbf{p}\| < \rho$. Show that the solution of the initial value problem $\mathbf{w}' = \mathbf{f}(z, \mathbf{w}, \mathbf{p})$, $\mathbf{w}(\hat{z}) = \hat{\mathbf{w}}$, is an analytic function of z and $\mathbf{p}$ for $|z - \hat{z}| < \gamma$, $\|\mathbf{p}\| < \rho$. (Successive approximations.)
3. Formulate and prove a result to the effect that the solution $\mathbf{w}$ of the initial value problem

$$\mathbf{w}' = \mathbf{f}(z, \mathbf{w}), \qquad \mathbf{w}(\hat{z}) = \hat{\mathbf{w}} \tag{9.1-9}$$

   depends analytically on the initial vector $\hat{\mathbf{w}}$. (Hint: The solution of (9.1-9) equals $\hat{\mathbf{w}} + \mathbf{v}$, where $\mathbf{v}$ is the solution of the initial value problem involving the parameter $\hat{\mathbf{w}}$,

$$\mathbf{v}' = \mathbf{g}(z, \mathbf{v}, \hat{\mathbf{w}}), \qquad \mathbf{v}(\hat{z}) = \mathbf{0},$$

   $\mathbf{g}(z, \mathbf{v}, \hat{\mathbf{w}}) := \mathbf{f}(z, \hat{\mathbf{w}} + \mathbf{v})$. Now apply problem 2.)

§9.2. POWER SERIES METHOD

Although the method of successive approximations used in the proof of the basic existence theorem by common consensus is called constructive, it is hard to conceive of any concrete problem where its application would be feasible if numerical approximations to the solution are desired. Because analytic functions can be expanded in Taylor series, it is reasonable to ask whether it might not be possible to construct directly the Taylor expansion of the solution from the Taylor expansion of the given function f. This construction is, indeed, possible and forms the basis of a solution algorithm that is frequently used in practice.

We begin with some preliminaries on double series (see e.g., Kamke [1947], pp. 66–68).

(a) Let a_{mn} $(m, n = 0, 1, 2, \ldots)$ be complex. By the **double series**

$$\sum_{m,n=0}^{\infty} a_{mn} \tag{9.2-1}$$

we mean the pair of double arrays of complex numbers

$$
\begin{array}{cccc}
a_{00} & a_{01} & a_{02} & \cdots \\
a_{10} & a_{11} & a_{12} & \cdots \\
a_{20} & a_{21} & a_{22} & \cdots \\
& \cdot & \cdot & \cdot \cdot
\end{array}
\quad \text{and} \quad
\begin{array}{cccc}
s_{00} & s_{01} & s_{02} & \cdots \\
s_{10} & s_{11} & s_{12} & \cdots \\
s_{20} & s_{21} & s_{22} & \cdots \\
& \cdot & \cdot & \cdot \cdot
\end{array}
$$

where[1]

$$s_{mn} := \sum_{k=0}^{m} \sum_{l=0}^{n} a_{kl}, \qquad m, n = 0, 1, 2, \ldots.$$

s_{mn} is called the (m, n)th partial sum of the double series, a_{mn} its (m, n)th term. The terms of a double series can be recovered from its partial sums by the formula

$$a_{mn} = s_{mn} - s_{m-1,n} - s_{m,n-1} + s_{m-1,n-1}.$$

(b) The double series (9.2-1) is called **convergent** if there exists a number s such that for every $\epsilon > 0$ there exists an integer $n_0 = n_0(\epsilon)$ with the property that (see Fig. 9.2a)

$$|s_{mn} - s| < \epsilon$$

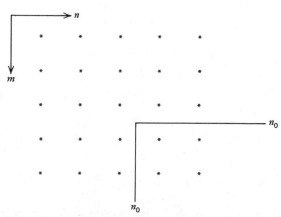

Fig. 9.2a. The s_{mn} in the lower right corner differ from s by less than ε.

---

[1] We avoid the definition of a series as a "formal sum," because it is devoid of meaning. See also §8.1 and §12.1.

for all m and n such that $m > n_0$ and $n > n_0$. The number s, if it exists, is uniquely determined, and is called the **value** or the **sum** of the double series.

To assert that the double series (9.2-1) is convergent is not the same as to assert that the **iterated series**

$$\sum_{m=0}^{\infty} \sum_{n=0}^{\infty} a_{mn} \qquad (9.2\text{-}2)$$

is convergent. Consider, for example, the double series with the partial sums

$$s_{mn} := \frac{(-1)^n}{m+1}.$$

It is convergent with the value 0. However, the series

$$\sum_{n=0}^{\infty} a_{0n}$$

has the nth partial sum $s_{0,n} = (-1)^n$ and is therefore divergent. Thus the iterated series (9.2-2) cannot even be formed in this case.

(c) *Cauchy criterion.* The series (9.2-1) is convergent if and only if for every $\epsilon > 0$ there exists $n_0 = n_0(\epsilon)$ such that

$$|s_{m_1,n_1} - s_{m_2,n_2}| < \epsilon$$

whenever $m_1, n_1, m_2, n_2 > n_0$.

(d) *Series with positive terms.* Let $a_{mn} \geq 0$, m, $n = 0, 1, \ldots$. The series (9.2-1) is convergent if and only if there exists a constant σ such that

$$s_{mn} < \sigma$$

for all m and n, and the value of the series in that case is sup s_{mn}. Under the same hypothesis, if any of the four series

$$\sum_{m,n=0}^{\infty} a_{mn}, \quad \sum_{m=0}^{\infty} \sum_{n=0}^{\infty} a_{mn}, \quad \sum_{n=0}^{\infty} \sum_{m=0}^{\infty} a_{mn}, \quad \sum_{k=0}^{\infty} \sum_{m+n=k} a_{mn}$$

$$(9.2\text{-}3)$$

is convergent, the remaining three are convergent, and the values of the four series are equal.

(e) For arbitrary a_{mn}, any of the four series (9.2-3) is called **absolutely convergent** if the series formed with $|a_{mn}|$ in place of a_{mn} is convergent.

(f) The absolute convergence of any of the series (9.2-3) implies its convergence.

(g) If any of the series (9.2-3) converges absolutely, then the remaining series are likewise absolutely convergent (and therefore convergent), and the values of the four series are the same.

(h) A power series in two variables is a double series of the form

$$\sum_{m,n=0}^{\infty} a_{mn} h^m k^n. \tag{9.2-4}$$

Unlike to what we know about power series in one variable, convergence of the series (9.2-4) for $h = a \neq 0$, $k = b \neq 0$ does not imply the convergence of the series for $|h| < |a|$, $|k| < |b|$. Consider, for instance, the series

$$\sum_{m=0}^{\infty} m! h^m (1+k)$$

formed with the coefficients $a_{m0} = a_{m1} = m!$, $a_{mn} = 0$ for $n \geq 2$. It converges for $k = -1$ and arbitrary h, but is clearly divergent for all (h, k) such that $h \neq 0$, $k \neq -1$. However, if we suppose the series (9.2-4) to converge *absolutely* for $h = a$, $k = b$, then the series

$$\sum_{m,n=0}^{\infty} |a_{mn}| |a|^m |b|^n$$

is a convergent majorant of (9.2-4) for all (h, k) such that $|h| \leq |a|$, $|k| \leq |b|$. By virtue of (d) and (e) the series thus converges absolutely for $|h| \leq |a|$, $|k| \leq |b|$ and by (g) it can be summed in any of the three modes

$$\sum_{m=0}^{\infty} h^m \sum_{n=0}^{\infty} a_{mn} k^n, \qquad \sum_{n=0}^{\infty} k^n \sum_{m=0}^{\infty} a_{mn} h^m,$$

$$\sum_{j=0}^{\infty} \sum_{m+n=j} a_{mn} h^m k^n.$$

We have already mentioned that a function of two complex variables that in a neighborhood of a point $(\hat{z}, \hat{w})$ can be represented by a series of the form (9.2-4), where $h := z - \hat{z}$, $k := w - \hat{w}$, is called **analytic** at that point.

We are now ready to prove an equivalent formulation of the existence theorem 9.1 based on power series. To avoid cumbersome notation, we consider only the case of a single equation. (See §17.1 for how to deal elegantly with power series in n variables.) Thus in the present section **w** is one-dimensional.

THEOREM 9.2

Let $\alpha > 0$, $\beta > 0$, and let f be an analytic function of two complex variables represented by

$$f(z, w) = \sum_{m,n=0}^{\infty} a_{mn} h^m k^n$$

for $|h| \leq \alpha$, $|k| \leq \beta$, where $h := z - \hat{z}$, $k := w - \hat{w}$, and let

$$\mu := \sum_{m,n=0}^{\infty} |a_{mn}| \alpha^m \beta^n < \infty.$$

Then there exist coefficients c_k $(k = 1, 2, \ldots)$, that are polynomials in the coefficients a_{mn} $(m, n = 0, 1, \ldots, k-1)$, such that the power series

$$w(z) := \hat{w} + \sum_{m=1}^{\infty} c_m h^m \tag{9.2-5}$$

has radius of convergence $\geq \gamma := \min(\alpha, \beta/\mu)$, and the function w represented by it for $|h| < \gamma$ is the unique analytic solution of the scalar differential equation

$$w' = f(z, w)$$

satisfying $w(\hat{z}) = \hat{w}$.

Proof. We begin the proof by examining what the coefficients of the Taylor series at $\hat{z}$ of a solution of the initial value problem (whose existence is not assumed) would have to look like. Let

$$g(z) = \hat{w} + \sum_{m=1}^{\infty} c_m h^m \qquad (h := z - \hat{z}) \tag{9.2-6}$$

be the Taylor series representation of a function g analytic at $\hat{z}$ and satisfying $g(\hat{z}) = \hat{w}$. For $n = 0, 1, 2, \ldots$, we write

$$(g(z) - \hat{w})^n = \sum_{m=n}^{\infty} c_m^{(n)} h^m.$$

Clearly, $c_0^{(0)} = 1$, $c_m^{(0)} = 0$ for $m > 0$. For $n > 0$, the coefficients $c_m^{(n)}$ are polynomials with positive coefficients in $c_1, c_2, \ldots, c_{m-n+1}$. Now let $\rho > 0$ be less than the radius of convergence of the series (9.2-6) and also such that $|h| \leq \rho$ implies

$$|g(z) - \hat{w}| \leq \beta. \tag{9.2-7}$$

For $|h| \leq \rho$, the double series

$$f(z, g(z)) = \sum_{m,n=0}^{\infty} a_{mn} h^m (g(z) - \hat{w})^n$$

is then absolutely convergent. By (h) above, we may sum it by columns, obtaining

$$f(z, g(z)) = \sum_{n=0}^{\infty} (g(z) - \hat{w})^n \sum_{m=0}^{\infty} a_{mn} h^m,$$

where the sum with respect to n is uniformly convergent for $|h| \leqslant \rho$. The nth term in that sum can be written as a product of two simple power series and hence can be evaluated by Cauchy multiplication:

$$(g(z) - \hat{w})^n \sum_{m=0}^{\infty} a_{mn} h^m = \sum_{l=n}^{\infty} c_l^{(n)} h^l \sum_{m=0}^{\infty} a_{mn} h^m = \sum_{p=n}^{\infty} c_{pn} h^p,$$

where

$$c_{p,n} := \sum_{l=n}^{p} c_l^{(n)} a_{p-l,n}. \tag{9.2-8}$$

It was shown above that the series with the nth term

$$\sum_{p=n}^{\infty} c_{p,n} h^p$$

converges uniformly for $|h| \leqslant \rho$. Hence by the Weierstrass double series theorem (Corollary 3.4c) the coefficients of each power of z may be summed separately, and we get

$$f(z, g(z)) = \sum_{p=0}^{\infty} \left(\sum_{n=0}^{p} c_{p,n} \right) h^p.$$

If g is a solution of the differential equation, this must equal

$$g'(z) = \sum_{p=0}^{\infty} (p+1) c_{p+1} h^p.$$

Thus by comparing coefficients we get

$$(p+1) c_{p+1} = \sum_{n=0}^{p} c_{p,n}, \qquad p = 0, 1, 2, \ldots.$$

Written out explicitly, using (9.2-8), this becomes

$$(p+1) c_{p+1} = a_{p,0}$$
$$+ c_1^{(1)} a_{p-1,1} + c_2^{(1)} a_{p-2,1} + \cdots + c_p^{(1)} a_{0,1}$$
$$+ c_2^{(2)} a_{p-2,2} + \cdots + c_p^{(2)} a_{0,2} \tag{9.2-9}$$
$$+ \cdots$$
$$+ c_p^{(p)} a_{0,p}.$$

Because the coefficients $c_m^{(n)}$ appearing on the right depend only on c_1, $c_2, \ldots, c_p$, the foregoing is a recurrence relation that determines the coefficients $c_1, c_2, \ldots$, uniquely. This already proves the uniqueness assertion of the theorem.

We now prove the statement that each c_p is a polynomial with nonnegative coefficients in the variables a_{mn} where $m < p$, $n < p$. Because $c_1 = a_{00}$, this assertion is clearly true for $p = 1$. Assuming its truth for some $p \geqslant 1$, it first follows that all $c_m^{(n)}$ appearing in (9.2-8) are such polynomials, and the assertion now follows immediately for c_{p+1}.

To complete the proof of the theorem, we shall show the following: if the coefficients c_p are determined recursively by (9.2-9), then the power series (9.2-6) has radius of convergence $\geqslant \gamma$, and the inequality (9.2-7) holds for $|h| \leqslant \gamma$. If this is shown, the formal operations leading to (9.2-9) are legitimate and show that g, indeed, is a solution function.

To estimate the radius of convergence, we use the *method of majorants* due to Cauchy. It is based on the following observation: we replace the coefficients a_{mn} by positive numbers $\beta_{mn} \geqslant |a_{mn}|$ and solve the resulting recurrence relations (9.2-9). In place of the coefficients c_p we obtain certain nonnegative coefficients δ_p, $p = 1, 2, \ldots$. Because the c_p are polynomials in the a_{mn} with nonnegative coefficients, it is clear that

$$|c_p| \leqslant \delta_p, \qquad p = 1, 2, \ldots. \tag{9.2-10}$$

The success of the method now depends on choosing the β_{mn} in such a manner that the δ_p can be estimated easily. Two methods for accomplishing this are now sketched.

(I) The first, due to Cauchy himself, is based on the analog for double power series of Cauchy's inequality for the coefficients of a simple power series. For each fixed value of k such that $|k| \leqslant \beta$, the function

$$f(z, w) = \sum_{m=0}^{\infty} h^m \sum_{n=0}^{\infty} a_{mn} k^n$$

is analytic in z for $|h| = |z - \hat{z}| < \alpha$. Hence by the Cauchy coefficient estimate (Theorem 2.2f)

$$\left| \sum_{n=0}^{\infty} a_{mn} k^n \right| \leqslant \frac{\mu}{\alpha^m}, \qquad m = 0, 1, 2, \ldots.$$

Applying Cauchy's estimate to the function

$$\sum_{n=0}^{\infty} a_{mn} k^n$$

analytic for $|k| < \beta$ and continuous in $|k| \le \beta$, we get

$$|a_{mn}| \le \frac{\mu}{\alpha^m \beta^n}, \qquad m, n = 0, 1, 2, \ldots . \qquad (9.2\text{-}11)$$

Thus the absolute values of the c_p do not exceed the Taylor coefficients at $\hat{z}$ of the solution $w(z)$ of the differential equation

$$w' = f_1(z, w),$$

where

$$f_1(z, w) := \sum_{m,n=0}^{\infty} \frac{\mu}{\alpha^m \beta^n} h^m k^n = \frac{\alpha \beta \mu}{(\alpha - h)(\beta - k)}.$$

By separating variables we find (using $k = 0$ for $h = 0$)

$$2\beta k - k^2 = -2\alpha \beta \mu \{ \text{Log}(\alpha - h) - \text{Log}\, \alpha \}$$

and solving for k yields

$$k = w(z) - \hat{w} = \beta - \sqrt{\beta^2 + 2\alpha \beta \mu \, \text{Log} \frac{\alpha - h}{\alpha}}. \qquad (9.2\text{-}12)$$

Clearly, the function w is analytic at $z = \hat{z}$, and its Taylor series, say, $\hat{w} + \sum \delta_p h^p$ thus has a positive radius of convergence ρ_0. The series $\hat{w} + \sum c_p h^p$, whose coefficients are not larger, thus likewise has a positive radius of convergence.

From (9.2-12) we can easily compute ρ_0. Because $w(z)$ has a singularity for $h = \alpha$ and also where the square root is zero, there follows

$$\rho_0 = \min(\alpha, \xi), \qquad (9.2\text{-}13)$$

where ξ is the solution of

$$\beta^2 + 2\alpha \beta \mu \, \text{Log} \frac{\alpha - \xi}{\alpha} = 0,$$

that is,

$$\xi := \alpha \left\{ 1 - \exp\left(-\frac{\beta}{2\alpha \mu} \right) \right\}.$$

Because $1 - e^{-x} \le x$, we have

$$\xi \le \frac{\beta}{2\mu}.$$

Thus the estimate (9.2-13) for the radius of convergence is considerably weaker than that given by the method of successive approximations. The reason lies in the crudeness of Cauchy's estimate for the coefficients. Whereas the function f is assumed to be bounded for $|h| \leq \alpha$, $|k| \leq \beta$, the majorizing function f_1 is clearly unbounded.

(II) The following method, due to Lindelöf (1870–1946), yields the estimate mentioned in the theorem. We replace a_{mn} by $\beta_{mn} := |a_{mn}|$ and obtain certain coefficients $\delta_p \geq 0$ as before. Let

$$g_p(h) := \hat{w} + \sum_{n=1}^{p} \delta_n h^n.$$

By the construction of the recurrence relations (9.2-9) the sum

$$\sum_{m=0}^{p-1} \sum_{n=1}^{p-1} |a_{mn}| h^m (g_{p-1}(h) - \hat{w})^n \qquad (9.2\text{-}14)$$

will, if rearranged in powers of h, agree with $g_p'(h)$ through the terms in h^{p-1}. Now let h be real, $0 \leq h \leq \gamma = \min(\alpha, \beta/\mu)$. Then $g_{p-1}(h) - \hat{w} \geq 0$, and the terms in (9.2-14) involving higher powers than h^{p-1} will be positive. Thus

$$g_p'(h) \leq \sum_{m=0}^{p-1} \sum_{n=0}^{p-1} |a_{mn}| h^m (g_{p-1}(h) - \hat{w})^n. \qquad (9.2\text{-}15)$$

We now show that for $0 \leq h \leq \gamma$ and $p = 0, 1, 2, \ldots$, we have $0 \leq g_p(h) - \hat{w} \leq \beta$. This is clearly true for $p = 0$. If it is true for some $p - 1 \geq 0$, then the expression on the right of (9.2-15) is $\leq \beta$ by hypothesis, and it follows on integration that

$$0 \leq g_p(h) - \hat{w} \leq \gamma\mu \leq \beta.$$

Thus for each $h \in [0, \gamma]$, the sequence $\{g_p(h) - \hat{w}\}$ is monotonically increasing and bounded and, therefore, convergent. In particular, it is convergent for $h = \gamma$, which means that the radius of convergence of the series $\sum \delta_p h^p$, and thus also of the series $\sum c_p h^p$, is at least γ, as asserted. ∎

PROBLEMS

1. The following scalar initial value problems are to be solved by power series and the solutions are to be compared with the elementary "closed-form" solution:

   (a) $w' = w$, $w(0) = 1$;

   (b) $w' = w^2$, $w(0) = 1$;

   (c) $w' = 1 + w^2$, $w(0) = 0$;

   (d) $w' = \dfrac{z}{2w}$, $w(0) = 1$.

2. To obtain a numerical approximation to the solution $w(z)$ of the initial value problem

$$w' = f(w), \qquad w(\hat{z}) = \hat{w}$$

in cases in which the derivatives of the function f are not easily obtainable, one frequently uses formulas of the type

$$\tilde{w}(z) = \hat{w} + h \sum_{i=1}^{m} a_i k_i,$$

where $z = \hat{z} + h$, $k_1 := f(\hat{w})$,

$$k_i := f\left(\hat{w} + h \sum_{j=1}^{i-1} p_{ij} k_j\right), \qquad i = 2, \ldots, m,$$

and where a_i and p_{ij} $(i = 1, \ldots, m; j = 1, \ldots, i-1)$ denote suitable constants.
(a) Show that for $m = 2, 3, 4$ there exist systems of values a_i and p_{ij} such that

$$\tilde{w}(z) - w(z) = 0(h^{m+1}), \qquad h \to 0. \tag{9.2-16}$$

[The special system where $m = 4$, $a_1 = a_4 = \frac{1}{6}$, $a_2 = a_3 = \frac{1}{3}$, $p_{21} = p_{32} = \frac{1}{2}$, $p_{43} = 1$, all other $p_{ij} = 0$, defines what is known as the **formula of Runge and Kutta**.]
(b) Show that for $m = 5$ no system of constants a_i and p_{ij} exists such that (9.2-16) holds (Butcher).
3. Let the series

$$f(z) := z + a_2 z^2 + a_3 z^3 + \cdots$$

have a positive radius of convergence. Use Cauchy's method of majorants to show that there exists a power series

$$g(w) := w + b_2 w^2 + b_3 w^3 + \cdots$$

of positive radius of convergence such that

$$g(f(z)) = z$$

for all sufficiently small $|z|$.

§9.3. LINEAR SYSTEMS

Let $\mathbf{A} = \mathbf{A}(z)$ be a $n \times n$ matrix and $\mathbf{b} = \mathbf{b}(z)$ an n vector whose elements are analytic functions in a region S (not necessarily simply connected) of the complex plane. The special case of (9.1-1) where $\mathbf{f}(z, \mathbf{w}) := \mathbf{A}(z)\mathbf{w} + \mathbf{b}(z)$ is called a **linear system of differential equations**. Thus such a system has the form

$$\mathbf{w}' = \mathbf{A}(z)\mathbf{w} + \mathbf{b}(z). \tag{L}$$

The system is called **homogeneous** if $\mathbf{b}(z) \equiv \mathbf{0}$ and **nonhomogeneous** otherwise.

As before, we consider the *initial value problem* of finding a solution $\mathbf{w} = \mathbf{w}(z)$ of (L) satisfying a given initial condition

$$\mathbf{w}(\hat{z}) = \hat{\mathbf{w}}, \qquad\qquad (9.3\text{-}1)$$

where $\hat{z} \in S$ and $\hat{\mathbf{w}}$ is a given vector. The theorems proved in §9.1 and §9.2 already guarantee the local existence and uniqueness of such a solution. However, they do not yield the best possible result.

THEOREM 9.3a

Let the functions $\mathbf{A}(z)$ and $\mathbf{b}(z)$ be analytic in the disk $|z - \hat{z}| < \gamma$, and let $\hat{\mathbf{w}}$ be an arbitrary vector. Then the linear system (L) possesses a solution $\mathbf{w}$ satisfying (9.3-1) which is likewise analytic in $|z - \hat{z}| < \gamma$.

Thus, in the case of a *linear* system, the maximal disk of analyticity of the solution depends only on the maximal disk of analyticity of the coefficient matrix $\mathbf{A}$ and of the nonhomogeneous term $\mathbf{b}$. Unlike the nonlinear case, a solution cannot stop being analytic unless the right-hand term of the differential equation stops being analytic.

Proof of Theorem 9.3a. Without loss of generality we may assume $\hat{z} = 0$, for this can be achieved by introducing the new variable $h := z - \hat{z}$. Let

$$\mathbf{A}(z) = \sum_{n=0}^{\infty} \mathbf{A}_n z^n, \qquad \mathbf{b}(z) = \sum_{n=0}^{\infty} \mathbf{b}_n z^n,$$

$|z| < \gamma$. The coefficients of the Taylor expansion

$$\mathbf{w}(z) = \sum_{n=0}^{\infty} \mathbf{w}_n z^n \qquad\qquad (9.3\text{-}2)$$

of the desired solution satisfy $\mathbf{w}_0 = \hat{\mathbf{w}}$,

$$(n+1)\mathbf{w}_{n+1} = \sum_{k=0}^{n} \mathbf{A}_k \mathbf{w}_{n-k} + \mathbf{b}_n, \qquad n = 0, 1, 2, \ldots .$$

By the triangle inequality,

$$(n+1)\|\mathbf{w}_{n+1}\| \leq \sum_{k=0}^{n} \|\mathbf{A}_k\| \, \|\mathbf{w}_{n-k}\| + \|\mathbf{b}_n\|. \qquad\qquad (9.3\text{-}3)$$

Let ρ be any number such that $\rho < \gamma$. By Cauchy's estimate for the coefficients of a power series, there exist constants α and β such that

$$\|\mathbf{A}_n\| \leq \frac{\alpha}{\rho^{n+1}}, \|\mathbf{b}_n\| \leq \frac{\beta}{\rho^{n+1}}, \qquad n = 0, 1, 2, \ldots . \qquad\qquad (9.3\text{-}4)$$

It follows from (9.3-3) and (9.3-4) that $\|\mathbf{w}_n\| \leqslant \omega_n$, where $\{\omega_n\}$ is the solution of the difference equation

$$(n+1)\omega_{n+1} = \sum_{k=0}^{n} \frac{\alpha}{\rho^{k+1}} \omega_{n-k} + \frac{\beta}{\rho^{n+1}}, \qquad n = 0, 1, \ldots,$$

satisfying $\omega_0 = \|\hat{\mathbf{w}}\|$. We now proceed as in the proof of Theorem 9.2 by Cauchy's method of majorants. The series

$$\omega(z) := \sum_{n=0}^{\infty} \omega_n z^n$$

is a formal solution of the scalar initial value problem $\omega(0) = \omega_0$,

$$\omega' = \sum_{n=0}^{\infty} \frac{\alpha}{\rho^{n+1}} z^n + \sum_{n=0}^{\infty} \frac{\beta}{\rho^{n+1}} z^n$$

$$= \frac{\alpha}{\rho - z} + \frac{\beta}{\rho - z} \quad (|z| < \rho)$$

whose solution

$$\omega(z) = \left(\omega_0 + \frac{\beta}{\alpha} \right) \left(1 - \frac{z}{\rho} \right)^{-\alpha} - \frac{\beta}{\alpha}$$

is analytic in $|z| < \rho$. Hence the power series $\sum_{n=0}^{\infty} \omega_n z^n$ has radius of convergence $\geqslant \rho$. Because $\|\mathbf{w}_n\| \leqslant \omega_n$ the same holds for the series (9.3-2). Because this is true for any $\rho < \gamma$, the radius of convergence cannot be less than γ. Because the series formally satisfies the differential equation, it defines an actual solution in the disk $|z| < \gamma$. ■

COROLLARY 9.3b

Let the matrix **A** *and the vector* **b** *be analytic in some region S. Then any vector-valued function* **w** *that in the neighborhood of some point* $\hat{z} \in S$ *is a solution of* (L) *can be continued analytically from* $\hat{z}$ *along any path in S.*

Proof. Let Γ be any path with starting point $\hat{z}$ lying in S, and let z_e be the terminal point of Γ. As in §3.6, we can construct a finite sequence of disks $D_i \subset S$, $i = 0, 1, \ldots, n$, whose centers z_i lie on Γ and have the property that $z_i \in D_{i-1}$, $i = 1, 2, \ldots, n$; $z_0 = \hat{z}$, $z_n = z_e$. By Theorem 9.3a, the solution **w** considered in the corollary either is analytic in D_0, or can be continued to a solution that is analytic in D_0. We let $\mathbf{w} := \mathbf{w}_0$, and for $k = 1, 2, \ldots, n$ define $\mathbf{w}_k$ as the solution of (L) analytic in D_k and satisfying $\mathbf{w}_k(z_k) = \mathbf{w}_{k-1}(z_k)$. These functions are analytic continuations of each other, because at each z_k the values and all derivatives of the functions $\mathbf{w}_k$ and $\mathbf{w}_{k-1}$ coincide. ■

No assumption was made in Corollary 9.3b concerning the connectivity of S. If S is multiply connected, it may happen that the function $\mathbf{w}$ will not return to its initial value if continued along a closed curve Γ. This is a basic phenomenon in the theory of linear differential equations with isolated singular points, which is studied in great detail from §9.4 onward. However, in the case in which S is simply connected, the phenomenon cannot occur.

THEOREM 9.3c

Let the matrix $\mathbf{A}$ *and the vector* $\mathbf{b}$ *be analytic in a simply connected region* S, *let* $\hat{z} \in S$, *and let* $\hat{\mathbf{w}}$ *be any vector. Then the initial value problem for* (L) *possesses a solution that is analytic in* S.

Proof. Certainly there exists a solution in a neighborhood of $\hat{z}$, by Theorem 9.3a. By the corollary, this solution can be continued along any path emanating from $\hat{z}$. Because S is simply connected, the monodromy theorem (Theorem 3.5c) implies that at any point $z \in S$ the values obtained by all these analytic continuations coincide. All analytic continuations thus define a single analytic solution $\mathbf{w}$ in S. ∎

We now state some properties of solutions of the system (L) in simply connected regions. These properties are completely analogous to what is known for systems of *real* differential equations.

We begin with the study of the *homogeneous system*

$$\mathbf{w}' = \mathbf{A}(z)\mathbf{w}. \tag{LH}$$

It is clear that the zero vector $\mathbf{w} = \mathbf{0}$ is a solution of the homogeneous system. We shall call it the **trivial solution.** If any solution $\mathbf{w}$ of the homogeneous system is zero at any point $z \in S$, it must be the trivial solution by virtue of the uniqueness statement of the existence theorem.

THEOREM 9.3d

The set of all solutions of the homogeneous equation (LH) *(where* $\mathbf{w}$ *is* n dimensional) *forms an* n-*dimensional vector space over the complex field.*

Proof. If $\mathbf{w}_1$ and $\mathbf{w}_2$ are any two solutions of (LH) and if c_1, c_2 are any two complex numbers, then simple algebra shows that $\mathbf{w} := c_1\mathbf{w}_1 + c_2\mathbf{w}_2$ is again a solution of (LH). This shows that the solutions form a vector space over the complex field.

To show that the vector space is n dimensional, we must find n linearly independent elements $\mathbf{w}_i$ $(i = 1, \ldots, n)$ in it such that every element in the space can be represented as a linear combination of the $\mathbf{w}_i$. Let $\hat{z} \in S$, and let $\hat{\mathbf{w}}_1, \ldots, \hat{\mathbf{w}}_n$ be n linearly independent vectors (for instance, the n coordinate unit vectors). Let $\mathbf{w}_i$ be the solution of (LH) satisfying $\mathbf{w}_i(\hat{z}) = \hat{\mathbf{w}}_i$. These solutions are linearly independent, for if not, then there would exist n constants $c_1, \ldots, c_n$, not all zero, such that

$$\sum_{i=1}^{n} c_i \mathbf{w}_i(z) = \mathbf{0}, z \in S.$$

This would imply in particular that

$$\sum_{i=1}^{n} c_i \mathbf{w}_i(\hat{z}) = \sum_{i=1}^{n} c_i \hat{\mathbf{w}}_i = \mathbf{0},$$

contrary to our assumption that the vectors $\hat{\mathbf{w}}_i$ are linearly independent. Now let $\mathbf{w}$ be any solution of (LH), and let $\mathbf{w}(\hat{z}) = \hat{\mathbf{w}}$. Because the $\hat{\mathbf{w}}_i$ form a basis in complex n space, there exist constants $c_1, \ldots, c_n$ such that

$$\hat{\mathbf{w}} = \sum_{i=1}^{n} c_i \hat{\mathbf{w}}_i.$$

Consider the function $\mathbf{w}_0 := \mathbf{w} - \sum_{i=1}^{n} c_i \mathbf{w}_i$. It is a solution of (LH) that vanishes at $\hat{z}$ and hence vanishes identically. Hence $\mathbf{w} = \sum c_i \mathbf{w}_i$, as was to be shown. ∎

Any set $\mathbf{w}_1, \ldots, \mathbf{w}_n$ of n linearly independent solutions of (LH) is called a **fundamental system** of solutions. A matrix $\mathbf{W}$ whose columns form a fundamental system of solutions is called a **fundamental matrix** of (LH). A fundamental matrix satisfies the identity

$$\mathbf{W}'(z) = \mathbf{A}(z)\mathbf{W}(z),$$

because each column on the left agrees with the corresponding column on the right. A fundamental matrix thus may be called a solution of the **matrix differential equation**

$$\mathbf{W}' = \mathbf{A}(z)\mathbf{W}. \tag{9.3-5}$$

If $\mathbf{G}$ is any matrix, any linear combination of the columns of $\mathbf{G}$ can be written in the form $\mathbf{Gc}$, where $\mathbf{c}$ is an n vector. Thus if $\mathbf{W}$ is a fundamental matrix of (LH), any solution $\mathbf{w}$ of (LH) can be written as

$$\mathbf{w}(z) = \mathbf{W}(z)\mathbf{c},$$

where $\mathbf{c}$ is a certain constant vector. It follows immediately that a solution matrix $\mathbf{W}$ of (9.3-5) is a fundamental matrix of (LH) if and only if $\det \mathbf{W}(z) \neq 0$ for all $z \in S$, because one must be able to determine $\mathbf{c}$ from the equation $\hat{\mathbf{w}} = \mathbf{W}(z)\mathbf{c}$ for arbitrary $\hat{z} \in S$ and arbitrary $\hat{\mathbf{w}}$. Actually, for $\mathbf{W}$ to be a fundamental matrix it suffices that $\det \mathbf{W}(z) \neq 0$ at a single point $z \in S$, because if $\det \mathbf{W}(z_0)$ for some $z_0 \in S$, then the equation $\mathbf{W}(z_0)\mathbf{c} = \mathbf{0}$ has a nontrivial solution c; the solution $w(z) := W(z) c$, which vanishes at z_0, is the trivial solution, and thus the columns of $W(z)$ are linearly dependent for *all* $z \in S$.

We deduce from the foregoing that if W is a solution matrix of (9.3-5) such that $\det W(z_0) \neq 0$ for some $z_0 \in S$, then $\det W(z) \neq 0$ for all $z \in S$. This fact can be given a quantitative formulation. We use the notation

$$\text{tr } A := \sum_{i=1}^{n} a_{ii}$$

for the **trace** of an $n \times n$ matrix $A = (a_{ij})$.

THEOREM 9.3e

If W is any solution matrix of (9.3-5), then for all $z \in S$ and all $z_0 \in S$

$$\det W(z) = \det W(z_0) \exp \int_{z_0}^{z} \text{tr } A(t) \, dt. \tag{9.3-6}$$

Proof. Let the *rows* of $W = (w_{ij})$ be denoted by $r_i := (w_{i1}, w_{i2}, \ldots, w_{in})$, so that

$$W = \begin{pmatrix} r_1 \\ r_2 \\ \vdots \\ r_n \end{pmatrix}.$$

By a well-known rule for differentiating determinants we then have

$$(\det W)' = \det \begin{pmatrix} r'_1 \\ r_2 \\ \vdots \\ r_n \end{pmatrix} + \det \begin{pmatrix} r_1 \\ r'_2 \\ \vdots \\ r_n \end{pmatrix} + \cdots + \det \begin{pmatrix} r_1 \\ r_2 \\ \vdots \\ r'_n \end{pmatrix}.$$

Because the columns of W satisfy (LH),

$$w'_{ij} = \sum_{k=1}^{n} a_{ik} w_{kj}, \qquad i, j = 1, \ldots, n.$$

Thus the derivative of each row can be written as a linear combination of all rows:

$$r'_i = \sum_{k=1}^{n} a_{ik} r_k, \qquad i = 1, \ldots, n,$$

and by elementary transformations on rows we find

$$\det \begin{pmatrix} \mathbf{r}_1' \\ \mathbf{r}_2 \\ \vdots \\ \mathbf{r}_n \end{pmatrix} = a_{11} \det \begin{pmatrix} \mathbf{r}_1 \\ \mathbf{r}_2 \\ \vdots \\ \mathbf{r}_n \end{pmatrix} = a_{11} \det \mathbf{W},$$

and similar expressions for the other determinants. Thus we obtain for $\det \mathbf{W}$ the scalar first-order differential equation

$$(\det \mathbf{W})' = \operatorname{tr} \mathbf{A} \det \mathbf{W},$$

which on integration yields the statement of the theorem. ∎

If $\mathbf{W}$ is any fundamental matrix of (LH), and if $\mathbf{C}$ is any nonsingular constant matrix, then each column of the matrix $\mathbf{WC}$ is a linear combination of solutions of (LH), hence $\mathbf{WC}$ is a matrix solution. Because $\det \mathbf{WC} = \det \mathbf{W} \det \mathbf{C} \neq 0$, the matrix $\mathbf{WC}$ is likewise a fundamental matrix. The question is whether all fundamental matrices can be obtained in this way. Let $\mathbf{W}$ be the given fundamental matrix, and let $\mathbf{V}$ be any other fundamental matrix. We form $\mathbf{Q} := \mathbf{W}^{-1}\mathbf{V}$. Differentiation yields $\mathbf{Q}' = -\mathbf{W}^{-1}\mathbf{W}'\mathbf{W}^{-1}\mathbf{V} + \mathbf{W}^{-1}\mathbf{V}'$ which on using the differential equation simplifies to $-\mathbf{W}^{-1}\mathbf{A}\mathbf{W}\mathbf{W}^{-1}\mathbf{V} + \mathbf{W}^{-1}\mathbf{A}\mathbf{V} = \mathbf{0}$. Thus $\mathbf{Q}$ is constant, $\mathbf{Q} = \mathbf{C}$, and it follows that $\mathbf{V} = \mathbf{WC}$. We thus have proved:

THEOREM 9.3f

The set of all fundamental matrices of (LH) *coincides with the set of all matrices* $\mathbf{WC}$, *where* $\mathbf{W}$ *is a fixed fundamental matrix and* $\mathbf{C}$ *is an arbitrary nonsingular matrix.*

It should be noted that if $\mathbf{W}$ is a fundamental matrix and $\mathbf{C}$ is a nonsingular matrix, then $\mathbf{CW}$ need not be a fundamental matrix.

There are not many cases in which a fundamental matrix can be exhibited in terms of elementary matrix functions. Two such cases are noted below.

EXAMPLE 1

Let $\mathbf{A}$ be a *constant* matrix. Applying successive approximations to the equation

$$\mathbf{W}' = \mathbf{A}\mathbf{W}$$

under the initial condition $\mathbf{W}(0) = \mathbf{I}$ we find the solution matrix

$$\mathbf{W}(z) := \mathbf{I} + z\mathbf{A} + \frac{z^2}{2!}\mathbf{A}^2 + \cdots = e^{z\mathbf{A}},$$

which is fundamental because $\det \mathbf{W}(0) = 1$.

EXAMPLE **2**

Again let **A** be a constant matrix, and let D be a simply connected region not containing the origin. Then $\log z$ can be defined as a single-valued analytic function in D. We assert that the differential equation

$$\mathbf{w}' = \frac{1}{z}\mathbf{A}\mathbf{w}$$

possesses in D the fundamental matrix

$$\mathbf{W}(z) := z^{\mathbf{A}} = e^{\mathbf{A}\log z}.$$

This follows because

$$(e^{\mathbf{A}\log z})' = \frac{1}{z}\mathbf{A}\,e^{\mathbf{A}\log z},$$

and because $\det(1^{\mathbf{A}}) = \det e^{\mathbf{A}\log 1} = \det \mathbf{I} = 1$.

As is seen in later sections, fundamental matrices can be constructed by series expansions in many other cases. However, the reader is warned against inferring from the one-dimensional case that the matrix

$$\mathbf{M}(z) := \exp\!\left(\int_{\hat{z}}^{z} \mathbf{A}(t)\,dt\right)$$

is a fundamental matrix of (LH) for all functions $\mathbf{A}(z)$. The matrix **M** is a solution of (LH) only if $\mathbf{A}(t)$ and $\mathbf{A}(s)$ commute for all $t, s \in S$, which is the case, for instance, if there exists a constant matrix **T** such that $\mathbf{T}\mathbf{A}(z)\mathbf{T}^{-1}$ is diagonal for all $z \in S$.

The knowledge of a fundamental matrix **W** enables one to write a simple formula for the solution **w** of the nonhomogeneous equation

$$\mathbf{w}' = \mathbf{A}(z)\mathbf{w} + \mathbf{b}(z) \tag{LNH}$$

satisfying a given initial condition

$$\mathbf{w}(\hat{z}) = \hat{\mathbf{w}}.$$

Because **W** is nonsingular, **w** can in any case be written in the form $\mathbf{w} = \mathbf{W}\mathbf{c}$, where **c**, instead of being constant, now is an analytic function we wish to determine. Inserting into the differential equation yields

$$\mathbf{W}'\mathbf{c} + \mathbf{W}\mathbf{c}' = \mathbf{A}\mathbf{W}\mathbf{c} + \mathbf{b}$$

or, since $\mathbf{W}' = \mathbf{A}\mathbf{W}$,

$$\mathbf{c}' = \mathbf{W}^{-1}\mathbf{b}.$$

Hence

$$\mathbf{c}(z) = \int_{\hat{z}}^{z} \mathbf{W}^{-1}(t)\mathbf{b}(t)\, dt + \mathbf{c}_0,$$

where $\mathbf{c}_0$ is constant. The initial condition yields $\mathbf{c}_0 = \mathbf{W}^{-1}(\hat{z})\hat{\mathbf{w}}$. We thus have obtained the so-called **variation-of-constants formula**,

$$\mathbf{w}(z) = \mathbf{W}(z)\mathbf{W}^{-1}(\hat{z})\hat{\mathbf{w}} + \mathbf{W}(z)\int_{\hat{z}}^{z} \mathbf{W}^{-1}(t)\mathbf{b}(t)\, dt. \qquad (9.3\text{-}7)$$

It exhibits $\mathbf{w}$ as the sum of the solution of (LH) satisfying the given initial condition and of the solution of (LNH) vanishing at $\hat{z}$.

It is remarkable that the matrix $\mathbf{W}^{-1}$, the inverse of the fundamental matrix, can be found as the solution of a differential equation closely related to (LH). Indeed,

$$(\mathbf{W}^{-1})' = -\mathbf{W}^{-1}\mathbf{W}'\mathbf{W}^{-1} = -\mathbf{W}^{-1}\mathbf{A}.$$

Denoting the transposed matrix with the superscript T, this may be written

$$((\mathbf{W}^{-1})^{T})' = -\mathbf{A}^{T}(z)(\mathbf{W}^{-1})^{T}.$$

We thus see that the matrix $\mathbf{V} := (\mathbf{W}^{-1})^{T}$ is a fundamental matrix of the system

$$\mathbf{v}' = -\mathbf{A}^{T}(z)\mathbf{v}.$$

This makes it possible to determine $\mathbf{W}^{-1}$ by solving a differential equation instead of computing an inverse.

An important application of the theory of linear systems of differential equations is the theory of the linear differential equation of order n for a single unknown function u,

$$p_0(z)u^{(n)} + p_1(z)u^{(n-1)} + \cdots + p_n(z)u = q(z),$$

where the functions $p_0, p_1, \ldots, p_n, q$ are analytic in some simply connected region S. In applications, the function p_0 frequently has zeros. This leads to analytical complications that will be dealt with in detail in §§9.4–9.12. Here we assume that $p_0(z) \neq 0$, $z \in S$. We then may divide by the leading coefficient and, on renaming the coefficients, write the equation in the form

$$u^{(n)} + p_1(z)u^{(n-1)} + \cdots + p_n(z)u = q(z), \qquad (\mathrm{L_n})$$

where $p_1, \ldots, p_n, q$ are again analytic in S.

With the equation (L_n) we associate the system

$$\mathbf{w}' = \mathbf{A}(z)\mathbf{w} + \mathbf{b}(z), \qquad\qquad\qquad\text{(NH)}$$

where

$$\mathbf{A}(z) := \begin{pmatrix} 0 & 1 & 0 & \cdots & 0 \\ 0 & 0 & 1 & \cdots & 0 \\ & \cdots\cdots\cdots & & \\ 0 & 0 & 0 & \cdots & 1 \\ -p_n(z) & -p_{n-1}(z) & -p_{n-2}(z) & \cdots & -p_1(z) \end{pmatrix}, \qquad (9.3\text{-}8)$$

$$\mathbf{b}(z) := \begin{pmatrix} 0 \\ 0 \\ \vdots \\ 0 \\ q(z) \end{pmatrix}.$$

It is seen immediately that if u is a solution of (L_n), then the vector

$$\mathbf{w} := \begin{pmatrix} u \\ u' \\ u'' \\ \vdots \\ u^{(n-1)} \end{pmatrix}$$

is a solution of (NH). Conversely, if $\mathbf{w}$ is a solution of (NH) satisfying the initial condition $\mathbf{w}(\hat{z}) = \hat{\mathbf{w}}$, where

$$\hat{\mathbf{w}} = \begin{pmatrix} \hat{w}_1 \\ \hat{w}_2 \\ \vdots \\ \hat{w}_n \end{pmatrix},$$

then the first component $w_1 = u$ of the vector $\mathbf{w}$ is a solution of (L_n) satisfying the initial conditions

$$u(\hat{z}) = \hat{w}_1, \, u'(\hat{z}) = \hat{w}_2, \ldots, u^{(n-1)}(\hat{z}) = \hat{w}_n. \qquad (9.3\text{-}9)$$

All the facts proved earlier about linear systems can thus be reinterpreted as facts about (L_n). In particular, for arbitrarily given complex numbers

$\hat{w}_1, \ldots, \hat{w}_n$ and arbitrary $\hat{z} \in S$ there exists a unique solution u of (L_n) that is analytic in S and satisfies the initial conditions (9.3-9). The totality of all solutions of the homogeneous equation (L_n) ($q(z) = 0$) forms an n-dimensional vector space; every solution of the homogeneous equation can be written as a linear combination of a **fundamental system** of solutions, that is, a set of n linearly independent solutions. If the functions $u_1, u_2, \ldots, u_n$ are solutions of the homogeneous equation (L_n), the determinant of the corresponding solution matrix of (LH) becomes the **Wronskian determinant**

$$w(z) = w(z; u_1, \ldots, u_n) := \begin{vmatrix} u_1 & u_2 & \cdots & u_n \\ u_1' & u_2' & \cdots & u_n' \\ u_1^{(n-1)} & u_2^{(n-1)} & \cdots & u_n^{(n-1)} \end{vmatrix} \quad (9.3\text{-}10)$$

Thus the system $(u_1, \ldots, u_n)$ is fundamental for (L_n) if and only if its Wronskian determinant $w(z) \neq 0$ for at least one $z \in S$. If $w(z) \neq 0$ for some z, then $w(z) \neq 0$ for all $z \in S$, as can also be seen from the formula

$$w(z) = w(\hat{z}) \exp\left(-\int_{\hat{z}}^{z} p_1(t) \, dt \right)$$

following from Theorem 9.3e in view of tr $\mathbf{A} = -p_1$. Finally, the variation-of-constants formula (9.3-7) leads to the following formula for the solution u of the nonhomogeneous equation (L_n) satisfying $u(\hat{z}) = u'(\hat{z}) = \cdots = u^{(n-1)}(\hat{z}) = 0$:

$$u(z) = \int_{\hat{z}}^{z} \frac{\begin{vmatrix} u_1(t) & u_2(t) & \cdots & u_n(t) \\ u_1'(t) & u_2'(t) & \cdots & u_n'(t) \\ u_1^{(n-2)}(t) & u_2^{(n-2)}(t) & \cdots & u_n^{(n-2)}(t) \\ u_1(z) & u_2(z) & & u_n(z) \end{vmatrix}}{w(t)} q(t) \, dt. \quad (9.3\text{-}11)$$

Transformation of Variable

For purposes of formal simplification or for other reasons, it is often necessary to introduce new variables in a differential equation. For later reference we state the basic facts concerning such transformations.

THEOREM 9.3g (transformation of independent variable in linear systems)

Let $\mathbf{w}$ be a solution of $\mathbf{w}' = \mathbf{A}(z)\mathbf{w}$, where $\mathbf{A}$ is analytic in some region S, and let $z = g(t)$, where g is analytic in a region T, $g(T) \subset S$, $g'(t) \neq 0$, $t \in T$. Then *the vector* $\mathbf{v} := \mathbf{w} \circ g$ *is a solution of*

$$\mathbf{v}' = g'(t)\mathbf{A}(g(t))\mathbf{v}.$$

Remark.   It is not required that g be one to one. This must be assumed only if $\mathbf{w}$ is to be recovered from $\mathbf{v}$.

Proof. $\mathbf{w}$ is analytic by hypothesis. Thus by the chain rule

$$\mathbf{v}'(t) = \mathbf{w}'(g(t))g'(t) = g'(t)\mathbf{A}(g(t))\mathbf{w}(g(t))$$

$$= g'(t)\mathbf{A}(g(t))\mathbf{v}(t). \quad \blacksquare$$

THEOREM 9.3h (transformation of dependent variable in linear systems)

Let $\mathbf{w}$ be a solution of $\mathbf{w}' = \mathbf{A}(z)\mathbf{w}$, where $\mathbf{A}$ is analytic in a region S, and let $\mathbf{B}(z)$ be an analytic and nonsingular matrix in S. If $\mathbf{w} = \mathbf{B}\mathbf{v}$, then $\mathbf{v}$ satisfies the equation

$$\mathbf{v}' = \mathbf{B}^{-1}(z)[\mathbf{A}(z)\mathbf{B}(z) - \mathbf{B}'(z)]\mathbf{v}.$$

The proof is by straightforward substitution.

For convenience we state the analogous results for scalar second-order equations.

THEOREM 9.3i (transformation of variables in scalar second-order equation)

Let u be a solution of

$$u'' + p(z)u' + q(z)u = 0, \tag{9.3-12}$$

where p and q are analytic in S.

(a) *If g is as in Theorem 9.3g, then the function $v := u \circ g$ (i.e., $v(t) := u(g(t))$, $t \in T$) satisfies the differential equation*

$$v'' + \left[p(g(t))g'(t) - \frac{g''(t)}{g'(t)} \right] v' + q(g(t))\{g'(t)\}^2 v = 0.$$

(b) *If the function h is analytic and different from zero in S, and if $u = hw$, $l := h'/h$, then w satisfies*

$$w'' + [2l(z) + p(z)]w' + [l^2(z) + l'(z) + p(z)l(z) + q(z)]w = 0.$$

The proofs are again by straightforward verification.

In dealing with scalar second-order equations, it is often convenient to assume that the coefficient of the first derivative is zero. The question arises whether the multiplier h can be chosen in such a way that this is the case. By

Theorem 9.3i, it is only necessary to chose h such that

$$\frac{h'}{h} = -\tfrac{1}{2}p.$$

We thus obtain at once

THEOREM 9.3j (reduction of scalar second order equation to special form)

Let the coefficient p in (9.3-12) possess a primitive in S; that is, let there exist an analytic function r such that $r' = p$. (This is always the case if S is simply connected.) *If u is a solution of (9.3-12), then the function*

$$w(z) := e^{r(z)/2} u(z)$$

satisfies the special second-order equation

$$w'' + [q(z) - \tfrac{1}{2}p'(z) - \tfrac{1}{4}\{p(z)\}^2]w = 0.$$

PROBLEMS

1. Prove the Theorems 9.3a and 9.3c by the method of successive approximations.
2. Find the general solution of the system

$$w_1' = w_2 + w_3$$
$$w_2' = w_3 + w_1$$
$$w_3' = w_1 + w_2$$

    (a) by matrix methods;
    (b) by converting the system to scalar equations for single unknown functions, using methods of elimination.
3. Let the matrices $\mathbf{A}$ and $\mathbf{B}$ be analytic in some simply connected region S containing the origin, and let $\mathbf{W}$ be a fundamental matrix of the system

$$\mathbf{w}' = \mathbf{A}(z)\mathbf{w}.$$

    Show that the solution of the matrix initial value problem

$$\mathbf{Z}' = \mathbf{A}(z)\mathbf{Z} + \mathbf{Z}\mathbf{A}^T(z) + \mathbf{B}(z), \qquad \mathbf{Z}(0) = \mathbf{0},$$

    can be represented in the form

$$\mathbf{Z}(z) = \mathbf{W}(z)\int_0^z \mathbf{W}^{-1}(t)\mathbf{B}(t)(\mathbf{W}^{-1})^T(t)\, dt\; \mathbf{W}^T(z).$$

4. Let the scalar function

$$p(z) := \sum_{n=0}^{\infty} a_n z^n$$

be analytic in $|z| < \alpha$, and let

$$w(z) = \sum_{n=0}^{\infty} b_n z^n$$

be a nontrivial solution of the scalar equation

$$w' = p(z)w.$$

If $\mathbf{A}$ is any constant matrix, show that the series

$$\mathbf{W}(z) := \sum_{n=0}^{\infty} b_n \mathbf{A}^n z^n$$

represents for $|z| < \alpha \|\mathbf{A}\|^{-1}$ a fundamental matrix for the system

$$w' = \mathbf{P}(z)\mathbf{w},$$

where

$$\mathbf{P}(z) := \mathbf{A}p(\mathbf{A}z) = \sum_{n=0}^{\infty} a_n \mathbf{A}^{n+1} z^n.$$

5. Show that the system

$$w' = \alpha \mathbf{A}(\mathbf{I} - z\mathbf{A})^{-1}\mathbf{w}$$

$(|z| < \|\mathbf{A}\|^{-1})$ possesses the fundamental matrix

$$\mathbf{W}(z) := (\mathbf{I} - z\mathbf{A})^{-\alpha} = \sum_{n=0}^{\infty} \frac{(\alpha)_n}{n!} \mathbf{A}^n z^n.$$

6. Let $\mathbf{A}$ be a matrix of order n with n distinct eigenvalues λ_i and corresponding eigenvectors $\mathbf{v}_i$ $(i = 1, 2, \ldots, n)$, and let all $\frac{1}{2}n(n+1)$ sums $\lambda_i + \lambda_j$ $(i \leqslant j)$ be different. Show that the matrix differential equation

$$\mathbf{W}' = \mathbf{A}\mathbf{W} + \mathbf{W}\mathbf{A}^T$$

has the linearly independent solutions

$$\mathbf{W}_{ij}(z) := \mathbf{v}_i \mathbf{v}_j^T \exp(\lambda_i + \lambda_j)z, \qquad 1 \leqslant i \leqslant j \leqslant n.$$

7. *The Wronskian determinant.* Prove: Any system of n functions u_i $(i = 1, \ldots, n)$ analytic in some region S whose Wronskian determinant is different from zero is a fundamental system of some linear differential equation of type $(\mathrm{L_n})$. [If u is any linear combination of the u_i, then

$$\begin{vmatrix} u_1 & u_2 & \cdots & u_n & u \\ u_1' & u_2' & \cdots & u_n' & u' \\ \vdots & \vdots & & \vdots & \vdots \\ u_1^{(n)} & u_2^{(n)} & \cdots & u_n^{(n)} & u^{(n)} \end{vmatrix} = 0.]$$

8. Let the functions u_i $(i = 1, 2, \ldots, n)$ in some region S form a system of linearly independent solutions of the differential equation

$$u^{(n)} + p_1(z)u^{(n-1)} + \cdots + p_n(z)u = 0$$

and also of the equation

$$u^{(n)} + q_1(z)u^{(n-1)} + \cdots + q_n(z)u = 0.$$

Show that $q_j = p_j$, $j = 1, 2, \ldots, n$.

9. Can a scalar second order equation be reduced to special form by a transformation of the *independent* variable?

§9.4. LINEAR SYSTEMS WITH ISOLATED SINGULARITIES

The theory developed in the preceding section satisfactorily describes the structure of the solutions of a linear differential equation in the neighborhood of a point where the matrix **A** is analytic. However, many important problems in mathematical physics and applied mathematics (for instance, all problems involving the Laplacian differential operator under the assumption of spherical or cylindrical symmetry) give rise to differential equations in which the matrix **A** arising in the systems formulation has an isolated singularity, and it is precisely the behavior of the solutions near the singularity that is of interest. The singularity can be located either at a finite point $z = z_0$ or at the point $z = \infty$ of the extended complex plane. However, by the trivial substitutions $z' := z - z_0$ or $z' := z^{-1}$, which do not alter the analytic character of **A** in the neighborhood of the singularity, we can always achieve that the singularity is located at $z = 0$.

Consequently, we here consider the homogeneous linear system

$$\mathbf{w}' = \mathbf{A}(z)\mathbf{w}, \tag{LH$_0$}$$

where $\mathbf{A}(z)$ is analytic in a punctured disk $D : 0 < |z| < \gamma$ ($\gamma > 0$ or $\gamma = \infty$). Generally, we are interested in the case in which the singularity is not removable, because otherwise the results of §9.3 are applicable. Then **A** will either have a pole (i.e., all elements of the matrix **A** have at most poles) or will have an essential singularity. It will be seen that already pole-type singularities can lead to considerable complications.

The fundamental fact about differential equations with isolated singularities is the occurrence of multivalued solutions. This can already be seen from the scalar differential equation

$$w' = \frac{r}{z}w, \tag{9.4-1}$$

where r is not an integer. Here the solutions are $w = cz^r$, where c is a constant. Although a single-valued branch of the solution can be defined in a neighborhood of any point $z_0 \neq 0$, this solution will not return to its initial value if continued analytically along a circle around the origin. However, because $z^r = e^{r \log z}$, we can say with some vagueness that "the multivaluedness of the solution is no worse than that of the function $\log z$."

A similar situation prevails, in a sense to be made precise, in the general case of equation (LH$_0$), with $\mathbf{A}(z)$ analytic in the punctured disk $D: 0 < |z| < \gamma$. Let z_0 be an arbitrary point of D, and let $N: |z - z_0| < \delta$ be a neighborhood of z_0, $N \subset D$. By Theorem 9.3a, there exists a unique solution $\mathbf{w}_0$ of (LH$_0$) taking a preassigned value $\hat{\mathbf{w}}$ at z_0. By Corollary 9.3b, this solution can be continued analytically along an arbitrary path in D that emanates from z_0 (see Fig. 9.4). The problem is to find an analytical representation for all these continuations.

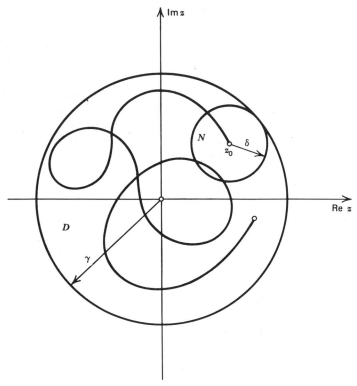

Fig. 9.4.

We recall from §4.6 the notion of the *continuous argument*. If $\Gamma: z = z(\tau)$, $\alpha \leqslant \tau \leqslant \beta$, is any arc not passing through O, then a continuous argument of z on Γ is any continuous function ϕ defined on $[\alpha, \beta]$ such that, for every $\tau \in [\alpha, \beta]$, $\phi(\tau)$ equals one of the values of $\arg z(\tau)$. According to Theorem 4.6a such a continuous argument can always be defined, and any two continuous arguments differ by a constant integral multiple of 2π.

Using continuous arguments, the multiple-valued solutions of (LH_0) can be described as follows.

THEOREM 9.4a

Let z_0 be a point of the annulus $D : 0 < |z| < \gamma$, let $|z_0| = \rho_0$, let θ_0 be a value of arg z_0, and let $\hat{\mathbf{w}}$ be a given vector. There exists a function $\tilde{\mathbf{w}}(\rho, \theta)$ of two real variables ρ and θ, defined for $0 < \rho < \gamma$ and $-\infty < \theta < \infty$, with the following properties:

(i) $\hat{\mathbf{w}}(\rho_0, \theta_0) = \hat{\mathbf{w}}$;

(ii) In a neighborhood of any point (ρ_1, θ_1), $0 < \rho_1 < \gamma$, $\hat{\mathbf{w}}(\rho, \theta)$ is an analytic function of the single complex variable $z := \rho e^{i\theta}$;

(iii) The analytic functions defined in (ii) are solutions of (LH_0);

(iv) Let $\Gamma : z = z(\tau), \alpha \leqslant \tau \leqslant \beta$, be a path in D emanating from z_0. If $\theta(\tau)$ is the continuous argument of z on Γ with initial value θ_0, the values on Γ of the analytic continuation of the solution of (LH_0) along Γ assuming the initial value $\hat{\mathbf{w}}$ at z_0 are given by

$$\tilde{\mathbf{w}}(|z(\tau)|, \theta(\tau)), \qquad \alpha \leqslant \tau \leqslant \beta.$$

Proof. Let $N : |z - z_0| < \epsilon$ be a neighborhood of z_0, $N \subset D$, and let $\mathbf{w}(z)$ be the unique solution of (LH_0) in N assuming the value $\hat{\mathbf{w}}$ at z_0. For $z \in N$, let $t := \log z$ denote that branch of the logarithm that is analytic in N and takes the value $t_0 := \mathrm{Log}\, \rho_0 + i\theta_0$ at z_0. The function $t = \log z$ defines an analytic one-to-one mapping of N onto a certain neighborhood N^* of t_0. For $t \in N^*$ we define $\mathbf{v}(t) := \mathbf{w}(e^t)$. The function $\mathbf{v}$ satisfies

$$\mathbf{v}'(t) = \mathbf{w}'(e^t)e^t = e^t \mathbf{A}(e^t)\mathbf{v}(t)$$

and thus a solution of the differential equation

$$\mathbf{v}' = \mathbf{B}(t)\mathbf{v}, \tag{9.4-2}$$

where

$$\mathbf{B}(t) := e^t \mathbf{A}(e^t), \ t \in N^*. \tag{9.4-3}$$

The relation (9.4-3) can be used to define $\mathbf{B}(t)$ in the entire half-plane $D^* : -\infty < \mathrm{Re}\, t < \mathrm{Log}\, \gamma$ as an analytic function, and with this definition the differential equation (9.4-2) has a sense in the whole region D^*. It follows by Theorem 9.3b that $\mathbf{v}$ can be continued analytically from t_0 along any path Γ^* in D^*. Because D^* is simply connected, the monodromy theorem (Theorem 3.5c) implies that the values obtained by all these continuations are the values of a single analytic function $\mathbf{v}$ in D^*.

We now define $\tilde{\mathbf{w}}$:

$$\tilde{\mathbf{w}}(\rho, \theta) := \mathbf{v}(\mathrm{Log}\, \rho + i\theta). \tag{9.4-4}$$

It is clear that $\tilde{\mathbf{w}}$ satisfies (i). To verify (ii), let N_1 be a neighborhood of (ρ_1, θ_1) such that $|\theta - \theta_1| < \pi$ for $(\rho, \theta) \in N_1$. Then if $z := \rho e^{i\theta}$, $\mathrm{Log}\, \rho + i\theta$ is a well-defined analytic branch of $\log z$. Hence $\tilde{\mathbf{w}}(\rho, \theta) = \mathbf{v}(\log z)$ is a composition of analytic functions, and therefore an analytic function of z. Calling this function $\mathbf{w}_1(z)$, we have, using (9.4-2) and (9.4-3),

$$\mathbf{w}_1'(z) = \mathbf{v}'(\log z)z^{-1} = \mathbf{B}(\log z)z^{-1}\mathbf{v}(\log z) = \mathbf{A}(z)\mathbf{w}_1(z),$$

proving (iii). To prove (iv), we have to show that for any point $z_1 := z(\tau_1)$ of Γ the values of $\tilde{\mathbf{w}}(|z(\tau)|, \theta(\tau))$ agree for $|\theta - \theta_1|$ sufficiently small with the values on Γ of a solution of (LH$_0$) which is analytic at z_1. This is a direct consequence of (ii) and (iii) above. ∎

For convenience, a function of (ρ, θ) defined on a set $\tilde{D}: 0 < \rho < \gamma$, $-\infty < \theta < \infty$ and enjoying property (ii) above is called **log-holomorphic** on the punctured disk D. We adhere to the established, although ambiguous, practice of denoting the pair of arguments (ρ, θ) of a log-holomorphic function by the symbol z, where it is understood that $z = \rho e^{i\theta}$. If z denotes the point (ρ, θ), the symbol $z e^{2i\pi}$ is used to denote the point $(\rho, \theta + 2\pi)$.

A log-holomorphic function is analytic if and only if it is periodic in θ with period 2π, for in that case and in that case only it always returns to its initial value if continued along a circle that winds around the origin. In general, however, a log-holomorphic function must not be thought of as a function defined on a set of points in the complex plane, because to each of the infinitely many values of $\theta = \arg z$ there may correspond different values of a log-holomorphic function $\mathbf{w}$. It may help the imagination to think of $\mathbf{w}$ as a single-valued function defined on a surface winding around the origin like a circular staircase pressed flat. An increase of θ by 2π then amounts to going around one turn of the staircase, which does not bring us back to the starting point. Somewhat vulgarizing Riemann's idea, this surface is called the **Riemann surface** of the function $\log z$.

It has been shown above that if $\mathbf{w}_0$ is an analytic solution of (LH$_0$) defined in a neighborhood of a point $z_0 \in D$, all analytic continuations of $\mathbf{w}_0$ can be obtained from a single log-holomorphic solution of (LH$_0$). An analogous statement also applies to the continuations of any fundamental matrix $\mathbf{W}_0$ of (LH$_0$). We now show that these log-holomorphic fundamental matrices have a surprisingly simple structure:

THEOREM 9.4b

Let $\mathbf{A}$ be analytic in the punctured disk $D: 0 < |z| < \gamma$, and let $\mathbf{W}$ be a log-holomorphic fundamental matrix of (LH$_0$) in $\tilde{D}$. Then there exist a constant matrix $\mathbf{R}$ and a matrix $\mathbf{P}$ that is analytic in D such that

$$\mathbf{W}(z) = \mathbf{P}(z)z^{\mathbf{R}} \quad \text{for all } z \in \tilde{D}. \tag{9.4-5}$$

Here $z^{\mathbf{R}} := e^{\mathbf{R}\log z}$; see §2.6. We thus see that the multiple valuedness of

W can be absorbed into the simple factor $z^{\mathbf{R}}$, entirely as in the case of the simple model equation $w' = (r/z)w$.

Proof of Theorem 9.4b. By the definition of a fundamental matrix,

$$\mathbf{W}'(z) = \mathbf{A}(z)\mathbf{W}(z), \, z \in \tilde{D}.$$

Because $\mathbf{A}(z\,e^{2\pi i}) = \mathbf{A}(z)$, there follows

$$\mathbf{W}'(z\,e^{2\pi i}) = \mathbf{A}(z)\mathbf{W}(z\,e^{2\pi i}), z \in \tilde{\mathbf{D}}.$$

Hence $\mathbf{V}(z) := \mathbf{W}(z\,e^{2\pi i})$ is a fundamental matrix, and thus, by Theorem 9.3f,

$$\mathbf{W}(z\,e^{2\pi i}) = \mathbf{W}(z)\mathbf{C},$$

where **C** is constant and nonsingular. Thus by Theorem 2.6h, there exists a matrix **R** (not uniquely determined) such that

$$\mathbf{C} = e^{2\pi i \mathbf{R}}. \tag{9.4-6}$$

Thus we also may write

$$\mathbf{W}(z\,e^{2\pi i}) = \mathbf{W}(z)\,e^{2\pi i \mathbf{R}}, \, z \in \tilde{D}. \tag{9.4-7}$$

Now let

$$\mathbf{P}(z) := \mathbf{W}(z)z^{-\mathbf{R}}. \tag{9.4-8}$$

Clearly, **P** is log-holomorphic in D. We shall show that it is even analytic in D by showing that **P** is periodic in θ with period 2π. In fact, because the matrices $z^{-\mathbf{R}}$ and $e^{2\pi i \mathbf{R}}$ commute,

$$\mathbf{P}(z\,e^{2\pi i}) = \mathbf{W}(z\,e^{2\pi i})(z\,e^{2\pi i})^{-\mathbf{R}} = \mathbf{W}(z)\,e^{2\pi i \mathbf{R}}\,e^{-2\pi i \mathbf{R}}$$

$$= \mathbf{W}(z)z^{-\mathbf{R}} = \mathbf{P}(z). \quad \blacksquare$$

The matrix **P** occurring in the statement of Theorem 9.4b is not uniquely determined by the differential equation. It depends on the choice of the fundamental solution **W** and of the matrix **R** in (9.4-6). If **W** and **R** can be chosen such that the matrix **P** (which has an isolated singularity at $z = 0$) has at most a pole at $z = 0$, then we shall call $z = 0$ a **regular singular point** of the differential equation (German terminology: *Stelle der Bestimmtheit* or *ausserwesentlich singuläre Stelle*). If the matrix **P** has an essential singularity for all choices of **W** and of **R**, $z = 0$ is called an **irregular singular point** (*Stelle der Unbestimmtheit, wesentlich singuläre Stelle*). The same terminology is used if the matrix $\mathbf{A}(z)$ has an isolated singularity at an arbitrary point $z = z_0$.

The German terms for regular and irregular singular point (whose literal translation is point of determinacy and point of indeterminacy, respectively) are suggested by the following fact.

THEOREM 9.4c

The point $z = 0$ is a regular singular point of (LH$_0$) *if and only if for every solution* **w** *there exists a real number* κ *such that*

$$\rho^\kappa \|\mathbf{w}(\rho\, e^{i\theta})\| \to 0 \qquad (9.4\text{-}9)$$

if $\rho \to 0$ *and* θ *is bounded.*

Thus, roughly speaking, $z = 0$ is a regular singular point if and only if no solution grows more rapidly than a fixed power of $|z|$ as $|z| \to 0$.

Proof of Theorem 9.4c.   (a) Let $z = 0$ be a regular singular point, and let $\mathbf{W}(z) = \mathbf{P}(z)\, z^{\mathbf{R}}$ be a log-holomorphic fundamental matrix such that $\mathbf{P}$ has at most a pole. Because every solution can be represented in the form $\mathbf{W}(z)\mathbf{c}$, where $\mathbf{c}$ is a fixed vector, and because

$$\|\mathbf{W}(z)\mathbf{c}\| \leqslant \|\mathbf{W}(z)\|\, \|\mathbf{c}\|,$$

it suffices to estimate the norm of $\mathbf{W}$. We make use of the fact that

$$\|\mathbf{W}(z)\| = \|\mathbf{P}(z) z^{\mathbf{R}}\| \leqslant \|\mathbf{P}(z)\|\, \|z^{\mathbf{R}}\|.$$

Because $\mathbf{P}$ has at most a pole at $z = 0$, there exists a real number μ such that

$$|z|^\mu \|\mathbf{P}(z)\| \to 0$$

as $z \to 0$. To estimate $\|z^{\mathbf{R}}\|$ we use

$$z^{\mathbf{R}} = e^{\mathbf{R} \log z} = e^{\mathbf{R} \operatorname{Log} \rho}\, e^{i\mathbf{R}\theta},$$

hence

$$\|z^{\mathbf{R}}\| \leqslant \|e^{\mathbf{R} \operatorname{Log} \rho}\|\, \|e^{i\mathbf{R}\theta}\|. \qquad (9.4\text{-}10)$$

By the power series definition of the exponential, $\|e^{\mathbf{A}}\| \leqslant e^{\|\mathbf{A}\|}$ for any matrix $\mathbf{A}$, hence

$$\|e^{\mathbf{R} \operatorname{Log} \rho}\| \leqslant e^{|\operatorname{Log} \rho|\, \|\mathbf{R}\|} = \rho^{-\|\mathbf{R}\|}$$

if $0 < \rho \leqslant 1$. By the same token, if $|\theta| \leqslant \theta_0$,

$$\|e^{i\mathbf{R}\theta}\| \leqslant e^{\theta_0 \|\mathbf{R}\|}.$$

It follows that for $z = \rho\, e^{i\theta}$ where $0 < \rho \leqslant 1$ and $|\theta| \leqslant \theta_0$,

$$\|z^{\mathbf{R}}\| \leqslant e^{\theta_0 \|\mathbf{R}\|} \rho^{-\|\mathbf{R}\|}. \qquad (9.4\text{-}11)$$

By (9.4-11), (9.4-9) is now seen to be true with $\kappa := \mu + \|\mathbf{R}\|$.

   (b) Let every solution satisfy (9.4-9). Then a similar relation must also hold for every fundamental matrix $\mathbf{W}$ of (LH$_0$). (For if not, the relation could not hold for at least one element $w_{ij}(z)$ of $\mathbf{W}$, and thus it could not hold for the solution column containing that element.) Now from (9.4-8),

$$\|\mathbf{P}(z)\| \leqslant \|\mathbf{W}(z)\| \|z^{-\mathbf{R}}\|$$

and (9.4-11) shows that

$$|z|^{\mu} \|\mathbf{P}(z)\| \to 0 \qquad (9.4\text{-}12)$$

for $|z| \to 0$, where $\mu := \kappa + \|\mathbf{R}\|$. Because $\mathbf{P}$ has an isolated singularity at $z = 0$, Theorem 4.4e now shows that this singularity is at most a pole. ∎

Theorem 9.4c implies that in the representation (9.4-8) the matrix $\mathbf{P}$, although not uniquely determined, either always has at most a pole or always has an essential singularity for a given differential equation.

The question may be asked whether it is possible to determine the character of a singular point directly from the differential equation, without first constructing a fundamental solution. There indeed exists a simple sufficient condition for a singular point to be regular.

THEOREM 9.4d

Let the matrix $\mathbf{A}(z)$ have a pole of order 1 at $z = 0$. Then $z = 0$ is a regular singular point of (LH_0).

Proof. Let $\mathbf{w}$ be a nonzero, log-holomorphic vector solution of (LH_0), and let $\mathbf{h}(\rho, \theta) = \mathbf{w}(\rho\, e^{i\theta})$. Then, using the differential equation,

$$\frac{\partial \mathbf{h}}{\partial \rho}(\rho, \theta) = \frac{d\mathbf{w}}{dz}(\rho\, e^{i\theta})\, e^{i\theta} = \mathbf{A}(\rho\, e^{i\theta})\mathbf{w}(\rho\, e^{i\theta})\, e^{i\theta}$$

and hence

$$\left\|\frac{\partial \mathbf{h}}{\partial \rho}(\rho, \theta)\right\| \leqslant \|\mathbf{A}(\rho\, e^{i\theta})\| \, \|\mathbf{h}(\rho, \theta)\|.$$

Let $0 < \rho_1 < \gamma$. By the hypothesis on $\mathbf{A}$ there exists a constant κ such that

$$\|\mathbf{A}(\rho\, e^{i\theta})\| \leqslant \frac{\kappa}{\rho}, \quad 0 < \rho \leqslant \rho_1.$$

We thus have

$$\left\|\frac{\partial \mathbf{h}}{\partial \rho}(\rho, \theta)\right\| \leqslant \frac{\kappa}{\rho}\|\mathbf{h}(\rho, \theta)\|. \qquad (9.4\text{-}13)$$

We now require the fact that if $\|\mathbf{h}(\rho, \theta)\| \neq 0$, then

$$\left|\frac{\partial}{\partial \rho}\|\mathbf{h}(\rho, \theta)\|\right| \leqslant \left\|\frac{\partial \mathbf{h}}{\partial \rho}(\rho, \theta)\right\|. \qquad (9.4\text{-}14)$$

This follows in view of

$$\frac{\partial}{\partial\rho}\|\mathbf{h}\| = \frac{\partial}{\partial\rho}\left[\sum_{i=1}^{n}|h_i|^2\right]^{1/2}$$

$$= \frac{1}{2\|\mathbf{h}\|}\sum_{i=1}^{n}\left(\frac{\partial h_i}{\partial\rho}\overline{h_i} + h_i\frac{\overline{\partial h_i}}{\partial\rho}\right)$$

$$\leqslant \frac{1}{\|\mathbf{h}\|}\sum_{i=1}^{n}\left|\frac{\partial h_i}{\partial\rho}\right||h_i|$$

$$\leqslant \frac{1}{\|\mathbf{h}\|}\left[\sum_{i=1}^{n}\left|\frac{\partial h_i}{\partial\rho}\right|^2\right]^{1/2}\left[\sum_{j=1}^{n}|h_j|^2\right]^{1/2}$$

$$= \left[\sum_{i=1}^{n}\left|\frac{\partial h_i}{\partial\rho}\right|^2\right]^{1/2}$$

$$= \left\|\frac{\partial\mathbf{h}}{\partial\rho}\right\|,$$

where we have used Cauchy's inequality. Writing $\chi(\rho, \theta) := \|\mathbf{h}(\rho, \theta)\|$, we thus have from (9.4-13) and (9.4-14)

$$\left|\frac{\partial\chi}{\partial\rho}(\rho, \theta)\right| \leqslant \frac{\kappa}{\rho}\chi(\rho, \theta)$$

and thus

$$\frac{\partial\chi}{\partial\rho} + \frac{\kappa}{\rho}\chi \geqslant 0$$

or

$$\frac{\partial}{\partial\rho}(\rho^\kappa\chi) \geqslant 0, \qquad 0 < \rho \leqslant \rho_1.$$

This implies

$$\rho_1^\kappa\chi(\rho_1, \theta) \geqslant \rho^\kappa\chi(\rho, \theta), \qquad 0 < \rho \leqslant \rho_1,$$

and, if for given $\theta_1 > 0$ we define

$$\mu := \max_{|\theta|\leqslant\theta_1}\chi(\rho_1, \theta),$$

we see that

$$\chi(\rho, \theta) \leqslant \frac{\rho_1^\kappa\mu}{\rho^\kappa}$$

for $0 < \rho \leqslant \rho_1$, $|\theta| \leqslant \theta_1$. Because $\chi(\rho, \theta) = \|\mathbf{w}(\rho\, e^{i\theta})\|$ there follows

$$\rho^{\kappa+1}\|\mathbf{w}(\rho\, e^{i\theta})\| \to 0$$

as $\rho \to 0$, $|\theta| \leqslant \theta_1$. Because $\mathbf{w}$ is an arbitrary solution, the differential equation thus satisfies the criterion of Theorem 9.4c, and it follows that the singular point $z = 0$ is regular. ■

The converse of Theorem 9.4d is not true for $n \geqslant 2$, as can be seen from Problem 3.

PROBLEMS

1. The system $\mathbf{w}' = \mathbf{A}(z)\mathbf{w}$, where

$$\mathbf{A}(z) := \begin{vmatrix} 0 & -iz^{-1} & 0 & 0 \\ \dfrac{3}{4i}(z+z^{-3})+\dfrac{1}{2i}z^{-1} & 0 & -\dfrac{3}{4}(z-z^{-3}) & 0 \\ 0 & 0 & 0 & -iz^{-1} \\ -\dfrac{3}{4}(z-z^{-3}) & 0 & -\dfrac{3}{4i}(z+z^{-3})+\dfrac{1}{2i}z^{-1} & 0 \end{vmatrix}$$

occurs in celestial mechanics in the theory of perturbed Keplerian motion.
(a) Show that the system has the four linearly independent solutions

$$\mathbf{w}_1(z) := \begin{vmatrix} \dfrac{1}{2i}(z-z^{-1}) \\[2mm] \dfrac{1}{2}(z+z^{-1}) \\[2mm] -\dfrac{1}{2}(z+z^{-1}) \\[2mm] \dfrac{1}{2i}(z-z^{-1}) \end{vmatrix}, \qquad \mathbf{w}_2(z) := \begin{vmatrix} 3-\dfrac{1}{2}(z^2+z^{-2}) \\[2mm] \dfrac{1}{2i}(z^2-z^{-2}) \\[2mm] -\dfrac{1}{2i}(z^2-z^{-2}) \\[2mm] -(z^2+z^{-2}) \end{vmatrix}$$

$$\mathbf{w}_3(z) := \begin{vmatrix} -\dfrac{1}{2i}(z^2-z^{-2}) \\[2mm] -(z^2+z^{-2}) \\[2mm] 3+\dfrac{1}{2}(z^2+z^{-2}) \\[2mm] -\dfrac{1}{i}(z^2-z^{-2}) \end{vmatrix}, \qquad \mathbf{w}_4(z) := \begin{vmatrix} z+z^{-1}-\dfrac{3}{2}(z-z^{-1})\log z \;\cdot \\[2mm] \dfrac{1}{2i}(z-z^{-1})+\dfrac{3}{2i}(z+z^{-1})\log z \\[2mm] \dfrac{1}{i}(z-z^{-1})-\dfrac{3}{2i}(z+z^{-1})\log z \\[2mm] -\dfrac{1}{2}(z+z^{-1})-\dfrac{3}{2}(z-z^{-1})\log z \end{vmatrix}$$

(b) Construct a fundamental matrix satisfying $\mathbf{W}(1) = \mathbf{I}$.

(c) Determine a matrix $\mathbf{R}$, and represent $\mathbf{W}$ in the form (9.4-5).

2.   Let the distinct eigenvalues of the matrix $\mathbf{R}$ occurring in Theorem 9.4b be $\lambda_1, \lambda_2, \ldots, \lambda_m$ $(m \leqslant n)$. Show that $(\mathrm{LH_0})$ has m solutions of the form

$$\mathbf{w}_k(z) = z^{\lambda_k} \mathbf{u}_k(z),$$

where the functions $\mathbf{u}_k$ are analytic in $|z| < \gamma$.

3.   Show that the converse of Theorem 9.4d is untrue by considering the system

$$\mathbf{w}' = (z^{-2}\mathbf{A} + \mathbf{B})\mathbf{w},$$

where

$$\mathbf{A} := \begin{pmatrix} 0 & 0 \\ -\dfrac{3}{16} & 0 \end{pmatrix}, \qquad \mathbf{B} := \begin{pmatrix} 0 & 1 \\ 0 & 0 \end{pmatrix}.$$

[Show that $w_2 = w_1'$ and find solutions of the form $w_1(z) = z^{\alpha}$.]

4.   If $z = 0$ is a singular point of $(\mathrm{LH_0})$, must every fundamental matrix necessarily be singular? Examine the scalar differential equation

$$w' = \frac{1}{z} w$$

to show that the answer is negative.

5.   Let $z = 0$ be a nonremovable singular point of $\mathbf{A}(z)$, and let $\mathbf{W}(z)$ be a fundamental matrix of $(\mathrm{LH_0})$ which is analytic at $z = 0$. Prove that $\det \mathbf{W}(0) = 0$. [For the sake of illustration, compare the example in Problem 3.]

6.   Let the functions u_i $(i = 1, \ldots, n)$ be analytic in a region S and form a system of linearly independent solutions of the differential equation

$$u^{(n)} + p_1(z)u^{(n-1)} + \cdots + p_n(z)u = 0$$

in $S - \{z_0\}$, where at least one of the coefficients p_i has a nonremovable singularity at z_0. Prove that the Wronskian determinant of the functions $u_1, \ldots, u_n$ (which is known to be different from zero in $S - \{z_0\}$) *vanishes* at z_0.

§9.5. SINGULARITIES OF THE FIRST KIND: FORMAL SOLU-TIONS

If in the differential equation

$$\mathbf{w}' = \mathbf{A}(z)\mathbf{w} \tag{9.5-1}$$

the matrix $\mathbf{A}$ is analytic in a punctured disk $0 < |z - z_0| < \gamma$ and has a pole of order 1 at z_0, then z_0 is called a **singularity of the first kind**. Theorem 9.4c shows that in a neighborhood of a singularity of the first kind no solution of the differential equation can grow more rapidly than a fixed power of $|z - z_0|$ if $z \to z_0$ in such a manner that $\arg(z - z_0)$ is bounded. This was shown to be equivalent to the statement that any fundamental matrix has the form $\mathbf{P}(z)(z-z_0)^{\mathbf{R}}$, where $\mathbf{R}$ is a constant matrix and $\mathbf{P}$ has at most a pole at z_0. It is the purpose of the present section to derive an explicit algorithm for constructing the matrices $\mathbf{P}$ and $\mathbf{R}$. A discourse on certain formal series is required.

We recall that a **formal Laurent series** is a sequence of complex numbers $\{c_m\}_{m=-\infty}^{\infty}$ written in the form

$$F = \sum_{m=-\infty}^{\infty} c_m z^m, \tag{9.5-2}$$

where all but finitely many of the c_m with negative indices are zero. No convergence is assumed, and no value is ascribed to a formal series. However, as was seen in §1.8, the operations of addition and multiplication can be defined for such formal series as if they were convergent, and under these operations the totality of all formal series form a field. The zero element is the formal series with all its coefficients zero, and the unit element is the formal series with $c_0 = 1$, $c_m = 0$ for $m \neq 0$. We recall that the derivative of the formal series (9.5-2) is the formal series

$$F' := \sum_{m=-\infty}^{\infty} (m+1)c_{m+1} z^m.$$

Let $\{F_{jk}\}$, $j, k = 0, 1, \ldots$, $F_{jk} = 0$ for $j+k$ large, be a finite set of formal Laurent series, and let μ_j, $j = 0, 1, \ldots$, be a sequence of complex numbers. The finite sum

$$P = \sum_{j,k=0}^{\infty} F_{jk} z^{\mu_j} (\log z)^k \tag{9.5-3}$$

is called a **formal logarithmic sum.** If

$$Q = \sum_{j,k=0}^{\infty} G_{jk} z^{\nu_j} (\log z)^k$$

also is a formal logarithmic sum, the sum $P+Q$ and the product PQ are defined as though the coefficients F_{jk} and G_{jk} were scalars. With this definition, the set of all formal logarithmic sums is closed under addition and multiplication. The derivative of a formal logarithmic sum P is defined by

$$P' := \sum_{j,k=0}^{\infty} [F'_{jk} + \mu_j F_{jk} z^{-1} + (k+1)F_{j,k+1} z^{-1}] z^{\mu_j} (\log z)^k. \tag{9.5-4}$$

and thus is again a formal logarithmic sum.

If two of the numbers μ_j, say μ_1 and μ_2, differ by an integer, the corresponding terms in the formal logarithmic sum (9.5-3) can be combined, yielding the simplified sum

$$\sum_{k=0}^{\infty} [F_{1k}z^{\mu_1} + F_{2k}z^{\mu_2}](\log z)^k = \sum_{k=0}^{\infty} G_{1k}z^{\mu_1}(\log z)^k,$$

where G_{1k} is again a formal Laurent series,

$$G_{1k} := F_{1k} + F_{2k}z^{\mu_2 - \mu_1}.$$

By performing this operation repeatedly, if necessary, we can always bring a formal logarithmic sum into a form where no two μ_j differ by an integer. Such a sum is called **reduced**. Note that even the reduced form of a formal logarithmic sum is not uniquely determined, since the μ_j are only determined modulo 1. A reduced sum is called zero if and only if all its coefficients F_{jk} are zero. A logarithmic sum is said to be zero if and only if one of its reduced sums is zero. (It is clear that if one reduced sum is zero, all reduced sums are zero.) Two formal logarithmic sums are considered equal if their difference is zero.

A **formal logarithmic matrix** (flm) **L** is a matrix whose elements are formal logarithmic sums. Because the formal logarithmic sums form a commutative ring, sums and products of formal logarithmic matrices can be defined in the usual way. The derivative of the flm $\mathbf{L} = (L_{ij})$ is the matrix with the elements L'_{ij}.

What is the connection of all this with differential equations? Let us consider the case where $z = 0$ is a singularity of the first kind of the differential equation (9.5-1). This means that the matrix $\mathbf{A}(z)$ is analytic in a punctured disk $0 < |z| < \gamma$ and has a pole of order 1 at $z = 0$. It thus can be expanded in a Laurent series

$$\mathbf{A}(z) = \sum_{m=-1}^{\infty} \mathbf{A}_m z^m$$

convergent for $0 < |z| < \gamma$. Evidently, the elements of this matrix can be regarded as formal Laurent series, and thus the matrix itself as an (especially simple) flm. Moreover, if **W** is a fundamental matrix of (9.5-1), then it follows from §9.4 that **W** can be represented in the form

$$\mathbf{W}(z) = \mathbf{P}(z)z^{\mathbf{R}},$$

where **P** is analytic in $0 < |z| < \gamma$ and has at most a pole at $z = 0$, and where **R** is a constant matrix. It is clear that **P** is an flm, and the explicit form given earlier shows that $z^{\mathbf{R}}$ likewise may be regarded an flm. An flm **W** will be called a **formal solution** of the differential equation (9.5-1), if it satisfies the relation

$$\mathbf{W}' = \mathbf{A}\mathbf{W}$$

considered as an identity between formal logarithmic matrices.

THEOREM 9.5a

Every actual solution matrix of (9.5-1) is a formal solution, and every formal solution is an actual solution.

Proof.   (a) Let $\mathbf{W}$ be an actual solution matrix of the differential equation. This means that the elements of the matrix

$$\mathbf{H}(z) := \mathbf{W}'(z) - \mathbf{A}(z)\mathbf{W}(z)$$

are identically zero for $0 < |z| < \gamma$. Now every element of $\mathbf{H}$ is a formal logarithmic sum, which in reduced form may be written

$$H = \sum_{j=1}^{J} z^{\mu_j} \sum_{k=0}^{K} F_{jk} (\log z)^k,$$

where the F_{jk} are formal Laurent series in the variable z that are convergent for $0 < |z| < \gamma$, and where no two μ_j differ by an integer.

We know that H is identically zero and wish to conclude that all F_{jk} vanish identically. To this end we introduce the variable $t := \log z$ and write

$$f_{jk}(t) := F_{jk}(e^t).$$

As a Laurent series F_{jk} is single-valued and f_{jk} thus is periodic with period $2\pi i$. Our hypothesis now says that the function

$$h(t) := \sum_{j=1}^{J} e^{\mu_j t} \sum_{k=0}^{K} f_{jk}(t) t^k$$

vanishes identically in the strip $D^*: -\infty < \operatorname{Re} t < \log \gamma$. Let t_0 be an arbitrary point of D^*. Then the sequence $\{x_n\}_{n=0}^{\infty}$, where

$$x_n := h(t_0 + 2\pi i n), \qquad n = 0, 1, \ldots,$$

is the zero sequence. Letting

$$c_j := e^{\mu_j t_0} \quad \text{and} \quad \omega_j := e^{2\pi i \mu_j}$$

we have

$$x_n = \sum_{j=1}^{J} \omega_j^n \sum_{k=0}^{K} a_{jk}(t_0 + 2\pi i n)^k, \tag{9.5-5}$$

where

$$a_{jk} := c_j f_{jk}(t_0),$$

and where the ω_j are distinct because the sum H is reduced. By Theorem 7.2b, the x_n are the coefficients of the Taylor series at 0 of a certain rational function r having poles of order $\leq K$ at the points $\omega_j^{-1}, j = 1, \ldots, J$. Because the x_n are zero, r is the zero function, which implies that all $a_{jk} = 0$. Because

$c_j \neq 0$, this implies that all $f_{jk}(t_0) = 0$. Since t_0 was arbitrary, it follows that the functions f_{jk} vanish identically. By the uniqueness of the Laurent series, the series F_{jk}, considered as formal series, thus are all zero.

(b) Let, on the other hand, $\mathbf{W}$ be a formal solution of (9.5-1). By §9.4, there exists an actual fundamental solution $\mathbf{W}_0$ of (9.5-1) which is of the form $\mathbf{P}(z)z^{\mathbf{R}}$, where $\mathbf{P}$ is meromorphic at $z = 0$, hence may be regarded as a flm. Also the matrix $\mathbf{W}_0^{-1} = z^{-\mathbf{R}}\mathbf{P}^{-1}$ is a flm. By (a) $\mathbf{W}_0$ is not only an actual, but also a formal solution of the differential equation. Hence if all identities are considered as formal,

$$(\mathbf{W}_0^{-1}\mathbf{W})' = -\mathbf{W}_0^{-1}\mathbf{W}_0'\mathbf{W}_0^{-1}\mathbf{W} + \mathbf{W}_0^{-1}\mathbf{W}'$$
$$= -\mathbf{W}_0^{-1}\mathbf{A}\mathbf{W}_0\mathbf{W}_0^{-1}\mathbf{W} + \mathbf{W}_0^{-1}\mathbf{A}\mathbf{W}$$

and thus

$$(\mathbf{W}_0^{-1}\mathbf{W})' = \mathbf{0}. \tag{9.5-6}$$

It will now be shown that if $\mathbf{H}$ is any flm such that $\mathbf{H}' = \mathbf{0}$, then $\mathbf{H}$ must be a matrix of constants. This is a consequence of

LEMMA 9.5b

Let P be a formal logarithmic sum such that $P' = 0$. Then P is a constant.

Taking the lemma for granted, we conclude from (9.5-6) that $\mathbf{W}_0^{-1}\mathbf{W} = \mathbf{C}$. Hence each column of $\mathbf{W}$ is a linear combination of actual solutions of the differential equation, and all formal series in $\mathbf{W}$ are convergent for $0 < |z| < \gamma$.

Proof of Lemma 9.5b. Let P be written in the form (9.5-3), and let P be reduced. The hypothesis then implies

$$F_{jk}' + \mu_j F_{jk}z^{-1} + (k+1)F_{j,k+1}z^{-1} = 0 \tag{9.5-7}$$

for all j and k. If P is not zero, let K be the highest power of $\log z$ with a nonzero coefficient, and let

$$F_{1K}z^{\mu_1} + F_{2K}z^{\mu_2} + \cdots + F_{JK}z^{\mu_J}$$

be that coefficient. For $k = K$, (9.5-7) yields

$$F_{jK}' + \mu_j F_{jK}z^{-1} = 0, \qquad j = 1, \ldots, J,$$

because $F_{j,K+1} = 0$. If $F_{jK} = \sum_{m=-\infty}^{\infty} c_{jm}^{(K)}z^m$, the last relation implies that

$$(m + \mu_j)c_{jm}^{(K)} = 0$$

for all m and for $j = 1, \ldots, J$. This in turn implies that μ_j is an integer for some j, for otherwise $c_{jm}^{(K)} = 0$ for all m and for all $j = 1, \ldots, J$, contradicting the assumption that the coefficient of $(\log z)^K$ is not identically zero. There is at most one μ_j which is an integer, for P is reduced. Let it be μ_1; we may assume that $\mu_1 = 0$ without loss of generality. Then it follows that

$$F_{1K} = c_{10}^{(K)} \neq 0;$$

$$F_{jk} = 0, j = 2, \ldots, J.$$

(9.5-8)

Now assume that $K \geq 1$. Consider (9.5-7) where $k = K - 1$. If $j = 2, \ldots, J$, (9.5-7) yields

$$F'_{j,K-1} + \mu_j F_{j,K-1} z^{-1} = 0$$

or, looked at coefficientwise,

$$(m + \mu_j) c_{jm}^{(K-1)} = 0, \qquad m = 0, \pm 1, \pm 2, \ldots.$$

Because μ_j is not an integer, this implies

$$F_{j,K-1} = 0, \qquad j = 2, 3, \ldots, J.$$

For $j = 1$, (9.5–7) yields in view of $\mu_1 = 0$

$$F'_{1,K-1} + K c_{10}^{(K)} z^{-1} = 0.$$

Because the derivative of a formal Laurent series has no term in z^{-1}, this relation is impossible unless $K c_{10}^{(K)} = 0$. Because F_{1K} was assumed nonzero, we must have $K = 0$. This contradicts our assumption that $K \geq 1$. Hence $K = 0$, and by (9.5-8), $P = c_{10}^{(0)}$, proving the lemma. ∎

We now are ready to tackle the problem of actually constructing (and not merely prove the existence of) the matrices $\mathbf{P}$ and $\mathbf{R}$ occurring in the representation $\mathbf{W}(z) = \mathbf{P}(z) z^{\mathbf{R}}$ of a fundamental matrix of (9.5-1). According to Theorem 9.5a, it suffices to construct these matrices in such a manner that the resulting $\mathbf{W}$ satisfies the differential equation formally.

We assume that the coefficient matrix $\mathbf{A}(z)$ is given as a Laurent series,

$$\mathbf{A}(z) = z^{-1} \mathbf{R} + \sum_{m=0}^{\infty} \mathbf{A}_m z^m,$$

(9.5-9)

where the power series is convergent for $|z| < \gamma$. The following heuristic consideration will motivate our construction. If all $\mathbf{A}_m = \mathbf{0}$, the differential equation reduces to the system $\mathbf{w}' = z^{-1} \mathbf{R} \mathbf{w}$. By §9.3, this has the fundamental matrix $\mathbf{W}(z) = z^{\mathbf{R}}$. Looking at the general equation (9.5-1) as a perturbation of $\mathbf{w}' = z^{-1} \mathbf{R} \mathbf{w}$, we suspect the existence of a fundamental matrix $\mathbf{W}$ of the form

$$\mathbf{W}(z) = \mathbf{P}(z) z^{\mathbf{R}},$$

(9.5-10)

where $\mathbf{P}$ is analytic (and not merely meromorphic) at $z = 0$, $\mathbf{P}(0) = \mathbf{I}$.

It will now be proved that under one additional hypothesis such a formal solution indeed exists. We shall say that a matrix satisfies hypothesis $\mathscr{H}$, if no two of its eigenvalues differ by an integer $\neq 0$. We then have:

THEOREM 9.5c

Let the matrix $\mathbf{R}$ *in* (9.5-9) *satisfy hypothesis* $\mathscr{H}$. *Then the system* (9.5-1) *possesses a fundamental matrix of the form* (9.5-10), *where* $\mathbf{P}$ *is analytic at* $z = 0$,

$$\mathbf{P}(z) = \sum_{m=0}^{\infty} \mathbf{P}_m z^m, \qquad \mathbf{P}_0 = I. \tag{9.5-11}$$

If

$$\mathbf{R} = \mathbf{T}\mathbf{S}\mathbf{T}^{-1} \tag{9.5-12}$$

where $\mathbf{T}$ *is nonsingular and* $\mathbf{S}$ *is upper triangular,*[2] *then* $\mathbf{P}_m = \mathbf{Q}_m \mathbf{T}^{-1}$, *where* $\mathbf{Q}_0 = \mathbf{T}$ *and for* $m > 0$, $\mathbf{Q}_m$ *is a rational function of the elements of the matrices* $\mathbf{A}_0, \mathbf{A}_1, \ldots, \mathbf{A}_{m-1}$ *and* $\mathbf{S}$.

Proof. If $\mathbf{W}$ is a fundamental matrix, then also $\mathbf{W}\mathbf{T}$ is one. Because $z^{\mathbf{R}} = \mathbf{T} z^{\mathbf{S}} \mathbf{T}^{-1}$, we may seek $\mathbf{W}$ in the form

$$\mathbf{W}(z) = \mathbf{Q}(z) z^{\mathbf{S}}, \tag{9.5-13}$$

where

$$\mathbf{Q}(z) = \sum_{m=0}^{\infty} \mathbf{Q}_m z^m, \qquad \mathbf{Q}_0 = \mathbf{T}$$

We have formally

$$\mathbf{W}'(z) = \mathbf{Q}'(z) z^{\mathbf{S}} + \mathbf{Q}(z) \mathbf{S} z^{-1} z^{\mathbf{S}}$$

$$= \sum_{m=0}^{\infty} [m\mathbf{Q}_m + \mathbf{Q}_m \mathbf{S}] z^{m-1} z^{\mathbf{S}},$$

and

$$\mathbf{A}(z)\mathbf{W}(z) = \left[\sum_{m=0}^{\infty} \mathbf{R}\mathbf{Q}_m z^{m-1} + \sum_{m=0}^{\infty} \mathbf{R}_m z^m \right] z^{\mathbf{S}}$$

where

$$\mathbf{C}_m := \sum_{k=0}^{m} \mathbf{A}_k \mathbf{Q}_{m-k}, \qquad m = 0, 1, 2, \ldots.$$

---

[2] $\mathbf{S}$ may be a Jordan canonical form of $\mathbf{R}$, or it may be a Schur form, in which case $\mathbf{T}$ may be taken unitary.

On substituting into the differential equation, we find that (9.5-13) is a formal solution if and only if the following relations hold:

$$\mathbf{Q}_0 \mathbf{S} = \mathbf{R} \mathbf{Q}_0; \tag{9.5-14}$$

$$\mathbf{Q}_m(m\mathbf{I}+\mathbf{S}) = \mathbf{R}\mathbf{Q}_m + \mathbf{C}_{m-1}, \qquad m = 1, 2, \dots. \tag{9.5-15}$$

Relation (9.5-14) is satisfied by taking $\mathbf{Q}_0 = \mathbf{T}$, as indicated. To deal with (9.5-15), assume that the matrices $\mathbf{Q}_0, \mathbf{Q}_1, \dots, \mathbf{Q}_{m-1}$ have been determined. Then $\mathbf{C}_{m-1}$ is a known matrix, and the problem is to solve for $\mathbf{Q}_m$. The problem is somewhat unorthodox, because $\mathbf{Q}_m$ occurs both as a right and as a left factor. Some special notation is required to deal with this problem.

We denote the columns of $\mathbf{Q}_m$ and of $\mathbf{C}_{m-1}$ by $\mathbf{q}_k$ and $\mathbf{c}_k$, respectively ($k = 1, \dots, n$). We furthermore let

$$\mathbf{S} = \mathbf{\Lambda} + \mathbf{M},$$

where $\mathbf{\Lambda} := \mathrm{diag}(\lambda_1, \lambda_2, \dots, \lambda_n)$ is the diagonal matrix whose diagonal elements coincide with those of $\mathbf{S}$ (and thus with the eigenvalues of $\mathbf{R}$), and where $\mathbf{M}$ has zero elements on and below the main diagonal. For $j = 1, \dots, n$, the jth column of the matrix equation (9.5-15) now can be written

$$(m+\lambda_j)\mathbf{q}_j + \mathbf{p}_j = \mathbf{R}\mathbf{q}_j + \mathbf{c}_j, \tag{9.5-16}$$

where $\mathbf{p}_j$ denotes the jth column of $\mathbf{Q}_m\mathbf{M}$. Because of the special form of the matrix $\mathbf{M}$, $\mathbf{p}_1 = \mathbf{0}$ and for $j > 1$ $\mathbf{p}_j$ involves elements of $\mathbf{q}_1, \mathbf{q}_2, \dots, \mathbf{q}_{j-1}$ only. This circumstance enables us to determine the columns $\mathbf{q}_1, \dots, \mathbf{q}_n$ recursively as follows. For $j = 1$, (9.5-16) is

$$[(m+\lambda_1)\mathbf{I} - \mathbf{R}]\mathbf{q}_1 = \mathbf{c}_1.$$

In view of hypothesis $\mathcal{H}$, $m + \lambda_1$ is not an eigenvalue of $\mathbf{R}$, and thus the matrix in brackets is nonsingular. Thus the equation can be solved for $\mathbf{q}_1$. Assuming that $\mathbf{q}_1, \dots, \mathbf{q}_{j-1}$ have been determined, $\mathbf{p}_j$ is known, and (9.5-16) may be written

$$[(m+\lambda_j)\mathbf{I} - \mathbf{R}]\mathbf{q}_j = \mathbf{c}_j - \mathbf{p}_j.$$

Again by virtue of hypothesis $\mathcal{H}$ this can be solved for $\mathbf{q}_j$. Thus the complete matrix $\mathbf{Q}_m$ can be determined under the assumption that $\mathbf{Q}_0, \dots, \mathbf{Q}_{m-1}$ have been determined, and because $\mathbf{Q}_0$ has been found, the construction of a formal fundamental solution has thus been completed for differential equations satisfying hypothesis $\mathcal{H}$. ■

It remains to deal with equations that do not satisfy hypothesis $\mathcal{H}$, which include many equations of practical importance. (Among others, the differential equations satisfied by the Bessel and the Legendre functions are in this category for special values of the parameters.) This is accomplished by repeated applications of the following fact:

THEOREM 9.5d

Let the distinct eigenvalues of $\mathbf{R}$ be $\lambda_1, \ldots, \lambda_m (m \leq n)$. There exists a matrix $\mathbf{V}(z)$, analytic at $z = 0$ and nonsingular for $z \neq 0$, such that the function $\hat{\mathbf{w}}$ defined by $\mathbf{w} = \mathbf{V}\hat{\mathbf{w}}$ satisfies a differential equation of the form (9.5-1),

$$\hat{\mathbf{w}}' = \left(z^{-1}\hat{\mathbf{R}} + \sum_{m=0}^{\infty} z^m \hat{\mathbf{A}} \right)\hat{\mathbf{w}},$$

where $\hat{\mathbf{R}}$ has the eigenvalues $\lambda_1 - 1, \lambda_2, \ldots, \lambda_m$.

Proof. Let $\mathbf{T}$ be any nonsingular constant matrix. If $\mathbf{w}$ satisfies (9.5-1), the function $\tilde{\mathbf{w}} := \mathbf{T}\mathbf{w}$ then will satisfy

$$\tilde{\mathbf{w}}' = \mathbf{T}\mathbf{w}' = \mathbf{T}\mathbf{A}(z)\mathbf{w} = \mathbf{T}\mathbf{A}(z)\mathbf{T}^{-1}\tilde{\mathbf{w}}.$$

It thus satisfies a differential equation of the same form, where $\mathbf{A}$ is replaced by the similar matrix $\mathbf{T}\mathbf{A}\mathbf{T}^{-1}$. Now let the multiplicity of the eigenvalue λ_1 of $\mathbf{R}$ be p, and let $\mathbf{T}$ be so chosen that

$$\tilde{\mathbf{R}} := \mathbf{T}\mathbf{R}\mathbf{T}^{-1} = \begin{pmatrix} \mathbf{R}_1 & \mathbf{0} \\ \mathbf{0} & \mathbf{R}_2 \end{pmatrix},$$

where $\mathbf{R}_1$ is a $p \times p$ upper triangular matrix with the p-fold eigenvalue λ_1, and where $\mathbf{R}_2$ is a square matrix of order $n - p$ with the eigenvalues $\lambda_2, \ldots, \lambda_m$. By the fact just mentioned, the function $\tilde{\mathbf{w}} := \mathbf{T}\mathbf{w}$ then satisfies

$$\tilde{\mathbf{w}}' = (z^{-1}\tilde{\mathbf{R}} + \mathbf{H}(z))\tilde{\mathbf{w}},$$

where $\mathbf{H}$ is analytic at $z = 0$. Now let

$$\mathbf{U} := \begin{pmatrix} z^{-1}\mathbf{I} & \mathbf{0} \\ \mathbf{0} & \mathbf{I} \end{pmatrix},$$

where the partitioning is similar to that of $\tilde{\mathbf{R}}$. (The two unit matrices $\mathbf{I}$ appearing here do not necessarily have the same dimension.) Clearly, $\mathbf{U}$ is nonsingular for $z \neq 0$, and

$$\mathbf{U}^{-1} = \begin{pmatrix} z\mathbf{I} & \mathbf{0} \\ \mathbf{0} & \mathbf{I} \end{pmatrix}.$$

is analytic at $z = 0$. Now let $\hat{\mathbf{w}} := \mathbf{U}\tilde{\mathbf{w}}$. For $z \neq 0$ this satisfies

$$\hat{\mathbf{w}}' = \mathbf{U}'\tilde{\mathbf{w}} + \mathbf{U}\tilde{\mathbf{w}}'$$

$$= \mathbf{U}'\tilde{\mathbf{w}} + \mathbf{U}(z^{-1}\tilde{\mathbf{R}} + \mathbf{H}(z))\tilde{\mathbf{w}}$$

$$= (\mathbf{U}'\mathbf{U}^{-1} + z^{-1}\mathbf{U}\tilde{\mathbf{R}}\mathbf{U}^{-1} + \mathbf{U}\mathbf{H}(z)\mathbf{U}^{-1})\hat{\mathbf{w}}.$$

Now

$$\mathbf{U}'\mathbf{U}^{-1} = \begin{pmatrix} -z^{-1}\mathbf{I} & \mathbf{0} \\ \mathbf{0} & \mathbf{0} \end{pmatrix}, \qquad \mathbf{U}\check{\mathbf{R}}\mathbf{U}^{-1} = \tilde{\mathbf{R}}.$$

Furthermore, if $\mathbf{H}(z) = \sum_{m=0}^{\infty} \mathbf{H}_m z^m$ and

$$\mathbf{H}_0 = \begin{pmatrix} \mathbf{H}_{11} & \mathbf{H}_{12} \\ \mathbf{H}_{21} & \mathbf{H}_{22} \end{pmatrix},$$

then

$$\mathbf{U}\mathbf{H}(z)\mathbf{U}^{-1} = z^{-1} \begin{pmatrix} \mathbf{0} & \mathbf{H}_{12} \\ \mathbf{0} & \mathbf{0} \end{pmatrix} + \text{analytic matrix.}$$

Thus $\hat{\mathbf{w}}$ satisfies a differential equation of the form (9.5-1), where $\mathbf{R}$ is replaced by the matrix

$$\hat{\mathbf{R}} := \begin{pmatrix} \mathbf{R}_1 - \mathbf{I} & \mathbf{H}_{12} \\ \mathbf{0} & \mathbf{R}_2 \end{pmatrix}$$

which has the eigenvalues $\lambda_1 - 1, \lambda_2, \ldots, \lambda_m$. Composing the foregoing transformations, the statement of the theorem is proved for $\mathbf{V} := \mathbf{T}^{-1}\mathbf{U}^{-1}$. ∎

By repeated applications of the construction of Theorem 9.5d, a differential equation with a singularity of the first kind at $z = 0$ can be transformed into a differential equation satisfying hypothesis $\mathscr{H}$ by a substitution of the form $\mathbf{w} = \mathbf{V}\hat{\mathbf{w}}$, where $\mathbf{V}$ is nonsingular for $z \neq 0$. For the transformed system a fundamental matrix $\hat{\mathbf{W}}$ can be constructed by the algorithm given in Theorem 9.5c. A fundamental matrix of the original system is then given by $\mathbf{W} := \mathbf{V}\hat{\mathbf{W}}$. It has the form $\mathbf{W}(z) = \mathbf{P}(z)z^{\hat{\mathbf{R}}}$ where $\mathbf{P}$ is a power series and $\hat{\mathbf{R}}$ is a matrix whose eigenvalues coincide with those of $\mathbf{R}$ modulo one. More precisely, if the eigenvalues of $\mathbf{R}$ are divided into sets of eigenvalues differing by an integer, the eigenvalues of $\hat{\mathbf{R}}$ consist of the lowest representatives of each set. The multiplicity of each such eigenvalue equals the number of eigenvalues in the set to which it belongs.

The construction of a fundamental system is simple enough for differential equations satisfying hypothesis $\mathscr{H}$, but it can be laborious for equations that do not satisfy this hypothesis, especially if two eigenvalues differ by a large integer. For systems that arise from a single nth order equation there exists a somewhat simpler method for constructing a fundamental system in the exceptional cases. This method, which is due to Frobenius, is discussed in §9.6.

PROBLEMS

1. Use the heuristic method sketched after the proof of Theorem 9.5a to construct a fundamental system for the equation

$$\mathbf{w}' = (z^{-1}\alpha\mathbf{I} + \mathbf{A})\mathbf{w},$$

where α is a complex number and $\mathbf{A}$ a matrix of constants. Interpret the result.

2. *Transformation of variable at singular point of first kind.* In the system

$$\mathbf{w}' = (z^{-1}\mathbf{R} + \mathbf{H}(z))\mathbf{w},$$

where $\mathbf{H}$ is analytic at $z = 0$, a new variable t is introduced by setting $z = g(t)$, where g is analytic at $t = t_0$, $g(t_0) = 0$, $g'(t_0) \neq 0$. Show that $\mathbf{u}(t) := \mathbf{w}(g(t))$ satisfies a linear system of differential equations with t_0 as a singular point of the first kind, and that the residue of its coefficient matrix at t_0 is again $\mathbf{R}$.

3. Let $\mathbf{W}$ be a fundamental matrix near $z = 0$ of the system

$$\mathbf{w}' = (z^{-1}\mathbf{R} + \mathbf{H}(z))\mathbf{w},$$

where $\mathbf{H}$ is analytic at $z = 0$, and let the matrix $\mathbf{C}$ (as in the proof of Theorem 9.4b) be defined by

$$\mathbf{W}(z\, e^{2\pi i}) = \mathbf{W}(z)\mathbf{C}.$$

If $\mathbf{R}$ satisfies hypothesis $\mathcal{H}$, show that the matrices $\mathbf{C}$ and $\mathbf{V} := e^{2\pi i \mathbf{R}}$ are similar.

4. (Continuation). By considering the example

$$\mathbf{R} := \begin{pmatrix} 0 & 0 \\ 0 & -1 \end{pmatrix}, \qquad \mathbf{H}(z) := \begin{pmatrix} 0 & 1 \\ 0 & 0 \end{pmatrix}$$

show that $\mathbf{C}$ and $\mathbf{V}$ need not be similar if $\mathbf{R}$ does not satisfy hypothesis $\mathcal{H}$ (Gantmacher).

§9.6. SCALAR EQUATIONS OF HIGHER ORDER: METHOD OF FROBENIUS

Here we are concerned with the scalar equation of the nth order,

$$u^{(n)} + p_1(z)u^{(n-1)} + p_2(z)u^{(n-2)} + \cdots + p_n(z)u = 0, \qquad (9.6\text{-}1)$$

where the coefficients p_k are analytic in some punctured disk $0 < |z - z_0| < \gamma$, $\gamma > 0$. Without loss of generality we may assume $z_0 = 0$. As in §9.3 we can associate with (9.6-1) the first order system $\mathbf{w}' = \mathbf{A}(z)\mathbf{w}$, where $\mathbf{A}$ is given by (9.3-8), and where the vector $\mathbf{w}^T := (u, u', \ldots, u^{(n-1)})$. As the associated system has an isolated singularity at $z = 0$, we can apply the theory of §9.4 and find that any solution of (9.6-1) is a linear combination of functions of the form

$$z^\mu (\log z)^k s(z),$$

where s is analytic in $0<|z|<\gamma$. In accordance with the terminology introduced in §9.4 we shall call $z=0$ a **regular singular point** of (9.6-1) if and only if every solution of (9.6-1) has a representation of the foregoing form in which all functions s have at most a pole at $z=0$. If (9.6-1) has a solution for which no such representation exists, then $z=0$ is called an **irregular singular point**.

It follows from Theorem 9.4c that $z=0$ is a regular singular point if and only if for every solution u of (9.6-1) there exists a real number κ such that

$$\rho^\kappa |u(\rho\, e^{i\theta})| \to 0$$

if $\rho \to 0$ and θ is bounded.

Again there arises the problem of telling from the differential equation alone, without constructing the solutions, whether a given singular point is regular. By considering the associated system, it follows immediately from Theorem 9.4c that $z=0$ is a regular singular point if the coefficients p_k have at most a pole of order 1 at $z=0$.

However, this condition is not necessary. Assume that the coefficient p_k in (9.6-1) has a pole of order not exceeding k for $k=1,2,\ldots,n$, or in other words, that it can be represented in the form

$$p_k(z) = z^{-k} q_k(z), \qquad k=1,2,\ldots,n, \tag{9.6-2}$$

where q_k is analytic at $z=0$. (In each term of the differential equation, the order of the derivative and the order of the pole add up to at most n.) If the equation has this property, $z=0$ is called a **singular point of the first kind**.

THEOREM 9.6a

If $z=0$ is a singular point of the first kind of the scalar equation (9.6-1), then it is a regular singular point.

Proof. We associate with (9.6-1) a firstorder system different from that in §9.3. Let u be a solution of (9.6-1). Then the n functions

$$v_k(z) := z^{k-1} u^{(k-1)}(z), \qquad k=1,2,\ldots,n,$$

satisfy

$$v_k'(z) = (k-1) z^{k-2} u^{(k-1)}(z) + z^{k-1} u^{(k)}(z)$$

$$= (k-1) z^{-1} v_k(z) + z^{-1} v_{k+1}(z), \; k=1,2,\ldots,n-1,$$

and

$$v_n'(z) = (n-1) z^{-1} v_n(z) + z^{n-1} u^{(n)}(z)$$

$$= (n-1) z^{-1} v_n(z) - z^{n-1} \sum_{k=1}^{n} p_k(z) u^{(n-k)}(z)$$

or, using (9.6-2),

$$v'_n(z) = (n-1)z^{-1}v_n(z) - z^{-1}\sum_{k=1}^{n} q_k(z)v_{n-k+1}(z).$$

Thus the vector $\mathbf{v} := (v_1, v_2, \ldots, v_n)^T$ is a solution of the system

$$\mathbf{v}' = z^{-1}\mathbf{B}(z)\mathbf{v}, \tag{9.6-3}$$

where

$$\mathbf{B}(z) := \begin{pmatrix} 0 & 1 & 0 & \cdots & 0 \\ 0 & 1 & 1 & \cdots & 0 \\ 0 & 0 & 2 & \cdots & -0 \\ & \cdot \quad \cdot \quad \cdot \quad \cdot \quad \cdot \quad \cdot & & \\ 0 & 0 & 0 & \cdots & 1 \\ -q_n(z) & -q_{n-1}(z) & -q_{n-2}(z) & \cdots & n-1-q_1(z) \end{pmatrix}$$

The matrix $\mathbf{B}$ is analytic at $z = 0$; hence for the system (9.6-3) $z = 0$ is a singularity of the first kind, in the sense defined in §9.5. It follows from Theorem 9.4d that $z = 0$ is a regular singular point, and hence that every solution $\mathbf{v}$ satisfies the limit relation (9.4-9) for a suitable value of κ. An analogous relation thus must hold for the first component $v_1 = u$ of $\mathbf{v}$. Hence $z = 0$ is a regular point for the nth order scalar equation (9.6-1). ∎

It should be remembered that the converse of Theorem 9.4d does not hold. That is, not every regular singular point of a system is a singular point of the first kind. However, we mention without proof that the converse of Theorem 9.6a is true: Every regular singular point of an nth order scalar equation is necessarily a singular point of the first kind.

The transformation used in the proof of Theorem 9.6a can also be utilized to construct a fundamental system of solutions for equation (9.6-1) in the manner indicated in §9.5. Writing

$$z^{-1}\mathbf{B}(z) = z^{-1}\mathbf{R} + \sum_{m=0}^{\infty} \mathbf{B}_m z^m,$$

we must first of all compute the eigenvalues of the matrix $\mathbf{R} = \mathbf{B}(0)$. Writing $q_k(0) =: b_k$, $k = 1, \ldots, n$, and expanding the characteristic determinant $\pi(\lambda) := \det(\lambda\mathbf{I} - \mathbf{B}(0))$ in terms of the elements of the last row, we find

$$\pi(\lambda) = \lambda(\lambda - 1)(\lambda - 2) \cdots (\lambda - n + 1)$$
$$+ b_1\lambda(\lambda - 1) \cdots (\lambda - n + 2)$$
$$+ b_2\lambda(\lambda - 1) \cdots (\lambda - n + 3)$$
$$+ \cdots$$
$$+ b_{n-1}\lambda + b_n.$$

The polynomial π is called the **indicial polynomial** belonging to the singularity $z = 0$ of the equation (9.6-1), and its zeros are called **characteristic exponents**. If the zeros $\lambda_1, \ldots, \lambda_n$ are distinct, and if no two zeros differ by an integer, it follows from §9.5 that (9.6-1) has n linearly independent solutions of the form $z^{\lambda_i} s_i(z)$, where the functions s_i are analytic at $z = 0$. Their power series can be computed by the method of undetermined coefficients; the formal solutions thus obtained are actual solutions by Theorem 9.5a.

Complications arise if several zeros of the indicial polynomial coincide, or if they differ by an integer. These difficulties can be overcome in a systematic manner by a method due to Frobenius (1849–1917), which is based on a generalization of the concept of formal logarithmic series.

Frobenius' Method

In §9.5 we considered formal logarithmic sums,

$$P = \sum_{j,k=0}^{\infty} F_{jk} z^{\mu_j} (\log z)^k, \tag{9.6-4}$$

where the symbols F_{jk} denoted formal Laurent series whose coefficients were complex numbers, and where the exponents μ_j likewise were complex numbers. However, in defining the algebraic structure of these formal logarithmic sums no special properties of the complex numbers were used beyond the fact that they form a field. We thus may consider formal logarithmic sums with exponents and Laurent series coefficients taken from an arbitrary field $\mathcal{F}$. For our present purposes it is convenient to assume that $\mathcal{F}$ is the field of functions that are meromorphic in some region S, usually the complex plane. The totality of formal logarithmic sums whose exponents and Laurent series coefficients are meromorphic in the complex plane are denoted by $\mathcal{L}$. It is convenient to think of the elements of $\mathcal{L}$ as formal logarithmic sums that depend on a parameter λ. By evaluating the exponents and coefficients at a point at which they are all finite, an ordinary formal logarithmic sum is obtained.

We use a backward prime (`) to indicate differentiation with respect to the parameter. If the sum (9.6-4) is in $\mathcal{L}$, its derivative with respect to λ is defined as

$$P^` := \sum_{j,k=0}^{\infty} [F_{jk}^` z^{\mu_j} (\log z)^k + F_{jk} \mu_j^` z^{\mu_j} (\log z)^{k+1}],$$

where $F_{jk}^`$ denotes the formal Laurent series obtained by differentiating each coefficient with respect to the parameter. It is evident that $P^`$ is again a formal logarithmic sum in $\mathcal{L}$. If the μ_j are not constant, differentiation with respect to the parameter brings in additional powers of $\log z$. It is clear that the usual rules hold for differentiation with respect to the parameter; in particular, $P = O$ implies $P\grave{} = O$. Moreover, it can be verified that differentiation with respect to the parameter commutes with ordinary differentiation, that is, we have $P\grave{}\,' = P'\grave{}$.

In the following, repeated use is made of

OBSERVATION 9.6b

Let $R_0, R_1, \ldots, R_n$, and S be formal logarithmic sums in $\mathscr{L}$, and let, for any $U \in \mathscr{L}$,

$$L(U) := R_0 U^{(n)} + R_1 U^{(n-1)} + \cdots + R_n U. \qquad (9.6\text{-}5)$$

If U is a formal solution of

$$L(U) = S, \qquad (9.6\text{-}6)$$

then $U\grave{}$ is a formal solution of

$$L(U\grave{}) = S\grave{} - L\grave{}(U), \qquad (9.6\text{-}7)$$

where

$$L\grave{}(U) := R_0\grave{}\, U^{(n)} + R_1\grave{}\, U^{(n-1)} + \cdots + R_n\grave{}\, U.$$

The truth of the observation follows at once by differentiating the formal identity (9.6-5) with respect to the parameter.

We now return to our problem of finding formal solutions of (9.6-1). We let

$$L(U) := z^n U^{(n)} + z^{n-1} q_1 U^{(n-1)} + \cdots + q_n U, \qquad (9.6\text{-}8)$$

where $q_1, \ldots, q_n$ are the power series (and thus special logarithmic sums) representing the functions defined by (9.6-2). These series do not depend on a parameter, and we thus have $L\grave{} = 0$.

To motivate what follows, we first consider the case in which the functions q_k are constant, $q_k(z) = b_k$, $k = 1, \ldots, n$. It is then verified directly that for any complex λ

$$\dot{L}(z^\lambda) = \pi(\lambda) z^\lambda.$$

Thus z^λ is a solution of $L(U) = 0$ if and only if λ is a zero of the indicial polynomial. [The case in which $p_k(z) = z^{-k} b_k$ is known as *Euler's differential equation*, and the foregoing is merely the textbook method for solving this special equation.] By differentiating this identity with respect to the parameter λ, we obtain, because $L\grave{} = 0$,

$$L(z^\lambda \log z) = [\pi\grave{}(\lambda) + \pi(\lambda) \log z] z^\lambda.$$

If λ_0 is a zero of multiplicity two of π, then $\pi(\lambda_0) = \pi\grave{}(\lambda_0) = 0$, and we see that $z^{\lambda_0} \log z$ is a second solution of the equation.

In the general case in which the q_k are not constant, we determine a formal series

$$U = z^\lambda \sum_{m=0}^{\infty} c_m z^m, \tag{9.6-9}$$

whose coefficients are rational functions of λ, in such a way that

$$L(U) = \pi(\lambda) z^\lambda \tag{9.6-10}$$

holds as a formal identity between logarithmic series in $\mathscr{L}$. Substituting U into L and collecting coefficients of like powers of z, we obtain, in a simplified notation,

$$L(U) = z^\lambda \{\pi(\lambda)c_0 + [\pi(\lambda+1)c_1 - g_1]z + [\pi(\lambda+2)c_2 - g_2]z^2 + \cdots\}.$$

Each bracket is a sum of convolution products of the sequence $\{c_m\}$ with the sequences of Taylor coefficients of the functions q_k. We have singled out the products involving the highest c_m and the lowest Taylor coefficients b_k. The remaining terms g_j are linear homogeneous expressions in $c_0, c_1, \cdots, c_{j-1}$ with coefficients that are polynomials in λ. We now wish to determine the c_m such that (9.6-10) holds. To this end we must set $c_0 := 1$ and

$$c_m := \frac{g_m}{\pi(\lambda+m)}, \quad m = 1, 2, \ldots. \tag{9.6-11}$$

An induction argument shows that (9.6-10) determines the c_m as rational functions with denominator

$$\prod_{k=1}^{m} \pi(\lambda+k);$$

they thus may have poles at those points λ for which $\pi(\lambda+k) = 0$ for some integer k such that $0 < k \leqslant m$. With this choice of the c_m, the series (9.6-9) satisfies (9.6-10) as an identity between formal logarithmic sums in $\mathscr{L}$.

Let now λ_1 be a zero of the indicial polynomial π, and let $\pi(\lambda_1 + k) \neq 0$ for all positive integers k. Then all coefficients c_m have finite values at $\lambda = \lambda_1$, and the series U evaluated at λ_1 is a formal solution of $L(U) = 0$, and hence is an actual solution of (9.6-1). We denote this solution by U_1.

We now consider the case in which the zero λ_1 has a multiplicity $m_1 > 1$. We differentiate the formal identity (9.6-10) with respect to the parameter λ. In view of $L' = 0$, observation 9.6b yields

$$L(U') = [\pi'(\lambda) + \pi(\lambda) \log z] z^\lambda.$$

If $m_1 = 2$, then $\pi(\lambda_1) = \pi'(\lambda_1) = 0$, and the foregoing relation shows that U' evaluated at λ_1 is another formal solution of $L(U) = 0$, and hence another actual solution of (9.6-1). Evidently,

$$U'(\lambda_1) = U_1 \log z + U_2, \tag{9.6-12}$$

where

$$U_2 := z^{\lambda_1} \sum_{m=0}^{\infty} c_m'(\lambda_1) z^m.$$

The presence of a logarithmic term shows that $U'(\lambda_1)$ is linearly independent from U_1. If $m_1 > 2$, it is clear that $m_1 - 1$ differentiations of (9.6-10) would yield the required $m_1 - 1$ additional linearly independent solutions.

We next consider the situation in which the indicial polynomial π has two simple zeros λ_1 and λ_2 such that $\lambda_1 - \lambda_2 =: k$ is a positive integer. We assume that $\pi(\lambda_2 + j) \neq 0$ for integral j such that $0 < j < k$ and $j > k$. A solution U_1 belonging to the zero λ_1 can be found as indicated above. Again, the problem is to find a second, linearly independent solution. The method outlined above cannot be applied because $\pi(\lambda_2 + k) = 0$. However, the method can be modified as follows. Instead of choosing $c_0 := 1$ as our initial coefficient, we now choose $c_0 := \lambda - \lambda_2$. The coefficients $c_1, c_2, \ldots, c_{k-1}$ then likewise will have the factor $\lambda - \lambda_2$. Therefore in the equation for c_k,

$$\pi(\lambda + k)c_k = g_k,$$

not only the left side vanishes for $\lambda = \lambda_2$ of order 1 but also the right. Thus c_k is determined as a rational function of λ which does not have λ_2 as a pole. The same holds for the coefficients c_m where $m > k$. The series thus determined formally satisfies

$$L(U) = \pi(\lambda)(\lambda - \lambda_2) z^\lambda \tag{9.6-13}$$

identically in λ.

If we now would set $\lambda = \lambda_2$, the first k terms in the series U would vanish, because they have the factor $\lambda - \lambda_2$. Thus we would obtain a series with leading term $z^{\lambda_2 + k} = z^{\lambda_1}$, and because the series is uniquely determined by its leading coefficient, this series would merely be a multiple of the series U_1 found above. To get a linearly independent solution it is necessary to differentiate the identity (9.6-13) formally with respect to λ before putting $\lambda = \lambda_2$. Doing so we obtain

$$L(U') = \pi(\lambda)z^\lambda + (\lambda - \lambda_2)[\pi'(\lambda) + \pi(\lambda) \log z]z^\lambda.$$

Letting $\lambda = \lambda_2$ now yields

$$L(U_2) = 0,$$

where

$$U_2 := U'(\lambda_2).$$

Thus U_2 is a formal solution of $L(U) = 0$, and hence an actual solution of (9.6-1). The leading term of U_2 is z^{λ_2}, hence U_2 is linearly independent from the solution U_1 found before. The first k coefficients of U_2 are what one would obtain by straightforward application of the method of undetermined coefficients; logarithmic terms enter from the term in z^{λ_1} onward.

If λ_1 has multiplicity m_1 and λ_2 has multiplicity m_2, it is readily seen that the same process carried out with $c_0 := (\lambda - \lambda_2)^{m_1}$ will yield a formal series U whose first $m_1 - 1$ λ-derivatives evaluated at λ_2 yield m_1 formal solutions associated with λ_1, and whose next m_2 derivatives yield m_2 formal solutions associated with λ_2. The case of several zeros differing by integers and having arbitrary multiplicities can be dealt with in a similar manner. Because the differential equations that are most studied are of the second order, the two special instances of the method of Frobenius given above suffice for most cases of interest. Some applications of the method are discussed in §9.7.

PROBLEMS

1. Let $z = 0$ be a singular point of the first kind of equation (9.6-1). Show that for each characteristic exponent λ there exists at most one solution u of the differential equation that can be represented in the form

$$u(z) = z^\lambda h(z),$$

   where h is analytic at $z = 0$, $h(0) = 1$. State a sufficient condition for the existence of such a solution.

2. The equation

$$u^{(n)} + q_1(z)u^{(n-1)} + z^{-1}q_2(z)u^{(n-2)} + \cdots + z^{-n+1}q_n(z)u = 0,$$

   where the functions q_k are analytic at $z = 0$, has at $z = 0$ the characteristic exponents $0, 1, 2, \ldots, n - 1$.
   (a) Prove that the equation has at least one solution that is analytic at $z = 0$.
   (b) By considering the example $u'' + z^{-1}au = 0$, show that for $n \geq 2$ all solutions need not be analytic at $z = 0$.

3. Let $z = 0$ be an isolated singular point of (9.6-1), and let the equation possess a system of solutions $u_1, u_2, \ldots, u_n$, all analytic at $z = 0$, whose Wronskian determinant at 0 is different from zero. Show that $z = 0$ is, in fact, a regular point of the differential equation. [Compare problem 6, §9.4.]

4. Let $z = 0$ be a singular point of the first kind of (9.6-1), with characteristic exponents $\lambda_1, \lambda_2, \ldots, \lambda_n$. Let $z = g(t)$, where g is analytic at $t = 0$, $g(0) = 0$, $g'(0) \neq 0$. If u is any solution of (9.6-1), show that the functions $v := u \circ g$ satisfy a differential equation that again has $t = 0$ as a singular point with the characteristic exponents $\lambda_1, \lambda_2, \ldots, \lambda_n$.

5. In the setting of Problem 4, assume that $z = t^k g_1(t)$, where k is a positive integer and g_1 is analytic at $t = 0$, $g_1(0) \neq 0$. Show that the differential equation satisfied by $v = u \circ g$ now at $t = 0$ has the charactistic exponents $k\lambda_j$, $j = 1, \ldots, n$.

6. Let $z = 0$ be a singular point of the first kind of (9.6-1), with characteristic exponents $\lambda_1, \lambda_2, \ldots, \lambda_n$. Let the function h be analytic at $z = 0$, $h(0) \neq 0$. If u is any solution of (9.6-1), $u = hw$, then the functions w again satisfy a differential equation that has $z = 0$ as a singular point of the first kind with the characteristic exponents $\lambda_1, \ldots, \lambda_n$.

7. In the setting of Problem 6, let h be a log-holomorphic function at $z = 0$ that has the special form $h(z) = z^\alpha h_1(z)$, where α is a complex number and where h_1 is analytic $z = 0$, $h_1(0) \neq 0$. If $u = hw$, show that the differential equation satisfied by w has at $z = 0$ a singular point of the first kind with characteristic exponents $\lambda_j - \alpha$, $j = 1, \ldots, n$. (The nontrivial fact is that the coefficients of the transformed equation are analytic, and not merely log-holomorphic, in some punctured vicinity of $z = 0$.)

§9.7. TWO EXAMPLES: THE EQUATIONS OF KUMMER AND BESSEL

In this section we apply the algorithm of Frobenius outlined in §9.6 to the solution of two classical differential equations, those of Kummer and Bessel.

EXAMPLE 1 **Kummer's equation**

A linear differential equation of the second order that is regular at all points except at $z = 0$, where it has a regular singular point, is necessarily of the form

$$z^2 u'' + z p(z) u' + q(z) u = 0,$$

where the functions p and q are entire. As a simple paradigm of the Frobenius method, we consider the case where p and q are polynomials of degree one:

$$p(z) := a_0 + a_1 z, \qquad q(z) = b_0 + b_1 z.$$

The indicial polynomial of the resulting equation is

$$\pi(\lambda) := \lambda(\lambda - 1) + a_0 \lambda + b_0.$$

Let its zeros be λ_1 and λ_2. Putting $u(z) = z^{\lambda_1} w(z)$ we obtain a differential equation of similar type for w (see Theorem 9.3i or problem 7, §9.6) where $b_0 = 0$, because one of the characteristic exponents now is zero. Assuming $a_1 \neq 0$, we can further standardize the equation by letting $z = -t/a_1$. This again leads to an equation of the same type where $a_1 = -1$. (The negative sign is chosen for later convenience.) We thus are led to consider a differential equation depending on two parameters only, which may be written

$$zu'' + (\gamma - z)u' - \alpha u = 0. \tag{9.7-1}$$

This is the equation known as **Kummer's differential equation** or also as the **confluent hypergeometric equation**. The equation occurs, among other places, in the quantum theory of the hydrogen atom and in the theory of diffraction of electromagnetic waves at a parabolic surface.

The indical equation of Kummer's equation is

$$\pi(\lambda) = \lambda(\lambda - 1) + \gamma\lambda = \lambda(\lambda - 1 + \gamma) = 0;$$

consequently, the exponents are 0 and $1 - \gamma$. To find a system of linearly independent solutions by the Frobenius method, we first determine a formal solution

$$U = z^\lambda \sum_{n=0}^{\infty} c_n z^n \tag{9.7-2}$$

of the nonhomogeneous equation

$$L(U) = \pi(\lambda)z^\lambda, \tag{9.7-3}$$

where, in accordance with (9.6-8),

$$L(U) := z^2 U'' + (\gamma z - z^2)U' - z\alpha U,$$

and where λ is a complex parameter. Substituting (9.7-2) yields

$$L(U) = z^\lambda \{\pi(\lambda)c_0 + \sum_{n=1}^{\infty} [\pi(\lambda + n)c_n - (\lambda + n - 1 + \alpha)c_{n-1}]z^n\};$$

consequently, (9.7-3) is satisfied if $c_0 = 1$ and

$$\pi(\lambda + n)c_n = (\lambda + n - 1 + \alpha)c_{n-1};$$

that is, for

$$c_n = c_n(\lambda) := \frac{(\lambda + \alpha)_n}{(\lambda + 1)_n (\lambda + \gamma)_n}, \qquad n = 0, 1, 2, \ldots. \tag{9.7-4}$$

If γ is not an integer, we can set λ equal to zero or to $1 - \gamma$ and thus obtain two formal solutions of $L(U) = 0$ which automatically are actual solutions of Kummer's equation. The resulting series are confluent hypergeometric series of type ${}_1F_1$ and are given by

$$\left.\begin{array}{l} u_1(z) = {}_1F_1(\alpha; \gamma; z), \\[2mm] u_2(z) = z^{1-\gamma}{}_1F_1(1 - \gamma + \alpha; 2 - \gamma; z). \end{array}\right\} \tag{9.7-5}$$

The radius of convergence of these series is infinite, as predicted by the theory.

If $\gamma = 1$, the two characteristic exponents coincide, and the two solutions given above are identical. According to Frobenius' theory, a second linearly independent solution is given by $\partial U/\partial \lambda$ evaluated at $\lambda = 0$, where U is given by (9.7-2) with coefficients c_n as determined in (9.7-4). The differentiation is carried out easily if α is not zero or a negative integer and is seen to yield

$$v(z) := \log z {}_1F_1(\alpha; 1; z) + \sum_{n=1}^{\infty} \frac{(\alpha)_n}{(n!)^2}[h_n(\alpha) - 2h_n(1)]z^n,$$

where, for $n = 1, 2, \ldots$ and $\zeta \neq 0, -1, \ldots, -n + 1$

$$h_n(\zeta) := \frac{1}{\zeta} + \frac{1}{\zeta + 1} + \cdots + \frac{1}{\zeta + n - 1}. \tag{9.7-6}$$

If α is a negative integer, $\alpha = -m$, the process of differentiation yields

$$v(z) = \log z \, {}_1F_1(-m; 1; z) + \sum_{n=1}^{m} \frac{(-m)_n}{(n!)^2} [h_n(-m) - 2h_n(1)] z^n$$

$$+ \sum_{n=m+1}^{\infty} \frac{(-1)^m m!(n-m-1)!}{(n!)^2} z^n.$$

We next consider the case where $\gamma = 1 + k$, where k is a positive integer. Here the characteristic exponents are 0 and $-k$. Of the two solutions (9.7-5), u_1 is still meaningful, but u_2 (as predicted by the theory) is meaningless because of a nonpositive integer as denominator parameter. By Frobenius' prescription, we now have to construct a formal solution of

$$L(U) = \pi(\lambda)(\lambda + k) z^{\lambda}.$$

Proceeding as before we find

$$U = z^{\lambda} \sum_{n=0}^{\infty} \frac{(\lambda + \alpha)_n (\lambda + k)}{(\lambda + 1)_n (\lambda + 1 + k)_n} z^n,$$

which may be written

$$U = z^{\lambda} (\lambda + k) \sum_{n=0}^{k-1} \frac{(\lambda + \alpha)_n}{(\lambda + 1)_n (\lambda + 1 + k)_n} z^n$$

$$+ z^{\lambda} \frac{1}{(\lambda + 1)_{k-1}} \sum_{n=k}^{\infty} \frac{(\lambda + \alpha)_n}{(\lambda + k + 1)_{n-k}(\lambda + k + 1)_n} z^n.$$

The second solution is now obtained by differentiating with respect to λ and setting $\lambda = -k$. If α is not a negative integer, this rather cumbersome process yields

$$v(z) = z^{-k} \sum_{n=0}^{k-1} \frac{(\alpha - k)_n}{(1-k)_n n!} z^n$$

$$- \frac{(1-\alpha)_k}{(k-1)!k!} \sum_{m=0}^{\infty} \frac{(\alpha)_m}{m!(k+1)_m} \{\log z + h_{k+m}(\alpha - k) \qquad (9.7\text{-}7)$$

$$- h_m(1) - h_{m+1}(k)\} z^m.$$

These examples seem sufficient to illustrate the method. We leave it to the reader to treat the cases in which α or $\gamma - 1$ are negative integers.

We now utilize the analytic theory of differential equations to derive a nontrivial identity between certain solutions of Kummer's equation. The function

$$u(z) := {}_1F_1(\gamma - \alpha; \gamma; z)$$

is a solution of

$$zu'' + (\gamma - z)u' - (\gamma - \alpha)u = 0.$$

Consequently, $v(z) := u(-z)$ solves

$$zv'' + (\gamma + z)v' + (\gamma - \alpha)v = 0.$$

It follows that $w(z) := e^z v(z)$ satisfies

$$zw'' + (-2z + \gamma + z)w' + [z - (\gamma + z) + (\gamma - \alpha)]w = 0$$

(compare problem 9, §9.3), which is identical with (9.7-1). We thus have found two solution functions of that differential equation, both of which are analytic at $z = 0$ and take the value 1 there. Because there can be only one such solution, the two functions are identical. This proves the so-called **first identity of Kummer**,

$$_1F_1(\alpha; \gamma; z) = e^z {}_1F_1(\gamma - \alpha; \gamma; -z). \tag{9.7-8}$$

This identity was proved in §1.5 by Cauchy multiplication. It can be established also by transforming certain integrals representing the function $_1F_1$. The foregoing method of transforming the differential equation will prove to be of great value in the theory of the Gaussian hypergeometric series $_2F_1$.

EXAMPLE **2**   **Bessel's equation**

Such is the name given to the equation

$$u'' + \frac{1}{z}u' + \left(1 - \frac{\nu^2}{z^2}\right)u = 0, \tag{9.7-9}$$

where ν is a complex parameter. Bessel's equation occurs in many problems of mathematical physics, especially those involving rotational symmetry (see Chap. 18). We show, first of all, that Bessel's equation can be reduced to Kummer's equation. Set $u(z) = z^\nu e^{iz} w(z)$. By Theorem 9.3i the function w then satisfies

$$w'' + \left(\frac{2\nu + 1}{z} + 2i\right)w' + \frac{(2\nu + 1)i}{z}w = 0. \tag{9.7-10}$$

This is an equation of the form studied at the beginning of the present section and thus can be reduced to Kummer's equation. Letting $v(t) := w(z)$ where $z = -t/2i$ we obtain

$$v'' + \left(\frac{2\nu + 1}{t} - 1\right)v' - \frac{\nu + \frac{1}{2}}{t}v = 0,$$

which is equivalent to Kummer's equation with parameters $\gamma = 2\nu + 1$, $\alpha = \nu + \frac{1}{2}$. If $2\nu + 1$ is not an integer, the last equation has by (9.7-5) the two linearly independent solutions

$$v_1(t) = {}_1F_1(\nu + \tfrac{1}{2}; 2\nu + 1; t),$$

$$v_2(t) = t^{-2\nu} {}_1F_1(-\nu + \tfrac{1}{2}; -2\nu + 1; t).$$

It follows that Bessel's equation has the two solutions

$$u_1(z) := z^\nu e^{iz} {}_1F_1(\nu + \tfrac{1}{2}; 2\nu + 1; -2iz), \\ u_2(z) := z^{-\nu} e^{iz} {}_1F_1(-\nu + \tfrac{1}{2}; -2\nu + 1; -2iz). \Bigg\}$$ (9.7-11)

By Kummer's first identity, these functions may also be written

$$u_1(z) = z^\nu e^{-iz} {}_1F_1(\nu + \tfrac{1}{2}; 2\nu + 1; 2iz), \\ u_2(z) = z^{-\nu} e^{-iz} {}_1F_1(-\nu + \tfrac{1}{2}; -2\nu + 1; 2iz). \Bigg\}$$ (9.7-12)

One of these solutions remains valid even if $2\nu + 1$ is an integer. The second solution can then be found by the Frobenius method as indicated under **1**.

Naturally, Bessel's equation can also be handled directly. The indicial polynomial is readily seen to be $\pi(\lambda) := \lambda^2 - \nu^2$ [this follows also from (9.7-11)], showing the exponents at 0 to be $\pm\nu$. Following the method of Frobenius, we seek a formal solution

$$U = z^\lambda \sum_{n=0}^{\infty} c_n z^n$$ (9.7-13)

of the nonhomogeneous equation

$$L_\nu(U) = \pi(\lambda) z^\lambda,$$ (9.7-14)

where

$$L_\nu(U) := z^2 U'' + z U' + (z^2 - \nu^2) U.$$

Substituting (9.7-13) into (9.7-14) yields

$$L_\nu(U) = z^\lambda \left\{ \pi(\lambda) c_0 + \sum_{n=1}^{\infty} [\pi(\lambda + n) c_n + c_{n-2}] z^n \right\},$$

where $c_{-1} := 0$. It follows that (9.7-13) is a solution of (9.7-14) if and only if $c_0 = 1$ and

$$\pi(\lambda + n) c_n + c_{n-2} = 0, \qquad n = 1, 2, \dots,$$

which implies

$$c_{2k+1} = 0, \qquad k = 0, 1, \dots,$$

and

$$c_{2k} = -\frac{1}{(\lambda + \nu + 2k)(\lambda - \nu + 2k)} c_{2k-2}, \qquad k = 1, 2, \dots.$$

The recurrence can be solved to yield

$$c_{2k} = c_{2k}(\lambda) = \frac{(-1)^k}{2^{2k} \left(\dfrac{\lambda + \nu}{2} + 1 \right)_k \left(\dfrac{\lambda - \nu}{2} + 1 \right)_k}.$$ (9.7-15)

According to Frobenius we get solutions of Bessel's equation if 2ν, the difference of the exponents, is not an integer by setting λ equal to one of the characteristic exponents $\pm\nu$. In hypergeometric notation this yields the two solutions

$$u_3(z) := z^\nu {}_0F_1\left(\nu+1; -\frac{z^2}{4}\right),$$

$$u_4(z) := z^{-\nu} {}_0F_1\left(-\nu+1; -\frac{z^2}{4}\right). \tag{9.7-16}$$

A comparison of (9.7-11) and (9.7-16) shows that both functions u_1 and u_3 are solutions of Bessel's equation enjoying the following properties:
  (i) $z^{-\nu}u(z)$ is single-valued in a neighborhood of $z = 0$;

 (ii) $\lim_{z \to 0} z^{-\nu}u(z) = 1$.

Because there can be only one solution having these properties, $u_3 = u_1$. We thus obtain the identity

$$ {}_1F_1(\nu+\tfrac{1}{2}; 2\nu+1; -2iz) = e^{-iz} {}_0F_1\left(\nu+1; -\frac{z^2}{4}\right)$$

or, naming the parameters differently,

$$ {}_1F_1(\alpha; 2\alpha; 2z) = e^z {}_0F_1\left(\alpha+\frac{1}{2}; \frac{z^2}{4}\right). \tag{9.7-17}$$

This is known as **Kummer's second identity**. The remarks made on occasion of Kummer's first identity likewise apply here.

In theory as well as in applications, none of the functions introduced thus far is used as standard solution of Bessel's equation. Instead, it has become customary to introduce the function

$$J_\nu(z) := \frac{1}{2^\nu \Gamma(\nu+1)} u_1(z),$$

where Γ denotes the gamma function. In hypergeometric notation,

$$J_\nu(z) = \frac{(z/2)^\nu}{\Gamma(\nu+1)} {}_0F_1\left(\nu+1; \frac{-z^2}{4}\right). \tag{9.7-18}$$

J_ν is the **Bessel function of order** ν. For integers ν this function has already made its appearance in §4.5. For arbitrary ν the series

$$\sum_{m=0}^{\infty} \frac{(-z^2/4)^m}{\Gamma(\nu+1+m)m!} = \left(\frac{2}{z}\right)^\nu J_\nu(z)$$

evidently converges for all complex z and thus defines an entire function of z, as predicted by the analytic theory. Less obviously, Stirling's formula shows that for every fixed z the series also converges uniformly with respect to ν on every compact set of the ν plane. Thus $J_\nu(z)$ is an entire function of ν for every fixed z with the possible exception of $z = 0$.

If 2ν is not an integer, the function $J_{-\nu}$ is proportional to the solution u_2 introduced above, and the functions

$$J_\nu(z) \quad \text{and} \quad J_{-\nu}(z)$$

thus form a system of two linearly independent solutions of Bessel's equation. Because the coefficients c_m where m is odd vanish identically, these functions are defined also when 2ν is an odd integer, and by comparing the leading terms it is easily seen that they remain linearly independent in that case. In the remaining case, when ν is an integer, both functions J_ν and $J_{-\nu}$ remain meaningful due to the Γ factor in the denominator. But it is easy to see that the two functions are no longer independent. Because $1/\Gamma(m) = 0$ for $m = 0, -1, -2, \ldots$ we have if ν equals a nonnegative integer n,

$$J_{-n}(z) = \left(\frac{z}{2}\right)^{-n} \sum_{m=0}^{\infty} \frac{(-z^2/4)^m}{\Gamma(-n+m+1)m!} = (-1)^n \left(\frac{z}{2}\right)^n \sum_{r=0}^{\infty} \frac{(-z^2/4)}{r!(n+r)!}$$

and thus

$$J_{-n}(z) = (-1)^n J_n(z). \tag{9.7-19}$$

It thus becomes necessary to search for another solution if $\nu = n$.

It is quite feasible to construct the missing solution by Frobenius' method; in fact C. Neumann constructed a solution exactly along these lines. Here we adopt a different procedure that leads directly to a function which today is firmly established as second solution of Bessel's equation for all values, integral or nonintegral, of the parameter ν.

It is clear that for nonintegral ν the function

$$v_\nu(z) := J_\nu(z) \cos \nu\pi - J_{-\nu}(z)$$

is a solution of Bessel's equation that is independent from J_ν. For $\nu = n$ this solution becomes trivial by virtue of (9.7-19). However, for each fixed $z = 0$, v_ν is an entire function of ν, because this is the case for both J_ν and $\cos \nu\pi$. We may thus hope to cancel the triviality by dividing by another entire function of ν (independent of z) which likewise vanishes at the integers, such as $\sin \nu\pi$. Thus we are led to consider the function

$$Y_\nu(z) := \frac{J_\nu(z) \cos \nu\pi - J_{-\nu}(z)}{\sin \nu\pi}, \tag{9.7-20}$$

called **Bessel function of the second kind** of order ν.

At the moment, Y_ν is defined only for nonintegral ν. But because the zeros of the denominator are simple, the singularities at the integers are removable, and by defining

$$Y_n(z) := \lim_{\nu \to n} Y_\nu(z), \qquad n = 0, \pm 1, \pm 2, \ldots, \tag{9.7-21}$$

Y_ν becomes an entire function of ν for each fixed $z \neq 0$.

We now show that Y_n is a solution of Bessel's equation for $\nu = n$. Defining v_ν as before we have by L'Hopital's rule

$$Y_n(z) = \frac{(-1)^n}{\pi} v_\nu'(z)\Big|_{\nu=n},$$

where $' := \partial/\partial\nu$. The formal logarithmic sum V_ν representing v_ν formally satisfies

$$L_\nu(V_\nu) := z^2 V_\nu'' + z V_\nu' + (z^2 - \nu^2) V_\nu = 0$$

identically in ν. Thus by differentiating with respect to the parameter ν and applying Observation 9.6b we obtain, writing $W_\nu := V_\nu'$,

$$z^2 W_\nu'' + z W_\nu' + (z^2 - \nu^2) W_\nu = 2\nu V_\nu$$

for all values of ν. If $\nu = n$, the function on the right vanishes, and we have $L_n(W_n) = 0$. By Theorem 9.5a, the series $W_n = V_n'$ thus defines an actual solution of Bessel's equation for $\nu = n$.

It remains to calculate the explicit form of the series representation of Y_ν when $\nu = n$. To this end we must evaluate J_ν' and $J_{-\nu}'$ at $\nu = n$. We recall the relation (8.5-16),

$$\frac{\Gamma'(z)}{\Gamma(z)} = -\frac{1}{z} - \gamma + \sum_{n=1}^{\infty} \left(\frac{1}{n} - \frac{1}{z+n}\right) \tag{9.7-22}$$

and the formula following from it,

$$\Gamma'(m+1) = m!\{h_m(1) - \gamma\},$$

where γ denotes Euler's constant and

$$h_m(\zeta) := \frac{1}{\zeta} + \frac{1}{\zeta+1} + \cdots + \frac{1}{\zeta+m-1}.$$

Because for fixed z the convergence of the series for J_ν is uniform with respect to ν in every compact set of the ν plane, we may differentiate term by term to obtain

$$\frac{\partial J_\nu(z)}{\partial \nu} = \left(\frac{z}{2}\right)^\nu \sum_{m=0}^{\infty} \frac{(-z^2/4)^m}{m!\,\Gamma(\nu+m+1)} \left[\log\frac{z}{2} - \frac{\Gamma'(\nu+m+1)}{\Gamma(\nu+m+1)}\right]$$

and thus, if $\nu = n \geqslant 0$,

$$\frac{\partial J_\nu}{\partial \nu}\Big|_{\nu=n} = \left(\frac{z}{2}\right)^n \sum_{m=0}^{\infty} \frac{(-z^2/4)^m}{m!(n+m)!} \left[\log\frac{z}{2} - h_{n+m}(1) + \gamma\right]. \tag{9.7-23}$$

In a like manner,

$$\frac{\partial J_{-\nu}(z)}{\partial \nu} = \left(\frac{z}{2}\right)^{-\nu} \sum_{m=0}^{\infty} \frac{(-z^2/4)^m}{m!\,\Gamma(-\nu+m+1)} \left[-\log\frac{z}{2} + \frac{\Gamma'(-\nu+m+1)}{\Gamma(-\nu+m+1)}\right];$$

however, if $m = 0, 1, \ldots, n-1$, some care is now required when $\nu \to n$. From (9.7-22) we have

$$\frac{\Gamma'(z)}{[\Gamma(z)]^2} = -\frac{1}{z\Gamma(z)} - \frac{\gamma}{\Gamma(z)} + \sum_{n=1}^{\infty} \frac{1}{\Gamma(z)}\left(\frac{1}{n} - \frac{1}{z+n}\right).$$

Using the fact that $(z + k)\Gamma(z) \to (-1)^k/k!$ for $z \to -k$ by (8.4-15), it easily follows that

$$\lim_{z \to -k} \frac{\Gamma'(z)}{[\Gamma(z)]^2} = (-1)^k k!, \quad k = 0, 1, 2, \dots.$$

We thus find, after simplification,

$$\left. \frac{\partial J_{-\nu}(z)}{\partial \nu} \right|_{\nu = n} = \left(\frac{-z}{2} \right)^{-n} \sum_{m=0}^{n-1} \frac{(n - m - 1)!}{m!} \left(\frac{z^2}{4} \right)^m$$

$$+ \left(\frac{-z}{2} \right)^n \sum_{r=0}^{\infty} \frac{(-z^2/4)^r}{r!(n+r)!} \left[-\log \frac{z}{2} + h_r(1) - \gamma \right]. \tag{9.7-24}$$

From (9.7-23) and (9.7-24) we now get

$$Y_n(z) = \frac{(z/2)^n}{\pi} \sum_{r=0}^{\infty} \frac{(-z^2/4)^r}{r!(n+r)!} \left[2 \log \frac{z}{2} - h_r(1) - h_{n+r}(1) + 2\gamma \right]$$

$$- \frac{(z/2)^{-n}}{\pi} \sum_{m=0}^{n-1} \frac{(n + m - 1)!}{m!} \left(\frac{z^2}{4} \right)^m. \tag{9.7-25}$$

The formula holds for $n = 1, 2, \dots$ and also for $n = 0$, in which case the second sum is to be omitted. For negative integers n we have

$$Y_n(z) = (-1)^n Y_{-n}(z),$$

as in the case of the Bessel functions J_n. Because the series for Y_n always contains a logarithmic term, the functions J_n and Y_n are linearly independent. It follows that J_ν and Y_ν form a fundamental system of solutions of Bessel's equation for all values of ν.

A discussion of the descriptive properties of the Bessel functions is postponed to §11.5 and §11.8 in which it will find a natural place in the context of the theory of asymptotic methods. Here we merely point out a remarkable fact concerning the algebraic nature of certain Bessel functions. From the definition (9.7-18) we see in view of $\Gamma(\frac{1}{2}) = \sqrt{\pi}$, $\Gamma(\frac{3}{2}) = \frac{1}{2}\sqrt{\pi}$ that

$$J_{1/2}(z) = \frac{(z/2)^{1/2}}{\frac{1}{2}\sqrt{\pi}} \sum_{m=0}^{\infty} \frac{(-z^2/4)^m}{(\frac{3}{2})_m m!} = \sqrt{\frac{2z}{\pi}} \sum_{m=0}^{\infty} \frac{(-z^2)^m}{(2m + 1)!}$$

$$= \sqrt{\frac{2}{\pi z}} \sin z, \tag{9.7-26a}$$

and

$$J_{-1/2}(z) = \frac{(z/2)^{-1/2}}{\sqrt{\pi}} \sum_{m=0}^{\infty} \frac{(-z^2/4)^m}{(\frac{1}{2})_m m!} = \sqrt{\frac{2}{\pi z}} \sum_{m=0}^{\infty} \frac{(-z^2)^m}{(2m)!}$$

$$= \sqrt{\frac{2}{\pi z}} \cos z. \tag{9.7-26b}$$

Thus, the functions $J_{1/2}$ and $J_{-1/2}$ are elementary in the sense that they can be represented as a composition of a finite number of rational or exponential functions or inverse functions thereof. We now show that, more generally, all functions J_ν where $\nu - \frac{1}{2}$ is an integer are elementary. This is a consequence of the relations, valid for arbitrary ν,

$$\frac{1}{z}\frac{d}{dz}\{z^\nu J_\nu(z)\} = z^{\nu-1}J_{\nu-1}(z),$$

$$\frac{1}{z}\frac{d}{dz}\{z^{-\nu}J_\nu(z)\} = -z^{-\nu-1}J_{\nu+1}(z),$$

(9.7-27)

which are easily proved by term-by-term differentiation of the series expansions. By iterating the relations (9.7-27), we have in an obvious symbolic notation

$$\left(\frac{d}{zdz}\right)^n \{z^\nu J_\nu(z)\} = z^{\nu-n}J_{\nu-n}(z),$$

$$\left(\frac{d}{zdz}\right)^n \{z^{-\nu}J_\nu(z)\} = (-1)^n z^{-\nu-n}J_{\nu+n}(z),$$

$n = 0, 1, 2, \ldots$. For $\nu = \pm\frac{1}{2}$ this yields in view of (9.7-26)

$$J_{1/2+n}(z) = (-z)^n\sqrt{2z/\pi}\left(\frac{d}{zdz}\right)^n\left(\frac{\sin z}{z}\right),$$

$$J_{-1/2-n}(z) = z^n\sqrt{2z/\pi}\left(\frac{d}{zdz}\right)^n\left(\frac{\cos z}{z}\right),$$

(9.7-28)

$n = 0, 1, 2, \ldots$. Because the derivatives of an elementary function are elementary, these relations indeed establish the elementary character of the Bessel functions involved. It can be shown—but this is much more difficult—that the cases where $\nu - \frac{1}{2}$ is an integer are the only ones where J_ν is an elementary function in the sense defined above.

PROBLEMS

1. If γ is not an integer, show that a fundamental system for the equation

$$zu'' + \gamma u' - ku = 0$$

is given by the functions

$$u_1(z) := {}_0F_1(\gamma; kz),$$
$$u_2(z) := z^{1-\gamma}{}_0F_1(2-\gamma; kz).$$

Use the method of Frobenius to obtain a fundamental system for $\gamma = 1$.

2. Let a, b, c, d, e be arbitrary complex numbers, and let k be an integer. All solutions of the differential equation

$$u'' + \frac{1}{z}(a + bz^k)u' + \frac{1}{z^2}(c + dz^k + ez^{2k})u = 0$$

can be represented in the form

$$u(z) = z^\alpha e^{\beta z^k} v(\gamma z^k),$$

where α, β, γ are suitable constants, and where v is a solution of Kummer's equation.

3. Let λ be a complex number. The differential equation

$$z^2 u'' + zu' - u = \lambda z^2 (1 - z^2)u$$

possesses a solution analytic at $z = 0$ which is given by

$$u(z) = z\, e^{-2\beta z^2} {}_1F_1(1 - \beta; 2; 4\beta z^2),$$

where β satisfies $-16\beta^2 = \lambda$. Confirm by means of Kummer's first transformation that u is unchanged if β is replaced by $-\beta$.

4. Let q be analytic in some region S. If u_1 and u_2 both are solutions of the differential equation

$$u'' + q(z)u = 0,$$

then the function $v := u_1 u_2$ satisfies

$$v''' + 4q(z)v' + 2q'(z)v = 0.$$

Use this result to obtain representations of the products

$$\{{}_0F_1(1 + \gamma; z)\}^2, \qquad {}_0F_1(1 + \gamma; z){}_0F_1(1 - \gamma; z)$$

in terms of single hypergeometric series.

5. Let ν and $z \neq 0$ be two complex numbers. The *difference* equation in the variable n,

$$y_{n+1} - \frac{2(\nu + n)}{z}y_n + y_{n-1} = 0$$

possesses the linearly independent solutions

$$y_n^{(1)} := J_{\nu+n}(z), \qquad y_n^{(2)} := Y_{\nu+n}(z).$$

6. Bessel's equation may be written

$$z(zu')' = (\nu^2 - z^2)u.$$

Conclude that for $\nu > 0$ and z real, $0 < z < \nu$, the function J_ν is increasing and hence different from zero.

7. By expanding the integral in series and interchanging summation and integration, establish **Poisson's integral representation**

$$J_\nu(z) = \frac{(z/2)^\nu}{\Gamma(\tfrac{1}{2})\Gamma(\nu + \tfrac{1}{2})} \int_0^\pi \cos(z \cos \theta)(\sin \theta)^{2\nu}\, d\theta,$$

valid for $\mathrm{Re}\, \nu > -\tfrac{1}{2}$.

§9.8. THE INFINITE POINT: EQUATIONS OF FUCHSIAN TYPE

Let the function f have an isolated singularity at $z = \infty$ (see §4.5). We recall that f is said to be analytic at $z = \infty$ if and only if the function $\hat{f}$ defined by $\hat{f}(z) := f(1/z)$ has a removable singularity at $z = 0$. Furthermore, $z = \infty$ is said to be a zero or pole of f if $z = 0$ is a zero or pole of $\hat{f}$. The order of the zero or pole of f at ∞ is defined to be the order of the zero or pole of $\hat{f}$ at 0.

Let the matrix $\mathbf{A}$ and the coefficients $p_1, \ldots, p_n$ have isolated singularities at $z = \infty$. To study the behavior of the solutions of the system

$$\mathbf{w}' = \mathbf{A}(z)\mathbf{w} \qquad (9.8\text{-}1)$$

or of the nth order scalar equation

$$u^{(n)} + p_1(z)u^{(n-1)} + \cdots + p_n(z)u = 0 \qquad (9.8\text{-}2)$$

in the neighborhood of $z = \infty$, we make the substitution $z \to 1/z$ in the differential equation and study the behavior of their solutions near $z = 0$. Substitution means, for instance in the case of the system (9.8-1), the construction of a new system that has the vector $\hat{\mathbf{w}}$ as a solution if and only if the vector $\mathbf{w}(z) := \hat{\mathbf{w}}(z^{-1})$ is a solution of (9.8-1). The new system is called the system **induced** by the transformation $z \to 1/z$. The point $z = \infty$ is said to be a singularity of a given type if and only if $z = 0$ is a singularity of the same type for the induced system.

If $\mathbf{w}$ is a solution of (9.8-1), then for the vector $\hat{\mathbf{w}}(z) := \mathbf{w}(z^{-1})$ there holds

$$\hat{\mathbf{w}}'(z) = \mathbf{w}'\left(\frac{1}{z}\right)\left(-\frac{1}{z^2}\right) = -\frac{1}{z^2}\mathbf{A}\left(\frac{1}{z}\right)\mathbf{w}\left(\frac{1}{z}\right).$$

Consequently, $\hat{\mathbf{w}}$ is a solution of

$$\hat{\mathbf{w}}' = -\frac{1}{z^2}\hat{\mathbf{A}}(z)\hat{\mathbf{w}}. \qquad (9.8\text{-}3)$$

Conversely, if $\hat{\mathbf{w}}$ is any solution of (9.8-3), then $\mathbf{w}(z) := \hat{\mathbf{w}}(z^{-1})$ will be a solution of (9.8-1). Thus the induced system is given by (9.8-3). Evidently, $z = 0$ is a singular point of the first kind of (9.8-3) if and only if $\hat{\mathbf{A}}$ is analytic at $z = 0$ and $\hat{\mathbf{A}}(0) = \mathbf{0}$. Consequently we have:

THEOREM 9.8a

The point $z = \infty$ is at most a singular point of the first kind of (9.8-1) if and only if $\mathbf{A}$ is analytic at ∞ and $\mathbf{A}(\infty) = \mathbf{0}$.

For the equation (9.8-2) the substitution is somewhat more difficult. We write the equation as

$$z^n u^{(n)} + z^{n-1}q_1(z)u^{(n-1)} + z^{n-2}q_2(z)u^{(n-2)} + \cdots + q_n(z)u = 0, \qquad (9.8\text{-}4)$$

where $q_k(z) := z^k p_k(z)$, $k = 1, \ldots, n$. If $u(z) = \hat{u}(1/z)$, we have

$$u'(z) = -\frac{1}{z^2}\hat{u}'\left(\frac{1}{z}\right), \qquad \text{hence } zu'(z) = -\frac{1}{z}\hat{u}'\left(\frac{1}{z}\right),$$

$$u''(z) = \frac{2}{z^3}\hat{u}'\left(\frac{1}{z}\right) + \frac{1}{z^4}\hat{u}''\left(\frac{1}{z}\right), \qquad \text{hence } z^2 u''(z) = \frac{1}{z^2}\hat{u}''\left(\frac{1}{z}\right) + \frac{2}{z}\hat{u}'\left(\frac{1}{z}\right),$$

and an easy induction will show that

$$z^m u^{(m)}(z) = \frac{(-1)^m}{z^m}\hat{u}^{(m)}\left(\frac{1}{z}\right) + \sum_{j=1}^{m-1} \alpha_{jm}\frac{1}{z^j}\hat{u}^{(j)}\left(\frac{1}{z}\right),$$

where the α_{jm} are numerical constants. Thus if u satisfies (9.8-4), then $\hat{u}$ is a solution of the differential equation

$$z^n \hat{u}^{(n)} + r_1(z)z^{n-1}\hat{u}^{(n-1)} + r_2(z)z^{n-2}\hat{u}^{(n-2)} + \cdots + r_n(z)u = 0, \qquad (9.8\text{-}5)$$

where, letting $q_0(z) := 1$,

$$r_k(z) := (-1)^{n-k}q_k\left(\frac{1}{z}\right) + \sum_{j=n-k+1}^{n} \alpha_{n-k,j}q_{n-j}\left(\frac{1}{z}\right), \quad k = 1, \ldots, n-1, \qquad (9.8\text{-}6)$$

and

$$r_n(z) := q_n\left(\frac{1}{z}\right). \qquad (9.8\text{-}7)$$

Conversely, if $\hat{u}$ is a solution of (9.8-5), then $u(z) := \hat{u}(1/z)$ is a solution of (9.8-4). According to the definition given in §9.6, $z = 0$ is a singular point of the first kind for equation (9.8-5) if and only if all r_k are analytic at $z = 0$. By (9.8-6) it follows inductively that this is the case if and only if the functions q_k are analytic at ∞ for $k = 1, 2, \ldots, n-1$, and (9.8-7) shows that, moreover, q_n must be analytic at ∞. Reverting to the notation used in (9.8-2), we have proved:

THEOREM 9.8b

The point $z = \infty$ is at most a singular point of the first kind of (9.8.2) if and only if each coefficient p_k is analytic at ∞ and has a zero there of order at least k.

If $z = \infty$ is a singular point of the first kind of a system of first-order equations or of a scalar equation of the nth order, a fundamental matrix or fundamental system can be constructed by the methods of §9.5 and §9.6, either by making the substitution $z \to 1/z$, or by working directly with formal series in $1/z$.

It is of interest to determine the structure of those differential equations that in the extended complex plane (including the point $z = \infty$) have singular points of the first kind only. Such equations (or first-order systems) are called of **Fuchsian type**. The singularities of a system or of an equation of Fuchsian type are necessarily finite in number; otherwise they would have a point of accumulation on the Riemann sphere that could not be an isolated singularity.

Considering the system (9.8-1) first, let the singularities be located at the points $z_1, z_2, \ldots, z_k$, and possibly ∞. If z_m is a singular point of the first kind, the matrix $\mathbf{A}(z)$ has a pole of order 1 at z_m. Consequently, if $\mathbf{R}_m$ denotes the residue at z_m, the function

$$\mathbf{F}(z) := \mathbf{A}(z) - \sum_{m=1}^{k} \frac{\mathbf{R}_m}{z - z_m}$$

is entire. Because $\mathbf{A}$ is analytic at ∞, $\mathbf{F}$ must be bounded, and hence is constant by Liouville's theorem (Theorem 3.3b). Letting $z \to \infty$ and observing that $\mathbf{A}(\infty) = \mathbf{0}$, the value of the constant is determined as zero. Thus $\mathbf{A}$ must have the form

$$\mathbf{A}(z) = \sum_{m=1}^{k} \frac{\mathbf{R}_m}{z - z_m}, \tag{9.8-8}$$

where the $\mathbf{R}_m$ are constant nonzero matrices. Conversely, it is immediately clear that every $\mathbf{A}$ of the foregoing form yields a Fuchsian system with singularities at $z_1, \ldots, z_k$. To study the singularity at ∞ we expand in powers of z^{-1}:

$$A(z) = \frac{1}{z} \sum_{m=1}^{k} \mathbf{R}_m + O(z^{-2}).$$

It follows that $z = \infty$ is a singular point if and only if the sum of the residues is not zero. We thus have proved:

THEOREM 9.8c

The system (9.8-1) is of Fuchsian type, with singularities at the points $z_1, \ldots, z_k$ and possibly ∞, if and only if the matrix A is of the form (9.8-8), where $\mathbf{R}_m \neq \mathbf{0}$, $m = 1, 2, \ldots, k$. The point ∞ is a singular point if and only if $\sum \mathbf{R}_m \neq \mathbf{0}$.

The nth order scalar equation (9.8-2) has singular points of the first kind at $z_1, z_2, \ldots, z_k$ if and only if the functions

$$q_h(z) := \prod_{m=1}^{k} (z - z_m)^h p_h(z), \qquad h = 1, 2, \ldots, n,$$

are analytic at all points z_m, that is, if they are entire. The point $z = \infty$ is a singular point of the first kind if and only if each function p_h is analytic at $z = \infty$ and has a zero of order h there. Because

$$p_h(z) = \prod_{m=1}^{k} (z - z_m)^{-h} q_h(z),$$

this is the case if and only if q_h is a polynomial of degree $\leq kh - h$. Thus we have:

THEOREM 9.8d

The nth order scalar equation (9.8-2) is of Fuchsian type, with singularities at the points $z_1, z_2, \ldots, z_k$ and possibly ∞, if and only if the coefficients have the form

$$p_h(z) = q_h(z) \prod_{m=1}^{k} (z - z_m)^{-h}, \qquad h = 1, \ldots, n,$$

where q_h is a polynomial of degree $\leq kh - h$.

The necessary and sufficient condition for $z = \infty$ to be an actual singular point is not as simple as in the case of systems and is therefore omitted.

We pursue the theory of Fuchsian equations further by considering in more detail the case of a scalar equation of the second order. Let

$$u'' + p(z)u' + q(z)u = 0 \qquad (9.8-9)$$

be a Fuchsian equation with singularities at the points $z_1, z_2, \ldots, z_k$ and possibly ∞. By Theorem 9.8d, the coefficients are rational functions of the form

$$p(z) = p^*(z) \prod_{m=1}^{k} (z - z_m)^{-1},$$

$$q(z) = q^*(z) \prod_{m=1}^{k} (z - z_m)^{-2},$$

where p^* and q^* are polynomials of degrees not exceeding $k - 1$ and $2k - 2$, respectively, and can therefore be represented by partial fractions as follows:

$$p(z) = \sum_{m=1}^{k} \frac{a_m}{z - z_m}, \qquad q(z) = \sum_{m=1}^{k} \left[\frac{b_m}{(z - z_m)^2} + \frac{c_m}{z - z_m} \right]. \qquad (9.8-10)$$

Because q has a zero of order 2 at ∞, we have

$$\sum_{m=1}^{k} c_m = 0. \qquad (9.8-11)$$

Also, no point z_m is regular if and only if at least one of the three numbers a_m, b_m, c_m is different from zero for each value of m.

Here we derive the necessary and sufficient condition in order that $z = \infty$ is a regular point of the differential equation. Making the substitution $z \to 1/z$ in (9.8-9) yields the following equation for the function $\hat{u}(z) := u(1/z)$:

$$\hat{u}'' + \left(\frac{2}{z} - \frac{1}{z^2}\hat{p}(z)\right)\hat{u}' + \frac{1}{z^4}\hat{q}(z)\hat{u} = 0,$$

where $\hat{p}(z) := p(z^{-1})$, $\hat{q}(z) := q(z^{-1})$. Evidently $z = 0$ is a regular point if and only if $\hat{p}(z) - 2z$ has a zero of order 2 at $z = 0$, and $\hat{q}(z)$ has a zero of order 4. In view of

$$\hat{p}(z) = \sum_{m=1}^{k} \frac{a_m z}{1 - zz_m} = z \sum_{m=1}^{k} a_m + O(z^2),$$

$$\hat{q}(z) = \sum_{m=1}^{k} \left[\frac{b_m z^2}{(1 - z_m z)^2} + \frac{c_m z}{1 - z_m z}\right]$$

$$= z^2 \sum_{m=1}^{k} (b_m + z_m c_m) + z^3 \sum_{m=1}^{k} (2b_m z_m + c_m z_m^2) + O(z^4),$$

this is the case if and only if in addition to (9.8-11) the three following relations are satisfied:

$$\sum_{m=1}^{k} a_m = 2, \qquad \sum_{m=1}^{k} (b_m + c_m z_m) = 0,$$

$$\sum_{m=1}^{k} (2b_m z_m + c_m z_m^2) = 0.$$
(9.8-12)

As the principal terms in the Laurent series of the coefficients are evident from (9.8-10), it is easy to write down the indicial polynomial belonging to the point $z = z_m$. It is given by

$$\pi_m(\lambda) := \lambda(\lambda - 1) + a_m \lambda + b_m.$$

Denoting its zeros by α_m, β_m, we evidently have

$$a_m = 1 - \alpha_m - \beta_m, \qquad b_m = \alpha_m \beta_m.$$
(9.8-13)

If all singular points are finite, then it follows from (9.8-12) that

$$\sum_{m=1}^{k} (\alpha_m + \beta_m) = k - 2.$$
(9.8-14)

This formula is called the **relation of Fuchs**. It shows that the characteristic exponents in a Fuchsian equation cannot be prescribed arbitrarily.

We specialize further by restricting the number of singular points. It is an easy matter to write down the general Fuchsian equation with one or two singular points, either finite or infinite. The solutions of these equations are all elementary. A highly structured theory is obtained, however, by considering the case of three singular points z_1, z_2, z_3. We first assume that these are finite. The coefficients p and q are then given by (9.8-10) where $k = 3$, with constants a_m, b_m, c_m satisfying (9.8-11) and (9.8-12). Because the three conditions to be satisfied by the three constants c_m are linear, it is to be expected that the c_m can be expressed by the a_m and the b_m. If so, then by (9.8-13) all constants a_m, b_m, c_m can be expressed in terms of the characteristic exponents, which by (9.8-14) must satisfy

$$\sum_{m=1}^{3} (\alpha_m + \beta_m) = 1.$$

Thus the equation, for given singularities z_1, z_2, z_3, would depend on only five essential parameters.

The equations to be satisfied by the c_m are

$$c_1 + c_2 + c_3 = 0,$$

$$z_1 c_z + z_2 c_2 + z_3 c_3 = b,$$

$$z_1^2 c_1 + z_2^2 c_2 + z_3^2 c_3 = b^*,$$

where $b := -\sum b_m$, $b^* := -2 \sum z_m b_m$. The determinant of the system is the Vandermondian

$$\Delta := (z_2 - z_1)(z_3 - z_1)(z_3 - z_2)$$

and thus is different from zero. Thus the system can be solved for every choice of the b_m. Expressing the solution in terms of the characteristic exponents, we obtain the following general form of the equation with three singular points of the first kind, all finite:

$$u'' + \left(\frac{1 - \alpha_1 - \beta_1}{z - z_1} + \frac{1 - \alpha_2 - \beta_2}{z - z_2} + \frac{1 - \alpha_3 - \beta_3}{z - z_3} \right) u'$$

$$- \left(\frac{\alpha_1 \beta_1}{(z - z_1)(z_2 - z_3)} + \frac{\alpha_2 \beta_2}{(z - z_2)(z_3 - z_1)} + \frac{\alpha_3 \beta_3}{(z - z_3)(z_1 - z_2)} \right)$$

$$\times \frac{(z_1 - z_2)(z_2 - z_3)(z_3 - z_1)}{(z - z_1)(z - z_2)(z - z_3)} u = 0. \tag{9.8-15a}$$

If one of the three singular points, say z_3, lies at ∞, the result is what one would obtain by letting $z_3 \to \infty$ in (9.8-15a) formally:

$$u'' + \left(\frac{1-\alpha_1-\beta_1}{z-z_1} + \frac{1-\alpha_2-\beta_2}{z-z_2} \right) u'$$

$$+ \left(\frac{\alpha_1\beta_1}{z-z_1} - \frac{\alpha_2\beta_2}{z-z_2} + \frac{\alpha_3\beta_3}{z_1-z_2} \right) \frac{z_1-z_2}{(z-z_1)(z-z_2)} u = 0. \quad (9.8\text{-}15b)$$

For reasons that will become apparent, the equation (9.18-15a) and its limiting case (9.8-15b) are called the **general hypergeometric equation**. This equation is now studied in detail in §9.9.

PROBLEMS

1. Let $z = \infty$ be a singular point of the first kind of equation (9.8-2), with characteristic exponents $\lambda_1, \ldots, \lambda_n$. Let $z = g(t)$, where g has a pole of order 1 at t_0. If u is any solution of (9.8-2), then the function $v := u \circ g$ satisfies an equation with t_0 as a regular singular point with the exponents $\lambda_1, \ldots, \lambda_n$.

2. In the setting of Problem 1, let the function g now have a pole of order k at t_0. Show that $v = u \circ g$ now satisfies a differential equation that has t_0 as a singular point of the first kind with characteristic exponents $k\lambda_j, j = 1, 2, \ldots, n$.

3. Let $z = \infty$ be a singular point of the first kind of equation (9.8-2), with the characteristic exponents $\lambda_1, \ldots \lambda_n$. If u denotes any solution, and if h is analytic at ∞, $h(\infty) \neq 0$, then the functions w defined by $u = hw$ satisfy a differential equation that again has ∞ as a singular point of the first kind with the same characteristic exponents.

4. In the setting of Problem 3, let h now be a log–holomorphic function of the special form $h(z) = z^\alpha h_1(z)$, where h_1 is analytic, $h_1(\infty) \neq 0$. Show that the differential equation satisfied by w at ∞ now has the characteristic exponents $\lambda_j + \alpha, j = 1, \ldots, n$.

5. Show that $\mathbf{w}' = \mathbf{0}$ is the only Fuchsian system with no singular point.

6. Prove that there exists no Fuchsian system with precisely one singular point.

7. Prove that there exists no second-order Fuchsian differential equation with no singular point.

8. Show that $z = \infty$ is a singular point of the differential equation $u'' = 0$, and determine the corresponding fundamental system.

9. Show that a Fuchsian second-order differential equation with a single singular point $z = a \neq \infty$ is necessarily of the form

$$u'' + \frac{2}{z-a} u' = 0,$$

and construct a fundamental system.

10. The totality of all second-order Fuchsian equations with precisely two singular

points z_1 and z_2, both finite, is identical with the totality of all equations

$$u'' + \frac{2z + a}{(z - z_1)(z - z_2)} u' + \frac{b}{(z - z_1)^2 (z - z_2)^2} u = 0,$$

where a and b are arbitrary constants. Determine all solutions of this equation by introducing the new variable

$$t := \frac{z - z_1}{z - z_2}.$$

§9.9. THE HYPERGEOMETRIC DIFFERENTIAL EQUATION

In this section we dwell on the theory of the Fuchsian differential equation of the second order with three singular points. This theory was developed by Riemann [1857] on an almost purely conceptual basis, with a minimum amount of analytic computation. Riemann based his work on ideas of analytic continuation which today are not easily accounted for without the appropriate mathematical machinery. The following presentation, although rigorous, tries to preserve some of the spirit of Riemann's work.

The general form of the equation under consideration is given near the end of §9.8. The equation is fully determined by the location of its singular points, which we now denote by a, b, c, and by the three pairs of characteristic exponents, now denoted by (α, α'), (β, β'), (γ, γ'), which must satisfy the Fuchsian relation

$$\alpha + \alpha' + \beta + \beta' + \gamma + \gamma' = 1. \tag{9.9-1}$$

In this notation equation (9.8-15a) appears in the form

$$u'' + \left(\frac{1 - \alpha - \alpha'}{z - a} + \frac{1 - \beta - \beta'}{z - b} + \frac{1 - \gamma - \gamma'}{z - c} \right) u'$$

$$- \left(\frac{\alpha\alpha'}{(z - a)(b - c)} + \frac{\beta\beta'}{(z - b)(c - a)} + \frac{\gamma\gamma'}{(z - c)(a - b)} \right) \frac{(a - b)(b - c)(c - a)}{(z - a)(z - b)(z - c)} u = 0. \tag{9.9-2}$$

It is our aim to construct for every point of the extended complex plane a system of solutions that is fundamental at that point and to show how these solutions can be continued analytically along an arbitrary path avoiding the singular points. The method consists in reducing, by suitable transformations of both the dependent and the independent variable, the number of parameters in the foregoing equation. It will turn out that its solutions depend in a nontrivial way on only three parameters.

Following Riemann, the set of all solutions of (9.9-2) are denoted by **Riemann's P symbol**

$$P\left\{\begin{matrix} a & b & c & \\ \alpha & \beta & \gamma & z \\ \alpha' & \beta' & \gamma' & \end{matrix}\right\}. \qquad (9.9\text{-}3)$$

It is clear that in this symbol the first three columns may be interchanged arbitrarily. Also, the two characteristic exponents in each column may be interchanged.

The symbol z in (9.9-3) does not represent a complex number, but it should be regarded as representing the unit function $i(z) := z$. We write

$$v \in P\left\{\begin{matrix} \cdot & \cdot & \cdot & \\ \cdot & \cdot & \cdot & g(z) \\ \cdot & \cdot & \cdot & \end{matrix}\right\}$$

to indicate that $v = u \circ g$, where u is a member of (9.9-3) with identical parameters. Similarly, the notation

$$w \in h(z)P\left\{\begin{matrix} \cdot & \cdot & \cdot & \\ \cdot & \cdot & \cdot & z \\ \cdot & \cdot & \cdot & \end{matrix}\right\}$$

indicates that $w = hu$, where u again is a member of (9.9-3). These notations are especially convenient when, as will usually be the case, the functions g and h are given by simple explicit formulas.

We recall the following corollaries of the Theorems 9.3g, 9.3h, and 9.3i. Let (L) be a scalar linear differential equation, and let z_0 be a finite singular point of the first kind of (L) with the characteristic exponents $\alpha_1, \ldots, \alpha_m$. Let the function g either be analytic at z_0 and satisfy $g'(z_0) \neq 0$, or let it be meromorphic and have a simple pole. Then every solution u of (L) can locally be represented in the form $u = v \circ g$, where v satisfies a differential equation (L′) which at $g(z_0)$ has a singular point of the first kind, again with the characteristic exponents $\alpha_1, \ldots, \alpha_m$. The same holds if $z_0 = \infty$, provided g either has a pole of order 1 at ∞ or the derivative of $g(1/z)$ does not vanish at 0. Furthermore, if z_0 is a regular point of (L), then $g(z_0)$ is a regular point of (L′) under the same conditions on g.

If (L) is Fuchsian, it makes sense to ask for those functions g for which the transformed equation is again Fuchsian. Such functions must define a one-to-one mapping of the extended complex plane onto itself. As follows from Theorem 5.2a and Theorem 5.10a, the only such functions are the Moebius transformations. It is readily verified that these functions also have the other properties required. In the special case under consideration, we thus obtain

THEOREM 9.9a

Let a, b, c be any three distinct points of the extended complex plane, and let α, α', β, β', γ, γ' be any six complex numbers satisfying the Fuchsian relation (9.9-1). Then for any Moebius transformation t,

$$P\left\{\begin{matrix} a & b & c \\ \alpha & \beta & \gamma & z \\ \alpha' & \beta' & \gamma' \end{matrix}\right\} = P\left\{\begin{matrix} t(a) & t(b) & t(c) \\ \alpha & \beta & \gamma & t(z) \\ \alpha' & \beta' & \gamma' \end{matrix}\right\}.$$

We next transform the dependent variable. Again let (L) be a linear differential equation that at the finite singular point z_0 has the characteristic exponents $\alpha_1, \ldots, \alpha_m$, and let h be a log–holomorphic function at z_0 of the special form $h(z) = (z - z_0)^\beta h_0(z)$, where h_0 is analytic at z_0, $h_0(z_0) \neq 0$. It is then easily verified by computation that if u is a solution of (L) and $u = hw$, the function w satisfies a differential equation (L*) which again has z_0 as a singular point, but now with the exponents $\alpha_1 - \beta, \ldots, \alpha_m - \beta$. If z_0 is regular and $\beta = 0$, z_0 remains regular. If $z_0 = \infty$ and $h(z) = z^\beta h_0(z)$, where h_0 is analytic at ∞, $h_0(\infty) \neq 0$, the characteristic exponents will become $\alpha_1 + \beta, \ldots, \alpha_m + \beta$.

Let (L) now be Fuchsian. We wish to determine those h for which the transformed equation is again Fuchsian, with the same singular points. If the singular points $z_1, \ldots, z_k$ are all finite, h may have at most a singularity of the form $(z - z_j)^{\beta_j}$ at each z_j, and must therefore be of the form

$$\prod_{j=1}^{k} (z - z_j)^{\beta_j} h_0(z),$$

where h_0 is a nonvanishing entire function. However, because h must be analytic at ∞, h_0 must be constant. Furthermore, because the modified characteristic exponents again must satisfy the Fuchsian relation (9.8-14), the sum of the β_j must be zero. The most general multiplier thus is of the form

$$h(z) = \prod_{j=1}^{k} (z - z_j)^{\beta_j}, \qquad \sum_{j=1}^{k} \beta_j = 0.$$

If $z_k = \infty$, h may be taken as

$$h(z) = \prod_{j=1}^{k-1} (z - z_j)^{\beta_j},$$

where the β_j are arbitrary. The exponents at ∞ will then be increased by $\sum \beta_j$. Applying the foregoing to the hypergeometric equation ($k = 3$), we obtain

THEOREM 9.9b

If a, b, c are distinct points in the (unextended) *complex plane, then for arbitrary exponents satisfying* (9.9-1) *and for arbitrary complex δ and ϵ*

$$P\left\{\begin{matrix} a & b & c & \\ \alpha & \beta & \gamma & z \\ \alpha' & \beta' & \gamma' & \end{matrix}\right\}$$

$$= \left(\frac{z-a}{z-b}\right)^{\delta}\left(\frac{z-b}{z-c}\right)^{\epsilon} P\left\{\begin{matrix} a & b & c & \\ \alpha-\delta & \beta+\delta-\epsilon & \gamma+\epsilon & z \\ \alpha'-\delta & \beta'+\delta-\epsilon & \gamma'+\epsilon & \end{matrix}\right\};$$

$$P\left\{\begin{matrix} a & \infty & c & \\ \alpha & \beta & \gamma & z \\ \alpha' & \beta' & \gamma' & \end{matrix}\right\}$$

$$= (z-a)^{\delta}(z-c)^{\epsilon} P\left\{\begin{matrix} a & \infty & c & \\ \alpha-\delta & \beta+\delta+\epsilon & \gamma-\epsilon & z \\ \alpha'-\delta & \beta'+\delta+\epsilon & \gamma'-\epsilon & \end{matrix}\right\}.$$

These theorems are now used to standardize the hypergeometric equation. Because in a Moebius transformation three image points may be prescribed arbitrarily, we can choose the transformation in Theorem 9.9a such that $t(a)=0$, $t(b)=\infty$, and $t(c)=1$. If a, b, c are finite, the explicit form of t is

$$t(z) = \frac{(z-a)(c-b)}{(c-a)(z-b)}. \qquad (9.9\text{-}4)$$

This yields the identity

$$P\left\{\begin{matrix} a & b & c & \\ \alpha & \beta & \gamma & z \\ \alpha' & \beta' & \gamma' & \end{matrix}\right\} = P\left\{\begin{matrix} 0 & \infty & 1 & \\ \alpha & \beta & \gamma & \dfrac{(z-a)(c-b)}{(c-a)(z-b)} \\ \alpha' & \beta' & \gamma' & \end{matrix}\right\}.$$

Following Riemann, we write

$$P\left\{\begin{matrix} 0 & \infty & 1 & \\ \alpha & \beta & \gamma & z \\ \alpha' & \beta' & \gamma' & \end{matrix}\right\} =: P\left\{\begin{matrix} \alpha & \beta & \gamma & \\ \alpha' & \beta' & \gamma' & z \end{matrix}\right\}$$

The set of solutions is further standardized by means of

$$P\left\{\begin{matrix} \alpha & \beta & \gamma & \\ \alpha' & \beta' & \gamma' & z \end{matrix}\right\} = z^{\alpha}(1-z)^{\gamma} P\left\{\begin{matrix} 0 & \beta+\alpha+\gamma & 0 & \\ \alpha'-\alpha & \beta'+\alpha+\gamma & \gamma'-\gamma & z \end{matrix}\right\},$$

which is a special case of Theorem 9.9b. We thus have found

THEOREM 9.9c

If a, b, c are finite, every member u of (9.9-3) can be represented in the form
$u = (hv) \circ t$, *where t is given by (9.9-4), $h(z) := z^\alpha (1-z)^\gamma$, and*

$$v \in P \left\{ \begin{matrix} 0 & \alpha + \beta + \gamma & 0 \\ \alpha' - \alpha & \alpha + \beta' + \gamma & \gamma' - \gamma \end{matrix} \; z \right\}.$$

Thus the totality of solutions of (9.9-2) can be obtained by elementary transformations from the solutions of the equation with singular points 0, ∞, 1 and one of the exponents at 0 and 1 equaling zero. Because the parameters still must satisfy the Fuchsian relation, this set depends on only three parameters. For formal convenience we choose as two of the parameters the two exponents at ∞ and as the third parameter 1 minus the nonzero exponent at 0. The reduced P set is then given by

$$P \left\{ \begin{matrix} 0 & \alpha & 0 \\ 1 - \gamma & \beta & \gamma - \alpha - \beta \end{matrix} \; z \right\}. \tag{9.9-5}$$

It consists, by definition, of all functions that are solutions of the equation

$$u'' + \left(\frac{\gamma}{z} + \frac{1 - \gamma + \alpha + \beta}{z - 1} \right) u' + \frac{\alpha \beta}{z(z-1)} u = 0, \tag{9.9-6}$$

called the **reduced hypergeometric equation**.

We now determine fundamental systems of solutions of the reduced hypergeometric equation under the assumption that none of the exponent differences $1 - \gamma$, $\alpha - \beta$, $\gamma - \alpha - \beta$ is an integer. The exceptional cases can be dealt with without difficulty by the method of §9.6.

It will be seen that the only point where it is necessary to determine a fundamental system by power series computation is $z = 0$. Here the indicial polynomial is $\pi(\lambda) := \lambda(\lambda - 1 + \gamma)$, by construction. Writing

$$L(U) := z(z - 1)U'' + [(1 + \alpha + \beta)z - \gamma]U' + \alpha \beta U$$

we shall, following Frobenius' method, determine rational functions $c_n = c_n(\lambda)$, $n = 0, 1, 2, \ldots$, such that the formal logarithmic series

$$U = z^\lambda \sum_{n=0}^{\infty} c_n z^n$$

formally satisfies

$$L(U) = z^\lambda \pi(\lambda), \tag{9.9-7}$$

identically in λ. Substituting U into L and comparing coefficients, we find that this is the case if and only if $c_0 = 1$ and

$$(\lambda + n + 1)(\lambda + n + \gamma)c_{n+1} = (\lambda + \alpha + n)(\lambda + \beta + n), \qquad n = 0, 1, 2, \ldots.$$

The recurrence relation can be solved immediately to yield

$$c_n = \frac{(\lambda + \alpha)_n (\lambda + \beta)_n}{(\lambda + 1)_n (\lambda + \gamma)_n}, \qquad n = 0, 1, \ldots . \tag{9.9-8}$$

The resulting series U is a formal solution of (9.9-6) and not merely of (9.9-7) if we set λ equal to one of the characteristic exponents 0 and $1 - \gamma$. The resulting series are (ordinary) hypergeometric series of type $_2F_1$. Because formal solutions are actual solutions, we have found

THEOREM 9.9d

If γ is not an integer, the functions

$$u_1(z) := F(\alpha, \beta; \gamma; z)$$
$$u_2(z) := z^{1-\gamma} F(1 + \alpha - \gamma, 1 + \beta - \gamma; 2 - \gamma; z) \tag{9.9-9}$$

form for $|z| < 1$ a fundamental system of the set (9.9-5), that is, a fundamental system for the reduced hypergeometric equation (9.9-6).

It is now easy to write down fundamental systems for the other singular points. The transformation

$$t_2(z) := \frac{1}{z}$$

interchanges 0 and ∞ and leaves 1 fixed. Thus by Theorem 9.9a,

$$P \begin{Bmatrix} 0 & \alpha & 0 & \\ 1 - \gamma & \beta & \gamma - \alpha - \beta & z \end{Bmatrix} = P \begin{Bmatrix} \alpha & 0 & 0 & 1 \\ \beta & 1 - \gamma & \gamma - \alpha - \beta & z \end{Bmatrix}.$$

By Theorem 9.9b, the last set is identical with

$$\left(\frac{1}{z} \right)^\alpha P \begin{Bmatrix} 0 & \alpha & 0 & 1 \\ \beta - \alpha & 1 - \gamma + \alpha & \gamma - \beta - \alpha & z \end{Bmatrix}$$

An application of Theorem 9.9d now yields the following fundamental system valid near $z = \infty$:

$$u_3(z) := (-z)^{-\alpha} F\left(\alpha, 1 + \alpha - \gamma; 1 + \alpha - \beta; \frac{1}{z} \right)$$
$$u_4(z) := (-z)^{-\beta} F\left(\beta, 1 + \beta - \gamma; 1 + \beta - \alpha; \frac{1}{z} \right) \tag{9.9-10}$$

The powers of -1 are introduced for later convenience.

To obtain a fundamental system valid near $z = 1$, we interchange the singular points 0 and 1 by means of

$$t_3(z) := 1 - z$$

which leaves ∞ fixed. Theorem 9.9a yields

$$P\left\{\begin{matrix} 0 & \alpha & 0 \\ 1-\gamma & \beta & \gamma-\alpha-\beta \end{matrix}\ z\right\} = P\left\{\begin{matrix} 0 & \alpha & 0 \\ \gamma-\alpha-\beta & \beta & 1-\gamma \end{matrix}\ 1-z\right\}.$$

By Theorem 9.9d, the second set has the fundamental system

$$u_5(z) := F(\alpha, \beta; 1-\gamma+\alpha+\beta; 1-z)$$
$$u_6(z) := (1-z)^{\gamma-\alpha-\beta}F(\gamma-\alpha, \gamma-\beta; 1+\gamma-\alpha-\beta; 1-z) \tag{9.9-11}$$

These six solutions are not the only ones that can be constructed by Riemann's method. At each singular point, a new set of fundamental solutions is obtained by permuting the two other singular points. At $z = 0$ this permutation is effected by the transformation

$$t_4(z) := \frac{z}{z-1}.$$

By Theorem 9.9a, we thus have

$$P\left\{\begin{matrix} 0 & \alpha & 0 \\ 1-\gamma & \beta & \gamma-\alpha-\beta \end{matrix}\ z\right\} = P\left\{\begin{matrix} 0 & 0 & \alpha \\ 1-\gamma & \gamma-\alpha-\beta & \beta \end{matrix}\ \frac{z}{z-1}\right\}.$$

Adjusting the exponents in the third column by means of Theorem 9.9b, the latter set in view of

$$1 - \frac{z}{z-1} = t_3 \circ t_4(z) = \frac{1}{1-z}$$

is identical with

$$(1-z)^{-\alpha}P\left\{\begin{matrix} 0 & \alpha & 0 \\ 1-\gamma & \gamma-\beta & \beta-\alpha \end{matrix}\ \frac{z}{z-1}\right\}.$$

This has the fundamental system

$$u_7(z) := (1-z)^{-\alpha}F\left(\alpha, \gamma-\beta; \gamma; \frac{z}{z-1}\right)$$
$$u_8(z) := z^{1-\gamma}(1-z)^{\gamma-\alpha-1}F\left(1+\alpha-\gamma, 1-\beta; 2-\gamma; \frac{z}{z-1}\right). \tag{9.9-12}$$

In an analogous manner we can find fundamental systems at $z = \infty$ and at $z = 1$ in terms of the variables

$$t_5(z) := t_4 \circ t_2(z) = \frac{1}{1-z}$$

and

$$t_6(z) := t_4 \circ t_3(z) = 1 - \frac{1}{z}.$$

There result the systems

$$u_9(z) := (1-z)^{-\alpha} F\left(\alpha, \gamma - \beta; 1 + \alpha - \beta; \frac{1}{1-z}\right)$$

$$u_{10}(z) := (1-z)^{-\beta} F\left(\beta, \gamma - \alpha; 1 + \beta - \alpha; \frac{1}{1-z}\right)$$

(9.9-13)

valid near $z = \infty$, and

$$u_{11}(z) := z^{-\alpha} F\left(\alpha, 1 + \alpha - \gamma; 1 + \alpha + \beta - \gamma; 1 - \frac{1}{z}\right)$$

$$u_{12}(z) := z^{\alpha - \gamma}(1-z)^{\gamma - \alpha - \beta} F\left(\gamma - \alpha, 1 - \alpha; 1 + \gamma - \alpha - \beta; 1 - \frac{1}{z}\right)$$

(9.9-14)

valid near $z = 1$.

A further set of 12 solutions can be obtained by interchanging the role of the two exponents in the last column of the P symbol. By Theorem 9.9b we have, for instance,

$$P\left\{\begin{matrix} 0 & \alpha & 0 \\ 1 - \gamma & \beta & \gamma - \alpha - \beta \end{matrix}\ z\right\} = (1-z)^{\gamma - \alpha - \beta} P\left\{\begin{matrix} 0 & \gamma - \beta & 0 \\ 1 - \gamma & \gamma - \alpha & \alpha + \beta - \gamma \end{matrix}\ z\right\}.$$

An application of Theorem 9.9d now yields the following new fundamental system valid near $z = 0$:

$$u_{13}(z) := (1-z)^{\gamma - \alpha - \beta} F(\gamma - \alpha, \gamma - \beta; \gamma; z)$$

$$u_{14}(z) := z^{1-\gamma}(1-z)^{\gamma - \alpha - \beta} F(1 - \alpha, 1 - \beta; 2 - \gamma; z)$$

(9.9-15)

By a similar procedure we find the following fundamental systems in terms of the variables $t_2, \ldots, t_6$:

$$u_{15}(z) := (-z)^{\beta - \gamma}(1-z)^{\gamma - \alpha - \beta} F\left(1 - \beta, \gamma - \beta; 1 + \alpha - \beta; \frac{1}{z}\right)$$

$$u_{16}(z) := (-z)^{\alpha - \gamma}(1-z)^{\gamma - \alpha - \beta} F\left(1 - \alpha, \gamma - \alpha; 1 + \beta - \alpha; \frac{1}{z}\right)$$

(9.9-16)

$$u_{17}(z) := z^{1-\gamma} F(1 + \alpha - \gamma, 1 + \beta - \gamma; 1 + \alpha + \beta - \gamma; 1 - z)$$

$$u_{18}(z) := z^{1-\gamma}(1-z)^{\gamma - \alpha - \beta} F(1 - \alpha, 1 - \beta; 1 - \alpha - \beta + \gamma; 1 - z)$$

(9.9-17)

$$u_{19}(z) := (1-z)^{-\beta} F\left(\gamma - \alpha, \beta; \gamma; \frac{z}{z-1}\right)$$

$$(9.9\text{-}18)$$

$$u_{20}(z) := z^{1-\gamma}(1-z)^{\gamma-\beta-1} F\left(1+\beta-\gamma, 1-\alpha; 2-\gamma; \frac{z}{z-1}\right)$$

$$u_{21}(z) := (-z)^{1-\gamma}(1-z)^{\gamma-\alpha-1} F\left(1-\beta, 1+\alpha-\gamma; 1+\alpha-\beta; \frac{1}{1-z}\right)$$

$$(9.9\text{-}19)$$

$$u_{22}(z) := (-z)^{1-\gamma}(1-z)^{\gamma-\beta-1} F\left(1-\alpha, 1+\beta-\gamma; 1+\beta-\alpha; \frac{1}{1-z}\right)$$

$$u_{23}(z) := z^{-\beta} F\left(1+\beta-\gamma, \beta; 1+\alpha+\beta-\gamma; 1-\frac{1}{z}\right)$$

$$(9.9\text{-}20)$$

$$u_{24}(z) := z^{\beta-\gamma}(1-z)^{\gamma-\alpha-\beta} F\left(\gamma-\beta, 1-\beta; 1+\gamma-\alpha-\beta; 1-\frac{1}{z}\right)$$

We thus have obtained Kummer's table of 24 solutions of the hypergeometric equation. If we write

$$t_1(z) := z,$$

for symmetry, four solutions have been obtained in each of the arguments $t_k(z)$, $k = 1, 2, \ldots, 6$. The solutions with argument t_k were denoted by $u_{2k-1}, u_{2k}, u_{2k+11}, u_{2k+12}$. Taken in a literal sense as series expansions, these solutions are valid in the sets $S_k := \{z : t_k(z)| < 1\}$. Figure 9.9a shows the sets S_1, S_3, and S_4. Because $t_2 = 1/t_1$, $t_5 = 1/t_3$, $t_6 = 1/t_4$, the sets S_2, S_5, and S_6 are complementary to the sets S_1, S_3, S_4. It will be noted that $z = e^{\pm i\pi/3}$ are the only points that are not interior points of some set S_k.

By the general theory of analytic linear differential equations, any of the 24 solutions u_k can be continued analytically along any path Γ that does not pass through one of the singular points $0, \infty, 1$. Naturally, the values of the functions obtained by this process of continuation depend on the path Γ; if a solution is continued along a closed path around a singular point, its values will be changed, unless it happens to be analytic at the singular point. Thus no single-valued analytic solutions can be defined in $\mathbb{C} - \{0, 1\}$ by the process of continuation. Nevertheless, it is customary to define such solutions artificially by erecting barriers in the complex plane across which they must not be continued. These barriers are placed on the real axis between the points $-\infty$ and 0 and between 1 and $+\infty$. It is evident that in this "cut" complex plane all paths of continuation are homotopic and thus define the same analytic function. *We shall normalize the solutions* $u_1, \ldots, u_{24}$ *by giving the powers of* $z, 1-z,$ *and* (near $z = \infty$) $-1/z$ *their principal values and use the symbols* $u_1, \ldots, u_{24}$ *also to denote the functions obtained from the*

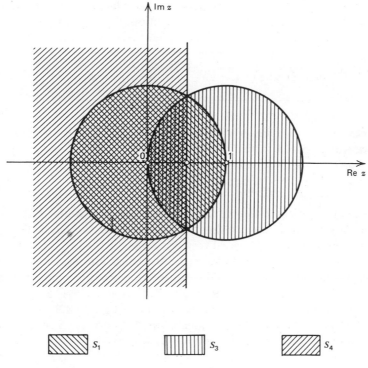

Fig. 9.9a.

original series definition by analytic continuation in the cut plane. If a solution is analytic at a finite singular point (such as u_1 at 0 or u_5 at 1) the cut joining that point to ∞ may be removed, because the values of the function are not changed by continuing it around the singular point.

Because the functions thus defined are all solutions of one and the same differential equation, any three of them must be linearly dependent. A full knowledge of these dependencies will enable us to exhibit the value of any u_k at any point z of the complex plane, $z \neq e^{\pm i\pi/3}$, in terms of series with argument $t_k(z)$ where $z \in S_k$. It turns out that all these dependencies can be mastered explicitly. A number of these formulas are trivial. Of the remaining formulas, we shall state those giving the full analytic continuation for the function u_0, from which all others can be obtained by trivial manipulation. (For a complete listing of the dependencies, see Erdélyi *et al.* [1953], vol. 1, pp. 106–108.)

We begin with the dependencies between the four solutions whose original domain of definition is the same set S_k. As to S_0, both functions u_1

and u_{13} are analytic at $z = 0$ and assume the value 1 there. Because the differential equation has only one such solution, there follows $u_1 = u_{13}$ or

$$F(\alpha, \beta; \gamma; z) = (1-z)^{\gamma-\alpha-\beta} F(\gamma-\alpha, \gamma-\beta; \gamma; z) \qquad (9.9\text{-}21)$$

This formula was called **Euler's first identity** in §1.5. There it was established as an identity between formal power series.

In a similar manner one can prove that

$$u_k = u_{k+12}, \qquad k = 1, 2, \ldots, 12. \qquad (9.9\text{-}22)$$

This relation enables us to confine our attention to the first 12 solutions defined. Any relation between these functions remains valid if the indices are increased by 12.

We next turn to the dependencies between solutions whose original domain of definition has a nonempty intersection. For instance, the original domain of definition of u_1, u_2, u_7, u_8 includes the set $S_1 \cap S_4$. Again the dependencies between these solutions turn out to be trivial. Both u_1 and u_7 are analytic at $z = 0$ and take the value 1 there. Because, as already noted, there is only one such solution, we have $u_1 = u_7$ or

$$F(\alpha, \beta; \gamma; z) = (1-z)^{-\alpha} F\left(\alpha, \gamma-\beta; \gamma; \frac{z}{z-1}\right) \qquad (9.9\text{-}23)$$

This formula is known as **Euler's second identity**. Because the series on the right converges for $\text{Re } z < \frac{1}{2}$, the formula gives an explicit representation of the analytic continuation of the hypergeometric series into the half plane $\text{Re} < \frac{1}{2}$. (The solutions involved being analytic at $z = 0$, it is not necessary here to cut the plane between $-\infty$ and 0.) As before, the formula can also be obtained in a more pedestrian way by explicit computation of the Taylor coefficients of u_7 at $z = 0$ (see §1.6).

In an analogous manner one can prove that

$$u_k = u_{k+6}, \qquad k = 1, 2, \ldots, 6. \qquad (9.9\text{-}24)$$

Because the original domains of definition of the two functions are not the same, each of these formulas provides an explicit representation of the analytic continuation of some u_{2k-1} or u_{2k} into a region exterior to S_k.

We next determine the dependence between u_1, u_5, u_6. Because the functions u_5 and u_6 form a fundamental system in $S_1 \cap S_3$, there exist constants a and b such that $u_1 = a u_5 + b u_6$ or, written out in full,

$$F(\alpha, \beta; \gamma; z) = aF(\alpha, \beta; 1 + \alpha + \beta - \gamma; 1 - z)$$

$$+ b(1 - z)^{\gamma - \alpha - \beta} F(\gamma - \alpha, \gamma - \beta; 1 + \gamma - \alpha - \beta; 1 - z).$$

$$(9.9\text{-}25)$$

To calculate a and b, we temporarily assume $\operatorname{Re}(\gamma - \alpha - \beta) > 0$ and make use of the following facts:

(i) The series $F(\alpha, \beta; \gamma; z)$ converges for $z = 1$. We use the ad hoc notation

$$F(\alpha, \beta; \gamma; 1) =: s(\alpha, \beta, \gamma).$$

(ii) By Abel's theorem (Theorem 2.2e)

$$\lim_{z \to 1-} F(\alpha, \beta; \gamma; z) = s(\alpha, \beta, \gamma).$$

Letting $z \to 1$ from the left in (9.9-25) we thus find

$$s(\alpha, \beta, \gamma) = a.$$

In addition to $\operatorname{Re}(\gamma - \alpha - \beta) > 0$ we now also assume that $\operatorname{Re}(1 - \gamma) > 0$. (The set of triples (α, β, γ) in $\mathbb{C}^3$ satisfying both conditions is open and connected.) Then both series on the right of (9.9-25) converge for $z = 0$, and letting $z \to 0$ from the right we obtain

$$1 = as(\alpha, \beta, 1 + \alpha + \beta - \gamma) + bs(\gamma - \alpha, \gamma - \beta, 1 + \gamma - \alpha - \beta)$$

which permits the determination of b.

By the Gauss formula (Theorem 8.6b)

$$a = s(\alpha, \beta, \gamma) = \frac{\Gamma(\gamma)\Gamma(\gamma - \alpha - \beta)}{\Gamma(\gamma - \alpha)\Gamma(\gamma - \beta)},$$

and using $\Gamma(\gamma)\Gamma(1 - \gamma) = \pi(\sin \pi\gamma)^{-1}$ we get

$$b = \frac{\sin \pi(\gamma - \alpha) \sin \pi(\gamma - \beta) - \sin \pi\gamma \sin \pi(\gamma - \alpha - \beta)}{\sin \pi\alpha \sin \pi\beta} \frac{\Gamma(\gamma)\Gamma(\alpha + \beta - \gamma)}{\Gamma(\alpha)\Gamma(\beta)}.$$

It is easily seen that the value of the first fraction is 1. We thus have found:

$$
\begin{aligned}
F(\alpha, \beta; \gamma; z) \\
= \frac{\Gamma(\gamma - \alpha - \beta)\Gamma(\gamma)}{\Gamma(\gamma - \alpha)\Gamma(\gamma - \beta)} F(\alpha, \beta; 1 + \alpha + \beta - \gamma; 1 - z) \\
+ \frac{\Gamma(\gamma)\Gamma(\alpha + \beta - \gamma)}{\Gamma(\alpha)\Gamma(\beta)} (1 - z)^{\gamma - \alpha - \beta} \\
\cdot F(\gamma - \alpha, \gamma - \beta; 1 + \gamma - \alpha - \beta; 1 - z)
\end{aligned}
\qquad (9.9\text{-}26)
$$

At the moment, this formula is established only in a connected open subset of the parameter space. However, because for any fixed $z \in S_1 \cap S_3$ both sides of (9.9-26) are meromorphic functions in each of the parameters, the formula remains valid for any set of parameters for which both sides are meaningful.

The formulas (9.9-23) and (9.9-26) enable us to express any hypergeometric series of argument z in terms of series with argument $t_3(z)$ or $t_4(z)$. In view of the relations

$$
t_2 = t_3 \circ t_4 \circ t_3 = t_4 \circ t_3 \circ t_4, \qquad t_5 = t_3 \circ t_4, \qquad t_6 = t_4 \circ t_3,
$$

the transformations t_3 and t_4 can be used to generate the full set of transformations $t_2, \ldots, t_6$. We thus can express the function u_1 in terms of any of the pairs (u_{2k-1}, u_{2k}). For instance, to obtain a representation in terms of series of argument t_6, we apply (9.9-24) to each of the functions on the right of (9.9-26). Similarly, to find a representation in terms of series of argument t_5, we apply (9.9-26) to the function u_7 appearing in the relation $u_1 = u_7$. A further application of (9.9-24) will then yield a representation in terms of $t_4 \circ t_3 \circ t_4 = t_2$. This last representation is especially interesting, because it connects functions with the nonintersecting original domains of definition S_1 and S_2.

In Table 9.9a we have listed the explicit representations of u_1 in terms of series of argument $t_k(z)$, $k = 2, 3, \ldots, 6$. Because u_1 is analytic at $z = 0$, these representations are valid in the plane cut from 1 to ∞. The powers of $1 - z$ and $-z$ have their principal values. A further set of representations is obtained by applying Euler's first transformation to each of the functions on the right.

Table 9.9a. Linear transforms and analytic continuation of the hypergeometric series

---

$F(\alpha, \beta; \gamma; z)$

$$= \frac{\Gamma(\beta-\alpha)\Gamma(\gamma)}{\Gamma(\gamma-\alpha)\Gamma(\beta)}(-z)^{-\alpha}F\left(\alpha, 1+\alpha-\gamma; 1+\alpha-\beta; \frac{1}{z}\right)$$

$$+ \frac{\Gamma(\alpha-\beta)\Gamma(\gamma)}{\Gamma(\gamma-\beta)\Gamma(\alpha)}(-z)^{-\beta}F\left(\beta, 1+\beta-\gamma; 1+\beta-\alpha; \frac{1}{z}\right)$$

$$= \frac{\Gamma(\gamma-\alpha-\beta)\Gamma(\gamma)}{\Gamma(\gamma-\alpha)\Gamma(\gamma-\beta)}F(\alpha, \beta; 1+\alpha+\beta-\gamma; 1-z)$$

$$+ \frac{\Gamma(\alpha+\beta-\gamma)\Gamma(\gamma)}{\Gamma(\alpha)\Gamma(\beta)}(1-z)^{\gamma-\alpha-\beta}F(\gamma-\alpha, \gamma-\beta; 1+\gamma-\alpha-\beta; 1-z)$$

$$= (1-z)^{-\alpha}F\left(\alpha, \gamma-\beta; \gamma; \frac{z}{z-1}\right)$$

$$= \frac{\Gamma(\beta-\alpha)\Gamma(\gamma)}{\Gamma(\gamma-\alpha)\Gamma(\beta)}(1-z)^{-\alpha}F\left(\alpha, \gamma-\beta; 1+\alpha-\beta; \frac{1}{1-z}\right)$$

$$+ \frac{\Gamma(\alpha-\beta)\Gamma(\gamma)}{\Gamma(\gamma-\beta)\Gamma(\alpha)}(1-z)^{-\beta}F\left(\beta; \gamma-\alpha; 1+\beta-\alpha; \frac{1}{1-z}\right)$$

$$= \frac{\Gamma(\gamma-\alpha-\beta)\Gamma(\gamma)}{\Gamma(\gamma-\alpha)\Gamma(\gamma-\beta)}z^{-\alpha}F\left(\alpha, 1-\gamma+\alpha; 1+\alpha+\beta-\gamma; 1-\frac{1}{z}\right)$$

$$+ \frac{\Gamma(\alpha+\beta-\gamma)\Gamma(\gamma)}{\Gamma(\alpha)\Gamma(\beta)}z^{\alpha-\gamma}(1-z)^{\gamma-\alpha-\beta}$$

$$\cdot F\left(\gamma-\alpha, 1-\alpha; 1+\gamma-\alpha-\beta; 1-\frac{1}{z}\right)$$

PROBLEMS

1. The Jacobi polynomials $P_n^{(\alpha, \beta)}(z)$ can be defined by

$$P_n^{(\alpha, \beta)}(z) := \frac{(1+\alpha)_n}{n!}F\left(-n, 1+\alpha+\beta+n; 1+\alpha; \frac{1-z}{2}\right).$$

Establish the following alternate representations in terms of hypergeometric series:

$$P_n^{(\alpha,\beta)}(z) = \frac{(1+\alpha+\beta+n)_n}{n!}\left(\frac{z-1}{2}\right)^n F\left(-n,-\alpha-n;-\alpha-\beta-2n;\frac{2}{1-z}\right)$$

$$= (-1)^n \frac{(1+\beta)_n}{n!} F\left(-n,1+\alpha+\beta+n;1+\beta;\frac{1+z}{2}\right)$$

$$= \frac{(1+\alpha)_n}{n!}\left(\frac{1+z}{2}\right)^n F\left(-n,-\beta-n;1+\alpha;\frac{z-1}{z+1}\right)$$

$$= (-1)^n \frac{(1+\alpha+\beta+n)}{n!}\left(\frac{z+1}{2}\right)^n \; F\left(-n,-\beta-n;-\alpha-\beta-2n;\frac{2}{1+z}\right)$$

$$= \frac{(1+\beta)_n}{n!}\left(\frac{z-1}{2}\right)^n F\left(-n,-\alpha-n;1+\beta;\frac{z+1}{z-1}\right).$$

2. Find, by Frobenius' method, a second solution of the hypergeometric equation when $\gamma = 1$.

3. The **complete elliptic integrals** can for $|k^2|<1$ be expressed in terms of hypergeometric series as follows:

$$K(k) := \int_0^{\pi/2} \frac{1}{\sqrt{1-k^2(\sin\phi)^2}}\,d\phi = \frac{\pi}{2}F(\tfrac{1}{2},\tfrac{1}{2};1;k^2),$$

$$E(k) := \int_0^{\pi/2} \sqrt{1-k^2(\sin\phi)^2}\,d\phi = \frac{\pi}{2}F(-\tfrac{1}{2},\tfrac{1}{2};1;k^2).$$

Show that for $|1-k^2|<1$

$$K(k) = \sum_{n=0}^{\infty} \frac{(\tfrac{1}{2})_n(\tfrac{1}{2})_n}{n!n!}\{h_n(1)-h_n(\tfrac{1}{2})+2\log 2-\tfrac{1}{2}\log(1-k^2)\}(1-k^2)^n,$$

$$E(k) = 1+\tfrac{1}{4}\sum_{n=0}^{\infty} \frac{(\tfrac{1}{2})_n(\tfrac{3}{2})_n}{n!(n+1)!}\{h_{n+1}(1)+h_n(1)-h_{n+1}(\tfrac{1}{2})-h_n(\tfrac{1}{2})$$

$$-4\log 2 - \log(1-k^2)\}(1-k^2)^n.$$

4. Find the analytic continuation of u_1 into the region S_2 by means of the relation $t_2 = t_3 \circ t_4 \circ t_3$.

5. Show that a second-order Fuchsian equation with the four singular points a, b, c, ∞ and corresponding exponent pairs $(\alpha, 0), (\beta, 0), (\gamma, 0), (\delta, \delta')$ is of the form

$$u'' + \left(\frac{1-\alpha}{z-a}+\frac{1-\beta}{z-b}+\frac{1-\gamma}{z-c}\right)u' + \frac{\delta\delta'z+p}{(z-a)(z-b)(z-c)}u = 0,$$

where p may have any value. (**Equation of Heun.** The constant p is called the **accessory parameter.**)

6. Prove Elliot's formula

$$F(\tfrac{1}{2}+\alpha, -\tfrac{1}{2}-\gamma; 1+\alpha+\beta; z)F(\tfrac{1}{2}-\alpha, \tfrac{1}{2}+\gamma; 1+\beta+\gamma; 1-z)$$
$$+F(\tfrac{1}{2}+\alpha, \tfrac{1}{2}-\gamma; 1+\alpha+\beta; z)F(-\tfrac{1}{2}-\alpha, \tfrac{1}{2}+\gamma; 1+\beta+\gamma; 1-z)$$
$$-F(\tfrac{1}{2}+\alpha, \tfrac{1}{2}-\gamma; 1+\alpha+\beta; z)F(\tfrac{1}{2}-\alpha, \tfrac{1}{2}+\gamma; 1+\beta+\gamma; 1-z)$$
$$=\frac{\Gamma(1+\alpha+\beta)\Gamma(1+\beta+\gamma)}{\Gamma(\tfrac{3}{2}+\alpha+\beta+\gamma)\Gamma(\tfrac{1}{2}+\beta)},$$

and obtain as a special case Legendre's relation

$$E(k)K(k')+E(k')K(k)-K(k)K(k')=\frac{\pi}{2},$$

where E and K denote complete elliptic integrals, and $k'^2 := 1-k^2$.

§9.10.   QUADRATIC TRANSFORMS: LEGENDRE FUNCTIONS

In this section we continue our investigation of the hypergeometric differential equation by studying certain of its special cases whose solutions admit transformations where the variables are linked not by Moebius transformations, as in §9.9, but by certain rational functions whose numerators and denominators are of degree two.

Let us consider a linear differential equation of order two that at $z = 0$ has a singular point of the first kind with exponents 0 and $\tfrac{1}{2}$. Of the two functions making up the fundamental system at this point, one can be chosen analytic, and the other as $z^{1/2}$ times an analytic function. By subjecting the independent variable to the transformation $s := z^{1/2}$, both functions will appear as analytic functions of s. Thus the singularity at $z = 0$ can be removed. On the other hand, new singularities will be introduced; generally, if $z = z_0$ were a singular point, both points $s = \pm z_0^{1/2}$ will now be singular. If there is only one finite singular point besides $z = 0$, the total number of singular points remains the same, and if the given equation was hypergeometric, we may expect it to transform into another hypergeometric equation by the process.

Consequently, let us consider the hypergeometric differential equation with singular points at $0, \infty, 1$ and corresponding exponents $0, \tfrac{1}{2}; \alpha, \beta; \gamma, \delta$, where $\alpha+\beta+\gamma+\delta=\tfrac{1}{2}$. By (9.8-15b), this equation is of the form

$$u''+\left(\frac{1}{2z}+\frac{1-\gamma-\delta}{z-1}\right)u'+\left(\alpha\beta+\frac{\gamma\delta}{z-1}\right)\frac{1}{z(z-1)}u=0.$$

If $u(z)=v(s)$, $s=z^{1/2}$, then v is readily seen to satisfy

$$v''+(1-\gamma-\delta)\frac{2s}{s^2-1}v'+\left(\alpha\beta+\frac{\gamma\delta}{s^2-1}\right)\frac{4}{s^2-1}v=0. \qquad (9.10\text{-}1)$$

This may be written

$$v'' + (1 - \gamma - \delta)\left(\frac{1}{s-1} + \frac{1}{s+1}\right)v'$$

$$+ \left(\frac{\gamma\delta}{s-1} - \frac{\gamma\delta}{s+1} + \frac{4\alpha\beta}{2}\right)\frac{2}{(s-1)(s+1)}v = 0,$$

and a comparison with (9.8-15b) shows that this equation is hypergeometric with the singular points -1, ∞, 1 and the corresponding exponents γ, δ; 2α, 2β; γ, δ whose sum is 1 as required. In terms of Riemann's P symbol we thus have proved:

$$P\left\{\begin{matrix} 0 & \infty & 1 \\ 0 & \alpha & \gamma & z^2 \\ \frac{1}{2} & \beta & \delta \end{matrix}\right\} = P\left\{\begin{matrix} -1 & \infty & 1 \\ \gamma & 2\alpha & \gamma & z \\ \delta & 2\beta & \delta \end{matrix}\right\},$$

where for consistency we have written z in place of s. The singular points -1, ∞, 1 can be brought back to 0, ∞, 1 in two different ways. Using the simplified version of the P symbol, we thus get

$$P\left\{\begin{matrix} 0 & \alpha & \gamma \\ \frac{1}{2} & \beta & \delta \end{matrix} \; z^2\right\} = P\left\{\begin{matrix} \gamma & 2\alpha & \gamma & \frac{1+z}{2} \\ \delta & 2\beta & \delta \end{matrix}\right\}$$

$$= P\left\{\begin{matrix} \gamma & 2\alpha & \gamma & \frac{1-z}{2} \\ \delta & 2\beta & \delta \end{matrix}\right\}. \qquad (9.10\text{-}2)$$

As an application we derive a quadratic transformation due to Goursat. If $\gamma = 0$ and hence $\delta = \frac{1}{2} - \alpha - \beta$, the set on the left by Theorem 9.9d has the fundamental system

$$v_1(z) = F(\alpha, \beta; \tfrac{1}{2}; z^2)$$

$$v_2(z) = zF(\alpha + \tfrac{1}{2}, \beta + \tfrac{1}{2}; \tfrac{3}{2}; z^2). \qquad (9.10\text{-}3)$$

On the other hand, the two functions

$$v_{3,4}(z) = F\left(2\alpha, 2\beta; \alpha + \beta + \tfrac{1}{2}; \frac{1 \pm z}{2}\right)$$

are both members of the set on the right. They are linearly independent, and hence likewise form a fundamental system. Hence it is possible, for instance, to find constants a and b such that $v_1 = av_3 + bv_4$. Because v_1 is unchanged if z is replaced by $-z$, $b = a$. To determine a, we set $z = 0$ to find

$$1 = 2aF(2\alpha, 2\beta; \alpha + \beta + \tfrac{1}{2}; \tfrac{1}{2}).$$

The hypergeometric series has the sum

$$\frac{\Gamma(\tfrac{1}{2})\Gamma(\alpha+\beta+\tfrac{1}{2})}{\Gamma(\alpha+\tfrac{1}{2})\Gamma(\beta+\tfrac{1}{2})},$$

as shown in Problem 2, §8.6. We thus may calculate a to find

$$
\frac{2\Gamma(\tfrac{1}{2})\Gamma(\alpha+\beta+\tfrac{1}{2})}{\Gamma(\alpha+\tfrac{1}{2})\Gamma(\beta+\tfrac{1}{2})}F(\alpha,\beta;\tfrac{1}{2};z^2)
$$
$$
=F\!\left(2\alpha,2\beta;\alpha+\beta+\frac{1}{2};\frac{1+z}{2}\right)+F\!\left(2\alpha,2\beta;\alpha+\beta+\frac{1}{2};\frac{1-z}{2}\right)
$$

$$(9.10\text{-}4)$$

In a similar manner we find

$$
\frac{4\Gamma(\tfrac{1}{2})\Gamma(\alpha+\beta+\tfrac{1}{2})}{\Gamma(\alpha)\Gamma(\beta)}zF(\alpha+\tfrac{1}{2},\beta+\tfrac{1}{2};\tfrac{3}{2};z^2)
$$
$$
=F\!\left(2\alpha,2\beta;\alpha+\beta+\frac{1}{2};\frac{1+z}{2}\right)-F\!\left(2\alpha,2\beta;\alpha+\beta+\frac{1}{2};\frac{1-z}{2}\right)
$$

$$(9.10\text{-}5)$$

A number of transforms can be derived from the following general principle. Suppose we have an identity of the form

$$
P\begin{Bmatrix} 0 & * & 0 \\ * & * & * \end{Bmatrix} z \end{Bmatrix} = P\begin{Bmatrix} 0 & * & 0 \\ * & * & * \end{Bmatrix} g(z) \end{Bmatrix} \tag{9.10-6}
$$

where the corresponding exponents on either side are not all equal in pairs, and where g is analytic at 0 and satisfies

$$g(0)=0. \tag{9.10-7}$$

Each of the sets (9.10-6) contains precisely one function that is analytic and equal to 1 at $z=0$. This function can be represented by a hypergeometric series as indicated in Theorem 9.9d. Because the sets (9.10-6) are identical, the two representations are identical. The method will generally yield identities of the form $F(*,*;*;z)=h(z)F(*,*;*;g(z))$, where h is an elementary factor.

A first set of transforms is obtained by writing (9.10-2) as follows:

$$
P\begin{Bmatrix} \gamma & 2\alpha & \gamma \\ \delta & 2\beta & \delta \end{Bmatrix} z \end{Bmatrix} = P\begin{Bmatrix} 0 & \alpha & \gamma \\ \tfrac{1}{2} & \beta & \delta \end{Bmatrix} (2z-1)^2 \end{Bmatrix}. \tag{9.10-8}
$$

Here we may replace z by any of the functions $t_k(z)$; the resulting P set on the left can then be written as a P set with argument z by suitably permuting the exponents. The arguments on the right become

$$q_k(z) := (2t_k(z) - 1)^2.$$

It is readily seen that $q_1 = q_3$, $q_4 = q_5$, $q_2 = q_6$. Because the P symbol on the left has the same form for each pair of transformations (t_1, t_3), (t_4, t_5), (t_2, t_6), because two pairs of exponents are equal, nothing is lost by considering the cases $k = 1, 4, 6$ only. The functions q_k do not satisfy (9.10-7), because their values at $z = 0$ are 1 for $k = 1, 4$ and ∞ for $k = 6$. Therefore, we must subject the P set on the right to one more Moebius transformation, which brings $q_k(0)$ to 0. For $k = 1, 4$ this can be done by either t_3 or t_6, for $k = 6$ by t_2 or t_5. We thus arrive at a total of six transformations of type (9.10-6), where

$$g(z) = t_j(q_k(z)).$$

for suitable values of j and k. The six resulting identities are presented in Table 9.10a.

Table 9.10a

k		j
1	$P\left\{\begin{matrix} \gamma & 2\alpha & \gamma \\ \delta & 2\beta & \delta \end{matrix}\ z\right\} = \begin{cases} P\left\{\begin{matrix} \gamma & \alpha & 0 \\ \delta & \beta & \frac{1}{2} \end{matrix}\ 4z(1-z)\right\} \\[2ex] P\left\{\begin{matrix} \gamma & 0 & \alpha \\ \delta & \frac{1}{2} & \beta \end{matrix}\ \dfrac{4z(z-1)}{(2z-1)^2}\right\} \end{cases}$	$\begin{matrix} 3 \\[2ex] 6 \end{matrix}$
4	$P\left\{\begin{matrix} \gamma & \gamma & 2\alpha \\ \delta & \delta & 2\beta \end{matrix}\ z\right\} = \begin{cases} P\left\{\begin{matrix} \gamma & \alpha & 0 \\ \delta & \beta & \frac{1}{2} \end{matrix}\ -\dfrac{4z}{(1-z)^2}\right\} \\[2ex] P\left\{\begin{matrix} \gamma & 0 & \alpha \\ \delta & \frac{1}{2} & \beta \end{matrix}\ \dfrac{4z}{(1+z)^2}\right\} \end{cases}$	$\begin{matrix} 3 \\[2ex] 6 \end{matrix}$
6	$P\left\{\begin{matrix} 2\alpha & \gamma & \gamma \\ 2\beta & \delta & \delta \end{matrix}\ z\right\} = \begin{cases} P\left\{\begin{matrix} \alpha & \gamma & 0 \\ \beta & \delta & \frac{1}{2} \end{matrix}\ \dfrac{z^2}{4(z-1)}\right\} \\[2ex] P\left\{\begin{matrix} \alpha & 0 & \gamma \\ \beta & \frac{1}{2} & \delta \end{matrix}\ \dfrac{z^2}{(2-z)^2}\right\} \end{cases}$	$\begin{matrix} 5 \\[2ex] 2 \end{matrix}$

To obtain the desired identities between hypergeometric series, we now must in each case choose either α or γ such that one of the exponents at 0 becomes 0, and identify the analytic solutions by Theorem 9.9d. We thus obtain the six identities given in Table 9.10b, where the parameters α and β are not necessarily equal to those in Table 9.10a.

Table 9.10b

$$F\left(\alpha,\beta;\frac{\alpha}{2}+\frac{\beta}{2}+\frac{1}{2};z\right)=\begin{cases} F\left(\dfrac{\alpha}{2},\dfrac{\beta}{2};\dfrac{\alpha}{2}+\dfrac{\beta}{2}+\dfrac{1}{2};4z(1-z)\right) \\[1.2em] (1-2z)^{-\alpha}F\left(\dfrac{\alpha}{2},\dfrac{\alpha}{2}+\dfrac{1}{2};\dfrac{\alpha}{2}+\dfrac{\beta}{2}+\dfrac{1}{2};\dfrac{4z(z-1)}{(2z-1)^2}\right) \end{cases}$$

$$F(\alpha,\beta;\alpha-\beta+1;z)=\begin{cases} (1-z)^{-\alpha}F\left(\dfrac{\alpha}{2},\dfrac{\alpha}{2}-\beta+\dfrac{1}{2};\alpha-\beta+1;-\dfrac{4z}{(1-z)^2}\right) \\[1.2em] (1-z)^{-\alpha}F\left(\dfrac{\alpha}{2},\dfrac{\alpha}{2}+\dfrac{1}{2};\alpha-\beta+1;\dfrac{4z}{(1+z)^2}\right) \end{cases}$$

$$F(\alpha,\beta;2\beta;z)=\begin{cases} (1-z)^{-\alpha/2}F\left(\dfrac{\alpha}{2},\beta-\dfrac{\alpha}{2};\beta+\dfrac{1}{2};\dfrac{z^2}{4(z-1)}\right) \\[1.2em] \left(1-\dfrac{z}{2}\right)^{-\alpha}F\left(\dfrac{\alpha}{2},\dfrac{\alpha}{2}+\dfrac{1}{2};\beta+\dfrac{1}{2};\dfrac{z^2}{(2-z)^2}\right) \end{cases}$$

Another set of six identities is obtained by interchanging the roles of z and $g(z)$. This leads to expressions involving $\sqrt{1-z}$, $\sqrt{z}$, and $\sqrt{-z}$. The root $\sqrt{1-z}$ must be so defined as to become $+1$ for $z=0$. The roots $\sqrt{z}$ and $\sqrt{-z}$ can be chosen arbitrarily, but the same determination must be chosen in all places where it appears in a formula. The resulting formulas are listed in Table 9.10c, p. 170.

All these identities are based on relations between two different kinds of P sets, namely (i) P sets in which one set of exponents has difference $\frac{1}{2}$, and (ii) P sets in which two sets of exponents are identical. A nontrivial identity between two P sets of type (ii) is obtained as follows. As shown, any P set of type (ii), where (γ,δ) is the recurring pair of exponents, is equal to a set of type (i) of the form

$$P\left\{\begin{matrix} 0 & \alpha & \gamma \\ \frac{1}{2} & \beta & \delta \end{matrix} \quad q_k(z)\right\},$$

Table 9.10c

	k	j
$F(\alpha, \beta; \alpha+\beta+\frac{1}{2}; z) = \Bigg\{$ $F\left(2\alpha, 2\beta; \alpha+\beta+\frac{1}{2}; \frac{1-\sqrt{1-z}}{2}\right)$	1	3
$\left(\frac{1+\sqrt{1-z}}{2}\right)^{-2\alpha} F\left(2\alpha, \alpha-\beta+\frac{1}{2}; \alpha+\beta+\frac{1}{2}; \frac{\sqrt{1-z}-1}{\sqrt{1-z}+1}\right)$	4	3
$\left(\sqrt{1-z}+\sqrt{-z}\right)^{-2\alpha} F\left(2\alpha, \alpha+\beta; 2\alpha+2\beta; \frac{2\sqrt{-z}}{\sqrt{1-z}+\sqrt{-z}}\right)$	6	5
$F(\alpha, \alpha+\frac{1}{2}; \beta; z) = \Bigg\{$ $(1-z)^{-\alpha} F\left(2\alpha, 2\beta-2\alpha-1; \beta; \frac{\sqrt{1-z}-1}{2\sqrt{1-z}}\right)$	1	6
$\left(\frac{1+\sqrt{1-z}}{2}\right)^{-2\alpha} F\left(2\alpha, 2\alpha-\beta+1; \beta; \frac{1-\sqrt{1-z}}{1+\sqrt{1-z}}\right)$	4	6
$(1+\sqrt{z})^{-2\alpha} F\left(2\alpha, \beta-\frac{1}{2}; 2\beta-1; \frac{2\sqrt{z}}{1+\sqrt{z}}\right)$	6	2

where k is 1, 4, or 6. We interchange the second and third pair of exponents and obtain

$$P\left\{\begin{matrix} 0 & \gamma & \alpha & \dfrac{q_k(z)}{q_k(z)-1} \\ \frac{1}{2} & \delta & \beta & \end{matrix}\right\}.$$

Applying (9.10-2) and thereby going back to a set of type (ii), this becomes

$$P\left\{\begin{matrix} \alpha & 2\gamma & \alpha & \\ \beta & 2\delta & \beta & r_k(z) \end{matrix}\right\},$$

where

$$r_k(z) := \frac{1+\sqrt{\dfrac{q_k(z)}{q_k(z)-1}}}{2},$$

either value of the square root being acceptable. The resulting identity is of the form (9.10-6) only if $r_k(0) = 0$, which is the case only if $q_k(0) = \infty$, that is, for $k = 6$, if the root is chosen appropriately. In the remaining cases $k = 1, 4$ we have $q_k(0) = 1$ and consequently $r_k(0) = \infty$, and we must apply t_2 or t_5 to bring the infinite point to zero. In this manner we obtain identities where a P set of type (ii) of argument z is expressed as another P set of type (ii) with argument

$$g(z) = t_j(r_k(z)),$$

where j and k are chosen as indicated previously. The results are listed in Table 9.10d.

Table 9.10d

k			j
1	$P\left\{\begin{matrix} \gamma & 2\alpha & \gamma \\ \delta & 2\beta & \delta \end{matrix}\ z\right\}=$	$\begin{cases} P\left\{\begin{matrix} 2\gamma & \alpha & \alpha & \dfrac{4\sqrt{-z}\sqrt{1-z}}{(\sqrt{1-z}+\sqrt{z})^2} \\ 2\delta & \beta & \beta \end{matrix}\right\} \\[4ex] P\left\{\begin{matrix} 2\gamma & \alpha & \alpha & \dfrac{-4\sqrt{-z}\sqrt{1-z}}{(\sqrt{1-z}-\sqrt{-z})^2} \\ 2\delta & \beta & \beta \end{matrix}\right\} \end{cases}$	5 2
4	$P\left\{\begin{matrix} \gamma & \gamma & 2\alpha \\ \delta & \delta & 2\beta \end{matrix}\ z\right\}=$	$\begin{cases} P\left\{\begin{matrix} 2\gamma & \alpha & \alpha & -\dfrac{2\sqrt{z}}{(1-\sqrt{z})^2} \\ 2\delta & \beta & \beta \end{matrix}\right\} \\[4ex] P\left\{\begin{matrix} 2\gamma & \alpha & \alpha & \dfrac{2\sqrt{z}}{(1+\sqrt{z})^2} \\ 2\delta & \beta & \beta \end{matrix}\right\} \end{cases}$	5 2

Table 9.10d—*continued*

k		j
6	$$P\left\{\begin{matrix} 2\alpha & \gamma & \gamma \\ 2\beta & \delta & \delta \end{matrix}\; z\right\} = \begin{cases} P\left\{\begin{matrix} \alpha & 2\gamma & \alpha & -\dfrac{(1-\sqrt{1-z})^2}{2\sqrt{1-z}} \\ \beta & 2\delta & \beta & \end{matrix}\right\} \\[6mm] P\left\{\begin{matrix} \alpha & 2\gamma & \alpha & \dfrac{(1-\sqrt{1-z})^2}{(1+\sqrt{1-z})^2} \\ \beta & 2\delta & \beta & \end{matrix}\right\} \end{cases}$$	1 4

To obtain identities between hypergeometric series, we again choose either α or γ such that one of the exponents at 0 becomes 0. Theorem 9.9d then enables us to identify the analytic solutions. (In the cases $k = 1, 4$, where the functions g have a factor $\sqrt{z}$ at 0, it is necessary to replace z by z^2 before identifying analytic solutions.) In this manner we obtain the six identities shown in Table 9.10e.

Because these transforms are based on identities between P sets of type (ii), the transforms obtained by interchanging the roles of z and $g(z)$ are already contained in Table 9.10e.

The 18 transforms listed in Tables 9.10b, 9.10c, and 9.10e are the basic quadratic transforms of Gauss and Kummer. From each formula given, three others can be obtained by subjecting one or both hypergeometric series to Euler's first identity. Thus a total number of 72 quadratic transforms can be obtained.

Legendre's equation

The theory of quadratic transforms is intimately linked with the theory of a differential equation that is of considerable importance in mathematical physics, called the **Legendre differential equation**. This equation depends on two complex parameters, commonly denoted by μ and ν, and is given by

$$(1-z^2)u'' - 2zu' + \left[\nu(\nu+1) - \frac{\mu^2}{1-z^2}\right]u = 0. \qquad (9.10\text{-}9)$$

Table 9.10e

$$F\left(\alpha, \beta; \frac{\alpha}{2}+\frac{\beta}{2}+\frac{1}{2}; z\right) = \begin{cases} (\sqrt{1-z}+\sqrt{-z})^{-2\alpha}\, F\left(\alpha, \frac{\alpha}{2}+\frac{\beta}{2}; \alpha+\beta; \dfrac{4\sqrt{-z}\sqrt{1-z}}{(\sqrt{1-z}+\sqrt{-z})^2}\right) \\[4mm] (\sqrt{1-z}-\sqrt{-z})^{-2\alpha}\, F\left(\alpha, \frac{\alpha}{2}+\frac{\beta}{2}; \alpha+\beta; \dfrac{-4\sqrt{-z}\sqrt{1-z}}{(\sqrt{1-z}-\sqrt{-z})^2}\right) \end{cases}$$

$$F(\alpha, \beta; \alpha-\beta+1; z) = \begin{cases} (1-\sqrt{z})^{-2\alpha}\, F\left(\alpha, \alpha-\beta+\frac{1}{2}; 2\alpha-2\beta+1; \dfrac{-4\sqrt{z}}{(1-\sqrt{z})^2}\right) \\[4mm] (1+\sqrt{z})^{-2\alpha}\, F\left(\alpha, \alpha-\beta+\frac{1}{2}; 2\alpha-2\beta+1; \dfrac{4\sqrt{z}}{(1+\sqrt{z})^2}\right) \end{cases}$$

$$F(\alpha, \beta; 2\beta; z) = \begin{cases} (1-z)^{-\alpha/2}\, F\left(\alpha, 2\beta-\alpha; \beta+\frac{1}{2}; -\dfrac{(1-\sqrt{1-z})^2}{4\sqrt{1-z}}\right) \\[4mm] \left(\dfrac{1+\sqrt{1-z}}{2}\right)^{-2\alpha} F\left(\alpha, \alpha-\beta+\frac{1}{2}; \beta+\frac{1}{2}; \dfrac{(1-\sqrt{1-z})^2}{(1+\sqrt{1-z})^2}\right) \end{cases}$$

The equation occurs, for instance, when Laplace's equation is separated in spherical, elliptical, or toroidal coordinates. A comparison with (9.10-1) shows that the set of solutions is

$$
P\left\{\begin{matrix} -1 & \infty & 1 \\ \dfrac{\mu}{2} & \nu+1 & \dfrac{\mu}{2} & z \\ -\dfrac{\mu}{2} & -\nu & -\dfrac{\mu}{2} \end{matrix}\right\},
$$

and hence is identical with

$$
P\left\{\begin{matrix} \dfrac{\mu}{2} & \nu+1 & \dfrac{\mu}{2} \\ & & & \dfrac{1-z}{2} \\ -\dfrac{\mu}{2} & -\nu & -\dfrac{\mu}{2} \end{matrix}\right\},
$$

Here $(1-z)/2$ can be replaced by $(1+z)/2$ because the exponent sets belonging to 0 and to 1 are identical. Reducing one of the exponents at 0 and 1 to zero to express the solutions by hypergeometric series, the set of solutions of (9.10-9) can also be written in the form

$$
\left(\frac{z+1}{z-1}\right)^{\mu/2} P\left\{\begin{matrix} 0 & \nu+1 & 0 & \dfrac{1-z}{2} \\ \mu & -\nu & -\mu \end{matrix}\right\}
$$

$$
= \left(\frac{z+1}{z-1}\right)^{-\mu/2} P\left\{\begin{matrix} 0 & \nu+1 & 0 & \dfrac{1+z}{2} \\ \mu & -\nu & -\mu \end{matrix}\right\}. \qquad (9.10\text{-}10)
$$

In these expressions it is permissible to replace μ by $-\mu$ and ν by $-\nu-1$ because this does not change the original set.

Numerous solutions of the Legendre equation now can be written down by specializing parameters in the solutions $u_1, \ldots, u_{24}$ obtained in §9.9. Since Hobson [1931] it has become traditional to employ as standard solutions the two functions

$$
P_\nu^\mu(z) := \frac{1}{\Gamma(1-\mu)}\left(\frac{z+1}{z-1}\right)^{\mu/2} F\left(-\nu, \nu+1; 1-\mu; \frac{1-z}{2}\right)
$$

$$(9.10\text{-}11)$$

and

$$
Q_\nu^\mu(z) := e^{i\mu\pi} 2^{-\nu-1}\sqrt{\pi}\frac{\Gamma(\nu+\mu+1)}{\Gamma(\nu+\frac{3}{2})} z^{-\nu-\mu-1}(z^2-1)^{\mu/2}
$$

$$(9.10\text{-}12)$$

$$
\cdot F\left(\frac{\nu}{2}+\frac{\mu}{2}+\frac{1}{2}, \frac{\nu}{2}+\frac{\mu}{2}+1; \nu+\frac{3}{2}; \frac{1}{z^2}\right),
$$

called **Legendre function** of the **first** and **second kind**, respectively. It is assumed that $|\arg z| < \pi$, $|\arg(z \pm 1)| < \pi$; consequently, these functions are undefined for real $z \leq 1$. (The convention to define them as the average of the limits of the values above and below the cut is frequently used.) The function P_ν^μ is proportional to u_1, and the function Q_ν^μ is derived from u_3 by means of the last quadratic transformation in Table 9.10b. Evidently, P_ν^μ has been selected so as to have a simple behavior at the points $z = \pm 1$, and Q_ν^μ has been selected so as to have a simple behavior at $z = \infty$. The normalizing factors have been chosen to give a simple appearance to some formulas occurring in the theory. The factors $\Gamma(1-\mu)$ and $\Gamma(\nu+\frac{3}{2})$ prevent the definitions from losing their meaning for $\mu = 1, 2, \ldots$ and $\nu = -\frac{3}{2}, -\frac{5}{2}, \ldots$, respectively. We note that for $\mu = 0$, $\nu = n = 0, 1, 2, \ldots$ the function P_ν^μ reduces to a polynomial of degree n. This is called the **Legendre polynomial**. It is commonly denoted by P_n and has the simple representation

$$P_n(z) = F\left(-n, n+1; 1; \frac{1-z}{2}\right). \tag{9.10-13}$$

It is obvious that the theory of linear and of quadratic transforms implies a large number of relations between Legendre functions. In fact, because the set of solutions of the Legendre equation is elementarily related to a P set of type (ii), every relation between hypergeometric series admitting quadratic transforms can be written as a relation between Legendre functions. We refer to Chapter III (written by F. Oberhettinger) of Erdélyi [1953] for a complete listing of these results.

PROBLEMS

1. Obtain the following expansions either directly or as special cases of quadratic transforms:

    (a) $\left(\dfrac{1+\sqrt{1-z}}{2}\right)^{-\alpha} = F\left(\dfrac{\alpha}{2}, \dfrac{\alpha}{2} + \dfrac{1}{2}; \alpha+1; z\right)$

    (b) $\dfrac{\left(\dfrac{1+\sqrt{1-z}}{2}\right)^{-\alpha}}{\sqrt{1-z}} = F\left(\dfrac{\alpha}{2} + \dfrac{1}{2}, \dfrac{\alpha}{2} + 1; \alpha+1; z\right)$

    (c) $\frac{1}{2}\{(1+z)^\alpha + (1-z)^\alpha\} = F\left(-\dfrac{\alpha}{2}, \dfrac{1}{2} - \dfrac{\alpha}{2}; \dfrac{1}{2}; z^2\right)$

    (d) $\dfrac{1}{2\alpha z}\{(1+z)^\alpha - (1-z)^\alpha\} = F\left(\dfrac{1}{2} - \dfrac{\alpha}{2}, 1 - \dfrac{\alpha}{2}; \dfrac{3}{2}; z^2\right)$

    (e) $\dfrac{1}{2}\{(\sqrt{1-z^2}+iz)^\alpha + (\sqrt{1-z^2}-iz)^\alpha\} = F\left(\dfrac{\alpha}{2}, -\dfrac{\alpha}{2}; \dfrac{1}{2}; z^2\right)$

(f) $\dfrac{1}{2\alpha i z}\{(\sqrt{1-z^2}+iz)^\alpha - (\sqrt{1-z^2}-iz)^\alpha\} = F\left(\dfrac{1}{2}-\dfrac{\alpha}{2}, \dfrac{1}{2}+\dfrac{\alpha}{2}; \dfrac{3}{2}; z^2\right)$

2. Show that for $-\pi/2 < \phi < \pi/2$ and arbitrary α,

$$\cos \alpha\phi = F\left(\dfrac{\alpha}{2}, -\dfrac{\alpha}{2}; \dfrac{1}{2}; (\sin \phi)^2\right),$$

$$\dfrac{\sin \alpha\phi}{\alpha \sin \phi} = F\left(\dfrac{1}{2}+\dfrac{\alpha}{2}, \dfrac{1}{2}-\dfrac{\alpha}{2}; \dfrac{3}{2}; (\sin \phi)^2\right).$$

3. The **Chebyshev polynomials** of the first kind are for $n = 0, 1, 2, \ldots$ defined by

$$T_n(x) := \cos(n \arccos x).$$

Show that

$$T_n(x) = F\left(n, -n; \dfrac{1}{2}; \dfrac{1-x}{2}\right).$$

4. If P_n denotes the Legendre polynomial, show that

$$P_{2n}(x) = (-1)^n \dfrac{(2n)!}{2^{2n}(n!)^2} F(-n, n+\tfrac{1}{2}; \tfrac{1}{2}; x^2),$$

$$P_{2n+1}(x) = (-1)^n \dfrac{(2n+1)!}{2^{2n}(n!)^2} xF(-n, n+\tfrac{3}{2}; \tfrac{3}{2}; x^2),$$

$n = 0, 1, 2, \ldots$.

5. *Cubic transforms.* If $\alpha + \beta = \tfrac{1}{3}$ and $w := e^{i\pi/3}$, show that

$$P\left\{\begin{matrix} 0 & \infty & 1 \\ 0 & 0 & \alpha \\ \tfrac{1}{3} & \tfrac{1}{3} & \beta \end{matrix} \; z^3\right\} = P\left\{\begin{matrix} 1 & w^2 & w^4 \\ \alpha & \alpha & \alpha \\ \beta & \beta & \beta \end{matrix} \; z\right\}$$

6. Deduce from problem 5 that

$$P\left\{\begin{matrix} \alpha & \alpha & \alpha \\ \beta & \beta & \beta \end{matrix} \; z\right\} = P\left\{\begin{matrix} \alpha & 0 & 0 & \dfrac{3i\sqrt{3}z(1-z)}{(z-w)^3} \\ \beta & \tfrac{1}{3} & \tfrac{1}{3} \end{matrix}\right\}$$

7. By identifying analytic members of the Riemann sets of problem 6, show that

$$F(3\alpha, \tfrac{1}{3}+\alpha; \tfrac{2}{3}+2\alpha; z) = \left(1-\dfrac{z}{w}\right)^3 F\left(\alpha, \dfrac{1}{3}+\alpha; \dfrac{2}{3}+2\alpha; \dfrac{3i\sqrt{3}z(1-z)}{(z-w)^3}\right).$$

8. By subjecting the Riemann sets of problem 7 to appropriate quadratic transformations, establish Goursat's cubic transforms:

(a) $F(3\alpha, 3\alpha +\tfrac{1}{2}; 4\alpha +\tfrac{2}{3}; z) = \left(1-\dfrac{9z}{8}\right)^{-2\alpha} F\left(\alpha, \alpha +\dfrac{1}{2}; \alpha +\dfrac{5}{6}; \dfrac{27z^2(z-1)}{(8-9z)^2}\right)$

(b) $F(3\alpha, 3\alpha + \frac{1}{2}; 2\alpha + \frac{5}{6}; z)$

$$= (1 - 9z)^{-2\alpha} F\left(\alpha, \alpha + \frac{1}{2}; 2\alpha + \frac{5}{6}; -\frac{27z(1 - z)^2}{(1 - 9z)^2}\right)$$

(c) $F(3\alpha, \alpha + \frac{1}{6}; 4\alpha + \frac{2}{3}; z) = \left(1 - \frac{z}{4}\right)^{-3\alpha} F\left(\alpha, \alpha + \frac{1}{3}; 2\alpha + \frac{5}{6}; \frac{27z^2}{(4 - z)^3}\right)$

(d) $F(3\alpha, \frac{1}{3} - \alpha; 2\alpha + \frac{5}{6}; z) = (1 - 4z)^{-3\alpha} F\left(\alpha, \alpha + \frac{1}{3}; 2\alpha + \frac{5}{6}; \frac{27z}{(4z - 1)^3}\right).$

Problems $9 \div 16$ deal with applications of the algorithm of the *arithmetic-geometric mean*, due to Gauss.

9. The complete elliptic integrals of the first and of the second kind, conventionally denoted by $K(k)$ and $E(k)$, were expressed in terms of hypergeometric series in problem 3, §9.9.
(a) If $0 \leqslant k < 1$, $k' := \sqrt{1 - k^2}$, show by means of a quadratic transform that

$$K(k) = \frac{2}{1 + k'} K\left(\frac{1 - k'}{1 + k'}\right). \tag{9.10-14}$$

(b) For $0 < k < 1$, let $k_0 := k$,

$$k'_m := \sqrt{1 - k_m^2}, \qquad k_{m+1} := \frac{1 - k'_m}{1 + k'_m}, \, m = 0, 1, 2, \ldots.$$

Show that $k_m \to 0$ as $m \to \infty$, and hence by iterating (9.10-14) and using $K(0) = \pi/2$ that

$$K(k) = \frac{\pi}{2} \prod_{m=0}^{\infty} \frac{2}{1 + k'_m}. \tag{9.10-15}$$

10. Let the sequences $\{\alpha_n\}$ and $\{\beta_n\}$ be defined by choosing two positive numbers α_0 and β_0 and forming

$$\alpha_{n+1} := \tfrac{1}{2}(\alpha_n + \beta_n), \qquad \beta_{n+1} := \sqrt{\alpha_n \beta_n}, \tag{9.10-16}$$

$n = 0, 1, 2, \ldots.$
(a) If $\alpha_0 > \beta_0 > 0$, show that the sequence $\{\alpha_n\}$ decreases monotonically and the sequence $\{\beta_n\}$ increases monotonically, and that both sequences have a common limit $\mu(\alpha_0, \beta_0)$, called the **arithmetic-geometric mean** of α_0 and β_0.
(b) If $\alpha_0 := 1$, $\beta_0 := k'$, show that, in the notation of problem 9,

$$\frac{\beta_n}{\alpha_n} = k'_n, \qquad \frac{\alpha_{n+1}}{\alpha_n} = \frac{1 + k'_n}{2},$$

hence

$$\alpha_n = \prod_{m=0}^{n-1} \frac{1 + k'_m}{2}, \, n = 1, 2, \ldots.$$

Conclude that

$$K(k) = \frac{\pi}{2\mu(1, k')}. \tag{9.10-17}$$

(c) Letting

$$\gamma_{n+1} := \tfrac{1}{2}(\alpha_n - \beta_n), \qquad n = 0, 1, 2, \ldots \tag{9.10-18}$$

show that under the initial conditions (b)

$$k_n = \frac{\gamma_n}{\alpha_n}, \; n = 1, 2, \ldots,$$

and establish the inequality

$$k_{n+1} \le \frac{1}{4k_0'} k_n^2.$$

Conclude that the sequences $\{\alpha_n\}$ and $\{\beta_n\}$ tend to their common limit with quadratic convergence.

11. If $0 < k < 1$, use quadratic and linear transforms to establish the identity

$$E(k) = (1 + k')E\left(\frac{1 - k'}{1 + k'}\right) - k'K(k). \tag{9.10-19}$$

12. Letting

$$Q(k) := \frac{K(k) - E(k)}{K(k)},$$

show that in the notation of problem 9

$$Q(k) = \frac{(1 + k')^2}{2} Q(k_1) + \tfrac{1}{2}k^2,$$

hence, using problem 10c, that

$$Q(k) = \frac{1}{2} \sum_{m=0}^{\infty} 2^m \prod_{i=0}^{m-1} \left[\frac{1 + k_i'}{2}\right]^2 k_m^2. \tag{9.10-20}$$

13. Using the notation of problem 10, show that the mth term in the sum (9.10-20) equals $2^m \gamma_m^2$, $m = 1, 2, \ldots$ Hence, putting $\gamma_0 := k_0 = k$,

$$Q(k) = \frac{1}{2} \sum_{m=0}^{\infty} 2^m \gamma_m^2,$$

and

$$E(k) = K(k)\left\{1 - \frac{1}{2} \sum_{m=0}^{\infty} 2^m \gamma_m^2\right\}. \tag{9.10-21}$$

[Note: The formulas (9.10-17) and (9.10-21) are very useful for the efficient numerical evaluation of $K(k)$ and $E(k)$.]

14. For $-1 < x < 1$, let

$$A(x) := \text{Artanh } x := \frac{1}{2} \text{Log} \frac{1+x}{1-x}.$$

(a) Applying a quadratic transform to the representation

$$A(x) = x_2 F_1(\tfrac{1}{2}, 1; \tfrac{3}{2}; x^2),$$

or by some other method, show that

$$A(x) = \frac{2}{1+x'} A\left(\sqrt{\frac{1-x'}{1+x'}}\right),$$

where $x' := \sqrt{1-x^2}$.

(b) Conclude that

$$A(x) = x \prod_{m=0}^{\infty} \frac{2}{1+x_m'}, \qquad (9.10\text{-}22)$$

where $x_0 := x$,

$$x_m' := \sqrt{1-x_m^2}, \; x_{m+1} := \left[\frac{1-x_m'}{1+x_m'}\right]^{1/2},$$

$m = 0, 1, 2, \ldots$.

15. Let the sequences $\{\alpha_n\}$ and $\{\beta_n\}$ now be defined by choosing two positive numbers α_0 and β_0 and letting

$$\alpha_{n+1} := \tfrac{1}{2}(\alpha_n + \beta_n), \qquad \beta_{n+1} := \sqrt{\alpha_{n+1}\beta_n}, \qquad (9.10\text{-}23)$$

$n = 0, 1, 2, \ldots$.

(a) Show that the sequences $\{\alpha_n\}$ and $\{\beta_n\}$ again have a common limit $\nu(\alpha_0, \beta_0)$, called the **modified arithmetic-geometric mean** of α_0 and β_0.

(b) If $\alpha_0 := 1$, $\beta_0 := x'$, show that in the notation of problem 14

$$\frac{\beta_n}{\alpha_n} = x_n', \frac{\alpha_{n+1}}{\alpha_n} = \frac{1+x_n'}{2},$$

hence

$$\alpha_n = \prod_{m=0}^{n-1} \frac{1+x_m'}{2}, \, n = 1, 2, \ldots.$$

Conclude that

$$A(x) = \frac{x}{\nu(1, x')}. \qquad (9.10\text{-}24)$$

(c) Show that, as $m \to \infty$,

$$\frac{x_{m+1}}{x_m} \to \frac{1}{4},$$

and conclude that the sequence $\{\alpha_n\}$ tends to its limit with *linear* convergence, with a convergence ratio that is independent of x.

16. For any real x, let

$$x'' := \sqrt{1 + x^2}.$$

Show that

$$\text{Arctan } x = \frac{x}{\nu(1, x'')}. \tag{9.10-25}$$

§9.11.   SINGULARITIES OF THE SECOND KIND: FORMAL SOLUTIONS

We return to the local treatment of a general system of linear differential equations with an isolated singularity at the point $z = \infty$. We are interested only in the case where the coefficient matrix has at most a pole at ∞. The system then can be written in the form

$$\mathbf{w}' = z^r \mathbf{A}(z)\mathbf{w}, \tag{9.11-1}$$

where r is an integer, and where $\mathbf{A}$ is analytic at ∞ and thus possesses for $|z|$ sufficiently large a representation

$$\mathbf{A}(z) = \mathbf{A}_0 + \mathbf{A}_1 z^{-1} + \mathbf{A}_2 z^{-2} + \cdots.$$

Excluding the trivial situation where $\mathbf{A} = \mathbf{0}$, we shall assume that the integer r is so chosen that $\mathbf{A}_0 \neq \mathbf{0}$.

The following was proved in §9.8: If $r \leq -2$, then $z = \infty$ is a regular point of the differential equation in the sense that the transformation $z \to 1/z$ will yield a regular point at $z = 0$. If $r = -1$, the same transformation will yield a singular point of the first kind at $z = 0$ which can be treated by the methods of §9.5. It thus remains to consider the case where $r \geq 0$. After Poincaré it is customary to call the integer $r + 1$ the **rank** of the singularity. We thus are interested in singularities of positive rank. For many equations of practical importance (Bessel, Kummer) the rank of the singularity at ∞ is 1. Because the discussion of the general case is very involved, for the most part we restrict our treatment to the case $r + 1 = 1$.

The one result at our disposal for singularities of arbitrary, even infinite, rank is contained in Theorem 9.4b, which, if adapted to an isolated singularity at ∞, states that there always exists a fundamental matrix of the form

$$\mathbf{W}(z) = \mathbf{P}(z)z^{\mathbf{R}},$$

where $\mathbf{P}$ is analytic in a punctured vicinity of ∞, and where $\mathbf{R}$ is a suitable matrix of constants. If the rank of the singularity is zero (i.e., if the singular point is regular), $\mathbf{P}$ may be assumed to be analytic at ∞, and there exist recursive algorithms for the determination of $\mathbf{R}$ and for the coefficients of the power series representing $\mathbf{P}$.

If the rank is positive, already the simplest example

$$\mathbf{w}' = \mathbf{A}\mathbf{w}$$

($\mathbf{A} = \text{const.}$), which has the fundamental matrix $e^{\mathbf{A}z}$, shows that the matrix $\mathbf{P}$ can have an essential singularity. No recursive schemes for the determination of $\mathbf{R}$ and of the coefficients of $\mathbf{P}$ are available in that case. Even if $\mathbf{R}$ is assumed diagonal (which is justifiable only if the original $\mathbf{R}$ is similar to a diagonal matrix), the substitution of a tentative solution vector

$$\mathbf{w}(z) = z^{\lambda} \sum_{n=-\infty}^{\infty} \mathbf{p}_n z^n \qquad (9.11\text{-}2)$$

into (9.11-1) leads to an infinite system of homogeneous equations, containing the parameter λ, which, in general, cannot be solved recursively. A theory for dealing with such systems has been developed by Helge von Koch [1892]; the theory does not make it possible, however, to determine λ or any of the coefficients $\mathbf{p}_n$ in (9.11-2) in a finite number of steps.

For these reasons it is all the more important that if $z = \infty$ is a singularity of *finite* rank, the equation (9.11-1) possesses a different kind of formal solution constructed recursively. These solutions were discovered by Thomé [1872], who called them *normal solutions*. Because exponential functions already occur in the solutions of the simplest equations with singularities of positive rank, it seems reasonable to include them as a basic element of construction in the formal solutions. Indeed, it turns out that Thomé's solutions are made up of so-called *log–exponential sums*, which we define presently.

A **formal log–exponential sum** is a finite expansion of the form

$$U = \sum_{j=1}^{k} P_j e^{\mu_j}, \qquad (9.11\text{-}3)$$

where the P_j are formal logarithmic sums in powers of z^{-1}, and where the μ_j are distinct polynomials in the variable z whose constant term is zero. (In most later applications, the μ_j are *linear* polynomials and thus of the form γz, where γ is a constant.) The ordering of the terms in (9.11-3) is considered irrelevant; that is, U is identified with any sum obtained by rearranging the terms in (9.11-3). If

$$V = \sum_{j=1}^{m} Q_j e^{\nu_j}$$

is another log–exponential sum, then V is defined to be equal to U if and only if $k = m$ and $Q_{j_i} = P_i$, $\nu_{j_i} = \mu_i$, $i = 1, \ldots, m$, for some permutation $(j_1, j_2, \ldots, j_m)$ of $(1, 2, \ldots, m)$.

If we denote by $\omega_1, \ldots, \omega_n$ the distinct polynomials in the set $\{\mu_1, \ldots, \mu_k, \nu_1, \ldots, \nu_m\}$, then U and V may be written as

$$U = \sum_{j=1}^{n} P_j e^{\omega_j}, \qquad V = \sum_{j=1}^{n} Q_j e^{\omega_j},$$

where any coefficient P_j or Q_j may be zero. The sum $U + V$ is then defined to be the formal log–exponential sum

$$U + V := \sum_{j=1}^{n} (P_j + Q_j)\, e^{\omega_j}.$$

Similarly, if we denote by $\sigma_1, \sigma_2, \ldots, \sigma_r$ the set of all distinct polynomials of the form $\mu_i + \nu_j$ $(i = 1, \ldots, k; j = 1, \ldots, m)$, then the product UV is defined to be the formal log–exponential sum

$$UV := \sum_{k=1}^{r} \left(\sum_{\mu_i + \nu_j = \sigma_k} P_i Q_j \right) e^{\sigma_k}.$$

The derivative U' of (9.11-3) is defined to be the formal log–exponential sum

$$U' := \sum_{j=1}^{k} (P_j' + P_j \mu_j')\, e^{\mu_j},$$

where the derivatives of the P_j are to be formed as if the series were convergent. It can readily be verified that with the foregoing definitions the formal log–exponential series form a commutative ring, and that the usual rules for the differentiation of sums and of products are satisfied.

A **formal log–exponential matrix** is defined to be a matrix $\mathbf{U} = (U_{ij})$ whose elements are formal log–exponential sums. The sum and the product of two such matrices are defined in the usual manner. The derivative $\mathbf{U}'$ is defined to be the matrix with the elements U_{ij}'. The set of formal log–exponential matrices is closed under the operations of addition, multiplication, and differentiation. Two formal log–exponential matrices are considered equal if and only if they are equal element by element, in the sense of equality as defined for formal log–exponential sums.

We now return to the differential equation (9.11-1). Its coefficient, $z^r \mathbf{A}(z)$, may clearly be regarded as a formal log–exponential matrix. It is, in fact, a matrix whose elements are formal Laurent series in $1/z$ (involving, as they must, only finitely many negative powers of $1/z$, i.e., positive powers of z.) A **formal log–exponential solution matrix** of (9.11-1) thus may be defined as a formal log–exponential matrix $\mathbf{W}$ that satisfies

$$\mathbf{W}' = z^r \mathbf{A}(z)\mathbf{W}$$

in the sense of equality for such matrices.

Concerning equations of rank 1, we prove:

THEOREM 9.11a

Let $A(z)$ be analytic at ∞, and let

$$A(z) = \sum_{k=0}^{\infty} A_k z^{-k}$$

in some neighborhood of ∞, where A_0 is similar to a diagonal matrix Λ with n distinct eigenvalues $\lambda_1, \ldots, \lambda_n$, $A_0 = T\Lambda T^{-1}$. Then the system

$$w' = A(z)w$$

possesses a formal log–exponential solution matrix

$$W = Pz^{\Delta} e^{\Lambda z}, \tag{9.11-4}$$

where P is a formal power series in $1/z$,

$$P = \sum_{n=0}^{\infty} P_n z^{-n}, \tag{9.11-5}$$

where Δ is a diagonal matrix that depends only on T and A_1, and where the matrices P_n can be constructed rationally from T, Λ, $A_1, \ldots, A_{n+1}$ ($n = 0, 1, \ldots$).

Proof. The following Lemma is used repeatedly.

LEMMA 9.11b

Let $\Gamma := \mathrm{diag}(\gamma_1, \ldots, \gamma_n)$, where $\gamma_i \neq \gamma_j$ for $i \neq j$, and let $C = (c_{ij})$ be any matrix. Then the equation for the matrix X,

$$X\Gamma - \Gamma X = C, \tag{9.11-6}$$

has a solution if and only if all diagonal elements of C are zero. If this condition is satisfied, the off-diagonal elements of a solution matrix X are uniquely determined, whereas the diagonal elements remain arbitrary.

Proof of the lemma. The matrix $X = (x_{ij})$ is a solution of (9.11-6) if and only if

$$x_{ij}(\gamma_j - \gamma_i) = c_{ij} \tag{9.11-7}$$

for $i, j = 1, 2, \ldots, n$. For $i = j$ this relation is possible only if $c_{ii} = 0$, and then it is satisfied for arbitrary x_{ii}. For $i \neq j$, x_{ij} is uniquely determined by (9.11-7) by virtue of $\gamma_j - \gamma_i \neq 0$. ∎

We now turn to the proof of the theorem. Suppose $\mathbf{W}$ is a matrix of the form (9.11-4), $\mathbf{W} = \mathbf{P}z^{\Delta} e^{\Gamma z}$, where Γ is diagonal. Then

$$\mathbf{W}' = \mathbf{P}'z^{\Delta} e^{\Gamma z} + \mathbf{P}\Delta z^{-1}z^{\Delta} e^{\Gamma z} + \mathbf{P}z^{\Delta}\Gamma e^{\Gamma z}$$
$$= [\mathbf{P}' + z^{-1}\mathbf{P}\Delta + \mathbf{P}\Gamma]z^{\Delta} e^{\Gamma z},$$

because the diagonal matrices z^{Δ} and Γ commute. The $\mathbf{W}$ is a solution matrix if and only if the above equals $\mathbf{AW} = \mathbf{AP}z^{\Delta} e^{\Gamma z}$, which is the case if

$$\mathbf{P}' + z^{-1}\mathbf{P}\Delta + \mathbf{P}\Gamma = \mathbf{AP},$$

or, substituting the series expansions for $\mathbf{P}$ and $\mathbf{A}$,

$$\sum_{k=0}^{\infty} z^{-k-1}\mathbf{P}_k(\Delta - k\mathbf{I}) + \sum_{k=0}^{\infty} z^{-k}\mathbf{P}_k\Gamma = \sum_{k=0}^{\infty} z^{-k}\mathbf{A}_k \sum_{m=0}^{\infty} z^{-m}\mathbf{P}_m.$$

Comparing coefficients, this is equivalent to the relations

$$\mathbf{P}_0\Gamma - \mathbf{A}_0\mathbf{P}_0 = \mathbf{0}, \tag{9.11-8a}$$

$$\mathbf{P}_{k-1}[\Delta - (k-1)\mathbf{I}] + \mathbf{P}_k\Gamma = \sum_{m=0}^{k} \mathbf{A}_m\mathbf{P}_{k-m}, \qquad k = 1, 2, \ldots. \tag{9.11-8b}$$

Without loss of generality we may assume that $\mathbf{A}_0$ is diagonal, $\mathbf{A}_0 = \Lambda$. If $\mathbf{W}$ is a formal solution for this special case, and if $\mathbf{T}$ is any non-singular matrix, then in view of

$$(\mathbf{TW})' = \mathbf{TW}' = \mathbf{TAW} = \mathbf{TAT}^{-1}(\mathbf{TW})$$

the matrix $\mathbf{TW}$ is a formal solution of the same type for the differential equation where $\mathbf{A}$ has been replaced by $\mathbf{TAT}^{-1}$, and thus Λ by $\mathbf{T\Lambda T}^{-1}$.

If $\mathbf{A}_0$ is diagonal, $\mathbf{A}_0 = \Lambda$, relation (9.11-8a) is satisfied if

$$\mathbf{P}_0 = \mathbf{I}, \qquad \Gamma = \Lambda. \tag{9.11-9}$$

Equation (9.11-8b) reads for $k = 1$

$$\mathbf{P}_0\Delta + \mathbf{P}_1\Gamma = \mathbf{A}_0\mathbf{P}_1 + \mathbf{A}_1\mathbf{P}_0$$

which in view of (9.11-9) yields

$$\mathbf{P}_1\Lambda - \Lambda\mathbf{P}_1 = \mathbf{A}_1 - \Delta.$$

This equation for $\mathbf{P}_1$ is of the type covered by Lemma 9.11b. In order that it has a solution, it is necessary that the diagonal elements of the matrix on the right vanish. This condition determines the diagonal matrix Δ. Its diagonal elements must agree with the corresponding elements of $\mathbf{A}_1$. We express this by writing

$$\Delta := \mathrm{diag}(\mathbf{A}_1). \tag{9.11-10}$$

If this is satisfied, the off-diagonal elements of $\mathbf{P}_1$ are uniquely determined, and the diagonal elements remain arbitrary. We thus have

$$\mathbf{P}_1 = \tilde{\mathbf{P}}_1 + \mathbf{\Pi}_1,$$

where $\tilde{\mathbf{P}}_1$ is a matrix with zeros on the main diagonal, which is uniquely determined, and where $\mathbf{\Pi}_1$ is an undetermined diagonal matrix.

Suppose now that $k-1 > 0$, and that we have determined the matrices $\mathbf{P}_0 = \mathbf{I}, \mathbf{P}_1, \ldots, \mathbf{P}_{k-2}$, and that

$$\mathbf{P}_{k-1} = \tilde{\mathbf{P}}_{k-1} + \mathbf{\Pi}_{k-1},$$

where $\tilde{\mathbf{P}}_{k-1}$ is a matrix with zeros on the main diagonal, and the diagonal matrix $\mathbf{\Pi}_{k-1}$ is as yet undetermined. Then relation (9.11-8b), written out in full, reads

$$\mathbf{P}_{k-1}[\mathbf{\Delta} - (k-1)\mathbf{I}] + \mathbf{P}_k \mathbf{\Lambda} = \mathbf{S}_k + \mathbf{A}_1 \mathbf{P}_{k-1} + \mathbf{\Lambda} \mathbf{P}_k,$$

where

$$\mathbf{S}_k := \mathbf{A}_k \mathbf{P}_0 + \mathbf{A}_{k-1} \mathbf{P}_1 + \cdots + \mathbf{A}_2 \mathbf{P}_{k-2}$$

is fully determined. Separating the determined and undetermined part of $\mathbf{P}_{k-1}$, this may be written

$$\mathbf{P}_k \mathbf{\Lambda} - \mathbf{\Lambda} \mathbf{P}_k = \mathbf{S}_k + \mathbf{A}_1 \tilde{\mathbf{P}}_{k-1} + \tilde{\mathbf{A}}_1 \mathbf{\Pi}_{k-1} + \mathbf{\Delta} \mathbf{\Pi}_{k-1}$$
$$+ (k-1)\tilde{\mathbf{P}}_{k-1} + (k-1)\mathbf{\Pi}_{k-1} - \tilde{\mathbf{P}}_{k-1}\mathbf{\Delta} - \mathbf{\Pi}_{k-1}\mathbf{\Delta}, \qquad (9.11\text{-}11)$$

where $\tilde{\mathbf{A}}_1 := \mathbf{A}_1 - \mathbf{\Delta}$ is a matrix with zeros on the main diagonal. Because $\mathbf{\Delta}$ and $\mathbf{\Pi}_{k-1}$ are both diagonal, $\mathbf{\Delta}\mathbf{\Pi}_{k-1} - \mathbf{\Pi}_{k-1}\mathbf{\Delta} = \mathbf{0}$. The expression on the right now may be regarded as the sum of three matrices $\mathbf{M}_k^{(i)}$, $i = 1, 2, 3$, namely of the matrix

$$\mathbf{M}_k^{(1)} := \mathbf{S}_k + \tilde{\mathbf{A}}_1 \mathbf{P}_{k-1} + (k-1)\tilde{\mathbf{P}}_{k-1},$$

which is completely determined, of the matrix

$$\mathbf{M}_k^{(2)} := \tilde{\mathbf{A}}_1 \mathbf{\Pi}_{k-1} - \tilde{\mathbf{P}}_{k-1}\mathbf{\Delta},$$

which has zeros on the main diagonal, and of the diagonal matrix

$$\mathbf{M}_k^{(3)} := (k-1)\mathbf{\Pi}_{k-1},$$

which is undetermined. In order that (9.11-11) has a solution $\mathbf{P}_k$, it is necessary by Lemma 9.11b that

$$\mathbf{\Pi}_{k-1} := \frac{1}{k-1} \operatorname{diag}(\mathbf{M}_k^{(1)}).$$

With this choice of $\mathbf{\Pi}_{k-1}$, $\mathbf{M}_k^{(2)}$ is determined, and equation (9.11-11) now completely determines the off-diagonal elements of $\mathbf{P}_k$, while the diagonal elements remain undetermined. The situation which was assumed to hold at the index $k-1$ thus prevails also at k. This shows that with our initial choice of $\mathbf{P}_0$ and $\mathbf{\Gamma}$ all $\mathbf{P}_k$ are ultimately determined uniquely. Conversely, the $\mathbf{P}_k$ thus determined satisfy the relations (9.11-9), and the matrix $\mathbf{W} := \mathbf{P}z^{\mathbf{\Delta}} e^{\mathbf{\Lambda}z}$, where $\mathbf{P} := \sum \mathbf{P}_k z^{-k}$, is a formal solution matrix of the desired type. The construction of the $\mathbf{P}_k$ indicated above verifies the final sentence of the theorem, which thus is fully proved. ■

This theorem makes no statement concerning the analytic properties of the formal solutions constructed. It is shown in §9.12 that, in general, the formal series in (9.11-4) is not convergent. However, the expression (9.11-5) always has a meaning as an *asymptotic expansion* of a suitable solution of the differential equation. This is proved in §11.4.

PROBLEMS

1. To motivate the structure of the formal solutions (9.11-4), consider the scalar first-order equation

$$w' = a(z)w,$$

where $a(z)$ is analytic at ∞,

$$a(z) = \lambda + \frac{\delta}{z} + \frac{a_2}{z^2} + \frac{a_3}{z^3} + \cdots, \qquad |z| > \rho.$$

Verify that this equation possesses the solution

$$w(z) = z^{\delta} e^{\lambda z} p(z),$$

where p is analytic at ∞,

$$p(z) := \exp\left(-\frac{a_2}{z} - \frac{a_3}{2z^3} - \frac{a_4}{3z^3} - \cdots\right).$$

and that the series $\mathbf{P}$ constructed in the proof of Theorem 9.11a in this case represents p.

2. The first-order system obtained from the scalar second-order equation

$$u'' - \frac{1}{z}u' - \frac{1}{z}u = 0$$

by letting $\mathbf{w}^T := (u, u')$ does not satisfy the hypotheses of Theorem 9.11a, because the matrix $\mathbf{A}_0$ has a nonlinear elementary divisor. Show that by introducing the new variable $t := z^{1/2}$ one can obtain a system satisfying the hypotheses, and establish the existence of formal solutions of the form

$$u = e^{\pm 2iz^{1/2}} z^{-1/4} p(z^{1/2}),$$

where p denotes a formal series in z^{-1}.

3. Show that the system of rank $r > 1$,

$$\mathbf{w}' = z^r \mathbf{A}(z)\mathbf{w},$$

where the matrix $\mathbf{A}_0$ has n distinct eigenvalues, possesses a formal solution of the form

$$\mathbf{W} = \mathbf{P}z^{\Delta} e^{\Gamma},$$

where $\mathbf{P}$ is a formal series in z^{-1}, Δ is a constant diagonal matrix, and

$$\Gamma := z\Gamma_r + \frac{z^2}{2}\Gamma_{r-1} + \cdots + \frac{z^{r+1}}{r+1}\Gamma_0,$$

with diagonal matrices $\Gamma_0, \ldots, \Gamma_r$. Formulate an algorithm for determining the Γ_i and the coefficients of $\mathbf{P}$. [It is now necessary to carry along r undetermined diagonal matrices.]

§9.12. SINGULARITIES OF THE SECOND KIND OF SPECIAL SECOND-ORDER EQUATIONS

In this section we adapt the construction of the preceding section to obtain formal solutions of the special scalar second-order equation

$$u'' - r(z)u = 0, \tag{9.12-1}$$

where, for $|z|$ sufficiently large,

$$r(z) = \sum_{n=0}^{\infty} r_n z^{-n}.$$

It is assumed that $r_0 \neq 0$.

By writing (9.12-1) as a system of first-order equations,

$$\mathbf{w}' = \mathbf{A}(z)\mathbf{w}, \tag{9.12-2}$$

where

$$\mathbf{w} := \begin{pmatrix} u \\ u' \end{pmatrix}, \qquad \mathbf{A}(z) := \begin{pmatrix} 0 & 1 \\ r(z) & 0 \end{pmatrix},$$

we see that $z = \infty$ is a singular point of the second kind whose rank is 1. We have

$$\mathbf{A}(z) = \sum_{n=0}^{\infty} \mathbf{A}_n z^{-n},$$

with

$$\mathbf{A}_0 := \begin{pmatrix} 0 & 1 \\ r_0 & 0 \end{pmatrix}, \qquad \mathbf{A}_n := \begin{pmatrix} 0 & 0 \\ r_n & 0 \end{pmatrix}, \qquad n = 1, 2, \ldots.$$

The eigenvalues of $\mathbf{A}_0$ are $\pm\lambda$, where λ is a solution of

$$\lambda^2 = r_0, \tag{9.12-3}$$

and thus, by our assumption on r_0, distinct. It follows from Theorem 9.11a that the system (9.12-2) has a formal log–exponential solution of the form $\mathbf{P}z^{\Delta} e^{\Lambda z}$, where $\mathbf{P}$ is a formal series in powers of z^{-1}, and where the matrices Δ and Λ are diagonal.

We begin by determining the matrices Δ and Λ. The matrix Λ is a Jordan canonical form of $\mathbf{A}_0$ and thus may be assumed as

$$\Lambda = \begin{pmatrix} \lambda & 0 \\ 0 & -\lambda \end{pmatrix}, \tag{9.12-4}$$

where λ satisfies (9.12-3). To find Δ, we must determine $\mathbf{T}$ such that

$$\mathbf{T}^{-1}\mathbf{A}_0\mathbf{T} = \Lambda; \tag{9.12-5}$$

by (9.11-10), we then have

$$\Delta := \text{diag}(\mathbf{T}^{-1}\mathbf{A}_1\mathbf{T}). \tag{9.12-6}$$

By computing the eigenvectors of $\mathbf{A}_0$, it is seen that a matrix $\mathbf{T}$ satisfying (9.12-5) is given by

$$\mathbf{T} := \begin{pmatrix} 1 & -1 \\ \lambda & \lambda \end{pmatrix}.$$

We have

$$\mathbf{T}^{-1} = \frac{1}{2\lambda}\begin{pmatrix} \lambda & 1 \\ -\lambda & 1 \end{pmatrix};$$

consequently

$$\mathbf{T}^{-1}\mathbf{A}_1\mathbf{T} = \frac{1}{2\lambda}\begin{pmatrix} r_1 & -r_1 \\ r_1 & -r_1 \end{pmatrix};$$

hence

$$\Delta = \begin{pmatrix} \delta & 0 \\ 0 & -\delta \end{pmatrix},$$

where

$$\delta := \frac{1}{2\lambda}r_1. \tag{9.12-7}$$

It remains to determine the matrix $\mathbf{P}$. However, rather than determining this matrix completely, we find the elements in its first row. (Those in the second row are then obtained by differentiation.) These are log–exponential series of the form

$$U = z^\delta e^{\lambda z} \sum_{n=0}^{\infty} u_n z^{-n},$$

and a similar series with δ and λ replaced by $-\delta$ and $-\lambda$, respectively. For the purpose of formal differentiation we write this in the form

$$U = e^{\lambda z} \sum_{n=-\infty}^{\infty} u_n z^{\delta-n},$$

where it is understood that $u_n = 0$ for $n < 0$. We now find easily

$$U' = e^{\lambda z} \sum_{n=-\infty}^{\infty} [\lambda u_n + (\delta - n + 1)u_{n-1}] z^{\delta-n},$$

$$U'' = e^{\lambda z} \sum_{n=-\infty}^{\infty} [\lambda^2 u_n + 2\lambda(\delta - n + 1)u_{n-1} + (\delta - n + 1)(\delta - n + 2)u_{n-2}] z^{\delta-n}.$$

Substituting into (9.12-1) we get, after cancelling the common factor $e^{\lambda z} z^\delta$ (which is permissible also in the formal sense)

$$\sum_{n=-\infty}^{\infty} [\lambda^2 u_n + 2\lambda(\delta - n + 1)u_{n-1} + (\delta - n + 1)(\delta - n + 2)u_{n-2}] z^{-n}$$

$$- \sum_{n=-\infty}^{\infty} u_n z^{-n} \sum_{m=0}^{\infty} r_m z^{-m} = 0.$$

Considering $u_n = 0$ for $n < 0$ and comparing coefficients, we find that the u_n must satisfy

$$\lambda^2 u_n + 2\lambda(\delta - n + 1)u_{n-1} + (\delta - n + 1)(\delta - n + 2)u_{n-2}$$

$$- [u_n r_0 + u_{n-1} r_1 + u_{n-2} r_2 + \cdots + u_0 r_n] = 0$$

for $n = 0, 1, 2, \ldots$. In view of $\lambda^2 = r_0$, $2\lambda\delta = r_1$ these relations simplify to

$$2\lambda(1 - n)u_{n-1} + (\delta - n + 1)(\delta - n + 2)u_{n-2} - \sum_{k=2}^{n} u_{n-k} r_k = 0,$$

or after replacing n by $n + 1$,

$$2\lambda n u_n - (\delta - n)(\delta - n + 1)u_{n-1} + \sum_{k=1}^{n} u_{n-k} r_{k+1} = 0. \qquad (9.12\text{-}8)$$

These relations must hold for $n = 0, 1, 2, \ldots$. It turns out that u_0 is arbitrary. Once u_0 is chosen, the coefficients $u_1, u_2, \ldots$ are determined uniquely, and a formal solution of the differential equation is thus obtained. A second formal solution results by replacing λ and δ by $-\lambda$ and $-\delta$, respectively.

Before discussing special cases, we recall that the general second-order equation

$$u'' + p(z)u' + q(z)u = 0 \tag{9.12-9}$$

can always be reduced to the form (9.12-1) by setting

$$u = hv, \tag{9.12-10a}$$

where

$$h(z) := \exp\left(-\frac{1}{2}\int_{z_0}^{z} p(t)\,dt\right) \tag{9.12-10b}$$

(see Theorem 9.3j). There results the equation

$$v'' - r(z)v = 0, \tag{9.12-11}$$

where

$$r := \tfrac{1}{4}p^2 + \tfrac{1}{2}p' - q. \tag{9.12-12}$$

We now carry out the construction of formal solutions for the two special differential equations considered in §9.7.

EXAMPLE 1 Kummer's equation

If u is a solution of Kummer's differential equation (9.7-1),

$$u'' + \left(\frac{\gamma}{z} - 1\right)u' - \frac{\alpha}{z}u = 0,$$

we find by carrying out the substitution (9.12-10) that

$$u(z) = z^{-\gamma/2}\, e^{z/2} v(z),$$

where

$$v'' + \left(-\frac{1}{4} + \frac{\gamma - 2\alpha}{2z} + \frac{1 - (\gamma - 1)^2}{4z^2}\right)v = 0.$$

It is customary to introduce the new parameters

$$\mu := \tfrac{1}{2}(\gamma - 1), \qquad \kappa := \tfrac{1}{2}\gamma - \alpha,$$

which is to say

$$\alpha = \mu + \tfrac{1}{2} - \kappa, \qquad \gamma = 2\mu + 1,$$

so that the equation appears in the form

$$v'' - \left(\frac{1}{4} - \frac{\kappa}{z} + \frac{\mu^2 - 1/4}{z^2}\right)v = 0. \tag{9.12-13}$$

This equation is frequently called **Whittaker's differential equation**. It follows from the results of §9.7 that for $2\mu + 1$ not an integer a system of linearly independent solutions of (9.12-13) is given by the functions

$$M_{\kappa,\mu}(z) \quad \text{and} \quad M_{\kappa,-\mu}(z),$$

where

$$M_{\kappa,\mu}(z) := z^{\mu+1/2} \, e^{-z/2} \, {}_1F_1(\mu - \kappa + \tfrac{1}{2}; 2\mu + 1; z) \tag{9.12-14}$$

is called the **Whittaker function of the first kind**.

Whittaker's equation (9.12-13) is of the form (9.12-1), with the coefficients of the series representing $r(z)$ given by

$$r_0 := \tfrac{1}{4}, \qquad r_1 := -\kappa, \qquad r_2 := \mu^2 - \tfrac{1}{4},$$

and $r_k := 0$ for $k \geqslant 3$. The recurrence relation (9.12-8) thus reduces to

$$2\lambda n u_n = (-\delta - \mu + \tfrac{1}{2} + n)(-\delta + \mu + \tfrac{1}{2} + n)u_{n-1}, \qquad n = 1, 2, \ldots.$$

Assuming $u_0 = 1$, this yields

$$u_n = \frac{(1/2 - \delta - \mu)_n (1/2 - \delta + \mu)_n}{(2\lambda)^n n!}, \qquad n = 0, 1, \ldots.$$

Relation (9.12-3) yields $\lambda^2 = \tfrac{1}{4}$, thus $\lambda = \pm\tfrac{1}{2}$, and from (9.12-7) it follows that $\delta = \mp\kappa$, where the upper and the lower signs go together. We thus obtain the following two formal solutions of Whittaker's equation:

$$U^{(I)} := e^{-z/2} z^\kappa \, {}_2F_0\left(\frac{1}{2} - \kappa - \mu, \frac{1}{2} - \kappa + \mu; -\frac{1}{z}\right),$$

$$\tag{9.12-15}$$

$$U^{(II)} := e^{z/2} z^{-\kappa} \, {}_2F_0\left(\frac{1}{2} + \kappa - \mu, \frac{1}{2} + \kappa + \mu; \frac{1}{z}\right).$$

In §11.5 $U^{(I)}$ is shown to be an asymptotic expansion for $W_{\kappa,\mu}(z)$, the *Whittaker function of the second kind* which is defined by eq. (10.5-14).

EXAMPLE 2   **Bessel's equation**

Let u now be a solution of Bessel's equation

$$u'' + \frac{1}{z}u' + \left(1 - \frac{\nu^2}{z^2}\right)u = 0. \tag{9.12-16}$$

Carrying out the substitution (9.12-10) yields $u(z) = z^{-1/2}v(z)$, where

$$v'' + \left(1 - \frac{\nu^2 - 1/4}{z^2}\right)v = 0.$$

This equation is of the form (9.12-1). The coefficients of the series representing r here are given by

$$r_0 = -1, \qquad r_1 = 0, \qquad r_2 = v^2 - \tfrac{1}{4}, \tag{9.12-17}$$

and $r_k = 0$ for $k \geqslant 3$. The recurrence relation (9.12-8) for the coefficients of the formal solution thus reduces to

$$2\lambda n v_n = (-\delta - v - \tfrac{1}{2} + n)(-\delta + v - \tfrac{1}{2} + n)v_{n-1}.$$

Assuming $v_0 = 1$ this yields

$$v_n = \frac{(1/2 - \delta - v)_n (1/2 - \delta + v)_n}{(2\lambda)^n n!},$$

$n = 0, 1, 2, \ldots$. Relation (9.12-3) furnishes $\lambda^2 = -1$, thus $\lambda = \pm i$, and (9.12-7) shows that $\delta = 0$. Thus the following two formal solutions of Bessel's equation are obtained:

$$U^{(I)} := e^{iz} z^{-1/2} \, {}_2F_0\left(\frac{1}{2} - v, \frac{1}{2} + v; \frac{1}{2iz}\right),$$

$$\tag{9.12-18}$$

$$U^{(II)} := e^{-iz} z^{-1/2} \, {}_2F_0\left(\frac{1}{2} - v, \frac{1}{2} + v; -\frac{1}{2iz}\right).$$

The series appearing here, like the series in (9.12-15), have radius of convergence zero for general values of the parameters. Thus they cannot represent actual solutions. Again, it is shown in §11.5 that they represent asymptotic expansions for certain solutions of Bessel's equation and as such can even be used to approximate these solutions numerically.

We note that for $v = n + \tfrac{1}{2}$, where n is an integer, both series in (9.12-18) terminate. Thus for these special values of the parameter the formal series define functions, which by construction are solutions of Bessel's equation. Because the two solutions are clearly independent, they form a fundamental system. It follows that all solutions of Bessel's equation of order $n + \tfrac{1}{2}$ are elementary functions. This confirms a result that was already obtained in §9.7, and provides us with a new representation of these elementary solutions.

PROBLEMS

1. Representing Y_v in terms of the functions (9.7-11) and evaluating the limit as $v \to n + \tfrac{1}{2}$, show that for $n = 0, 1, 2, \ldots$

$$Y_{n+1/2}(z) = -(-i)^n \sqrt{\frac{2}{\pi}} \, u^{(I)}(z) + iJ_{n+1/2}(z),$$

$$= -i^n \sqrt{\frac{2}{\pi}} \, u^{(II)}(z) - iJ_{n+1/2}(z),$$

where $u^{(I)}$ and $u^{(II)}$ are the formal solutions (9.12-18) evaluated for $\nu = n + \frac{1}{2}$,

$$u^{(I)}(z) := e^{iz}z^{-1/2} \, {}_2F_0\left(-n, n+1; \frac{1}{2iz}\right),$$

$$u^{(II)}(z) := e^{-iz}z^{-1/2} \, {}_2F_0\left(-n, n+1; -\frac{1}{2iz}\right).$$

Conclude that

$$J_{n+1/2}(z) + iY_{n+1/2}(z) = (-i)^{n+1} \sqrt{\frac{2}{\pi}} u^{(I)}(z),$$

$$J_{n+1/2}(z) - iY_{n+1/2}(z) = i^{n+1} \sqrt{\frac{2}{\pi}} u^{(II)}(z).$$

2. Construct formal log–exponential solutions for the differential equations satisfied by the functions zJ_ν^2, $zJ_\nu Y_\nu$, zY_ν^2,

$$u''' + \left(4 - \frac{4\nu^2 - 1}{z^2}\right)u' + \frac{4\nu^2 - 1}{z^3}u = 0$$

(cf. Problem 4, §9.7).

SEMINAR ASSIGNMENTS

1. Write a program to solve differential equations $w' = f(z, w)$, where f is rational, by the power series method. The following outline is suggested:

   (a) Letting $f = g/h$, find an algorithm that expands the polynomials g and h at any point $(\hat{z}, \hat{w})$ in powers of $z - \hat{z}$ and $w - \hat{w}$;

   (b) Find the power series $w(z) = \sum a_n(z - \hat{z})^n$ by the method of undetermined coefficients;

   (c) Estimate the radius of convergence by the Cauchy–Hadamard formula, and choose your integration step accordingly.

   Observe how your program behaves near singularities. Use it to make a numerical study of Painlevé's equation (see Davis [1962]).

2. Create an analog of Riemann's theory for the differential equation satisfied by the generalized hypergeometric series ${}_3F_2$ (see §1.5). Obtain Whipple's formula (Problem 4, §1.6) as an example of a quadratic transform.

3. Study the form of the most general linear second-order differential equation with regular singular points at the fourth roots of unity, and nowhere else (a special case of Heun's equation). Obtain the Taylor coefficients of a fundamental system at the origin.

4. Determine closed trajectories for the **restricted three-body problem** (mass of third body negligeable compared to the mass of the two other bodies). We assume that the two heavy bodies are at a constant distance 1, that they rotate around their common center of gravity with constant angular velocity 1, and that the motion of the third body takes place in the plane of rotation. Placing the origin of the coordinate system at the center of gravity and the x axis through the heavy bodies, the motion of the small body is governed by the system of differential equations

$$x'' = x + 2y' - \nu \frac{x + \mu}{[(x + \mu)^2 + y^2]^{3/2}} - \mu \frac{x - \nu}{[(x - \nu)^2 + y^2]^{3/2}},$$

$$y'' = y - 2x' - \nu \frac{y}{[(x + \mu)^2 + y^2]^{3/2}} - \mu \frac{y}{[(x - \nu)^2 + y^2]^{3/2}}.$$

Here μ and $\nu := 1 - \mu$ are the ratios of the masses of the bodies to the right and to the left of the origin to the total mass. For the system earth–moon, $\mu = 0.01213$ (see Walter [1972], p. 5).

NOTES

Standard references on the analytic theory of ordinary differential equations include Ince [1926], Poole [1936], Coddington and Levinson [1955]. Our treatment of the general theory generally follows Coddington and Levinson. Walter [1972] gives an elegant and concise treatment of some aspects of the theory.

§9.2.   For a detailed treatment of the numerical solution of the initial value problem for analytic equations see Henrici [1962].

§9.3.   In addition to Coddington and Levinson, Gantmacher [1959] in Chapter 14 offers a thorough treatment of linear equations from the matrix point of view.

§9.4.   Log–holomorphic functions (that is, analytic functions of log z) are introduced here because the usual treatment of multivalued solutions in texts on *differential equations* seem unnecessarily vague. On the other hand, we wished to avoid the machinery that goes with the rigorous presentation as given, for instance, by Ahlfors [1966], Chapter 8. The classical reference on singular points is Birkhoff [1909].

§9.5.   In his lectures, G. D. Birkhoff strongly emphasized the difference between formal and actual solutions (G. Birkhoff, private communication).

§9.6.   For Frobenius' method, see Frobenius [1873].

§9.7.   For an exhaustive treatment of Bessel functions, Watson [1944] is unsurpassed. Confluent hypergeometric functions are discussed thoroughly by Buchholz [1953].

§9.8.   The Fuchsian theory dates back to Fuchs [1866–1868].

§9.9.   See also Chapter 2 (by W. Magnus) of Erdelyi [1953].

§9.10.   For complete treatments of Legendre functions, see Hobson [1931], MacRobert [1967], Robin [1957–1959]. Chapter 3 of Erdelyi [1953] contains an excellent collection of formulas.

10
INTEGRAL TRANSFORMS

The prominence of integral transforms in applied analysis has one of its roots in the attempts of certain nineteenth century mathematicians to algebraize the analytical problem of solving differential equations. Let D denote the differentiation operator, $D := d/d\tau$, $D^2 := d/d\tau(d/d\tau) = d^2/d\tau^2$, and so on. The linear differential equation of order n with constant coefficients may then be written

$$(D^n + a_{n-1}D^{n-1} + \cdots + a_1 D + a_0)X = F(\tau), \qquad (10.0\text{-}1)$$

where X is the unknown function. Treating D as if it were a complex number, we would get

$$X = \frac{1}{D^n + a_{n-1}D^{n-1} + \cdots + a_0} F(\tau). \qquad (10.0\text{-}2)$$

But what should be the meaning of the expression on the right?

Among the early proponents of symbolic methods, Oliver Heaviside[1] (1850–1925) was perhaps the most original, and it is highly provocative to study his interpretation of (10.0-2). Let $s_1, \ldots, s_n$ be the zeros of the characteristic polynomial of (10.0-1),

$$p(s) := s^n + a_{n-1}s^{n-1} + \cdots + a_0,$$

and assume that these zeros are all distinct. Then by partial fraction decomposition,

$$\frac{1}{D^n + a_{n-1}D^{n-1} + \cdots + a_0} = \sum_{k=1}^{n} \frac{1}{p'(s_k)(D - s_k)},$$

---

[1] Heaviside is an outstanding example of a genius who went unrecognized by the scientific establishment. See the obituary by E. T. Whittaker [1928].

and to interpret (10.0-2) it thus suffices to give a meaning to

$$\frac{1}{D-s_k}F(\tau).$$

Always treating D as a complex number (of large modulus) we may write

$$\frac{1}{D-s_k}=\frac{1}{D}\frac{1}{1-s_k/D}=\frac{1}{D}+\frac{s_k}{D^2}+\frac{s_k^2}{D^3}+\dots .$$

If D means differentiation, $1/D$ must mean integration, for instance between the limits zero and τ. This suggests the interpretation

$$\frac{1}{D-s_k}F(\tau):=X_k(\tau),$$

where

$$X_k(\tau):=\int_0^\tau F(\tau_1)\,d\tau_1+s_k\int_0^\tau d\tau_1\int_0^{\tau_1}F(\tau_2)\,d\tau_2+\dots . \qquad (10.0\text{-}3)$$

If F is continuous on the real line, the series on the right is easily shown to converge uniformly on every finite interval, and thus to define a function X_k. Could it be that, as suggested by the above formal work, the function

$$X(\tau):=\sum_{k=1}^n\frac{1}{p'(s_k)}X_k(\tau) \qquad (10.0\text{-}4)$$

is a solution of (10.0-1)?

It is not hard to see that this is indeed the case. From (10.0-3) there follows immediately

$$(D-s_k)X_k(\tau)=F(\tau), \qquad k=1,2,\dots,n.$$

In view of

$$p(D)=\prod_{k=1}^n(D-s_k)$$

we thus get

$$p(D)X(\tau)=\sum_{k=1}^n\frac{1}{p'(s_k)}\prod_{\substack{m=1\\m\neq k}}^n(D-s_m)F(\tau).$$

But in view of the Lagrangian interpolation formula

$$\sum_{k=1}^n\frac{1}{p'(s_k)}\prod_{\substack{m=1\\m\neq k}}^n(D-s_m)=1,$$

the foregoing expression reduces to $F(\tau)$, as desired. It can furthermore be verified that the X defined by (10.0-4) is the solution of (10.0-1) satisfying the initial conditions $X(0) = X'(0) = \cdots = X^{(n-1)}(0) = 0$.

It greatly irritated some of Heaviside's mathematical contemporaries that his "symbolic" methods, although devoid of any rational basis, seemed to produce correct results. Although it is not difficult to place the foregoing algorithm for solving linear differential equations with constant coefficients on a rigorous footing, no such simple explanation seemed available for some more spectacular successes of Heaviside's method, such as problems involving fractional powers of the operator D.

The present century has seen several attempts to rationalize Heaviside's technique and to broaden its applicability. (See Erdélyi [1962] for a brief survey.) One such attempt is the operational calculus of J. Mikusinski [1953]. Mikusinski's approach is algebraical: A certain family of functions defined on the positive half-line are shown to form an algebra if multiplication is defined by convolution (see §10.4). Because there are no divisors of zero, the quotient field can be formed. Certain elements of the quotient field represent ordinary functions, whereas others play the role of differentiation or integration operators (see also L. Schwartz [1950], and for expositions of related theories, L. Berg [1967], Krabbe [1970], Moore [1971]).

Another successful rationalization of the symbolic methods, instigated principally by Bromwich, Carson, and Doetsch, interprets Heaviside's calculus in terms of the Laplace transformation. This forms the main content of the present chapter. The transformation of Laplace associates with the elements X of a certain space Ω of functions defined on the real line and zero on the negative half-line the function $x = \mathscr{L}X$ of a complex variable s by virtue of

$$x(s) := \int_0^\infty e^{-s\tau} X(\tau) \, d\tau. \tag{10.0-5}$$

If the derivatives $X^{(m)}$ are in Ω and if $X^{(m)}(0) = 0$, $m = 0, 1, 2, \ldots$, then an integration by parts yields

$$\int_0^\infty e^{-s\tau} X^{(m)}(\tau) \, d\tau = s \int_0^\infty e^{-s\tau} X^{(m-1)}(\tau) \, d\tau,$$

$m = 1, 2, \ldots$. Thus induction yields

$$\mathscr{L}X^{(m)}(s) = s^m x(s), \qquad m = 0, 1, 2, \ldots, \tag{10.0-6}$$

and we see that differentiation in Ω is reflected in the space of image functions $x(s)$ simply by multiplication by s. This fact provides an evident link with the symbolic method, although it must be emphasized that (10.0-6) holds only if the initial values of X and of its derivatives are zero.

Before developing the theory of the Laplace transformation in a rigorous manner, let us briefly show how the method would proceed to solve (10.0-1), say in the special case $F(\tau) = 1$, assuming initial values zero. Assuming the existence of a solution and applying the transformation $\mathscr{L}$ to both sides of (10.0-1) we obtain in view of (10.0-6)

$$p(s)x(s) = \frac{1}{s};$$

hence

$$x(s) = \frac{1}{sp(s)}. \tag{10.0-7}$$

Assuming $\mathscr{L}$ to be one to one (which essentially is the case, see Theorem 10.1h), the whole problem now amounts to identifying the preimage of the function (10.0-7). If the zeros of p are simple and $\neq 0$, we find by partial fractions

$$\frac{1}{sp(s)} = \frac{1}{p(0)s} + \sum_{k=1}^{n} \frac{1}{s_k p'(s_k)(s - s_k)}$$

Because $e^{a\tau}$ $(a \in \mathbb{C})$ has the image $(s - a)^{-1}$, we find

$$X(\tau) = \mathscr{L}^{-1}x(\tau) = \frac{1}{p(0)} + \sum_{k=1}^{n} \frac{1}{s_k p'(s_k)} \exp(s_k \tau).$$

The idea underlying this method of solution may be described thus: The given problem is transformed into a different space where the solution can be found more easily (above, it was found by solving a linear algebraic equation). Then the solution is transformed back from the image space into the original space. The reader will recall that the same principle was used in the solution of two-dimensional boundary value problems by conformal transplantation. Another example on a more elementary level is the multiplication of numbers by taking logarithms.

The algebraization of differential equation problems by means of the Laplace transformation has become an all-pervasive mathematical tool in electric circuit theory and in (linear) control theory. The same method can be useful also for the solution of partial differential problems and of certain integral equations, especially those occurring in actuarial science. In addition, the theory of Laplace transforms has applications to topics in special functions, asymptotics, and even analytic number theory. We also deal with the so-called discrete Laplace transform, which recently has sprung into prominence in connection with digital techniques for transmitting information.

§10.1. DEFINITION AND BASIC PROPERTIES OF THE $\mathscr{L}$ TRANSFORMATION

The *Laplace transformation* (called $\mathscr{L}$ transformation for brevity) is a mapping of a set of functions $F, G, \ldots$ defined on the real line onto a set of functions $f, g, \ldots$ of a complex variable. The domain of definition of the transformation is called the *original space* and is denoted by Ω. The range $\mathscr{L}\Omega$ of the transformation is called the image space. The members of Ω are called *original functions*, and those of $\mathscr{L}\Omega$ *image functions*.

In the present work, the **original space** Ω is taken to consist of all complex valued functions F that satisfy the following conditions:

(i) $F(\tau)$ is defined for all real τ and identically zero for $\tau < 0$;

(ii) there exists an increasing sequence of points $0 = \tau_0 < \tau_1 < \tau_2 < \cdots$ with no finite point of accumulation such that F is continuous on each open interval (τ_{k-1}, τ_k) $(k = 1, 2, \ldots)$, and such that the one-sided limits

$$F(\tau_k -) := \lim_{\substack{\tau \to \tau_k \\ \tau < \tau_k}} F(\tau) \quad \text{and} \quad F(\tau_k +) := \lim_{\substack{\tau \to \tau_k \\ \tau > \tau_k}} F(\tau)$$

exist for $k = 1, 2, \ldots$;

(iii) the limit $F(0+)$ need not exist, but $|F(\tau)|$ is at least improperly integrable at $\tau = 0$.

Remarks. To assume that F is defined on the whole real line and is zero on the negative half-line is a matter of formal convenience. Whenever F is defined by some explicit formula, there is always the tacit understanding that the formula holds only for $\tau \geq 0$ and is to be replaced by 0 for $\tau < 0$. Condition (ii) does not prevent F from having only finitely many, or even no discontinuities. Condition (iii) enables us to consider functions such as $\tau^{-1/2}$ as original functions.

If F and G are in Ω, then so is $aF + bG$ for arbitrary complex numbers a, b [not, however, FG, because this does not necessarily satisfy (iii)]. Moreover, if $F \in \Omega$, then so is G where

$$G(\tau) := \int_0^\tau F(\sigma) \, d\sigma.$$

In addition, G is even continuous—a useful fact in many theoretical investigations.

DEFINITION

If F is in Ω, then the **Laplace transform** *$f = \mathscr{L}F$ of F is the function defined by*

$$f(s) := \int_0^\infty e^{-s\tau} F(\tau) \, d\tau \tag{10.1-1}$$

for all complex s for which the integral converges.

The integral on the right of (10.1-1) is called a **Laplace integral**. It is improper at the upper limit and may [by virtue of (iii)] be improper at the lower limit. It is therefore to be understood as

$$\lim_{\substack{\omega \to \infty \\ \delta \to 0}} \int_{\delta}^{\omega} e^{-s\tau} F(\tau) \, d\tau,$$

where $\omega \to \infty$ and $\delta \to 0$ independently from each other. For functions F in Ω, the integrals from δ to ω may be understood in the Riemann sense.

We now discuss some basic properties of Laplace transforms.

I. Set of Convergence

The set of all s such that the integral (10.1-1) converges is called the **set of (simple) convergence** of the integral and is denoted by $C(F)$. Our first task is the study of the set $C(F)$. In addition, we are interested in the **set of absolute convergence**, to be denoted by $A(F)$, which is the set of all s such that

$$\int_{0}^{\infty} |e^{-s\tau} F(\tau)| d\tau$$

converges. Because absolute convergence implies convergence, $A(F) \subset C(F)$.

Clearly there exist functions F (such as $F(\tau) := e^{\tau^2}$) such that $C(F)$ is empty. A sufficient condition in order that $C(F)$ is not empty is the existence of constants μ and γ such that

$$|F(\tau)| < \mu \, e^{\gamma \tau} \qquad (10.1\text{-}2)$$

for all $\tau > 1$, say. Functions F with this property are said to be of **bounded exponential growth**. The greatest lower bound of all γ such that μ exists such that (10.1-2) holds is called the **growth indicator** of F and is denoted by γ_F. If no such γ exists, we set $\gamma_F := \infty$. It may also happen that $\gamma_F = -\infty$, for instance if $F(\tau) = 0$ for all sufficiently large τ.

If $\gamma_F < \infty$, then for every $\gamma > \gamma_F$ there exists μ such that (10.1-2) holds. Consequently, if $s = \sigma + i\omega$ and $\sigma > \gamma$,

$$|e^{-s\tau} F(\tau)| < \mu \, e^{-(\sigma - \gamma)\tau},$$

and the integral (10.1-1) exists. Thus for functions $F \in \Omega$ such that $\gamma_F < \infty$, the domain of convergence $C(F)$ certainly contains the half-plane $\operatorname{Re} s > \gamma_F$. It may also contain some or all s such that $\operatorname{Re} s = \gamma_F$ (example: $F(\tau) := (1 + \tau^2)^{-1}$, $\gamma_F = 0$). It is also possible to construct examples where $C(F)$ contains s such that $\operatorname{Re} s < \gamma_F$. These examples involve functions F that are zero except on a sequence of short intervals where F is large (see §10.7V).

The following lemma plays a basic role in the determination of the set $C(F)$ and at other places.

LEMMA 10.1a

Let $F \in \Omega$, $0 \leqslant \beta < \pi/2$. If the point s_0 belongs to $C(F)$, then the Laplace integral (10.1-1) converges uniformly on the set $\hat{S}_\beta$ of all s such that $|\arg(s - s_0)| \leqslant \beta$.

Stated more explicitly, the lemma asserts that (i) for every $s \in \hat{S}_\beta$

$$f(s) := \lim_{\omega \to \infty} \int_0^\omega e^{-s\tau} F(\tau)\, d\tau \tag{10.1-3}$$

exists, and (ii) the convergence to the limit is uniform in the sense that, given any $\epsilon > 0$, there exists $\omega_0 = \omega_0(\epsilon)$ such that $\omega > \omega_0$ implies

$$\left| f(s) - \int_0^\omega e^{-s\tau} F(\tau)\, d\tau \right| < \epsilon$$

for all $s \in \hat{S}_\beta$.

Proof. We prove the assertions (i) and (ii) simultaneously by showing the existence of the limit (10.1-3) by means of Cauchy's convergence criterion: For every $\epsilon > 0$ there exists $\omega_0 = \omega_0(\epsilon)$ such that for all ω and χ such that $\omega_0 < \omega \leqslant \chi$, and for all $s \in \hat{S}_\beta$,

$$\left| \int_\omega^\chi e^{-s\tau} F(\tau)\, d\tau \right| < \epsilon.$$

For $\tau \geqslant 0$ let

$$G(\tau) := -\int_\tau^\infty e^{-s_0\sigma} F(\sigma)\, d\sigma.$$

The integral exists because $s_0 \in C(F)$; for the same reason,

$$\lim_{\tau \to \infty} G(\tau) = 0. \tag{10.1-4}$$

Also, G is differentiable at all points of continuity of F, and $G'(\tau) = e^{-s_0\tau} F(\tau)$. Thus if $0 < \omega < \chi$ we get by integration by parts

$$\int_\omega^\chi e^{-s\tau} F(\tau)\, d\tau = \int_\omega^\chi e^{-(s-s_0)\tau} e^{-s_0\tau} F(\tau)\, d\tau$$

$$= [e^{-(s-s_0)\tau} G(\tau)]_\omega^\chi + (s - s_0) \int_\omega^\chi e^{-(s-s_0)\tau} G(\tau)\, d\tau.$$

Let $\epsilon > 0$ be given. By (10.1-4) there exists ω_0 such that $\tau > \omega_0$ implies $|G(\tau)| < \epsilon$. Thus if $\chi \geq \omega > \omega_0$ and $\mathrm{Re}(s - s_0) > 0$, the first term on the right is bounded by 2ϵ. The second term is bounded by

$$\left| (s - s_0)\epsilon \int_\omega^\chi e^{-\mathrm{Re}(s-s_0)\tau} \, d\tau \right| < \frac{|s - s_0|}{\mathrm{Re}(s - s_0)} \epsilon.$$

If $|\arg(s - s_0)| \leq \beta$, where $\beta < \pi/2$, then $\mathrm{Re}(s - s_0) \geq |s - s_0| \cos \beta$. Hence the last bound may be replaced by $\epsilon/\cos\beta$, and it follows that for $\chi \geq \omega > \omega_0$ and $s \in \hat{S}_\beta$,

$$\left| \int_\omega^\chi e^{-s\tau} F(\tau) \, d\tau \right| \leq \left(2 + \frac{1}{\cos\beta} \right)\epsilon.$$

Because the bound on the right does not depend on s and tends to zero with ϵ, the uniformity of the convergence is proved. ∎

Lemma 10.1a implies, in particular, that if $s_0 \in C(F)$, then any s such that $\mathrm{Re}\, s > \mathrm{Re}\, s_0$ also belongs to $C(F)$, for any such s is contained in an angular domain $|\arg(s - s_0)| \leq \beta$ where $\beta < \pi/2$. The infimum α_F of all real numbers α such that $C(F)$ contains an s with $\mathrm{Re}\, s = \alpha$ is called the **abscissa of simple convergence** of the Laplace integral (10.1-1). (If $C(F)$ is empty, we set $\alpha_F := \infty$.) If α_F is finite, then $C(F)$ contains all s such that $\mathrm{Re}\, s > \alpha_F$, and no s such that $\mathrm{Re}\, s < \alpha_F$. No general statement can be made concerning the convergence of the integral for $\mathrm{Re}\, s = \alpha_F$. We summarize:

THEOREM 10.1b

The set of simple convergence of a Laplacian integral, if not empty, is either the full plane or a right half-plane, possibly including some or all of its boundary points.

A similar but simpler result holds for $A(F)$, the set of absolute convergence. Let $s_0 \in A(F)$, and let $\mathrm{Re}\, s_1 \geq \mathrm{Re}\, s_0$. Then for all $\tau > 0$,

$$\left| e^{-s_1\tau} F(\tau) \right| \leq \left| e^{-s_0\tau} F(\tau) \right|,$$

and since

$$\int_0^\infty \left| e^{-s\tau} F(\tau) \right| d\tau < \infty \tag{10.1-5}$$

holds for $s = s_0$ it also holds for $s = s_1$. There follows

THEOREM 10.1c

The set of absolute convergence of a Laplacian integral, if not empty, is either the full plane or an open or closed right half-plane.

The infimum of all numbers Re s such that s satisfies (10.1-5) is called the **abscissa of absolute convergence** of the integral (10.1-1) and is denoted by β_F. We put $\beta_F = \infty$ if (10.1-5) holds for no s.

If γ_F denotes the growth indicator of F, then evidently $\alpha_F \leq \beta_F \leq \gamma_F$. The literature on the $\mathscr{L}$ transformation attaches some weight to the fact that in any of the statements $-\infty \leq \alpha_F$, $\alpha_F \leq \beta_F$, $\beta_F \leq \gamma_F$, $\gamma_F \leq \infty$ either equality or inequality can actually occur. Nevertheless, as we shall see later, the numbers α_F, β_F, γ_F coincide in most cases of practical interest. They always coincide if F is defined by a power series.

II. $\mathscr{L}$ Transforms are Analytic

THEOREM 10.1d

Let $F \in \Omega$, $\alpha_F < \infty$. Then $f := \mathscr{L}F$ is analytic in the interior of $C(F)$, that is, at all points s such that Re $s > \alpha_F$.

This is one of the basic facts about the $\mathscr{L}$ transformation. It makes it possible to apply the powerful tools of complex analysis to the solution of real variable problems.

Proof of Theorem 10.1d.   For Re $s > \alpha_F$,

$$f(s) = \lim_{n \to \infty} f_n(s), \qquad (10.1\text{-}6)$$

where

$$f_n(s) := \int_0^n e^{-s\tau} F(\tau) \, d\tau, \qquad n = 0, 1, 2, \ldots.$$

By Theorem 4.1a, each f_n is analytic in Re $s > \alpha_F$ (in fact, entire). By Theorem 3.4b, the limit function f is analytic if the convergence of the limit (10.1-6) is uniform on every compact subset T of Re $s > \alpha_F$. This, however, follows at once from Lemma 10.1a, for every such compact subset is contained in some angular domain $|\arg(s - \alpha_F)| \leq \beta$, where $\beta < \pi/2$, and the convergence has been shown to be uniform on such angular domains.   ■

COROLLARY 10.1e

Let $F \in \Omega$, $\alpha_F < \infty$, $f := \mathscr{L}F$. Then for $n = 1, 2, \ldots$ the functions $\tau^n F(\tau)$ have the Laplace transforms $(-1)^n f^{(n)}(s)$; their abscissa of convergence is $\leq \alpha_F$.

Proof.   It suffices to establish the Corollary for $n = 1$. We first show that the Laplacian integral

$$h(s) := \int_0^\infty e^{-s\tau} \tau F(\tau) \, d\tau$$

exists for Re $s > \alpha_F$. Let $\delta > 0$ be such that Re $(s - \delta) > \alpha_F$. Then the function G defined by

$$G(\tau) := -\int_\tau^\infty e^{-(s-\delta)\sigma} F(\sigma) \, d\sigma$$

tends to zero for $\tau \to \infty$. Using integration by parts,

$$\int_\omega^X e^{-s\tau} \tau F(\tau) \, d\tau = \int_\omega^X \tau e^{-\delta\tau} e^{-(s-\delta)\tau} F(\tau) \, d\tau$$

$$= [\tau e^{-\delta\tau} G(\tau)]_\omega^X + \int_\omega^X (1 - \delta\tau) e^{-\delta\tau} G(\tau) \, d\tau.$$

Given $\epsilon > 0$, this can be made $< \epsilon$ by choosing ω, $\chi > \tau_0$ because $\int_0^\infty |1 - \delta\tau| e^{-\delta\tau} \, d\tau$ exists. Thus $h(s)$ exists.

Using the notation of the preceding proof, the fact that $f_n \to f$ uniformly on T by the Weierstrass double series theorem implies that $f_n' \to f'$. By Theorem 4.1a,

$$f_n'(s) = -\int_0^n \tau e^{-s\tau} F(\tau) \, d\tau.$$

Because the limit on the left exists as $n \to \infty$ and equals $f'(s)$, the limit on the right exists and has the same value. But the limit on the right has been shown to equal $-h(s)$. ∎

III.   Behavior of $\mathscr{L}F$ on the Boundary of $C(F)$

Let $F \in \Omega$, and let $\alpha_F < \infty$. It is clear that $f := \mathscr{L}F$, as an analytic function, is continuous at all interior points of the set of convergence $C(F)$. This means that

$$\lim_{s \to s_0} f(s) = f(s_0)$$

holds at all interior points of $C(F)$. It is less trivial that a similar result also holds at the boundary points of $C(F)$.

THEOREM 10.1f (Abelian Theorem)

Let $F \in \Omega$, $f := \mathscr{L}F$. If $C(F)$ contains the boundary point s_0, then for every β such that $0 \leq \beta < \pi/2$,

$$\lim_{s \to s_0} f(s) = f(s_0),$$

where the approach of s to s_0 is restricted to the angular domain $\hat{S}_\beta$: $|\arg(s - s_0)| \leqslant \beta$.

Proof. Let $\epsilon > 0$ be given. By Lemma 10.1a, ω can be selected such that

$$\left| \int_\omega^\infty e^{-s\tau} F(\tau) \, d\tau \right| < \frac{\epsilon}{3}$$

for all $s \in \hat{S}_\beta$, including s_0. Then, letting

$$f_\omega(s) := \int_0^\omega e^{-s\tau} F(\tau) \, d\tau,$$

we have

$$|f(s) - f(s_0)| \leqslant |f(s) - f_\omega(s)| + |f_\omega(s) - f_\omega(s_0)| + |f_\omega(s_0) - f(s_0)|$$

$$\leqslant \frac{2\epsilon}{3} + |f_\omega(s) - f_\omega(s_0)|.$$

Because f_ω is an entire analytic function of s (the interval of integration being finite), there exists $\delta > 0$ such that $|s - s_0| < \delta$ implies

$$|f_\omega(s) - f_\omega(s_0)| < \frac{\epsilon}{3},$$

and there follows for $|s - s_0| < \delta$, $s \in \hat{S}_\beta$

$$|f(s) - f(s_0)| < \epsilon,$$

proving the assertion of the theorem. ■

Several applications of Theorem 10.1f are given in §10.2, examples **4** and **5**.

IV. Behavior of $\mathscr{L} F$ at Infinity

The basic fact here is that Laplace transforms tend to zero if s tends to infinity in an appropriate manner: see Theorem 10.7a. Here we state an incomplete form of this result, which suffices for many applications.

THEOREM 10.1g

Let $F \in \Omega$, $f := \mathscr{L} F$, and let s_0 be any complex number. Then for any β such that $0 \leqslant \beta < \pi/2$,

$$\lim_{s \to \infty} f(s) = 0,$$

provided that s tends to infinity in the set $\hat{S}_\beta$: $|\arg(s - s_0)| \leqslant \beta$.

Proof. Without loss of generality one may assume that $s_0 \in C(F)$, because the faraway points of any set $\hat{S}_\beta$ are contained in any other such set, perhaps after enlarging β slightly.

Let $\epsilon > 0$ be given. We write

$$f(s) = \int_0^\infty e^{-s\tau} F(\tau)\, d\tau = \int_0^\omega + \int_\omega^\eta + \int_\eta^\infty.$$

In view of Lemma 10.1a we can choose η such that

$$\left| \int_\eta^\infty e^{-s\tau} F(\tau)\, d\tau \right| < \frac{\epsilon}{3}$$

for all $s \in \hat{S}_\beta$. If $\mathrm{Re}(s - s_0) > 0$, $\omega > 0$ can be chosen such that

$$\left| \int_0^\omega e^{-s\tau} F(\tau)\, d\tau \right| \le e^{|s_0|\omega} \int_0^\omega |F(\tau)|\, d\tau < \frac{\epsilon}{3}.$$

We then have, if $s \in \hat{S}_\beta$,

$$\left| \int_\omega^\eta e^{-s\tau} F(\tau)\, d\tau \right| \le e^{-\omega|s-s_0|\cos\beta} \int_\omega^\eta |e^{-s_0\tau} F(\tau)|\, d\tau,$$

and this becomes $< \epsilon/3$ if $|s - s_0|$ is sufficiently large, $|s - s_0| > \rho$ say. Thus for all $s \in \hat{S}_\beta$ such that $|s - s_0| > \rho$, we find $|f(s)| < \epsilon$, establishing the assertion of the theorem. ∎

V.   Existence of the Inverse Transformation

The program for solving problems described in the introduction to this chapter can be carried out only if the $\mathscr{L}$ transformation is *invertible*, which is to say that the *inverse transformation* exists.

Let T be a transformation from a space X to a space Y, that is, with any $x \in X$ let there be associated an element $y = Tx \in Y$. If $Tx_1 = Tx_2$ implies that $x_1 = x_2$, then T is called one-to-one. If T is one-to-one, then the transformation T^{-1}, called the inverse of T, is defined by associating with any $y \in TX$ the unique $x \in X$ such that $Tx = y$. If the spaces X and Y are linear, and if T is a linear transformation, then T is one-to-one if and only if $Tx = O_y$ implies $x = O_x$. Here O_x and O_y denote the zero elements in the spaces X and Y, respectively.

In the case of the $\mathscr{L}$ transformation, the linear spaces X and Y are Ω and $\mathscr{L}\Omega$, respectively, and the zero elements in these spaces are the functions that vanish identically. Yet the function $F(\tau) \equiv 0$ is not the only function that has the image function $f(s) \equiv 0$, because, by the definition of the Riemann integral, any function $F \in \Omega$ that is different from zero only on a sequence of points $\{\tau_k\}$ such that $\tau_k \to \infty$ also has the image function zero.

The difficulty can be resolved in a simple manner. We call two functions F_1 and $F_2 \in \Omega$ **equivalent** and write $F_1 \sim F_2$ if they have identical values at all $\tau > 0$ except at most on a set of points with no finite point of accumulation. By this equivalence relation the set Ω is divided up into **equivalence classes**. Two functions are in the same equivalence class if and only if they are equivalent to each other. The class of all F equivalent to a given $F_0 \in \Omega$ is denoted by $[F_0]$.

By the foregoing, all functions in one and the same equivalence class have the same $\mathscr{L}$ transform. We now show that, conversely, functions in different equivalence classes have different $\mathscr{L}$ transforms.

THEOREM 10.1h

Let $F_k \in \Omega$, $f_k := \mathscr{L} F_k$ $(k = 1, 2)$. If $f_1(s) = f_2(s)$ for all s in some right half-plane, then F_1 and F_2 belong to the same equivalence class.

Actually a stronger result will be proved; see Corollary 10.1j. The proof is based on the following

LEMMA 10.1i

Let ϕ be a complex valued, continuous function on the interval $[0, 1]$ such that

$$\int_0^1 \phi(x) x^n \, dx = 0, \qquad n = 0, 1, 2, \ldots. \qquad (10.1\text{-}7)$$

Then ϕ vanishes identically.

Proof of the Lemma.   Without loss of generality we may assume ϕ real. For ϕ real the proof is indirect. Suppose ϕ does not vanish identically. Then by the continuity of ϕ,

$$\alpha := \int_0^1 [\phi(x)]^2 \, dx > 0, \beta := \int_0^1 |\phi(x)| dx > 0.$$

Let $\epsilon := \alpha/\beta$. Then $\epsilon > 0$. By the Weierstrass approximation theorem, there exists a polynomial π such that

$$|\phi(x) - \pi(x)| < \frac{\epsilon}{2} \quad \text{for all} \quad x \in [0, 1].$$

Then $\phi(x) = \pi(x) + (\epsilon/2)\theta(x)$ where θ is continuous, $|\theta(x)| \leq 1$. Now

$$\alpha = \int_0^1 [\phi(x)]^2 \, dx = \int_0^1 \phi(x)\pi(x) \, dx + \frac{\epsilon}{2} \int_0^1 \phi(x)\theta(x) \, dx.$$

The first integral on the right vanishes in view of (10.1-7), and the second is bounded by β. We thus get $\alpha \leq (\epsilon/2)\beta = \alpha/2$, a contradiction. ∎

Proof of Theorem 10.1*h.*   Let $F := F_1 - F_2$. Then the hypothesis implies that $f := f_1 - f_2$ vanishes identically in some right half-plane $H \subset C(F)$. Let $s_0 \in H$, then by integration by parts

$$f(s) = (s - s_0) \int_0^\infty e^{-(s-s_0)\tau} G(\tau) \, d\tau$$

for Re $s >$ Re s_0, where

$$G(\tau) := \int_0^\tau e^{-s_0\sigma} F(\sigma) \, d\sigma$$

is continuous. If f vanishes identically, then it certainly vanishes for $s = s_0 + n\lambda$, where $n = 1, 2, \ldots$ and λ is any fixed positive number. Thus

$$\int_0^\infty e^{-\tau n\lambda} G(\tau) \, d\tau = 0, \qquad n = 1, 2, \ldots.$$

This is reduced to the situation of Lemma 10.1i by putting $x := e^{-\tau\lambda}$,

$$\phi(x) := \begin{cases} G\left(-\dfrac{\text{Log } x}{\lambda}\right), & 0 < x \leq 1, \\[2ex] \lim_{\tau \to \infty} G(\tau), & x = 0. \end{cases}$$

We get

$$\int_0^1 x^{n-1} \phi(x) \, dx = 0, \qquad n = 1, 2, \ldots,$$

and because ϕ is continuous, it vanishes identically, by the lemma. Thus $G(\tau) \equiv 0$, and by differentiation

$$e^{-s_0\tau} F(\tau) = 0$$

for all τ, except perhaps on a set with no finite point of accumulation. Thus $F \in [0]$.   ∎

What we have actually proved is

COROLLARY 10.1j (Theorem of Lerch)

The conclusion of Theorem 10.1*h is already true if $f_1(s) = f_2(s)$ merely holds for $s = s_0 + n\lambda$, where $n = 1, 2, \ldots$, and s_0 and $\lambda > 0$ are arbitrary.*

The fact that only the equivalence class $[F]$ to which a function $F \in \Omega$ belongs, but not the individual function F itself can be identified from its transform $f := \mathcal{L}F$ need not bother us. In many applications it is known a priori that F is continuous. Because the class $[F]$ cannot contain two different continuous functions, F then is fully determined by $[F]$. In many applications where F is discontinuous (for instance, in electrical engineering), the definition of F at the points of discontinuity is physically irrelevant.

PROBLEMS

1. Prove that if an image function is a polynomial, then it vanishes identically.
2. Let f be an image function having a real period $\lambda > 0$, $f(s + \lambda) = f(s)$ for all s. Prove that f vanishes identically.
3. Let $F \in \Omega$, and let there exist $\omega > 0$ such that $F(\tau) = 0$ for $\tau > \omega$. Prove that $f := \mathcal{L}F$ is an entire analytic function.
4. Show that the growth indicator γ_F of a function $F \in \Omega$ is given by the formula

$$\gamma_F = \limsup_{\tau \to \infty} \frac{\text{Log} |F(\tau)|}{\tau}. \tag{10.1-8}$$

5. Find the growth indicators γ_F of the functions $F(\tau) =$

    (a) $e^{-\tau^2}$,

    (b) $\cosh\sqrt{\tau}$,

    (c) $\tau^{1/\text{Log}\,\tau}$,

    (d) $(\tau|1 - \tau|)^{-1/2}$.

    Which of these functions (supposed to be 0 for $\tau < 0$) are original functions?

§10.2. OPERATIONAL RULES; BASIC CORRESPONDENCES

To apply the $\mathcal{L}$ transformation to specific problems, it is indispensible to know explicitly the function $f := \mathcal{L}F$ for a large variety of original functions F. If F and f are given by explicit formulas it is often convenient to express the fact that $f = \mathcal{L}F$ by the **Doetsch symbol** ○———● in one of the following ways:

$$F(\tau) \circ\!\!-\!\!-\!\!-\!\!\bullet f(s), \qquad f(s) \bullet\!\!-\!\!-\!\!-\!\!\circ F(\tau).$$

Such an expression is called a **correspondence** for the $\mathcal{L}$ transformation (mnemotechnic device: in ○———● the ○ stands for "original," the ● for "image"). In this section, starting with some basic correspondences, we present eight operational rules (numbered I to VIII) by means of which a large number of further correspondences can be derived. These are illustrated by examples that at the same time serve as illustrations for the theorems of §10.1.

EXAMPLE 1   The exponential function

Let a be a complex number, and let

$$F(\tau) := \begin{cases} e^{a\tau}, & \tau \geq 0, \\ 0, & \tau < 0. \end{cases}$$

The resulting Laplacian integral

$$\int_0^\infty e^{-s\tau} e^{a\tau} d\tau = \int_0^\infty e^{-(s-a)\tau} d\tau$$

converges precisely for $\operatorname{Re} s > \operatorname{Re} a$. Its value in case of convergence is $(s-a)^{-1}$. We thus have obtained the basic correspondence

$$e^{a\tau} \circ\!\!-\!\!\!-\!\!\!-\!\bullet \frac{1}{s-a}, \qquad (10.2\text{-}1)$$

where it is understood that the original function is zero for $\tau < 0$.

For $a = 0$, F is the **Heaviside unit function** H defined by

$$H(\tau) := \begin{cases} 1, & \tau \geq 0, \\ 0, & \tau < 0. \end{cases}$$

Evidently,

$$H(\tau) \circ\!\!-\!\!\!-\!\!\!-\!\bullet \frac{1}{s}. \qquad (10.2\text{-}2)$$

In the following $F, G, \ldots$ always denote original functions, $f = \mathscr{L}F$, $g = \mathscr{L}G, \ldots$ their corresponding image functions. The definition of the operation $\mathscr{L}$ then obviously implies

RULE I: Linearity

For arbitrary complex numbers a and b,

$$aF(\tau) + bG(\tau) \circ\!\!-\!\!\!-\!\!\!-\!\bullet af(s) + bg(s),$$

where $C(aF + bG) \supset (C(F) \cap C(G))$. The inclusion can be proper.

EXAMPLE **2**

If a is any complex number, then for $\operatorname{Re} s > |\operatorname{Im} a|$

$$\cos a\tau \circ\!\!-\!\!\!-\!\!\!-\!\bullet \frac{s}{s^2 + a^2},$$

$$\qquad (10.2\text{-}3)$$

$$\sin a\tau \circ\!\!-\!\!\!-\!\!\!-\!\bullet \frac{a}{s^2 + a^2}.$$

Indeed we have by (10.2-1), for instance,

$$\sin a\tau = \frac{1}{2i}[e^{ia\tau} - e^{-ia\tau}] \circ\!\!-\!\!\!-\!\!\!-\!\bullet \frac{1}{2i}\left[\frac{1}{s-ia} - \frac{1}{s+ia}\right] = \frac{a}{s^2 + a^2}.$$

RULE II: Similarity

For every positive real number ρ,

$$F(\rho\tau) \circ\!\!-\!\!-\!\!-\!\!\bullet \frac{1}{\rho}f\left(\frac{s}{\rho}\right), \qquad s \in \rho C(F).$$

This follows by substituting $\rho\tau =: \sigma$ in the Laplacian integral (10.1-1).

Our next rule expresses the image of F' in terms of $\mathscr{L}F$. Naturally, not every $F \in \Omega$ is differentiable, and even if $F'(\tau)$ exists for all $\tau > 0$, F' need not be in Ω. We also do not wish to exclude the possibility that F' fails to exist at isolated points, which already happens, for instance, if F is continuous and piecewise linear. We therefore make the following assumptions.

Let $F \in \Omega$, and let

$$F(0+) := \lim_{\substack{\tau \to 0 \\ \tau > 0}} F(\tau) \qquad (10.2\text{-}4)$$

exist. Furthermore, let there exist a function $E \in \Omega$ such that for $\tau > 0$

$$F(\tau) - F(0+) = \int_0^\tau E(\sigma)\, d\sigma. \qquad (10.2\text{-}5)$$

This implies that F is continuous for $\tau > 0$, and that

$$F'(\tau) = E(\tau)$$

for all $\tau > 0$, with the possible exception of a set with no point of accumulation. We call any $E \in \Omega$ satisfying (10.2-5) a **generalized derivative** of F. If the ordinary derivative F' exists for all $\tau > 0$, then any generalized derivative belongs to the equivalence class $[F']$. Thus, for simplicity, we denote a generalized derivative of F likewise by F'.

RULE III: Differentiation of an original function.

Let $F \in \Omega$ satisfy (10.2-4), and let it possess a (possibly generalized) *deriva-tive $F' \in \Omega$. If $\sigma_0 \in C(F')$, $\sigma_0 \geqslant 0$, then $\alpha_F \leqslant \sigma_0$ and*

$$F'(\tau) \circ\!\!-\!\!-\!\!-\!\!\bullet sf(s) - F(0+) \qquad (10.2\text{-}6)$$

for all s such that $\operatorname{Re} s > \sigma_0$.

Proof. Let $g(s) \bullet\!\!-\!\!-\!\!-\!\!\circ F'(\tau)$. Then for $\operatorname{Re} s > \sigma_0$

$$g(s) = \lim_{\omega \to \infty} g_\omega(s), \text{ where } g_\omega(s) := \int_0^\omega e^{-s\tau} F'(\tau)\, d\tau.$$

Integrating by parts yields

$$g_\omega(s) = e^{-s\omega}F(\omega) - F(0+) + s \int_0^\omega e^{-s\tau}F(\tau)\,d\tau. \qquad (10.2\text{-}7)$$

Relation (10.2-6) is thus implied by the following proposition: *If* Re $s >$ $\sigma_0 \geq 0$, *then*

$$\lim_{\omega \to \infty} e^{-s\omega}F(\omega) = 0. \qquad (10.2\text{-}8)$$

Because

$$e^{-s\omega}F(\omega) = e^{-s\omega}F(0+) + e^{-s\omega} \int_0^\omega F'(\tau)\,d\tau,$$

it suffices in view of Re $s > \sigma_0 \geq 0$ to show that

$$\lim_{\omega \to \infty} e^{-s\omega} \int_0^\omega F'(\tau)\,d\tau = 0. \qquad (10.2\text{-}9)$$

Letting

$$G(\tau) = \int_0^\tau e^{-\sigma_0\alpha}F'(\alpha)\,d\alpha,$$

integration by parts yields

$$e^{-s\omega} \int_0^\omega F'(\tau)\,d\tau = e^{-s\omega} \int_0^\omega e^{\sigma_0\tau}G'(\tau)\,d\tau$$

$$= e^{-(s-\sigma_0)\omega}G(\omega) - e^{-s\omega}\sigma_0 \int_0^\omega e^{\sigma_0\tau}G(\tau)\,d\tau.$$

Because $\sigma_0 \in C(F')$, $\lim_{\omega \to \infty} G(\omega)$ exists; thus there exists μ such that $|G(\tau)| < \mu$ for all $\tau > 0$. Hence

$$\lim_{\omega \to \infty} e^{-(s-\sigma_0)\omega}G(\omega) = 0,$$

and

$$\left| e^{-s\omega}\sigma_0 \int_0^\omega e^{\sigma_0\tau}G(\tau)\,d\tau \right| \leq \mu\,e^{-\text{Re}(s-\sigma_0)\omega} \to 0.$$

Thus (10.2-9) follows, implying (10.2-8) and rule III. ■

Rule III provides one of the links between the operational calculus of Heaviside and the $\mathscr{L}$ transformation. If $F(0+) = 0$, rule III simply reads

$$F'(\tau) \circ\!\!\!-\!\!\!-\!\!\!-\!\!\bullet sf(s).$$

The operation "differentiation with respect to τ" thus is translated into the operation "multiplication by s". It also becomes now clear why Heaviside's calculus generally breaks down for functions F such that $F(0+) \neq 0$. Even in this more general case, however, rule III transforms differentiation, a transcendental operation requiring a limit process, into a simple algebraic operation.

Repeated application of rule III under suitable conditions yields:

RULE III': Repeated differentiation

Let $F \in \Omega$ be such that the derivatives $F', \ldots, F^{(n-1)}$ exist for $\tau > 0$, let the limits

$$F^{(k)}(0+) := \lim_{\substack{\tau \to 0 \\ \tau > 0}} F^{(k)}(\tau), \qquad k = 0, 1, \ldots, n-1$$

exist, and let F possess a (possibly generalized) nth derivative $F^{(n)} \in \Omega$. Then if $\sigma_0 \in C(F^{(n)})$, $\sigma_0 \geq 0$, then for all s such that $\text{Re } s > \sigma_0$

$$F^{(n)}(\tau) \circ\!\!\!-\!\!\!-\!\!\!-\!\!\bullet s^n f(s) - s^{n-1} F(0+) - s^{n-2} F'(0+)$$

$$- \cdots - F^{(n-1)}(0+).$$

Rule III, read backwards, translates the operation "multiplication by s" from the image space to the original space. We already know from Corollary 10.1e how to translate the operation "multiplication by τ" from the original space to the image space. For completeness, we state the result as:

RULE IV: Multiplication by τ of an original function

Let $F \in \Omega$, $\alpha_F < \infty$. Then if $\text{Re } s > \alpha_F$,

$$\tau^n F(\tau) \circ\!\!\!-\!\!\!-\!\!\!-\!\!\bullet (-1)^n f^{(n)}(s), \qquad n = 1, 2, \ldots. \tag{10.2-10}$$

EXAMPLE **3**

From (10.2-2) there follows

$$\tau^n \circ\!\!\!-\!\!\!-\!\!\!-\!\!\bullet \frac{n!}{s^{n+1}}, \qquad n = 0, 1, 2, \ldots. \tag{10.2-11}$$

More generally (10.2-1) implies

$$\tau^n e^{a\tau} \circ\!\!\!-\!\!\!-\!\!\!-\!\!\bullet \frac{n!}{(s-a)^{n+1}}, \qquad n = 0, 1, 2, \ldots. \tag{10.2-12}$$

RULE V: Integration of an original function

Let $F \in \Omega$, $\sigma_0 \in C(F)$, $\sigma_0 \geqslant 0$. Then if $\mathrm{Re}\ s > \sigma_0$,

$$\int_0^\tau F(\sigma)\, d\sigma \quad \circ\!\!-\!\!\bullet \quad \frac{f(s)}{s}. \tag{10.2-13}$$

This readily follows by applying rule III to the original function

$$G(\tau) := \int_0^\tau F(\sigma)\, d\sigma,$$

whose generalized derivative is F.

From rule IV we can obtain a result concerning integration in the image space. Let F be a function in Ω such that

$$G(\tau) := \frac{1}{\tau} F(\tau)$$

is likewise in Ω, $\alpha_G < \infty$, and let $G(\tau) \circ\!\!-\!\!\bullet g(s)$. Then by (10.2-10),

$$F(\tau) = \tau G(\tau) \quad \circ\!\!-\!\!\bullet \quad -g'(s);$$

hence $g'(s) = -f(s)$. Because f is analytic we have for arbitrary s and t such that $\mathrm{Re}\ s > \alpha_G$, $\mathrm{Re}\ t > \alpha_G$,

$$g(s) - g(t) = \int_s^t f(v)\, dv.$$

Letting $t \to \infty$ along an arbitrary ray making an acute angle with the positive real axis, $g(t) \to 0$ by Theorem 10.1g. Thus the limit of the integral on the right also exists, and we find

$$g(s) = \int_s^\infty f(v)\, dv.$$

We thus have obtained:

RULE VI: Division by τ of an original function

Let F be an original function such that $\tau^{-1}F(\tau)$ is an original function with finite abscissa of convergence α_G. Then

$$\frac{F(\tau)}{\tau} \quad \circ\!\!-\!\!\bullet \quad \int_s^\infty f(v)\, dv, \qquad \mathrm{Re}\ s > \alpha_G, \tag{10.2-14}$$

where the integral on the right may be extended along any ray $\arg(v - s) = \phi_0$ where $|\phi_0| < \pi/2$.

EXAMPLE **4**

Let $\omega > 0$. Then $\tau^{-1} \sin \omega\tau$ is an original function with $\alpha_G = 0$. From

$$\sin \omega\tau \; \circ\!\!-\!\!-\!\!\bullet \; \frac{\omega}{s^2 + \omega^2}$$

there follows

$$\frac{\sin \omega\tau}{\tau} \; \circ\!\!-\!\!-\!\!\bullet \; \int_s^\infty \frac{\omega}{v^2 + \omega^2} \, dv = \frac{\pi}{2} - \operatorname{Arctan} \frac{s}{\omega}.$$

This correspondence permits an interesting application of Theorem 10.1f. By integration by parts it is easily seen that

$$\int_0^\infty \frac{\sin \omega\tau}{\tau} \, d\tau$$

exists. Hence by the theorem quoted,

$$\int_0^\infty \frac{\sin \omega\tau}{\tau} \, d\tau = \lim_{s \to 0+} \int_0^\infty e^{-s\tau} \frac{\sin \omega\tau}{\tau} \, d\tau = \lim_{s \to 0+} \left[\frac{\pi}{2} - \operatorname{Arctan} \frac{s}{\omega} \right],$$

yielding the nonelementary result

$$\int_0^\infty \frac{\sin \omega\tau}{\tau} \, d\tau = \frac{\pi}{2}. \tag{10.2-15}$$

EXAMPLE **5**

Let α, β be real. By (10.2-3),

$$\cos \alpha\tau - \cos \beta\tau \; \circ\!\!-\!\!-\!\!\bullet \; \frac{s}{s^2 + \alpha^2} - \frac{s}{s^2 + \beta^2}.$$

The function

$$F(\tau) := \frac{\cos \alpha\tau - \cos \beta\tau}{\tau}$$

is in Ω (its limit as $\tau \to 0+$ exists). Thus by rule VI, if $\operatorname{Re} s > 0$,

$$\frac{\cos \alpha\tau - \cos \beta\tau}{\tau} \; \circ\!\!-\!\!-\!\!\bullet \; \lim_{t \to \infty} \int_s^t \left(\frac{v}{v^2 + \alpha^2} - \frac{v}{v^2 + \beta^2} \right) dv$$

$$= \frac{1}{2} \lim_{t \to \infty} \left[\operatorname{Log} \frac{t^2 + \alpha^2}{t^2 + \beta^2} - \operatorname{Log} \frac{s^2 + \alpha^2}{s^2 + \beta^2} \right]$$

$$= \frac{1}{2} \operatorname{Log} \frac{s^2 + \alpha^2}{s^2 + \beta^2}.$$

For $\alpha, \beta > 0$, $\int_0^\infty F(\tau) \, d\tau$ exists. Theorem 10.1f thus yields the result (unobtainable by elementary methods)

$$\int_0^\infty \frac{\cos \alpha\tau - \cos \beta\tau}{\tau} \, d\tau = \operatorname{Log} \frac{\beta}{\alpha} \quad (\alpha, \beta > 0).$$

RULE VII: Translation of an original function

Let $\lambda > 0$. Then

$$F(\tau - \lambda) \circ\!\!-\!\!-\!\!-\!\!-\bullet\, e^{-s\lambda} f(s).$$

Proof.   Because $F(\tau - \lambda) = 0$ for $\tau < \lambda$, the substitution $\tau - \lambda =: \sigma$ yields

$$F(\tau - \lambda) \circ\!\!-\!\!-\!\!-\!\!-\bullet \int_{\lambda}^{\infty} e^{-s\tau} F(t - \lambda)\, d\tau$$

$$= e^{-s\lambda} \int_{0}^{\infty} e^{-s\sigma} F(\sigma)\, d\sigma = e^{-s\lambda} f(s). \quad \blacksquare$$

EXAMPLE **6**

Let $I(\tau)$ represent an impulse of strength α and duration δ, beginning at $\tau = 0$ (see Fig. 10.2a). The function $I(\tau)$ may be expressed in terms of the Heaviside unit function $H(\tau)$:

$$I(\tau) = \alpha[H(\tau) - H(\tau - \delta)].$$

In view of (10.2-2) we thus have

$$I(\tau) \circ\!\!-\!\!-\!\!-\!\!-\bullet\, \alpha \frac{1 - e^{-s\delta}}{s}. \qquad\qquad (10.2\text{-}16)$$

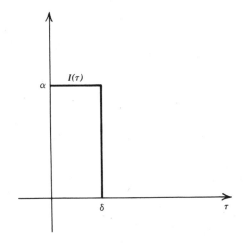

Fig. 10.2a.

Laplace Transform of Periodic Function

For $\tau \geq 0$, let the original function F be periodic with period $\lambda > 0$, that is, let

$$F(\tau) = F(\tau + \lambda) \quad \text{for all } \tau \geq 0.$$

If F_0 equals F in the interval $[0, \lambda)$ and is zero outside this interval, then

$$F(\tau) = F_0(\tau) + F_0(\tau - \lambda) + F_0(\tau - 2\lambda) + \cdots,$$

thus by rule VII the image function of F is

$$f(s) := (1 + e^{-s\lambda} + e^{-2s\lambda} + \cdots)f_0(s),$$

where $f_0 \bullet\!\!-\!\!-\!\!\circ F_0$, that is,

$$f_0(s) := \int_0^\lambda e^{-s\tau} F(\tau)\, d\tau.$$

By summing a geometric series we thus get for Re $s > 0$

$$F(\tau) \circ\!\!-\!\!-\!\!\bullet f(s) = \frac{1}{1 - e^{-s\lambda}} f_0(s). \tag{10.2-17}$$

To compute the Laplace transform of a periodic function, it thus suffices to know the Laplacian integral for the function restricted to one period.

EXAMPLE **7**

Let the impulse of Example **6** be repeated periodically with period $\lambda > \delta$. (10.2-17) yields

$$I(\tau) + I(\tau - \lambda) + I(\tau - 2\lambda) + \cdots \circ\!\!-\!\!-\!\!\bullet \frac{\alpha}{s} \frac{1 - e^{-s\delta}}{1 - e^{-s\lambda}}.$$

In §10.7 it is shown how to obtain the *Fourier series* of a periodic function very easily from its Laplace transform.

RULE VIII. Damping of original function

If s_0 is any complex number, then for all s such that $s - s_0 \in C(F)$

$$e^{s_0\tau} F(\tau) \circ\!\!-\!\!-\!\!\bullet f(s - s_0). \tag{10.2-18}$$

Indeed, if $s - s_0 \in C(F)$,

$$e^{s_0\tau} F(\tau) \circ\!\!-\!\!-\!\!\bullet \int_0^\infty e^{-s\tau} e^{s_0\tau} F(\tau)\, d\tau = f(s - s_0).$$

The word damping used in naming Rule VIII is really appropriate only if Re $s_0 < 0$. The rule permits to write down the transform of an original function F multiplied by $e^{s_0\tau}$ if only the transform of F is known. Thus the correspondences (10.2-3) imply

$$e^{-b\tau} \sin a\tau \circ\!\!-\!\!-\!\!\bullet \frac{a}{(s + b)^2 + a^2}, \qquad e^{-b\tau} \cos a\tau \circ\!\!-\!\!-\!\!\bullet \frac{s + b}{(s + b)^2 + a^2}$$

for arbitrary complex a and b.

Several of our examples illustrate a tendency of the $\mathscr{L}$ transformation to make functions "simpler." The images of transcendental functions such as the exponential function and of the trigonometric functions are rational. The images of functions such as $I(\tau)$ that have no elementary representation are simple transcendental functions. A similar tendency is shown by further, more complicated correspondences that are discussed later.

More important still is another feature of the $\mathscr{L}$ transformation that becomes apparent from the foregoing examples. Although originally defined only on $C(F)$, the set of convergence of the Laplace integral, all transforms f that we have encountered have the property that they can be continued into larger domains of the complex plane. In fact, in all the examples given, f can be continued as a rational or meromorphic function into the whole plane. Although this last statement does not hold for all Laplace transforms, the fact that some analytic continuation is usually possible will prove to be of great value when working in the image space.

PROBLEMS

1. Let F satisfy the hypotheses of Rule III (Differentiation). Show that

$$F(0+) = \lim_{s \to \infty} sf(s). \qquad (10.2\text{-}19)$$

More generally, if F satisfies the hypotheses of Rule III$'$, then for $k \geq 0$

$$F^{(k)}(0+) = \lim_{s \to \infty}[s^{k+1}f(s) - s^k F(0+) - \cdots - sF^{(k-1)}(0+)]. \qquad (10.2\text{-}20)$$

[These formulas are useful for checking correspondences. In addition, they show how the asymptotic behavior of the image function for $s \to \infty$ is linked to the asymptotic behavior of the original function for $\tau \to 0$. See §11.5 for a systematic discussion.]

2. Let $\xi > 0$. Establish the correspondences

$$\tau^{-3/2} \exp\left(-\frac{\xi^2}{4\tau}\right) \circ\!\!-\!\!-\!\!\bullet \frac{2\sqrt{\pi}}{\xi} e^{-\xi\sqrt{s}}, \qquad (10.2\text{-}21a)$$

$$\tau^{-1/2} \exp\left(-\frac{\xi^2}{4\tau}\right) \circ\!\!-\!\!-\!\!\bullet \frac{\sqrt{\pi}}{\sqrt{s}} e^{-\xi\sqrt{s}}. \qquad (10.2\text{-}21b)$$

(principal values of square root).
[Both correspondences are consequences of the identity

$$\int_0^\infty \exp\left[-\left(\tau - \frac{\alpha}{\tau}\right)^2\right] d\tau = \alpha \int_0^\infty \tau^{-2} \exp\left[-\left(\tau - \frac{\alpha}{\tau}\right)^2\right] d\tau = \frac{\sqrt{\pi}}{2},$$

which follows because

$$\int_0^\infty \left(1+\frac{\alpha}{\tau^2}\right) \exp\left[-\left(\tau-\frac{\alpha}{\tau}\right)^2\right] d\tau = \int_0^\infty \exp\left[-\left(\tau-\frac{\alpha}{\tau}\right)^2\right] d\left(\tau-\frac{\alpha}{\tau}\right) = \sqrt{\pi}.]$$

3. Establish the correspondence

$$e^{-\tau^2} \circ\!\!\longrightarrow\!\!\bullet \frac{\sqrt{\pi}}{2} e^{s^2/4} \operatorname{erfc}\left(\frac{s}{2}\right).$$

$$\left[\operatorname{erfc}(z) := \frac{2}{\sqrt{\pi}} \int_0^\infty e^{-t^2} dt.\right]$$

4. Let $F \in \Omega$, $F(\tau)$ bounded for $\tau \geqslant 0$, and let

$$G(\tau) := \int_\tau^\infty \frac{F(\sigma)}{\sigma} d\sigma$$

exist for $\tau > 0$. Show that $G \in \Omega$, and if $F \circ\!\!\longrightarrow\!\!\bullet f$, then

$$G(\tau) \circ\!\!\longrightarrow\!\!\bullet \frac{1}{s} \int_0^s f(t) dt.$$

5. Establish the correspondences

(a) $E(\tau) := \displaystyle\int_\tau^\infty \frac{e^{-\sigma}}{\sigma} d\sigma \circ\!\!\longrightarrow\!\!\bullet \frac{1}{s} \operatorname{Log}(1+s),$

(b) $\operatorname{Si}(\tau) := \displaystyle\int_\tau^\infty \frac{\sin \sigma}{\sigma} d\sigma \circ\!\!\longrightarrow\!\!\bullet \frac{1}{s} \operatorname{Arctg} s,$

(c) $\operatorname{Ci}(\tau) := \displaystyle\int_\tau^\infty \frac{\cos \sigma}{\sigma} d\sigma \circ\!\!\longrightarrow\!\!\bullet \frac{1}{s} \operatorname{Log}\sqrt{1+s^2}.$

6. Not all elementary functions have elementary images. Prove that

$$\frac{1}{1+\tau} \circ\!\!\longrightarrow\!\!\bullet e^s E(s),$$

where $E(s)$ is the *exponential integral* defined in Problem 5.
7. Let k be an integer, $k \geqslant 0$. Show that the function

$$F(\tau) := \begin{cases} \tau^k, & 0 \leqslant \tau \leqslant 1, \\ 0, & \tau > 1, \end{cases}$$

has the Laplace transform

$$f(s) := \frac{k!}{s^{k+1}} \left\{ 1 - e^{-s}\left(1 + \frac{s}{1!} + \frac{s^2}{2!} + \cdots + \frac{s^k}{k!}\right) \right\}.$$

Verify that f is entire (compare Problem 3, §10.1).

8. Prove

$$(-1)^{[\tau]} \circ\!\!-\!\!\bullet \frac{1}{s}\frac{1-e^{-s}}{1+e^{-s}}.$$

9. For $\omega > 0$, show that

$$|\sin \omega\tau| \circ\!\!-\!\!\bullet \frac{\omega}{s^2+\omega^2}\frac{1+e^{-s\pi/\omega}}{1-e^{-s\pi/\omega}}.$$

10. Let $B_0, B_1, B_2, \ldots$ denote the Bernoulli numbers,

$$B_n(\tau) := \tau^n + \binom{n}{1}B_1\tau^{n-1} + \cdots + \binom{n}{n}B_n$$

the nth Bernoulli polynomial, $B_n^*(\tau)$ the periodic continuation of the restriction of $B_n(\tau)$ to the interval $[0, 1]$. Show that for $n \geq 1$,

$$B_n^*(\tau) \circ\!\!-\!\!\bullet \frac{n!}{s^{n+1}}\left\{B_0 + \frac{B_1}{1!}s + \cdots + \frac{B_n}{n!}s^n - \frac{s}{e^s-1}\right\}.$$

11. Let α, β be real. Show that for $\text{Re } s > \max(-\alpha, -\beta)$

$$\frac{e^{-\alpha\tau}-e^{-\beta\tau}}{\tau} \circ\!\!-\!\!\bullet \text{Log} \frac{s+\beta}{s+\alpha}$$

and for $\alpha, \beta > 0$ deduce the value of the integral

$$\int_0^\infty \frac{e^{-\alpha\tau}-e^{-\beta\tau}}{\tau} \, d\tau.$$

12. Let $\alpha, \beta > 0$. Show that

$$\frac{\cos \alpha\tau - \cos \beta\tau}{\tau^2} \circ\!\!-\!\!\bullet \frac{s}{2}\text{Log}\frac{s^2+\alpha^2}{s^2+\beta^2} + \beta \, \text{Arctg}\frac{\beta}{s} - \alpha \, \text{Arctg}\frac{\alpha}{s}$$

and deduce the value of the integral

$$\int_0^\infty \frac{(\sin \alpha\tau)^2}{\tau^2} \, d\tau.$$

§10.3. ORDINARY DIFFERENTIAL EQUATIONS. SYSTEMS

The stage is now set for the presentation of an area of engineering science in which the Laplace transformation is applied with consistent success. This concerns the solution of initial value problems for ordinary linear differential equations with constant coefficients. The contribution of the $\mathscr{L}$ transformation is directed less toward problems of existence and uniqueness—these questions are easily settled in a more general context—than toward the construction, in an algorithmic fashion, of explicit representations of solutions whose existence and uniqueness are known. In this latter respect the $\mathscr{L}$ method can be successful even in cases in which classical methods fail.

I. Scalar Equations

Let $n > 0$, let $a_0, a_1, \ldots, a_{n-1}$ and $b_0, b_1, \ldots, b_{n-1}$ be complex constants, and let F be an original function. Roughly speaking, we wish to determine—for $\tau > 0$ only—the solution of the linear scalar nth order equation

$$X^{(n)} + a_{n-1} X^{(n-1)} + \cdots + a_0 X = F(\tau) \qquad (10.3\text{-}1a)$$

satisfying the initial conditions

$$X^{(k)}(0) = b_k, \qquad k = 0, 1, \ldots, n-1. \qquad (10.3\text{-}1b)$$

Because F need not be continuous, it must be explained what we mean by a solution of the differential equation. The function X is called a **solution** of (10.3-1a), if X and its derivatives $X', \ldots, X^{(n-1)}$ are continuous for $\tau > 0$, and if $X^{(n-1)}$ possesses a generalized derivative (see p. 211) which, if denoted by $X^{(n)}$, satisfies (10.3-1a) at all points of continuity of F. The solution is said to satisfy the initial conditions (10.3-1b) if

$$X^{(k)}(0+) = b_k, \qquad k = 0, 1, \ldots, n-1.$$

These definitions having been made, it follows from the standard existence theorem for linear scalar nth-order equations that for every $F \in \Omega$ the initial value problem (10.3-1) possesses a unique solution X.

Let us now assume that $\gamma_F < \infty$, that is, that F is of bounded exponential growth. The classical variations-of-constants formula (proved in §9.3 for F analytic, but valid for arbitrary $F \in \Omega$) then shows that the solution X likewise is of bounded exponential growth. Thus its Laplace transform possesses a nonempty half-plane of convergence.

Let us apply the operation $\mathscr{L}$ to both sides of (10.3-1a), considered as an identity in τ. Letting

$$x := \mathscr{L}X, \qquad f := \mathscr{L}L$$

we first have in view of (10.3-1b), using the operational rule III′ (Differentiation of original function)

$$X^{(k)}(\tau) \circ\!\!-\!\!\bullet\, s^k x(s) - s^{k-1} b_0 - s^{k-2} b_1 - \cdots - b_{k-1}, \qquad k = 1, 2, \ldots, n.$$

In view of Rule I (Linearity) (10.3-1a) thus yields

$$(s^n x(s) - s^{n-1} b_0 - s^{n-2} b_1 - \cdots - b_{n-1})$$
$$+ a_{n-1}(s^{n-1} x(s) - s^{n-2} b_0 - \cdots - b_{n-2})$$
$$+ \cdots + a_1(sx(s) - b_0) + a_0 x(s) = f(s).$$

Introducing the characteristic polynomial of the homogeneous equation (10.3-1a),

$$p(s) := s^n + a_{n-1} s^{n-1} + \cdots + a_0, \qquad (10.3\text{-}2)$$

and the polynomials

$$p_k(s) := s^k + a_{n-1}s^{k-1} + \cdots + a_{n-k} \tag{10.3-3}$$

$(k = 0, 1, \ldots, n-1)$ the last result may be written

$$p(s)x(s) = f(s) + b_0 p_{n-1}(s) + b_1 p_{n-2}(s) + \cdots + b_{n-1} p_0(s).$$

This is an algebraic equation for $x(s)$ whose solution is immediate:

$$x(s) = \frac{f(s) + b_0 p_{n-1}(s) + \cdots + b_{n-1} p_0(s)}{p(s)} \tag{10.3-4}$$

The task of determining the solution of the initial value problem (10.3-1) is thus reduced to the problem of identifying the function X whose Laplace transform is (10.3-4). In other words, the problem now is one of *inverting* the Laplace transformation. We know from Theorem 10.1h that this inversion is, in principle, always possible, that is, that a continuous original function is completely determined by its image function. Schematically, the method of the Laplace transformation in the solution of a differential equation problem may be illustrated by the following diagram:

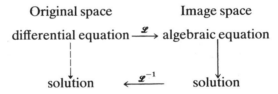

By means of the $\mathscr{L}$ transformation, the problem is translated into the image space where the solution is easy. In the original space the solution is then found by inverting the $\mathscr{L}$ transformation. In principle, the same scheme can be applied to the solution of functional equations other than ordinary linear differential equations with constant coefficients. Examples are given in subsequent paragraphs. In the present context, it is an advantage of the method that it permits the determination of a special solution of the differential equation, satisfying specified initial conditions, without making it necessary first to construct the "complete," or general, solution of the differential equation.

It is not claimed that the last step of the method (that of inverting the transformation) is always easy. In fact, the general theory of the inverse Laplace transformation requires a form of the Fourier integral theorem and is not presented until §10.7. Another method of inversion, which is applicable to image functions that are analytic at $s = \infty$, is given in §10.5. In still other cases the inversion may be accomplished by inspecting a table of explicit correspondences.

In addition to these possibilities there exists a large class of original functions F for which the problem of finding original functions belonging to (10.3-4) turns out to be elementary.

Let us call an original function F an **exponential polynomial** if for $\tau > 0$ it can be represented in the form

$$F(\tau) = \sum_{k=1}^{m} P_k(\tau) e^{c_k \tau},$$

where m is a positive integer, the c_k are complex numbers, and the P_k are polynomials with complex coefficients. It follows from the basic correspondence (10.2-12) that *if F is an exponential polynomial, then $f := \mathscr{L}F$ is a rational function vanishing at infinity. The converse is also true.* If f is a rational function vanishing at ∞, then by partial fraction decomposition it can be written as a finite sum of functions of the form

$$\sum_{j=1}^{k} \frac{a_j}{(s-c)^j},$$

where c and the a_j are complex numbers. By the correspondence just quoted, each such function is the image of an original function of the form $P(\tau) e^{c\tau}$, where P is a polynomial. There follows $f = \mathscr{L}F$, where F is an exponential polynomial. The proof given also provides a method to construct F.

Thus if the nonhomogeneous term F in (10.3-1a) is an exponential polynomial, then $f = \mathscr{L}F$ is a rational function vanishing at infinity, and by (10.3-4) x is a rational function of the same type. Thus $X = \mathscr{L}^{-1}x$ is an exponential polynomial, which can be constructed by decomposing $x(s)$ into partial fractions. The fact that X is an exponential polynomial could have been anticipated from the elementary theory of differential equations, but the $\mathscr{L}$ method provides an algorithm for the direct and explicit construction of the desired special solution.

Fortunately, in many technical applications the nonhomogeneous terms F are either constants or trigonometric functions, both of which are special cases of exponential polynomials. In actual applications it is often preferable to follow the *derivation* of equation (10.3-4) rather than to use the formula itself.

EXAMPLE **1**

To determine the solution of

$$X'' - X' - 2X = 20 \cos 2\tau$$

satisfying the initial conditions $X(0) = 1$, $X'(0) = 0$. If $X(\tau) \circ\!\!-\!\!-\!\!\bullet x(s)$, then

$$x'(\tau) \circ\!\!-\!\!-\!\!\bullet sx(s) - X(0) = sx(s) - 1,$$
$$X''(\tau) \circ\!\!-\!\!-\!\!\bullet s^2 x(s) - sX(0) - X'(0) = s^2 x(s) - s.$$

Thus applying $\mathscr{L}$ yields

$$(s^2 - s - 2)x(s) - s + 1 = \frac{20s}{s^2 + 4},$$

$$x(s) = \frac{20s + (s^2 + 4)(s - 1)}{(s^2 + 4)(s^2 - s - 2)}.$$

Expansion in partial fractions yields

$$x(s) = \frac{(-3 + i)/2}{s - 2i} + \frac{(-3 - i)/2}{s + 2i} + \frac{2}{s - 2} + \frac{2}{s + 1};$$

whence there follows

$$X(\tau) = \frac{-3 + i}{2} e^{2i} + \frac{-3 - i}{2} e^{-2i\tau} + 2 e^{2\tau} + 2 e^{-\tau}$$

$$= -3 \cos 2\tau - \sin 2\tau + 2 e^{2\tau} + 2 e^{-\tau}.$$

EXAMPLE 2

Let $\alpha > 0$. To solve

$$X'' + \alpha^2 X = \cos \alpha\tau$$

under the initial conditions $X(0) = X'(0) = 0$. In the elementary theory this problem presents an exceptional situation because the nonhomogeneous term is a solution of the homogeneous equation. To solve the problem by the $\mathscr{L}$ method, let $X(\tau) \circ\!\!-\!\!-\!\!\bullet x(s)$. Applying $\mathscr{L}$ to the equation then yields in view of the vanishing initial values

$$(s^2 + \alpha^2)x(s) = \frac{s}{s^2 + \alpha^2}, \qquad x(s) = \frac{s}{(s^2 + \alpha^2)^2}.$$

The inverse transformation could be achieved by decomposition in partial fractions, but it is somewhat simpler to note that

$$x(s) = -\frac{1}{2} \frac{d}{ds} \frac{1}{s^2 + \alpha^2}.$$

By virtue of

$$\frac{1}{s^2 + \alpha^2} \bullet\!\!-\!\!-\!\!\circ \frac{1}{\alpha} \sin \alpha\tau$$

Rule IV (differentiation in image space) implies

$$-\frac{d}{ds}\frac{1}{s^2+\alpha^2} \quad\bullet\!\!-\!\!-\!\!\circ\quad \frac{\tau}{\alpha}\sin\alpha\tau;$$

hence

$$X(\tau)=\frac{\tau}{2\alpha}\sin\alpha\tau.$$

II. Systems of Differential Equations

A similar calculus can be applied to systems of nonhomogeneous linear differential equations with constant coefficients. Let $\mathbf{A}$ be a square matrix of order n with complex elements, let $\mathbf{F}(\tau)$ be a vector whose components are functions in Ω with bounded exponential growth, and let $\mathbf{b}$ be a given vector. We consider the problem of finding the vector solution $\mathbf{X}$ of

$$\mathbf{X}'=\mathbf{AX}+\mathbf{F}(\tau) \tag{10.3-5a}$$

satisfying

$$\mathbf{X}(0+)=\mathbf{b}. \tag{10.3-5b}$$

It is shown in the theory of ordinary differential equations (see §9.3 for $\mathbf{F}$ analytic) that the solution can be represented in the form

$$\mathbf{X}(\tau)=e^{\tau\mathbf{A}}\mathbf{b}+\int_0^\tau e^{(\tau-\sigma)\mathbf{A}}\mathbf{F}(\sigma)\,d\sigma. \tag{10.3-6}$$

The solution $\mathbf{X}$ can also be identified by means of the $\mathscr{L}$ transformation. We let

$$\mathbf{F}(\tau)\;\circ\!\!-\!\!-\!\!\bullet\;\mathbf{f}(s),\qquad \mathbf{X}(\tau)\;\circ\!\!-\!\!-\!\!\bullet\;\mathbf{x}(s),$$

where the correspondences are to be understood componentwise. Applying $\mathscr{L}$ to (10.3-5a) then yields, considering (10.3-5b),

$$s\mathbf{x}(s)-\mathbf{b}=\mathbf{Ax}(s)+\mathbf{f}(s)$$

or

$$(s\mathbf{I}-\mathbf{A})\mathbf{x}(s)=\mathbf{f}(s)+\mathbf{b}.$$

The matrix $\mathbf{A}-s\mathbf{I}$ is nonsingular for all s with Re s sufficiently large, and for such s we have

$$\mathbf{x}(s)=(s\mathbf{I}-\mathbf{A})^{-1}(\mathbf{f}(s)+\mathbf{b}). \tag{10.3-7}$$

Again the problem has been reduced to a problem of computing inverse Laplace transforms.

The elements of the matrix $(s\mathbf{I} - \mathbf{A})^{-1}$ are rational functions of s that vanish at infinity. If the components of $\mathbf{F}$ are exponential polynomials, then the components of $\mathbf{f}$ are of the same kind, hence also the components of $\mathbf{x}(s)$. Thus the transformation $\mathscr{L}^{-1}$ can be effected by partial fraction decomposition. If the linear system of equations for the components of $\mathbf{x}$ is solved by elimination, only those components whose originals are wanted need to be computed.

EXAMPLE **3**

Two flywheels of moments of inertia θ_1 and θ_2 are connected by a shaft of negligeable moment of inertia. The shaft and the two wheels are rotating with constant angular velocity ω. At time $\tau = 0$ a constant breaking moment μ is applied to the first wheel. It is required to find the subsequent angular velocity of the second wheel.

We let Φ_1 and Φ_2 denote the angular displacements of the two wheels and may assume that $\Phi_1(0) = \Phi_2(0) = 0$. The assumption of angular velocity ω at $\tau = 0$ requires $\Phi_1'(0) = \Phi_2'(0) = \omega$. If the stiffness of the shaft is denoted by λ, the equations of motion are for $\tau > 0$

$$\theta_1 \Phi_1'' - \lambda(\Phi_2 - \Phi_1) = -\mu,$$

$$\theta_2 \Phi_2'' + \lambda(\Phi_2 - \Phi_1) = 0.$$

The function Φ_2' is desired. If $\phi_1 := \mathscr{L}\Phi_1$, $\phi_2 := \mathscr{L}\Phi_2$, the corresponding equations in the image space are

$$\theta_1 s^2 \phi_1 - \lambda(\phi_2 - \phi_1) = \theta_1 \omega - \frac{1}{s}\mu,$$

$$\theta_2 s^2 \phi_2 + \lambda(\phi_2 - \phi_1) = \theta_2 \omega.$$

Eliminating ϕ_1 yields

$$\phi_2(s) = \frac{\omega}{s^2} - \frac{\lambda\mu}{s^3[\theta_1 \theta_2 s^2 + \lambda(\theta_1 + \theta_2)]}.$$

Hence

$$\Phi_2'(\tau) \circ\!\!-\!\!\bullet \; s\phi_2(s) - \Phi_2(0) = s\phi_2(s)$$

or, if $\alpha^2 := \lambda(\theta_1 + \theta_2)/\theta_1\theta_2$,

$$\Phi_2'(\tau) \circ\!\!-\!\!\bullet \; \frac{\omega}{s} - \frac{\lambda\mu}{\theta_1 \theta_2 s^2 (s^2 + \alpha^2)}$$

$$= \frac{\omega}{s} - \frac{\lambda\mu}{\theta_1 \theta_2 \alpha^2}\left\{\frac{1}{s^2} - \frac{1}{s^2 + \alpha^2}\right\}.$$

Consulting a table of correspondences, there follows

$$\Phi_2'(\tau) = \omega - \frac{\mu}{\theta_1 + \theta_2}\tau + \frac{\mu}{\alpha(\theta_1 + \theta_2)}\sin \alpha\tau.$$

III.   Systems Engineering

In engineering science the word **system** has come to describe any device that effectively realizes a transformation T from one function space Ξ_1 to another function space Ξ_2. If $X_1 \in \Xi_1$ and $X_2 := TX_1$, then X_1 is called the **input** or **driving function** of the system, and X_2 is the **output** or **response** caused by X_1 (see Fig. 10.3a).

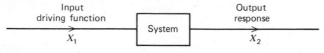

Fig. 10.3a.

Frequently in applications, input and output are functions of time, and the spaces Ξ_1 and Ξ_2 may both be taken as Ω, the space of original functions. It is then often possible to describe the system by a linear differential or integro-differential equation with constant coefficients. In this setting the $\mathscr{L}$ transformation proves to be a tool whose usefulness goes beyond that of a mere computational device. By shifting the analysis from the original space to the image space, the theory of systems attains a transparency which would not be available otherwise.

For the sake of illustration, consider the electric circuit depicted in Fig. 10.3b.

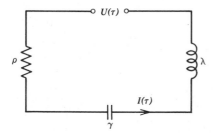

Fig. 10.3b.

The meaning of the symbols is as follows:

ρ Ohmian resistance
γ capacity of condenser
λ inductivity of coil
$U(\tau)$ imposed voltage[2]
$I(\tau)$ current[2]

---

[2] Use of capital letters for voltage and current does not imply (as it does in engineering literature) that we are dealing with direct current.

We conceive the circuit as a system with input U and output I. The working of the system is described by Kirchhoff's law stating that the sum of all voltage drops equals the imposed voltage or electromotive force. The various voltage drops are as follows:

$$\text{Resistance: } \rho I \text{ (Ohm's law)}$$

$$\text{Condenser: } \frac{\text{charge}}{\text{capacity}} = \frac{1}{\gamma} \int_0^\tau I(\sigma)\, d\sigma$$

$$\text{Coil: } \lambda \frac{dI}{d\tau} \text{ (Faraday's law)}$$

Kirchhoff's law thus yields

$$\rho I + \frac{1}{\gamma} \int_0^\tau I(\sigma)\, d\sigma + \lambda \frac{dI}{d\tau} = U(\tau). \tag{10.3-8}$$

We assume the system to be at rest initially: $I(0) = 0$. Letting $U(\tau) \circ\!\!-\!\!-\!\!-\!\bullet\, u(s)$, $I(\tau) \circ\!\!-\!\!-\!\!-\!\bullet\, i(s)$, we then have

$$\frac{dI}{d\tau} \circ\!\!-\!\!-\!\!-\!\bullet\, si(s), \quad \int_0^\tau I(\sigma)\, d\sigma \circ\!\!-\!\!-\!\!-\!\bullet\, \frac{1}{s} i(s).$$

Applying $\mathscr{L}$ to (10.3-8), there follows

$$\rho i + \frac{1}{s\gamma} i + s\lambda i = u \tag{10.3-9}$$

or

$$\left(\rho + \frac{1}{s\gamma} + s\lambda \right) i(s) = u(s).$$

Thus

$$i(s) = g(s)u(s), \tag{10.3-10}$$

where

$$g(s) := \frac{1}{\rho + \dfrac{1}{s\gamma} + s\lambda}.$$

The function g is called the **transfer function** of the system.

Thus in the image space the working of our system is described in a very simple manner, as follows: *The image function of the output is the product of the image function of the input and of the transfer function of the system* (see Fig. 10.3c).

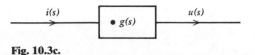

i(s)   • g(s)   u(s)

Fig. 10.3c.

We call a **linear system** any engineering system that maps Ω into Ω in such a manner that the images of input and output are connected by a homogeneous relation of the form (10.3-10). In addition to electric circuits such as the one considered above, typical examples of linear systems are mechanical vibrating systems subject to outside excitation.

Simple rules hold if several linear systems are connected. Consider, for instance, two linear systems with transfer functions g_1 and g_2. Let F_0 be the input of the first sytem, and let its output F_1 be used as input for the second system, which then in turn produces the output F_2 (see Fig. 10.3d).

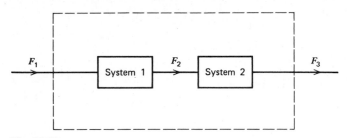

F_1 System 1 F_2 System 2 F_3

Fig. 10.3d.

If the respective image functions are denoted by f_0, f_1, f_2, we have

$$f_1(s) = g_1(s)f_0(s), \qquad f_2(s) = g_2(s)f_1(s);$$

hence

$$f_2(s) = g_1(s)g_2(s)f_0(s).$$

We conclude: *If several systems are connected serially, the transfer function of the combined system is the product of the transfer functions of the individual systems* (see Fig. 10.3e).

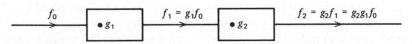

f_0 • g_1 $f_1 = g_1 f_0$ • g_2 $f_2 = g_2 f_1 = g_2 g_1 f_0$

Fig. 10.3e.

An important application of the foregoing arises in the case of **control systems**. Here the output of a system is used to drive a *control unit*. The output of the control unit is then subtracted algebraically from the input of the system and thereby modifies this input, as shown in Fig. 10.3*f*. If g_1 and

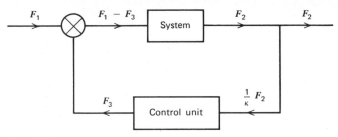

Fig. 10.3f.

g_2, respectively, denote the transfer functions of the system and of the control unit, then the following relations hold in the image space:

$$f_2 = g_1(f_1 - f_3),$$

$$f_3 = \frac{1}{\kappa} g_2 f_2$$

(κ = gain factor). There follows

$$f_2 = g_1 f_1 - \frac{1}{\kappa} g_1 g_2 f_2;$$

hence

$$f_2 = \frac{g_1}{1 + (1/\kappa) g_1 g_2} f_1.$$

The transfer function of the system including the control unit thus is

$$g := \frac{g_1}{1 + (1/\kappa) g_1 g_2}. \tag{10.3-11}$$

Stability. Generally, a system is called **stable** if a bounded input always produces a bounded output. It is easy to find a necessary and sufficient condition for the stability of a system whose transfer function g is rational, $g(\infty) = 0$. For simplicity, we restrict ourselves to bounded input functions of the form $F_1(\tau) := e^{i\omega\tau}$; in principle, any bounded input function may be synthesized from input functions of this particular form in view of Fourier's integral formula (see §10.6). Because $e^{i\omega\tau} \circ\!\!-\!\!\bullet (s - i\omega)^{-1}$, the image function of the output F_2 is

$$f_2(s) := \frac{1}{s - i\omega} g(s).$$

Let the poles of g be $s_1, \ldots, s_m$, and let the order of s_k be j_k. Then if $s_k \neq i\omega$ for all k, f_2 has a partial fraction decomposition

$$f_2(s) = \frac{g(i\omega)}{s - i\omega} + \sum_{k=1}^m r_k(s),$$

where r_k is a principal part of degree j_k. We have

$$r_k(s) \bullet\!\!\!-\!\!\!-\!\!\!\circ P_k(\tau) e^{s_k \tau},$$

where P_k is a polynomial of degree $j_k - 1$. Thus the response of the system is

$$F_2(\tau) = g(i\omega) e^{i\omega\tau} + \sum_{k=1}^m P_k(\tau) e^{s_k \tau}. \tag{10.3-12}$$

Clearly, this is bounded if and only if all Re $s_k \leq 0$, and if the poles such that Re $s_k = 0$ have order one. However, if g has a purely imaginary pole of order one, say $s_k = i\omega_0$, then if $\omega = \omega_0$ the partial fraction expansion of f_2 contains a quadratic polynomial in $(s - i\omega_0)^{-1}$, which again leads to an unbounded output. Thus we find that *a necessary and sufficient condition for g to be the transfer function of a stable system is that all poles of g have negative real parts.*

Assume now that the system described by g is stable. Then in view of Re $s_k < 0$, $k = 1, \ldots, m$, all terms in the sum (10.3-12) except the first tend to zero as $\tau \to \infty$, and after a sufficiently long time (which in practice may be a matter of microseconds) all that remains is the **steady-state response**

$$F_2(\tau) = g(i\omega) e^{i\omega\tau}.$$

This simply is the input, multiplied by a constant factor that depends on ω. The function $\omega \to g(i\omega)$ is called the **frequency admittance** of the system. It is customary to write $g(i\omega) = a(\omega) e^{i\psi(\omega)}$. The functions $a(\omega)$ and $\psi(\omega)$ indicate the changes in amplitude and phase of the incoming signal caused by the system. For graphical representation one plots $20 \log_{10} a(\omega)$ and $\psi(\omega)$ against $\log_{10}\omega$. This yields what is called the **Bode plot** of the system. Alternatively, one may plot the curve $(20 \log_{10} a(\omega), \psi(\omega))$ represented parametrically by means of ω (**Nichols plot**).

For most engineering applications, only stable systems are useful. The stability of a given system with a rational transfer function can be tested *algebraically* by means of the Routh–Hurwitz algorithm or one of its adaptions (see §6.7 and §12.7), or *graphically* by means of the **Nyquist diagram** (§4.10), which may be looked at as a direct application of the principle of the argument.

Shock Response. Let the transfer function g of a system be the image function of an original function G, $g(s)$ ●———○ $G(\tau)$. What is the physical meaning of G? If there existed an input function F_0 such that $F_0(\tau)$ ○———● 1, then G would be the response to that input, because

$$f_1(s) = g(s) \cdot 1 = g(s).$$

No F_0 with the image function $f_0(s) = 1$ exists, because its existence would contradict Theorem 10.1g. However, we can easily construct a sequence of original functions whose image functions *approach* the function $f_0(s) = 1$. For $\epsilon > 0$, let

$$F_\epsilon(\tau) := \begin{cases} \dfrac{1}{\epsilon}, & 0 \leqslant \tau < \epsilon, \\ \\ 0, & \text{otherwise.} \end{cases}$$

F_ϵ is a lone rectangular impulse of height ϵ^{-1} and duration ϵ. The integrated strength of the impulse is

$$\int_{-\infty}^{\infty} F_\epsilon(\tau)\, d\tau = 1. \tag{10.3-13}$$

By (10.2-16),

$$F_\epsilon(\tau) \; ○———● \; f_\epsilon(s) := \frac{1 - e^{-s\epsilon}}{\epsilon s},$$

and it is obvious that

$$\lim_{\epsilon \to 0} f_\epsilon(s) = 1,$$

uniformly on every bounded set of the s-plane. Now if the inverse Laplace transformation were continuous in the sense that $f_\epsilon \to f$ would imply $F_\epsilon \to F$ where $f = \mathscr{L}F$ then G would be the response to F. However, the limit

$$\lim_{\epsilon \to 0} F_\epsilon(\tau) = \begin{cases} \infty, & \tau = 0 \\ 0, & \tau \neq 0 \end{cases}$$

is not in Ω; it does not define a function in any conventional sense. It nevertheless helps our physical intuition to consider $G(\tau)$ as response to a "pseudofunction" δ such that

$$\delta(0) = \infty; \qquad \delta(\tau) = 0, \qquad \tau \neq 0$$

and enjoying the additional property that

$$\int_{-\infty}^{\infty} \delta(\tau)\, d\tau = 1. \tag{10.3-14}$$

The pseudofunction δ thus introduced is called the **Dirac δ function**. The function G, which is the response to the shock $\delta(\tau)$, may be called **shock response**. The term **weighing function** is also used.

Many physical situations involving short impulses are reasonably described by the δ function. This description has the advantage that it is independent of the actual duration of the impulse (which is usually accidental) and that it involves only the integrated strength of the impulse. If $\mathscr{L}$ transforms are taken, it is reasonable to postulate the correspondence

$$\delta(\tau) \circ\!\!-\!\!-\!\!\bullet 1. \qquad (10.3\text{-}15)$$

EXAMPLE **4**

A force $F \in \Omega$ acts on a mathematical pendulum of mass μ and length λ that is at rest for $\tau < 0$ (see Fig. 10.3g). To determine the motion of the pendulum for $\tau > 0$. If the

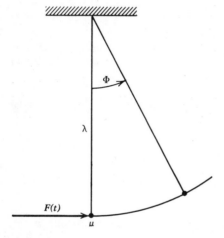

F(t)

Fig. 10.3g.

gravitational constant is denoted by γ, the differential equation of the motion is

$$\mu\lambda \Phi'' + \mu\gamma \sin \Phi = F(\tau),$$

which for small $|\Phi|$ is approximated by

$$\mu\lambda \Phi'' + \mu\gamma\Phi = F(\tau). \qquad (10.3\text{-}16)$$

Letting $\Phi \circ\!\!-\!\!-\!\!\bullet \phi$, $F \circ\!\!-\!\!-\!\!\bullet f$ and taking transforms, we get

$$\mu\lambda s^2\phi + \mu\gamma\phi = f(s),$$

$$\phi(s) = \frac{1}{\mu\lambda (s^2 + \gamma/\lambda)} f(s). \qquad (10.3\text{-}17)$$

Thus

$$g(s) := \frac{1}{\mu\lambda} \frac{1}{s^2 + \gamma/\lambda}$$

is the transfer function of the system. Thus the shock response of the pendulum, that is, the response to a force of short duration at $\tau = 0$ and integrated strength 1, is

$$G(\tau) := \frac{1}{\mu\sqrt{\gamma\lambda}} \sin\sqrt{\frac{\gamma}{\lambda}}\tau. \tag{10.3-18}$$

This result can be confirmed as follows: Let the impulse have duration ϵ. Because force equals mass times acceleration,

$$1 = \int_0^\epsilon F(\tau)\, d\tau = \mu\lambda\, \Phi'(\epsilon).$$

Thus as $\epsilon \to 0$,

$$\Phi'(0+) = \frac{1}{\mu\lambda}. \tag{10.3-19}$$

This indeed is the initial condition satisfied by (10.3-18). If the homogeneous pendulum equation is solved by means of Laplace transforms under the initial conditions $\Phi(0) = 0$ and (10.3-19), there results

$$\mu\lambda\left(s^2\phi(s) - \frac{1}{\mu\lambda}\right) + \mu\gamma\phi(s) = 0,$$

which yields

$$\phi(s) = \frac{1}{\mu\lambda s^2 + \mu\gamma},$$

as above.

We have introduced the δ function only as a means to promote the intuitive understanding of impulse situations; our presentation is devoid of any logical basis. It is not difficult to put the $\mathscr{L}$ theory of the δ function on sound logical principles. This can be done either by considering Laplace transforms on the space of tempered distributions, or by Mikusinski's operational calculus.

PROBLEMS

1. Assuming vanishing initial values, solve

    (a) $X''' + 3X'' + 3X' + X = 1,$

    (b) $X''' + X = 1,$

    (c) $X^{(4)} + 2X'' + X = \sin\tau.$

Problems 2–4 deal with electric circuits. The translation of a circuit problem into image space is facilitated by the observation that, because of the linearity of the transformation, Kirchhoff's laws remain valid in image space. Moreover in image space, an inductance of inductivity λ and a condenser of capacity γ may formally be treated like Ohmain resistances of strength $s\lambda$ and $1/s\gamma$, respectively.

2. Referring to Fig. 10.3h, can the inductive effect of a coil (inductivity λ, resistance ρ_1) be offset by shunting it with a condenser (capacity γ) and a resistance ρ_2 in series, if the initial charge is assumed to be zero?

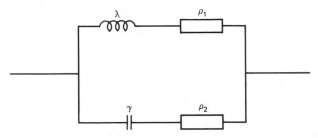

Fig. 10.3h.

[The symbolic resistances (or "impedances") of the two branches are

$$z_1 := s\lambda + \rho_1, \qquad z_2 := \frac{1}{s\gamma} + \rho_2,$$

The reciprocal of an impedance is called an admittance. If the branches are switched in parallel, admittances are added, thus the combined impedance z satisfies

$$\frac{1}{z} = \frac{1}{z_1} + \frac{1}{z_2}, \quad \text{i.e., } z = \frac{z_1 z_2}{z_1 + z_2}.$$

We want this to be real and independent of s.]

3. Let $\lambda > 4\rho^2\gamma$. Determine the shock response of the circuit shown in Fig. 10.3i.

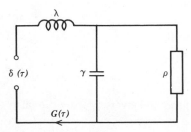

Fig. 10.3i.

4. A weakly damped ($\rho^2 < 4\lambda/\gamma$) electric circuit, after being at rest, is excited at time $\tau = 0$ by a very short voltage shock (see Fig. 10.3j). Determine the first time instant $\tau_0 > 0$ at which the current in the circuit passes through zero.

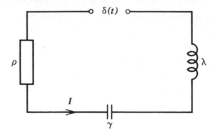

Fig. 10.3j.

5. The system

$$X'' - X + Y + Z = 0,$$
$$X + Y'' - Y + Z = 0,$$
$$X + Y + Z'' - Z = 0$$

is to be solved under the initial conditions $X(0) = 1$, $Y(0) = Z(0) = X'(0) = Y'(0) = Z'(0) = 0$.

6. Two identical pendula (mass μ, length λ) are coupled by means of a spring of strength κ. Assuming the angular displacements Φ_1, Φ_2 to be small, their motion is governed by the differential equations

$$\mu\lambda \Phi_1'' = -\mu\gamma\Phi_1 - \kappa\lambda (\Phi_1 - \Phi_2),$$
$$\mu\lambda \Phi_2'' = -\mu\gamma\Phi_2 + \kappa\lambda (\Phi_1 - \Phi_2)$$

($\gamma :=$ gravitational constant). Determine the motion of the pendula when the second pendulum is at rest initially and the first is released from the amplitude α with initial velocity zero. (For compact notation, let $\omega^2 := \gamma/\lambda$, $\beta^2 := \kappa/\lambda$.)

7. A number of $n - 1$ beads of equal mass are attached to a taut elastic string of length 1 at equidistant points. The beads are initially at rest. At time $\tau = 0$ the first bead is given an impulse perpendicular to the string. Determine the subsequent motion of all beads.

[The differential equations of motion are

$$X_1'' + \lambda (2X_1 - X_2) = 0,$$
$$X_k'' + \lambda (2X_k - X_{k-1} - X_{k+1}) = 0,$$
$$k = 2, 3, \ldots, n-2,$$
$$X_{n-1}'' + \lambda (2X_{n-1} - X_{n-2}) = 0,$$

where λ is proportional to the tension of the string and inversely proportional to the mass of a bead.]

8. Let F be a continuous bounded function on the real line. It is well known that the solution of the initial value problem for the heat equation in an unbounded one-dimensional medium,

$$\frac{\partial^2 U}{\partial x^2} = \frac{\partial U}{\partial \tau} \qquad (\tau > 0, -\infty < x < \infty) \qquad (10.3\text{-}20\text{a})$$

$$\lim_{\tau \to 0+} U(\tau, x) = F(x) \qquad (-\infty < x < \infty) \qquad (10.3\text{-}20\text{b})$$

is given by

$$U(\tau, x) = \int_{-\infty}^{\infty} \frac{1}{\sqrt{4\pi\tau}} \exp\left(-\frac{(x-\xi)^2}{4\tau}\right) F(\xi)\, d\xi. \qquad (10.3\text{-}21)$$

Derive this formula by applying the $\mathscr{L}$ transformation with respect to τ, assuming interchangeability of improper integrations wherever necessary. [If $U(\tau, x) \circ\!\!-\!\!\bullet\, u(s, x)$, applying $\mathscr{L}$ to (10.3-20a) yields an ordinary differential equation for $x \to u(s, x)$,

$$u'' - su = -F(x). \qquad (10.3\text{-}22)$$

The functions $x \to e^{\pm\sqrt{s}\,x}$ are linearly independent solutions of the homogeneous equation. By a version of the variation of constants formula, the solution of (10.3-22) that stays bounded for $x \to \pm\infty$ is

$$u(s, x) := \frac{1}{2\sqrt{s}} \int_{-\infty}^{x} e^{-(x-\xi)\sqrt{s}} F(\xi)\, d\xi + \frac{1}{2\sqrt{s}} \int_{x}^{\infty} e^{-(\xi-x)\sqrt{s}} F(\xi)\, d\xi$$

$$= \frac{1}{2\sqrt{s}} \int_{-\infty}^{\infty} e^{-|x-\xi|\sqrt{s}} F(\xi)\, d\xi.$$

To transform back we require the correspondence

$$\frac{1}{\sqrt{\pi\tau}} \exp\left(-\frac{\xi^2}{4\tau}\right) \circ\!\!-\!\!\bullet\, \frac{1}{\sqrt{s}} e^{-|\xi|\sqrt{s}}.]$$

§10.4. CONVOLUTION

Here we consider the following problem: Suppose F and G are in Ω, and $f := \mathscr{L}F$, $g := \mathscr{L}G$. How is the original function of the product fg (if it exists) related to F and G?

It is clear that, in general, FG does not correspond to fg. Let, for instance, H denote the Heaviside unit function. Then $H(\tau) \circ\!\!-\!\!\bullet\, s^{-1}$, and $[H(\tau)]^2 = H(\tau)$ does not correspond to $s^{-1} \cdot s^{-1} = s^{-2}$.

To discover the correct original function empirically, let $F(\tau)$ be the input of a linear system, and let $G(\tau)$ be the shock response of the system. Then

$f(s)g(s)$ is the image of the output, and it is our task to express the output $X(\tau)$ in terms of F and G.

To this end, we first approximate the input F by a step function (see Fig. 10.4a). Let $\eta > 0$, $\tau_n := n\eta$ $(n = 0, 1, 2, \ldots)$, and define

$$F_\eta(\tau) := F(\tau_n), \qquad \tau_n \le \tau < \tau_{n+1}, \qquad n = 0, 1, 2, \ldots.$$

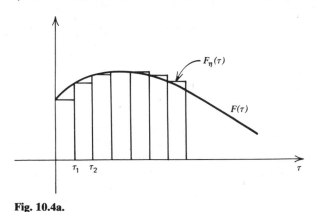

Fig. 10.4a.

If the input consisted only of a single impulse of height $F(\tau_0)$ and duration η at $\tau = 0$, then by the end of §10.3 the response of the system would be approximated by

$$\eta F(\tau_0) G(\tau),$$

because the integrated strength of the impulse is $\eta F(\tau_0)$. More generally, the response of a single impulse of height $F(\tau_k)$ between τ_k and τ_{k+1} is approximately

$$\eta F(\tau_k) G(\tau - \tau_k).$$

In view of the linearity of the system, the response at time τ to the entire step function input is the sum of the foregoing contributions from $k = 0$ to $k = n^*$ where n^* is the largest n such that $\tau_n < \tau$:

$$\eta \sum_{k=0}^{n^*} F(\tau_k) G(\tau - \tau_k).$$

If $\eta \to 0$, then $F_\eta \to F$, and the output caused by the step function can be expected to tend to the output caused by F. At the same time, the given sum tends to the Riemann integral of the function $F(\sigma) G(\tau - \sigma)$ between the limits 0 and τ. Thus we are led to conjecture that the original function corresponding to $f(s)g(s)$ is

$$X(\tau) := \int_0^\tau F(\sigma) G(\tau - \sigma) \, d\sigma. \qquad (10.4\text{-}1)$$

This is made precise presently.

The function X defined by (10.4-1) for all real τ (which implies that $X(\tau) = 0$ for $\tau \leq 0$) is called the **convolution** of F and G. The customary notation for the convolution is

$$X = F * G.$$

Clearly, the convolution of two functions is commutative:

$$F * G = G * F. \tag{10.4-2}$$

We next establish some general properties of the convolution of functions in Ω.

THEOREM 10.4a

Let $F, G \in \Omega$. *Then* $X := F * G$, *in addition to being in* Ω, *is continuous at all* $\tau > 0$.

Proof. We have $X = Y + Z$, where, for $\tau > 0$,

$$Y(\tau) := \int_0^{\tau/2} F(\sigma) G(\tau - \sigma) \, d\sigma, \qquad Z(\tau) := \int_0^{\tau/2} G(\sigma) F(\tau - \sigma) \, d\sigma.$$

Because the hypotheses on F and G are the same, it suffices to prove the assertion of the Theorem for Y.

We first show that Y exists. Let $\tau > 0$, and let $|G(\sigma)| \leq \mu$ for $\frac{1}{2}\tau \leq \sigma \leq \tau$. Then

$$|Y(\tau)| \leq \mu \int_0^{\tau/2} |F(\sigma)| \, d\sigma.$$

The last integral exists because, by the definition of Ω, F is absolutely integrable at $\tau = 0$.

We next show that Y is continuous at a fixed $\tau_0 > 0$. Let $\epsilon > 0$ be given. We have

$$Y(\tau_0 + \delta) - Y(\tau_0) = \int_0^{\tau_0/2} F(\sigma)[G(\tau_0 - \sigma + \delta) - G(\tau_0 - \sigma)] \, d\sigma$$

$$+ \int_{\tau_0/2}^{\tau_0/2 + \delta/2} F(\sigma) G(\tau_0 - \sigma + \delta) \, d\sigma.$$

Let $|\delta| < \frac{1}{4}\tau_0$, and let $|G(\sigma)| \leq \mu$ for $\frac{1}{4}\tau_0 \leq \sigma \leq \frac{3}{4}\tau_0$. Then the second integral is bounded by

$$\mu \int_{\tau_0/2}^{\tau_0/2 + \delta/2} |F(\sigma)| \, |d\sigma|,$$

which is $< \frac{1}{2}\epsilon$ for all sufficiently small δ. The first integral can be written as a sum of finitely many, say n, integrals over intervals $I_1, \ldots, I_n$ such that

$G(\tau_0 - \sigma)$ is continuous in the interior of each I_k and has finite one-sided limits at the endpoints. Let $[\tau_1, \tau_2]$ be a typical I_k. We write

$$\int_{\tau_1}^{\tau_2} = \int_{\tau_1}^{\tau_1 + \delta_1} + \int_{\tau_1 + \delta_1}^{\tau_2 - \delta_2} + \int_{\tau_2 - \delta_2}^{\tau_2}$$

and can make the first and the third integral less than $\epsilon/8n$ (in spite of the possible jump of G) by making δ_1 sufficiently small and letting $\delta < \delta_1$; the middle integral can then be made $< \epsilon/4n$ by the uniform continuity of G in (τ_1, τ_2).

It can be shown by means of examples that the limit $X(0+)$ need not exist. To prove that $X \in \Omega$ it thus remains to be shown that Y is absolutely integrable at $\tau = 0$. Let $\rho > 0$ be so small that both F and G are continuous in $(0, \rho)$, and let $0 < \delta < \frac{1}{2}\rho$. We then have

$$\int_{\delta}^{\rho} |Y(\tau)| \, d\tau \leq \int_{\delta}^{\rho} \left\{ \int_{0}^{\tau/2} |F(\sigma) G(\tau - \sigma)| \, d\sigma \right\} d\tau.$$

The domain of integration in the (τ, σ) plane consists of a rectangle R and a triangle S as shown in Fig. 10.4b. Changing the order of integrations, we have

$$\int_{R} = \int_{0}^{\delta/2} |F(\sigma)| \left\{ \int_{\delta}^{\rho} |G(\tau - \sigma)| \, d\tau \right\} d\sigma,$$

$$\int_{S} = \int_{\delta/2}^{\rho/2} |F(\sigma)| \left\{ \int_{2\sigma}^{\rho} |G(\tau - \sigma)| \, d\tau \right\} d\sigma,$$

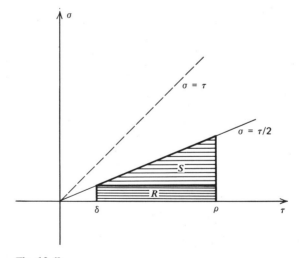

Fig. 10.4b.

and the assertion follows from the fact that the limits of these integrals as $\delta \to 0$ obviously exist. ∎

We now are able to state the so-called **convolution theorem**:

THEOREM 10.4b

Suppose F, $G \in \Omega$, and let $X := F * G$. If the Laplace transforms $f := \mathscr{L}F$ and $g := \mathscr{L}G$ converge absolutely for $s = s_0$, then so does $x := \mathscr{L}X$, and

$$x(s) = f(s)g(s) \tag{10.4-3}$$

holds for all s such that $\mathrm{Re}\, s \geq \mathrm{Re}\, s_0$.

Proof. If $\mathrm{Re}\, s \geq \mathrm{Re}\, s_0$,

$$f(s)g(s) = \lim_{\omega \to \infty} \int_0^\omega e^{-s\rho} F(\rho)\, d\rho \int_0^\omega e^{-s\sigma} G(\sigma)\, d\sigma$$

$$= \lim_{\omega \to \infty} \iint_{R_\omega} e^{-s(\rho+\sigma)} F(\rho)G(\sigma)\, d\rho\, d\sigma,$$

where R_ω denotes the rectangle $0 < \rho, \sigma < \omega$. Because the double integral converges absolutely, we may integrate over the triangle $S_\omega : \rho > 0$, $\sigma > 0$, $\rho + \sigma < \omega$ instead and let $\omega \to \infty$. Using the variables $\tau := \rho + \sigma$ and ρ in place of ρ and σ, we have

$$\iint_{S_\omega} e^{-(\rho+\sigma)s} F(\rho)G(\sigma)\, d\rho\, d\sigma$$

$$= \int_0^\omega e^{-s\tau} \left\{ \int_0^\tau F(\rho)G(\tau-\rho)\, d\rho \right\} d\tau$$

$$= \int_0^\omega e^{-s\tau} X(\tau)\, d\tau,$$

and the limit as $\omega \to \infty$ (which is already known to exist) by definition of the Laplace transform equals $x(s)$. ∎

A modification of this theorem, which is somewhat more difficult to prove, states that if $f(s)$ converges absolutely and $g(s)$ converges simply for $s = s_0$, then $x(s)$ converges at least simply for $s = s_0$, and (10.4-3) holds for $s = s_0$ and for $\mathrm{Re}\, s > \mathrm{Re}\, s_0$.

The convolution theorem must be regarded as yet another operational rule for the $\mathscr{L}$ transformation, which should be added to the eight rules given in §10.2. The dual rule, which concerns the image of the ordinary product FG in the image space, may be established as a by-product of results on the Fourier integral theorem, see §10.6.

It has already been mentioned that the convolution of two functions in Ω obeys the commutative law. It is trivial that the distributive law holds, and it is easy to see that the associative law is likewise valid. That is, if $F, G, H \in \Omega$, then

$$F * (G * H) = (F * G) * H.$$

This can be proved elementarily by an appropriate change of variables. If the transforms $f \bullet\!\!-\!\!-\!\!\circ F$, $g \bullet\!\!-\!\!-\!\!\circ G$, $h \bullet\!\!-\!\!-\!\!\circ H$ have a finite abscissa of absolute convergence, then the following short proof based on the foregoing theorems is available: Both functions $F * (G * H)$ and $(F * G) * H$ have the Laplace transform fgh. By Theorem 10.4a they are both continuous for $\tau > 0$, thus by Theorem 10.1h they are identical.

We turn to applications of the convolution theorem.

The first application expresses the theorem in the language of system theory. A system with transfer function g and input $F \circ\!\!-\!\!-\!\!\bullet f$ produces an output that in the image space is represented by $f(s)g(s)$. If g is an $\mathscr{L}$ transform, then the corresponding original function is the *shock response* of the system. Theorem 10.4b thus yields:

THEOREM 10.4c

The output of a linear system is the convolution of the input with the shock response.

EXAMPLE **1**

The pendulum considered in Example **4** of §10.3 has the shock response

$$G(\tau) := \frac{1}{\mu\sqrt{\gamma\lambda}} \sin \sqrt{\frac{\gamma}{\lambda}} \tau.$$

Consequently, the motion of the pendulum caused by an arbitrary forcing function $F \in \Omega$ is given by

$$\Phi(\tau) = \frac{1}{\mu\sqrt{\gamma\lambda}} \int_0^\tau F(\sigma) \sin \sqrt{\frac{\gamma}{\lambda}} (\tau - \sigma) \, d\sigma.$$

This result agrees with what one obtains classically by the method of variation of constants.

The following illustrations of the convolution theorem are more special.

EXAMPLE **2**

Which original function corresponds to $f(s) := s^{-1/2}$ (principal value)? Because

$$s^{-n} \;\bullet\!\!-\!\!-\!\!\circ\; \frac{1}{(n-1)!}\tau^{n-1}$$

for positive integers n, it seems reasonable to try $F(\tau) := c\tau^{-1/2}$, where the constant c is yet to be determined. Under this assumption the following chain of correspondences holds:

$$1\;\circ\!\!-\!\!-\!\!\bullet\; \frac{1}{s} = s^{-1/2} \cdot s^{-1/2} \;\bullet\!\!-\!\!-\!\!\circ\; F * F(\tau) = c^2 \int_0^\tau \frac{1}{\sqrt{\sigma(\tau-\sigma)}}\,d\sigma = c^2\pi.$$

Because $c < 0$ would yield a negative Laplace transform there follows $c = \pi^{-1/2}$; thus

$$\frac{1}{\sqrt{\tau}}\;\circ\!\!-\!\!-\!\!\bullet\;\frac{\sqrt{\pi}}{\sqrt{s}} \quad \text{(principal value)}, \tag{10.4-4}$$

an important correspondence that is generalized in §10.5. By Rule V (Integration) we have for $n = 1, 2, \dots$

$$\tau^{n-1/2}\;\circ\!\!-\!\!-\!\!\bullet\;\sqrt{\pi}(\tfrac{1}{2})_n s^{-1/2-n}. \tag{10.4-5}$$

EXAMPLE **3**

Which original function corresponds to

$$f(s) := \frac{1}{\sqrt{s^2+1}}?$$

(The square root for which $f(s) \sim s^{-1}$ for $|s|$ large is to be taken.) We have

$$f(s) = (s-i)^{-1/2}(s+i)^{-1/2},$$

where the roots have their principal values. By (10.4-4) and Rule VIII (Damping),

$$(s-i)^{-1/2}\;\bullet\!\!-\!\!-\!\!\circ\;\frac{1}{\sqrt{\pi}}e^{i\tau}\tau^{-1/2},$$

$$(s+i)^{-1/2}\;\bullet\!\!-\!\!-\!\!\circ\;\frac{1}{\sqrt{\pi}}e^{-i\tau}\tau^{-1/2}.$$

Thus by the convolution theorem the original function corresponding to f is

$$F(\tau) := \frac{1}{\pi}\int_0^\tau \frac{e^{-i\sigma}}{\sqrt{\sigma}}\frac{e^{i(\tau-\sigma)}}{\sqrt{\tau-\sigma}}\,d\sigma.$$

Letting $\sigma := [(1+\xi)/2]\tau$ we obtain

$$F(\tau) = \frac{1}{\pi}\int_{-1}^1 \frac{e^{-i\xi\tau}}{\sqrt{1-\xi^2}}\,d\xi.$$

The integral on the right is recognized as Poisson's (or Parseval's) integral for the Bessel function J_0 (problem 3, §4.5). We thus have found the important correspondence

$$J_0(\tau) \circ\!\!\!-\!\!\!-\!\!\!\bullet \frac{1}{\sqrt{s^2+1}}. \tag{10.4-6}$$

The images of Bessel functions of arbitrary order, and of various combinations thereof, are computed in §10.5.

EXAMPLE 4

Many curious integrals can be evaluated by means of the convolution theorem. From (10.4-6) and

$$\left(\frac{1}{\sqrt{s^2+1}}\right)^2 = \frac{1}{s^2+1} \bullet\!\!\!-\!\!\!-\!\!\!\circ \sin \tau$$

it follows that the convolution of J_0 with itself equals $\sin \tau$,

$$\int_0^\tau J_0(\sigma)J_0(\tau-\sigma)\, d\sigma = \sin \tau. \tag{10.4-7}$$

EXAMPLE 5   Abel's integral equation

Let F be continuous for $\tau \geq 0$ and differentiable for $\tau > 0$, $F(0) = 0$, and let $F' \in \Omega$. We wish to solve **Abel's integral equation** for an unknown function G,

$$\int_0^\tau \frac{G(\sigma)}{\sqrt{\tau-\sigma}}\, d\sigma = F(\tau), \qquad \tau \geq 0. \tag{10.4-8}$$

Assuming the existence of a solution $G \in \Omega$, let $F \circ\!\!\!-\!\!\!-\!\!\!\bullet f$, $G \circ\!\!\!-\!\!\!-\!\!\!\bullet g$, and apply $\mathcal{L}$ to (10.4-8). The expression on the left may be regarded as the convolution of $G(\tau)$ with $\tau^{-1/2}$. Thus the image of (10.4-8) is

$$g(s)\frac{\sqrt{\pi}}{\sqrt{s}} = f(s),$$

which is readily solved for g:

$$g(s) = \frac{1}{\sqrt{\pi}} \sqrt{s}\, f(s) = \frac{1}{\pi} \frac{\sqrt{\pi}}{\sqrt{s}}\, sf(s).$$

Because $sf(s) \bullet\!\!\!-\!\!\!-\!\!\!\circ F'(\tau)+F(0) = F'(\tau)$, transforming back yields

$$G(\tau) = \frac{1}{\pi}\int_0^\tau \frac{F'(\sigma)}{\sqrt{\tau-\sigma}}\, d\sigma. \tag{10.4-9}$$

This formula was derived under the hypothesis that a solution exists in Ω. It thus remains to be verified that the function G defined by (10.4-8) actually is a solution. The verification is simple: Being the convolution of two functions in Ω, G is in Ω. We thus may take Laplace transforms. Retracing the foregoing steps, we find that G actually solves (10.4-7).

PROBLEMS

1. For nonnegative integers m and n, let

    $$I(m, n) := \int_0^1 \tau^m (1-\tau)^n \, d\tau.$$

    The integral $I(m, n)$ may be regarded as a convolution. Use this fact to evaluate $I(m, n)$.

2. Let $F \in \Omega$. A form of Taylor's theorem implies that the n-fold indefinite integral of F,

    $$F^{(-n)}(\tau) := \int_0^\tau \int_0^{\tau_1} \cdots \int_0^{\tau_{n-1}} F(\tau_n) \, d\tau_n \cdots d\tau_1$$

    ($n = 1, 2, \ldots$) may be written as

    $$F^{(-n)}(\tau) = \frac{1}{(n-1)!} \int_0^\tau (\tau - \sigma)^{n-1} F(\sigma) \, d\sigma.$$

    Prove this fact by means of the convolution theorem.

3. Let α, β be positive, $\alpha \neq \beta$. Suppose the integrals

    $$\int_0^\infty J_0(\alpha \tau) \cos \beta \tau \, d\tau, \qquad \int_0^\infty J_0(\alpha \tau) \sin \beta \tau \, d\tau$$

    converge. What is their value?

4. Let $G \in \Omega$, $\alpha_G < \infty$. Show that the integral equation for an unknown function X,

    $$G(\tau) = X(\tau) + \lambda \int_0^\tau e^{\lambda(\tau - \sigma)} X(\sigma) \, d\sigma$$

    has the solution

    $$X(\tau) := G(\tau) - \lambda \int_0^\tau G(\sigma) \, d\sigma.$$

5. The equation

    $$X(\tau) = F(\tau) + \int_0^\tau X(\sigma) G(\tau - \sigma) \, d\sigma,$$

    where F and G are given and X is sought, is called a **renewal equation**. Assuming F, $G \in \Omega$ and α_F, $\alpha_G < \infty$, express the Laplace transform of X in terms of the transforms of F and G.

6. With the notation of the preceding problem, show that the solution obtained by the Laplace transformation is identical with that obtained by successive approximations as

$$\lim_{n \to \infty} X_n(\tau),$$

where $X_0(\tau) := F(\tau)$,

$$X_{n+1}(\tau) := F(\tau) + \int_0^\tau X_n(\sigma) G(\tau - \sigma) \, d\sigma,$$

$n = 0, 1, 2, \ldots$.

7. Suppose all bulbs from a shipment of light bulbs have a common continuous life-length distribution, with the probability of a light bulb burning out between the times τ and $\tau + d\tau$ being $F(\tau) \, d\tau$ ($\int_0^\infty F(\tau) \, d\tau = 1$). What is the expected number $X(\tau)$ of bulbs required to keep a lamp in service for a time interval of length τ, if a bulb is replaced by a new one as soon as it burns out? Consider the distribution

$$F(\tau) \, d\tau := \frac{1}{n!} \tau^n e^{-\tau} \, d\tau,$$

and relate $X(\tau)$ to the life expectancy of a bulb,

$$\mu := \int_0^\infty \tau F(\tau) \, d\tau.$$

[$X(\tau)$ satisfies the renewal equation

$$X(\tau) = 1 + \int_0^\tau X(\sigma) F(\tau - \sigma) \, d\sigma.]$$

8. Let $F \in \Omega$, $\alpha_F < \infty$. Consider the problem of the conduction of heat in a semi-infinite bar, when the temperature is $F(\tau)$ at the end of the bar and the whole bar is initially cold. The mathematical model is

$$\frac{\partial^2 U}{\partial x^2} = \frac{\partial U}{\partial \tau}, \qquad \tau > 0, x > 0;$$

$$U(0, x) = 0, \qquad x > 0;$$

$$U(\tau, 0) = F(\tau), \qquad \tau > 0;$$

$$U(x, \tau) \to 0 \quad \text{as } x \to \infty \text{ for all } \tau > 0.$$

Show as in Problem 8, §10.3, that for each $x > 0$, the solution $U(\tau, x)$ is the convolution of F with

$$G(\tau) := \frac{x}{2\sqrt{\pi}} \tau^{-3/2} e^{-x^2/4\tau} .$$

§10.5. SOME NONELEMENTARY CORRESPONDENCES

In this section we establish some results that permit us to greatly expand our repertory of correspondences. Use is made of the theory of the Γ function (cf. §§8.4–8.6) and of the hypergeometric function (cf. §§9.9–9.10).

We begin by computing the image of the unrestricted **fractional power function**. Let ν be a complex number, Re $\nu > -1$. The function

$$F(\tau) := \tau^\nu := e^{\nu \operatorname{Log} \tau}, \qquad \tau > 0$$

is an original function with growth indicator zero. Its $\mathscr{L}$ transform $f(s)$ thus exists at least for Re $s > 0$. Let Re $s > 0$. We have

$$f(s) = \lim_{\omega \to \infty} f_\omega(s),$$

where

$$f_\omega(s) := \int_0^\omega e^{-s\tau} \tau^\nu \, d\tau.$$

To evaluate the limit, we substitute $u := s\tau$. There results

$$f_\omega(s) = s^{-\nu-1} \int_\Lambda e^{-u} u^\nu \, du,$$

where the powers have their principal values and where Λ, the path of integration, now is the straight-line segment from $u = 0$ to $u = \rho e^{i\phi}$ where $\rho := |s|\omega$, $\phi := \arg s$ $(-\pi/2 < \phi < \pi/2)$. Because the integrand is an analytic function of u in the right half plane, we may by Cauchy's theorem integrate from $u = 0$ to $u = \rho$ and then along the circular arc of radius ρ to $u = \rho e^{i\phi}$. The integral along the circular arc is bounded by

$$\frac{\pi}{2} \rho e^{-\rho \cos \phi} \rho^{\operatorname{Re} \nu} e^{\pi/2 |\operatorname{Im} \nu|}$$

and thus tends to zero for $\rho \to \infty$. The integral along the real axis tends to Euler's integral for the Γ function. We thus have obtained

THEOREM 10.5a

If Re $\nu > -1$, the correspondence

$$\tau^\nu \;\circ\!\!-\!\!-\!\!\bullet\; \frac{\Gamma(\nu+1)}{s^{\nu+1}} \tag{10.5-1}$$

holds for Re $s > 0$, where $s^{\nu+1}$ has its principal value.

If $\nu = n$, $n = 0, 1, 2, \ldots$, then in view of $\Gamma(n+1) = n!$ (10.5-1) is in accordance with the elementary correspondence (10.2-11). For $\nu = n + \frac{1}{2}$, $\Gamma(n + \frac{1}{2}) = \sqrt{\pi}(\frac{1}{2})_n$, and we obtain (10.4-5).

We now shall make an algorithmically potent application of Theorem 10.5a. Still assuming $\mathrm{Re}\,\nu > -1$, suppose a function $F \in \Omega$ is for $\tau > 0$ represented by the infinite series

$$F(\tau) = \tau^\nu \sum_{n=0}^{\infty} a_n \tau^n \qquad (10.5\text{-}2)$$

with complex coefficients a_n, and suppose we wish to calculate $f := \mathcal{L}F$. Applying the $\mathcal{L}$ transformation term by term and using Theorem 10.5a formally yields

$$g(s) = \sum_{n=0}^{\infty} a_n \Gamma(\nu + n + 1) s^{-\nu - n - 1}. \qquad (10.5\text{-}3)$$

For what s (if any) does $g(s)$ represent $f(s)$? The answer is not clear, because (1) the Laplace integral formed with F is improper, both at its upper and (if $\mathrm{Re}\,\nu < 0$) at its lower limit, and (2) the convergence of the series in the integrand (unless the series terminates) is not uniform over $(0, \infty)$. However, the following result grants everything that can be hoped for.

THEOREM 10.5b   (Hardy's theorem)

Let the series $g(s)$ be convergent for some $s = s_0 > 0$. Then the series (10.5-2) converges for all $\tau > 0$, and $g(s) = f(s) := \mathcal{L}F(s)$ for $s = s_0$ and for all s such that $\mathrm{Re}\,s > s_0$.

Proof. It is convenient to write

$$b_n := a_n \Gamma(1 + \nu + n)$$

throughout the proof. The series $g(s)$, apart from the factor $s^{-1-\nu}$, is a power series in s^{-1}. Because it converges for $s = s_0$, it has a positive radius of convergence,

$$\limsup_{n \to \infty} |b_n|^{1/n} < \infty.$$

Stirling's formula now shows that

$$\limsup_{n \to \infty} |a_n|^{1/n} = 0,$$

that is, the radius of convergence of (10.5-2) is infinite. Because $\mathrm{Re}\,\nu > -1$, F is integrable at $\tau = 0$.

Because $g(s)$ converges for $s = s_0$, it converges for all $s > s_0$. Thus if it is shown that for any $s > 0$ such that $g(s)$ converges we have $s \in C(F)$ and $g(s) = f(s)$, it follows that $g(s) = f(s)$ for all real $s \geqslant s_0$. Because both g and (in view of Theorem 10.1d) f are analytic for $\mathrm{Re}\ s > s_0$, the principle of analytic continuation implies that $g(s) = f(s)$ for $\mathrm{Re}\ s > s_0$.

We thus must show only that convergence of $g(s)$ for $s = s_0 > 0$ implies $s_0 \in C(F)$ and $g(s_0) = f(s_0)$. Without loss of generality we may assume $s_0 = 1$, for this can be achieved by an appropriate change of variables. To simplify the presentation of the proof we also assume that ν is real. (The modifications necessary for complex ν are minor.) Without further loss of generality, we then may assume

$$-1 < \nu \leqslant 0, \tag{10.5-4}$$

for this can be accomplished by inserting some zero coefficients, if necessary.

For every $\omega > 0$ the series

$$\sum_{n=1}^{\infty} a_n \tau^{\nu+n}$$

(omitting the zeroth term) converges uniformly in $[0, \omega]$. Hence integration and summation may be interchanged to yield

$$\int_0^\omega e^{-\tau} \sum_{n=1}^{\infty} a_n \tau^{\nu+n}\, d\tau = \sum_{n=1}^{\infty} \frac{b_n}{\Gamma(\nu+n+1)} \int_0^\omega e^{-\tau} \tau^{\nu+n}\, d\tau.$$

Adding the convergent integral $a_0 \int_0^\omega e^{-\tau} \tau^\nu\, d\tau$, we get

$$\int_0^\omega e^{-\tau} F(\tau)\, d\tau = \sum_{n=0}^{\infty} \frac{b_n}{\Gamma(\nu+n+1)} \int_0^\omega e^{-\tau} \tau^{\nu+n}\, d\tau.$$

From this we subtract

$$g(1) = \sum_{n=0}^{\infty} \frac{b_n}{\Gamma(\nu+n+1)} \int_0^\infty e^{-\tau} \tau^{\nu+n}\, d\tau,$$

where the series on the right is convergent by hypothesis. There results

$$\int_0^\omega e^{-\tau} F(\tau)\, d\tau - g(1) = -\sum_{n=0}^{\infty} \frac{b_n}{\Gamma(\nu+n+1)} \int_\omega^\infty e^{-\tau} \tau^{\nu+n}\, d\tau,$$

and it only remains to be shown that the series on the right tends to zero for $\omega \to \infty$.

We make use of the following identity, established by repeated integration by parts:

$$\int_{\omega}^{\infty} e^{-\tau} \tau^{\nu+n} \, d\tau = e^{-\omega} \omega^{\nu+n} + (\nu+n) e^{-\omega} \omega^{\nu+n-1} + \cdots$$

$$+ (\nu+n)(\nu+n-1) \cdots (\nu+1) e^{-\omega} \omega^{\nu}$$

$$+ (\nu+n)(\nu+n-1) \cdots (\nu+1)\nu \int_{\omega}^{\infty} e^{-\tau} \tau^{\nu-1} \, d\tau.$$

This may be written

$$\int_{\omega}^{\infty} e^{-\tau} \tau^{\nu+n} \, d\tau = e^{-\omega} \omega^{\nu} (\nu+1)_n \left\{ \sum_{m=0}^{n} \frac{\omega^m}{(\nu+1)_m} + \epsilon_\omega \right\},$$

where

$$e^{-\omega} \omega^{\nu} \epsilon_\omega = \nu \int_{\omega}^{\infty} e^{-\tau} \tau^{\nu-1} \, d\tau \to 0$$

as $\omega \to \infty$. Using $\Gamma(\nu+1+n) = \Gamma(\nu+1)(\nu+1)_n$, we thus get

$$\sum_{n=0}^{\infty} \frac{b_n}{\Gamma(\nu+1+n)} \int_{\omega}^{\infty} e^{-\tau} \tau^{\nu+n} \, d\tau = \frac{e^{-\omega} \omega^{\nu}}{\Gamma(\nu+1)} \sum_{n=0}^{\infty} b_n \left\{ \sum_{m=0}^{n} \frac{\omega^m}{(\nu+1)_m} + \epsilon_\omega \right\}.$$

Since Σb_n is convergent by hypothesis, it is clear that

$$\epsilon_\omega e^{-\omega} \omega^{\nu} \sum_{n=0}^{\infty} b_n \to 0 \quad \text{for } \omega \to \infty,$$

and it only remains to be shown that also

$$\lim_{\omega \to \infty} e^{-\omega} \omega^{\nu} \sum_{n=0}^{\infty} b_n \sum_{m=0}^{n} \frac{\omega^m}{(\nu+1)_m} = 0. \qquad (10.5\text{-}5)$$

We let

$$r_n := \sum_{m=n}^{\infty} b_m, \qquad c_n := \sum_{m=0}^{n} \frac{\omega^m}{(\nu+1)_m},$$

so that

$$b_n = r_n - r_{n+1}, \qquad c_n - c_{n-1} = \frac{\omega^n}{(\nu+1)_n},$$

and furthermore

$$\lim_{n \to \infty} r_n = 0. \tag{10.5-6}$$

If n is any integer we then have, using summation by parts,

$$\sum_{k=0}^{n} b_k \sum_{m=0}^{k} \frac{\omega^m}{(\nu+1)_m} = \sum_{k=0}^{n} (r_k - r_{k+1})c_k$$

$$= \sum_{k=0}^{n} r_k(c_k - c_{k-1}) - r_{n+1}c_n$$

$$= \sum_{k=0}^{n} r_k \frac{\omega^k}{(\nu+1)_k} - r_{n+1}c_n.$$

Letting $n \to \infty$ the second term tends to zero in view of (10.5-6) because $\lim_{n \to \infty} c_n$ exists, and we get

$$\sum_{k=0}^{\infty} b_k \sum_{m=0}^{k} \frac{\omega^m}{(\nu+1)_m} = \sum_{k=0}^{\infty} r_k \frac{\omega^k}{(\nu+1)_k}.$$

We thus must only show that the expression on the right, when multiplied by $\omega^\nu e^{-\omega}$, tends to zero for $\omega \to \infty$.

Let $\alpha := \sup|r_n|$, and let $\epsilon > 0$ be given. Choosing n such that $|r_k| < \epsilon$ for $k > n$, we have

$$\left| \omega^\nu e^{-\omega} \sum_{k=0}^{\infty} r_k \frac{\omega^k}{(\nu+1)_k} \right| \leq \alpha \omega^\nu e^{-\omega} \sum_{k=0}^{n} \frac{\omega^k}{(\nu+1)_k} + \epsilon \omega^\nu e^{-\omega} \sum_{k=n+1}^{\infty} \frac{\omega^k}{(\nu+1)_k}.$$

Because

$$\sum_{k=n+1}^{\infty} \frac{\omega^k}{(\nu+1)_k} \leq \frac{1}{\nu+1} e^\omega,$$

the second term is $\leq \epsilon/(\nu+1)$ for all ω. The first term tends to zero for $\omega \to \infty$. This proves (10.5-5), and hence the assertion of the theorem. ■

Hardy's theorem provides a key to a large number of important correspondences. The following examples represent only a small sample.

EXAMPLE **1**

To compute the $\mathcal{L}$ transform of

$$F(\tau) := \frac{1}{\sqrt{\tau}} \cos(2\sqrt{\alpha \tau}),$$

where $\alpha > 0$. Applying $\mathscr{L}$ termwise we get

$$F(\tau) = \sum_{n=0}^{\infty} \frac{(-1)^n (4\alpha)^n \tau^{n-1/2}}{(2n)!}$$

$$\circ\!\!-\!\!\bullet \sum_{n=0}^{\infty} \frac{(-1)^n (4\alpha)^n}{(2n)!} \frac{\Gamma(n+1/2)}{s^{n+1/2}}$$

$$= \sqrt{\pi} \frac{(-\alpha)^n}{n!} \frac{1}{s^{n+1/2}} = \frac{\sqrt{\pi}}{\sqrt{s}} e^{-\alpha/s},$$

where we have used $\Gamma(n+1/2) = \sqrt{\pi} (\tfrac{1}{2})_n$ and $(2n)! = 4^n n! (\tfrac{1}{2})_n$. We thus have

$$\frac{1}{\sqrt{\tau}} \cos(2\sqrt{\alpha\tau}) \circ\!\!-\!\!\bullet \frac{\sqrt{\pi}}{\sqrt{s}} e^{-\alpha/s}. \qquad (10.5\text{-}7)$$

EXAMPLE 2

To compute the $\mathscr{L}$ transform of

$$F(\tau) := \tau^\mu J_\nu(\tau),$$

where J_ν is the Bessel function. The series

$$\tau^\mu J_\nu(\tau) = \frac{\tau^{\mu+\nu}}{2^\nu} \sum_{n=0}^{\infty} \frac{(-\tau^2/4)^n}{n!\Gamma(\nu+1+n)}$$

shows that $F \in \Omega$ for $\mathrm{Re}\,(\mu+\nu) > -1$. Hardy's theorem yields the transform

$$f(s) = \frac{1}{2^\nu} \sum_{n=0}^{\infty} \frac{(-1/4)^n \Gamma(\mu+\nu+2n+1)}{n!\Gamma(\nu+1+n)s^{\mu+\nu+1+2n}}.$$

We use the relations $\Gamma(\nu+1+n) = \Gamma(\nu+1)(\nu+1)_n$, $\Gamma(\mu+\nu+1+2n) = \Gamma(\mu+\nu+1)(\mu+\nu+1)_{2n}$ and

$$(\mu+\nu+1)_{2n} = 2^{2n} \left(\frac{\mu+\nu+1}{2}\right)_n \left(\frac{\mu+\nu}{2}+1\right)_n.$$

It follows that f can be expressed as a hypergeometric function,

$$f(s) = \frac{\Gamma(\mu+\nu+1)}{2^\nu \Gamma(\nu+1)s^{\mu+\nu+1}} {}_2F_1\left(\frac{\mu+\nu+1}{2}, \frac{\mu+\nu}{2}+1; \nu+1; -\frac{1}{s^2}\right).$$

The series converges at least for $|s| > 1$, and thus represents the $\mathscr{L}$ transform at least for $\mathrm{Re}\,s > 1$. However, the known asymptotic behavior of the Bessel function shows that the growth indicator γ_F of F is zero. Thus $f(s)$ is defined and analytic for $\mathrm{Re}\,s > 0$. Hence if by the symbol ${}_2F_1(\ldots, z)$ we understand not merely the hypergeometric series but also the analytic continuation of that series into the z plane cut between $z = 1$ and $z = \infty$, then it represents $f(s)$ in the whole half plane of convergence $\operatorname{Re} s > 0$. Applying Euler's transformation (9.9-23) to the series, we may put the correspondence in the form

$$\tau^\mu J_\nu(\tau) \circ\!\!-\!\!-\!\!\bullet \frac{\Gamma(\mu+\nu+1)}{2^\nu \Gamma(\nu+1)} (s^2+1)^{-(\mu+\nu+1)/2} \, _2F_1\left(\frac{\mu+\nu+1}{2}, \frac{\mu+\nu+2}{2}; \nu+1; \frac{1}{1+s^2}\right).$$
$$(10.5\text{-}8)$$

Simplifications occur for special values of μ.

(a) $\mu = \nu$. The hypergeometric series reduces to 1, and using the duplication formula of the Γ function we obtain:

$$\tau^\nu J_\nu(\tau) \circ\!\!-\!\!-\!\!\bullet \frac{2^\nu \Gamma(\nu+1/2)}{\sqrt{\pi}} (s^2+1)^{-\nu-1/2}.$$
$$(10.5\text{-}9)$$

For $\nu = 0$ we obtain (10.4-6).

(b) $\mu = 0$. By a quadratic transformation of the hypergeometric series,

$$_2F_1\left(\frac{\nu}{2}, \frac{\nu+1}{2}; \nu+1; \frac{1}{1+s^2}\right) = \left[\frac{\sqrt{1+s^2}+s}{2\sqrt{1+s^2}}\right]^{-\nu},$$

and we obtain the important correspondence

$$J_\nu(\tau) \circ\!\!-\!\!-\!\!\bullet \frac{(\sqrt{1+s^2}-s)^\nu}{\sqrt{1+s^2}}, \quad \operatorname{Re} \nu > -1.$$
$$(10.5\text{-}10)$$

(c) $\mu = -1$. By the same quadratic transformation, there now results

$$\tau^{-1} J_\nu(\tau) \circ\!\!-\!\!-\!\!\bullet \frac{1}{\nu(s+\sqrt{1+s^2})^\nu}, \quad \operatorname{Re} \nu > 0.$$
$$(10.5\text{-}11)$$

EXAMPLE 3

Let α, γ, ν be complex numbers, $\operatorname{Re} \nu > -1$, $\gamma \neq 0, -1, -2, \ldots$. Applying Hardy's theorem to the confluent hypergeometric function

$$F(\tau) := \tau^\nu \, _1F_1(\alpha; \gamma; \tau) = \sum_{n=0}^{\infty} \frac{(\alpha)_n}{(\gamma)_n} \frac{\tau^{n+\nu}}{n!},$$

we obtain the following representation for $f := \mathscr{L}F$:

$$f(s) = \sum_{n=0}^{\infty} \frac{(\alpha)_n}{(\gamma)_n} \frac{\Gamma(\nu+1+n)}{n! s^{\nu+1+n}}$$
$$= \frac{\Gamma(\nu+1)}{s^{\nu+1}} \, _2F_1\left(\alpha, \nu+1; \gamma; \frac{1}{s}\right).$$
$$(10.5\text{-}12)$$

The series converges for $|s| > 1$ and (unless it terminates) is singular at $s = 1$. Thus f, in general, cannot be continued into any open right half plane containing $\operatorname{Re} s = 1$ and is represented by the series in the whole set $C(F)$.

Using Rule VIII (Damping), we find for **Whittaker's function** (9.12-14)

$$M_{\kappa,\mu}(\tau) = \tau^{\mu+1/2} e^{-\tau/2} \,_1F_1(\mu - \kappa + 1/2; 2\mu + 1; \tau)$$

the correspondence

$$\tau^{\nu-1} M_{\kappa,\mu}(\tau) \circ\!\!-\!\!-\!\!\bullet \frac{\Gamma(\nu + \mu + 1/2)}{(s + 1/2)^{\nu+\mu+1/2}}$$

$$\cdot \,_2F_1\!\left(\mu - \kappa + 1/2; \nu + \mu + 1/2; 2\mu + 1; \frac{1}{s+1/2}\right). \tag{10.5-13}$$

The function

$$W_{\kappa,\mu}(\tau) := \frac{\Gamma(-2\mu)}{\Gamma(-\mu - \kappa + 1/2)} M_{\kappa,\mu}(\tau) + \frac{\Gamma(2\mu)}{\Gamma(\mu - \kappa + 1/2)} M_{\kappa,-\mu}(\tau),$$

$$\tag{10.5-14}$$

likewise a solution of the differential equation (9.12-13), is known as the **Whittaker function of the second kind**. If $\mathrm{Re}(\nu \pm \mu) > -\frac{1}{2}$, then $\tau^{\nu-1} W_{\kappa,\mu}(\tau)$ is in Ω and has the image

$$\frac{\Gamma(-2\mu)\Gamma(\mu + \nu + 1/2)}{2\Gamma(-\mu - \kappa + 1/2)} (s + 1/2)^{-\mu-\nu-1/2}$$

$$\cdot \,_2F_1\!\left(\mu - \kappa + 1/2, \mu + \nu + 1/2; 2\mu + 1; \frac{1}{s+1/2}\right)$$

$+$ a similar term with μ replaced by $-\mu$.

This simplifies by the linear transformation (9.9-26), and there results

$$\tau^{\nu-1} W_{\kappa,\mu}(\tau) \circ\!\!-\!\!-\!\!\bullet \frac{\Gamma(\mu + \nu + 1/2)\Gamma(-\mu + \nu + 1/2)}{\Gamma(\nu - \kappa + 1)} (s + 1/2)^{-\mu-\nu-1/2}$$

$$\cdot \,_2F_1\!\left(\mu - \kappa + 1/2, \mu + \nu + 1/2; \nu - \kappa + 1; \frac{s-1/2}{s+1/2}\right). \tag{10.5-15}$$

The image function is elementary for numerous special combinations of values of the parameters.

The foregoing list of correspondences could be augmented almost indefinitely. We refer the reader to a table of $\mathcal{L}$ transforms such as Oberhettinger and Badii [1973] for further examples.

Several interesting integral relations can be derived from these correspondences. For example, from (10.5-10) we know that for $\mathrm{Re}\,\nu > -1$

$$\int_0^\infty e^{-s\tau} J_\nu(\tau)\, d\tau = \frac{(\sqrt{1+s^2} - s)^\nu}{\sqrt{1+s^2}}$$

for all $s > 0$. From the known asymptotic behavior of $J_\nu(\tau)$ (see §11.5) we know that the integral exists for $s = 0$. By the Abelian theorem 10.1f we conclude that its value equals its limit as $s \to 0+$; that is, we have

$$\int_0^\infty J_\nu(\tau)\, d\tau = 1, \qquad \text{Re } \nu > -1. \tag{10.5-16}$$

From (10.5-8) we deduce in a similar manner, using the Gauss formula (8.6-6) to evaluate the hypergeometric series at $s = 0$, the more general formula

$$\int_0^\infty \tau^\mu J_\nu(\tau)\, d\tau = \frac{2^\mu \Gamma\left(\dfrac{\nu + \mu + 1}{2}\right)}{\Gamma\left(\dfrac{\nu - \mu + 1}{2}\right)}, \tag{10.5-17}$$

valid for $\text{Re } \mu < \tfrac{1}{2}$ and $\text{Re }(\nu + \mu) > -1$. The so-called **discontinuous integrals of Weber and Schafheitlin**,

$$\int_0^\infty \tau^\lambda J_\mu(\alpha\tau) J_\nu(\beta\tau)\, d\tau,$$

although more difficult, can be evaluated in a similar manner; see Problems 10–13.

The convolution theorem (Theorem 10.4b) is a further source of integral relations. For instance, let

$$F_\nu(\tau) := \frac{\nu}{\tau} J_\nu(\tau) \qquad (\text{Re } \nu > 0).$$

Then by (10.5-11) the transform $f_\nu \bullet\!\!-\!\!\circ F_\nu$ satisfies

$$f_\nu(s) f_\mu(s) = f_{\nu+\mu}(s).$$

There follows for all $\tau > 0$

$$\nu\mu \int_0^\tau \frac{J_\nu(\sigma)}{\sigma} \frac{J_\mu(\tau - \sigma)}{\tau - \sigma}\, d\sigma = (\nu + \mu) \frac{J_{\nu+\mu}(\tau)}{\tau} \tag{10.5-18}$$

$(\text{Re } \mu, \text{Re } \nu > 0)$. From (10.5-10) there follows in a similar manner

$$\int_0^\tau J_\nu(\sigma) J_\mu(\tau - \sigma)\, d\sigma = \int_0^\tau J_0(\sigma) J_{\mu+\nu}(\tau - \sigma)\, d\sigma \tag{10.5-19}$$

$(\text{Re } \mu, \text{Re } \nu, \text{Re}(\nu + \mu) > -1)$, and from (10.5-9)

$$\int_0^\tau \sigma^\nu J_\nu(\sigma)(\tau - \sigma)^\mu J_\mu(\tau - \sigma)\, d\sigma$$

$$= \frac{\Gamma(\nu + 1/2)\Gamma(\mu + 1/2)}{\Gamma(\nu + \mu + 1/2)\Gamma(1/2)} \tau^{\nu+\mu} J_{\nu+\mu}(\tau). \tag{10.5-20}$$

Hardy's theorem permits the calculation not only of Laplace transforms, but also of inverse Laplace transforms. Whenever a Laplace transform can be extended to a function of the form $s^{-\nu}g(s)$, where Re $\nu>0$ and g is analytic at infinity, the following result, which is a mere reformulation of Theorem 10.5b, permits to find the original function.

COROLLARY 10.5c

Let Re $\nu>0$ *and let*

$$f(s) = s^{-\nu} \sum_{n=0}^{\infty} c_n s^{-n}, \tag{10.5-21}$$

where the series converges for $|s|>\rho$, and where (if ν is not an integer) $|\arg s|<\pi$. Then f is an analytic continuation of $\mathscr{L}F$, where F is the original function defined for all $\tau>0$ by

$$F(\tau) := \sum_{n=0}^{\infty} \frac{c_n}{\Gamma(\nu+n)} \tau^{\nu+n-1}. \tag{10.5-22}$$

EXAMPLE **4**

Let $\alpha>0$, and let

$$f(s) := s^{-1} e^{-\alpha/s} = s^{-1} \sum_{n=0}^{\infty} \frac{(-\alpha)^n}{n!} s^{-n}.$$

Then $f(s) \bullet\!\!-\!\!\!-\!\!\!\circ F(\tau)$ where

$$F(\tau) = \sum_{n=0}^{\infty} \frac{(-\alpha\tau)^n}{(n!)^2} = J_0(2\sqrt{\alpha\tau}).$$

More generally, if Re $\nu>-1$,

$$s^{-\nu-1} e^{-\alpha/s} \bullet\!\!-\!\!\!-\!\!\!\circ \left(\frac{\tau}{\alpha}\right)^{\nu/2} J_\nu(2\sqrt{\alpha\tau}). \tag{10.5-23}$$

The method of inversion implied by Corollary 10.5c requires $f(s)$ to have a very special behavior at infinity. We therefore mention another method of inversion by series expansion that does not require any special behavior at ∞ and which is suggested by certain results of this section.

Assume the Laplace transform f is analytic in Re $s>0$ (this can always be achieved by a translation $s \to s+a$). The Moebius map

$$t:s \to z := \frac{s-1/2}{s+1/2}$$

maps $\operatorname{Re} s > 0$ onto $|z| < 1$. Therefore the function

$$g(z) := f(t^{[-1]}(z)) = f\left(\frac{1}{2}\frac{1+z}{1-z}\right)$$

is analytic in $|z| < 1$ and can be expanded in powers of z:

$$g(z) = \sum_{n=0}^{\infty} a_n z^n.$$

There follows

$$f(s) = g(t(s)) = \sum_{n=0}^{\infty} a_n \left(\frac{s-1/2}{s+1/2}\right)^n, \qquad \operatorname{Re} s > 0.$$

If original functions F_n such that

$$F_n(\tau) \circ\!\!-\!\!-\!\!\bullet \left(\frac{s-1/2}{s+1/2}\right)^n$$

were known, and if (as in the case of Hardy's theorem) term-by-term application of $\mathcal{L}$ were permissible, the function

$$F(\tau) := \sum_{n=0}^{\infty} a_n F_n(\tau)$$

would have the Laplace transform f, and thus would be the desired original function.

Clearly, the functions F_n do not exist, because their presumed image functions do not vanish at infinity. However, by specializing parameters in (10.5-13) we find

$$\tau^{-1/2} M_{n+1/2,0}(\tau) \circ\!\!-\!\!-\!\!\bullet \frac{1}{s+1/2} \, {}_2F_1\left(-n, 1; \, 1 \, ; \frac{1}{s+1/2}\right)$$

$$= \frac{(s-1/2)^n}{(s+1/2)^{n+1}}.$$

The function on the left is related to the **Laguerre polynomial** L_n (see §2.5 for definition and other properties); more precisely,

$$\tau^{-1/2} M_{n+1/2,0}(\tau) = e^{-\tau/2} L_n(\tau).$$

We thus have found: If

$$(s+1/2)f(s) = \sum_{n=0}^{\infty} a_n \left(\frac{s-1/2}{s+1/2}\right)^n, \qquad (10.5\text{-}24)$$

then $f = \mathcal{L}F$, at least formally, where

$$F(\tau) := e^{-\tau/2} \sum_{n=0}^{\infty} a_n L_n(\tau). \tag{10.5-25}$$

The following result concerning the validity of the foregoing operations is proved in Doetsch [1950], p. 301:

THEOREM 10.5d

Let f be analytic in $\mathrm{Re}\, s > 0$ and be represented there by (10.5-24), where $\sum |a_n|^2 < \infty$. Then f is the Laplace transform of a Lebesgue-integrable function F that is represented (in the $L_2(0, \infty)$ metric) *by the series* (10.5-25).

EXAMPLE 5

Let us pretend to be ignorant of the original function corresponding to

$$f(s) := \frac{1}{s+1/2} \exp\left(-\frac{\alpha}{s+1/2}\right) (\alpha > 0).$$

To expand $(s + 1/2)f(s)$ in powers of

$$z := \frac{s - 1/2}{s + 1/2},$$

we note that $(s + 1/2)^{-1} = 1 - z$. Consequently,

$$(s + 1/2)f(s) = e^{-\alpha(1-z)} = e^{-\alpha} \sum_{n=0}^{\infty} \frac{(\alpha z)^n}{n!}.$$

Clearly, the coefficients in this expansion satisfy the conditions of Theorem 10.5d. Thus $f = \mathcal{L}F$ where

$$F(\tau) := e^{-\alpha} \sum_{n=0}^{\infty} \frac{\alpha^n}{n!} e^{-\tau/2} L_n(\tau).$$

On the other hand, by (10.5-23) and Rule VIII (Damping) $f(s)$ corresponds to $e^{-\tau/2} J_0(2\sqrt{\alpha\tau})$. We thus have proved **Hille's expansion**

$$J_0(2\sqrt{\alpha\tau}) = e^{-\alpha} \sum_{n=0}^{\infty} \frac{\alpha^n}{n!} L_n(\tau). \tag{10.5-26}$$

PROBLEMS

1. Let $\mathrm{Re}\, \mu > 0$, $\mathrm{Re}\, \nu > 0$. Evaluate

$$\int_0^1 \tau^{\mu-1}(1-\tau)^{\nu-1}\, d\tau$$

by convolution and thus obtain a new proof of Theorem 8.7a.

2. For Re $\nu > 0$ and $\tau > 0$, let

$$\gamma(\nu, \tau) := \int_0^\tau \sigma^{\nu-1} e^{-\sigma} \, d\sigma,$$

an **incomplete Γ-function**. Prove

$$\gamma(\nu, \tau) \circ\!\!-\!\!-\!\!-\!\!\bullet \Gamma(\nu) \frac{(1+s)^{-\nu}}{s}.$$

3. Let $F \in \Omega$ possess a generalized derivative $F' \in \Omega$ (see §10.2). Then the integral equation

$$\int_0^\tau J_0(\tau - \sigma) X(\sigma) \, d\sigma = F(\tau)$$

is solved by

$$X(\tau) := \int_0^\tau \frac{J_1(\tau - \sigma)}{\tau - \sigma} F(\sigma) \, d\sigma + F'(\tau).$$

4. Let $F(\tau) \circ\!\!-\!\!-\!\!-\!\!\bullet f(s)$. Show that

$$G(\tau) := \int_0^\tau F(\sigma) J_0(2\sqrt{\sigma(\tau - \sigma)}) \, d\sigma \circ\!\!-\!\!-\!\!-\!\!\bullet \frac{1}{s} f\left(s + \frac{1}{s}\right).$$

[Interchange order of integrations and use (10.5-23).]
5. Establish the correspondence

$$J_\nu(2\sqrt{\alpha\tau}) J_\nu(2\sqrt{\beta\tau}) \circ\!\!-\!\!-\!\!-\!\!\bullet \frac{1}{s} e^{-(\alpha+\beta)/s} I_\nu\left(\frac{2\sqrt{\alpha\beta}}{s}\right),$$

where $I_\nu(z)$ denotes the modified Bessel function of the first kind,

$$I_\nu(z) := e^{-i\nu\pi/2} J_\nu(iz).$$

6. For $\tau > 0$ and Re $\mu > -\frac{1}{2}$,

$$\tau^\mu K_\mu(\tau) = 2^\mu \Gamma(\mu + 1) e^{-\tau} \sum_{n=0}^{\infty} \frac{(1/2 - \mu)_n}{(1/2 + \mu)_{n+1}} L_n(2\tau).$$

[$K_\mu(z) =$ modified Bessel function of the second kind,

$$K_\mu(z) := \frac{\pi}{2} \frac{I_{-\mu}(z) - I_\mu(z)}{\sin \mu\pi}.]$$

7. If L_n denotes the Laguerre polynomial, prove by means of Laplace transforms

$$\int_0^\tau L_n(\sigma) \, d\sigma = L_n(\tau) - L_{n+1}(\tau), \qquad n = 0, 1, 2, \ldots.$$

8. Establish the generating function of the Laguerre polynomials,

$$\frac{\exp\left(-\dfrac{z\tau}{1-z}\right)}{1-z} = \sum_{n=0}^{\infty} z^n L_n(\tau), \qquad |z| < 1,$$

by means of Theorem 10.5d.

9. The generalized Laguerre polynomial $L_n^{(\alpha)}$ is defined by

$$L_n^{(\alpha)}(\tau) := \frac{(\alpha+1)_n}{n!} {}_1F_1(-n; \alpha+1; \tau),$$

$n = 0, 1, 2, \ldots$; $\operatorname{Re} \alpha > -1$ $(L_n = L_n^{(0)})$. Prove that

(a) $L_n^{(\alpha)}(\tau) \; \circ\!\!-\!\!-\!\!-\!\!\bullet \; \dfrac{\Gamma(\alpha+1+n)}{s^{\alpha+1}} \left(\dfrac{s-1}{s}\right)^n,$

(b) $\displaystyle\sum_{n=0}^{\infty} L_n^{(\alpha)}(\tau) z^n = (1-z)^{-\alpha-1} \exp\left(-\dfrac{z\tau}{1-z}\right), \qquad |z| < 1,$

(c) $\displaystyle\sum_{n=0}^{\infty} \dfrac{L_n^{(\alpha)}(\tau)}{\Gamma(1+\alpha+n)} \xi^n = e^{\xi} (\xi\tau)^{-\alpha/2} J_\alpha(2\sqrt{\xi\tau}).$

10. *Discontinuous integrals of Weber and Schafheitlin.* Let $\alpha > \beta > 0$, and let $\operatorname{Re}(\mu+\nu+1) > \operatorname{Re}\lambda > -1$. Show that

$$\int_0^{\infty} \tau^{-\lambda} J_\mu(\alpha\tau) J_\nu(\beta\tau) \, d\tau$$

$$= \frac{\beta^\nu \Gamma((\mu+\nu-\lambda+1)/2)}{2^\lambda \alpha^{\nu-\lambda+1} \Gamma(\nu+1) \Gamma((\lambda+\mu-\nu+1)/2)}$$

$$\cdot {}_2F_1\left(\frac{\mu+\nu-\lambda+1}{2}, \frac{\nu-\lambda-\mu+1}{2}; \nu+1; \frac{\beta^2}{\alpha^2}\right).$$

11. (Continuation) If $\alpha > 0$ and $\operatorname{Re}(\mu+\nu+1) > \operatorname{Re}\lambda > 0$, show that

$$\int_0^{\infty} \tau^{-\lambda} J_\mu(\alpha\tau) J_\nu(\beta\tau) \, d\tau$$

$$= \frac{(\alpha/2)^{\lambda-1} \Gamma(\lambda) \Gamma((\mu+\nu-\lambda+1)/2)}{2\Gamma((\lambda+\nu-\mu+1)/2) \Gamma((\lambda+\nu+\mu+1)/2) \Gamma((\lambda+\mu-\nu+1)/2)}$$

12. (Continuation) As a special case, show that for $\alpha > 0$, $\beta > 0$, $\operatorname{Re}\mu > 0$

$$\int_0^{\infty} \frac{1}{\tau} J_\mu(\alpha\tau) J_\mu(\beta\tau) \, d\tau = \begin{cases} \dfrac{1}{2\mu}\left(\dfrac{\beta}{\alpha}\right)^\mu, & \beta \leq \alpha, \\[3mm] \dfrac{1}{2\mu}\left(\dfrac{\alpha}{\beta}\right)^\mu, & \beta \geq \alpha. \end{cases}$$

13. (Continuation) Deduce Weber's formulae

$$\int_0^\infty J_0(\alpha\tau) \sin \beta\tau \, d\tau = \begin{cases} 0, & \alpha > \beta > 0 \\ \\ \infty, & \alpha = \beta > 0 \\ \\ \dfrac{1}{\sqrt{\beta^2 - \alpha^2}}, & \beta > \alpha > 0 \end{cases}$$

$$\int_0^\infty J_0(\alpha\tau) \cos \beta\tau \, d\tau = \begin{cases} \dfrac{1}{\sqrt{\alpha^2 - \beta^2}}, & \alpha > \beta > 0 \\ \\ \infty, & \alpha = \beta > 0 \\ \\ 0, & \beta > \alpha > 0 \end{cases}$$

as special cases of the Weber–Schafheitlin integral.

§10.6. THE FOURIER INTEGRAL

Of the two general methods for inverting the $\mathscr{L}$ transformation we have discussed, the first (Corollary 10.5c) is valid only under the stringent hypothesis that $\mathscr{L}F$ (apart from a factor $s^{-\nu}$) is analytic at infinity. The second method (Theorem 10.5d) is subject to conditions that are not easily checked; also, it is too specialized to allow much flexibility. In §10.7 we derive an inversion formula that is valid under the sole assumption that the Laplace integral converges absolutely for some s, that is, that $\beta_F < \infty$. This includes all cases of practical interest.

The inversion formula mentioned is based on a form of the *Fourier integral theorem*, the discussion of which is the main goal of the present section. We begin by a brief informal discussion of the theorem. After a short discourse on Fourier series, we then prove a version of the integral theorem that is sufficient for inverting all absolutely convergent $\mathscr{L}$ transforms of functions in Ω.

I. Fourier theory: Informal discussion

Let F be a complex-valued function of period 2λ defined on the real line. Assuming that $|F|$ is (possibly improperly) Riemann integrable on $[-\lambda, \lambda]$, the numbers

$$a_n := \frac{1}{2\lambda} \int_{-\lambda}^{\lambda} F(\tau) \exp\left(-\frac{in\pi\tau}{\lambda}\right) d\tau, \qquad n = 0, \pm 1, \pm 2, \ldots, \quad (10.6\text{-}1)$$

called **Fourier coefficients** of F, can be defined, and the series

$$\sum_{n=-\infty}^{\infty} a_n \exp\left(\frac{in\pi\tau}{\lambda}\right),$$

(10.6-2)

called the **Fourier series** of F, can be formed. The theory of Fourier series is concerned with the question under what conditions the Fourier series of a function F converges, and to what sum. It is known that the convergence of the series at a point τ depends only on the behavior of F in the immediate vicinity of the point τ. If F meets a certain condition (C) at τ (see II below), then the series converges, and its value is $F(\tau)$:

$$F(\tau) = \sum_{n=-\infty}^{\infty} a_n \exp\left(\frac{in\pi\tau}{\lambda}\right).$$

(10.6-3)

We now drop the condition that F be periodic. Instead, we demand

$$\int_{-\infty}^{\infty} |F(\tau)| \, d\tau < \infty.$$

(10.6-4)

For $\lambda > 0$ and for real ω we define

$$G_\lambda(\omega) := \int_{-\lambda}^{\lambda} e^{-i\omega\sigma} F(\sigma) \, d\sigma,$$

(10.6-5)

so that

$$a_n = \frac{1}{2\lambda} G_\lambda\left(\frac{n\pi}{\lambda}\right), \qquad n = 0, \pm 1, \pm 2, \ldots.$$

If $-\lambda < \tau < \lambda$, and if F meets condition (C) at τ, then (10.6-3) still holds (the series does not depend on values of F outside $[-\lambda, \lambda]$), and we thus have

$$F(\tau) = \frac{1}{2\pi} \sum_{n=-\infty}^{\infty} \exp\left(\frac{in\pi\tau}{\lambda}\right) G_\lambda\left(\frac{n\pi}{\lambda}\right)\frac{\pi}{\lambda}.$$

(10.6-6)

We now let $\lambda \to \infty$, proceeding heuristically. The foregoing sum may be looked at as a numerical approximation to the integral

$$\int_{-\infty}^{\infty} e^{i\omega\tau} G_\lambda(\omega) \, d\omega,$$

evaluated with the step π/λ. As $\lambda \to \infty$, the step tends to zero, and the sum may be expected to converge to the integral. At the same time,

$$G_\lambda(\omega) \to G(\omega) := \int_{-\infty}^{\infty} e^{-i\omega\sigma} F(\sigma) \, d\sigma,$$

where the improper integral exists in view of hypothesis (10.6-4). Performing the two passages to the limit simultaneously, we obtain the identity

$$F(\tau) = \frac{1}{2\pi} \int_{-\infty}^{\infty} e^{i\omega\tau} \left\{ \int_{-\infty}^{\infty} e^{-i\omega\sigma} F(\sigma) \, d\sigma \right\} d\omega \qquad (10.6\text{-}7)$$

known as the **Fourier integral formula**. With a minor modification, this is proved in III below.

II. A Short Theory of Fourier Series

The basic tool we require (not only in the theory of Fourier series, but also for the Fourier and Laplace transforms) is the following result, known as the **Riemann–Lebesgue Lemma**.

THEOREM 10.6a

If H is a complex-valued function defined on the finite interval $[\alpha, \beta]$ such that $|H|$ is (possibly improperly) *integrable on $[\alpha, \beta]$, then*

$$\lim_{\omega \to \infty} \int_{\alpha}^{\beta} H(\tau) \sin \omega\tau \, d\tau = 0. \qquad (10.6\text{-}8)$$

The statement of the lemma is intuitively obvious: For large ω, the contributions of the factor $H(\tau)$ to the integral are cancelled out by the rapid oscillations of the factor $\sin \omega\tau$. If H has a continuous derivative, the lemma may be proved very simply by integration by parts. The following proof for the general case is due to G. N. Hardy.

Proof. Without loss of generality we may assume H to be real. It is convenient to first consider the case where H is bounded on $[\alpha, \beta]$. Let $|H(\tau)| \leqslant \kappa$ for $\tau \in [\alpha, \beta]$, and let $\epsilon > 0$ be given. Let $\Delta : \alpha = \tau_0 < \tau_1 < \cdots < \tau_{n-1} < \tau_n = \beta$ be any subdivision of $[\alpha, \beta]$, and let

$$\lambda_k := \inf_{\tau \in I_k} H(\tau), \qquad \mu_k := \sup_{\tau \in I_k} H(\tau),$$

where I_k denotes the interval $[\tau_{k-1}, \tau_k]$ $(k = 1, 2, \ldots, n)$. Let

$$\lambda_\Delta := \sum_{k=1}^{n} \lambda_k (\tau_k - \tau_{k-1}), \qquad \mu_\Delta := \sum_{k=1}^{n} \mu_k (\tau_k - \tau_{k-1});$$

because H is integrable, it is possible to choose a subdivision such that $\mu_\Delta - \lambda_\Delta < \varepsilon$.

In I_k, let

$$H_k(\tau) := H(\tau) - H(\tau_{k-1}),$$

so that $|H_k(\tau)| \leq \mu_k - \lambda_k$, $\tau \in I_k$. We then have

$$\left| \int_\alpha^\beta H(\tau) \sin \omega\tau \, d\tau \right|$$

$$= \left| \sum_{k=1}^n H(\tau_{k-1}) \int_{\tau_{k-1}}^{\tau_k} \sin \omega\tau \, d\tau + \sum_{k=1}^n \int_{\tau_{k-1}}^{\tau_k} H_k(\tau) \sin \omega\tau \, d\tau \right|$$

$$\leq \sum_{k=1}^n |H(\tau_{k-1})| \left| \int_{\tau_{k-1}}^{\tau_k} \sin \omega\tau \, d\tau \right| + \sum_{k=1}^n \int_{\tau_{k-1}}^{\tau_k} |H_k(\tau)| \, d\tau$$

$$\leq n\kappa \frac{2}{\omega} + (\mu_\Delta - \lambda_\Delta).$$

By taking ω sufficiently large (n remaining fixed after ϵ has been chosen), the last expression can be made less than 2ϵ, which implies (10.6-8).

Let now H be unbounded. Because H has an absolutely convergent integral, the points near which it is unbounded may be enclosed in a finite number of intervals $J_1, J_2, \ldots, J_p$ such that

$$\sum_{k=1}^p \int_{J_k} |H(\tau)| \, d\tau < \varepsilon.$$

It now suffices to apply the Lemma as proved above for bounded H to the finitely many intervals left over after removing from $[\alpha, \beta]$ the intervals $J_1, \ldots, J_p$. ∎

It is clear that (10.6-8) also holds when $\sin \omega\tau$ is replaced by $\cos \omega\tau$, or by $e^{i\omega\tau}$.

Let now F be periodic with period 2λ, and let $|F|$ be (possibly improperly) Riemann integrable over every finite interval, so that the Fourier coefficients of F can be defined by (10.6-1). We denote by

$$F_n(\tau) := \sum_{k=-n}^n a_k \exp\left(\frac{ik\pi\tau}{\lambda}\right),$$

the nth (symmetric) partial sum of the Fourier series of F. Under what conditions can we assert that $F_n(\tau) \to F(\tau)$?

By the definition of the Fourier coefficients we have, interchanging (finite) summation and integration, summing a terminating geometric series and making use of the periodicity of F,

$$F_n(\tau) = \frac{1}{2\lambda} \sum_{k=-n}^n \exp\left(\frac{ik\pi\tau}{\lambda}\right) \int_{-\lambda}^\lambda \exp\left(-\frac{ik\pi\sigma}{\lambda}\right) F(\sigma) \, d\sigma$$

$$= \frac{1}{2\lambda} \int_{-\lambda}^\lambda F(\sigma) \left\{ \sum_{k=-n}^n \exp\left(\frac{ik\pi(\tau-\sigma)}{\lambda}\right) \right\} d\sigma$$

$$= \frac{1}{2\lambda} \int_{-\lambda}^{\lambda} F(\sigma) \frac{\sin\{[(2n+1)\pi(\tau-\sigma)]/2\lambda\}}{\sin\{[\pi(\tau-\sigma)]/2\lambda\}} \, d\sigma$$

$$= \frac{1}{2\lambda} \int_{-\lambda}^{\lambda} F(\tau+\sigma) \frac{\sin\{[(2n+1)\pi\sigma]/2\lambda\}}{\sin(\pi\sigma/2\lambda)} \, d\sigma$$

$$= \frac{1}{2\lambda} \int_{0}^{\lambda} \{F(\tau+\sigma)+F(\tau-\sigma)\} \frac{\sin\{[(2n+1)\pi\sigma]/2\lambda\}}{\sin(\pi\sigma/2\lambda)} \, d\sigma.$$

Subtracting

$$F(\tau) = \frac{1}{2\lambda} \int_{0}^{\lambda} 2F(\tau) \frac{\sin\{[2n+1)\pi\sigma]/2\lambda\}}{\sin(\pi\sigma/2\lambda)} \, d\sigma,$$

there follows

$$F_n(\tau) - F(\tau) = \frac{1}{2\lambda} \int_{0}^{\lambda} S(\tau,\sigma) \frac{\sin\{[(2n+1)\pi\sigma]/2\lambda\}}{\sin(\pi\sigma/2\lambda)} \, d\sigma, \quad (10.6\text{-}9)$$

where

$$S(\tau,\sigma) := F(\tau+\sigma) - 2F(\tau) + F(\tau-\sigma) \qquad (10.6\text{-}10)$$

is a symmetric second difference of values of F.

THEOREM 10.6b

Let F be a complex-valued, 2λ-periodic function such that $|F|$ is (possibly improperly) Riemann integrable over every finite interval. Then the Fourier series of F converges (in the sense that $F_n \to F$) at every point τ such that for some α, $0 < \alpha \leq \lambda$,

$$\lim_{\omega \to \infty} \int_{0}^{\alpha} S(\tau,\sigma) \frac{\sin \omega\sigma}{\sigma} \, d\sigma = 0, \qquad (C)$$

where S is defined by (10.6-10).

Proof. It is to be shown that the integral on the right of (10.6-9) tends to zero for $n \to \infty$. We write

$$\int_{0}^{\lambda} = \int_{0}^{\alpha} + \int_{\alpha}^{\lambda}$$

and consider the two integrals on the right separately.

To show that the second integral tends to zero, we apply Theorem 10.6a where $[\alpha, \beta] = [\alpha, \lambda]$ and

$$H(\sigma) := \frac{S(\tau,\sigma)}{\sin(\pi\sigma/2\lambda)}.$$

Because $S(\tau, \sigma)$ as a function of σ is integrable on $[\alpha, \lambda]$, so is $H(\sigma)$, and the desired relation

$$\lim_{n \to \infty} \int_\alpha^\lambda \frac{S(\tau, \sigma)}{\sin(\pi\sigma/2\lambda)} \sin\left(\frac{2n+1}{2\lambda}\sigma\pi\right) d\sigma = 0$$

follows at once from the stronger assertion (10.6–8). To deal with the first integral, we let $\omega := [(2n+1)/2\lambda]\pi$ and consider

$$\int_0^\alpha S(\tau, \sigma)\frac{\sin \omega\sigma}{\sin(\pi\sigma/2\lambda)} d\sigma = \frac{2\lambda}{\pi} \int_0^\alpha S(\tau, \sigma)\frac{\sin \omega\sigma}{\sigma} d\sigma$$

$$+ \int_0^\alpha S(\tau, \sigma)\left\{\frac{1}{\sin(\pi\sigma/2\lambda)} - \frac{2\lambda}{\pi\sigma}\right\} \sin \omega\sigma \, d\sigma.$$

The first integral on the right tends to zero for $\omega \to \infty$ by hypothesis, and the second by the Riemann–Lebesgue lemma, because the function

$$G(\sigma) := \begin{cases} \dfrac{1}{\sin(\pi\sigma/2\lambda)} - \dfrac{2\lambda}{\pi\sigma}, & \sigma \neq 0 \\ 0 & , \quad \sigma = 0 \end{cases}$$

is continuous, hence $S(\tau, \sigma)G(\sigma)$ is integrable, on $[0, \lambda]$. ■

Condition (C), which if satisfied at a point τ ensures the convergence of the Fourier series at that point, occupies a central position in the theory but is not easily verified directly. It should be pointed out, first of all, that mere continuity of F at τ does not imply (C), for it can be shown that there exist continuous functions whose Fourier series diverge at certain points. On the other hand, it follows at once from the Riemann–Lebesgue lemma that (C) holds if $S(\tau, \sigma)\sigma^{-1}$ is integrable on $[0, \alpha]$. (This is known as **Dini's test** for convergence.) This is the case, for instance, if $F'(\tau)$ exists or, more generally, if there exist constants μ and $\eta > 0$ such that

$$|F(\tau+\sigma) - F(\tau)| \leq \mu|\sigma|^\eta \quad \text{for} \quad |\sigma| \leq \alpha.$$

We mention without proof that (C) also holds if F is of bounded variation[3]

---

[3] A function F is said to be of **bounded variation** on an interval $[\alpha, \beta]$ if there exists a constant μ such that, whatever subdivision $\alpha = \tau_0 < \tau_1 < \cdots < \tau_{n-1} < \tau_n = \beta$ of $[\alpha, \beta]$ is chosen,

$$\sum_{k=1}^n |F(\tau_k) - F(\tau_{k-1})| < \mu.$$

A function is of bounded variation on $[\alpha, \beta]$ if and only if it is the difference of two monotonic functions on $[\alpha, \beta]$. Hence the one-sided limits $F(\tau+)$ and $F(\tau-)$ such as those occurring in (10.6–11) exist at every interior point τ of $[\alpha, \beta]$.

on the interval $[\tau - \alpha, \tau + \alpha]$, and if

$$F(\tau) = \tfrac{1}{2}\{F(\tau+) + F(\tau-)\} \tag{10.6-11}$$

(**Jordan's test** for convergence). For this to be the case it is not necessary that F be continuous at τ.

III. The Fourier Integral Theorem

Let F now be a complex-valued function defined on the real line such that $|F|$ is (at least improperly) integrable on every finite interval, and

$$\int_{-\infty}^{\infty} |F(\tau)| \, d\tau < \infty. \tag{10.6-12}$$

The integral

$$G(\omega) := \int_{-\infty}^{\infty} e^{-i\omega\tau} F(\tau) \, d\tau \tag{10.6-13}$$

then exists for all real ω. The function G is called the **Fourier transform** of F. We note the following properties of G:

THEOREM 10.6c

Under the hypotheses on F stated above, its Fourier transform G is bounded and continuous, and it satisfies

$$\lim_{\omega \to \pm\infty} G(\omega) = 0. \tag{10.6-14}$$

Proof. G is bounded, because

$$|G(\omega)| \leq \int_{-\infty}^{\infty} |F(\tau)| \, d\tau.$$

G is continuous because it is the uniform limit of the continuous functions $\int_{-n}^{n} e^{-i\omega\tau} F(\tau) \, d\tau$, $n = 1, 2, \ldots$. To prove (10.6-14), let $\epsilon > 0$ be given. We write

$$G(\omega) = \left(\int_{-\infty}^{-\eta} + \int_{-\eta}^{\eta} + \int_{\eta}^{\infty} \right) e^{-i\omega\tau} F(\tau) \, d\tau \tag{10.6-15}$$

and choose η such that

$$\int_{\eta}^{\infty} |F(\pm\tau)| \, d\tau < \epsilon.$$

This makes the first and the third integral in (10.6-15) less than ϵ for all values of ω. The middle integral becomes less than ϵ for all sufficiently large $|\omega|$ by the Riemann–Lebesgue lemma. This shows that $|G(\omega)| < 3\epsilon$ for all sufficiently large ω, and thus proves (10.6-14). ∎

We now are able to state the following, not quite perfect, form of the Fourier integral theorem.

THEOREM 10.6d

Let F satisfy the conditions stated previously and let G be its Fourier transform. If τ is any point such that F satisfies condition (C) at τ for some $\alpha > 0$, then

$$F(\tau) = \frac{1}{2\pi} \lim_{\eta \to \infty} \int_{-\eta}^{\eta} e^{i\omega\tau} G(\omega)\, d\omega. \qquad (10.6\text{-}16)$$

The theorem is not perfect because the roles played by F and G are not symmetric. Whereas G exists as an ordinary improper integral,

$$G(\omega) = \lim_{\substack{\eta_1 \to -\infty \\ \eta_2 \to \infty}} \int_{\eta_1}^{\eta_2} e^{-i\omega\sigma} F(\sigma)\, d\sigma$$

(where η_1 and η_2 approach their limits independently), the integral in (10.6–16) exists only as a **principal value integral**, where the upper and the lower limit tend to $\pm\infty$ symmetrically. One also writes (see §4.8)

$$\lim_{\eta \to \infty} \int_{-\eta}^{\eta} =: PV \int_{-\infty}^{\infty}.$$

It can be shown by means of examples that under the conditions stated the ordinary integral

$$\int_{-\infty}^{\infty} e^{i\omega\tau} G(\tau)\, d\tau$$

need not exist.

Proof of Theorem 10.6d.   Let

$$F_\eta(\tau) := \frac{1}{2\pi} \int_{-\eta}^{\eta} e^{i\omega\tau} G(\omega)\, d\omega.$$

In view of (10.6-12) the improper integral defining G converges uniformly with respect to ω. Thus we may interchange integrations to obtain

$$F_\eta(\tau) = \frac{1}{2\pi} \int_{-\eta}^{\eta} e^{i\omega\tau} \left\{ \int_{-\infty}^{\infty} e^{-i\omega\sigma} F(\sigma)\, d\sigma \right\} d\omega$$

$$= \frac{1}{2\pi} \int_{-\infty}^{\infty} F(\sigma) \left\{ \int_{-\eta}^{\eta} e^{i\omega(\tau-\sigma)} \, d\omega \right\} d\sigma$$

$$= \frac{1}{\pi} \int_{-\infty}^{\infty} F(\sigma) \frac{\sin[\eta(\tau-\sigma)]}{\tau-\sigma} d\sigma$$

$$= \frac{1}{\pi} \int_{0}^{\infty} \{F(\tau+\sigma) + F(\tau-\sigma)\} \frac{\sin \eta\sigma}{\sigma} d\sigma.$$

From this we subtract the relation

$$F(\tau) = \frac{1}{\pi} \int_{0}^{\infty} 2F(\tau) \frac{\sin \eta\sigma}{\sigma} d\sigma,$$

which is a consequence of (10.2-15). This yields

$$F_\eta(\tau) - F(\tau) = \frac{1}{\pi} \int_{0}^{\infty} S(\tau, \sigma) \frac{\sin \eta\sigma}{\sigma} d\sigma,$$

where S is the integrable function defined by (10.6-10).

To show that $F_\eta(\tau) \to F(\tau)$ for $\eta \to \infty$, let $\epsilon > 0$ be given. We write

$$\int_{0}^{\infty} = \int_{0}^{\alpha} + \int_{\alpha}^{\chi} + \int_{\chi}^{\infty}, \tag{10.6-17}$$

and choose $\chi > 1$ such that

$$\int_{\chi}^{\infty} |F(\tau+\sigma) + F(\tau-\sigma)| \, d\sigma < \epsilon$$

[which is possible in view of (10.6-12)] and also such that

$$\left| 2F(\tau) \int_{\chi}^{\infty} \frac{\sin \eta\sigma}{\sigma} d\sigma \right| < \epsilon$$

for all $\eta > 1$ (which is possible because the improper integral converges uniformly for $\eta > 1$). This makes

$$\left| \int_{\chi}^{\infty} S(\tau, \sigma) \frac{\sin \eta\sigma}{\sigma} d\sigma \right|$$

$$\leqslant \left| \int_{\chi}^{\infty} \{F(\tau+\sigma) + F(\tau-\sigma)\} \frac{\sin \eta\sigma}{\sigma} d\sigma \right| + \left| 2F(\tau) \int_{\chi}^{\infty} \frac{\sin \eta\sigma}{\sigma} d\sigma \right|$$

$$\leqslant 2\epsilon.$$

The middle integral in (10.6-17) tends to zero for $\eta \to \infty$ by the Riemann–Lebesgue Lemma, and the first tends to zero by hypothesis (C). This shows that the sum of all three integrals is $< 4\epsilon$ for all sufficiently large η, proving (10.6-16). ∎

We have already commented on the lack of symmetry between the roles of F and G in the statement of Theorem 10.6d. A perfectly symmetric theory of the Fourier integral (due to M. Plancherel) is possible in the space $L^2(-\infty, \infty)$ of functions F satisfying

$$\int_{-\infty}^{\infty} |F(\tau)|^2 \, d\tau < \infty, \tag{10.6-18}$$

where the integral is a Lebesgue integral. It then turns out that the Fourier transform of F (if defined appropriately) is a member of the same space; moreover, there holds the relation

$$\int_{-\infty}^{\infty} |F(\tau)|^2 \, d\tau = \frac{1}{2\pi} \int_{-\infty}^{\infty} |G(\omega)|^2 \, d\omega \tag{10.6-19}$$

which is known as **Parseval's formula**.

IV. The Poisson Summation Formula

Let $\eta > 0$, and let F be an integrable function defined on the real line which, in addition to satisfying (10.6-12), is such that the series

$$f(\tau) := \sum_{n=-\infty}^{\infty} F(n\eta + \tau) \tag{10.6-20}$$

converges uniformly for $0 \leq \tau \leq \eta$. Then f is periodic with period η, and its Fourier coefficients can be formed. The nth coefficient is

$$a_n := \frac{1}{\eta} \int_0^{\eta} f(\tau) \exp\left(-\frac{2\pi i n \tau}{\eta}\right) d\tau.$$

Inserting the series (10.6-20) for f and interchanging summation and integration (permissible by uniform convergence) we obtain

$$a_n = \frac{1}{\eta} \sum_{k=-\infty}^{\infty} \int_0^{\eta} F(k\eta + \tau) \exp\left(-\frac{2\pi i n \tau}{\eta}\right) d\tau$$

$$= \frac{1}{\eta} \sum_{k=-\infty}^{\infty} \int_0^{\eta} F(k\eta + \tau) \exp\left(-\frac{2\pi i n (k\eta + \tau)}{\eta}\right) d\tau$$

$$= \frac{1}{\eta} G\left(\frac{2\pi n}{\eta}\right),$$

where G is the Fourier transform of F. Thus if f satisfies condition (C) at τ, there holds the formula

$$\sum_{k=-\infty}^{\infty} F(k\eta + \tau) = \frac{1}{\eta} PV \sum_{n=-\infty}^{\infty} G\left(\frac{2\pi n}{\eta}\right) \exp\left(\frac{2\pi i n \tau}{\eta}\right), \tag{10.6-21}$$

known as the **Poisson summation formula**. The symbol PV indicates that the sum is to be regarded as a *symmetric* limit of its partial sums.

The summation formula expresses a series of values at equidistant points of F as a series of values at equidistant points of G. If the step η on the τ axis becomes small, the corresponding step $2\pi/\eta$ on the ω axis becomes large—which may have a desirable effect on the speed of convergence.

THEOREM 10.6e

The summation formula (10.6-21) *holds for all real τ if F [in addition to (10.6-12)] satisfies either of the following conditions:*

(a) *F can be extended to a function of the complex variable t that is analytic in a strip $|\text{Im } t| < \theta$ such that the series* (10.6-20) *converges uniformly in that strip;*

(b) *F is of bounded variation on $(-\infty, \infty)$, and at points of discontinuity F is normalized such that*

$$F(\tau) = \tfrac{1}{2}\{F(\tau+) + F(\tau-)\}.$$

Proof. (a) If the convergence of the series (10.6-20) is uniform for $|\text{Im } t| < \theta$, then f is an analytic function of period η, and its Fourier series converges automatically by Theorem 4.5b. The formula (10.6-21) then actually holds for $|\text{Im } t| < \theta$.

(b) Let τ be an arbitrary point on the real line, and let I be an interval of length $\leq \eta$ containing τ in its interior. We shall show that f is of bounded variation on I. If $\tau_0 < \tau_1 < \cdots < \tau_n$ is any subdivision of I, then

$$\sum_{i=1}^{n} |f(\tau_i) - f(\tau_{i-1})| = \sum_{i=1}^{n} \left| \sum_{k=-\infty}^{\infty} \{F(k\eta + \tau_i) - F(k\eta + \tau_{i-1})\} \right|$$

$$\leq \sum_{i=1}^{n} \sum_{k=-\infty}^{\infty} |F(k\eta + \tau_i) - F(k\eta + \tau_{i-1})|$$

$$\leq \sum_{k=-\infty}^{\infty} \sum_{i=1}^{n} |F(k\eta + \tau_i) - F(k\eta + \tau_{i-1})|.$$

The last series is bounded by μ, the total variation of F on $(-\infty, \infty)$, and it follows that the variation of f on I is likewise bounded by μ. Thus f satisfies (C) at τ, and (10.6-21) follows. ∎

EXAMPLE

Let $z \in \mathbb{C}$ and $\lambda > 0$ be two parameters, and let

$$F(\tau) := \exp(-\pi\lambda\tau^2 + 2iz\tau).$$

This clearly satisfies (10.6-12) and condition (a) of Theorem 10.6e. For the Fourier transform G of F we find by completing the square in the exponent

$$G(\omega) = \int_{-\infty}^{\infty} \exp(-\pi\lambda\tau^2 + 2iz\tau - i\omega\tau)\, d\tau$$

$$= \exp\left[-\frac{1}{\pi\lambda}\left(z - \frac{\omega}{2}\right)^2\right] \int_{-\infty}^{\infty} \exp\left[-\pi\lambda\left(\tau - \frac{iz}{\pi\lambda} + \frac{i\omega}{2\pi\lambda}\right)^2\right] d\tau.$$

By Cauchy's theorem, the path of integration may be shifted to $\operatorname{Im}\tau = \operatorname{Re}(z/\pi\lambda - \omega/2\pi\lambda)$, and we get

$$G(\omega) = \exp\left[-\frac{1}{\pi\lambda}\left(z - \frac{\omega}{2}\right)^2\right] \int_{-\infty}^{\infty} \exp(-\pi\lambda s^2)\, ds$$

$$= \exp\left[-\frac{1}{\pi\lambda}\left(z - \frac{\omega}{2}\right)^2\right] \frac{1}{\sqrt{\lambda}}.$$

We apply the summation formula (10.6-21) where $\eta = 1$ and $\tau = 0$. The resulting series on the left,

$$\theta(z, \lambda) := \sum_{k=-\infty}^{\infty} \exp(-\pi\lambda k^2 + 2ikz), \tag{10.6-22}$$

is known as a θ function. (The parameter λ is usually denoted by $-i\tau$.) The series on the right can likewise be expressed by the above θ function, and we obtain **Jacobi's identity**

$$\theta(z, \lambda) = \frac{1}{\sqrt{\lambda}} \exp\left(-\frac{z^2}{\pi\lambda}\right) \theta\left(\frac{z}{i\lambda}, \frac{1}{\lambda}\right). \tag{10.6-23}$$

For $z = 0$ there results in particular $\theta(0, \lambda) = (1/\sqrt{\lambda})\theta(0, 1/\lambda)$ or

$$\sum_{n=-\infty}^{\infty} \exp(-\pi\lambda n^2) = \frac{1}{\sqrt{\lambda}} \sum_{n=-\infty}^{\infty} \exp\left(-\frac{\pi n^2}{\lambda}\right). \tag{10.6-24}$$

Setting $q := e^{-\pi\lambda}$, this relation expresses the value of the series

$$f(q) := \sum_{n=-\infty}^{\infty} q^{n^2} \tag{10.6-25}$$

(where $|q| < 1$) in terms of $f(q_1)$ where $\log q \log q_1 = \pi^2$. Thus if q is real, the series $f(q)$ (which is of importance in the theory of elliptic functions, and in the theory of partitions, see §8.2) can always be evaluated numerically with a value of q such that $|q| < e^{-\pi} = 0.04321 \ldots$, where the convergence is very rapid indeed.

PROBLEMS

1. Find the Fourier transforms of the functions

(a) $F(\tau) := \begin{cases} \alpha, & |\tau| < \beta \\ 0, & |\tau| \geq \beta \end{cases}$

(b) $\quad F(\tau) := \begin{cases} \alpha\left(1 - \dfrac{|\tau|}{\beta}\right), & |\tau| < \beta \\ 0, & |\tau| \geqslant \beta \end{cases}$

2. The relation (10.6-13) holds for

$$F(\tau) := \frac{1}{a^2 + \tau^2}, \qquad G(\omega) := \frac{\pi}{a} e^{-|\omega|a}$$

where $a > 0$ (see example 4, §4.8). Use this result to express

$$\sum_{k=-\infty}^{\infty} \frac{1}{a^2 + (k\eta + \tau)^2}$$

by a rapidly converging series when $\eta > 0$ is small. (Perform numerical tests for $a = 1$, $\tau = 0$, $\eta = 2^{-m}$, $m = 0, 1, 2, \ldots$).

3. For real $\alpha, \beta > 0$ show that if $\beta/2\pi$ is not an integer,

$$PV \sum_{n=-\infty}^{\infty} \frac{\sin[(\alpha + n)\beta]}{\alpha + n} = \pi \frac{\sin[(2m+1)\pi\alpha]}{\sin \pi\alpha}$$

where $m := [\beta/2\pi]$. Deduce that for $\beta < 2\pi$

$$PV \sum_{n=-\infty}^{\infty} \frac{\sin[(\alpha + n)\beta]}{\alpha + n} = \int_{-\infty}^{\infty} \frac{\sin \beta\tau}{\tau} d\tau = \pi.$$

[Compare problem 1.]

4. Show that for $\eta > 0$

$$\sum_{k=-\infty}^{\infty} \left(\frac{\sin k\eta}{k\eta}\right)^2 = \frac{\pi}{\eta}\left\{1 + 2m - \frac{\pi}{\eta}m(m+1)\right\},$$

where $m := [\eta/\pi]$. Deduce that for $\eta < \pi$

$$\sum_{k=-\infty}^{\infty} \left(\frac{\sin k\eta}{k\eta}\right)^2 = \int_{-\infty}^{\infty} \left(\frac{\sin \tau}{\tau}\right)^2 d\tau.$$

Also check the cases $\eta = \pi$, $\eta = \pi/2$. [Compare problem 1.]

5. Poisson's integral (9.7-29) shows that

$$G(\omega) := J_0(\omega)$$

is the Fourier transform of

$$F(\tau) := \begin{cases} \dfrac{1}{\pi} \dfrac{1}{\sqrt{1 - \tau^2}}, & |\tau| < 1, \\ 0, & |\tau| \geqslant 1. \end{cases}$$

Conclude that for $\alpha > 0$, $\tau \geqslant 0$

$$1 + 2 \sum_{n=1}^{\infty} J_0(\alpha n) \cos(\alpha n \tau) = 2 \sum_k \frac{1}{\sqrt{\alpha^2 - (2\pi k + \alpha\tau)^2}},$$

where the sum on the right involves those integers k for which the square root is real.

6. For some functions F satisfying the hypotheses of Theorem 10.6e it may happen that for certain $\eta > 0$

$$\eta \sum_{k=-\infty}^{\infty} F(k\eta) = \int_{-\infty}^{\infty} F(\tau)\, d\tau,$$

compare the problems above. Explain, and state a general result to this effect.

7. Let $\alpha > 0$. From the fact that

$$\int_{-\infty}^{\infty} \frac{1}{\sqrt{\alpha^2 + \tau^2}} e^{-i\omega\tau}\, d\tau = 2K_0(\alpha|\omega|)$$

(K_0 = modified Bessel function of second kind), deduce that for γ real, $\gamma \neq 0$, $\pm 2\pi$, $\pm 4\pi$, ...

$$\sum_{k=-\infty}^{\infty} \frac{e^{ik\gamma}}{\sqrt{\alpha^2 + k^2\eta^2}} = \frac{2}{\eta} \sum_{n=-\infty}^{\infty} K_0\left(\frac{\alpha}{\eta}|2\pi n - \gamma|\right).$$

The following problems are concerned with the approximation of the integral

$$I := \int_{\alpha}^{\beta} F(\tau)\, d\tau$$

$(-\infty < \alpha < \beta < \infty$, F continuous on $[\alpha, \beta]$) by the **trapezoidal values**

$$T(\eta) := \frac{\eta}{2}\{F(\tau_0) + 2F(\tau_1) + 2F(\tau_2) + \cdots + 2F(\tau_{N-1}) + F(\tau_N)\}$$

of the integral, where $N = 1, 2, \ldots$, $\eta := (\beta - \alpha)/N$, $\tau_k := \alpha + k\eta$. If F is analytic on $[\alpha, \beta]$, it can be shown (see §11.12) that there exist constants $c_1, c_2, \ldots$, such that

$$T(\eta) \approx I + \sum_{k=1}^{\infty} c_k \eta^{2k}$$

asympotically as $\eta \to 0$, which makes it possible to speed up the convergence of a sequence of trapezoidal values by means of the *Romberg algorithm*. If F is not analytic, Poisson's sum formula may be used to study the asymptotic behavior of $T(\eta)$, which then can be more complicated.

8. Show that the trapezoidal values for the integral

$$I := \int_{-1}^{1} \sqrt{1-\tau^2}\, d\tau$$

satisfy

$$T(\eta) = \frac{\pi}{2} + 2\pi \sum_{n=1}^{\infty} \frac{\eta}{2\pi n} J_1\left(\frac{2\pi n}{\eta}\right)$$

and use the known asymptotic behavior of the Bessel function (see §11.5) to show that for a certain constant c,

$$T(\eta) = \frac{\pi}{2} + c\eta^{3/2} + O(\eta^{5/2}).$$

9. If

$$I := -\int_{0}^{1} \tau \operatorname{Log} \tau\, d\tau,$$

show that there exist constants c_1, c_2 such that

$$T(\eta) = \tfrac{1}{2} + c_1\eta^2 + c_2\eta^2 \operatorname{Log} \eta + O(\eta^4).$$

How would the values $T(2^{-n})$ have to be treated by the Romberg method?

10. Show that for $\eta \to 0$ the trapezoidal values

$$T(\eta) := \eta \sum_{k=-\infty}^{\infty} e^{-(k\eta)^2}$$

for $\int_{-\infty}^{\infty} e^{-\tau^2}\, d\tau$ converge to $\sqrt{\pi}$ with an error that is of smaller order than any power of η.

11. Let F be a periodic analytic function on the real line with period $\lambda > 0$. Show that the trapezoidal values $T(\eta)$ for the integral

$$I := \int_{0}^{\lambda} F(\tau)\, d\tau$$

converge to I with an error that is of smaller order than any power of η. [Consider the Theorems 4.5b and 4.4c.]

12. Prove Parseval's relation (10.6-19) by writing

$$\int_{-\infty}^{\infty} |F(\tau)|^2\, d\tau = \int_{-\infty}^{\infty} F(\tau)\overline{F(\tau)}\, d\tau$$

and expressing $F(\tau)$ in terms of G, assuming that integrations may be interchanged and that $\int_{-\infty}^{\infty} |G(\omega)|\, d\omega$ converges.

13. Let G_k denote the Fourier transform of F_k, $k = 1, 2$. Under appropriate hypotheses prove the following generalization of Parseval's formula:

$$\int_{-\infty}^{\infty} F_1(\tau)\overline{F_2(\tau)}\, d\tau = \frac{1}{2\pi} \int_{-\infty}^{\infty} G_1(\omega)\overline{G_2(\omega)}\, d\omega.$$

14.   Let F be such that $\int_{-\infty}^{\infty} |F(\tau)| \, d\tau < \infty$, and let F satisfy condition (C) everywhere. In addition we assume that F is *band-limited*, that is, that its Fourier transform $G := \mathscr{F}F$ is zero outside a finite interval, which we take to be $[-\pi, \pi]$. By the Fourier integral theorem,

$$F(\tau) = \frac{1}{2\pi} \int_{-\pi}^{\pi} G(\omega) \, e^{i\omega\tau} \, d\tau, \quad -\infty < \tau < \infty.$$

The **sampling theorem** states that there also holds

$$F(\xi) = \sum_{n=-\infty}^{\infty} F(n) \frac{\sin \pi(\xi - n)}{\pi(\xi - n)}, \quad -\infty < \xi < \infty.$$

(a) Prove the sampling theorem by noting that the series on the right is $\sum a_{-n} b_n$, where $a_{-n} := F(n)$ is the $(-n)$th Fourier coefficient of the function G (continued periodically), and

$$b_n := \frac{\sin \pi(\xi - n)}{\pi(\xi - n)}$$

is the nth Fourier coefficient of $G_1(\omega) := e^{i\xi\omega}$. Thus the series equals the zeroth Fourier coefficient of GG_1, which can also be calculated directly.

(b) Prove the sampling theorem as a special case of the Poisson summation formula by noting that the product FF_1, where

$$F_1(\tau) := \frac{\sin \pi(\xi - \tau)}{\pi(\xi - \tau)},$$

has as its Fourier transform the convolution (appropriately defined) of G and G_1.

§10.7. THE LAPLACE TRANSFORM AS A FOURIER TRANSFORM

Let F be an original function with abscissa of absolute convergence $\beta_F < \infty$. Then for every $\sigma_0 > \beta_F$ (and possibly also for $\sigma_0 = \beta_F$), because $F(\tau) = 0$ for $\tau < 0$,

$$\int_{-\infty}^{\infty} |e^{-\sigma_0\tau} F(\tau)| \, d\tau < \infty. \tag{10.7-1}$$

If $s = \sigma + i\omega$ and $\sigma \geq \sigma_0$, the Laplace transform $f = \mathscr{L}F$ may be written

$$f(\sigma + i\omega) = \int_{-\infty}^{\infty} e^{-\sigma\tau} F(\tau) \, e^{-i\omega\tau} \, d\tau \tag{10.7-2}$$

and thus as a function of ω may be looked at as the *Fourier* transform of a function whose integral over $(-\infty, \infty)$ converges absolutely. The fact that

Laplace transforms are also Fourier transforms opens the arsenal of Fourier analysis to Laplace transform theory. Among other things, an entirely new look at the inversion problem is thus provided. This and some other consequences are discussed below.

I. Behavior of an $\mathscr{L}$ Transform at Infinity

The following is a more complete form of Theorem 10.1g.

THEOREM 10.7a

Let $F \in \Omega$, and let σ_0 be such that (10.7-1) holds. Then $f := \mathscr{L}F$ satisfies

$$\lim_{s \to \infty} f(s) = 0, \tag{10.7-3}$$

where s tends to infinity in the half plane $\mathrm{Re}\, s \geq \sigma_0$.

Proof.   We already know from Theorem 10.1g that (10.7-3) holds if $s \to \infty$ in any sector $|\arg s| \leq \beta$ where $0 \leq \beta < \pi/2$. Thus for fixed β, if $\epsilon > 0$ is given, there exists ρ such that $|\arg s| \leq \beta$ and $|s| > \rho$ implies $|f(s)| < \epsilon$. On the other hand, Theorem 10.6c implies that for every $\sigma \geq \sigma_0$

$$\lim_{\omega \to \pm\infty} f(\sigma + i\omega) = 0 \tag{10.7-4}$$

An examination of the proof of Theorem 10.6c and of the Riemann–Lebesgue lemma on which it is based shows that (10.7-4) holds uniformly for $\sigma \geq \sigma_0$. Thus for the ϵ given previously, there exists ω_0 such that $|f(\sigma + i\omega)| < \epsilon$ for all $\sigma \geq \sigma_0$ if $|\omega| > \omega_0$. Let now

$$\rho_1 := \max\left(\rho, \frac{\omega_0}{\sin \beta}\right).$$

A simple drawing then will convince the reader that $|f(s)| < \epsilon$ for all s such that $\mathrm{Re}\, s \geq \sigma_0$ and $|s| > \rho_1$, which amounts to the statement of the theorem.   ∎

II. The Complex Inversion Formula

If (10.7-1) holds, then for every $\sigma \geq \sigma_0$ the Fourier integral theorem permits us to recover $e^{-\sigma\tau}F(\tau)$ from the values $f(\sigma + i\omega)$ at all points τ where $e^{-\sigma\tau}F(\tau)$ [or $F(\tau)$] satisfies condition (C). If τ is any such point,

$$e^{-\sigma\tau}F(\tau) = \frac{1}{2\pi} PV \int_{-\infty}^{\infty} f(\sigma + i\omega)\, e^{i\omega\tau}\, d\omega$$

or equivalently

$$F(\tau) = \frac{1}{2\pi} PV \int_{-\infty}^{\infty} f(\sigma + i\omega) \, e^{(\sigma + i\omega)\tau} \, d\omega.$$

The integral may be interpreted as (the principal value of) the complex line integral of $f(s) \, e^{s\tau}$ taken along the vertical straight line $s = \sigma + i\omega$, $-\infty < \omega < \infty$. Because $ds = i \, d\omega$, there follows

$$F(\tau) = \frac{1}{2\pi i} PV \int_{\sigma - i\infty}^{\sigma + i\infty} e^{s\tau} f(s) \, ds. \qquad (10.7\text{-}5)$$

It is understood that

$$PV \int_{\sigma - i\infty}^{\sigma + i\infty} := \lim_{\eta \to \infty} \int_{\sigma - i\eta}^{\sigma + i\eta},$$

and that the integral on the right is taken along the vertical straight line segment joining the points $\sigma \pm i\eta$. The formula (10.7-5) is known as the **complex inversion formula** for the Laplace transformation. We summarize the conditions of its validity.

THEOREM 10.7b

Let $F \in \Omega$, $f := \mathcal{L}F$, and let σ_0 be such that (10.7-1) holds. Then the complex inversion formula (10.7-5) holds for all $\sigma \geq \sigma_0$ at all points τ where F satisfies condition (C).

Thus for instance, if τ is any point such that F has bounded variation in $[\tau - \alpha, \tau + \alpha]$ for some (small) $\alpha > 0$, then the integral on the right of (10.7-5) converges to

$$\tfrac{1}{2}[F(\tau+) + F(\tau-)].$$

EXAMPLE **1**

Let $F(\tau) := H(\tau)$, the Heaviside unit function. Then $f(s) = s^{-1}$, and the inversion formula yields for $\sigma > 0$

$$\frac{1}{2\pi i} \int_{\sigma - i\infty}^{\sigma + i\infty} \frac{e^{s\tau}}{s} \, ds = \begin{cases} 1, & \tau > 0, \\ \tfrac{1}{2}, & \tau = 0, \\ 0, & \tau < 0. \end{cases} \qquad (10.7\text{-}6)$$

These results can be verified directly, using the residue theorem.

Remark 1. At first sight it seems surprising that the principal value integral on the right of (10.7-5) does not depend on σ. This, however, is an immediate consequence of Cauchy's theorem. Let $\sigma_0 \leq \sigma_1 < \sigma_2$, and consider the rectangular Jordan curve with the four corners $\sigma_1 \pm i\eta$, $\sigma_2 \pm i\eta$.

Because $e^{s\tau}f(s)$ is analytic for $\operatorname{Re} s > \sigma_0$, the integral around the curve is zero. By Theorem 10.7a, the contributions of the horizontal pieces tend to zero for $\eta \to \infty$, and it thus follows that

$$PV \int_{\sigma_1 - i\infty}^{\sigma_1 + i\infty} = PV \int_{\sigma_2 - i\infty}^{\sigma_2 + i\infty}.$$

Remark 2. By virtue of its derivation, the inversion formula also holds for $\tau < 0$ and then must yield the value zero. This also can be seen directly, as follows. By Cauchy's theorem,

$$\int_{\sigma - i\rho}^{\sigma + i\rho} e^{s\tau}f(s)\, ds = \int_{\Gamma_\rho} e^{s\tau}f(s)\, ds,$$

where Γ_ρ denotes the circular arc $s = \sigma + \rho\, e^{i\phi}$, $-\pi/2 \le \phi \le \pi/2$. Let

$$\mu_\rho := \max_{s \in \Gamma_\rho} |f(s)|.$$

If $\tau < 0$, the integral on the right as in Jordan's lemma is estimated by

$$\mu_\rho \int_{\Gamma_\rho} |e^{s\tau}||ds| = 2\mu_\rho\rho\, e^{\sigma\tau} \int_0^{\pi/2} e^{\tau\rho\, \cos\phi}\, d\phi \le \frac{\pi}{(-\tau)}\mu_\rho\, e^{\sigma\tau}.$$

By virtue of Theorem 10.7a, $\mu_\rho \to 0$ as $\rho \to \infty$. Thus for $\tau < 0$,

$$PV \int_{\sigma - i\infty}^{\sigma + i\infty} e^{s\tau}f(s)\, ds = 0.$$

The complex inversion formula features, in a sense, an explicit representation of the original function F in terms of its image function f. No undue confidence should be placed, however, in the immediate usefulness of this formula for the purpose of determining the original function by means of numerical integration, because

(i) a separate integration would be required for each value of τ;

(ii) for large τ, the integrand oscillates rapidly, thus the integral is difficult to evaluate numerically;

(iii) the convergence of the improper integral is determined by the speed with which $f(\sigma \pm i\omega)$ tends to zero for $\omega \to \infty$, which may be low.

In the following subsections the inversion formula is applied to derive—under suitable additional conditions on f—several algorithmically more useful representations of F in terms of its image. For the first time in this chapter, use is made of the fact that because $\mathscr{L}$ transforms are analytic, their analytic continuation into any given region, if it exists, is unique. Some additional inversion formulas based on this fact are derived in §10.9.

III. The Heaviside Expansion Theorem

Let $F \in \Omega$, $\beta_F < \infty$, and let $f := \mathcal{L}F$ satisfy the following conditions:

(I) f can be continued analytically into the whole complex plane, with the exception of finitely or infinitely many isolated singularities, say at the points $s_1, s_2, \ldots$.

(II) There exists a sequence of circles $\Gamma_n : |s| = \rho_n$, where $\rho_n \to \infty$, such that the quantities

$$\mu_n := \sup_{s \in \Gamma_n} |f(s)| \tag{10.7-7}$$

tend to zero for $n \to \infty$. (The fact that f is a Laplace transform merely guarantees that $f(s) \to 0$ if s tends to infinity in some right half plane; we now postulate that $f(s) \to 0$ on a certain sequence of *whole* circles.)

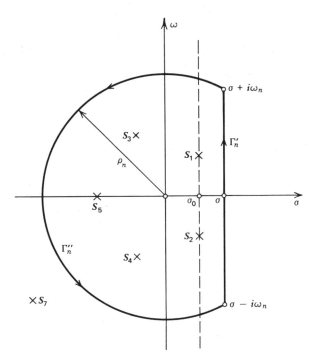

Fig. 10.7a.

Let $\sigma > \beta_F$, and let Γ_n intersect the straight line $\mathrm{Re}\, s = \sigma$ in the points $\sigma \pm i\omega_n$. We apply the residue theorem to the closed curve consisting of the straight line segment Γ_n' joining $\sigma \pm i\omega_n$ and of the part Γ_n'' of Γ_n lying to the left of Re $s = \sigma$. This yields

$$\frac{1}{2\pi i} \int_{\Gamma'_n + \Gamma''_n} e^{s\tau} f(s) \, ds = \sum_{|s_k| < \rho_n} \text{res}(e^{s\tau} f(s))_{s=s_k}. \qquad (10.7\text{-}8)$$

Here we let $n \to \infty$. By Jordan's lemma (Lemma 4.8b) relation (10.7-7) implies that

$$\lim_{n \to \infty} \int_{\Gamma''_n} e^{s\tau} f(s) \, ds = 0.$$

By definition of the principal value,

$$\lim_{n \to \infty} \frac{1}{2\pi i} \int_{\Gamma'_n} e^{s\tau} f(s) \, ds = \frac{1}{2\pi i} PV \int_{\sigma - i\infty}^{\sigma + i\infty} e^{s\tau} f(s) \, ds$$

whenever the latter exists. By Theorem 10.7b the principal value does exist and equals

$$\tfrac{1}{2}[F(\tau+) + F(\tau-)]$$

at every τ where F satisfies condition (C). Thus for every such τ the limit as $n \to \infty$ of the expression on the left of (10.7-8) exists. There follows the existence of the limit on the right, and we get

$$\tfrac{1}{2}[F(\tau+) + F(\tau-)] = \sum_{k=1}^{\infty} \text{res}(e^{s\tau} f(s))_{s=s_k}. \qquad (10.7\text{-}9)$$

This formula expresses the original function in terms of the singularities of its $\mathscr{L}$ transform. We summarize:

THEOREM 10.7c (Heaviside Expansion Theorem)

Let $F \in \Omega$, $\beta_F < \infty$, let $f := \mathscr{L}F$ be analytically continuable into the complex plane with the exception of finitely or infinitely many isolated singularities at the points $s_1, s_2, \ldots$, and let

$$\lim_{n \to \infty} \sup_{s \in \Gamma_n} |f(s)| = 0 \qquad (10.7\text{-}10)$$

for a suitable sequence of circles Γ_n whose radii tend to infinity as $n \to \infty$. Then (10.7-9) holds at every $\tau > 0$ where F satisfies condition (C).

EXAMPLE **2**

Let f be a rational function vanishing at ∞ with poles at $s_1, \ldots, s_n$. Then f trivially satisfies the hypotheses of Theorem 10.7c. If the principal part of f at s_k is

$$\sum_{m=1}^{m_k} \frac{a_{k,m}}{(s - s_k)^m},$$

then the residue of $e^{s\tau}f(s)$ at s_k equals

$$\text{res}\left\{e^{s_k\tau}\sum_{l=0}^{\infty}\frac{(s-s_k)^l\tau^l}{l!}\sum_{m=1}^{m_k}\frac{a_{k,m}}{(s-s_k)^m}\right\}_{s=s_k}=e^{s_k\tau}\sum_{m=1}^{m_k}\frac{a_{k,m}}{(m-1)!}\tau^{m-1}.$$

Thus $F=\mathcal{L}^{-1}f$ is the exponential polynomial already determined in §10.3.

In the proof of formula (10.7-9) it is not essential that the curves Γ_n are circles. What matters is only that these curves tend to infinity, and that f tends to zero on them. In some cases these properties are easier to verify for rectangular or even for parabolic curves.

EXAMPLE **3**

Here we show how to use the expansion theorem for obtaining the Fourier series of nonanalytic periodic functions. (For analytic functions, see §4.5.) Let $\omega > 0$,

$$F(\tau) := |\sin \omega\tau|, \qquad \tau \geqslant 0.$$

This satisfies (C) at all τ. We know (cf. Problem 9, §10.2) that

$$F(\tau) \circ\!\!-\!\!-\!\!\bullet \frac{\omega}{s^2+\omega^2}\frac{1+e^{-s\pi/\omega}}{1-e^{-s\pi/\omega}}=: f(s).$$

The poles of f are at

$$s = 2ik\omega, \qquad k = 0, \pm 1, \pm 2, \dots.$$

(The apparent poles at $s = \pm i\omega$ are cancelled by the corresponding zeros of the numerator.) Much as in the proof of the summation formulas of §4.9, one can show that

$$\frac{1+e^{-s\pi/\omega}}{1-e^{-s\pi/\omega}}$$

is bounded on the squares with corners $(\pm 1 \pm i)s_k$, where $s_k := (2k+1)\omega$. In view of the factor $(s^2+\omega^2)^{-1}$, $|f(s)|$ tends to zero on these squares. The same factor causes $|f|$ to be integrable on all vertical lines $\text{Re } s = \sigma > 0$. Thus the hypotheses of the expansion theorem are satisfied. In view of

$$\text{res}\left\{\frac{\omega}{s^2+\omega^2}\frac{1+e^{-s\pi/\omega}}{1-e^{-s\pi/\omega}}\right\}_{s=2i\omega k}=\frac{1}{\pi}\frac{2}{1-4k^2}$$

(10.7-9) yields after combining conjugate terms

$$|\sin \omega\tau| = \frac{2}{\pi} - \frac{4}{\pi}\sum_{k=1}^{\infty}\frac{1}{4k^2-1}\cos 2\omega k\tau.$$

IV.   A Real Inversion Formula

Here we assume that the $\mathcal{L}$ transform $f = \mathcal{L}F$ of an original function F $(\beta_F < \infty)$ can be continued analytically into the complex plane cut along a half line in the direction of the negative real axis, that is, into a set $|\arg(s - s_0)| < \pi$. Without loss of generality we may assume that $s_0 = 0$, because by rule VIII (damping) this can be achieved by multiplying F by $e^{-s_0\tau}$. Under certain additional hypotheses it will be shown that the original function F can, in turn, be represented by a Laplacian integral involving f.

The additional hypotheses are as follows:

(i) Near $s = 0$, f is such that

$$\lim_{s \to 0} sf(s) = 0, \tag{10.7-11}$$

without restriction on the manner of approach.

(ii) Without restriction on the manner of approach to ∞ in the cut plane $|\arg s| < \pi$, there holds

$$\lim_{s \to \infty} f(s) = 0. \tag{10.7-12}$$

(iii) On the upper edge of the cut, f can be defined in such a manner that f is continuous in the upper half plane Im $s \geq 0$, with the possible exception of $s = 0$, and similarly on the lower edge of the cut. The values of f on the upper and lower edges of the cut are denoted by $f(e^{i\pi}\sigma)$ and $f(e^{-i\pi}\sigma)$, respectively, where $\sigma > 0$.

We integrate $e^{s\tau}f(s)$, where $\tau > 0$, along the simple closed curve Γ shown in Fig. 10.7b. Because the integrand is analytic inside Γ, the value of the

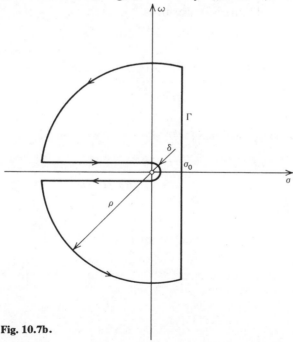

Fig. 10.7b.

integral is zero by Cauchy's theorem for all $\delta > 0$ and all $\rho > 0$. Letting $\delta \to 0$, the integral along the small semicircle tends to zero by (10.7-11), and the integrals along the horizontal parts by hypothesis (iii) tend to

$$\int_0^\rho e^{-\tau\sigma}\{f(e^{i\pi}\sigma) - f(e^{-i\pi}\sigma)\}\, d\sigma.$$

As $\rho \to \infty$, the integrals along the outer circular arcs tend to zero by (10.7-12), using Jordan's lemma. The integral along the vertical part of Γ tends to $2\pi i F(\tau)$ at every point τ where F satisfies condition (C). Thus in the limit we get

THEOREM 10.7d

Let $F \in \Omega$, $\beta_F < \infty$, and let $f := \mathcal{L}F$ be continuable into the cut plane $|\arg s| < \pi$ such that it satisfies the conditions (i), (ii), (iii) mentioned above. Then at every $\tau > 0$ where F satisfies condition (C),

$$F(\tau) = \frac{1}{2\pi i} \int_0^\infty e^{-\tau\sigma}\{f(e^{-i\pi}\sigma) - f(e^{i\pi}\sigma)\}\, d\sigma. \qquad (10.7\text{-}13)$$

If $f(s)$ is real for real $s > 0$, then by the symmetry principle $1/(2\pi i)$ $\{f(e^{-i\pi}\sigma) - f(e^{i\pi}\sigma)\}$ is real for $\sigma > 0$. Thus the above formula may be regarded as a "real" inversion formula.

EXAMPLE **4**

For Re $\alpha > 0$, the function $F(\tau) := \tau^{\alpha-1}$ is in Ω. Its image

$$f(s) = \frac{\Gamma(\alpha)}{s^\alpha}$$

satisfies the foregoing conditions for $0 < \text{Re } \alpha < 1$. We have

$$f(e^{-i\pi}\sigma) - f(e^{i\pi}\sigma) = \{e^{i\pi\alpha} - e^{-i\pi\alpha}\}\Gamma(\alpha)\sigma^{-\alpha}$$
$$= 2i \sin \pi\alpha\, \Gamma(\alpha)\sigma^{-\alpha};$$

thus (10.7-13) here implies

$$\tau^{\alpha-1} = \frac{1}{\pi}\Gamma(\alpha) \sin \pi\alpha \int_0^\infty e^{-\tau\sigma}\sigma^{-\alpha}\, d\sigma.$$

By (III), §8.4, the integral equals $\tau^{\alpha-1}\Gamma(1-\alpha)$. As a byproduct, we thus have obtained the well-known formula

$$\Gamma(\alpha)\Gamma(1-\alpha) = \frac{\pi}{\sin \pi\alpha}. \qquad (10.7\text{-}14)$$

EXAMPLE 5

The hypergeometric function is the classical example of a function that is analytic in a cut plane and behaves like a power at the endpoints of the cut. Thus we may hope to be able to apply (10.7-13) to the correspondence (10.5-15). By the damping rule there follows

$$\tau^{\nu-1} e^{\tau/2} W_{\kappa,\mu}(\tau) \circ\!\!\!-\!\!\!-\!\!\!-\!\!\bullet \frac{\Gamma(\mu+\nu+1/2)\Gamma(\nu-\mu+1/2)}{\Gamma(\nu-\kappa+1)}$$

$$\cdot s^{-\mu-\nu-1/2} F\left(\mu+\nu+\tfrac{1}{2}, \mu-\kappa+\tfrac{1}{2}; \nu-\kappa+1; 1-\frac{1}{s}\right).$$

This correspondence is valid under the condition $\mathrm{Re}(\nu\pm\mu)>-\tfrac{1}{2}$, and for such ν and μ the above $f(s)$ satisfies (10.7-12). To study the behavior at $s=0$, we apply a linear transformation to find

$$f(s) = \mathrm{const}\, F(\mu+\nu+\tfrac{1}{2}, \nu-\mu+\tfrac{1}{2}; \nu+\kappa+1; s)$$
$$+\Gamma(\kappa+\nu)s^{-\kappa-\nu}F(\mu-\kappa+\tfrac{1}{2}, -\mu-\kappa+\tfrac{1}{2}; 1-\nu-\kappa; s).$$

This shows that (10.7-11) is satisfied if $\mathrm{Re}(\nu+\kappa)<1$. Since the first term is analytic at $s=0$, this representation permits us to conclude that

$$f(e^{-i\pi}\sigma)-f(e^{i\pi}\sigma)=\Gamma(\kappa+\nu)[e^{i\pi(\kappa+\nu)}-e^{-i\pi(\kappa+\nu)}]$$
$$\sigma^{-\kappa-\nu}F(\mu-\kappa+\tfrac{1}{2}, -\mu-\kappa+\tfrac{1}{2}; 1-\kappa-\nu; -\sigma).$$

Applying (10.7-14) before inserting into (10.7-13), we obtain

$$\tau^{\nu-1} e^{\tau/2} W_{\kappa,\mu}(\tau)$$

$$=\frac{1}{\Gamma(1-\kappa-\nu)}\int_0^\infty e^{-\tau\sigma}\sigma^{-\kappa-\nu}F(\mu-\kappa+\tfrac{1}{2}, -\mu-\kappa+\tfrac{1}{2}; 1-\kappa-\nu; -\sigma)\, d\sigma.$$

$$(10.7\text{-}15)$$

Although this result was derived under certain restrictive conditions on μ, the formula actually holds for arbitrary values of μ, because the integral has a meaning for arbitrary μ and is an analytic function of μ. The condition $\mathrm{Re}(\nu+\kappa)<1$, however, is necessary to ensure the convergence of the integral.

Formula (10.7-15) is important for the study of the asymptotic behavior of the Whittaker functions (see §11.5), and also for the representation of these functions in terms of continued fractions (see §12.13). Moreover, it furnishes the following basic correspondence:

$$\frac{1}{\Gamma(1-\nu-\kappa)}\tau^{-\kappa-\nu}F(\mu-\kappa+\tfrac{1}{2}, -\mu-\kappa+\tfrac{1}{2}; 1-\nu-\kappa; -\tau)$$

$$\circ\!\!\!-\!\!\!-\!\!\!-\!\!\bullet s^{\nu-1} e^{s/2} W_{\kappa,\mu}(s). \qquad (10.7\text{-}16)$$

V.   Order of Magnitude of the Original Function

Here we are concerned with the problem of determining the order of magnitude and the asymptotic behavior of an original function F from a knowledge of the singularities of the image function $f := \mathcal{L}F$, without explicitly calculating F. It is convenient here to recall the definition of Landau's O symbols: If two functions F and G are defined for $\tau \geq \tau_0$, and if $G(\tau) \neq 0$ for τ sufficiently large, then

$$F(\tau) = o(G(\tau)) \text{ as } \tau \to \infty \text{ means that } \lim_{\tau \to \infty} \frac{F(\tau)}{G(\tau)} = 0;$$

$$F(\tau) = O(G(\tau)) \text{ as } \tau \to \infty \text{ means that } \frac{F(\tau)}{G(\tau)} \text{ is bounded as } \tau \to \infty.$$

We also use the symbol $\sim$ to denote asymptotic equivalence:

$$F(\tau) \sim G(\tau) \text{ as } \tau \to \infty \text{ means that } \lim_{\tau \to \infty} \frac{F(\tau)}{G(\tau)} = 1.$$

For a first orientation we consider the case in which the image function is *rational*. If the poles of f are located at the points $s_1, \ldots, s_n$, we know from §10.3 that the corresponding F has the form

$$F(\tau) = \sum_{k=1}^{n} P_k(\tau) \exp(s_k \tau), \tag{10.7-17}$$

where the P_k are polynomials (of degree zero if the poles are simple). There follows immediately

$$F(\tau) = o(e^{\sigma \tau}), \tag{10.7-18}$$

for every $\sigma > \max \operatorname{Re} s_k$. Moreover, for $\tau \to \infty$ the dominating terms in the sum (10.7-17) are clearly those for which $\operatorname{Re} s_k$ is largest. If there is only one such term, say, the first, then it describes the asymptotic behavior of F as $\tau \to \infty$ in the sense that

$$F(\tau) \sim P_1(\tau) \exp(s_1 \tau), \qquad \tau \to \infty. \tag{10.7-19}$$

Thus in the special case under discussion, f being analytic for $\operatorname{Re} s > \sigma_0$ implies that F is $o(e^{\sigma \tau})$ for every $\sigma > \sigma_0$; moreover, the asymptotic behavior of F is determined by the rightmost singularities of its Laplace transform. Do such statements also hold for original functions with nonrational transforms? The answer is, only under additional hypotheses on f or on F.

To show that additional hypotheses are necessary, let σ be any real number, and let G be any positive, monotonically increasing function defined for $\tau > 0$. G may tend to infinity with arbitrary rapidity. We shall construct a function $F \in \Omega$ such that $f := \mathcal{L}F$ is analytic for $\operatorname{Re} s > \sigma$, although the statement $F(\tau) = o(G(\tau))$ is not true (nor is it true for any function equivalent to F in the sense of §10.1). This shows that there is, in general, no connection between the growth of an original function and the abscissa of analyticity of its $\mathscr{L}$ transform.

The trick of our construction consists in letting F be of the order of G merely on a sequence of short intervals. We may assume $\sigma < 0$ without weakening our assertion. Let $\gamma_n := G(n)$,

$$\delta_n := \min\{1, \gamma_n^{-1} 2^{-n} e^{\sigma n}\}, \qquad n = 1, 2, \ldots,$$

and define

$$F(\tau) := \begin{cases} \gamma_n, & n \leq \tau < n + \delta_n, \qquad n = 1, 2, \ldots, \\ 0, & \text{otherwise.} \end{cases}$$

If $f := \mathscr{L}F$, then

$$f(\sigma) = \sum_{n=1}^{\infty} \gamma_n \int_n^{n+\delta_n} e^{-\sigma\tau} \, d\tau = \sum_{n=1}^{\infty} \gamma_n (-\sigma)^{-1} e^{-\sigma n} (e^{-\sigma \delta_n} - 1).$$

By the mean value theorem $e^{-\sigma\delta_n} - 1 \leq (-\sigma) e^{-\sigma\delta_n}$, thus

$$f(\sigma) \leq e^{-\sigma} \sum_{n=1}^{\infty} \gamma_n \delta_n e^{-\sigma n} \leq e^{-\sigma} \sum_{n=1}^{\infty} 2^{-n} = e^{-\sigma}.$$

Thus the Laplace integral of F converges for $s = \sigma$, and by Theorem 10.1d $f = \mathscr{L}F$ is analytic for $\operatorname{Re} s > \sigma$.

The theorems concerning the asymptotic behavior of F that can be obtained under additional hypotheses are of one of two types. In the first type, the additional hypotheses concern the behavior of f. Such theorems are called of **Abelian type**. In the second type, nontrivial hypotheses are made on F as well as on f. Such theorems are said to be of **Tauberian type**. We now require a lemma on Fourier transforms that slightly generalizes Theorem 10.6c.

LEMMA 10.7e (Haar's lemma)

Let G be a continuous, complex-valued function on $(-\infty, \infty)$, and let the principal value integral

$$F(\tau) := \frac{1}{2\pi} PV \int_{-\infty}^{\infty} e^{i\omega\tau} G(\omega) \, d\omega \tag{10.7-20}$$

converge uniformly with respect to τ for $\tau > \tau_0$. Then $\lim_{\tau \to \infty} F(\tau) = 0$.

Proof. Let $\epsilon > 0$ be given. By the hypothesis of uniform convergence, there exists η such that

$$\left| F(\tau) - \frac{1}{2\pi} \int_{-\eta}^{\eta} e^{i\omega\tau} G(\omega) \, d\omega \right| < \frac{\epsilon}{2}$$

for all $\tau > \tau_0$. By the Riemann–Lebesgue lemma there exists τ_1 (depending on ϵ) such that

$$\left| \frac{1}{2\pi} \int_{-\eta}^{\eta} e^{i\omega\tau} G(\omega) \, d\omega \right| < \frac{\epsilon}{2}$$

for all $\tau > \tau_1$. This implies that $|F(\tau)| < \epsilon$ for $\tau > \max(\tau_0, \tau_1)$. ■

It should be noted that the hypothesis of uniform convergence is certainly satisfied if, as hypothesized in Theorem 10.6c,

$$\int_{-\infty}^{\infty} |G(\omega)| \, d\omega < \infty.$$

This condition is violated in many cases of practical importance, for instance, if $0 < \lambda \leq 1$ for

$$G(\omega) := (\alpha + i\omega)^{-\lambda},$$

where $\alpha > 0$ and the power has its principal value. On the other hand, this function satisfies the conditions of Haar's lemma, for by integration by parts we have

$$\int_{-\eta}^{\eta} e^{i\omega\tau} (\alpha + i\omega)^{-\lambda} \, d\omega = \frac{1}{i\tau} [e^{i\omega\tau} (\alpha + i\omega)^{-\lambda}]_{-\eta}^{\eta} + \frac{\lambda}{\tau} \int_{-\eta}^{\eta} e^{i\omega\tau} (\alpha + i\omega)^{-\lambda - 1} \, d\omega,$$

and the limits of the two terms on the right as $\eta \to \infty$ exist uniformly for $\tau > 1$, say.

The following applications of Lemma 10.7e to $\mathscr{L}$ transforms yield two results of Abelian type.

THEOREM 10.7f

Let $F \in \Omega$ satisfy condition (C) at all large τ, and let

$$\int_{-\infty}^{\infty} |e^{-\sigma\tau} F(\tau)| \, d\tau < \infty$$

so that $f(s) := \mathscr{L}F(s)$ is defined for $\mathrm{Re}\, s \geq \sigma$. If the principal value integral

$$PV \int_{-\infty}^{\infty} e^{i\omega\tau} f(\sigma + i\omega) \, d\omega$$

converges uniformly for large values of τ, then

$$F(\tau) = o(e^{\sigma\tau}), \qquad \tau \to \infty. \qquad (10.7\text{-}21)$$

Proof. By the complex inversion formula,

$$e^{-\sigma\tau} F(\tau) = \frac{1}{2\pi} PV \int_{-\infty}^{\infty} e^{i\omega\tau} f(\sigma + i\omega) \, d\omega.$$

Applying Lemma 10.7e with $G(\omega) := f(\sigma + i\omega)$, there follows $e^{-\sigma\tau}F(\tau) = o(1)$, $\tau \to \infty$, which is equivalent to (10.7-21). ∎

If, for some positive integer n, $f^{(k)}(\sigma + i\omega) \to 0$ as $\omega \to \pm\infty$ for $k = 0, 1, \ldots, n-1$, and

$$PV \int_{-\infty}^{\infty} e^{i\omega\tau} f^{(n)}(\sigma + i\omega) \, d\omega$$

converges uniformly, then using rule (IV) (multiplication) it can be shown in a similar manner that

$$F(\tau) = o(\tau^{-n} e^{\sigma\tau}), \qquad \tau \to \infty.$$

Whereas Theorem 10.7f describes the order of magnitude of F as $\tau \to \infty$, the following theorem describes its asymptotic behavior.

THEOREM 10.7g

Let F satisfy the conditions of the preceding theorem, and in addition assume that $f = \mathcal{L}F$ can be continued analytically into a half plane Re $s > \sigma_0$ where $\sigma_0 < \sigma$, with the exception of a pole at $s = s_0$ ($\sigma_0 < $ Re $s_0 < \sigma$) where f has the principal part

$$\frac{a_1}{s - s_0} + \frac{a_2}{(s - s_0)^2} + \cdots + \frac{a_k}{(s - s_0)^k}.$$

If $f(s) \to 0$ for $s \to \infty$, Re $s > \sigma_0$, and if there exists σ_1, $\sigma_0 < \sigma_1 <$ Re s_0, such that

$$PV \int_{-\infty}^{\infty} e^{i\omega\tau} f(\sigma_1 + i\omega) \, d\omega$$

converges uniformly for $\tau \geq \tau_0$, then

$$F(\tau) = P(\tau) e^{s_0\tau} + o(e^{\sigma_1\tau}), \tag{10.7-22}$$

where

$$P(\tau) := a_1 + \frac{1}{1!} a_2 \tau + \cdots + \frac{1}{(k-1)!} a_k \tau^{k-1}.$$

Proof. By the complex inversion formula, if (C) is satisfied at τ,

$$F(\tau) = \frac{1}{2\pi i} PV \int_{\sigma - i\infty}^{\sigma + i\infty} e^{s\tau} f(s) \, ds.$$

Because $f(s) \to 0$ for $s \to \infty$, Re $s > \sigma_0$, we have by the residue theorem

$$F(\tau) = \frac{1}{2\pi i} PV \int_{\sigma_1 - i\infty}^{\sigma_1 + i\infty} e^{s\tau} f(s) \, ds + \mathrm{res}[e^{s\tau} f(s)]_{s = s_0}.$$

The residue equals $P(\tau)\, e^{s_0\tau}$, and we thus obtain

$$e^{-\sigma_1\tau}F(\tau) - P(\tau)\, e^{(s_0-\sigma_1)\tau} = \frac{1}{2\pi}PV\int_{-\infty}^{\infty} e^{i\omega\tau}f(\sigma_1+i\omega)\,d\omega.$$

The last term is $o(1)$ by Lemma 10.7e, which yields (10.7-22). ∎

Formula (10.7-22) describes the asymptotic behavior of F in terms of the rightmost singularity of its $\mathscr{L}$ transform. A completely analogous statement can be made if f has a finite number of singularities in the half plane $\operatorname{Re} s > \sigma_1$.

A difficulty in the application of the foregoing theorems in concrete situations arises because there are conditions imposed on the behavior of $f(\sigma + i\omega)$ for $\omega \to \pm\infty$ even for values of $\sigma < \operatorname{Re} s_0$. The following theorem of Tauberian type, which apparently concerns a much more special situation, avoids this difficulty. The theorem is of a more recondite character than Theorem 10.7g; we quote it without proof from Doetsch [1950], p. 524.

THEOREM 10.7h (Ikehara–Wiener theorem)

Let $F \in \Omega$ be nonnegative and monotonically (but not necessarily strictly) *increasing, and let $C(F)$ contain the half plane $\operatorname{Re} s > \sigma_0 > 0$. If $f := \mathscr{L}F$ can be extended to a function analytic for $\operatorname{Re} s \geq \sigma_0$ save for a pole of order one at $s = \sigma_0$ with residue $\alpha > 0$, then*

$$F(\tau) \sim \alpha\, e^{\sigma_0\tau}, \qquad \tau \to \infty. \tag{10.7-23}$$

The theorems of the present section draw conclusions about the asymptotic behavior of F from certain assumptions on $f := \mathscr{L}F$. This is the situation that occurs when problems concerning functions in Ω are attacked by means of the $\mathscr{L}$ transformation; see the examples given in §10.9 and §10.10.

The converse problem of describing the asymptotic behavior of $f = \mathscr{L}F$ for $s \to \infty$ in terms of properties of F is also of frequent interest. This may be solved by means of the Watson–Doetsch lemma discussed in §11.5.

PROBLEMS

1. *When is an f an $\mathscr{L}F$?* Let f be analytic in a half plane $H: \operatorname{Re} s > \sigma_0$, and let

   (i) $\displaystyle \lim_{\substack{s\to\infty \\ s\in H}} f(s) = 0;$

   (ii) $\displaystyle \int_{-\infty}^{\infty} |f(\sigma+i\omega)|\, d\omega < \infty \quad$ for all $\sigma > \sigma_0$.

Then for all real τ and for $\sigma > \sigma_0$ the integral

$$F(\tau) := \frac{1}{2\pi i} \int_{\sigma - i\infty}^{\sigma + i\infty} e^{s\tau} f(s) \, ds \qquad (10.7\text{-}24)$$

exists and is independent of σ, $F \in \Omega$, and $f(s) = \mathscr{L}F(s)$ for $s \in H$.

2. Prove: If $F \in \Omega$ is a piecewise smooth function with discontinuities, and if $f := \mathscr{L}F$, then

$$\int_{-\infty}^{\infty} |f(\sigma + i\omega)| \, d\omega = \infty$$

for all σ. [Thus condition (ii) of problem 1 is not necessary in order that f be a Laplace transform.]

3. Find the Fourier series expansions of the following nonanalytic periodic functions by expanding their image functions in partial fractions:

(a) $F(\tau) := (-1)^{[\tau]}$

(b) $F(\tau) := B_n^*(\tau)$ (Bernoulli function)

[See Problems 8 and 10, §10.2, for image functions.]

4. Let $0 < \kappa < \lambda$. Show that the function

$$F(\tau) := \begin{cases} \alpha\dfrac{\tau}{\kappa}, & 0 \leqslant \tau \leqslant \kappa, \\[2mm] \alpha\dfrac{\lambda - \tau}{\lambda - \kappa}, & \kappa \leqslant \tau \leqslant \lambda, \end{cases}$$

is represented by a trigonometric series as follows:

$$F(\tau) = \frac{\alpha}{2} + \frac{\alpha\lambda^2}{\pi^2 \kappa(\lambda - \kappa)} \sum_{m=1}^{\infty} \frac{\sin(m\pi\kappa/\lambda)}{m^2} \sin\left[\frac{2m\pi}{\lambda}\left(\tau - \frac{\kappa}{2}\right)\right].$$

5. Determine the motion of the pendulum of Example 4, §10.3, when the pendulum is excited periodically (period η) by a short impulse of constant integrated strength.
[Differential equation of motion is

$$\Phi'' + \nu^2 \Phi = \kappa \sum_{k=0}^{\infty} \delta(\tau - k\eta);$$

result of applying expansion theorem if ν is not an integral multiple of $\nu_1 := 2\pi/\eta$:

$$\frac{1}{\kappa}\Phi(\tau) = \frac{1}{\nu^2\eta} - \frac{1}{2\nu}\frac{\cos[\nu(\tau + \eta/2)]}{\sin(\nu\pi/2)} + \frac{2}{\eta}\sum_{k=1}^{\infty}\frac{\cos(k\nu_1\tau)}{\nu^2 - k^2\nu_1^2}.$$

Show that resonance occurs for $\nu = m\nu_1$.]

6. Let $0 < \beta < \pi/2$. The function

$$F(\tau) := \begin{cases} \min(\sin\tau, \sin\beta), & \text{if } \sin\tau \geqslant 0, \\ \max(\sin\tau, -\sin\beta), & \text{if } \sin\tau \leqslant 0 \end{cases}$$

(alternating current cut off at height $\sin \beta$) has the following Fourier series representation:

$$F(\tau) = \frac{2\beta + \sin 2\beta}{\pi} \sin \tau + \sum_{k=3,5,\ldots} \frac{k \sin[(k+1)2\beta] - (k+1) \sin 2k\beta}{k(k+1)(2k+1)} \sin k\tau$$

7. Let $p(s) := \alpha_n s^n + \alpha_{n-1} s^{n-1} + \cdots + \alpha_0$ $(\alpha_n \neq 0)$ be a stable real polynomial (see §6.7, §12.7), and let X denote the solution of $p(D)X = 0$ $(D := d/d\tau)$ satisfying the initial conditions

$$X(0) = X'(0) = \cdots = X^{(n-2)}(0) = 0, \; X^{(n-1)}(0) = 1.$$

The integral

$$I(p) := \int_0^\infty [X(\tau)]^2 \, d\tau,$$

which exists by virtue of the stability of p, is of interest in control theory ("quadratic control area"). By applying the Laplace transform and using Parseval's relation, show that

$$I(p) = \frac{1}{2\pi} \int_{-\infty}^\infty \frac{1}{|p(i\omega)|^2} \, d\omega.$$

8. Let p be as in the preceding problem, and let its zeros $s_1, \ldots, s_n$ be simple. Let X be the solution of $p(D)X = 0$ such that

$$X^{(k)}(0) = \sum_{i=1}^n s_i^k, \; k = 0, 1, \ldots, n-1.$$

Show that

$$J := \int_0^\infty [X(\tau)]^2 \, d\tau = \frac{1}{2\pi} \int_{-\infty}^\infty \left| \frac{p'(i\omega)}{p(i\omega)} \right|^2 \, d\omega,$$

and also that

$$J = - \sum_{i,j=1}^n \frac{1}{s_i + s_j}.$$

9. Verify Theorem 10.7d for

$$F(\tau) := \frac{1 - e^{-\beta\tau}}{\tau}$$

where $\beta > 0$.

10. The modified Bessel function of the second kind may be defined by

$$K_\mu(\tau) := \sqrt{\frac{\pi}{2\tau}} \, W_{0,\mu}(2\tau).$$

Use (10.7-15) to obtain the integral representation

$$K_\mu(\tau) = \int_0^\infty e^{-\tau \cosh \lambda} \cos(\mu\lambda) \, d\lambda.$$

11. From the Laplace transform of the incomplete Γ function (see problem 2, §10.5) conclude that

$$\gamma(\nu, \tau) = \Gamma(\nu) + o(e^{-\lambda\tau})$$

for every $\lambda < 1$. Also verify this fact directly.

§10.8. DIRICHLET SERIES: PRIME NUMBER THEOREM

Here we consider a special type of series, called *Dirichlet series*. Some functions of importance in analytic number theory are represented by such series; a prominent example is the zeta function of Riemann. The elementary properties of Dirichlet series are quite analogous to those of Laplace integrals. Indeed, it is shown that every Dirichlet series can be represented as the Laplacian integral of a function in Ω. This makes available to Dirichlet series the advanced results of §10.7. Assuming the Tauberian theorem 10.7h, we thus can draw a conclusion concerning the asymptotic behavior of a certain number-theoretical function. This implies the celebrated prime number theorem as an elementary consequence.

I. Dirichlet Series

Let $A = \{a_1, a_2, \ldots\}$ be a sequence of complex numbers, and let s be a complex variable. The formal series

$$\sum_{n=1}^{\infty} \frac{a_n}{n^s}, \tag{10.8-1}$$

where $n^s := e^{s \operatorname{Log} n}$, is called a **Dirichlet series**. The elementary theory of Dirichlet series can be developed very much along the theory of Laplace integrals given in §10.1. Indeed, as we shall see, Dirichlet series are a special case of the latter. The following lemma is instrumental.

LEMMA 10.8a

Let $0 \leqslant \beta < \pi/2$. If the series (10.8-1) *converges for $s = s_0$, it converges uniformly in the angular domain $|\arg(s - s_0)| \leqslant \beta$.*

Proof. It suffices to consider the case where $s_0 = 0$, for

$$\sum_{n=1}^{\infty} \frac{a_n}{n^s} = \sum_{n=1}^{\infty} \frac{b_n}{n^{s-s_0}},$$

where $b_n := a_n n^{-s_0}$.

Suppose, then, that $\sum_{n=1}^{\infty} a_n$ is convergent. Let

$$r_n := a_{n+1} + a_{n+2} + \cdots,$$

so that $r_n \to 0$ $(n \to \infty)$. Using summation by parts,

$$\sum_{n=k}^{m} \frac{a_n}{n^s} = \sum_{n=k}^{m} \frac{r_{n-1} - r_n}{n^s}$$

$$= \sum_{n=k}^{m} r_n \left\{ \frac{1}{(n+1)^s} - \frac{1}{n^s} \right\} + \frac{r_{k-1}}{k^s} - \frac{r_m}{(m+1)^s}.$$

If $\sigma := \operatorname{Re} s > 0$,

$$\left| \frac{1}{(n+1)^s} - \frac{1}{n^s} \right| = \left| s \int_n^{n+1} \frac{1}{\tau^{s+1}} \, d\tau \right| \leq |s| \int_n^{n+1} \frac{1}{\tau^{\sigma+1}} \, d\tau$$

$$= \frac{|s|}{\sigma} \left\{ \frac{1}{n^\sigma} - \frac{1}{(n+1)^\sigma} \right\}.$$

Thus if $\epsilon > 0$ is given and n_0 is chosen such that $|r_n| < \epsilon$ for $n > n_0$, then if $k > n_0 + 1$

$$\left| \sum_{n=k}^{m} \frac{a_n}{n^s} \right| \leq \frac{\epsilon |s|}{\sigma} \sum_{n=k}^{m} \left\{ \frac{1}{n^\sigma} - \frac{1}{(n+1)^\sigma} \right\} + \frac{2\epsilon}{k^\sigma} = \frac{3\epsilon |s|}{\sigma}.$$

If $|\arg s| \leq \beta$, then $|s|/\sigma = 1/\cos \beta$, and it follows that the series (10.8-1) satisfies Cauchy's test for uniform convergence. ∎

If the series (10.8-1) converges for some s, let σ be the infimum of the real parts of all such s. If the series diverges for all s, set $\sigma := \infty$. The symbol σ is called the **abscissa of convergence** of the series. Lemma 10.8a implies:

THEOREM 10.8b

Either the series (10.8-1) *converges for no s, or it converges for all s, or there exists a real number σ such that the series converges for all s such that* $\operatorname{Re} s > \sigma$ *and diverges for all s such that* $\operatorname{Re} s < \sigma$.

The question of convergence for s such that $\operatorname{Re} s = \sigma$ remains open; various cases are possible.

If the series converges *absolutely* for some $s = s_0$, then it converges absolutely for all s such that $\operatorname{Re} s = \operatorname{Re} s_0$. Thus we have the following analog of Theorem 10.1c:

THEOREM 10.8c

The set of all s for which the series (10.8-1) *converges absolutely, if not empty, is either the full plane or an open or closed half plane.*

Let $\sigma < \infty$, and let T denote any compact subset of $\operatorname{Re} s > \sigma$. Because T can be enclosed in an angular domain of the type described in Lemma 10.8a, the series converges uniformly on T. Because each term of the series is an analytic function of s, there follows:

THEOREM 10.8d

The sum of a Dirichlet series represents an analytic function in the interior of its domain of convergence.

A uniformly convergent series of analytic functions may be differentiated term by term. We thus have:

COROLLARY 10.8e

Let (10.8-1) *have an abscissa of convergence* $\sigma < \infty$, *and let*

$$a(s) := \sum_{n=1}^{\infty} \frac{a_n}{n^s}, \quad \text{Re } s > \sigma.$$

Then

$$a'(s) = - \sum_{n=1}^{\infty} \frac{a_n \text{ Log } n}{n^s}.$$

In a similar manner one could deduce from Lemma 10.8a the analog for Dirichlet series of the Abelian theorem 10.1f and of Theorem 10.1g concerning the behavior for $s \to \infty$.

EXAMPLE **1**

The best-known example of a Dirichlet series is the ζ function of Riemann defined by

$$\zeta(s) := \sum_{n=1}^{\infty} \frac{1}{n^s}. \tag{10.8-2}$$

The abscissa of convergence is clearly 1. Thus $\zeta(s)$ is analytic for Re $s > 1$.

EXAMPLE **2**

Consider the Dirichlet series

$$f(s) := \sum_{n=1}^{\infty} \frac{(-1)^{n-1}}{n^s}. \tag{10.8-3}$$

For real $s > 0$ this is an alternating series with terms whose absolute values decrease monotonically to zero. Hence by the Leibniz criterion the series converges. For Re $s < 0$ the series clearly diverges. Hence the abscissa of convergence is 0, and the sum $f(s)$ is analytic for Re $s > 0$. The abscissa of absolute convergence, on the other hand, is 1.

We now show that the function (10.8-3) is closely related to the Riemann ζ function. Suppose Re $s > 1$, so that the series converges absolutely, and hence its terms can be arranged in arbitrary order. From

$$\zeta(s) = \left(1 + \frac{1}{3^s} + \frac{1}{5^s} + \cdots\right) + \left(\frac{1}{2^s} + \frac{1}{4^s} + \frac{1}{6^s} + \cdots\right),$$

$$f(s) = \left(1 + \frac{1}{3^s} + \frac{1}{5^s} + \cdots\right) - \left(\frac{1}{2^s} + \frac{1}{4^s} + \frac{1}{6^s} + \cdots\right)$$

there follows by subtraction

$$\zeta(s) - f(s) = 2\left(\frac{1}{2^s} + \frac{1}{4^s} + \frac{1}{6^s} + \cdots\right)$$

$$= 2^{1-s}\left(1 + \frac{1}{2^s} + \frac{1}{3^s} + \cdots\right)$$

$$= 2^{1-s}\zeta(s).$$

Thus if $2^{1-s} \neq 1$,

$$\zeta(s) = \frac{1}{1 - 2^{1-s}} f(s). \tag{10.8-4}$$

This formula continues $\zeta(s)$ analytically into the half plane Re $s > 0$, with the possible exception of those values of s for which $2^{1-s} = 1$. At $s = 1$, $f(1) = \text{Log } 2$. On the other hand,

$$1 - 2^{1-s} = 1 - e^{-(s-1)\text{Log } 2} = (s - 1) \text{ Log } 2\{1 + O(s - 1)\}.$$

We conclude that at $s = 1$ the function $\zeta(s)$ has a pole of order 1 with residue 1. We see later that, contrary to what one might expect, the remaining points s such that $2^{1-s} = 1$ are *not* poles of ζ.

Dirichlet's Rule for Multiplication.   Suppose the Dirichlet series

$$a(s) := \sum_{k=1}^{\infty} \frac{a_k}{k^s}, \quad b(s) := \sum_{m=1}^{\infty} \frac{b_m}{m^s}$$

are absolutely convergent. Multiplying the two series formally yields

$$\sum_{k,m=1}^{\infty} \frac{a_k b_m}{(km)^s}.$$

According to an elementary theorem, the formal product of two absolutely convergent series converges (absolutely) to the product of the two sums, independently of the order in which the terms of the formal product are summed. We order the terms of the above series according to the value of $n := km$. The coefficient of n^{-s} then becomes

$$\sum_{km=n} a_k b_m = \sum_{k|n} a_k b_{n/k}.$$

The symbol $k|n$ means "k divides n"; thus the sum on the right comprises all divisors of n, 1 and n included. The resulting formula,

$$\sum_{k=1}^{\infty} \frac{a_k}{k^s} \sum_{m=1}^{\infty} \frac{b_m}{m^s} = \sum_{n=1}^{\infty} \frac{1}{n^s} \sum_{k|n} a_k b_{n/k},$$

is known as **Dirichlet's rule** for multiplying two Dirichlet series.

Dirichlet's rule should be compared with Cauchy's rule for multiplying power series. The appearance of the divisor symbol hints at the importance of Dirichlet series in the multiplicative analytical theory of numbers.

EXAMPLE **3**

For positive integers n we define[4]

$\Lambda(n) := \mathrm{Log}\, p$, if $n = p^m$, where p is a prime and m is a positive integer, $\Lambda(n) := 0$, otherwise.

The Dirichlet series

$$\sum_{n=1}^{\infty} \frac{\Lambda(n)}{n^s}$$

converges absolutely for Re $s > 1$. We multiply the series by the series representing $\zeta(s)$. Dirichlet's rule yields

$$\zeta(s) \sum_{n=1}^{\infty} \frac{\Lambda(n)}{n^s} = \sum_{n=1}^{\infty} \frac{a}{n^s} \sum_{k|n} \Lambda(k).$$

Let the decomposition of n in prime factors be

$$n = p_1^{m_1} p_2^{m_2} \cdots p_r^{m_r}.$$

Then the only divisors k of n such that $\Lambda(k) \neq 0$ are the prime powers

$$p_1, p_1^2, \cdots, p_1^{m_1}; \qquad p_2, p_2^2, \cdots, p_2^{m_2}; \qquad p_r, p_r^2, \cdots, p_r^{m_r}.$$

For the first set of divisors, $\Lambda(k) = \mathrm{Log}\, p_1$, for the second, $\Lambda(k) = \mathrm{Log}\, p_2$, and so forth. Hence

$$\sum_{k|n} \Lambda(k) = m_1 \mathrm{Log}\, p_1 + m_2 \mathrm{Log}\, p_2 + \cdots + m_r \mathrm{Log}\, p_r$$

$$= \mathrm{Log}(p_1^{m_1} p_2^{m_2} \cdots p_r^{m_r}) = \mathrm{Log}\, n$$

We thus find

$$\zeta(s) \sum_{n=1}^{\infty} \frac{\Lambda(n)}{n^s} = \sum_{n=1}^{\infty} \frac{\mathrm{Log}\, n}{n^s}.$$

By Corollary 10.8e, the series on the right equals $-\zeta'(s)$. There follows

$$-\frac{\zeta'(s)}{\zeta(s)} = \sum_{n=1}^{\infty} \frac{\Lambda(n)}{n^s}, \qquad \mathrm{Re}\, s > 1. \tag{10.8-5}$$

---

[4] The symbol Λ, like ζ, is traditional.

We now show that every convergent Dirichlet series can be represented (up to a trivial factor) as the Laplace transform of a certain function in Ω. This result enables us to apply to Dirichlet series the tools of §10.7.

We first show how this representation formula comes about formally. Let

$$a(s) = \sum_{n=1}^{\infty} \frac{a_n}{n^s} = \sum_{n=1}^{\infty} a_n e^{-s \operatorname{Log} n} \tag{10.8-6}$$

be a Dirichlet series, assumed convergent for certain values of s. We multiply the foregoing by s^{-1}. Then, in view of the correspondence

$$\frac{1}{s} e^{-s \operatorname{Log} n} \bullet\!\!-\!\!\!-\!\!\!-\!\!o\; H(\tau - \operatorname{Log} n)$$

($H =$ Heaviside unit function), $s^{-1}a(s)$ may formally be regarded as the image of the function

$$A(\tau) := \sum_{n=1}^{\infty} a_n H(\tau - \operatorname{Log} n).$$

For any τ, the only nonzero terms of the sum are those for which $\operatorname{Log} n \leq \tau$, and for these H has the value 1. We thus may also write

$$A(\tau) = \sum_{\operatorname{Log} n \leq \tau} a_n. \tag{10.8-7}$$

Clearly, $A \in \Omega$. We shall prove:

THEOREM 10.8f

Let the series (10.8-6) *be convergent for* $s = \sigma \geq 0$. *Then for all* s *such that* $\operatorname{Re} s > \sigma$,

$$a(s) = s \int_0^{\infty} e^{-s\tau} A(\tau) \, d\tau. \tag{10.8-8}$$

Proof. We first establish a result concerning the order of magnitude of the sums $s_n := a_1 + a_2 + \cdots + a_n$, and hence of the function $A(\tau)$. Denoting the partial sums of the series $a(\sigma)$ by b_n,

$$b_n := \sum_{k=1}^{n} \frac{a_k}{k^{\sigma}}, \qquad n = 1, 2, \ldots,$$

$b_0 := 0$, we have

$$s_n = \sum_{k=1}^{n} \frac{a_k}{k^{\sigma}} k^{\sigma} = \sum_{k=1}^{n} (b_k - b_{k-1}) k^{\sigma}$$

$$= \sum_{k=1}^{n-1} b_k \{k^{\sigma} - (k+1)^{\sigma}\} + b_n n^{\sigma}.$$

The convergence of $a(\sigma)$ implies the existence of a constant γ such that $|b_k| < \gamma$ for all k. There follows

$$|s_n| \leqslant 2\gamma n^\sigma,$$

hence

$$|A(\tau)| \leqslant 2\gamma e^{\sigma\tau}. \tag{10.8-9}$$

The existence of the integral in (10.8-8) for $\mathrm{Re}\, s > \sigma$ follows at once.

To show the identity of the integral with the sum on the left, consider for $\eta > 0$

$$a_\eta(s) := s \int_0^\eta e^{-s\tau} A(\tau)\, d\tau$$

$$= s \sum_{\mathrm{Log}\, n \leqslant \eta} a_n \int_{\mathrm{Log}\, n}^\eta e^{-s\tau}\, d\tau$$

$$= \sum_{\mathrm{Log}\, n \leqslant \eta} a_n \{e^{-s\,\mathrm{Log}\, n} - e^{-s\eta}\}$$

$$= \sum_{\mathrm{Log}\, n \leqslant \eta} \frac{a_n}{n^s} - e^{-s\eta} \sum_{\mathrm{Log}\, n \leqslant \eta} a_n.$$

By the above, the last term is bounded by $2\gamma |e^{(\sigma-s)\eta}|$. If $\mathrm{Re}\, s > \sigma$ this tends to zero for $\eta \to \infty$. At the same time,

$$a_\eta(s) \to s \int_0^\infty e^{-s\tau} A(\tau)\, d\tau,$$

and the identity (10.8-8) follows. ∎

II. The Prime Number Theorem

For $\tau > 0$, let

$$L(\tau) := \sum_{\mathrm{Log}\, n \leqslant \tau} \Lambda(n). \tag{10.8-10}$$

By virtue of (10.8-5) and Theorem 10.8f, the logarithmic derivative of the ζ function then admits the representation

$$-\frac{\zeta'(s)}{\zeta(s)} = s \int_0^\infty e^{-s\tau} L(\tau)\, d\tau, \qquad \mathrm{Re}\, s > 1. \tag{10.8-11}$$

THEOREM 10.8g

For $\tau \to \infty$,

$$L(\tau) \sim e^\tau. \tag{10.8-12}$$

Proof. The proof is an application of the Wiener–Ikehara Tauberian theorem 10.7h. Clearly, the function L is nonnegative and nondecreasing. The domain of convergence of the Laplace integral formed with L includes the half plane Re $s > 1$. The Laplace transform $l := \mathscr{L}L$ by (10.8-11) is

$$l(s) = -\frac{\zeta'(s)}{s\zeta(s)}.$$

We already know that ζ at $s = 1$ has a pole of order 1. It follows that ζ'/ζ at the same point has a pole of order 1 with residue -1, and that $l(s)$ has a pole of order 1 with residue $+1$. Thus (10.8-12) follows from (10.7-23) if we can show that with the exception of the point $s = 1$, ζ'/ζ is analytic on Re $s = 1$. This is an immediate consequence of the following two lemmas:

LEMMA 10.8h

The function

$$\zeta(s) - \frac{1}{s-1}$$

is entire.

Proof. For every s such that Re $s > 1$, the function

$$f(z) := (1+z)^{-s} = e^{-s \operatorname{Log}(1+z)}$$

satisfies the hypotheses of the Plana summation formula (Theorem 4.9c). Because

$$\sum_{n=0}^{\infty} f(n) = \zeta(s), \quad \int_0^{\infty} f(x)\, dx = \frac{1}{s-1},$$

there follows

$$\zeta(s) = \frac{1}{2} + \frac{1}{s-1} + Z(s), \tag{10.8-13}$$

where

$$Z(s) := i \int_0^{\infty} [(1+iy)^{-s} - (1-iy)^{-s}] \frac{1}{e^{2\pi y} - 1}\, dy.$$

Because

$$(1+iy)^{-s} - (1-iy)^{-s} := e^{-s \operatorname{Log}(1+iy)} - e^{-s \operatorname{Log}(1-iy)}$$

$$= (1+y^2)^{-s/2}(-2i) \sin(s \operatorname{Arctan} y),$$

the integral representing Z may be written

$$Z(s) = 2 \int_0^\infty (1+y^2)^{-s/2} \sin(s \text{ Arctan } y) \frac{1}{e^{2\pi y} - 1} \, dy.$$

Here the integrand for each $y \geq 0$ is an entire analytic function of s, and because $|\sin(s \text{ Arctan } y)| \leq e^{|s|\pi/2}$ the integral converges uniformly with respect to s in every compact set of the s plane. Thus $Z(s)$ is an entire function, tantamount to the assertion of Lemma 10.8h. ∎

LEMMA 10.8i

For $\text{Re } s = 1, s \neq 1,$

$$\zeta(s) \neq 0. \tag{10.8-14}$$

Proof. Although somewhat artificial, the following proof, due to E. Landau, has the advantage of being very short. Let

$$\lambda(s) := \frac{\zeta'(s)}{\zeta(s)}.$$

We already know that $\zeta(s)$ has a pole of order 1 at $s = 1$. Thus the residue of the logarithmic derivative at $s = 1$ is -1, which implies

$$\lim_{\epsilon \to 0} \epsilon \lambda(1+\epsilon) = -1. \tag{10.8-15}$$

If ζ has a zero of order m at s_0, then

$$\lim_{s \to s_0} (s - s_0)\lambda(s) = m;$$

this also holds for $m = 0$ ($\zeta(s_0) \neq 0$). Assume now, contrary to the assertion of Lemma 10.8i, that ζ has a zero of order $m_1 \geq 1$ at a point $s = 1 + i\omega$ such that $\omega \neq 0$. Then by the above

$$\lim_{\epsilon \to 0} \epsilon \lambda(1 + i\omega + \epsilon) = m_1. \tag{10.8-16}$$

The point $1 + 2i\omega$ is no pole of ζ. It may or may not be a zero of ζ; in any case,

$$\lim_{\epsilon \to 0} \epsilon \lambda(1 + 2i\omega + \epsilon) =: m_2 \geq 0. \tag{10.8-17}$$

Confining ϵ to real values and taking real parts, equations (10.8-15), (10.8-16), and (10.8-17) imply

$$\lim_{\epsilon \to 0} \epsilon\{3\lambda(1+\epsilon) + 4 \text{ Re } \lambda(1 + i\omega + \epsilon) + \text{Re } \lambda(1 + 2i\omega + \epsilon)\}$$

$$\tag{10.8-18}$$

$$= -3 + 4m_1 + m_2 \geq 1.$$

On the other hand, the expansion (10.8-5) yields for $\epsilon > 0$

$$3\lambda(1+\epsilon)+4\operatorname{Re}\lambda(1+i\omega+\epsilon)+\operatorname{Re}\lambda(1+2i\omega+\epsilon)$$

$$= -\sum_{n=1}^{\infty}\frac{\Lambda(n)}{n^{1+\epsilon}}[3+4\operatorname{Re}n^{-i\omega}+\operatorname{Re}n^{-2i\omega}].$$

But

$$3+4\operatorname{Re}n^{-i\omega}+\operatorname{Re}n^{-2i\omega}$$

$$= 2+4\cos(\omega\operatorname{Log}n)+1+\cos(2\omega\operatorname{Log}n)$$

$$= 2[1+2\cos(\omega\operatorname{Log}n)+\cos^2(\omega\operatorname{Log}n)]$$

$$= 2[1+\cos(\omega\operatorname{Log}n)]^2 \geqslant 0;$$

hence $\epsilon[3\lambda(1+\epsilon)+4\operatorname{Re}\lambda(1+i\omega+\epsilon)+\operatorname{Re}\lambda(1+2i\omega+\epsilon)]\leqslant 0$, contradicting (10.8-18). Thus the assumption that $\zeta(1+i\omega)=0$ for a real $\omega\neq 0$ cannot be maintained, proving Lemma 10.8i, and Theorem 10.8g. ■

The interest of Theorem 10.8g lies in the fact that the function L is elementarily related to the distribution of primes.

For $\tau > 0$, let $\pi(\tau)$ denote the number of primes not exceeding τ. Examples: $\pi(2)=1$, $\pi(3.5)=2$, $\pi(12)=5$. Although the distribution of the primes themselves appears to be fairly irregular, the function $\pi(\tau)$ has a smooth asymptotic behavior. The following theorem was conjectured by Gauss and proved simultaneously by Hadamard and De la Vallée-Poussin in 1894:

THEOREM 10.8j (The prime number theorem)

For $\tau \to \infty$,

$$\pi(\tau) \sim \frac{\tau}{\operatorname{Log}\tau}. \tag{10.8-19}$$

Proof. We estimate $\pi(\tau)$ from below and above in terms of the function $L(\tau)$. In the sum

$$L(\tau)=\sum_{\operatorname{Log}n\leqslant\tau}\Lambda(n) \tag{10.8-20}$$

only the terms for which $n=p^m$ (p prime) yield a nonzero contribution. For a fixed p these are the powers $p, p^2, \ldots, p^r$, where r is the greatest integer such that $r\operatorname{Log}p\leqslant\tau$ or

$$r := \left[\frac{\tau}{\operatorname{Log}p}\right].$$

Thus

$$L(\tau) = \sum_{\text{Log } p \leqslant \tau} \left[\frac{\tau}{\text{Log } p} \right] \text{Log } p,$$

and by omitting the brackets there follows

$$L(\tau) \leqslant \sum_{\text{Log } p \leqslant \tau} \frac{\tau}{\text{Log } p} \text{Log } p = \tau \sum_{\text{Log } p \leqslant \tau} 1 = \tau \pi(e^{\tau}).$$

We thus have

$$\pi(e^{\tau}) \geqslant \frac{L(\tau)}{\tau}. \qquad (10.8\text{-}21)$$

On the other hand, if $1 < \sigma < \tau$,

$$\pi(e^{\tau}) = \pi(e^{\sigma}) + \sum_{\sigma < \text{Log } p \leqslant \tau} 1$$

$$= \pi(e^{\sigma}) + \sum_{\sigma < \text{Log } p \leqslant \tau} \frac{\text{Log } p}{\text{Log } p} < \pi(e^{\sigma}) + \frac{1}{\sigma} \cdot \sum_{\sigma < \text{Log } p \leqslant \tau} \text{Log } p$$

or, because obviously $\pi(e^{\sigma}) < e^{\sigma}$,

$$\pi(e^{\tau}) < e^{\sigma} + \frac{1}{\sigma} L(\tau).$$

Together with (10.8-21) this implies

$$e^{-\tau} L(\tau) \leqslant e^{-\tau} \tau \pi(e^{\tau}) \leqslant e^{\sigma - \tau} + \frac{\tau}{\sigma} e^{-\tau} L(\tau). \qquad (10.8\text{-}22)$$

As $\tau \to \infty$, the lower bound for $e^{-\tau} \tau \pi(e^{\tau})$ tends to 1 by Theorem 10.8g. The prime number theorem is established if we can choose σ as a function of τ such that the upper bound likewise tends to 1. Again by Theorem 10.8g, this is the case if $\sigma(\tau)$ satisfies

$$\frac{\tau}{\sigma(\tau)} \to 1 \quad \text{and} \quad \tau e^{\sigma(\tau) - \tau} \to 0$$

as $\tau \to \infty$. The function $\sigma(\tau) = \tau - 2 \text{Log } \tau$ satisfies both these requirements. ■

More precise statements on the function $\pi(\tau)$ could be made if more information concerning the asymptotic behavior of the function L were available. By the developments of §10.7, such information would require a knowledge of the poles of $l(s) := \mathscr{L}L(s)$, that is, of the zeros of $\zeta(s)$. Riemann conjectured that, apart from the so-called "trivial" zeros at the points $s = -2, -4, \ldots$, the only zeros of $\zeta(s)$ lie on the line Re $s = \frac{1}{2}$. Hardy proved that there indeed exists an infinite number of zeros on that line. By rigorous numerical computations, Rosser, Schoenfeld, and Yoke showed in 1968 that the 3,900,000 zeros with smallest imaginary part indeed lie on the critical line. But Riemann's conjecture itself is still unsettled.

PROBLEMS

1. Let the Dirichlet series

$$a(s) := \sum_{n=1}^{\infty} \frac{a_n}{n^s}$$

converge absolutely for $s = \sigma_0$. Prove *Perron's formula*,

$$\sum_{n \leq \xi} a_n = \frac{1}{2\pi i} PV \int_{\sigma - i\infty}^{\sigma + i\infty} \frac{a(s)}{s} \xi^s \, ds,$$

valid for $\sigma \geq \sigma_0$ and for nonintegral ξ. What is the value of the expression on the right if ξ is an integer?

2. The *Moebius symbol* $\mu(n)$ is defined by

$$\mu(n) := \begin{cases} 1, & \text{if } n = 1 \\ (-1)^r, & \text{if } n \text{ is the product of } r \text{ distinct primes} \\ 0, & \text{in all other cases} \end{cases}$$

Prove that

(a) $\displaystyle \sum_{n=1}^{\infty} \frac{\mu(n)}{n^s} = \frac{1}{\zeta(s)}$,

(b) $\displaystyle \Lambda(n) = -\sum_{k \mid n} \mu(k) \, \text{Log } k$.

(See Hardy and Wright [1954], Chapter 17, for these and similar identities.)

3. For $\lambda > 0$, let

$$\omega(\lambda) := \sum_{n=1}^{\infty} e^{-n^2 \pi \lambda} = \tfrac{1}{2}\{\theta(0, \lambda) - 1\},$$

where θ denotes the function introduced in the example of §10.6. For Re $s > 1$, prove

$$\pi^{-s/2} \Gamma\left(\frac{s}{2}\right) \zeta(s) = \int_0^{\infty} \omega(\lambda) \lambda^{s/2-1} \, d\lambda. \tag{10.8-23}$$

[Use (10.5-1) to express

$$\int_0^{\infty} \exp(-n^2 \pi \lambda) \lambda^{s/2-1} \, d\lambda,$$

and integrate term by term. Justification is by monotone convergence (Beppo Levi theorem), see §8.4.]

4. (Continuation) Establish Riemann's integral representation,

$$\pi^{-s/2}\Gamma\!\left(\frac{s}{2}\right)\zeta(s) = -\frac{1}{s} - \frac{1}{1-s}$$

$$+ \int_1^\infty \{\lambda^{(1-s)/2} + \lambda^{s/2}\}\lambda^{-1}\omega(\lambda)\,d\lambda. \tag{10.8-24}$$

[In (10.8-23), write $\int_0^\infty = \int_0^1 + \int_1^\infty$; substitute $\lambda \to 1/\lambda$ in the first integral; and use Jacobi's transformation (10.6-24).]

5. (Continuation) Prove Riemann's functional relation for the ζ function: If $f(s) := \pi^{-s/2}\Gamma(s/2)\zeta(s)$, then $f(s) = f(1-s)$.

6. (Continuation) Prove that for $n = 1, 2, \ldots$

$$\zeta(-2n) = 0, \ \zeta(-2n+1) = -\frac{B_{2n}}{2n},$$

where B_{2n} is the $2n$th Bernoulli number.

7. Let $s = \frac{1}{2} + it,$

$$\xi(t) := \pi^{-s/2}\frac{s(s-1)}{2}\Gamma\!\left(\frac{s}{2}\right)\zeta(s).$$

Show that the Riemann hypothesis is true if and only if the function ξ has real zeros only.

§10.9. FUNCTIONS OF EXPONENTIAL TYPE

An **entire function** of a complex variable by definition is a function that is defined and analytic in the entire complex plane. If F is entire, its Taylor series at 0 converges and represents $F(z)$ for all complex z,

$$F(z) = \sum_{n=0}^\infty a_n z^n, \ |z| < \infty. \tag{10.9-1}$$

Let F be entire, and let, for all $\rho > 0$,

$$\mu(\rho) := \max_{|z|=\rho} |F(z)|. \tag{10.9-2}$$

The function F is said to be of **exponential type** if there exists a number $\alpha > 0$ such that

$$\mu(\rho) < e^{\alpha\rho} \text{ for all sufficiently large } \rho. \tag{10.9-3}$$

Let γ denote the infimum of all numbers α such that (10.9-3) is true. Then γ will be called the **growth parameter** of F. We have $\gamma \geq 0$ unless F is identically zero.

EXAMPLES **1**

The function $\exp(z^2)$ is not of exponential type. The function $z^{100} e^{-2z}$ is a function of exponential type with growth parameter 2. The function $\cosh\sqrt{z}$ is a function of exponential type with growth parameter zero, as is any polynomial.

The totality of entire functions of exponential type with growth parameter γ is denoted by Ω_γ.[5] If $F \in \Omega_\gamma$, then the function G defined by

$$G(\tau) := \begin{cases} F(\tau), & \tau \geq 0 \\ 0, & \tau < 0 \end{cases} \tag{10.9-4}$$

evidently belongs to Ω, the class of original functions defined in §10.1, and its growth indicator γ_G (see §10.1) satisfies $\gamma_G \leq \gamma$. [There may be inequality, as shown by $F(z) := \cos z$, where $\gamma = 1$ and $\gamma_G = 0$.] Thus the $\mathcal{L}$ transform of G is defined, and analytic, at least for Re $s > \gamma$. *For simplicity in what follows, we call "Laplace transform of F" and denote by $\mathcal{L}F$ what is really the Laplace transform of the function G defined by* (10.9-4).

Hardy's theorem 10.5b taught us how to compute the Laplace transform of certain entire functions by term-by-term integration. In the present section we are generally concerned with the connection between the growth properties of functions $F \in \Omega_\gamma$ and the location of the singularities of $f = \mathcal{L}F$.

To state our first result we denote, for all $\sigma \geq 0$, by ω_σ the class of all· functions $f(s)$ that vanish at ∞ and are analytic outside the circle $|s| = \sigma$, and outside no smaller circle. By the Cauchy–Hadamard formula for the radius of convergence of a power series, a function

$$f(s) = \sum_{n=0}^{\infty} b_n s^{-n-1}$$

belongs to ω_σ if and only if

$$\limsup_{n \to \infty} |b_n|^{1/n} = \sigma.$$

THEOREM 10.9a

For every $\gamma \geq 0$, the $\mathcal{L}$ transform defines a one-to-one mapping of the set Ω_γ onto the set ω_γ.

In other words, every $F \in \Omega_\gamma$ possesses a Laplace transform that can be extended to a function $f \in \omega_\gamma$. Conversely, every $f \in \omega_\gamma$ is an analytic continuation of the Laplace transform of a certain $F \in \Omega_\gamma$.

[5] In the terminology that is used in the general theory of entire functions, the class Ω_γ for $\gamma > 0$ is identical with the entire functions of order 1 and type γ, and for $\gamma = 0$ with the entire functions that are either of order 1 and of minimal type or of order < 1.

Proof. (a) Let the entire function F, represented by the power series (10.9-1), belong to Ω_y. This means that given any $\epsilon > 0$, there exists $\kappa > 0$ such that the function $\mu(\rho)$ defined by (10.9-2) satisfies

$$\mu(\rho) < \kappa \, e^{(\gamma+\epsilon)\rho}$$

for all $\rho > 0$. By Cauchy's estimate for the coefficients of a power series there follows

$$|a_n| \leqslant \frac{\kappa \, e^{(\gamma+\epsilon)\rho}}{\rho^n}, \qquad n = 0, 1, 2, \ldots$$

for all $\rho > 0$. Letting $\rho := n/(\gamma+\epsilon)$, we obtain in particular

$$|a_n| \leqslant \kappa \left(\frac{e}{n}\right)^n (\gamma+\epsilon)^n, \qquad n = 1, 2, \ldots. \tag{10.9-5}$$

By Hardy's theorem 10.5b, $f := \mathscr{L}F$ is represented by the series

$$f(s) = \sum_{n=0}^{\infty} a_n \frac{n!}{s^{n+1}} \tag{10.9-6}$$

for all $s > 0$ for which the series converges. By the Cauchy–Hadamard formula the series converges for $s > \sigma$, where

$$\sigma := \limsup_{n \to \infty} |n! a_n|^{1/n}.$$

By (10.9-5) and Stirling's formula,

$$|n! a_n| \leqslant \kappa \sqrt{2\pi n} (\gamma+\epsilon)^n (1 + O(1/n)),$$

and it is obvious that $\sigma \leqslant \gamma + \epsilon$. Because $\epsilon > 0$ was arbitrary, we have $\sigma \leqslant \gamma$. Thus $f := \mathscr{L}F \in \omega_\sigma$ where $\sigma \leqslant \gamma$.

(b) Let $f \in \omega_\sigma$,

$$f(s) = \sum_{n=0}^{\infty} b_n s^{-n-1}, \quad |s| > \sigma. \tag{10.9-7}$$

By Hardy's theorem, the restriction of f to $\operatorname{Re} s > \sigma$ is the Laplace transform (in the manner of speech used here) of the entire function

$$G(z) := \sum_{n=0}^{\infty} \frac{b_n}{n!} z^n.$$

We shall show that $G \in \Omega_\tau$ where $\tau \leqslant \sigma$. Because

$$\limsup_{n \to \infty} |b_n|^{1/n} = \sigma,$$

there exists, for every $\epsilon > 0$, n_0 such that $n > n_0$ implies

$$|b_n| < (\sigma + \epsilon)^n.$$

There follows

$$|G(z)| \leqslant \sum_{n=0}^{n_0} \frac{|b_n|}{n!} |z|^n + \sum_{n>n_0} \frac{(\sigma + \epsilon)^n}{n!} |z|^n$$

$$\leqslant \sum_{n=0}^{n_0} \frac{|b_n|}{n!} |z|^n + e^{(\sigma+\epsilon)|z|} \leqslant e^{(\sigma+2\epsilon)|z|},$$

for sufficiently large $|z|$. It follows that G is a function of exponential type with growth parameter $\tau \leqslant \sigma + 2\epsilon$. Because $\epsilon > 0$ was arbitrary, $\tau \leqslant \sigma$.

(c) Let $F \in \Omega_\gamma$ for some $\gamma \geqslant 0$. By (a), $\mathcal{L}$ maps this onto some $f \in \omega_\sigma$ where $\sigma \leqslant \gamma$. By (b), f is the image of some $G \in \Omega_\tau$ where $\tau \leqslant \sigma$. Because f can be the image of only one analytic original function, $G = F$, hence $\gamma = \tau = \sigma$. ∎

Let $F \in \Omega_\gamma$. Then the Laplace transform of F converges absolutely at least for Re $s > \gamma$, and thus the complex inversion formula (10.7-5) holds for any vertical path of integration to the right of Re $s = \gamma$. However, for functions of exponential type there holds yet another inversion formula which is more convenient because the path of integration has finite length.

Let $\rho > \gamma$, and let Γ denote the circle $s = \rho e^{i\phi}$, $0 \leqslant \phi \leqslant 2\pi$. For arbitrary complex z we consider the integral

$$I(z) := \frac{1}{2\pi i} \int_\Gamma f(s) e^{sz} \, ds,$$

where $f := \mathcal{L}F$. On Γ, $f(s)$ is represented by the series (10.9-6) which converges uniformly. Substituting the series and integrating term by term, we get

$$I(z) = \frac{1}{2\pi i} \sum_{n=0}^{\infty} a_n \int_\Gamma \frac{n!}{s^{n+1}} e^{sz} \, ds = \sum_{n=0}^{\infty} a_n z^n = F(z),$$

because the residue of $n! s^{-n-1} e^{sz}$ at $s = 0$ is z^n. By Cauchy's theorem, Γ may be replaced by any closed curve having winding number $+1$ with respect to the set $|s| \leqslant \gamma$. We thus have:

THEOREM 10.9b (Pincherle's Theorem)

Let $F \in \Omega_\gamma$ ($\gamma \geqslant 0$), $f := \mathcal{L}F$, and let Γ be any simple closed curve encircling the set $|s| \leqslant \gamma$ in the positive sense. Then for all complex z,

$$F(z) = \frac{1}{2\pi i} \int_\Gamma f(s) e^{sz} \, ds. \qquad (10.9\text{-}8)$$

It should be noted that the inversion formula (10.9-8), in contrast to (10.7-5), does *not* yield the value zero for $z = \tau < 0$.

The growth parameter measures the overall growth of a function of exponential type. We now introduce a measure for the growth in particular directions. For real ϕ, let $\gamma(\phi)$ denote the infimum of all real numbers α such that for all sufficiently large $\tau > 0$,

$$|F(e^{i\phi}\tau)| < e^{\alpha\tau}.$$

The function $\gamma(\phi)$ thus defined is called the **indicator function** of F. If the growth parameter of F is γ, then clearly $\gamma(\phi) \leqslant \gamma$ for all ϕ. It can happen that $\gamma(\phi) < \gamma$, and even that $\gamma(\phi) < 0$, for certain values of ϕ.

EXAMPLE 2

If $F(z) = e^z$, then for $\tau \geqslant 0$, $|F(e^{i\phi}\tau)| = e^{\tau \cos\phi}$, thus $\gamma(\phi) = \cos\phi$.

The growth indicator γ_F defined in §10.1 for arbitrary functions $F \in \Omega$ in the case of a function of exponential type clearly equals $\gamma(0)$. Because we know already from Theorem 10.1d that $f := \mathscr{L}F$ is analytic for Re $s > \gamma_F$, Theorem 10.9a can be modified as follows:

THEOREM 10.9c

Let $F \in \Omega_\gamma$ where $\gamma \geqslant 0$, and let $\gamma(\phi)$ be the indicator function of F. Then $f := \mathscr{L}F$ can be extended to a function which is analytic in the union of the two sets $|s| > \gamma$ (including $s = \infty$) and Re $s > \gamma(0)$.

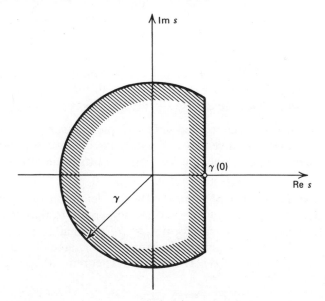

Fig. 10.9a.

Here we recall the following question on Laplace transforms. Let F be any function in Ω whose $\mathscr{L}$ transform has an abscissa of convergence $\alpha_F < \infty$. We already know from §10.1 that $f := \mathscr{L}F$ is analytic in the half plane Re $s > \alpha_F$, and in all examples that have been encountered this half plane was the maximal half-plane of analyticity; that is, it was not possible to continue f analytically into a half plane Re $s > \alpha_0$ where $\alpha_0 < \alpha_F$. However, we have not proved this impossibility. In fact, examples have been constructed (see Doetsch [1958], p. 38) in which the continuation is possible. On the other hand, it will now be shown that if F is of exponential type, $f := \mathscr{L}F$ *cannot* be continued into a half plane Re $s > \alpha_0$ where $\alpha_0 < \gamma_F = \gamma(0)$. Because f is analytic at ∞, this means that f has a singular point somewhere on the line Re $s = \gamma(0)$.

Let F satisfy the hypotheses of Theorem 10.9c, and let $f := \mathscr{L}F$. We express $F(\tau)$, where $\tau > 0$, in terms of f by means of the new inversion formula (10.9-8). Suppose f can be continued into a half plane Re $s > \alpha_0$ where $\alpha_0 < \gamma(0)$. Then for every $\epsilon > 0$, the path of integration Γ could be deformed into a path of finite length λ lying entirely in Re $s \leqslant \alpha_0 + \epsilon$. The maximum of $|e^{\tau s}|$ on the path of integration would then not exceed $e^{(\alpha_0 + \epsilon)\tau}$. Denoting the maximum of $|f(s)|$ on Γ by μ, (10.9-8) would then for every $\epsilon > 0$ yield the estimate

$$|F(\tau)| \leqslant \frac{1}{2\pi}\mu\lambda\, e^{(\alpha_0 + \epsilon)\tau}.$$

Because $\epsilon > 0$ was arbitrary, it would follow that $\gamma(0) = \alpha_0$, contrary to our assumption that $\alpha_0 < \gamma(0)$.

THEOREM 10.9d

Under the hypotheses of Theorem 10.9c, $f := \mathscr{L}F$ cannot be continued analytically into any half plane Re $s > \alpha_0$ *where* $\alpha_0 < \gamma(0)$; *that is, f has a singular point on the straight line segment* Re $s = \gamma(0)$, $|s| \leqslant \gamma$.

Thus far we have been concerned only with the Laplace transform of the restriction of F to the positive real axis. F is defined, however, for arbitrary complex values. We now apply the foregoing results to the function

$$F_\phi(\tau) := F(e^{i\phi}\tau), \qquad \tau \geqslant 0,$$

where ϕ is an arbitrary real number. By the definition of the indicator function, the growth of $F_\phi(\tau)$ as $\tau \to \infty$ is comparable to that of $e^{\gamma(\phi)\tau}$. By Theorem 10.9c the Laplace transform $f_\phi := \mathscr{L}F_\phi$ is analytic in the union of the sets $|s| > \gamma$, Re $s > \gamma(\phi)$, and by Theorem 10.9d f_ϕ actually has a singular point on the line Re $s = \gamma(\phi)$. By Hardy's Theorem 10.5b, f_ϕ is easily calculated. In view of

$$F_\phi(\tau) = \sum_{n=0}^{\infty} a_n e^{in\phi}\tau^n,$$

we have for $|s| > \gamma$

$$f_\phi(s) = \sum_{n=0}^{\infty} a_n e^{in\phi} \frac{n!}{s^{n+1}} = e^{-i\phi} f(e^{-i\phi} s).$$

[Thus rule II (similarity) in the case of functions of exponential type holds for arbitrary complex ρ.] We conclude from this that $f(e^{-i\phi} s)$ can be continued analytically into the half plane Re $s > \gamma(\phi)$, and not into any larger right half plane. This is the same as to say that $f(s)$ can be continued analytically into the half plane $H_\phi : \text{Re}(e^{i\phi} s) > \gamma(\phi)$, but not into any half plane $\text{Re}(e^{i\phi} s) > \alpha_0$ where $\alpha_0 < \gamma(\phi)$. Any point s on the boundary of H_ϕ such that f cannot be continued analytically into s is henceforth called a **singularity** of f.

The foregoing holds for every value of ϕ, $0 \leq \phi \leq 2\pi$. We conclude that f can be continued into the set

$$S := \bigcup_{0 \leq \phi \leq 2\pi} H_\phi,$$

the union of all half planes H_ϕ. The continuation is not possible into any larger set which is a union of half planes.

Let C denote the complement of S. The set C is contained in the disk $|s| \leq \gamma$, thus bounded. It is the intersection of the closed half planes $\hat{H}_\phi : \text{Re}(e^{i\phi} s) \leq \gamma(\phi)$, the complements of the open half planes H_ϕ. Each $\hat{H}_\phi$ contains all singular points of f. By Theorem 10.9d, the boundary of each $\hat{H}_\phi$ contains at least one singular point. Thus for fixed ϕ, $\hat{H}_\phi$ is the intersection of all closed half planes $\text{Re}(e^{i\phi} s) \leq \alpha$ containing all singular points of f. The set C thus may also be described as the intersection of *all* closed half planes containing all singular points of f. The intersection of all closed half planes containing a given set T is commonly described as the **convex hull** of T (see § 6.5 for another example). Thus C is the **convex hull of the singularities** of f.

We next show that the function $\gamma(\phi)$ is completely determined by C. Let ϕ be fixed. Then because C is the intersection of all closed half-planes $\hat{H}_\psi$, $0 \leq \psi \leq 2\pi$,

$$\sup_{s \in C} \text{Re}(e^{i\phi} s) = \inf_{0 \leq \psi \leq 2\pi} \sup_{s \in \hat{H}_\psi} \text{Re}(e^{i\phi} s).$$

The last supremum is $+\infty$ for all ψ except for $\psi = \phi$, where it has the value $\gamma(\phi)$. There follows

$$\gamma(\phi) = \sup_{s \in C} \text{Re}(e^{i\phi} s). \tag{10.9-9}$$

Now, if K is any bounded closed convex set in the plane, the function $\kappa(\phi)$ defined for all real ϕ by

$$\kappa(\phi) := \sup_{s \in K} \text{Re}(e^{-i\phi} s) \tag{10.9-10}$$

is called the **support function** of K (see Fig. 10.9b). We thus see that the support function of C is just $\gamma(-\phi)$. Altogether we have obtained the following striking result, due to Polya:

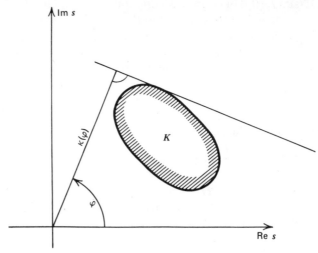

Fig. 10.9b.

THEOREM 10.9e

Let F be of exponential type, and let $\gamma(\phi)$ denote the indicator function of F. Let $f := \mathscr{L}F$, and let $\kappa(\phi)$ denote the support function of the convex hull of the singularities of f. Then for all ϕ,

$$\kappa(\phi) = \gamma(-\phi).$$

The value of this theorem lies in the fact that in many cases the function f is simple enough for its singularities, and hence the convex hull of the singularities and its support function, to be determined by inspection. The knowledge of $\kappa(\phi)$ then permits us to estimate the rate of growth of $|F(z)|$ along any ray $z = e^{i\phi}\tau$, $0 \le \tau < \infty$.

EXAMPLE **3**

If $F(z) := e^{az}$, where a is complex, then $f(s) := \mathscr{L}F(s) = (s - a)^{-1}$. The convex hull of the singularities of f reduces to the single point $s = a$. If $a \neq 0$, $a = |a|e^{i\alpha}$, the support function of this set is

$$\kappa(\phi) = |a| \cos(\phi - \alpha).$$

We conclude that $\gamma(\phi) = |a| \cos(-\phi - \alpha) = |a| \cos(\phi + \alpha)$, which can of course be verified immediately.

EXAMPLE **4**

Let

$$F(z) := \sum_{n=0}^{\infty} \frac{z^{3n}}{(3n)!}.$$

Here

$$f(s) := \mathscr{L}F(s) = \sum_{n=0}^{\infty} s^{-3n-1} = \frac{1}{s(1-s^{-3})} = \frac{s^2}{s^3-1}.$$

The singular points of f are 1, $w := e^{2\pi i/3}$, w^2, and the convex hull of the singularities is the triangle spanned by these points (see Fig. 10.9c). Its support function is

$$\kappa(\phi) = \max\left(\cos\phi, \cos\left(\phi - \frac{2\pi}{3}\right), \cos\left(\phi - \frac{4\pi}{3}\right)\right).$$

That $\gamma(\phi) = \kappa(-\phi)$ is clear in this case because

$$F(z) = \tfrac{1}{3}\{e^z + e^{wz} + e^{w^2z}\}.$$

EXAMPLE **5**

Let

$$F(z) := J_0(z) = \sum_{n=0}^{\infty} \frac{(-z^2/4)^n}{(n!)^2},$$

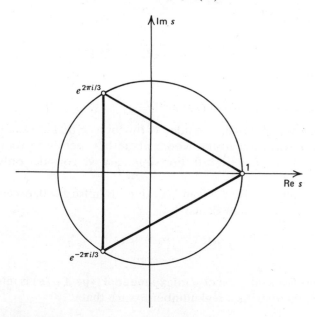

Fig. 10.9c.

the Bessel function of order zero. We already know

$$f(s) := \mathscr{L}F(s) = \frac{1}{\sqrt{s^2+1}}.$$

This function is in ω_1, thus $J_0(z)$ is a function of exponential type with growth parameter 1. The hull of the singularities of f is the straight line segment joining the points $\pm i$; its support function is $\kappa(\phi) = |\sin \phi|$. We conclude that the growth of $J_0(z)$ in the direction $\arg z = \phi$ is comparable to that of $e^{|\sin \phi| |z|}$. This is confirmed by more precise information about the asymptotic behavior of the Bessel functions, which is obtained in §11.5 and §11.8.

EXAMPLE **6**

Does there exist a function of exponential type such that its indicator function $\gamma(\phi)$ is identically equal to 1? If such a function exists, then the convex hull of the singularities of its Laplace transform must be the closed unit disk $|s| \leq 1$. This requires f to be a noncontinuable power series with radius of convergence unity. According to the Weierstrassian example **16** of §3.2,

$$f(s) := \sum_{n=1}^{\infty} s^{-n!-1}$$

is such a function. Thus

$$F(z) := \mathscr{L}^{-1}f(z) = \sum_{n=1}^{\infty} \frac{z^{n!}}{(n!)!}$$

$$= z + \frac{1}{2}z^2 + \frac{1}{720}z^6 + \frac{1}{620448'401733'239439'360000}z^{24} + \cdots$$

has the required property.

Functions of Semiexponential Type.

Some of these results can be extended to functions F that are analytic and of exponential growth not in the whole plane but merely in some angular domain $|\arg z| < \alpha$ where $\alpha > 0$. For simplicity we consider only the case $\alpha = \pi/2$.

Let F be analytic in Re $z > 0$ and continuous in Re $z \geq 0$, except possibly at $z = 0$. For $\rho > 0$ we now define

$$\mu(\rho) := \max_{\substack{|z|=\rho \\ \text{Re } z \geq 0}} |F(z)|.$$

The function F is said to be of **semiexponential type** if $\mu(\rho)$ is integrable at $\rho = 0$, and if there exists a real number α such that

$$\mu(\rho) < e^{\alpha\rho} \text{ for all sufficiently large } \rho. \tag{10.9-11}$$

The infimum γ of all α such that (10.9-11) holds is again called the **growth parameter** of F. Although γ measures the overall growth of F, the growth along the ray $\arg z = \phi$ is measured as before by the **indicator function** $\gamma(\phi)$ which is the infimum of all α such that

$$|F(e^{i\phi}\tau)| < e^{\alpha\tau} \text{ for all sufficiently large } \tau > 0.$$

The function $\gamma(\phi)$ is now defined only for $-\pi/2 \leq \phi \leq \pi/2$.

By Theorem 10.1d, $f := \mathscr{L}F$ is analytic for Re $s > \gamma(0)$. Similarly, if we let

$$F_\phi(\tau) := F(e^{i\phi}\tau), \qquad \tau > 0,$$

then $f_\phi := \mathscr{L}F_\phi$ is analytic for Re $s > \gamma(\phi)$, $-\pi/2 \leq \phi \leq \pi/2$. It will be shown that for all these ϕ, if Re $s > \gamma(0)$ and Re$(e^{i\phi}s) > \gamma(\phi)$,

$$f(s) = e^{i\phi}f_\phi(e^{i\phi}s). \tag{10.9-12}$$

When dealing with functions of exponential type, this relation could be read off the Laurent series of f at ∞. In the present case, there is no Laurent series, and a different approach is necessary. By Cauchy's theorem, the integral

$$\int_\Gamma e^{-sz}F(z)\,dz$$

has the same value whether extended along the straight line segment from $z = 0$ to $z = \rho > 0$ or along the straight line segment from $z = 0$ to $z = e^{i\phi}\rho$ and then along the circular arc centered at 0 to $z = \rho$. The integral along the circular arc is bounded by

$$\mu(\rho)\frac{\pi\rho}{2}e^{-|s|\rho\nu} < e^{-(|s|\nu - \gamma - \epsilon)\rho}$$

for all sufficiently large ρ, where ν is the minimum of $\cos(\arg s + \arg z)$ on the arc. Assume now $\phi > 0$ for definiteness. Then if s is restricted to a sector $\alpha < \arg s < \beta$ where $-\pi/2 - \phi < \alpha < \pi/2 - \phi$, $\nu > 0$ holds, and the integral along the circular arc tends to zero for $\rho \to \infty$ if $|s|$ is sufficiently large. There follows

$$\lim_{\rho\to\infty}\int_0^\rho e^{-s\tau}F(\tau)\,d\tau = \lim_{\rho\to\infty} e^{i\phi}\int_0^\rho e^{-se^{i\phi}\tau}F(e^{i\phi}\tau)\,d\tau,$$

which is to say, $f(s) = e^{i\phi}f_\phi(e^{i\phi}s)$. Because this identity between analytic functions holds in a nonempty open set, it holds throughout, and one function represents the analytic continuation of the other.

Thus (10.9-12) extends $f(s)$ analytically into the half plane $H_\phi : \text{Re}(e^{i\phi}s) > \gamma(\phi)$. Because this is true for every $\phi \in [-\pi/2, \pi/2]$, we may conclude the following:

THEOREM 10.9f

Let F be a function of semiexponential type, and let $\gamma(\phi)$ be its indicator function. Then $f := \mathscr{L}F$ can be extended to a function analytic in the union S of all half planes $H_\phi : \mathrm{Re}(e^{i\phi}s) > \gamma(\phi), -\pi/2 \leq \phi \leq \pi/2$.

Let C denote the complement of S. By Boolean algebra,

$$C = \bigcap_{-\pi/2 \leq \phi \leq \pi/2} \hat{H}_\phi,$$

where $\hat{H}_\phi$ is the complement of H_ϕ. To complete the analogy with Polya's Theorem 10.9e, we should like to show that the boundary of each $\hat{H}_\phi$ contains a singularity of f, and thus that C (in a sense) is again the convex hull of these singularities. To this end we require a modification of the complex inversion formula that is possible for functions of semi-exponential type.

Let $\tau > 0$, $-\pi/2 < \phi < \pi/2$, and let $\xi > \gamma$. Then by the ordinary inversion formula, using (10.9-12) and letting $w := e^{i\phi}$,

$$F(w\tau) = \frac{w^{-1}}{2\pi i} \lim_{\eta \to \infty} \int_{\xi - i\eta}^{\xi + i\eta} e^{t\tau} f(w^{-1}t) \, dt.$$

Substituting $s := w^{-1}t$,

$$F(w\tau) = \frac{1}{2\pi i} \lim_{\eta \to \infty} \int_{w^{-1}(\xi - i\eta)}^{w^{-1}(\xi + i\eta)} e^{sw\tau} f(s) \, ds.$$

Let now $\alpha > \gamma(\pi/2)$, $\beta > \gamma(-\pi/2)$. We close the path of integration as shown in Fig. 10.9d.

The circular arcs lie in the half plane $\mathrm{Re}(ws) \leq \sigma$, thus by Jordan's lemma, because $f(s) \to 0$ as $s \to \infty$ in the union of $\mathrm{Im}\, s \geq \beta$, $\mathrm{Im}\, s \leq -\alpha$, and $\mathrm{Re}\, s \geq \xi$, the integrals along the circular arcs tend to zero as $\eta \to \infty$. On $\mathrm{Im}\, s = -\alpha$, if $s = \sigma - i\alpha$, the integrand is bounded by

$$e^{\sigma \cos \phi + \alpha \sin \phi} |f(\sigma - i\alpha)|,$$

thus the integral for $\eta' \to \infty$ clearly exists; similarly on $\mathrm{Im}\, s = \beta$. Thus if $z := e^{i\phi}\tau$, we get for any z such that $\mathrm{Re}\, z > 0$

$$F(z) = \frac{1}{2\pi i} \int_{-\infty - i\alpha}^{-\infty + i\beta} e^{zs} f(s) \, ds; \tag{10.9-13}$$

the precise form of the path of integration does not matter as long as it starts out on $\mathrm{Im}\, s = -\alpha$, avoids C, and ends up on $\mathrm{Im}\, s = \beta$.

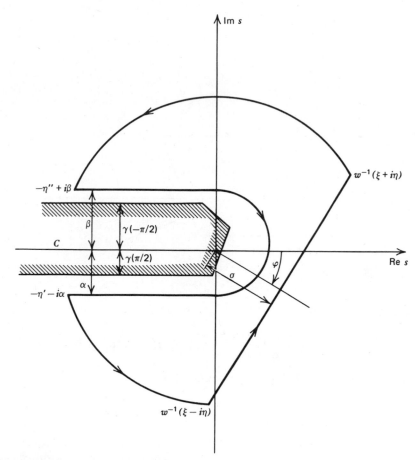

Fig. 10.9d.

We use (10.9-13) to prove the following analog of Theorem 10.9d:

THEOREM 10.9g

Under the hypotheses of Theorem 10.9f, if $-\pi/2 < \phi < \pi/2$, $f := \mathcal{L}F$ cannot be continued analytically into any half plane $\mathrm{Re}(e^{i\phi}s) > \theta$ where $\theta < \gamma(\phi)$.

The *proof* is indirect. Assume that f may be continued. We recover $F(e^{i\phi}\tau)$ from f by means of (10.9-13), choosing a path of integration consisting of three straight pieces Γ_1, Γ_2, Γ_3 lying, respectively, on $\mathrm{Im}\,s = -\alpha$, on $\mathrm{Re}(e^{i\phi}s) = \delta$, and on $\mathrm{Im}\,s = \beta$, where $\theta < \delta < \gamma(\phi)$ (see Fig. 10.9e). The straight pieces meet at the points

$$s_1 := \frac{\delta - \alpha \sin \phi}{\cos \phi} - i\alpha, \qquad s_2 := \frac{\delta + \beta \sin \phi}{\cos \phi} + i\beta.$$

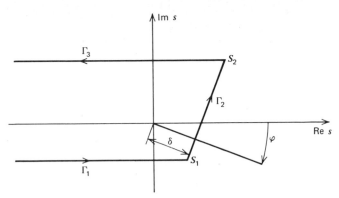

Fig. 10.9e.

Parametrizing $-\Gamma_1$ by

$$s = \frac{\delta - \lambda - \alpha \sin \phi}{\cos \phi} - i\alpha, \qquad 0 \leq \lambda < \infty,$$

we see that the integral along Γ_1 is bounded by

$$e^{\delta\tau} \int_0^\infty e^{-\lambda\tau} \left| f\left(\frac{\delta - \lambda - \alpha \sin \phi}{\cos \phi} - i\alpha\right) \right| d\lambda.$$

The integral may be regarded as a Laplacian integral of a bounded function, thus it tends to zero as $\tau \to \infty$. It follows that the integral along Γ_1 is $O(e^{\delta\tau})$ as $\tau \to \infty$. Similarly, the integral along Γ_3 may be shown to be $O(e^{\delta\tau})$. The integral along Γ_2 is bounded by

$$e^{\delta\tau} \int_{s_1}^{s_2} |f(s)| \, |ds|,$$

and thus likewise is $O(e^{\delta\tau})$. Under the assumptions made it thus follows that $F(e^{i\phi}\tau) = O(e^{\delta\tau})$ as $\tau \to \infty$, where $\delta < \gamma(\phi)$. This contradicts the definition of $\gamma(\phi)$. Thus the assumption that f can be continued into a half plane $\mathrm{Re}(e^{i\phi}s) > \theta$ where $\theta < \gamma(\phi)$ is untenable. ∎

As before we conclude that each line $\mathrm{Re}(e^{i\phi}s) = \gamma(\phi)$ $(-\pi/2 < \phi < \pi/2)$ actually contains a point into which f cannot be continued analytically. The totality of these points is again called the set of singular points of f. If $\phi = \pm\pi/2$, the lines $\mathrm{Re}(e^{i\phi}s) = \gamma(\phi)$ do not necessarily contain singular points. To state a result analogous to Theorem 10.9e, we enlarge the set of singular points by two "virtual" singular points $-\infty \mp i\gamma(\pm(\pi/2))$. A virtual point $-\infty + i\xi$ by definition is contained in a half plane $\mathrm{Re}(e^{i\phi}s) \leq \theta$ if and only if either $-\pi/2 < \phi < \pi/2$ or $\phi = -\pi/2$ and $\theta \geq \xi$ or $\phi = \pi/2$ and $-\theta \leq \xi$. Then the set C may again be described as the intersection of all closed half planes containing all singularities of f (including the virtual singularities), and thus as the convex hull of these singularities. Defining the support function of a not necessarily bounded convex set again by (10.9-10), we have

THEOREM 10.9h

Let F be a function of semiexponential type with indicator function $\gamma(\phi)$, let $f := \mathcal{L}F$, and let $\kappa(\phi)$ be the support function of the convex hull of the singularities of f, including the virtual singularities. Then $\kappa(\phi) = \gamma(-\phi)$ holds for $-\pi/2 \leq \phi \leq \pi/2$.

EXAMPLE 7

The function $F(z) := (1+z)^{-1}$ is of semiexponential type with indicator function $\gamma(\phi) = 0$. The virtual singularities are at $-\infty + i0$. Thus C consists of the negative real axis. Indeed $s = 0$ is known to be a singular point of

$$f(s) = \int_0^\infty \frac{e^{-s\tau}}{1+\tau} d\tau = e^s \int_s^\infty \frac{e^{-u}}{u} du.$$

EXAMPLE 8

Let

$$F(z) := e^{e^{-z}}.$$

F, though entire, is not of exponential type, because for z real, $z \to -\infty$ the exponent grows much more rapidly than linearly. But

$$\mathrm{Log}|F(e^{i\phi}\tau)| = e^{-\tau \cos\phi} \cos(\tau \sin \phi),$$

thus for $-\pi/2 \leq \phi \leq \pi/2$

$$\gamma(\phi) = \limsup_{\tau \to \infty} \frac{\mathrm{Log}|F(e^{i\phi}\tau)|}{\tau} = 0.$$

Thus F is of semiexponential type, and the set C again consists of the closed negative real axis. That this indeed is the convex hull of the singularities of the Laplace transform is confirmed by the explicit representation (obtained by substituting $\xi := e^{-\tau}$ in the Laplacian integral)

$$f(s) = \sum_{n=0}^\infty \frac{1}{n!(s+n)}$$

exhibiting its poles at $s = 0, -1, -2, \ldots$.

EXAMPLE **9**

Here we consider

$$F(z) := \frac{1}{\Gamma(1+z)}.$$

We know from §8.4 that F is entire. By Stirling's formula, if $z \to \infty$, $|\arg z| \le \alpha < \pi$,

$$\mathrm{Log}|\Gamma(1+e^{i\phi}\tau)| = \tau[(1-\mathrm{Log}\,\tau)\cos\phi + \phi\,\sin\phi] - \tfrac{1}{2}\mathrm{Log}\,\tau + O(1).$$

There follows

$$\gamma(\phi) = \lim_{\tau \to \infty}\sup \frac{\mathrm{Log}|\Gamma(1+e^{i\phi}\tau)|^{-1}}{\tau} = \begin{cases} -\infty, & -\dfrac{\pi}{2} < \phi < \dfrac{\pi}{2}, \\[2mm] \dfrac{\pi}{2}, & \phi = \pm\dfrac{\pi}{2}, \\[2mm] +\infty, & \dfrac{\pi}{2} < |\phi| < \pi. \end{cases}$$

Clearly, F is not of exponential type. It is, however, of semiexponential type. The virtual singularities are at $-\infty \pm i(\pi/2)$. However, because $\gamma(0) = -\infty$, $f := \mathscr{L}F$ is an entire function. The hull of its singularities is empty.

Problems

1. The indicator function of a function F of exponential or of semiexponential type satisfies

$$\gamma(\phi) = \lim_{\tau \to \infty}\sup \frac{\mathrm{Log}|F(e^{i\phi}\tau)|}{\tau}.$$

2. Let $\gamma \ge 0$, F not a polynomial. Prove: If F belongs to Ω_γ, then so do all derivatives of F.

3. Let α, β be complex numbers $\neq 0, -1, -2, \ldots$, and let

$$F(z) := {}_1F_1(\alpha; \beta; z).$$

Determine the convex hull of the singularities of $f := \mathscr{L}F$ and its support function, and compare the resulting statements on the growth of F with the results derived in §8.8 by means of integrals of the Mellin–Barnes type.

4. Deduce Poisson's integral for J_0 (see §10.4) from

$$J_0(\tau) \; \circ\!\!-\!\!-\!\!-\!\!\bullet \; \frac{1}{\sqrt{s^2+1}}$$

by means of Pincherle's inversion formula (10.9-8).

5. Prove: An entire function of exponential type with period 2π is necessarily a trigonometric polynomial. [If F is the function in question, then (*) $|F(z)| \le \alpha\,e^{\beta|z|}$ for all z. By the periodicity of F, the function $G(w) := F(-i \log w)$ is analytic for $0 < |w| < \infty$. Let n be an integer $> \beta$. Then by (*)

$$w^n G(w) = \begin{cases} O(w^{2n}), & w \to \infty, \\ O(1), & w \to 0. \end{cases}$$

Thus $w^n G(w)$ has a removable singularity at $w = 0$ and a pole at $w = \infty$ and hence is a polynomial. See Boas [1964].]

6. Show that the indicator function of an entire function of exponential type is continuous.

7. Let $\gamma(\phi)$ be the indicator function of an entire function of exponential type. If
$$\phi_1 < \phi_2 < \phi_3, \phi_2 - \phi_1 < \pi, \phi_3 - \phi_2 < \pi,$$
then
$$\gamma(\phi_1) \sin(\phi_3 - \phi_2) + \gamma(\phi_2) \sin(\phi_1 - \phi_3) + \gamma(\phi_3) \sin(\phi_2 - \phi_1) \geq 0.$$
[This is a purely geometric consequence of the fact that $\gamma(-\phi)$ is the support function of a convex set.]

8. Prove: An entire function of exponential type either tends to zero exponentially (i.e., has $\gamma(\phi) < 0$) in no direction ϕ, or the directions in which it tends to zero form a single open interval of length $\leq \pi$.
[Where is 0 located with respect to the set C?]

§10.10. THE DISCRETE LAPLACE TRANSFORMATION

I. General Theory

In some technical applications, the original function F can be sampled only at the times $\tau = n\eta$ ($n = 0, 1, 2, \ldots$), where η is a positive constant. At all other times τ, F is either undefined or unknown. In such cases it is possible to define a **discrete Laplace transform** $\mathscr{L}_\eta$ of F (evaluated with the step η) by

$$f_\eta(s) := \mathscr{L}_\eta F(s) := \eta \sum_{n=0}^{\infty} e^{-sn\eta} F(n\eta) \tag{10.10-1}$$

for all s for which the series is convergent. The series obviously may be regarded as a simple numerical approximation to the Laplacian integral (10.1-1).

Writing $z := e^{s\eta}$ we have $f_\eta(s) = p_\eta(z)$ where

$$p_\eta(z) := \eta \sum_{n=0}^{\infty} F(n\eta) z^{-n}. \tag{10.10-2}$$

Because p_η is a mere power series in z^{-1}, its domain of simple convergence by the Cauchy–Hadamard formula is the set $|z| > \rho$, where

$$\rho := \limsup_{n \to \infty} |F(n\eta)|^{1/n},$$

possibly including some or all of its boundary points. It follows that the domain of definition of $f_\eta(s)$ is given by

$$\operatorname{Re} s > \sigma_\eta := \frac{1}{\eta} \operatorname{Log} \rho,$$

boundary points possibly included, and that f_η is analytic and periodic with period $2\pi i/\eta$ in the interior of this set.

If $\eta = 1$, the function p_η defined by (10.10-2) is in the engineering literature referred to as the **z transform** of F. The z transform is essentially identical with what otherwise would be called the *generating function* of the sequence $\{F(n)\}$. All the many techniques for dealing with power series thus become available to the z transform, and no special theory is required for it as long as η is kept fixed.

If η is varied, however, it may be of interest to discuss the relationship of the $\mathcal{L}_\eta$ transform to the $\mathcal{L}$ transform. For this and other purposes the following result, in which f_η is expressed in terms of f, is useful.

THEOREM 10.10a

Let $F \in \Omega$, $\beta_F < \infty$, $f := \mathcal{L}F$, and let $e^{-s\tau}F(\tau)$ be of bounded variation on $(0, \infty)$ for $\text{Re } s \geq \sigma_0 \geq \beta_F$. If at points of discontinuity F is normalized such that

$$F(\tau) = \tfrac{1}{2}\{F(\tau+) + F(\tau-)\},$$

then for all s satisfying $\text{Re } s \geq \sigma_0$

$$f_\eta(s) = \eta \sum_{n=0}^{\infty} e^{-sn\eta} F(n\eta) = PV \sum_{n=-\infty}^{\infty} f\left(s + \frac{2\pi i n}{\eta}\right). \qquad (10.10\text{-}3)$$

Equation (10.10-3) is known as **Polya's formula**. It expresses $f_\eta(s)$ as a sum of values of f at equidistant points on the vertical line with abscissa $\text{Re } s$. The smaller η, the larger is the distance between these points.

Polya proved his formula for analytic functions of exponential type, using the residue theorem. However, the formula is an immediate corollary of the *Poisson summation formula* (Theorem 10.6e): For $\text{Re } s \geq \beta_F$, the Fourier transform of $e^{-s\tau}F(\tau)$ is $f(s + i\omega)$, and hypothesis (b) of Theorem 10.6e is satisfied for $\text{Re } s \geq \sigma_0$ because $F(\tau) = 0$ for $\tau < 0$. Transcribing (10.6-21) and setting $\tau = 0$ we obtain (10.10-3).

II.   Discrete $\mathcal{L}$ Transforms of Functions of Exponential Type

The theory of the discrete $\mathcal{L}$ transformation can be pushed farther if we assume that the original function belongs to some class Ω_γ of functions of the exponential type. We recall from §10.9 that the (continuous) Laplace transform $f := \mathcal{L}F$ then can be extended to a function analytic for $|s| > \gamma$, and vanishing and analytic at $s = \infty$. More precisely, f is analytic outside a certain compact convex set C, the hull of singularities of f, contained in the disk $|s| \leq \gamma$.

Because we do not wish to change the definition of $F(0)$ if $F(0) \neq 0$, Polya's formula must now be modified to read

$$\eta \sum_{n=0}^{\infty} F(n\eta) e^{-ns\eta} = \frac{\eta}{2} F(0) + PV \sum_{n=-\infty}^{\infty} f\left(s + \frac{2\pi i n}{\eta}\right). \qquad (10.10\text{-}4)$$

The variation of $G(\tau) := e^{-s\tau}F(\tau)$ on $(0, \infty)$ is bounded if

$$\int_0^\infty |G'(\tau)| \, d\tau < \infty. \tag{10.10-5}$$

Now if $F \in \Omega_\gamma$, F not constant, it easily follows from Cauchy's integral formula for the derivative that $F' \in \Omega_\gamma$. Thus (10.10-5) is satisfied, and (10.10-4) thus valid, if $\text{Re } s > \gamma$.

This range of validity will now be extended. Notice, however, that the expression on the right of (10.10-4) even ceases to have a meaning if s belongs to the set C_η of all s such that

$$s + \frac{2\pi i n}{\eta} \in C$$

for some integer n. (The set C_η consists of C and of all sets obtained by translating C by an integral multiple of $2\pi i/\eta$.) We denote by S_n the complement of C_η with respect to the unextended plane. Two cases are possible:

(a) The set S_η is connected. Because the diameter of C does not exceed 2γ, this certainly happens if $\eta\gamma < \pi$.

(b) The set S_η is not connected. Then S_η has a unique decomposition into components. Precisely one component contains the set $\text{Re } s > \gamma$. This is called the **right component** of S_η and is denoted by S_η^+.

In case (a) above, the right component of S_η equals S_η.

THEOREM 10.10b

Let $F \in \Omega_\gamma$, and let f denote the analytic continuation of $\mathscr{L}F$ into the exterior of C, the hull of singularities of f. Then $f_\eta := \mathscr{L}_\eta F$ can be extended analytically into S_η^+, the right component of the set S_η, and the analytic continuation is represented by

$$f_\eta(s) = \frac{\eta}{2}F(0) + PV \sum_{n=-\infty}^{\infty} f\left(s + \frac{2\pi i n}{\eta}\right), \tag{10.10-6}$$

where the series on the right converges uniformly on every compact subset of S_η^+.

Proof. We begin by proving the last statement. Let $s_0 \in C$. Then f can be expanded in a series of powers of $(s - s_0)^{-1}$ with no constant term and thus may be represented in the form

$$f(s) = \frac{a_0}{s - s_0} + \frac{1}{(s - s_0)^2} g(s),$$

where g is analytic in S and bounded for $|s| \geq \gamma + 1$, say. Let T be a compact subset of S_η^+. We know from a similar discussion in §7.10 that

$$PV \sum_{n=-\infty}^{\infty} \frac{1}{s + 2\pi i n/\eta - s_0}$$

converges uniformly in T and thus represents an analytic function there. The same is true of the series

$$\sum_{n=-\infty}^{\infty} \frac{1}{(s + 2\pi i n/\eta - s_0)^2} g\left(s + \frac{2\pi i n}{\eta}\right)$$

(even without the PV) because leaving aside the finitely many terms for which $|s + 2\pi i n/\eta| < \gamma + 1$, its terms are dominated by const $\cdot n^{-2}$.

We conclude that the series (10.10-6) converges uniformly on every compact subset T of S_η^+ and thus represents an analytic function in S_η^+. Because this function agrees with f_η for Re s sufficiently large, it represents the analytic continuation of f_η into S_η^+. ■

The series on the right of (10.10-6) converges merely like $\sum_n n^{-k}$ where $k \geq 2$, whereas the series defining f_η for Re $s > \gamma$ converges like $\sum_n q^n$ where $|q| < 1$. Thus the result can sometimes be used to convert slowly converging series into rapidly converging ones.

EXAMPLE **1**

Let $f(s) := s^{-k-1}$ (k a nonnegative integer). We know that f is tne image of $(1/k!)\tau^k$. Thus for $\eta = 1$ and $k = 0$ there follows

$$PV \sum_{n=-\infty}^{\infty} \frac{1}{s - 2\pi i n} = \frac{1}{2} + \sum_{n=1}^{\infty} e^{-sn}, \qquad (10.10\text{-}7)$$

and for $k = 1, 2, \ldots$

$$\sum_{n=-\infty}^{\infty} \frac{1}{(s - 2\pi i n)^{k+1}} = \frac{1}{k!} \sum_{n=1}^{\infty} n^k e^{-sn}. \qquad (10.10\text{-}8)$$

In both formulas the series on the right for Re $s > 0$ converge like a geometric series.

EXAMPLE **2**

Let $\alpha > 0$. Then from the correspondence

$$\frac{1}{\sqrt{s^2 + \alpha^2}} \quad\bullet\!\!-\!\!-\!\!\circ\quad J_0(\alpha\tau)$$

we obtain by (10.10-6) where $\eta = 1$ for Re $s > 0$

$$PV \sum_{n=-\infty}^{\infty} \frac{1}{\sqrt{(s + 2\pi i n)^2 + \alpha^2}} = \frac{1}{2} + \sum_{n=1}^{\infty} J_0(\alpha n) e^{-sn}.$$

If α is not an integral multiple of 2π, the series on the right still converges for $s = 0$. If the series is considered a series of powers of e^{-s}, it follows from Abel's theorem on power series (Theorem 2.2e) that its value for $s = 0$ equals the limit for $s = 0+$ of the expansion on the left. When $s = 0$, the terms of the series on the left where $n = \pm k$, $2\pi k > \alpha$ are purely imaginary and cancel each other, and there remains only

$$\frac{1}{2} + \sum_{n=1}^{\infty} J_0(\alpha n) = \frac{1}{\alpha} + 2 \sum_{n=1}^{r} \frac{1}{\sqrt{\alpha^2 - 4\pi^2 n^2}}, \tag{10.10-9}$$

where $r := [\alpha/2\pi]$.

From this point onward we assume that the set S_η is connected, which means that all translates of C, the hull of singularities, are disjoint. In this situation there is yet another possibility to express f_η in terms of f.

For $\epsilon > 0$, let C^ϵ denote the set of all points $s + d$ where $s \in C$ and $|d| \le \epsilon$, and let C_η^ϵ be the set of all s such that $s + 2\pi i k/\eta \in C^\epsilon$ for some integer k (see Fig. 10.10a). If ϵ is sufficiently small, the translates of C^ϵ are disjoint, and S_η^ϵ, the complement of C_η^ϵ, is still connected. Let Γ_ϵ denote the boundary curve of the convex set C^ϵ, oriented in the positive sense. For $s \in S_\eta^\epsilon$ we consider the integral

$$h(s) := \frac{1}{2\pi i} \int_{\Gamma_\epsilon} \frac{f(u)}{1 - e^{\eta(u-s)}} \, du.$$

The singularities of the integrand (considered as a function of s) occur only where

$$e^{\eta(u-s)} = 1, \qquad \text{i.e., for } s = u + \frac{2\pi i k}{\eta}$$

(k integer), that is, when s lies on Γ_ϵ or on one of its translates. Thus as a function of s the integrand is analytic in S_η^ϵ, and because the integrand also depends analytically on u, it is clear that $h(s)$ is analytic in S_η^ϵ. We now evaluate the integral. Let Re $\eta s \ge \gamma + 2\epsilon$. Then $|e^{\eta(u-s)}| \le e^{-\eta\epsilon} < 1$ on Γ_ϵ. Expanding into a geometric series and integrating term by term, we then get

$$h(s) = \frac{1}{2\pi i} \int_{\Gamma_\epsilon} \sum_{n=0}^{\infty} e^{-n(s-u)\eta} f(u) \, du$$

$$= \sum_{n=0}^{\infty} e^{-ns\eta} \frac{1}{2\pi i} \int_{\Gamma_\epsilon} e^{nu\eta} f(u) \, du.$$

But, by Theorem 10.9b,

$$\frac{1}{2\pi i} \int_{\Gamma_\epsilon} f(u) e^{nu\eta} \, du = F(n\eta);$$

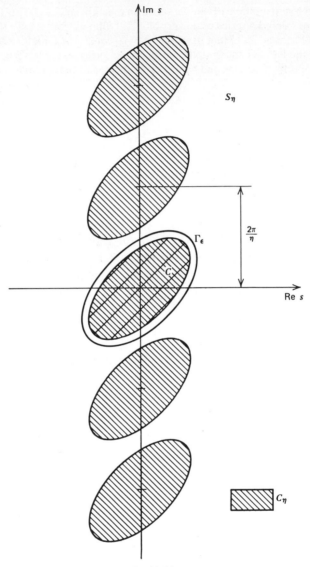

Fig. 10.10a.

therefore

$$h(s) = \sum_{n=0}^{\infty} e^{-sn\eta} F(n\eta) = f_\eta(s),$$

and we have obtained:

THEOREM 10.10c

Let $F \in \Omega_\gamma$ for some $\gamma \geqslant 0$, and let η be such that S_η is connected. Then for any $s \in S_\eta$ the continuation f_η of $\mathscr{L}_\eta F$ is represented by

$$f_\eta(s) = \frac{1}{2\pi i} \int_{\Gamma_\epsilon} \frac{f(u)}{1 - e^{-\eta(s-u)}} \, du, \qquad (10.10\text{-}10)$$

where Γ_ϵ runs parallel to the boundary of C at distance ϵ, and ϵ is chosen so small that s is exterior to Γ_ϵ and to all $(2\pi i/\eta)$ — translates of Γ_ϵ.

Now let s be such that Re $\eta s < -\gamma - 2\epsilon$. Then $|e^{\eta(s-u)}| < e^{-\eta\epsilon}$ at all points of Γ_ϵ. Expanding the kernel of the integral (10.10-10) in powers of $e^{\eta(s-u)}$ (in place of $e^{-\eta(s-u)}$ as before) by use of the identity $1/(1-z) = -z^{-1}(1-z^{-1})^{-1}$, we obtain

$$f_\eta(s) = -\frac{1}{2\pi i} \int_{\Gamma_\epsilon} \sum_{n=1}^{\infty} e^{n\eta(s-u)} f(u) \, du$$

$$= -\sum_{n=1}^{\infty} e^{ns\eta} \frac{1}{2\pi i} \int_{\Gamma_\epsilon} e^{-n\eta u} f(u) \, du$$

$$= -\sum_{n=1}^{\infty} e^{n\eta s} F(-n\eta),$$

where we have again used Theorem 10.9b. Thus *in the half-plane Re $\eta s < -\gamma$ the function f_η has the representation*

$$f_\eta(s) = -\sum_{n=1}^{\infty} e^{ns\eta} F(-n\eta). \qquad (10.10\text{-}11)$$

This result takes a particularly striking form when formulated in terms of the z transform. It suffices to state the case where $\eta = 1$.

THEOREM 10.10d (Leau–Wigert–Polya theorem)

Let F be a function of exponential type, let C be the convex hull of the singularities of $f := \mathscr{L}F$, and let the sets $C + 2\pi ik$ $(k = 0, \pm 1, \pm 2, \ldots)$ all be disjoint. Then the z transform of F,

$$p(z) := \sum_{n=0}^{\infty} F(n) z^{-n},$$

can be continued analytically to a function that is analytic at $z = 0$, and whose expansion in powers of z is

$$p(z) = -\sum_{n=1}^{\infty} F(-n) z^{n}. \qquad (10.10\text{-}12)$$

EXAMPLE 3

The hypotheses of Theorem 10.10c are satisfied if F is any polynomial P, for then C reduces to the single point $s = 0$. Thus if P is any polynomial, and

$$p(z) := \sum_{n=0}^{\infty} P(n)z^{-n}$$

for $|z| > 1$, then the analytic continuation of p into $|z| < 1$ is given by

$$p(z) = -\sum_{n=1}^{\infty} P(-n)z^{n}.$$

EXAMPLE 4

The function $F(\tau) := \operatorname{Cosh}\sqrt{\alpha\tau}(\alpha > 0)$ belongs to the class Ω_0. Thus C again consists of the sole point $s = 0$. The function

$$p(z) := \sum_{n=0}^{\infty} \operatorname{Cosh}\sqrt{n\alpha}\, z^{-n}$$

$(|z| > 1)$ can be continued into the set $|z| < 1$ and there has the representation

$$p(z) = -\sum_{n=1}^{\infty} \cos\sqrt{n\alpha}\, z^{n}.$$

We now turn to the problem of *inverting* the discrete Laplace transform. If by inversion we merely mean recovering the numbers $F(n\eta)$ from the function $f_{\eta}(s)$, then this is merely a matter of determining the coefficients of the Taylor series of an analytic function and need not be discussed any further. If by inversion we mean determining $F(\tau)$ from $f_{\eta}(s)$ also at points τ that are not of the form $n\eta$, then the inversion problem clearly has no unique solution, because for $\tau \neq n\eta$ the definition of $F(\tau)$ may be changed arbitrarily without influencing $f_{\eta}(s)$.

The following theorem shows that matters are different if F is an entire function of exponential type.

THEOREM 10.10e

Let F be an entire function of exponential type, let C be the hull of singularities of $f := \mathscr{L}F$, and let $\eta > 0$ be so small that the set S_{η} is connected. Then for arbitrary complex t, if Γ is any positively oriented simple closed curve in S_{η} encircling C but none of its translates,

$$F(t) = \frac{1}{2\pi i} \int_{\Gamma} e^{ts} f_{\eta}(s)\, ds. \qquad (10.10\text{-}13)$$

Proof. Using Polya's formula (10.10-6), we write

$$f_{\eta}(s) = f(s) + g_{\eta}(s),$$

where

$$g_\eta(s) := \frac{\eta}{2}F(0) + PV \sum_{\substack{n=-\infty \\ n \neq 0}}^{\infty} f\left(s + \frac{2\pi i n}{\eta}\right)$$

Because g_η is analytic on and in the interior of Γ,

$$\int_\Gamma e^{ts} g_\eta(s)\, ds = 0,$$

and by Theorem 10.9b we have

$$\frac{1}{2\pi i} \int_\Gamma e^{ts} f_\eta(s)\, ds = \frac{1}{2\pi i} \int_\Gamma e^{ts} f(s)\, ds = F(t). \quad \blacksquare$$

This result shows that a function of exponential type is fully determined by its values at the points $t = n\eta$, $n = 0, 1, 2, \ldots$, provided that η is chosen such that S_η is connected. As we have noted, this is surely so if $\eta\gamma < \pi$, where γ denotes the growth parameter of the function. We thus have:

COROLLARY 10.10f

Let $F \in \Omega_\gamma$ where $\gamma \geq 0$, and let $F(n\eta) = 0$, $n = 0, 1, 2, \ldots$, for some $\eta > 0$ such that $\eta\gamma < \pi$. Then F vanishes identically.

The bound given for η is best possible, because the function $F(t) := \sin t$ satisfies $|\sin t| \leq e^{|t|}$, $\sin n\pi = 0$, $n = 0, 1, 2, \ldots$, yet it does not vanish identically.

III. Functions of Semiexponential Type

The discrete Laplace transform $\mathscr{L}_\eta$ was defined by (10.10-1). An alternate representation of $f_\eta := \mathscr{L}_\eta F$ is given by Polya's formula (10.10-6). If F is of exponential type, and if the set S_η is connected, then f_η or its analytic continuation is also represented by the contour integral (10.10-10) or by the series (10.10-11).

To the foregoing representations we now add one that is based on the Plana summation formula (Theorem 4.9c). This representation generally holds for $\mathscr{L}_\eta$ transforms of semiexponential type, provided that η is sufficiently small. It permits us to draw conclusions similar to those of the Theorems 10.10c and 10.10d.

Let F be analytic for Re $t \geq 0$, let F be of semiexponential type, and let $\gamma(\phi)$ be the indicator function of F and

$$\gamma := \sup_{-\frac{\pi}{2} \leq \phi \leq \frac{\pi}{2}} \gamma(\phi)$$

its growth parameter. Let $\eta\gamma < 2\pi$. Then if s is real, $s > \gamma$, the function

$$t \to e^{-s\eta t}F(\eta t)$$

satisfies the conditions under which the Plana summation formula holds (see Theorem 4.9c), and there results

$$\sum_{n=0}^{\infty} e^{-s\eta n}F(n\eta) = \frac{1}{2}F(0) + \int_0^{\infty} e^{-s\eta\xi}F(\eta\xi)\,d\xi$$

$$+ i\int_0^{\infty} [e^{-is\eta\xi}F(i\eta\xi) - e^{is\eta\xi}F(-i\eta\xi)]\frac{1}{e^{2\pi\xi}-1}\,d\xi.$$

The sum on the left is $\eta^{-1}f_\eta(s)$, and the first integral on the right equals $\eta^{-1}f(s)$. Thus we have obtained

$$f_\eta(s) = f(s) + g_\eta(s), \tag{10.10-14}$$

where

$$g_\eta(s) := \frac{\eta}{2}F(0) + i\int_0^{\infty} \frac{e^{-is\tau}F(i\tau) - e^{is\tau}F(-i\tau)}{e^{2\pi\tau/\eta}-1}\,d\tau. \tag{10.10-15}$$

The representation (10.10-14) has been derived for s real, $s > \gamma$. It is now used to continue f_η analytically. We recall that in any case f_η is analytic in the set Re $s > \gamma$, and that it is periodic with period $2\pi i/\eta$. It thus suffices to study the analytic behavior of f_η in a horizontal strip of width $> 2\pi/\eta$.

By Theorem 10.9f, the function f can be continued to a function analytic in the complement of the unbounded convex set

$$C := \left\{ s : \text{Re}(e^{i\phi}s) \le \gamma(\phi),\ -\frac{\pi}{2} \le \phi \le \frac{\pi}{2} \right\},$$

the convex hull of the singularities of f (see Fig. 10.9d). It thus remains to continue g_η.

The integrand in the integral (10.10-15) representing g_η is an analytic function of s. If s is complex, $s = \sigma + i\omega$, then the integrand is of the order of the larger of the two functions

$$\exp\left[\left(\omega + \gamma\left(\frac{\pi}{2}\right) - \frac{2\pi}{\eta}\right)\tau\right] \quad \text{and} \quad \exp\left[\left(-\omega + \gamma\left(-\frac{\pi}{2}\right) - \frac{2\pi}{\eta}\right)\tau\right].$$

The integral thus converges whenever the exponents are both negative, which is the case if

$$\omega \in I_\eta := \left(-\frac{2\pi}{\eta} + \gamma\left(-\frac{\pi}{2}\right),\ \frac{2\pi}{\eta} - \gamma\left(\frac{\pi}{2}\right) \right).$$

(Because $\gamma(\pm\pi/2) \leq \gamma < 2\pi/\eta$ by hypothesis, the interval I_η has positive length.) It is obvious that the integral converges uniformly with respect to ω whenever ω is restricted to a closed subinterval of I_η. By the Theorems 4.1a and 3.4b it follows as usual that g_η is analytic in the strip

$$T_\eta := -\infty < \operatorname{Re} s < \infty, \qquad -\frac{2\pi}{\eta} + \gamma\left(-\frac{\pi}{2}\right) < \operatorname{Im} s < \frac{2\pi}{\eta} - \gamma\left(\frac{\pi}{2}\right).$$

We now assume that η is so small that

$$\eta\left[\gamma\left(-\frac{\pi}{2}\right) + \gamma\left(\frac{\pi}{2}\right)\right] < 2\pi. \qquad (10.10\text{-}16)$$

Then the width of the strip T_η exceeds $2\pi/\eta$, the width of a period strip of f_η; furthermore, T_η contains the set C defined above. It follows, first of all, that f_η can be continued into $T_\eta \backslash C$; however, in view of the periodicity of f_η we may conclude that f_η can be continued to a function that is analytic on the complement S_η of the set C_η of all s such that

$$s + \frac{2\pi i}{\eta} n \in C$$

for some integer n. The set S_η has infinitely many infinitely long arms extending to the left, as shown in Fig. 10.10b.

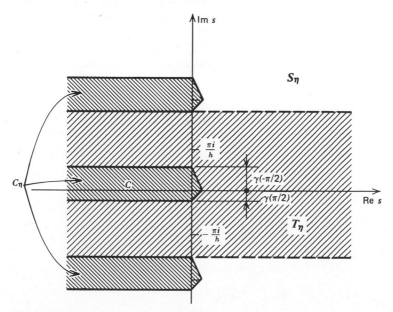

Fig. 10.10b.

Summarizing, we obtain

THEOREM 10.10g

Let F be a function of semiexponential type analytic for Re $t \geq 0$, *and let* $\gamma(\phi)$ *be the indicator function of* $F(-\pi/2 \leq \phi \leq \pi/2)$. *If* $\eta[\gamma(\pi/2) + \gamma(-\pi/2)] < 2\pi$, *then* $f_\eta := \mathcal{L}_\eta F$ *can be extended analytically into the set* S_η *described above. Furthermore, for* $-2\pi/\eta + \gamma(\pi/2) < \text{Im } s < 2\pi/\eta - \gamma(\pi/2)$ *there holds the representation* (10.10-14), *where the function* $g_\eta(s)$ *given by* (10.10-15) *is analytic.*

Let again $s = \sigma + i\omega$, where $\omega \in I_\eta := (-2\pi/\eta + \gamma(-\pi/2), 2\pi/\eta - \gamma(\pi/2))$. We contend that

$$\lim_{\sigma \to -\infty} g_\eta(\sigma + i\omega) = 0, \tag{10.10-17}$$

uniformly with respect to ω if ω is restricted to any closed subinterval of I_η. This follows because g_η is the sum of two integrals of the form

$$\int_0^\infty e^{\pm i\sigma\tau} \cdot \text{continuous function of } \tau,$$

which converge uniformly with respect to σ and ω. As in the proof, say, of Theorem 10.6d, we first may cut off a tail of the integrals which is $\leq \epsilon/2$. The assertion then follows by applying the Riemann–Lebesgue Lemma (Theorem 10.6a) to the remaining finite integrals.

We make two applications of (10.10-14) that correspond to the Theorems 10.10c and 10.10d on $\mathcal{L}_\eta$ transforms of functions of *exponential* type.

THEOREM 10.10h

Let F be a function of semiexponential type (analytic for Re $t \geq 0$) *with indicator function* $\gamma(\phi)$. *If* η *satisfies* (10.10-16), *then the z transform of F,*

$$p_\eta(z) := \eta \sum_{n=0}^\infty F(n\eta) z^{-n},$$

can be continued analytically into the sector

$$\eta\gamma\left(-\frac{\pi}{2}\right) < \arg z < 2\pi - \eta\gamma\left(\frac{\pi}{2}\right), \qquad |z| < 1;$$

moreover, $p_\eta(z) \to 0$ *when* $z \to 0$ *in any closed subsector of that sector.*

Proof. Let $f := \mathcal{L}F$. By (10.9-12),

$$f(s) = -if^*(-is),$$

where f^* likewise is a $\mathscr{L}$ transform (of $F(-i\tau)$). It thus follows from Theorem 10.7a that

$$\lim_{\sigma \to -\infty} f(\sigma + i\omega) = 0$$

uniformly in ω for $\omega \geq \omega_0 > \gamma(-\pi/2)$. Hence by (10.10-14) and (10.10-17) we have

$$\lim_{\sigma \to -\infty} f_\eta(\sigma + i\omega) = 0$$

uniformly in ω for $\gamma(-\pi/2) < \omega_0 \leq \omega \leq \omega_1 < 2\pi/\eta - \gamma(\pi/2)$. The result now follows since $p_\eta(e^{-\eta s}) = f_\eta(s)$. ∎

EXAMPLE 5

An interesting example is provided by the so-called **Mittag–Leffler function**

$$E_\eta(z) := \sum_{n=0}^{\infty} \frac{z^n}{\Gamma(1 + n\eta)}.$$

This is $1/\eta$ times the z_η transform of $F(t) := [\Gamma(1+t)]^{-1}$, evaluated at z^{-1} (see example **9** of §10.9). We have $\gamma(\pi/2) = \gamma(-\pi/2) = \pi/2$. Thus

$$E_\eta(z) \to 0 \text{ if } z \to \infty \text{ in the sector } |\arg(-z)| < \pi - \frac{\eta\pi}{2} - \epsilon$$

where $\epsilon > 0$. Two special cases may be noted:

$E_1(z) = e^z$, where the statement is obvious;
$E_2(z) = \mathrm{Cosh}\sqrt{z}$, where the statement just becomes vacuous.

THEOREM 10.10i

Let F and η satisfy the hypotheses of Theorem 10.10g. If α and β are such that

$$-\frac{2\pi}{\eta} + \gamma\left(-\frac{\pi}{2}\right) < -\alpha < -\gamma\left(\frac{\pi}{2}\right), \qquad \gamma\left(-\frac{\pi}{2}\right) < \beta < \frac{2\pi}{\eta} - \gamma\left(\frac{\pi}{2}\right),$$

then there holds for every $\tau > 0$ the inversion formula

$$F(\tau) = \frac{1}{2\pi i} \int_{-\infty-i\alpha}^{-\infty+i\beta} e^{s\tau} f_\eta(s) \, ds, \tag{10.10-18}$$

where the path of integration avoids the set C_η defined earlier.

Proof. In the inversion formula (10.9-13), replace f by $f_\eta - g_\eta$. In the integral of g_η the path of integration may by virtue of Theorem 10.10g be shifted to the left as much as we please. By (10.10-17), the integral of g_η thus has the value zero. There remains (10.10-18). ∎

The result shows that functions of semiexponential type likewise are fully determined by their values at the points $\tau = n\eta$, $n = 0, 1, 2, \ldots$, if η is small enough. Corollary 10.10f thus has the following analog:

COROLLARY 10.10j (Carlson's theorem)

Let F be a function of semiexponential type with indicator function $\gamma(\phi)$, and let $F(n\eta) = 0$, $n = 0, 1, 2, \ldots$, where η satisfies (10.10-16). Then F vanishes identically.

PROBLEMS

1. The function $F(t) := [\Gamma(-t)]^{-1}$ is entire, satisfies $\gamma(\pm\pi/2) = \pi/2$, vanishes for $t = 0, 1, 2, \ldots$, and yet does not vanish identically. How do you explain the apparent contradiction to Corollary 10.10j?

(a) t,   (b) $e^{\alpha t}$,   (c) $\dfrac{1}{\sqrt{5}}\{e^{t\,\text{Log}\,\omega} - e^{-t\,(\text{Log}\,\omega + i\pi)}\}$, $\omega := \dfrac{\sqrt{5}+1}{2}$.

The remaining problems deal with the z transform

$$a(z) := \sum_{n=0}^{\infty} a_n z^{-n} \qquad (10.10\text{-}19)$$

of a given sequence $A = \{a_0, a_1, a_2, \ldots\}$, without reference to the origin of A. We denote by l the linear space of all sequences $\{a_n\}_{n=0}^{\infty}$ with real or complex elements a_n, and we use the symbol

$$a(z) \bullet \!\!\!\overset{z}{-\!\!\!-}\!\!\!\circ A = \{a_0, a_1, a_2, \ldots\}$$

to indicate that $a(z)$ is the z transform of an element $A \in l$. Any z transform may be regarded a formal power series; it defines a function analytic at $z = \infty$ if and only if $\{a_n\} \in l^0$, where l^0 denotes the subspace of l consisting of all sequences $\{a_n\}$ such that

$$\limsup_{n \to \infty} |a_n|^{1/n} < \infty.$$

3. Establish the following pairs of z transforms:

(a) $\{1, 1, 1, \ldots\} \circ\!\!\!-\!\!\!\overset{z}{-\!\!\!-}\!\!\!\bullet \dfrac{z}{z-1}$,

(b) $\{1, 2, 3, \ldots\} \circ\!\!\!-\!\!\!\overset{z}{-\!\!\!-}\!\!\!\bullet \dfrac{z^2}{(z-1)^2}$,

(c) $\left\{\dbinom{n+k}{k}\right\} \circ\!\!\!-\!\!\!\overset{z}{-\!\!\!-}\!\!\!\bullet \dfrac{z^k}{(z-1)^k}$.

4. Which is the z transform of the sequence $\{1, \frac{1}{2}, \frac{1}{3}, \ldots\}$?

5. Show: If $a(z) \bullet\!\!\!-\!\!\!-\!\!\!-\!\!\!\circ \{a_n\}$, and if $\lim_{n\to\infty} a_n$ exists, then

$$\lim_{n\to\infty} a_n = \lim_{z\to 1+} (z-1)a(z).$$

6. Show: If $A \in l^0$, $B \in l^0$, and if $A \circ\!\!\!-\!\!\!-\!\!\!-\!\!\!\bullet a(z)$, $B \circ\!\!\!-\!\!\!-\!\!\!-\!\!\!\bullet b(z)$, then $a(z)b(z) \bullet\!\!\!-\!\!\!-\!\!\!-\!\!\!\circ C$, where the sequence $C = \{c_n\}$ is the *Cauchy product* of A and B,

$$c_n = \sum_{k=0}^{n} a_k b_{n-k}.$$

[This is essentially a restatement of Theorem 2.3e.]

7. Let $A, B \in l^0$, and let $C := \{a_n b_n\}$, the *Hadamard product* of A and B. Show that for $|z|$ sufficiently large

$$C \circ\!\!\!-\!\!\!-\!\!\!-\!\!\!\bullet c(z) := \frac{1}{2\pi i} \int_{\Gamma} a(t)b\left(\frac{1}{t}\right)\frac{1}{t}\,dt$$

where Γ is an appropriately large circle.

The following problems deal with *digital filters*. A **digital filter** is a device that maps sequences $X = \{x_n\} \in l$ onto sequences $Y = \{y_n\} \in l$ according to a formula

$$y_n + \sum_{k=1}^{N} b_k y_{n-k} = \sum_{k=0}^{M} a_k x_{n-k}, \qquad n = 0, 1, 2, \ldots, \qquad (10.10\text{-}20)$$

where $a_0, a_1, \ldots, a_M, b_1, \ldots, b_N$ are given constants, $a_M \neq 0$, $b_N \neq 0$, and $x_m = y_m := 0$ for $m < 0$. N may be zero, in which case the sum on the left of (10.10-20) is interpreted as zero, and the filter is called **nonrecursive**; otherwise the filter is called **recursive**. The sequences X and Y are, respectively, called the **input** and the **output** of the filter.

8. Show: If the input of a digital filter belongs to l^0, then so does the output.

9. Show: If $X \in l^0$, $X \circ\!\!\!-\!\!\!-\!\!\!-\!\!\!\bullet x(z)$, then the output $Y \circ\!\!\!-\!\!\!-\!\!\!-\!\!\!\bullet y(z) = h(z)x(z)$, where

$$h(z) := \frac{\sum_{k=0}^{M} a_k z^{-k}}{\sum_{k=0}^{N} b_k z^{-k}}.$$

(The rational function h is called the **transfer function** of the digital filter.)

10. If $h(z)$ is the transfer function of a digital filter, then the sequence $H = \{h_n\} \circ\!\!\!-\!\!\!-\!\!\!-\!\!\!\bullet h(z)$ is called the **shock response** of the filter. Show that the

output of a filter is the convolution of the input with the shock response,

$$y_n = \sum_{k=0}^{n} h_k x_{n-k}, \qquad n = 0, 1, 2, \ldots,$$

as in the case of a linear system.

11. Let $l_\infty \subset l$ denote the space of bounded sequences. A digital filter is called **stable** if it maps l_∞ into l_∞; that is, if a bounded input always produces a bounded output. Prove that the following three statements are equivalent:

   (i) The digital filter F is stable.
   (ii) The poles of the transfer function $h(z)$ of F all are located in the interior of the unit disk.
   (iii) If $h(z) \bullet\!\!\overset{z}{\underline{\qquad}}\!\!\circ \{h_n\}$, then $\sum_{n=0}^{\infty} |h_n| < \infty$.
   [M. Gutknecht]

§10.11. SOME INTEGRAL TRANSFORMS RELATED TO THE $\mathscr{L}$ TRANSFORM

Here we discuss very briefly certain integral transformations that are related to the Laplace transformation in an elementary manner.

I. The Two-sided Laplace Transformation

Let F be a function defined on the real line, $F(\tau) = F_1(\tau) + F_2(-\tau)$, where the functions F_1 and F_2 both are in Ω. We then may consider

$$f(s) := \mathscr{L}_{II} F(s) := \int_{-\infty}^{\infty} e^{-s\tau} F(\tau)\, d\tau, \qquad (10.11\text{-}1)$$

the so-called **two-sided Laplace transform** of F. To find the domain of convergence of the two-sided Laplace integral, we note that

$$\int_{-\infty}^{\infty} e^{-s\tau} F(\tau)\, d\tau = I_1 + I_2,$$

where

$$I_1 := \int_{0}^{\infty} e^{-s\tau} F_1(\tau)\, d\tau = \mathscr{L}F_1(s)$$

and

$$I_2 := \int_{-\infty}^{0} e^{-s\tau} F_2(-\tau)\, d\tau = \int_{0}^{\infty} e^{s\tau} F_2(\tau)\, d\tau = \mathscr{L}F_2(-s).$$

If α_1 and α_2 denote the abscissas of convergence of the Laplace integrals of

F_1 and F_2, respectively, then I_1 converges and is analytic for Re $s > \alpha_1$, and I_2 converges and is analytic for Re$(-s) > \alpha_2$; that is, for Re $s < -\alpha_2$. Thus provided that $\alpha_1 < -\alpha_2$, the two-sided transform converges and is an analytic function in the strip $\alpha_1 < $ Re $s < -\alpha_2$.

Because $\mathscr{L}_{II}F(s) = \mathscr{L}F_1(s) + \mathscr{L}F_2(-s)$, many of the results that were stated in earlier chapters for the one-sided $\mathscr{L}$ transform can be stated also for the two-sided transform. In particular, there holds the complex inversion formula. Let the integral (10.11-1) converge *absolutely* for $\alpha_1 < $ Re $s < -\alpha_2$. Then for every σ such that $\alpha_1 < \sigma < -\alpha_2$,

$$f(\sigma + i\omega) = \int_{-\infty}^{\infty} e^{-\sigma\tau} F(\tau) e^{-i\omega\tau} \, d\tau$$

as a function of ω is the Fourier transform of the function

$$G(\tau) := e^{-\sigma\tau} F(\tau), \qquad -\infty < \tau < \infty,$$

whose integral over $(-\infty, \infty)$ converges absolutely. Thus by the Fourier integral theorem (Theorem 10.6d), at every point where F satisfies condition (C),

$$e^{-\sigma\tau} F(\tau) = \frac{1}{2\pi} PV \int_{-\infty}^{\infty} e^{i\omega\tau} f(\sigma + i\omega) \, d\omega$$

or

$$F(\tau) = \frac{1}{2\pi i} PV \int_{\sigma - i\infty}^{\sigma + i\infty} e^{s\tau} f(s) \, ds. \tag{10.11-2}$$

II. The Mellin Transform

In (10.11-1), let $e^{-\tau} = x$, $\Phi(e^{-\tau})$, so that $\Phi(x) = F(-\text{Log } x)$. Then

$$f(s) = \int_0^{\infty} x^{s-1} \Phi(x) \, dx. \tag{10.11-3}$$

The function f is called the **Mellin transform** of Φ; we shall write $f = \mathscr{M}\Phi$. For convenience, we regard the Mellin transform as being defined on the space Ω; this space is contained in the space of all functions $F(-\text{Log } x) = F_1(-\text{Log } x) + F_2(\text{Log } x)$ where $F_1, F_2 \in \Omega$. Thus if the set of convergence of the integral (10.11-3) has a nonempty interior, this interior is a strip, $\alpha < $ Re $s < \beta$, and $f := \mathscr{M}\Phi$ is analytic in this strip.

If the integral (10.11-3) converges absolutely for $\alpha < \mathrm{Re}\, s < \beta$, then by (10.11-2) we have at every point x such that $F(\tau)$ satisfies (C) at $\tau = -\mathrm{Log}\, x$

$$\Phi(x) = \frac{1}{2\pi i} PV \int_{\sigma - i\infty}^{\sigma + i\infty} x^{-s} f(s)\, ds \qquad (10.11\text{-}4)$$

for all σ such that $\alpha < \sigma < \beta$. This is known as the **Mellin inversion formula.**

We use the symbol $\Phi \circ\!\!\!-\!\!\!-\!\!\!\bullet\, f$ to indicate the relationship $f = \mathcal{M}\Phi$. The Mellin transform obeys operational rules similar to those established for the Laplace transform. For instance, if $\Phi \circ\!\!\!-\!\!\!-\!\!\!\bullet\, f$ and $\alpha > 0$,

$$\Phi(\alpha x) \circ\!\!\!-\!\!\!-\!\!\!\bullet\, \frac{f(s)}{\alpha^{s}}, \qquad (10.11\text{-}5)$$

and for arbitrary complex a,

$$x^{a} \Phi(x) \circ\!\!\!-\!\!\!-\!\!\!\bullet\, f(s + a). \qquad (10.11\text{-}6)$$

Furthermore, if $\Phi' \in \Omega$ and

$$x^{s-1} \Phi(x) \to 0 \quad \text{for } x \to 0 \text{ and for } x \to \infty,$$

then it is easily shown by integration by parts that

$$\Phi'(x) \circ\!\!\!-\!\!\!-\!\!\!\bullet\, -(s - 1) f(s - 1). \qquad (10.11\text{-}7)$$

If

$$x^{s} \Phi(x) \to 0 \quad \text{for } x \to 0 \text{ and for } x \to \infty,$$

one can show in a similar manner that

$$x\Phi'(x) \circ\!\!\!-\!\!\!-\!\!\!\bullet\, -sf(s).$$

More generally, if

$$x^{s+k} \Phi(x) \to 0 \quad \text{for } x \to 0 \text{ and for } x \to \infty,$$

$k = 0, 1, \ldots, n - 1$, then

$$x^{k} \Phi^{(k)}(x) \circ\!\!\!-\!\!\!-\!\!\!\bullet\, (-1)^{k} (s)_{k} f(s), \qquad (10.11\text{-}8)$$

$k = 1, 2, \ldots, n$. This last relation is especially useful for solving so-called **Eulerian differential equations**

$$x^{n} \Phi^{(n)} + a_{1} x^{n-1} \Phi^{(n-1)} + \cdots + a_{n} \Phi = \Psi(x).$$

Certain well-known integral representations of higher transcendental functions can be interpreted as Mellin transforms, for example,

$$\Gamma(s) = \int_{0}^{\infty} x^{s-1} e^{-x}\, dx, \qquad \mathrm{Re}\, s > 0,$$

$$\Gamma(s)\zeta(s) = \int_{0}^{\infty} x^{s-1} (e^{x} - 1)^{-1}\, dx, \qquad \mathrm{Re}\, s > 1.$$

The application of the Mellin inversion formula (10.11-4) in these instances yields formulas that belong to the working kit of analytic number theory.

III.   The Stieltjes Transform

Suppose $F \in \Omega$, and

$$\int_0^\infty |F(\tau)| \, d\tau < \infty.$$

Then $f(s) := \mathcal{L}F(s)$ exists and is continuous for Re $s \geq 0$, and analytic for Re $s > 0$. Because $f(s) \to 0$ for $s \to \infty$, Re $s \geq 0$, f is a function of semiexponential type with growth parameter 0 (see §10.9). It follows from Theorem 10.9f that

$$g(z) := \mathcal{L}f(z) = \int_0^\infty e^{-zs} f(s) \, ds$$

exists and can be extended to a function that is analytic in the cut plane $|\arg z| < \pi$.

For $z > 0$ we have, interchanging the order of the integrations,

$$g(z) = \int_0^\infty e^{-zs} \int_0^\infty e^{-s\tau} F(\tau) \, d\tau \, ds = \int_0^\infty \left\{ F(\tau) \int_0^\infty e^{-s(z+\tau)} \, ds \right\} d\tau$$

$$= \int_0^\infty \frac{F(\tau)}{z+\tau} \, d\tau.$$

By analytic continuation, this representation evidently holds for all z such that $|\arg z| < \pi$.

The functional transformation

$$g(z) = \mathcal{S}F(z) := \int_0^\infty \frac{F(\tau)}{z+\tau} \, d\tau \qquad (10.11\text{-}9)$$

is called the **Stieltjes transformation**. We write $g \bullet\!\!-\!\!\circ F$ in order to indicate that $g = \mathcal{S}F$. The following operational rules for the Stieltjes transformation are easily established: If $F(\tau) \circ\!\!-\!\!\bullet g(z)$, then

$$\tau F(\tau) \circ\!\!-\!\!\bullet \int_0^\infty F(\tau) \, d\tau - z g(z), \qquad (10.11\text{-}10)$$

$$F(\sqrt{\tau}) \circ\!\!-\!\!\bullet g(i\sqrt{z}) + g(-i\sqrt{z}) \qquad (10.11\text{-}11)$$

(principal values). Moreover, if $\rho > 0$,

$$F(\rho\tau) \circ\!\!-\!\!\bullet g(\rho z), \qquad (10.11\text{-}12)$$

$$\frac{1}{\tau} F\left(\frac{\rho}{\tau}\right) \circ\!\!-\!\!\bullet \frac{1}{z} g\left(\frac{\rho}{z}\right), \qquad (10.11\text{-}13)$$

and for any complex a such that $|\arg a| < \pi$,

$$\frac{F(\tau)}{\tau + a} \circ\!\!-\!\!-\!\!-\!\!\bullet -\frac{g(z) - g(a)}{z - a}. \tag{10.11-14}$$

If F' exists and $F' \in \Omega$, then

$$F'(\tau) \circ\!\!-\!\!-\!\!-\!\!\bullet -\frac{1}{z}F(0) - g'(z). \tag{10.11-15}$$

An inversion formula for Stieltjes transforms of functions in Ω can be obtained by way of the theory of Cauchy integrals; see §13.5. In the later part of chapter 12 we encounter Stieltjes transforms defined in terms of Stieltjes integrals. The domain of definition of the Stieltjes transform as considered in Chapter 12 contains the nonnegative functions in Ω as a proper subset. It is seen there that an intimate connection exists between Stieltjes transforms and certain continued fractions. This forms the basis for numerically efficient algorithms for the evaluation of Laplace transforms.

§10.12. SOME APPLICATIONS TO PARTIAL DIFFERENTIAL EQUATIONS

In some treatments of integral transform techniques great emphasis is placed on applications of integral transforms to initial or boundary value problems involving partial differential operators. The method consists in applying the transformation to one selected variable and treating the other variables as parameters. The result in the image space is a boundary value problem whose dimension is reduced by one. The desired solution then is obtained by translating back into the original space.

To apply the method rigorously, it is usually necessary to make a number of assumptions, such as the assumption that differentiation with respect to the parameters may be interchanged with the application of the integral transformation. These assumptions can usually be verified only a posteriori. This means that after the solution has been found by formal methods, it must be verified that the expression found actually *is* a solution of the posed problem. This drawback is shared by the method of integral transforms with other methods for solving partial differential equations.

Another, more serious, criticism may be voiced. What is obtained by applying the integral transformation to a partial differential equation frequently is nothing more than what could have been found by the completely elementary method of separating variables and integrating with respect to the separation parameter. Indeed, the decision to use a certain integral transformation already prejudices the values (real or complex) that the separation parameter is permitted to assume. In our view greater freedom of action is obtained if variables are separated first, the method of integral transforms being used only afterwards to adjust the solution to the boundary conditions. We illustrate the method by three simple examples.

I. Heat Conduction

We wish to find a solution U of the heat equation in one spatial dimension,

$$\frac{\partial U}{\partial \tau} = \frac{\partial^2 U}{\partial \xi^2},$$

(10.12-1)

defined and continuous for $\tau \geq 0$, $-\infty < \xi < \infty$ and satisfying the conditions

$$\lim_{\tau \to \infty} U(\xi, \tau) = 0 \text{ for all } \xi,$$

(10.12-2)

$$U(\xi, 0) = F(\xi),$$

(10.12-3)

where F is a given function such that, say,

$$\int_{-\infty}^{\infty} |F(\xi)| \, d\xi < \infty.$$

Separating variables in (10.12-1) yields the solutions $e^{-\omega^2 \tau} e^{i\omega\xi}$, where ω is a separation parameter. In principle, this parameter may have arbitrary complex values; in view of (10.12-2), however, it seems reasonable to restrict ω to real values. We thus seek U in the form

$$U(\xi, \tau) = \int_{-\infty}^{\infty} G(\omega) \, e^{-\omega^2 \tau} \, e^{i\omega\xi} \, d\omega;$$

(10.12-4)

if G is sufficiently smooth this will automatically satisfy (10.12-1) and (10.12-2). The condition (10.12-3) requires

$$F(\xi) = \int_{-\infty}^{\infty} G(\omega) \, e^{i\omega\xi} \, d\omega.$$

This is satisfied at every point ξ where F satisfies condition (C) of §10.6 if the integral is taken as a principal value integral and

$$G(\omega) = \frac{1}{2\pi} \int_{-\infty}^{\infty} F(\eta) \, e^{-i\omega\eta} \, d\eta.$$

Inserting this in (10.12-4) and reversing the order of the integrations, there results for $\tau > 0$

$$U(\xi, \tau) = \frac{1}{2\pi} \int_{-\infty}^{\infty} F(\eta) \left\{ \int_{-\infty}^{\infty} e^{-\omega^2 \tau + i\omega(\xi - \eta)} \, d\omega \right\} d\eta.$$

By a well-known formula (see the last example in §10.6) the inner integral has the value

$$\sqrt{\frac{\pi}{\tau}} \exp\left(-\frac{(\xi-\eta)^2}{4\tau}\right).$$

We thus find

$$U(\xi, \tau) = \frac{1}{2\sqrt{\pi\tau}} \int_{-\infty}^{\infty} F(\eta) \exp\left(-\frac{(\xi-\eta)^2}{4\tau}\right) d\eta. \qquad (10.12\text{-}5)$$

As pointed out earlier, the foregoing method rests on some unverified assumptions such as the interchangeability of certain improper integrals. It can be shown, however, that the function defined by (10.12-5) for $\tau > 0$ satisfies (10.12-1) and (10.12-2); moreover,

$$\lim_{\tau \to 0+} U(\xi, \tau) = F(\xi)$$

at every ξ where F is continuous.

II. A Dirichlet Problem

In the (x, y) plane we introduce polar coordinates (ρ, ϕ) defined by $x = \rho \cos \phi$, $y = \rho \sin \phi$ and consider the angular domain $D : -\alpha < \phi < \alpha$, where $\alpha < \pi$. In this domain we seek a solution $U = U(\rho, \phi)$ of Laplace's equation

$$\Delta U = \frac{\partial^2 U}{\partial \rho^2} + \frac{1}{\rho} \frac{\partial U}{\partial \rho} + \frac{1}{\rho^2} \frac{\partial^2 U}{\partial \phi^2} = 0 \qquad (10.12\text{-}6)$$

satisfying the boundary conditions

$$\begin{aligned} U(\rho, \pm\alpha) &= U_0, & 0 < \rho < 1, \\ U(\rho, \pm\alpha) &= 0, & \rho > 1, \end{aligned} \qquad (10.12\text{-}7)$$

where U_0 is constant, and

$$\lim_{\rho \to \infty} U(\rho, \phi) = 0. \qquad (10.12\text{-}8)$$

In addition, we require U to be bounded in D.

This problem could be solved by the method of conformal mapping (see §5.6). Here we solve it by separating variables. It is well known that the functions

$$\rho^{\pm s}(A \cos s\phi + B \sin s\phi)$$

are solutions of (10.12-6) for every (real or complex) value of the separation parameter s. Because the desired solution will be an even function of ϕ, we drop the sine term and seek the solution in the form

$$U(\rho, \phi) = \int_\Gamma f(s)\rho^{-s} \cos s\phi \, ds. \qquad (10.12\text{-}9)$$

If f is suitably well behaved, this expression satisfies Laplace's equation for every choice of the path of integration Γ. The boundary condition (10.12-7) requires

$$\int_\Gamma f(s)\rho^{-s} \cos s\alpha \, ds = G(\rho), \qquad (10.12\text{-}10)$$

where

$$G(\rho) := \begin{cases} U_0, & 0 < \rho < 1, \\ 0, & \rho > 1. \end{cases}$$

The integral (10.12-10) acquires a familiar look if for Γ we select a path $s = \sigma + i\omega$, $-\infty < \omega < \infty$. It then becomes a case of the *Mellin inversion formula* (10.11-4) and will be satisfied if $2\pi i f(s) \cos s\alpha$ is the Mellin transform of $G(\rho)$, that is, if

$$f(s) \cos s\alpha = \frac{1}{2\pi i} \int_0^\infty \rho^{s-1} G(\rho) \, d\rho = \frac{1}{2\pi i} \frac{U_0}{s},$$

and if $\sigma > 0$. Thus the function

$$U(\rho, \phi) := \frac{U_0}{2\pi i} \int_{\sigma-i\infty}^{\sigma+i\infty} \rho^{-s} \frac{\cos s\phi}{\cos s\alpha} \frac{ds}{s} \qquad (10.12\text{-}11)$$

satisfies the conditions (10.12-6) and (10.12-7) for every $\sigma > 0$ such that the path of integration does not run through one of the poles of the integrand, that is, for $\sigma \neq \pi/2\alpha$, $3\pi/2\alpha$, ...

For $0 < \sigma < \pi/2\alpha$ the integral is independent of σ, because the integrand is analytic in the strip $0 < \operatorname{Re} s < \pi/2\alpha$ and vanishes sufficiently rapidly as $\operatorname{Im} s \to \pm\infty$. We let $\sigma \to 0$ and take into account the singularity at $s = 0$ by counting the residue with the factor $\frac{1}{2}$. In view of

$$\rho^{-i\omega} = e^{-i\omega \operatorname{Log}\rho} = \cos(\omega \operatorname{Log} \rho) - i \sin(\omega \operatorname{Log} \rho)$$

this yields, omitting the odd part of the integrand,

$$U(\rho, \phi) = \frac{1}{2} U_0 - \frac{U_0}{\pi} \int_0^\infty \frac{\sin(\omega \operatorname{Log} \rho)}{\omega} \frac{\cosh \omega\phi}{\cosh \omega\alpha} \, d\omega. \qquad (10.12\text{-}12)$$

Evidently, this solution is bounded. If in (10.12-11) we shifted the path of integration beyond the simple pole of the integrand at $s = \pi/2\alpha$, there would by calculating the residue arise a term

$$\rho^{-\pi/2\alpha} \cos \frac{\pi\phi}{2\alpha},$$

which satisfies the differential equation and the boundary condition, but not the condition of boundedness. Similar terms would arise from the other poles. Thus (10.12-12) is the desired solution of the problem.

III. Sommerfeld's Solution of the Helmholtz Equation

Here we consider the so-called **Helmholtz equation**

$$\Delta u + k^2 u = 0$$

where Δ denotes the Laplacian in two dimensions, and where k is a real constant. [This equation arises when we seek solutions of the wave equation $u_{\tau\tau} = a^2 \Delta u$ which are of the form $e^{i\omega\tau}u(x, y)$.] Without loss of generality we may assume that $k = 1$, so that the Helmholtz equation assumes the form

$$\Delta u + u = 0. \tag{10.12-13}$$

We wish to determine particular solutions of the Helmholtz equation that in polar coordinates (ρ, ϕ) have the special form $u(\rho, \phi) = e^{i\nu\phi}v(\rho)$, where ν is a given real number. Such solutions could, naturally, be obtained by separating variables. It would be found that the function v has to be a solution of Bessel's equation,

$$v'' + \frac{1}{\rho}v' + \left(1 - \frac{\nu^2}{\rho^2}\right)v = 0, \tag{10.12-14}$$

see §9.7. Here we pursue an entirely different approach, which will lead very directly to certain integral representations of the Bessel functions due to A. Sommerfeld.

It is clear that the functions

$$u(\rho, \phi) = e^{ix} = e^{i\rho\cos\phi} \quad \text{and} \quad u(\rho, \phi) = e^{iy} = e^{i\rho\cos(\phi - \pi/2)}$$

are both solutions of (10.12-13). So is, by virtue of the rotational invariance of Δ, the function

$$u(\rho, \phi) = e^{i(x\cos\alpha + y\sin\alpha)}$$
$$= e^{i\rho\cos(\phi - \alpha)}$$

for any fixed value of α. This property is preserved if we integrate with respect to α. Thus if $A(\alpha)$ is any sufficiently smooth function,

$$u(\rho, \phi) = \int_a^b A(\alpha) e^{i\rho\cos(\phi - \alpha)} d\alpha$$

solves (10.12-13). All the foregoing holds if α assumes complex values. In this case the integral is to be taken along some specified path of integration from a to b. If $A(\alpha)$ is analytic, only the endpoints of the path are relevant.

We now select $A(\alpha) := Ce^{i\nu\alpha}$, where ν is the parameter specified and C is a constant. Making the substitution $\beta := \alpha - \phi$ in the integral, this yields

$$u(\rho, \phi) = Ce^{i\nu\phi} \int_{a-\phi}^{b-\phi} e^{i\rho \cos \beta + i\nu\beta} \, d\beta.$$

This solution of (10.12-13) does not yet have the desired form $e^{i\nu\phi}v(\rho)$, because the integral (through the limits of integration) still depends on ϕ. We can force it to be independent of ϕ, however, by letting the path originate and terminate in an area where a parallel displacement of the path does not change the value of the integral. This is accomplished by letting the path extend to infinity in areas where the integrand is small.

If ν is real, $\rho > 0$, and $\beta = \xi + i\eta$, the real part of the exponent of the exponential function is

$$\text{Re}\{i\rho \cos \beta + i\nu\beta\} = \rho \sinh \eta \sin \xi - \nu\eta.$$

For fixed ξ, this tends to $-\infty$ exponentially

for $\eta \to -\infty$ if $2n\pi < \xi < (2n+1)\pi$,

for $\eta \to +\infty$ if $(2n+1)\pi < \xi < (2n+2)\pi$

$(n = 0, \pm 1, \pm 2, \ldots)$, that is, in the shaded areas of Fig. 10.12a.

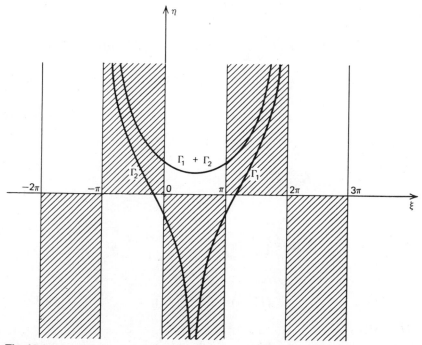

Fig. 10.12a.

We now select a path Γ_1 from $\xi_0 + i\infty$ to $\xi_1 - i\infty$ where $\xi_0 \in (-\pi, 0)$, $\xi_1 \in (0, \pi)$, and a path Γ_2 from $\xi_1 - i\infty$ to $\xi_2 + i\infty$ where $\xi_2 \in (\pi, 2\pi)$, the precise choices of ξ_0, ξ_1, ξ_2 being irrelevant, and define the functions

$$H_\nu^{(1)}(\rho) := C \int_{\Gamma_1} e^{i\rho \cos \beta + i\nu\beta} \, d\beta,$$

$$H_\nu^{(2)}(\rho) := C \int_{\Gamma_2} e^{i\rho \cos \beta + i\nu\beta} \, d\beta,$$

(10.12-15)

where

$$C := \frac{1}{\pi} e^{-i\nu\pi/2}.$$

(10.12-16)

The functions $H_\nu^{(1)}$ and $H_\nu^{(2)}$ are called the **first and second Hankel function of order** ν, respectively.

A number of elementary properties of the Hankel functions can be derived directly from these integrals. Making the substitution $\beta \to \pi - \beta$ in the integral

$$H_\nu^{(1)}(\rho) = \frac{1}{\pi} \int_{\Gamma_1} e^{i\rho \cos \beta + i\nu(\beta - \pi/2)} \, d\beta$$

(10.12-17)

yields

$$H_\nu^{(1)}(\rho) = -\frac{1}{\pi} \int_{\pi - \Gamma_1} e^{-i\rho \cos \beta - i\nu(\beta - \pi/2)} \, d\beta.$$

But $\pi - \Gamma_1$ is a path equivalent to $-\Gamma_2^-$, where the bar denotes complex conjugation. We thus have

$$H_\nu^{(1)}(\rho) = [H_\nu^{(2)}(\rho)]^-.$$

(10.12-18)

Substituting $\beta \to -\beta$ in

$$H_{-\nu}^{(1)}(\rho) = \frac{1}{\pi} \int_{\Gamma_1} e^{i\rho \cos \beta - i\nu(\beta - \pi/2)} \, d\beta$$

we find

$$H_{-\nu}^{(1)}(\rho) = -\frac{1}{\pi} \int_{-\Gamma_1} e^{i\rho \cos \beta + i\nu(\beta + \pi/2)} \, d\beta;$$

hence

$$H_{-\nu}^{(1)}(\rho) = e^{i\nu\pi} H_\nu^{(1)}(\rho),$$

(10.12-19a)

and by (10.12-18)

$$H_{-\nu}^{(2)}(\rho) = e^{-i\nu\pi} H_\nu^{(2)}(\rho).$$

(10.12-19b)

By virtue of their construction, the Hankel functions are solutions of the Bessel differential equation (10.12-14). It thus must be possible to express them linearly in terms of the functions J_ν and Y_ν, the Bessel functions of the first and second kind defined in §9.7.

THEOREM 10.12a

For all real ν and for all $\rho > 0$,

$$H_\nu^{(1)}(\rho) = J_\nu(\rho) + iY_\nu(\rho),$$
$$H_\nu^{(2)}(\rho) = J_\nu(\rho) - iY_\nu(\rho). \qquad (10.12\text{-}20)$$

Proof. We define (for the purpose of the present proof only)

$$S_\nu := \tfrac{1}{2}\{H_\nu^{(1)} + H_\nu^{(2)}\},$$

$$D_\nu := \frac{1}{2i}\{H_\nu^{(1)} - H_\nu^{(2)}\}.$$

We then have to show that $S_\nu = J_\nu$, $D_\nu = Y_\nu$. By (10.12-19),

$$S_{-\nu} = \tfrac{1}{2}\{e^{i\nu\pi}H_\nu^{(1)} + e^{-i\nu\pi}H_\nu^{(2)}\} = \cos \nu\pi S_\nu - \sin \nu\pi D_\nu.$$

Hence if ν is not an integer,

$$D_\nu = \frac{\cos \nu\pi S_\nu - S_{-\nu}}{\sin \nu\pi}.$$

In view of (9.7-20), Theorem 10.12a will be proved if we can show that $S_\nu = J_\nu$.

We have

$$S_\nu(\rho) = \frac{1}{2\pi} \int_{\Gamma_1 + \Gamma_2} e^{i\rho \cos \beta + i\nu(\beta - \pi/2)} \, d\beta. \qquad (10.12\text{-}21)$$

The path of integration is equivalent to the rectangular path shown in Fig. 10.12b. In the above integral, we substitute

$$u := e^{-i(\beta - \pi/2)}.$$

This makes

$$\frac{du}{u} = -i \, d\beta, \qquad \cos \beta = \sin\left(\frac{\pi}{2} - \beta\right) = \frac{1}{2i}\left(u - \frac{1}{u}\right);$$

hence after reversing the sense of integration the integral becomes

$$S_\nu(\rho) = \frac{1}{2\pi i} \int_\Gamma \exp\left(\frac{\rho}{2}\left(u - \frac{1}{u}\right)\right) u^{-\nu-1} \, du,$$

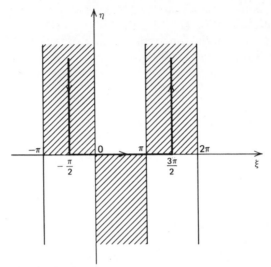

Fig. 10.12b.

where the path of integration in the u plane is shown in Fig. 10.12c. The principal value of $\arg u$ must be taken. To evaluate the integral, we substitute $\rho u = 2v$, obtaining

$$S_\nu(\rho) = \frac{1}{2\pi i} \left(\frac{\rho}{2}\right)^\nu \int_\Gamma \exp\left(v - \frac{\rho^2}{4v}\right) v^{-\nu-1} \, dv;$$

the same path of integration may be used. To compute the last integral, we use the expansion

$$\exp\left(-\frac{\rho^2}{4v}\right) = \sum_{n=0}^{\infty} \frac{(-\rho^2/4)^n}{n!} v^{-n},$$

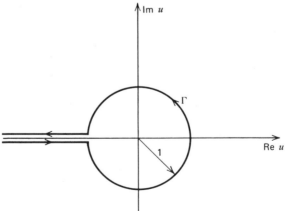

Fig. 10.12c.

which converges uniformly on Γ. Because of the presence of the factor e^v, summation and integration may be interchanged (inspite of the path of integration extending to infinity). Using Hankel's integral (Theorem 8.4b) valid for all complex α,

$$\frac{1}{2\pi i}\int_{\Gamma} e^v v^{-\alpha}\,dv = \frac{1}{\Gamma(\alpha)},$$

we obtain

$$S_\nu(\rho) = \left(\frac{\rho}{2}\right)^\nu \sum_{n=0}^{\infty} \frac{(-\rho^2/4)^n}{n!}\frac{1}{2\pi i}\int_{\Gamma} e^v v^{-1-\nu-n}\,dv$$

$$= \left(\frac{\rho}{2}\right)^\nu \sum_{n=0}^{\infty} \frac{(-\rho^2/4)^n}{n!\,\Gamma(1+\nu+n)}.$$

The last series by definition equals $J_\nu(\rho)$. ∎

Sommerfeld's integrals may be used to develop the theory of Bessel functions *ab ovo*, without making use of any of the results obtained previously. Naturally, there is no need here to dwell on this alternate development. However, we use the integrals to obtain certain important asymptotic expansions for Bessel functions of large order; see §11.8.

SEMINAR ASSIGNMENTS

1. Compute the actual distribution of primes up to 10^6, and compare it to the distribution predicted by the prime number theorem.
2. Develop the basic properties of the z transform along the lines of the Laplace transform. Establish a dictionary of correspondences. Consult the engineering literature for problems that have been attacked successfully by means of the z transform.

NOTES

The basic reference on the Laplace transformation is the "Handbuch" by Doetsch (3 vols., Doetsch [1950], [1955], [1956]). Shorter presentations are Doetsch [1958], 2nd ed. [1970] and, more practically oriented, Doetsch [1967]. See also the article by Doetsch in Sauer and Szabo [1967], pp. 232–484. Excellent introductions with many striking applications are given by Carlslaw and Jaeger [1947], and in Chapter 6 of Lawrentiew and Schabat [1967]. For a mathematical introduction with less emphasis on applications see Widder [1941].

§10.1.   All theorems in this section can be found in Doetsch [1950] or Doetsch [1958].

§10.2.   Arrangement of operational rules follows Lawrentiew and Schabat [1967]. Extensive tables of correspondences are given in Erdélyi [1954] and Oberhettinger and Badii [1973]. Shorter tables are included, for example, in Magnus, Oberhettinger, and Soni [1966] and in Lawrentiew and Schabat [1967].

§10.3.   For additional examples see Carslaw and Jaeger [1947] or Lawrentiew and Schabat [1967]. An introduction to system theory is given in Sauer and Szabo [1967]. For applications to control theory see Effertz and Kolberg [1963]. Rigorous but elementary introductions to the theory of distributions are given in Doetsch [1967] and Doetsch [1970].

§10.4.   See Doetsch [1950] and Doetsch [1958]. For the renewal equation considered in Problems 7–9, see Bellman and Cooke [1963]. For applications to actuarial mathematics consult Saxer [1958].

§10.5.   Theorem 10.5b is a slight generalization of Theorem 1.79 in Titchmarsh [1939], compare Hardy [1904]. For problems 10–12 (discontinuous integrals of Weber and Schafheitlin) see Watson [1944], pp. 398–415.

§10.6.   For the proof of the Riemann–Lebesgue lemma given here see Whittaker and Watson [1927], §9.41. Otherwise, the presentation of Fourier theory follows Titchmarsh [1939], chapter 13. For Poisson's summation formula see also Courant and Hilbert [1930], pp. 64; Titchmarsh [1937], Theorem 45; Boas [1946]; Bochner [1955], Theorem 2.4.2. For Problem 6, and the examples preceding it, see Boas and Pollard [1973]. For proofs of the sampling theorem under different conditions see Boas [1972]. Oberhettinger [1957] gives a collection of Fourier transforms.

§10.7.   See Doetsch [1950], except for the statement of the Heaviside expansion theorem, which is taken from Lawrentiew and Schabat [1967]. For a proof of the Wiener–Ikehara theorem see also Chandrasekharan [1968]. Problems 7 and 8 are from Henrici [1970], but see also Talbot [1959]. Concerning the numerical inversion of Laplace transforms see Bellmann et al. [1966], Dubner and Abate [1968], Stehfest [1970].

§10.8.   For an elementary account of Dirichlet series see Titchmarsh [1939], Chapter 9. Dirichlet's rule of multiplication is amply illustrated by Hardy and Wright [1954], who also sketch a formal theory of Dirichlet series similar to the theory of formal power series given in Chapter 1.

§10.9.   See Polya [1929]. For Problem 3 see Boas [1964].

§10.10.   The z transform is dealt with by Doetsch in Sauer and Szabo [1967], pp. 409–414, and by Doetsch [1967]. See also Jury [1964]. Theorems 10.10g–10.10j are based on communications by H. Rutishauser. For Problems 7–9 see Gutknecht [1973]. See also Kaiser [1966].

§10.11.   See Lawrentiew and Schabat [1967]. Tables and operational rules for the Mellin and the Stieltjes transforms are to be found in Erdelyi [1954] and, for the Mellin transform, in Oberhettinger [1974].

§10.12.   For Sommerfeld's integrals see Sommerfeld [1947].

11
ASYMPTOTIC METHODS

In the vaguest sense, the theory of asymptotics aims at describing the behavior of functions near the boundary of their domain of definition. When dealing with analytic functions, one frequently encounters a situation in which a function f is defined on some unbounded region of the complex plane, and it is the behavior of $f(z)$ as $z \to \infty$ in S that is of interest. This behavior is trivial if f has either a removable singularity or a pole at ∞; thus we concentrate on studying the asymptotic behavior at ∞ of functions that either have an essential singularity at ∞ or that are defined merely in some angular region extending to ∞.

Even with these restrictions, the question of asymptotics is too general to admit of any nontrivial, generally valid answers. It thus has become customary to study the asymptotic behavior of certain standardized classes of functions. One such class (dealt with in §11.4) is given by the solutions of differential equations where ∞ is a singular point of the second kind. Another large class (treated in §11.5–11.8) is given by the functions that are defined by definite integrals such as those occurring in the definition of integral transforms. A third standardized question in asymptotics concerns the behavior as $n \to \infty$ of the elements of a sequence whose generating function has well-understood singularities on the boundary of its disk of convergence (see §11.10). The chapter concludes with the application of asymptotic theory to the numerical evaluation of limits (§11.12).

§11.1. AN EXAMPLE; ASYMPTOTIC POWER SERIES

Leonhard Euler in 1739 considered the formal series in z^{-1},

$$F := \frac{0!}{z} - \frac{1!}{z^2} + \frac{2!}{z^3} - \cdots = \sum_{n=0}^{\infty} \frac{(-1)^n n!}{z^{n+1}} \qquad (11.1\text{-}1)$$

which is manifestly divergent for all values of z. Thus clearly, the series has no immediate analytic meaning. On the other hand, if $|z|$ is large, then from a

351

strictly numerical point of view the series behaves much like a convergent series. The terms first decrease steadily as long as $n < |z|$, until at $n = [|z|]$ we reach a smallest term whose absolute value, by Stirling's formula, approximately equals

$$\frac{n!}{n^{n+1}} \sim \sqrt{\frac{2\pi}{n}}\, e^{-n}.$$

For $n = 20$, this is about $1.2 \cdot 10^{-9}$. Thus, numerically speaking, it would seem reasonable to terminate the summation of the series at this point. It is only afterward that the terms of the series begin to increase without bounds, causing the ultimate divergence of the series.

Series of the above type have been called *semiconvergent* (Stieltjes) or *initially convergent* (Jahnke and Emde). Is it possible to express the phenomenon of semiconvergence in rigorous analytical language? As we shall see, the answer is yes. The first who succeeded in doing so was H. Poincaré, who formalized a concept which today we call *asymptotic series*.

Before giving definitions, let us return to our example. Formal differentiation yields

$$F' = -\frac{1!}{z^2} + \frac{2!}{z^3} - \frac{3!}{z^4} + \cdots ;$$

thus in the sense of a formal identity we have

$$F' = F - \frac{1}{z}.$$

The idea does not seem far fetched to somehow relate F to a solution of the differential equation

$$u' = u - \frac{1}{z}. \tag{11.1-2}$$

If the series F were convergent, its value at $z = \infty$ would be zero. It thus seems reasonable to relate F to a solution of the differential equation that becomes small if z tends to ∞, for instance, along the positive real axis. The function

$$f(z) := e^z \int_z^\infty \frac{e^{-t}}{t}\, dt \tag{11.1-3}$$

evidently is a solution of (11.1-2), and for $x > 0$ it satisfies

$$f(x) \leqslant \frac{e^x}{x} \int_x^\infty e^{-t}\, dt = \frac{1}{x}.$$

It thus has the required limiting behavior. No other solution has this behavior, because it would have to be obtained from f by adding a nonzero multiple of e^z, a solution of the homogeneous equation, which tends to ∞ as $z \to \infty$ through the positive reals.

But what is the analytical connection between F and f? A first connection can be established as follows. Letting $t = z + \tau$, where $\tau \geq 0$, we can write (11.1-3) as

$$f(z) = \int_0^\infty \frac{e^{-\tau}}{z + \tau} \, d\tau = \frac{1}{z} \int_0^\infty \left(1 + \frac{\tau}{z}\right)^{-1} e^{-\tau} \, d\tau. \tag{11.1-4}$$

Expanding $(1 + \tau/z)^{-1}$ in a geometric series and integrating term by term by use of the formula

$$\int_0^\infty \tau^n e^{-\tau} \, d\tau = n!,$$

we obtain

$$f(z) = \frac{1}{z} \int_0^\infty \sum_{n=0}^\infty \left(-\frac{\tau}{z}\right)^n e^{-\tau} \, d\tau = \sum_{n=0}^\infty (-1)^n \frac{n!}{z^{n+1}},$$

the series F evaluated at z. Naturally, the foregoing procedure is grossly invalid—the integrated series does not even converge on the whole interval of integration, let alone uniformly—and the punishment is immediate in view of the meaninglessness of the result.

Nevertheless, this rough-and-tumble approach holds the key to a rigorous analysis. Instead of expanding in a geometric series, we can use Taylor's formula with remainder, which in the present case has the very explicit form

$$\left(1 + \frac{\tau}{z}\right)^{-1} = 1 + \frac{-\tau}{z} + \frac{(-\tau)^2}{z^2} + \cdots + \frac{(-\tau)^{n-1}}{z^{n-1}} + \frac{(-\tau)^n}{z^n} \left(1 + \frac{\tau}{z}\right)^{-1},$$

where $n = 0, 1, 2, \ldots$. Substituting this into (11.1-4), we obtain a sum of finitely many terms that certainly may be integrated term by term. The result, valid for $n = 0, 1, 2, \ldots$, is

$$f(z) = F_n(z) + r_n(z), \tag{11.1-5}$$

where

$$F_n(z) := \frac{0!}{z} - \frac{1!}{z^2} + \frac{2!}{z^3} - \cdots + (-1)^{n-1} \frac{(n-1)!}{z^n}$$

is the nth partial sum of the formal series F evaluated at z, and

$$r_n(z) := \frac{(-1)^n}{z^n} \int_0^\infty \frac{\tau^n e^{-\tau}}{z + \tau} \, d\tau. \tag{11.1-6}$$

We proceed to estimate r_n. For $z > 0$, we evidently have

$$|r_n(z)| \leq \frac{1}{z^{n+1}} \int_0^\infty \tau^n e^{-\tau} d\tau = \frac{n!}{z^{n+1}}.$$

Thus, if n is fixed and $z \to \infty$ through positive values, $r_n(z)$ is "of the order of" z^{-n-1}, that is, there exists a real constant γ_n (here $\gamma_n = n!$) such that

$$|r_n(z)| \leq \gamma_n z^{-n-1}.$$

If the exact value of γ_n is irrelevant, it is convenient to express this fact by the **Landau O symbol** which is defined as follows. If h and g are two functions with an unbounded domain of definition S, one writes

$$g(z) = O(h(z)), \qquad z \to \infty, \qquad z \in S,$$

to indicate that there exists a constant γ such that

$$|g(z)| \leq \gamma |h(z)|$$

for all $z \in S$ with $|z|$ sufficiently large. In this notation our result simply reads

$$r_n(z) = O(z^{-n-1}), \qquad z \to \infty, \qquad z > 0.$$

Yet more is true. The integral (11.1-4) defines f for all values of z such that $|\arg z| < \pi$, and the decomposition (11.1-5), with the formula (11.1-6) for the remainder, likewise holds for all such values. Now if $0 < \alpha \leq \pi$, let us denote by S_α the open wedge-shaped region of all z such that $|\arg z| < \alpha$,

$$S_\alpha := \{z : |\arg z| < \alpha\}.$$

If $0 \leq \alpha < \pi$, we also define

$$\hat{S}_\alpha := \{z : |\arg z| \leq \alpha\}.$$

(These notations are used throughout the present chapter, and also in Chapter 12.) Let $0 < \alpha < \pi$; we shall estimate $r_n(z)$ if $z \in \hat{S}_\alpha$. If $\operatorname{Re} z \geq 0$, we clearly have $|z + \tau| \geq |z|$ for all $\tau \geq 0$; hence

$$|r_n(z)| \leq \frac{n!}{|z|^{n+1}},$$

as before. If $\operatorname{Re} z < 0$, then

$$|z + \tau| \geq |z| \sin \alpha;$$

hence

$$|r_n(z)| \leq \frac{n!}{\sin \alpha} \frac{1}{|z|^{n+1}}.$$

Thus in any case it is still true that for $n = 0, 1, 2, \ldots$.

$$r_n(z) = O(z^{-n-1}), \qquad z \to \infty, \qquad z \in \hat{S}_\alpha. \qquad (11.1\text{-}7)$$

To sum up, we see that the formal series F is related to the function f in the following way: If $0 \leqslant \alpha < \pi$, the partial sums F_n $(n = 0, 1, 2, \ldots)$ of F satisfy

$$f(z) = F_n(z) + O(z^{-n-1})$$

for $z \to \infty$, $z \in \hat{S}_\alpha$. Thus although it is not true for any fixed value of z that the sequence of partial sums converges, it is true that for any fixed n the function F_n approximates f arbitrarily well for $|z|$ sufficiently large, $z \in \hat{S}$. The approximation is so good, in fact, that the error tends to zero for $z \to \infty$ even if it is multiplied by z^n.

The following definition should now be well motivated. Let S be an unbounded set in the complex plane, let f be defined on S (it is not necessary that f be analytic, although this will be the main case of interest), and let

$$F = a_0 + a_1 z^{-1} + a_2 z^{-2} + \cdots$$

be a formal power series whose partial sums we denote by F_n:

$$F_n(z) := a_0 + a_1 z^{-1} + a_2 z^{-2} + \cdots + a_n z^{-n}.$$

We shall say that the series F **represents** f **asymptotically as** $z \to \infty$, $z \in S$ (or, equivalently, that f **admits the asymptotic power series** F as $z \to \infty$, $z \in S$) if the pair (f, F) has

PROPERTY (A)

For $n = 0, 1, 2, \ldots$,

$$f(z) - F_n(z) = O(z^{-n-1}), \qquad z \to \infty, \qquad z \in S. \qquad (11.1\text{-}8)$$

Written out in full, this means that for $n = 0, 1, 2, \ldots$ there exist positive real numbers γ_n and ρ_n such that

$$|f(z) - F_n(z)| \leqslant \gamma_n |z|^{-n-1} \qquad (11.1\text{-}9)$$

for all $z \in S$ such that $|z| > \rho_n$. Actually, we may assume that all ρ_n are the same, because (11.1-9) for $n = 0$ implies that

$$|f(z) - a_0| \leqslant \gamma_0 |z|^{-1} \quad \text{for} \quad |z| > \rho_0, \qquad z \in S,$$

and thus in particular that f is bounded for $|z| > \rho_0$, $z \in S$. Hence for any n such that $\rho_n > \rho_0$

$$z^{n+1}(f(z) - F_n(z))$$

is bounded, say by $\bar{\gamma}_n$, on the set $z \in S$, $\rho_0 < |z| \leqslant \rho_n$. Replacing γ_n by $\max(\gamma_n, \bar{\gamma}_n)$, it follows that (11.1-9) holds for $|z| > \rho_0$, $z \in S$.

Relation (11.1-9) clearly implies

$$|z^n[f(z)-F_n(z)]| \leq \gamma_n |z|^{-1}$$

for $|z| > \rho_n$, $z \in S$, and hence that f and F satisfy

PROPERTY (B)

For $n = 0, 1, 2, \ldots$,

$$\lim_{\substack{z \to \infty \\ z \in S}} \{z^n[f(z)-F_n(z)]\} = 0. \tag{11.1-10}$$

This may be written yet differently. Since

$$z^n F_n(z) = z^n(a_0 + a_1 z^{-1} + \cdots + a_{n-1} z^{-n+1}) + a_n$$
$$= z^n F_{n-1}(z) + a_n$$

(where $F_{-1} := 0$), and because $\lim a_n = a_n$, property (B) clearly implies

PROPERTY (C)

For $n = 0, 1, 2, \ldots$.

$$\lim_{\substack{z \to \infty \\ z \in S}} \{z^n[f(z)-F_{n-1}(z)]\} \text{ exists and equals } a_n. \tag{11.1-11}$$

THEOREM 11.1a

Properties A, B and C are equivalent.

Proof.   We have already seen that $(A) \Rightarrow (B) \Rightarrow (C)$; it thus remains to show that $(C) \Rightarrow (A)$. Let n be any nonnegative integer. Then (C) implies

$$a_{n+1} = \lim_{\substack{z \to \infty \\ z \in S}} \{z^{n+1}[f(z)-F_n(z)]\}.$$

Thus for $|z|$ sufficiently large, $|z| > \rho_n$ say,

$$|z^{n+1}[f(z)-F_n(z)]| \leq 1 + |a_{n+1}|$$

or

$$|f(z)-F_n(z)| \leq \frac{1 + |a_{n+1}|}{|z|^{n+1}}.$$

But this is precisely (11.1-9) where $\gamma_n = 1 + |a_{n+1}|$.   ∎

By virtue of Theorem 11.1a, we may henceforth use the Properties (A), (B) and (C) as equivalent conditions for f to be represented asymptotically by F as $z \to \infty$, $z \in S$. It is customary to use the notation

$$f \approx F, \qquad z \to \infty, \qquad z \in S$$

to indicate that these conditions are satisfied. If S is a full neighborhood of ∞, or if there is no ambiguity about the manner in which z is allowed to tend to ∞, the qualifier "$z \in S$" may be omitted.

An important conclusion may be drawn from property (C).

THEOREM 11.1b

A function f can in a given unbounded set S admit **at most one** *asymptotic power series as $z \to \infty$, $z \in S$.*

Proof by contradiction. Let

$$F = a_0 + a_1 z^{-1} + a_2 z^{-2} + \cdots,$$
$$G = b_0 + b_1 z^{-1} + b_2 z^{-2} + \cdots$$

be two distinct asymptotic power series representations of f as $z \to \infty$, $z \in S$. Let n be the smallest integer such that $a_n \neq b_n$. Then $F_{n-1} = G_{n-1}$. Hence (11.1-9) implies that $a_n = b_n$, which contradicts the definition of n. ∎

EXAMPLES of asymptotic series:

1 The computations carried out near the beginning of this section show that for every α such that $0 \leq \alpha < \pi$, the series (11.1-1) represents the function (11.1-3) asymptotically as $z \to \infty$, $z \in \hat{S}_\alpha$.

2 Theorem 8.8b shows that under appropriate conditions on a and c the divergent series

$$_2F_0(a, 1 + a - c; -z^{-1})$$

is an asymptotic power series for the function

$$\frac{\Gamma(c-a)}{\Gamma(c)} (-z)^a {}_1F_1(a; c; z)$$

if $z \to \infty$, $-z \in \hat{S}_\alpha$, for every α such that $0 < \alpha < \pi/2$.

3 By the method of example **1** an important asymptotic series can be obtained for the *Binet function* defined by (8.5-7),

$$J(z) = \frac{1}{\pi} \int_0^\infty \frac{z}{\eta^2 + z^2} \operatorname{Log} \frac{1}{1 - e^{-2\pi\eta}} d\eta, \qquad (11.1-12)$$

occurring in Stirling's formula for $\Gamma(z)$. Clearly the integrals

$$\beta_m := \frac{1}{\pi} \int_0^\infty \eta^{2m} \operatorname{Log} \frac{1}{1 - e^{-2\pi\eta}} d\eta,$$

$m = 0, 1, 2, \ldots,$ all exist, $\beta_m > 0$. Using the geometric series with the remainder term,

$$\frac{z}{z^2 + \eta^2} = \frac{1}{z}\left\{ 1 - \frac{\eta^2}{z^2} + \frac{\eta^4}{z^4} - \cdots + \frac{(-\eta^2)^{k-1}}{z^{2k-2}} + \frac{(-\eta^2)^k}{z^{2k-2}(z^2 + \eta^2)} \right\}$$

and integrating in (11.1-12) term by term, we obtain

$$J(z) = \frac{\beta_0}{z} - \frac{\beta_1}{z^3} + \frac{\beta_2}{z^5} - \cdots + (-1)^{k-1}\frac{\beta_{k-1}}{z^{2k-1}} + r_k(z),$$

where the remainder,

$$r_k(z) := \frac{(-1)^k}{\pi} \frac{1}{z^{2k-1}} \int_0^\infty \frac{\eta^{2k}}{z^2 + \eta^2} \mathrm{Log}\frac{1}{1 - 2^{-2\pi\eta}} \, d\eta, \qquad (11.1\text{-}13)$$

can be estimated as follows: If $\mathrm{Re}\, z^2 \geq 0$, then clearly

$$|z^2 + \eta^2| \geq |z^2|;$$

hence

$$|r_k(z)| \leq \frac{\beta_k}{|z|^{2k+1}}.$$

If $|\arg z| \leq \alpha$ where $\pi/4 < \alpha < \pi/2$, then $|z^2 + \eta^2| \geq |z^2|\sin 2\alpha$,

$$|r_k(z)| \leq \frac{\beta_k}{\sin 2\alpha |z|^{2k+1}}.$$

Thus for every α such that $0 < \alpha < \pi/2$ we have

$$J(z) \approx \sum_{k=0}^\infty (-1)^k\frac{\beta_k}{z^{2k+1}}, \qquad z \to \infty, \qquad z \in \hat{S}_\alpha. \qquad (11.1\text{-}14)$$

By rotating the path of integration as in the proof of Theorem 8.5a, one can show that this expansion holds for $\alpha < \pi$.

To compute the β_k, we expand the logarithm in powers of $e^{-2\pi\eta}$ to obtain for $k = 0, 1, 2, \ldots$

$$\beta_k = \frac{1}{\pi} \int_0^\infty \eta^{2k} \sum_{m=1}^\infty \frac{1}{m} e^{-2m\pi\eta} \, d\eta.$$

Although the convergence is not uniform near $\eta = 0$, summation and integration may be interchanged by Beppo Levi's theorem (see §8.4). Using Euler's integral in the form (10.5-1), this yields

$$\beta_k = \frac{1}{\pi} \sum_{m=1}^\infty \frac{1}{m} \int_0^\infty \eta^{2k} e^{-2m\pi\eta} \, d\eta$$

$$= \frac{1}{\pi} \sum_{m=1}^\infty \frac{1}{m} \frac{(2k)!}{(2\pi m)^{2k+1}}$$

$$= \frac{2(2k)!}{(2\pi)^{2k+2}} \sum_{m=1}^\infty \frac{1}{m^{2k+2}}.$$

The series can be summed by (4.9-5) or (7.10-13), and we find

$$\beta_k = \frac{(-1)^k}{(2k+1)(2k+2)} B_{2k+2}, \qquad k = 0, 1, 2, \ldots, \tag{11.1-15}$$

where B_{2k+2} is a *Bernoulli number*. Thus the asymptotic expansion for $J(z)$ takes the final form

$$J(z) \approx \sum_{k=1}^{\infty} \frac{B_{2k}}{2k(2k-1)} \frac{1}{z^{2k-1}}, \qquad z \to \infty, \qquad z \in \hat{S}_\alpha \tag{11.1-16}$$

where $\alpha < \pi$.

Other ways of identifying the β_k are discussed in §11.2 and §11.11.

The definition of an asymptotic series given is applicable when z approaches ∞, which is the situation that occurs most frequently. However, we also may consider the approach to a finite value. Let f be defined on a set with point of accumulation z_0, and let

$$F = a_0 + a_1(z - z_0) + a_2(z - z_0)^2 + \cdots$$

be a formal power series in $z - z_0$. We write

$$F_n(z) := a_0 + a_1(z - z_0) + \cdots + a_n(z - z_0)^n, \qquad n = 0, 1, 2, \ldots,$$

$F_{-1}(z) := 0$. We say that F is an asymptotic power series for f as $z \to z_0$, $z \in S$, if F has any of the following three equivalent properties:

(A') For $n = 0, 1, 2, \ldots$,
$$f(z) - F_n(z) = O((z - z_0)^{n+1}), \qquad z \to z_0, \qquad z \in S,$$

which is to say that for every $n = 0, 1, 2, \ldots$ there exists a real number γ_n such that

$$|f(z) - F_n(z)| \leq \gamma_n |z - z_0|^{n+1}$$

for $|z - z_0|$ sufficiently small, $z \in S$.

(B') For $n = 0, 1, 2, \ldots$,

$$\lim_{\substack{z \to z_0 \\ z \in S}} \{(z - z_0)^{-n}[f(z) - F_n(z)]\} = 0.$$

(C') For $n = 0, 1, 2, \ldots$,

$$\lim_{\substack{z \to z_0 \\ z \in S}} \{(z - z_0)^{-n}[f(z) - F_{n-1}(z)]\} = a_n.$$

We shall use the notation

$$f \approx F, \qquad z \to z_0, \qquad z \in S$$

to indicate that f meets these conditions. It is clear that the analog of Theorem 11.1b holds in this case.

We finally note some variation in the use of the symbol $\approx$. For simplicity we assume that S is unbounded, and that z approaches ∞. Analogous conventions may be made for $z \to z_0$. If h is a function defined on S and $\neq 0$ for $|z|$ sufficiently large, we write

$$f \approx hF, \qquad z \to \infty, \qquad z \in S$$

to mean

$$h^{-1}f \approx F, \qquad z \to \infty, \qquad z \in S.$$

Furthermore, if g has the range S,

$$\lim_{z \to \infty} g(z) = \infty,$$

and if the inverse function $g^{[-1]}$ exists, we shall write

$$f(z) \approx a_0 + a_1[g(z)]^{-1} + a_2[g(z)]^{-2} + \cdots$$

to indicate that

$$f \circ g^{[-1]} \approx F, \qquad z \to \infty, \qquad z \in g^{[-1]}(S).$$

PROBLEMS

1. The following rules for manipulating the O symbol hold for arbitrary integers m and n:
   (a) $O(z^{-m}) + O(z^{-n}) = O(z^{-\min(m,n)}), \qquad z \to \infty$;
   (b) $O(z^{-m})O(z^{-n}) = O(z^{-m-n}), \qquad z \to \infty.$

2. Let f be defined on an unbounded set S, let $z_0 \in \mathbb{C}$, and let, for $z \in T := z_0 + S^{-1}$,

$$g(z) := f\left(\frac{1}{z - z_0}\right).$$

   Then

$$f(z) \approx a_0 + a_1 z^{-1} + a_2 z^{-2} + \cdots, \qquad z \to \infty, \qquad z \in S$$

   if and only if

$$g(z) \approx a_0 + a_1(z - z_0) + a_2(z - z_0)^2 + \cdots, \qquad z \to z_0, \qquad z \in T.$$

3. Let

$$f(z) \approx a_0 + a_1 z^{-1} + a_2 z^{-2} + \cdots, \qquad z \to \infty, \qquad z \in S.$$

   Then for $n = 1, 2$,

$$z^n[f(z) - (a_0 + a_1 z^{-1} + \cdots + a_n z^{-n})]$$
$$\approx a_n + a_{n+1} z^{-1} + a_{n+2} z^{-2}, \qquad z \to \infty, \qquad z \in S.$$

4. Let h be a real, continuous function on $[0, \infty)$ such that all integrals

$$c_n := \int_0^\infty h(\tau)\tau^n \, d\tau, \qquad n = 0, 1, 2, \dots,$$

are absolutely convergent (example: $h(\tau) = e^{-\tau}$). Show that the function

$$f(z) := \int_0^\infty \frac{h(\tau)}{z + \tau} \, d\tau, \qquad z \in S_\pi,$$

admits for every $\alpha < \pi$ the asymptotic power series

$$f(z) \approx c_0 z^{-1} - c_1 z^{-2} + c_2 z^{-3} - \cdots, \qquad z \to \infty, \qquad z \in \hat{S}_\alpha.$$

5. Let $0 < \gamma < 1$. Then the function

$$f(z) := \sum_{k=1}^\infty \frac{\gamma^k}{z + k}$$

is defined for $z \in S_\pi$. Show that for every $\alpha < \pi$ it admits the asymptotic power series

$$f(z) \approx a_1 z^{-1} - a_2 z^{-2} + a_3 z^{-3} - \cdots, \qquad z \to \infty, \qquad z \in \hat{S}_\alpha,$$

where

$$a_n := \sum_{k=1}^\infty k^{n-1}\gamma^k, \qquad n = 1, 2, \dots.$$

6. Let $\beta > 0$, $\gamma > 0$. The function

$$f(z) := \int_0^\infty \frac{e^{-\tau^\beta}}{(1 + \tau z^{-1})^\gamma} \, d\tau, \qquad z \in S_\pi,$$

possesses, for every $\alpha < \pi$, the asymptotic power series

$$f(z) \approx \frac{1}{\beta} \sum_{n=0}^\infty (-1)^n \frac{(\gamma)_n \Gamma((n+1)/\beta)}{n!} z^{-n}, \qquad z \to \infty, \qquad z \in \hat{S}_\alpha.$$

7. In the definition of asymptotic power series it was required that one of the three properties (A), (B), (C) holds for all nonnegative integers n. Show that it actually suffices that any of these properties holds on some unbounded set of integers.

8. The formal series $F = \sum \alpha_k x^{-k}$ is said to **envelop** the function f for $x > 0$, if for each n and all $x > 0$

$$f(x) = F_n(x) + \theta_n(x)\alpha_{n+1} x^{-n-1},$$

where $0 \le \theta_n(x) \le 1$. (The remainder is a positive fraction of the first neglected term.) Show that if F envelops f, it represents f asymptotically as $x \to \infty$. Also show that the converse is not true.

9. Prove: If, in problem 4, h has constant sign, then the series $\sum (-1)^n c_n x^{-n-1}$ envelops $f(x)$ for $x > 0$. In particular, the series (11.1-1) envelops the exponential integral (11.1-4).

10. Prove: The series (11.1-14) (considered as z^{-1} times a series in z^{-2}) envelops the Binet function $J(z)$ for $z = x > 0$.
11. The Taylor series at 0 of the functions e^{-x}, $\text{Log}(1+x)$, $(1+x)^{-\alpha}$ where $d > 0$ (considered as series in x rather than x^{-1}) envelop these functions for $x > 0$. State a general theorem that contains these examples as special cases.

§11.2   THE ALGEBRA OF ASYMPTOTIC POWER SERIES

We recall some conventions on formal power series. If

$$F = a_0 + a_1 z^{-1} a_2 z^{-2} + \cdots,$$

$$G = b_0 + b_1 z^{-1} + b_2 z^{-2} + \cdots$$

are two formal power series, and if α and β are any two complex numbers, we denote by $\alpha F + \beta G$ the formal power series with nth coefficient $\alpha a_n + \beta b_n$, by FG the formal Cauchy product of the two series, and, if $a_0 \neq 0$, by F^{-1} the solution of the formal equation $FX = 1$. It is evident from §1.2 that with these definitions the formal power series form an *algebra* $\mathcal{A}$ (see §2.1). The *units* of $\mathcal{A}$, that is, the elements that possess reciprocals, are precisely the series F with $a_0 \neq 0$.

We have seen in Chapter 2 that the formal power series with positive radius of convergence form a subalgebra of $\mathcal{A}$. It will now be shown that the formal series that are *asymptotic* to some function in a given set S form another (larger) subalgebra.

THEOREM 11.2a

Let S be an unbounded set in $\mathbb{C}$, let the functions f and g be defined on S, and let F and G be two formal power series such that

$$f(z) \approx F, \qquad g(z) \approx G, \qquad z \to \infty, \qquad z \in S.$$

Then for arbitrary complex α and β,

(i) $\qquad\qquad\qquad\qquad \alpha f(z) + \beta g(z) \approx \alpha F + \beta G,$

furthermore

(ii) $\qquad\qquad\qquad\qquad f(z)g(z) \approx FG$

and, if $a_0 \neq 0$,

(iii) $\qquad\qquad\qquad\qquad f^{-1}(z) \approx F^{-1}$

as $z \to \infty$, $z \in S$.

Proof. As in §11.1 we use subscripts to denote partial sums. Since $(\alpha F + \beta G)_n = \alpha F_n + \beta G_n$, we have, using property (A),

$$(\alpha f + \beta g)(z) - (\alpha F + \beta G)_n(z) = \alpha[f(z) - F_n(z)] + \beta[g(z) - G_n(z)]$$

$$= O(z^{-n-1}) + O(z^{-n-1}) = O(z^{-n-1})$$

for $n = 0, 1, 2, \ldots$. Thus the series $\alpha F + \beta G$ has property (A) with regard to the function $\alpha f + \beta g$, proving (i).

Although it is not true in general that $(FG)_n = F_n G_n$, we have

$$(FG)_n(z) = (F_n G_n)(z) + O(z^{-n-1}).$$

Hence

$$f(z)g(z) - (FG)_n(z)$$
$$= f(z)g(z) - F_n(z)G_n(z) + O(z^{-n-1})$$
$$= [f(z) - F_n(z)]g(z) + [g(z) - G_n(z)]F_n(z) + O(z^{-n-1})$$
$$= O(z^{-n-1})g(z) + O(z^{-n-1})F_n(z) + O(z^{-n-1}),$$

and this is $O(z^{-n-1})$, in view of the fact that the limits of g and F_n exist as $z \to \infty$, $z \in S$. This establishes (ii).

To prove (iii), we first note that, since by property (C)

$$\lim_{\substack{z \to \infty \\ z \in S}} f(z) = a_0$$

exists and is $\neq 0$, f^{-1} is defined for $|z|$ sufficiently large, $z \in S$. To simplify the notation, let

$$G := F^{-1} = \frac{1}{a_0} - \frac{a_1}{a_0^2} z^{-1} + \cdots.$$

Then

$$\frac{1}{f(z)} - G_n(z) = \frac{1 - f(z)G_n(z)}{f(z)}$$

$$= \frac{1}{f(z)} \{1 - [F_n(z) + O(z^{-n-1})]G_n(z)\}.$$

By the definition of G,

$$F_n(z)G_n(z) = 1 + O(z^{-n-1});$$

hence the last expression equals

$$\frac{1}{f(z)} \{O(z^{-n-1}) + G_n(z)O(z^{-n-1})\} = O(z^{-n-1})$$

for $z \to \infty$, $z \in S$. Thus the series G has property (A) with regard to $1/f$, proving (iii). ■

Formal power series, as was pointed out in Chapter 1, in addition to forming an algebra admit in certain cases the operation of composition. If

$$F = a_1 z^{-1} + a_2 z^{-2} + \cdots$$

is a non-unit in $\mathcal{A}$ (leading coefficient zero) and if

$$G = b_0 + b_1 w + b_2 w^2 + \cdots$$

is any formal power series, we can substitute F into G. The result of this operation is known as the *composition* of G with F and denoted by $G \circ F$. The coefficients of

$$G \circ F = c_0 + c_1 z^{-1} + c_2 z^{-2} + \cdots$$

are calculated as follows: Let, for $k = 1, 2, \ldots$,

$$F^k =: a_k^{(k)} z^{-k} + a_{k+1}^{(k)} z^{-k-1} + \cdots,$$

then $c_0 = b_0$, and for $n > 0$, c_n is a finite sum,

$$c_n := b_1 a_n^{(1)} + b_2 a_n^{(2)} + \cdots + b_n a_n^{(n)},$$

$n = 1, 2, \ldots$. The question naturally arises whether the composition of functions possessing asymptotic power series mirrors itself in the composition of their asymptotic power series. The answer is affirmative.

THEOREM 11.2b

Let S be an unbounded set in $\mathbb{C}$, let f be defined on S and

$$f(z) \approx F = a_1 z^{-1} + a_2 z^{-2} + \cdots, \qquad z \to \infty, \qquad z \in S,$$

so that, in particular,

$$\lim_{\substack{z \to \infty \\ z \in S}} f(z) = 0.$$

Let g be defined on $T := f(S)$, and let

$$g(w) \approx G = b_0 + b_1 w + b_2 w^2 + \cdots, \qquad w \to 0, \qquad w \in T.$$

Then

$$g \circ f(z) \approx G \circ F, \qquad z \to \infty, \qquad z \in S.$$

Proof. We continue to denote partial sums of formal series by subscripts. Let n be a fixed nonnegative integer. Theorem 11.2a implies that for $k = 0, 1, 2, \ldots, n$

$$[f(z)]^k = [F_n(z)]^k + O(z^{-n-1}).$$

Multiplying these relations by b_k and summing, we get

$$G_n \circ f(z) = G_n \circ F_n(z) + O(z^{-n-1}).$$

But

$$g \circ f(z) = G_n \circ f(z) + O((f(z)^{n+1}));$$

hence by the above and because $f(z) = O(z^{-1})$,

$$g \circ f(z) = G_n \circ F_n(z) + O(z^{-n-1}).$$

The definition of composition implies

$$G_n \circ F_n(z) = (G \circ F)_n(z) + O(z^{-n-1});$$

thus

$$g \circ f(z) = (G \circ F)_n(z) + O(z^{-n-1}),$$

and $G \circ F$ is seen to possess property (A) with regard to $g \circ f$. ∎

EXAMPLE

We use Theorem 11.2b to identify once more the coefficients β_k occurring in the asymptotic expansion (11.1-14) of Binet's function $J(z)$. These coefficients were calculated in §11.1 using nontrivial tools, both from real analysis (Beppo Levi's theorem) and from complex analysis (the residue theorem). Such tools are avoided in the purely formal calculation that follows, which requires merely that some asymptotic expansion is known to exist, plus one of the functional relations satisfied by the gamma function.

The Binet function is connected with $\Gamma(z)$ by

$$\text{Log } \Gamma(z) = \tfrac{1}{2} \text{Log } 2\pi + (z - \tfrac{1}{2}) \text{Log } z - z + J(z), \tag{11.2-1}$$

where we suppose $z > 0$ to avoid any ambiguity in the choices of the logarithms. Taking logarithms in $\Gamma(z+1) = z\Gamma(z)$ we have

$$\text{Log } \Gamma(z+1) = \text{Log } z + \text{Log } \Gamma(z);$$

thus (11.2-1) after simplification implies

$$J(z+1) - J(z) = 1 - (z + \tfrac{1}{2}) \text{Log}\left(1 + \frac{1}{z}\right). \tag{11.2-2}$$

The method now consists in computing the asymptotic power series of the functions on either side and comparing coefficients. The function $J(z+1)$ may be viewed as the composition of $J(w^{-1})$ with $w := (1+z)^{-1}$. We have

$$w \approx z^{-1} - z^{-2} + z^{-3} - \cdots, \qquad z \to \infty$$

and more generally for $m = 1, 2, \ldots,$

$$w^m \approx z^{-m} \sum_{n=0}^{\infty} (-1)^n \frac{(m)_n}{n!} z^{-n}, \qquad z \to \infty$$

(these expansions are, in fact, convergent). Assuming that $J(z)$ admits an asymptotic expansion,

$$J(z) \approx \sum_{n=1}^{\infty} c_n z^{-n}, \qquad z > 0, \qquad z \to \infty,$$

we thus have

$$J(z+1) \approx \sum_{n=1}^{\infty} d_n z^{-n}, \qquad z > 0, \qquad z \to \infty,$$

where

$$d_n = \frac{(n)_0}{0!} c_n - \frac{(n-1)_1}{1!} c_{n-1} + \cdots + (-1)^{n+1} \frac{(1)_{n-1}}{(n-1)!} c_1,$$

$n = 1, 2, \ldots$. The coefficient of z^{-n} in the asymptotic power series for $J(z+1) - J(z)$ thus is

$$\sum_{k=1}^{n-1} (-1)^{n-k} \frac{(k)_{n-k}}{(n-k)!} c_k = (-1)^{n-1} \sum_{k=0}^{n-1} \frac{(-n+1)_{k-1}}{(k-1)!} c_k.$$

On the other hand, by Theorem 11.2a,

$$1 - \left(z + \frac{1}{2}\right) \mathrm{Log}\left(1 + \frac{1}{z}\right)$$

$$\approx 1 - (z + \tfrac{1}{2})(z^{-1} - \tfrac{1}{2} z^{-2} + \tfrac{1}{3} z^{-3} - \cdots)$$

$$= \sum_{n=2}^{\infty} (-1)^{n-1} \frac{n-1}{2n(n+1)} z^{-n}, \qquad z \to \infty.$$

Comparing coefficients there follows

$$\sum_{k=1}^{n-1} \frac{(-n+1)_{k-1}}{(k-1)!} c_k = \frac{n-1}{2n(n+1)}, \tag{11.2-3}$$

$n = 1, 2, \ldots$. To identify the c_k in terms of the Bernoulli numbers, it is only necessary to put

$$c_k =: \frac{(-1)^{k+1}}{k(k+1)} b_{k+1}, \qquad k = 1, 2, \ldots.$$

This yields

$$\sum_{k=1}^{n-1} \frac{(-n-1)_{k+1}}{(k+1)!} (-1)^{k+1} b_{k+1} = \frac{n-1}{2},$$

that is,

$$\sum_{k=2}^{n} \frac{(-n-1)_k}{k!} (-1)^k b_k = \frac{n-1}{2}$$

$(n = 1, 2, \ldots)$ or, replacing n by $n - 1$,

$$1 - n \cdot \frac{1}{2} + \sum_{k=2}^{n-1} \frac{(-n)_k(-1)^k}{k!} b_k = 0,$$

$n = 2, 3, \ldots$. If we define $b_0 := 1$, $b_1 := \frac{1}{2}$, then this may be written

$$\sum_{k=0}^{n-1} \binom{n}{k} b_k = 0,$$

$n = 2, 3, \ldots$, which is identical with the recurrence relation of the Bernoulli numbers B_n following directly from their generating function

$$\frac{z}{e^z - 1} = \sum_{n=0}^{\infty} \frac{B_n}{n!} z^n.$$

In view of $b_0 = B_0$ we thus have $b_k = B_k$ $(k = 1, 2, \ldots)$; hence

$$c_k = \frac{(-1)^{k+1}}{k(k+1)} B_{k+1}, \qquad k = 1, 2, \ldots,$$

in agreement with (11.1-16).

PROBLEMS

1. Let S be an unbounded set, and let

$$f(z) \approx F = a_0 + a_1 z^{-1} + a_2 z^{-2} + \cdots, \qquad z \to \infty, \qquad z \in S,$$

where $|\arg a_0| < \pi$. Show that $|\arg f(z)| < \pi$ for $|z|$ sufficiently large, and that for any complex number α the function $f^\alpha := e^{\alpha \, \text{Log} f}$ (principal value) satisfies

$$f^\alpha(z) \approx F^\alpha = a_0^\alpha \left\{ 1 + \frac{a_1}{a_0} z^{-1} + \cdots \right\}^\alpha$$

$$= a_0^\alpha + a_1 a_0^{\alpha-1} z^{-1} + \cdots, \qquad z \to \infty, \qquad z \in S.$$

2. Let S be an unbounded set, k a positive integer, and

$$f(z) \approx F = z^{-k} + a_{k+1} z^{-k-1} + \cdots, \qquad z \to \infty, \qquad z \in S.$$

Show that a branch of $f^{1/k}$ can be defined such that

$$f^{1/k}(z) \approx F^{1/k} = z^{-1} \{ 1 + a_{k+1} z^{-1} + \cdots \}^{1/k}, \qquad z \to \infty, \qquad z \in S.$$

3. Let the function f be one-to-one on a set S having 0 as a point of accumulation, and let

$$f(z) \approx F = a_1 z + a_2 z^2 + a_3 z^3 + \cdots, \qquad z \to 0, \qquad z \in S.$$

where $a_1 \neq 0$. If $T := f(S)$, and if $F^{[-1]}$ denotes the formal reversion of F, show that

$$f^{[-1]}(w) \approx F^{[-1]}, \qquad w \to 0, \qquad w \in T.$$

4. Deduce the recurrence relation (11.2-3) of the coefficients of the asymptotic expansion of Binet's function from the functional relation

$$\Gamma(z)\Gamma(1-z) = \frac{\pi}{\sin \pi z}.$$

5. If A is a matrix of functions having asymptotic expansions, we write

$$A(z) \approx A_0 + A_1 z^{-1} + A_2 z^{-2} + \cdots, \qquad z \to \infty, \qquad z \in S.$$

Show that if $\det A_0 \neq 0$, then the matrix A^{-1} has an asymptotic expansion with leading term A_0^{-1}.

6. If J denotes the Binet function, show that the asymptotic series for $f(z) := 2J(z) - J(2z)$ envelops f for $z = x > 0$ (see Problem 8, §11.1). Conclude that for all integers $n > 0$

$$2^{-2n}\sqrt{\pi n}\binom{2n}{n} = \exp\left(-\frac{1}{8n} + \frac{1}{192n^3} - \frac{1}{640n^5} + \frac{\theta_n}{840n^7}\right),$$

where $0 < \theta_n < 1$.

7. Show that for all positive integers n,

$$2^{-2n}\sqrt{\pi n}\binom{2n}{n} \leqslant \exp\left(-\frac{1}{2\sqrt{2}} \operatorname{Arctan}\frac{1}{2\sqrt{2n}}\right),$$

and that the error in the upper bound is $O(n^{-5})$. Verify that already for $n = 1$ the error is less than 0.1%.

§11.3. ANALYTIC PROPERTIES OF ASYMPTOTIC POWER SERIES

Let $\rho \geqslant 0$, and let f be analytic for $\rho < |z| < \infty$. We recall from §4.4 that f is represented by its Laurent series,

$$f(z) = \sum_{n=-\infty}^{\infty} c_n z^{-n}, \tag{11.3-1}$$

which converges for $|z| > \rho$ and whose coefficients are uniquely determined by f. It is possible that only finitely many or no coefficients c_n with negative indices (corresponding to positive powers of z) are different from zero. The vanishing of such c_n depends on the existence of $\lim_{z \to \infty} f(z)$. Three possibilities exist:

(i) $\lim_{z \to \infty} f(z)$ exists as a proper (finite) limit. In this case $c_n = 0$ for all $n < 0$; the isolated singularity at ∞ is removable in the sense that by defining $f(\infty) := c_0$, f becomes analytic at ∞.

(ii) $\lim_{z \to \infty} f(z) = \infty$. Here there exists a positive integer m such that $c_n = 0$ for $n < -m$, $c_{-m} \neq 0$. The function f is now said to have a pole at ∞, and m is called the order of the pole.

(iii) $\lim_{z \to \infty} f(z)$ does not exist either as a proper or as an improper limit. In this case $c_n \neq 0$ for infinitely many negative indices n. The function f is now said to have an essential singularity at ∞.

Under what circumstances does a function f that is analytic for $\rho < |z| < \infty$ possess an asymptotic power series for $z \to \infty$ if the approach is unrestricted?

THEOREM 11.3a

Let f be analytic for $\rho < |z| < \infty$. Then f admits an asymptotic power series for $z \to \infty$ (approach unrestricted) if and only if the singularity at ∞ is removable, in which case the asymptotic series is identical with the Laurent series of f at ∞, and hence converges at least for $|z| > \rho$.

Proof.   (a) If f is analytic at ∞, let its Laurent series be

$$f(z) = \sum_{n=0}^{\infty} c_n z^{-n}, \tag{11.3-2}$$

and let

$$F_n(z) := c_0 + c_1 z^{-1} + \cdots + c_n z^{-n}, \qquad n = 0, 1, 2, \ldots.$$

Then for every nonnegative integer n

$$z^n [f(z) - F_{n-1}(z)] = c_n + c_{n+1} z^{-1} + \cdots,$$

where the series on the right converges for $|z| > \rho$. There follows

$$\lim_{z \to \infty} z^n [f(z) - F_{n-1}(z)] = c_n$$

for $n = 0, 1, 2, \ldots$. Hence the formal series

$$F := c_0 + c_1 z^{-1} + c_2 z^{-2} + \cdots$$

with regard to f possesses property (C) of §11.1, and thus represents f asymptotically.

(b) If f admits an asymptotic power series $F = c_0 + c_1 z^{-1} + c_2 z^{-2} + \cdots$ as $z \to \infty$, where the approach is unrestricted, then by (C)

$$\lim_{z \to \infty} f(z) = c_0 \quad \text{(approach unrestricted)}$$

exists. By Theorem 4.4d (Riemann's theorem) $z = \infty$ is a removable singularity of f. The Laurent series thus involves no powers with positive exponents, and by (a) also is an asymptotic series. By the uniqueness of asymptotic series (Theorem 11.1b) F agrees with the Laurent series, and thus is convergent.   ∎

It is clear that a convergent power series is a more effective computational tool than a mere asymptotic series. A series such as (11.3-2) permits us (in principle) to compute the values of f at *any* point z such that $|z| > \rho$ with an *arbitrarily* small error. A nonconvergent asymptotic series (as the first example in §11.1 shows) merely may enable us to compute f at *any* point z with a *certain error*, or to achieve an *arbitrarily* small error at *some* points $z \neq \infty$.

Theorem 11.3a implies the following: If a function f that is analytic for $\rho < |z| < \infty$ admits an asymptotic power series for $z \to \infty$ that is not identical with its Laurent series, then $z = \infty$ cannot be a removable singularity of f. Neither can it be a pole of f, because if it were a pole, then $\lim_{z \to \infty} f(z) = \infty$, no matter whether or how the approach is restricted. Thus z must be an essential singularity.

The following example shows that a function with an essential singularity at ∞ may indeed admit an asymptotic series in powers of z^{-1} (which even may be convergent) if the approach to ∞ is suitably restricted. Let $0 \leq \alpha < \pi/2$,

$$f(z) := e^{-z}, \qquad z \in S_\alpha.$$

We assert that the zero series is an asymptotic power series for f, that is, that

$$e^{-z} \approx 0 + 0 z^{-1} + 0 z^{-2} + \cdots, \qquad z \to \infty, \qquad z \in S_\alpha.$$

This is immediately verified by means of property (C). Because $F = 0$, all partial sums F_n are zero, and it is to be shown that

$$\lim_{\substack{z \to \infty \\ z \in S_\alpha}} z^n e^{-z} = 0, \qquad n = 0, 1, 2, \ldots.$$

Setting $z = \rho e^{i\phi}$, $-\alpha \leq \phi \leq \alpha$, this is a consequence of

$$|z^n e^{-z}| = \rho^n e^{-\rho \cos \phi} \leq \rho^n e^{-\rho \cos \alpha}$$

and the fact that $\rho^n e^{-\gamma \rho} \to 0$ $(\rho \to \infty)$ for all integers n and any $\gamma > 0$.

An important conclusion to be drawn from our example is the fact that two different functions defined on the same unbounded set S may be admit the same asymptotic series. In fact, the functions $f(z) := 0$ and $f(z) := e^{-z}$ $(z \in S_\alpha)$ both possess the asymptotic power series representation 0.

The last remark is of interest in connection with the problem of *summing an asymptotic series*, which we state as follows: Given a formal power series

$$F = a_0 + a_1 z^{-1} + a_2 z^{-2} + \cdots \qquad (11.3\text{-}3)$$

and a sector

$$S : \alpha < \arg z < \beta$$

$(\beta - \alpha \leq 2\pi)$, does there exist a function f, analytic for $|z|$ sufficiently large, $z \in S$, such that F is asymptotic to f as $z \to \infty$, $z \in S$? Our example shows that if the problem has any solution at all, this solution cannot be unique. The question of the *existence* of a solution, however, is answered positively without restrictions by the following theorem due to J. F. Ritt, which even gives an explicit formula for the "sum" f. For notational simplicity we assume, clearly without loss of generality, that S is bisected by the positive real axis.

THEOREM 11.3b (Ritt's theorem)

Let $0 < \alpha \leq \pi$, and let (11.3-3) be any formal power series. If β is any real number such that $0 < \beta < \min(1, \pi/2\alpha)$, then the series

$$\sum_{\substack{n=0 \\ a_n \neq 0}}^{\infty} a_n \left[1 - \exp\left(-\frac{z^\beta}{|a_n|} \right) \right] z^{-n} \tag{11.3-4}$$

(where z^β has its principal value) converges in $S_\alpha \cap \{|z| > 1\}$, uniformly on every compact subset, and its sum $f(z)$ satisfies

$$f(z) \approx F, \qquad z \to \infty, \qquad z \in S_\alpha. \tag{11.3-5}$$

Proof. We require the inequality

$$|1 - e^{-t}| < |t|, \quad \operatorname{Re} t > 0. \tag{11.3-6}$$

This follows from

$$|1 - e^{-t}| = \left| \int_0^t e^{-s} \, ds \right| < \int_0^t e^{-\operatorname{Re} s} \, ds,$$

where the integral is taken along the straight line segment from 0 to t.

We first show that the series (11.3-4) converges as indicated. For $z \in S_\alpha$, $|\arg z^\beta| = \beta |\arg z| \leq \beta \alpha < \pi/2$, hence $\operatorname{Re} z^\beta > 0$. Thus (11.3-6) is applicable with $t := z^\beta |a_n|^{-1}$ and yields

$$\left| a_n \left[1 - \exp\left(-\frac{z^\beta}{|a_n|} \right) \right] \right| < |z^\beta|. \tag{11.3-7}$$

Thus the series is majorized by $\sum_{n=0}^{\infty} |z^{\beta-n}|$, which converges uniformly on every compact subset of $S_\alpha \cap \{|z| > 1\}$. It follows that f is analytic.

To prove the asymptotic relation (11.3-5), we establish property (B) of § 11.1. Let n be any nonnegative integer, and let F_n denote the nth partial sum of F. Then

$$z^n [f(z) - F_n(z)] = -z^n \sum_{k=0}^{n} a_k \exp\left(-\frac{z^\beta}{|a_k|} \right) z^{-k}$$

$$+ z^n \sum_{k=n+1}^{\infty} a_k \left[1 - \exp\left(-\frac{z^\beta}{|a_k|} \right) \right] z^{-k}.$$

If $z \in S_\alpha$, then $\operatorname{Re} z^\beta \geq |z|^\beta \cos(\beta\alpha)$. In view of $\cos(\beta\alpha) > 0$, every term in the first sum on the right is thus bounded by

$$|z|^{n-k}|a_k|\exp(-|z|^\beta \gamma_k),$$

where $\gamma_k > 0$, and thus tends to zero for $z \to \infty$, $z \in S_\alpha$. Using (11.3-7), the second sum is majorized by

$$|z|^{n+\beta} \sum_{k=n+1}^{\infty} |z|^{-k} = \frac{|z|^{\beta-1}}{1-|z|^{-1}}$$

which again tends to zero for $z \to \infty$ in view of $\beta < 1$. Thus property (B) is established, which completes the proof of Theorem 11.3b. ∎

We conclude with some remarks concerning integration and differentiation of asymptotic power series. Beginning with integration, we first consider a function that possesses an asymptotic power series at some finite point z_0 that without loss of generality we assume to be 0. For simplicity we assume that S is an angular set of type $\hat{S}_\alpha$ where $\alpha \geq 0$. We thus let f be defined on $\hat{S}_\alpha$, and

$$f(z) \approx a_0 + a_1 z + a_2 z^2 + \cdots, \qquad z \to 0, \qquad z \in \hat{S}_\alpha.$$

We call **formal integral** of the series on the right the series

$$a_0 z + \frac{a_1}{2} z^2 + \frac{a_2}{3} z^3 + \cdots .$$

The question is whether this formal integral is an asymptotic series for the indefinite integral $\int_0^z f(t)\, dt$.

Before attacking the problem, it is necessary to note that the hypothesis of f possessing an asymptotic expansion does not guarantee the integrability of f. This is already seen by the function defined for $x > 0$ by

$$f(x) := \begin{cases} e^{-1/x}, & x \text{ rational}, \\ 0, & x \text{ irrational}. \end{cases}$$

This function has the asymptotic expansion 0 for $x \to 0$, $x > 0$, but, being discontinuous at every $x > 0$, it is clearly not integrable. It is thus necessary to add integrability to our hypotheses.

THEOREM 11.3c

Let f be defined on some set $\hat{S}_\alpha$ ($\alpha \geq 0$), let

$$f(z) \approx a_0 + a_1 z + a_2 z^2 + \cdots, \qquad z \to 0, \qquad z \in \hat{S}_\alpha, \qquad (11.3\text{-}8)$$

and let

$$g(z) := \int_0^z f(t)\, dt$$

(the integral taken along the straight line segment from 0 to z) exist for every $z \in \hat{S}_\alpha$. Then

$$g(z) \approx a_0 z + \frac{a_1}{2} z^2 + \frac{a_2}{3} z^3 + \cdots, \qquad z \to 0, \qquad z \in \hat{S}_\alpha.$$

It is not necessary to assume that f is analytic, although if $\alpha > 0$ this obviously is the only case of interest.

Proof. We denote the nth partial sum of the series on the right of (11.3-8) by $F_n(z)$. By property (B') there exists, for every $\epsilon > 0$ and for every positive integer n, $\delta > 0$ such that

$$|f(z) - F_{n-1}(z)| \leq \epsilon |z|^{n-1}$$

if $|z| < \delta$, $z \in \hat{S}_\alpha$. Letting

$$G_n(z) := \int_0^z F_{n-1}(t)\, dt = a_0 z + \frac{a_1}{2} z^2 + \cdots + \frac{a_{n-1}}{n} z^n.$$

it follows that for the same values of z

$$|g(z) - G_n(z)| = \left| \int_0^z [f(t) - F_{n-1}(t)]\, dt \right|$$

$$\leq \int_0^z |f(t) - F_{n-1}(t)|\, |dt|$$

$$\leq \epsilon \int_0^z |t|^{n-1} |dt| = \frac{\epsilon}{n} |z|^n.$$

This implies

$$\lim_{\substack{z \to \infty \\ z \in \hat{S}_\alpha}} z^{-n} [g(z) - G_n(z)] = 0.$$

for $n = 1, 2, \ldots$. Clearly,

$$\lim_{\substack{z \to 0 \\ z \in \hat{S}_\alpha}} g(z) = 0.$$

Thus the series

$$G := a_0 z + \frac{a_1}{2} z^2 + \frac{a_2}{3} z^3 + \cdots$$

has property (B') with regard to the function g, establishing the assertion of the theorem. ∎

The case in which

$$f(z) \approx a_0 + a_1 z^{-1} + a_2 z^{-2} + \cdots, \qquad z \to \infty, \qquad z \in \hat{S}_\alpha \quad (11.3\text{-}9)$$

is more complicated because even for f continuous the improper integral

$$\int_z^\infty f(t)\, dt$$

does not exist unless $a_0 = a_1 = 0$. However, the following can be established:

THEOREM 11.3d

Let f be continuous on $\hat{S}_\alpha$, and let (11.3-9) hold. Then the improper integral

$$h(z) := \int_z^\infty \left[f(t) - a_0 - \frac{a_1}{t} \right] dt$$

(taken along the ray $\arg t = \arg z$) exists for $z \in \hat{S}_\alpha$ and possesses the asymptotic series

$$h(z) \approx a_2 z^{-1} + \frac{a_3}{2} z^{-2} + \frac{a_4}{3} z^{-3} + \cdots, \qquad z \to \infty, \qquad z \in \hat{S}_\alpha.$$

$$(11.3\text{-}10)$$

Proof. For $t \in \hat{S}_\alpha$ let $\hat{f}(t) := f(t^{-1})$. By hypothesis,

$$\hat{f}(t) \approx a_0 + a_1 t + a_2 t^2 + \cdots, \qquad t \to 0, \qquad t \in \hat{S}_\alpha;$$

hence

$$\frac{\hat{f}(t) - a_0 - a_1 t}{t^2} \approx a_2 + a_3 t + a_4 t^2 + \cdots, \qquad t \to 0, \qquad t \in \hat{S}_\alpha.$$

We conclude that

$$\hat{h}(z) := \lim_{\delta \to 0+} \int_{\delta z}^z \frac{\hat{f}(t) - a_0 - a_1 t}{t^2}\, dt$$

exists. By the preceding theorem,

$$\hat{h}(z) \approx a_2 z + \frac{a_3}{2} z^2 + \frac{a_4}{3} z^3 + \cdots, \qquad z \to 0, \qquad z \in \hat{S}_\alpha$$

or, what is the same,

$$\hat{h}(z^{-1}) \approx a_2 z^{-1} + \frac{a_3}{2} z^{-2} + \frac{a_4}{3} z^{-3} + \cdots, \qquad z \to \infty, \qquad z \in \hat{S}_\alpha.$$

However, by substituting $t \to t^{-1}$ in the integral,

$$\hat{h}(z^{-1}) = \lim_{\delta \to 0+} \int_z^{\delta^{-1}z} \left[f(t) - a_0 - \frac{a_1}{t} \right] dt.$$

Thus the improper integral defining $h(z)$ exists and equals $\hat{h}(z^{-1})$. Because $\hat{h}(z^{-1})$ admits the asymptotic power series on the right of (11.3-10), the same holds for $h(z)$. ∎

We next discuss differentiation. Here we must first appreciate the fact that if f possesses an asymptotic power series and is differentiable, then f' need not possess an asymptotic power series. For example, let

$$f(x) := e^{-x} \sin(e^x), \qquad x > 0.$$

Clearly f is differentiable, and $f \approx 0$, $x \to \infty$, $x > 0$. But $f'(x) = -e^{-x} \sin(e^x) + \cos(e^x)$, and this does not possess an asymptotic power series, because $\lim_{x \to \infty} f'(x)$ does not exist. To obtain results, the real and the complex cases must be discussed separately.

In the real case, only a weak result may be proved inasmuch as the existence of an asymptotic series for f' must be postulated.

THEOREM 11.3e

Let f be defined for $x > 0$, let it possess the asymptotic power series (11.3-9) where $\alpha = 0$, and let f' exist, be continuous, and possess an asymptotic power series. Then

$$f'(x) \approx -a_1 x^{-2} - 2a_2 x^{-3} - 3a_3 x^{-4} - \cdots, \qquad x \to \infty, \qquad x > 0.$$

Proof. By hypothesis, there exist $b_0, b_1, \ldots$ such that

$$f'(x) \approx b_0 + b_1 z^{-1} + b_2 z^{-2} + \cdots, \qquad x \to \infty, \qquad x > 0;$$

it is to be shown that $b_0 = b_1 = 0$, $b_{k+1} = -ka_k$, $k = 1, 2, \ldots$. Let $x_0 > 0$. Then

$$f(x) = f(x_0) + \int_{x_0}^{x} f'(\xi)\, d\xi$$

$$= f(x_0) + \int_{x_0}^{x} (b_0 + b_1 \xi^{-1})\, d\xi + \int_{x_0}^{x} [f'(\xi) - b_0 - b_1 \xi^{-1}]\, d\xi$$

$$= c + b_0 x + b_1 \operatorname{Log} x - \int_{x}^{\infty} [f'(\xi) - b_0 - b_1 \xi^{-1}]\, d\xi,$$

where c is a constant. The integral has, by Theorem 11.3d, the asymptotic power series

$$-b_2 x^{-1} - \frac{b_3}{2} x^{-2} - \frac{b_4}{3} x^{-3} - \cdots, \qquad x \to \infty, \qquad x > 0.$$

Because f admits an asymptotic power series, $\lim_{x \to \infty} f(x)$ exists, and it follows that $b_0 = b_1 = 0$. Because an asymptotic power series is unique, we may compare coefficients and find

$$b_k = -(k-1)a_{k-1}, \qquad k = 2, 3, \ldots,$$

establishing the result. ■

If f is analytic in a sector S_α where $\alpha > 0$, it is not necessary to assume that f' possesses an asymptotic power series.

THEOREM 11.3f

Let f be analytic in S_α where $\alpha > 0$, and let it possess the asymptotic power series (11.3-9). Then for every β such that $0 < \beta < \alpha$,

$$f'(z) \approx a_1 z^{-2} - 2a_2 z^{-3} - 3a_3 z^{-4} - \cdots, \qquad z \to \infty, \qquad z \in \hat{S}_\beta.$$

Proof. We use Cauchy's formula for the derivative. Let $0 < \epsilon < \sin(\alpha - \beta)$, $z \in \hat{S}_\beta$. Then the circle $\Gamma : t = z(1 + \epsilon\, e^{i\phi})$, $0 \leq \phi \leq 2\pi$, lies in S_α, and

$$f'(z) = \frac{1}{2\pi i} \int_\Gamma \frac{f(t)}{(t-z)^2}\, dt.$$

Also by Cauchy's formula,

$$-kz^{-k-1} = \frac{1}{2\pi i} \int_\Gamma \frac{t^{-k}}{(t-z)^2}\, dt, \qquad k = 0, 1, 2, \ldots,$$

hence for $n = 0, 1, 2, \ldots$

$$f'(z) + a_1 z^{-2} + 2a_2 z^{-3} + \cdots + na_n z^{-n-1}$$

$$= \frac{1}{2\pi i} \int_\Gamma \frac{1}{(t-z)^2}[f(t) - a_0 - a_1 t^{-1} - \cdots - a_n t^{-n}]\, dt \tag{11.3-11}$$

We now use property (A). For every $n = 0, 1, 2, \ldots$ there exist constants $\gamma_n > 0$ and $\rho_n > 0$ such that for $t \in S_\alpha$, $|t| \geq \rho_n$,

$$|f(t) - a_0 - a_1 t - \cdots - a_n t^{-n}| \leq \gamma_n |t|^{-n-1}.$$

If $|z| \geq (1-\epsilon)^{-1}\rho_n$, this estimate holds for all $t \in \Gamma$. Hence (11.3-11) implies for $n = 0, 1, 2, \ldots$ that

$$\left| f'(z) + \sum_{k=1}^n ka_k z^{-k-1} \right| \leq \frac{\gamma_n}{\epsilon(1-\epsilon)^{n+1}} |z|^{-n-2}, \qquad z \in \hat{S}_\beta; \qquad |z| \geq \frac{\rho_n}{1-\epsilon}.$$

Thus the formally differentiated series

$$F' = -a_1 z^{-2} - 2a_2 z^{-3} - 3a_3 z^{-4} - \cdots$$

possesses property (A) with regard to f', establishing the result. ■

EXAMPLE

In §11.1 we have obtained an asymptotic power series, valid for $z \to \infty$ in S_α for every $\alpha < \pi$, for Binet's function $J(z)$ appearing in Stirling's formula,

$$\Gamma(z) = \sqrt{2\pi} \exp[(z - \tfrac{1}{2}) \operatorname{Log} z - z] e^{J(z)}.$$

We now convert this into an asymptotic expansion for the Γ function itself. By Theorem 11.2b,

$$e^{J(z)} \approx H = 1 + h_1 z^{-1} + h_2 z^{-2} + \cdots, \qquad z \to \infty, \qquad z \in S_\alpha, \qquad (11.3\text{-}12)$$

where H is the composition of the exponential series with the asymptotic series for $J(z)$. Lacking a simple explicit formula for the coefficients h_n, a recurrence relation can be established as follows. Theorem 11.3f permits us to differentiate (11.3-12) formally to obtain

$$J'(z) e^{J(z)} \approx H' = -h_1 z^{-2} - 2h_2 z^{-3} + \cdots, \qquad z \to \infty, \qquad z \in S_\alpha.$$

Applying Theorem 11.3f to the asymptotic series for $J(z)$, H' by Theorem 11.2a equals

$$\left(-\frac{B_2}{2} z^{-2} - \frac{B_4}{4} z^{-4} - \cdots \right) (1 + h_1 z^{-1} + h_2 z^{-2} + \cdots).$$

Comparing coefficients yields

$$nh_n = \frac{B_2}{2} h_{n-1} + \frac{B_4}{4} h_{n-3} + \frac{B_6}{6} h_{n-5} + \cdots, \qquad n = 1, 2, \ldots.$$

The sum on the right terminates when the index of h becomes negative. From $h_0 = 1$ and from the known values of the Bernoulli numbers the h_n are now easily calculated. We obtain for $\alpha \in (0, \pi)$

$$e^{J(z)} \approx 1 + \frac{1}{12} z^{-1} + \frac{1}{288} z^{-2} - \frac{139}{51840} z^{-3} + \cdots, \qquad z \to \infty, \qquad z \in \hat{S}_\alpha. \qquad (11.3\text{-}13)$$

PROBLEMS

1. Let f be defined for $|z| > \rho$, and

$$f(z) \approx \sum_{n=0}^{\infty} c_n z^{-n}, \qquad z \to \infty,$$

    where the approach to ∞ is unrestricted. Is f analytic at ∞?

2. Let $\alpha < \pi/2$. Show that

$$\operatorname{Log}(1 - e^{-z}) \approx 0, \qquad z \to \infty, \qquad z \in S_\alpha.$$

3. Let $\alpha < \pi/4$, and for $z \in S_\alpha$ let

$$\mathrm{erfc}_0(z) := \frac{2}{\sqrt{\pi}} \int_z^\infty e^{-t^2} \, dt, \qquad \mathrm{erfc}_{n+1}(z) := \int_z^\infty \mathrm{erfc}_n(t) \, dt,$$

$n = 0, 1, \ldots$, the integrations being performed along the ray $\arg t = \arg z$. Using the known result (see §11.5)

$$\mathrm{erfc}_0(z) \approx \frac{1}{\sqrt{\pi}} \frac{e^{-z^2}}{z} {}_2F_0(\tfrac{1}{2}, 1; -z^{-2}), \qquad z \to \infty, \qquad z \in \hat{S}_\alpha,$$

verify that

$$\mathrm{erfc}_n(z) \approx \frac{2}{\sqrt{\pi}} \frac{e^{-z^2}}{(2z)^{n+1}} {}_2F_0\left(\frac{n+1}{2}, \frac{n+2}{2}; -z^{-2}\right), \qquad z \to \infty, \qquad z \in \hat{S}_\alpha.$$

§11.4. ASYMPTOTIC SOLUTIONS OF DIFFERENTIAL EQUATIONS

Here we apply the theory of asymptotic power series to the study of singular points of ordinary differential equations. The notations and concepts introduced in Chapter 9 are used. Our discussion concerns the general system of order n,

$$\mathbf{w}' = \mathbf{A}(z)\mathbf{w}, \qquad\qquad\qquad (11.4\text{-}1)$$

where $\mathbf{A}(z)$ is analytic at $z = \infty$, and hence representable by a convergent series in powers of z^{-1},

$$\mathbf{A}(z) = \sum_{m=0}^\infty \mathbf{A}_m z^{-m},$$

for $|z|$ sufficiently large. We shall assume that $\mathbf{A}_0 \neq \mathbf{0}$. The point $z = \infty$ then is an *irregular singular point of rank one* of the system (11.4-1).

As in §9.11 we assume that the matrix $\mathbf{A}_0$ has n distinct eigenvalues. Without further loss of generality it then may be taken to be in diagonal form, $\mathbf{A}_0 = \mathrm{diag}(\lambda_1, \lambda_2, \ldots, \lambda_n)$. Theorem 9.11a then stated the existence of a *formal solution* of (11.4-1) of the form

$$\mathbf{W} = \mathbf{P} z^\Delta e^{\Lambda z}, \qquad\qquad\qquad (11.4\text{-}2)$$

where $\mathbf{P}$ is a formal power series,

$$\mathbf{P} = \mathbf{P}_0 + \mathbf{P}_1 z^{-1} + \mathbf{P}_2 z^{-2} + \cdots,$$

and where $\Delta = \mathrm{diag}(\delta_1, \delta_2, \ldots, \delta_n)$ is a certain diagonal matrix depending on $\mathbf{A}_0$ and $\mathbf{A}_1$.

The main objective here is to show that each column of the formal solution $\mathbf{W}$ is, in suitable sectors of the complex plane, an asymptotic series to a certain actual solution of the system (11.4-1). More specifically, we shall prove:

THEOREM 11.4a

Let the columns of the formal solution matrix $\mathbf{W}$ *be denoted by* $\mathbf{w}_1, \mathbf{w}_2, \ldots, \mathbf{w}_n$. *Then for every* i, $i = 1, 2, \ldots, n$, *the following holds: If S is a closed sector of the z plane containing none of the straight lines* $\text{Re}(\lambda_i - \lambda_j)z = 0$ $(j = 1, 2, \ldots, n; j \neq i)$, *then there exists in S for $|z|$ sufficiently large an actual solution* $\mathbf{v}$ *of the system* (11.4-1) *such that*

$$\mathbf{v}(z) \approx \mathbf{w}_i, \qquad z \to \infty, \qquad z \in S. \tag{11.4-3}$$

Proof. Let i be one of the indices $1, 2, \ldots, n$. This index will be kept fixed throughout the proof.

We begin with some normalizations designed to simplify the notations of the proof. By a substitution of the form $\mathbf{w} \to \exp(\lambda_i z) z^{\delta_i} \mathbf{w}$ in (11.4-1) we can achieve that $\lambda_i = \delta_i = 0$ in the transformed differential equation. The formal solution $\mathbf{w}_i$ then is a power series in z^{-1},

$$\mathbf{w}_i = \mathbf{p}_{i0} + \mathbf{p}_{i1} z^{-1} + \mathbf{p}_{i2} z^{-2} + \cdots, \tag{11.4-4}$$

whose coefficients are complex n vectors. The condition on S now requires that it contains none of the straight lines $\text{Re} \lambda_j z = 0$ $(j \neq i)$. By a substitution of the form $\mathbf{w}(z) \to \mathbf{w}(\epsilon z)$ $(|\epsilon| = 1)$ we can achieve that S is a sector S_α with a suitable α, $0 < \alpha < \pi/2$.

We begin the proof proper by reducing the assertion to the assertion that a certain differential equation has a solution that is asymptotic to zero. Let $\beta > \alpha$. By Ritt's Theorem 11.3b there exists a vector $\mathbf{f}$ of functions analytic in S_β such that

$$\mathbf{f}(z) \approx \mathbf{w}_i, \qquad z \to \infty, \qquad z \in S_\beta. \tag{11.4-5}$$

(Here and subsequently, asymptotic relations between vector-valued functions are to be understood componentwise.) By virtue of Theorem 11.3f,

$$\mathbf{f}'(z) \approx \mathbf{w}_i', \qquad z \to \infty, \qquad z \in S_\alpha. \tag{11.4-6}$$

Because $\mathbf{w}_i$ is a formal solution, $\mathbf{w}_i' = \mathbf{A}\mathbf{w}_i$, where the product on the right is to be interpreted formally. Because the convergent series $\mathbf{A}(z)$ is asymptotic to its sum, it follows from (11.4-5) and from Theorem 11.2a that $\mathbf{A}(z)\mathbf{f}(z) \approx \mathbf{A}\mathbf{w}_i$ and hence that

$$\mathbf{f}'(z) - \mathbf{A}(z)\mathbf{f}(z) \approx \mathbf{w}_i' - \mathbf{A}\mathbf{w}_i = \mathbf{0}, \qquad z \to \infty, \qquad z \in S_\alpha. \tag{11.4-7}$$

A vector $\mathbf{v}$ is a solution of $\mathbf{w}' = \mathbf{A}(z)\mathbf{w}$ satisfying relation (11.4-3) as required by the theorem if and only if the function

$$\mathbf{h} := \mathbf{v} - \mathbf{f}$$

has the following properties:
   (i) $\mathbf{h}(z) \approx \mathbf{w}_i - \mathbf{w}_i$ and hence is asymptotic to the zero series,

$$\mathbf{h}(z) \approx \mathbf{0} + \mathbf{0}z^{-1} + \mathbf{0}z^{-2} + \cdots, \qquad z \to \infty, \qquad z \in S_\alpha; \quad (11.4\text{-}8a)$$

   (ii) $\mathbf{h}$ satisfies the differential equation

$$\mathbf{h}' = \mathbf{v}' - \mathbf{f}'$$
$$= \mathbf{A}(z)\mathbf{v} - \mathbf{f}'$$
$$= \mathbf{A}(z)\mathbf{h} - (\mathbf{f}' - \mathbf{A}(z)\mathbf{f})$$

or

$$\mathbf{h}' = \mathbf{A}(z)\mathbf{h} + \mathbf{b}(z), \qquad\qquad (11.4\text{-}8b)$$

where

$$\mathbf{b}(z) := \mathbf{A}(z)\mathbf{f}(z) - \mathbf{f}'(z). \qquad\qquad (11.4\text{-}8c)$$

By (11.4-7), $\mathbf{b}$ is asymptotic to zero,

$$\mathbf{b}(z) \approx \mathbf{0} + \mathbf{0}z^{-1} + \mathbf{0}z^{-2} + \cdots, \qquad z \to \infty, \qquad z \in S_\alpha.$$

To prove the existence of an $\mathbf{h}$ satisfying (11.4-8), we now set up an integral equation that is equivalent to (11.4-8b). The integral equation is solved by the method of successive approximations, beginning with the initial approximation $\mathbf{h}_0 := \mathbf{0}$. The solution is then shown to be asymptotic to zero.

Some preliminaries are required to set up the integral equation. The formal solution matrix $\mathbf{W} = \mathbf{P}z^{\mathbf{\Delta}} e^{\mathbf{\Lambda}z}$ satisfies

$$\mathbf{W}' = \mathbf{A}(z)\mathbf{W} \qquad\qquad (11.4\text{-}9)$$

as an identity between formal logarithmic series. Since $\det \mathbf{P}_0 \neq 0$, $\mathbf{P}^{-1}$, and hence $\mathbf{W}^{-1} := e^{-\mathbf{\Lambda}z} z^{-\mathbf{\Delta}} \mathbf{P}^{-1}$ exist formally, and (11.4-9) implies

$$\mathbf{W}'\mathbf{W}^{-1} = \mathbf{A}, \qquad\qquad (11.4\text{-}10)$$

which is the same as

$$(\mathbf{P}' + z^{-1}\mathbf{P}\mathbf{\Delta} + \mathbf{P}\mathbf{\Lambda})\mathbf{P}^{-1} = \mathbf{A}. \qquad\qquad (11.4\text{-}11)$$

Let

$$\mathbf{P}_{(m)} := \sum_{k=0}^{m} \mathbf{P}_k z^{-k}, \qquad m = 0, 1, 2, \ldots.$$

We define

$$\mathbf{W}_{(m)} := \mathbf{P}_{(m)} z^{\Delta} e^{\Lambda z}.$$

Clearly, $\mathbf{W}_{(m)}$ may be regarded as a function of z. Since $\det \mathbf{P}_0 \neq 0$, $\mathbf{P}_{(m)}^{-1}$ is defined as a function of z for $|z|$ sufficiently large, $|z| \geqslant \rho$, say. It follows that for such $|z|$ also

$$\mathbf{W}_{(m)}^{-1} = e^{-\Lambda z} z^{-\Delta} \mathbf{P}_{(m)}^{-1}$$

is defined. We set

$$\mathbf{A}_{(m)} := \mathbf{W}_{(m)}' \mathbf{W}_{(m)}^{-1} = (\mathbf{P}_{(m)}' + z^{-1}\mathbf{P}_{(m)}\Delta + \mathbf{P}_{(m)}\Lambda)\mathbf{P}_{(m)}^{-1}.$$

Evidently, $\mathbf{A}_{(m)}$ is analytic at ∞; moreover, it follows by comparison with (11.4-11) that

$$\mathbf{A}_{(m)}(z) = \mathbf{A}(z) + O(z^{-m-1}). \qquad (11.4\text{-}13)$$

By construction, the matrix $\mathbf{W}_{(m)}$ is a fundamental matrix of the differential equation

$$\mathbf{w}' = \mathbf{A}_{(m)}(z)\mathbf{w} \qquad (|z| \geqslant \rho) \qquad (11.4\text{-}14)$$

near $z = \infty$. By (11.4-13), this system can be made arbitrarily "close" to the given system $\mathbf{w}' = \mathbf{A}(z)\mathbf{w}$ by making m sufficiently large. This is used later on. By the general theory of linear analytic systems of differential equations (cf. (9.3-7)), a solution of the nonhomogeneous equation

$$\mathbf{w}' = \mathbf{A}_{(m)}(z)\mathbf{w} + \mathbf{a}(z) \qquad (11.4\text{-}15)$$

is given by the variation of constants formula,

$$\mathbf{w}(z) = \int_a^z \mathbf{K}(z, t)\mathbf{a}(t) \, dt,$$

where

$$\mathbf{K}(z, t) := \mathbf{W}_{(m)}(z)\mathbf{W}_{(m)}^{-1}(t). \qquad (11.4\text{-}16)$$

More generally, if the kernel $\mathbf{K}$ is decomposed as follows,

$$\mathbf{K}(z, t) = \mathbf{K}_0(z, t) + \mathbf{K}_1(z, t), \qquad (11.4\text{-}17a)$$

where each $\mathbf{K}_i$ is a solution of the homogeneous system,

$$\frac{\partial}{\partial z}\mathbf{K}_i(z, t) = \mathbf{A}_{(m)}(z)\mathbf{K}_i(z, t), \qquad i = 0, 1, \qquad (11.4\text{-}17b)$$

then the function

$$\mathbf{w}(z) := \int_a^z \mathbf{K}_0(z, t)\mathbf{a}(t)\, dt - \int_z^b \mathbf{K}_1(z, t)\mathbf{a}(t)\, dt,$$

where $|a| > \rho$, $|b| > \rho$, again is a solution of (11.4-15), as is immediately verified by differentiation. This statement also holds for $b = \infty$ provided that the improper integral converges locally uniformly with respect to z.

How does the foregoing relate to our task of finding a solution of (11.4-8b) that satisfies (11.4-8a)? Write (11.4-8b) in the form

$$\mathbf{h}' = \mathbf{A}_{(m)}(z)\mathbf{h} + (\mathbf{B}_{(m)}(z)\mathbf{h} + \mathbf{b}(z)), \qquad (11.4\text{-}18)$$

where

$$\mathbf{B}_{(m)}(z) := \mathbf{A}(z) - \mathbf{A}_{(m)}(z).$$

Considering this as a differential equation of the form (11.4-15) with nonhomogeneous term $\mathbf{a}(z) := \mathbf{B}_{(m)}(z)\mathbf{h}(z) + \mathbf{b}(z)$, it follows that if $\mathbf{h}$ is a solution of the integral equation

$$\mathbf{h}(z) = \int_a^z \mathbf{K}_0(z, t)[\mathbf{B}_{(m)}(t)\mathbf{h}(t) + \mathbf{b}(t)]\, dt$$

$$\qquad\qquad\qquad\qquad\qquad\qquad (11.4\text{-}19)$$

$$\quad - \int_z^\infty \mathbf{K}_1(z, t)[\mathbf{B}_{(m)}(t)\mathbf{h}(t) + \mathbf{b}(t)]\, dt,$$

then $\mathbf{h}$ is a solution of (11.4-18), and hence of (11.4-8b).

We now choose the integer m, the decomposition (11.4-17) of the kernel, and a, the lower limit of integration, in such a way that the process of successive approximations applied to (11.4-19) converges and furnishes a solution with the correct asymptotic behavior.

Let I_0 be the set of all integers $j \in \{1, 2, \ldots, n\}$ such that

$$\operatorname{Re} \lambda_j z \to -\infty, \qquad z \to \infty, \qquad z \in S_\alpha,$$

and let I_1 be the complementary set. We denote by $\mathbf{D}_0$ the diagonal matrix that has ones in the jth position where $j \in I_0$, and zeros elsewhere, and by $\mathbf{D}_1$ the matrix $\mathbf{I} - \mathbf{D}_0$. With this we define

$$\mathbf{K}_i(z, t) := \mathbf{W}_{(m)}(z)\mathbf{D}_i\mathbf{W}_{(m)}^{-1}(t)$$

$$\qquad = \mathbf{P}_{(m)}(z)z^{\Delta} e^{\Lambda z}\mathbf{D}_i e^{-\Lambda t} t^{-\Delta}\mathbf{P}_{(m)}^{-1}(t), \qquad (11.4\text{-}20)$$

$i = 0, 1$. The effect of these definitions is to make the real parts of the exponents in the exponential factors negative in $\mathbf{K}_0$ for $\operatorname{Re}(z - t) > 0$ and negative in $\mathbf{K}_1$ for $\operatorname{Re}(z - t) < 0$.

Bounds on the kernels $\mathbf{K}_i$ are required to prove existence and convergence of the successive approximations. In $\mathbf{K}_0$, we may assume $|z| \geq |t|$; in $\mathbf{K}_1$, $|z| \leq |t|$. Furthermore, $|\arg z| \leq \alpha$ and $|\arg t| \leq \alpha$. Because $\mathbf{P}_{(m)}(z)$ and $\mathbf{P}_{(m)}^{-1}(z)$ are analytic for $\rho \leq |z| \leq \infty$, the contributions of these factors to the norms $\|\mathbf{K}_i\|$ are bounded. Because S_α contains none of the rays $\operatorname{Re} \lambda_j z = 0$, there exist constants γ and $\mu > 0$ such that for z, $t \in S_\alpha$, writing $\zeta := |z|$, $\tau := |t|$,

$$\left|\exp(\lambda_j(z-t))\right| \leq \gamma e^{-\mu(\zeta-\tau)}, \qquad \zeta \geq \rho, \qquad j \in I_0,$$

and (because $i \in I_1$ and $\lambda_i = 0$)

$$\left|\exp(\lambda_j(z-t))\right| \leq \gamma, \qquad \zeta \leq \tau, \qquad j \in I_1.$$

Letting $\delta := \max |\operatorname{Re} \delta_j|$, it follows that for a suitable constant κ,

$$\begin{aligned}
\|\mathbf{K}_0(z, t)\| &\leq \kappa \zeta^\delta \tau^{-\delta} e^{-\mu(\zeta-\tau)}, \qquad \zeta \geq \tau, \\
\|\mathbf{K}_1(z, t)\| &\leq \kappa \zeta^{-\delta} \tau^\delta, \qquad \zeta \leq \tau.
\end{aligned} \qquad (11.4\text{-}21)$$

We recall that as a consequence of (11.4-13), there exists for each m a constant β_m such that

$$\|\mathbf{B}_{(m)}(t)\| \leq \frac{\beta_m}{\tau^{m+1}}, \qquad \tau \geq \rho. \qquad (11.4\text{-}22)$$

We now fix m to satisfy $m > 2\delta$, $m > 1$. With this choice of m we attempt to construct a sequence of successive approximations $\{\mathbf{h}_k\}$ to a solution of (11.4-19) in the following obvious way:

$$\mathbf{h}_0(z) := \mathbf{0};$$

$$\begin{aligned}
\mathbf{h}_{k+1}(z) &:= \int_a^z \mathbf{K}_0(z, t)[\mathbf{B}_{(m)}\mathbf{h}_k(t) + \mathbf{b}(t)]\, dt \\
&\quad - \int_z^\infty \mathbf{K}_1(z, t)[\mathbf{B}_{(m)}\mathbf{h}_k(t) + \mathbf{b}(t)]\, dt,
\end{aligned} \qquad (11.4\text{-}23)$$

$$k = 0, 1, 2, \ldots; \qquad |z| \geq |a|.$$

These approximations still depend on a, which is going to be chosen conveniently later. However, we assume a to lie on the axis of symmetry of S_α, and hence to be real. The paths of integration in (11.4-23) are chosen as shown in Fig. 11.4a.

In the first integral we first integrate from $t = a$ to $t = a_z := z|az^{-1}|$ along the circular segment $|t| = a$, then from a_z to z along the ray $\arg t = \arg z$. The integration from z to ∞ is again along the ray $\arg t = \arg z$.

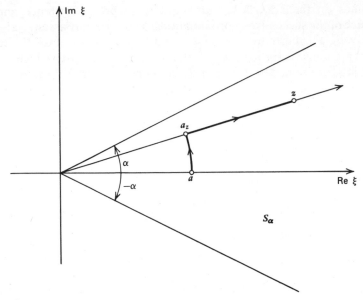

Fig. 11.4a.

LEMMA 11.4b

The successive approximations $\mathbf{h}_k$ exist and are analytic for $|z| \geq a, z \in S_\alpha$, and there exist constants χ_k such that

$$\|\mathbf{h}_k(t)\| \leq \chi_k \tau^\delta, \qquad \tau \geq a. \tag{11.4-24}$$

Proof. The assertion is clearly true for $k = 0$. Assume the assertion to be true for some $k \geq 0$. Let $\mathbf{g}_k := \mathbf{B}_{(m)}\mathbf{h}_k + \mathbf{b}$. Because $\mathbf{b}$ is asymptotic to zero, there exists by (11.4-22) a constant γ_k such that

$$\|\mathbf{g}_k(t)\| \leq \gamma_k \tau^{\delta - m - 1}, \qquad \tau \geq a.$$

Using the estimates (11.4-21), we now find

$$\left\| \int_a^{a_z} \mathbf{K}_0(z, t)\mathbf{g}_k(t)\, dt \right\| \leq \gamma_k \kappa \frac{\pi}{2} \zeta^\delta a^{-m},$$

$$\left\| \int_{a_z}^z \mathbf{K}_0(z, t)\mathbf{g}_k(t)\, dt \right\| \leq \gamma_k \kappa m^{-1} \zeta^\delta a^{-m},$$

and, since $m > 2\delta$,

$$\left\| \int_z^\infty \mathbf{K}_1(z, t)\mathbf{g}_k(t)\, dt \right\| \leq \gamma_k \kappa (m - 2\delta)^{-1} \zeta^\delta \zeta^{-m}.$$

Thus $\mathbf{h}_{k+1}$ exists and satisfies (11.4-24) where

$$\chi_{k+1} := \frac{\chi_k \kappa}{a^m}\left(\frac{\pi}{2} + \frac{1}{m} + \frac{1}{m-2\delta}\right).$$

Because the integrands in (11.4-23) depend analytically on z and because the improper integral converges uniformly with respect to z on compact subsets of S_α, $\mathbf{h}_{k+1}$ is analytic. The assertion of the lemma thus holds for $k+1$, and hence for all k. ■

LEMMA 11.4c

If a is chosen sufficiently large, the sequence $\{\mathbf{h}_k\}$ converges uniformly on every compact subset of $|z| \geq a$, $z \in S_\alpha$.

Proof. Let $\mathbf{d}_k := \mathbf{h}_k - \mathbf{h}_{k-1}$. It follows from (11.4-23) that

$$\mathbf{d}_{k+1}(z) = \int_a^z \mathbf{K}_0(z,t)\mathbf{B}_{(m)}(t)\mathbf{d}_k(t)\,dt$$

$$- \int_z^\infty \mathbf{K}_1(z,t)\mathbf{B}_{(m)}(t)\mathbf{d}_k(t)\,dt.$$

Now suppose that

$$|\mathbf{d}_k(t)| \leq \theta_k \tau^\delta, \qquad \tau \geq a. \tag{11.4-25}$$

Proceeding much as in the proof of the preceding lemma, we then have, letting $\beta := \beta_m$,

$$\left\| \int_a^{a_z} \mathbf{K}_0(z,t)\mathbf{B}_{(m)}(t)\mathbf{d}_k(t)\,dt \right\| \leq \kappa\beta\theta_k \frac{\pi}{2}\zeta^\delta a^{-m},$$

$$\left\| \int_{a_z}^z \mathbf{K}_0(z,t)\mathbf{B}_{(m)}(t)\mathbf{d}_k(t)\,dt \right\| \leq \kappa\beta\theta_k m^{-1}\zeta^\delta a^{-m},$$

and

$$\left\| \int_z^\infty \mathbf{K}_1(z,t)\mathbf{B}_{(m)}(t)\mathbf{d}_k(t)\,dt \right\| \leq \kappa\beta\theta_k(m-2\delta)^{-1}\zeta^\delta\zeta^{-m}.$$

Let now a be so chosen that

$$a^m \geq 2\kappa\beta\left(\frac{\pi}{2} + \frac{1}{m} + \frac{1}{m-2\delta}\right). \tag{11.4-26}$$

Then the foregoing estimates show that

$$\|\mathbf{d}_{k+1}(z)\| \leq \theta_{k+1}\zeta^\delta, \qquad \zeta \geq a,$$

where $\theta_{k+1} \leq \frac{1}{2}\theta_k$.

In view of $\mathbf{d}_1 = \mathbf{h}_1$, (11.4-25) is true for $k = 1$ by virtue of Lemma 11.4b. There follows

$$\|\mathbf{d}_k(z)\| \leq 2^{1-k} \theta_1 \zeta^\delta, \qquad \zeta \geq a, \qquad k = 1, 2, \dots,$$

which implies the uniform convergence of the series

$$\sum_{k=1}^{\infty} \mathbf{d}_k(z) = \sum_{k=1}^{\infty} (\mathbf{h}_k(z) - \mathbf{h}_{k-1}(z))$$

on every set $a \leq |z| \leq \rho_1$, and hence of the sequence $\{\mathbf{h}_k\}$. Our argument also shows that the limit function $\mathbf{h}$ satisfies

$$\|\mathbf{h}(z)\| \leq 2\theta_1 \zeta^\delta, \qquad \zeta \geq a. \tag{11.4-27}$$

By virtue of the uniform convergence, $\mathbf{h}$ is a solution of the integral equation (11.4-19), and hence of the differential equation (11.4-8b). It remains to prove:

LEMMA 11.4d

The limit function $\mathbf{h}$ is asymptotic to zero.

Proof. It suffices to show that for $p = 0, 1, 2, \dots$, there exist constants η_p such that

$$\|\mathbf{h}(z)\| \leq \eta_p \zeta^{\delta - p}, \qquad \zeta \geq a. \tag{11.4-28}$$

By (11.4-27) this is true for $p = 0$. Let $\mathbf{g} =: \mathbf{B}_{(m)} \mathbf{h} + \mathbf{b}$. If (11.4-28) is true for some $p \geq 0$, we have, since $\mathbf{b} \approx \mathbf{0}$,

$$\|\mathbf{g}(z)\| \leq \beta \eta_p \zeta^{\delta - p - m - 1}, \qquad \zeta \geq a.$$

Much as in the preceding lemmas we now estimate

$$\left\| \int_a^{a_z} \mathbf{K}_0(z, t) \mathbf{g}(t) \, dt \right\| \leq \kappa \beta \eta_p \frac{\pi}{2} a^{\delta - m - p} \zeta^\delta e^{-\mu(\zeta - a)}.$$

The integral from a_z to z is split in two by first integrating to the halfway point $z_1 := \frac{1}{2}(a_z + z)$. This yields

$$\left\| \int_{a_z}^{z_1} \mathbf{K}_0(z, t) \mathbf{g}(t) \, dt \right\| \leq \kappa \beta \eta_p (m + p)^{-1} a^{-m-p} \zeta^\delta \exp\left(-\frac{\mu}{2}(\zeta - a)\right).$$

$$\left\| \int_{z_1}^{z} \mathbf{K}_0(z, t) \mathbf{g}(t) \, dt \right\| \leq \kappa \beta \eta_p (m + p)^{-1} \left(\frac{\zeta + a}{2}\right)^{-m-p} \zeta^\delta$$

Finally,

$$\left\| \int_z^{\infty} \mathbf{K}_1(z, t) \mathbf{g}(t) \, dt \right\| \leq \kappa \beta \eta_p (m + p - 2\delta)^{-1} \zeta^{\delta - m - p}.$$

By virtue of $m > 1$, each of these estimates is $O(\zeta^{\delta-p-1})$ for $\zeta \to \infty$. This shows that (11.4-28) is true, with p increased by one. The induction step is thus complete, and with it the proof of Lemma 11.4d and of Theorem 11.4a. ∎

The next result now follows easily.

COROLLARY 11.4e

Let S be a closed sector of the complex plane containing none of the **critical lines** $\mathrm{Re}(\lambda_i - \lambda_j)z = 0$, $i, j = 1, 2, \ldots, n$, $i \neq j$. *Then equation* (11.4-1) *possesses in S a fundamental system of solutions* $v_1, v_2, \ldots, v_n$ *such that*

$$v_i(z) \approx w_i, \qquad z \to \infty, \qquad z \in S, \qquad i = 1, 2, \ldots, n. \quad (11.4\text{-}29)$$

Proof. The sector S satisfies the hypotheses of Theorem 11.4a for every i. It thus follows that n solutions v_i satisfying (11.4-29) exist, and it remains only to be shown that they form a fundamental system. This follows from the fact that for $z \to \infty$

$$\det(v_1, \ldots, v_n) = \det\{(\mathbf{P}_0 + O(z^{-1}))z^{\boldsymbol{\Delta}} e^{\boldsymbol{\Lambda}z}\}$$

$$= z^{\Sigma \delta_i} e^{z \Sigma \lambda_i}\{\det \mathbf{P}_0 + O(z^{-1})\}$$

$$\neq 0. \quad \blacksquare$$

By construction, the solutions v_i mentioned in the corollary depend on the sector S. By the general theory of analytic differential equations, each of these solutions can, of course, be continued analytically across the boundaries of S, and across the critical lines $\mathrm{Re}(\lambda_i - \lambda_j)z = 0$. However, it must not be assumed that the asymptotic relation (11.4-29) remains valid for the continued function. Typically, the function v_i will be represented asymptotically by a different linear combination of the formal solutions w_j in each sector bounded by the lines $\mathrm{Re}(\lambda_i - \lambda_j)z = 0$. After its discoverer, the discontinuous change in the asymptotic representations is called the **Stokes phenomenon**.

To illustrate the results of this section, we consider their implications for the special second-order equation

$$u'' - r(z)u = 0, \quad (11.4\text{-}30)$$

where

$$r(z) = \sum_{n=0}^{\infty} r_n z^{-n}$$

for $|z|$ sufficiently large, $r_0 \neq 0$. As shown in §9.12, the usual reduction to a first-order system here yields a system (11.4-1) with $n = 2$ where $\mathbf{A}_0$ has the eigenvalues $\pm r_0^{1/2}$. There is only one critical line, namely, the line

Re $r_0^{1/2} z = 0$. Thus in each of the two half planes separated by this line there exist two actual solutions that are asymptotic to the two formal solutions constructed in §9.12.

EXAMPLE

To study a specific case, we consider the equation

$$u'' + \left(1 - \frac{\nu^2 - 1/4}{z^2}\right) u = 0$$

satisfied by $z^{1/2} v$ where v is a solution of Bessel's equation (9.7-9). Since $r_0 = -1$, the critical line is given by Re $iz = 0$, and thus coincides with the real axis. The two formal solutions found by the algorithm of §9.12 were

$$V^{(\mathrm{I})} = e^{iz} {}_2F_0\left(\frac{1}{2} - \nu, \frac{1}{2} + \nu; \frac{1}{2iz}\right),$$

$$V^{(\mathrm{II})} = e^{-iz} {}_2F_0\left(\frac{1}{2} - \nu, \frac{1}{2} + \nu; -\frac{1}{2iz}\right).$$

The actual solutions having these formal solutions as asymptotic expansions are in §11.5 identified as certain Hankel functions. In fact, there it is shown that

$$\sqrt{\frac{\pi z}{2}} \exp\left[i\left(\frac{\nu\pi}{2} + \frac{\pi}{4}\right)\right] H_\nu^{(1)}(z) \approx V^{(\mathrm{I})}, \qquad z \to \infty, \qquad -iz \in S_{3\pi/2 - \epsilon},$$

$$\sqrt{\frac{\pi z}{2}} \exp\left[-i\left(\frac{\nu\pi}{2} + \frac{\pi}{4}\right)\right] H_\nu^{(2)}(z) \approx V^{(\mathrm{II})}, \qquad z \to \infty, \qquad iz \in S_{3\pi/2 - \epsilon}.$$

In this case the sectors in which the asymptotic expansions hold are larger than what could be inferred from the general theory. Nevertheless, the Stokes phenomenon is present. To observe it, consider the Bessel function $z^{1/2} J_\nu(z)$. In view of $J_\nu = \frac{1}{2}(H_\nu^{(1)} + H_\nu^{(2)})$ we have

$$z^{1/2} J(z) \approx \tfrac{1}{2}\{V^{(\mathrm{I})} + V^{(\mathrm{II})}\}, \qquad z \to \infty, \qquad z \in S_{\pi - \epsilon}. \tag{11.4-31}$$

If arg z is increased by 2π, the expression on the left is multiplied by $\exp((\nu + \frac{1}{2})2\pi i)$, whereas the expression on the right is unchanged. Thus (11.4-31) cannot hold in any larger sector.

PROBLEM

1. Show that Theorem 11.4a remains true even if $\mathbf{A}(z)$ is not represented by a convergent expansion in z^{-1}, but merely possesses the asymptotic expansion

$$\mathbf{A}(z) \approx \sum_{m=0}^{\infty} \mathbf{A}_m z^{-m}, \qquad z \to \infty$$

in a suitable sector S_α.

§11.5. THE WATSON–DOETSCH LEMMA

This deals with the asymptotic behavior as $z \to \infty$, $z \in S_\alpha$ of functions defined by a Laplacian integral,

$$f(z) := (\mathscr{L}\phi)(z) := \int_0^\infty e^{-z\tau}\phi(\tau)\, d\tau. \qquad (11.5\text{-}1)$$

The main point is that this behavior is fully determined by the behavior of ϕ near $\tau = 0$. It will be shown, for instance, that if ϕ is an original function of bounded exponential growth that possesses an asymptotic power series for $\tau \to 0$, $\tau > 0$, then its Laplace transform f for every $\epsilon > 0$ possesses an asymptotic power series as $z \to \infty$, $z \in \hat{S}_{\pi/2-\epsilon}$ whose coefficients are very easily calculated from the asymptotic series for ϕ. If ϕ is analytic in a sector S_α, then the asymptotic representation for f even holds for $z \in S_{\pi/2+\alpha-\epsilon}$. Because a large number of important special functions, notably solutions of ordinary differential equations, may be regarded as Laplace transforms, the method has numerous applications.

We recall from Chapter 10 the following facts about the Laplace transform:

(i) Let ϕ be an *original function* with growth indicator γ (see §10.1 for definitions). Then $f := \mathscr{L}\phi$ is analytic at least for Re $z > \gamma$.

(ii) Let $\phi = \phi(t)$ be analytic for $|\arg t| \leq \alpha$ where $0 < \alpha < \pi$, and let the function

$$\mu(\tau) := \sup_{\substack{|t|=\tau \\ |\arg t| \leq \alpha}} |\phi(t)|, \qquad \tau > 0$$

be integrable at $\tau = 0$ and satisfy

$$\gamma := \limsup_{\tau \to \infty} \frac{\operatorname{Log} \mu(\tau)}{\tau} < \infty.$$

(If $\alpha = \pi/2$, such functions were called of *semiexponential type* in §10.9.) Then $f := \mathscr{L}\phi$ can be continued analytically into every half plane Re $e^{-i\psi}z > \gamma$, where $-\alpha \leq \psi \leq \alpha$. If $\alpha > \pi/2$, this means that f can be continued across parts of the negative real axis. The continuations from the top down and from the bottom up will, in general, be different. We use the symbol f to denote the continued function, with the understanding that it must be thought of as a log-holomorphic function as defined in §9.4. The proof of these assertions is given explicitly in §10.9 for $\alpha = \pi/2$; the general case can be dealt with similarly.

The precise formulation of the result alluded to initially is as follows:

THEOREM 11.5 (Watson–Doetsch lemma)

Let ϕ satisfy either condition (i) *or condition* (ii), *and let*

$$\phi(t) \approx \Phi := t^b\{a_0 + a_1 t^c + a_2 t^{2c} + \cdots\} \qquad (11.5\text{-}2)$$

as $t \to 0$ and $t > 0$ [case (i)] or $t \to 0$ and $t \in \hat{S}_\alpha$ [case (ii)], where b and c are complex numbers such that Re $b > -1$, Re $c > 0$, and where the powers have their principal values. Then for every $\delta > 0$ the function $f := \mathcal{L}\phi$ satisfies

$$f(z) \approx F := \frac{1}{z^{b+1}} \sum_{m=0}^{\infty} a_m \frac{\Gamma(b+1+mc)}{z^{mc}} \tag{11.5-3}$$

as $z \to \infty$ and $z \in \hat{S}_{\pi/2-\delta}$ in case (i) and $z \to \infty$ and $z \in \hat{S}_{\pi/2+\alpha-\delta}$ in case (ii).

Proof. Before proceeding to the proof proper, it is instructive to see how the result (11.5-3) comes about in a purely formal way. Substituting the series (11.5-2) into (11.5-1) and integrating term by term as if the interval of integration were finite and the series uniformly convergent, we obtain

$$f(z) = \int_0^\infty e^{-z\tau} \tau^b \sum_{m=0}^\infty a_m \tau^{mc} \, d\tau = \sum_{m=0}^\infty a_m \int_0^\infty e^{-z\tau} \tau^{b+mc} \, d\tau$$

and this is formally identical with F in view of Euler's integral in the form (10.5-1).

It is obvious that the foregoing formal procedure is invalid, but it suggests how the result was found and how it might be proved. We now carry out the proof for case (i); the proof for the other case is obtained by rotating the path of integration exactly as done in §10.9.

Let n be any nonnegative integer, and let Φ_n and F_n have their usual meaning as partial sums of the series Φ and F. Let $\epsilon > 0$ be given, and let

$$\epsilon_1 := \frac{(\sin \delta)^{\beta+n\gamma+1}}{2\Gamma(\beta+n\gamma+1)} \epsilon,$$

where $\beta := $ Re b, $\gamma := $ Re c. By property (B') of §11.1 there exists $\eta > 0$ such that for $0 < \tau < \eta$

$$|\tau^{-b-nc}[\phi(\tau) - \Phi_n(\tau)]| \leq \epsilon_1,$$

or equivalently,

$$|\phi(\tau) - \Phi_n(\tau)| \leq \epsilon_1 \tau^{\beta+n\gamma}. \tag{11.5-4}$$

Observing that

$$f(z) - F_n(z) = \int_0^\infty e^{-z\tau}[\phi(\tau) - \Phi_n(\tau)] \, d\tau, \tag{11.5-5}$$

we estimate $f - F_n$ by writing

$$\int_0^\infty = \int_0^\eta + \int_\eta^\infty.$$

For the first integral we have from (11.5-4), letting $x := \operatorname{Re} z$ and using $x \geqslant |z| \sin \delta$,

$$\left| \int_0^\eta e^{-z\tau} [\phi(\tau) - \Phi_n(\tau)] \, d\tau \right| \leqslant \epsilon_1 \int_0^\eta e^{-x\tau} \tau^{\beta+n\gamma} \, d\tau < \epsilon_1 \frac{\Gamma(\beta + n\gamma + 1)}{x^{\beta+n\gamma+1}}$$

$$\leqslant \frac{\epsilon}{2} |z|^{-\beta-n\gamma-1}. \tag{11.5-6}$$

To estimate the second integral, let $\lambda > 0$ be such that

$$|\phi(\tau)| \leqslant e^{\lambda\tau}, \qquad \tau \geqslant \eta.$$

Then the function

$$\tau^{-\beta-n\gamma-1} e^{-\lambda\tau} |\phi(\tau) - \Phi_n(\tau)|$$

is piecewise continuous for $\tau \geqslant \eta$ and tends to zero for $\tau \to \infty$. It thus has a finite supremum, which we call κ. There follows

$$|\phi(\tau) - \Phi_n(\tau)| \leqslant \kappa \tau^{\beta+n\gamma+1} e^{\lambda\tau}, \qquad \tau \geqslant \eta.$$

Hence if z is such that $x = \operatorname{Re} z \geqslant 2\lambda$,

$$\left| \int_\eta^\infty e^{-z\tau} [\phi(\tau) - \Phi_n(\tau)] \, d\tau \right| \leqslant \kappa \int_\eta^\infty e^{-\tau(x-\lambda)} \tau^{\beta+n\gamma+1} \, d\tau$$

$$< \kappa \frac{\Gamma(\beta + n\gamma + 2)}{(x-\lambda)^{\beta+n\gamma+2}}$$

$$\leqslant \frac{\kappa_1}{|z|} \frac{1}{|z|^{\beta+n\gamma+1}},$$

where

$$\kappa_1 := \frac{2^{\beta+n\gamma+2} \Gamma(\beta + n\gamma + 2)}{(\sin \delta)^{\beta+n\gamma+2}} \kappa.$$

Thus if $|z| \geqslant 2\kappa_1/\epsilon$, $z \in \hat{S}_{\pi/2-\delta}$,

$$\left| \int_\eta^\infty e^{-z\tau} [\phi(\tau) - \Phi_n(\tau)] \, d\tau \right| \leqslant \frac{\epsilon}{2} |z|^{-\beta-n\gamma-1},$$

and by (11.5-5) and (11.5-6) it follows that for the same values of z

$$|z^{b+nc+1}[f(z) - F_n(z)]| < e^{\pi/2|\operatorname{Im}(b+nc)|} \epsilon.$$

Because this is true for all n and $\epsilon > 0$ was arbitrary, the foregoing shows that the series F possesses property (B) with regard to f, proving the assertion. ∎

In many applications of Theorem 11.5 the series Φ has a positive radius of convergence. For this special case, the theorem was proved by G. N. Watson, and the corresponding result is frequently referred to as **Watson's lemma.**

Several typical applications of the Watson–Doetsch lemma are presented. It can be seen that even in the simplest applications some preliminary transformations are usually required before the lemma can be applied successfully.

EXAMPLE **1**   **The incomplete Γ function.**

The classical Γ function can be defined by Euler's integral

$$\Gamma(a) := \int_0^\infty e^{-s}s^{a-1}\,ds, \qquad \text{Re } a > 0.$$

Here we consider an **incomplete Γ function** defined by

$$\Gamma(a, z) := \int_z^\infty e^{-s}s^{a-1}\,ds, \tag{11.5-7}$$

where $|\arg z| < \pi$, $|\arg a| < \pi$, and the path of integration runs parallel to the real axis. We shall determine a series asymptotic to $\Gamma(a, z)$ for $z \to \infty$, $z \in \hat{S}_{3\pi/2-\delta}$ $(\delta > 0)$, where a is fixed. (The behavior for $|a|$ large and z fixed, or when a and z tend to ∞ simultaneously, is more complicated.) We assume temporarily that $|\arg z| < \pi/2$. Rotating the path of integration, we then have

$$\Gamma(a, z) = \int_\Lambda e^{-s}s^{a-1}\,ds,$$

where the path Λ runs from z to ∞ in the direction $\arg z$. Letting $s = z(1+\tau)$, where $0 \le \tau < \infty$, we obtain

$$\Gamma(a, z) = z^a e^{-z} \int_0^\infty e^{-z\tau}(1+\tau)^{a-1}\,d\tau,$$

all powers having their principal values. The integral now has precisely the form considered in Theorem 11.5 where

$$\phi(t) := (1+t)^{a-1}.$$

Clearly, ϕ satisfies the conditions (ii) for every $\alpha < \pi$. Because ϕ is analytic at $t = 0$, it is asymptotic to its Taylor series:

$$\phi(t) \approx \sum_{n=0}^\infty \frac{(1-a)_n}{n!}(-t)^n, \qquad t \to 0.$$

Watson's lemma thus applies with $b = 0$, $c = 1$, and

$$a_n := (-1)^n \frac{(1-a)_n}{n!}, \qquad n = 0, 1, 2, \ldots.$$

In view of $\Gamma(n+1) = n!$ we thus find

$$\Gamma(a, z) \approx z^{a-1} e^{-z} \left\{ 1 - \frac{(1-a)_1}{z} + \frac{(1-a)_2}{z^2} - \cdots \right\}, \qquad z \to \infty, \qquad z \in \hat{S}_{3\pi/2-\delta}.$$

$$(11.5\text{-}8)$$

EXAMPLE 2   **The complex error integral.**

Let

$$\operatorname{erf}(z) := \frac{2}{\sqrt{\pi}} \int_0^z e^{-s^2} \, dx. \qquad (11.5\text{-}9)$$

It is well known that

$$\int_0^\infty e^{-s^2} \, ds = \frac{\sqrt{\pi}}{2};$$

hence by Cauchy's theorem

$$\operatorname{erf}(z) = 1 - \frac{2}{\sqrt{\pi}} \int_z^\infty e^{-s^2} \, ds,$$

where the path of integration is parallel to the real axis or, if $|\arg z| < \pi/4$, along the ray $\arg s = \arg z$. To reduce the integral to a form where the Watson–Doetsch lemma is applicable, set

$$s^2 = z^2(1+\tau), \qquad ds = z \frac{d\tau}{2\sqrt{1+\tau}}.$$

This yields

$$\operatorname{erf}(z) = 1 - \frac{z \, e^{-z^2}}{\sqrt{\pi}} \int_0^\infty e^{-z^2\tau} \frac{1}{\sqrt{1+\tau}} \, d\tau.$$

Again, the function

$$\phi(t) := (1+t)^{-1/2} \approx \sum_{n=0}^\infty \frac{(1/2)_n}{n!} (-t)^n, \qquad t \to 0,$$

satisfies the conditions (ii) of Theorem 11.5 for every $\alpha < \pi$. Replacing z by z^2 we infer

$$\operatorname{erf}(z) \approx 1 - \frac{e^{-z^2}}{\sqrt{\pi} z} \left\{ 1 - \frac{1/2}{z^2} + \frac{(1/2)_2}{z^4} - \cdots \right\}, \qquad z \to \infty, \qquad z \in \hat{S}_{3\pi/4-\delta}. \quad (11.5\text{-}10)$$

EXAMPLE 3   **Bessel Functions.**

In the first place, we consider the Hankel functions of the first and second kind and of unrestricted order ν that were defined in §10.12 and shown to be related to the Bessel functions as follows:

$$H_\nu^{(1)} = J_\nu + iY_\nu,$$
$$H_\nu^{(2)} = J_\nu - iY_\nu. \tag{11.5-11}$$

Watson's lemma enables us to study the asymptotic behavior of these functions, and thereby of the Bessel functions, as $z \to \infty$ in suitable sectors of the complex plane. Our first task is to represent these functions by Laplacian integrals. Among the means at our disposal, the simplest way is to express the Hankel functions by Whittaker functions and to make use of the key formula (10.7-15).

Using (9.7-20), $H_\nu^{(1)}$ can be expressed in terms of J_ν and $J_{-\nu}$. Expressing the Bessel functions by confluent hypergeometric series of argument $-2iz$ as in (9.7-11) and using (9.12-14), we find that $H_\nu^{(1)}$ is a linear combination of two Whittaker functions of the first kind that is proportional to a Whittaker function of the second kind. There results

$$H_\nu^{(1)}(z) = \exp\left[-\left(\nu + \frac{1}{2}\right)\frac{i\pi}{2}\right]\frac{1}{\sqrt{2\pi z}}W_{0,\nu}(-2iz). \tag{11.5-12}$$

Similarly, we find

$$H_\nu^{(2)}(z) = \exp\left[\left(\nu + \frac{1}{2}\right)\frac{i\pi}{2}\right]\frac{1}{\sqrt{2\pi z}}W_{0,\nu}(2iz). \tag{11.5-13}$$

Now the integral (10.7-15) can be used. Replacing z by iz in (11.5-12), we find

$$H_\nu^{(1)}(iz) = -i\exp\left(-\frac{i\nu\pi}{2}\right)\frac{1}{\sqrt{2\pi z}}W_{0,\nu}(2z)$$
$$= -i\exp\left(-\frac{i\nu\pi}{2}\right)e^{-z}\frac{2}{\pi}\int_0^\infty e^{-2z\sigma}\sigma^{-1/2}{}_2F_1(\tfrac{1}{2}+\nu,\tfrac{1}{2}-\nu;\tfrac{1}{2};-\sigma)\,d\sigma.$$

Watson's lemma is now applicable for every $\alpha < \pi$ and in view of $\Gamma(n+\tfrac{1}{2}) = \sqrt{\pi}(\tfrac{1}{2})_n$ yields

$$H_\nu^{(1)}(iz) \approx -i\exp\left(-\frac{i\nu\pi}{2}\right)e^{-z}\sqrt{\frac{2}{\pi z}}\,{}_2F_0\left(\frac{1}{2}+\nu,\frac{1}{2}-\nu;\ -\frac{1}{2z}\right),$$

$$z \to \infty, \qquad iz \in S_{3\pi/2-\delta},$$

or

$$H_\nu^{(1)}(z) \approx \sqrt{\frac{2}{\pi z}}\exp\left[i\left(z - \frac{\nu\pi}{2} - \frac{\pi}{4}\right)\right]{}_2F_0\left(\frac{1}{2}+\nu,\frac{1}{2}-\nu;\frac{1}{2iz}\right),$$
$$z \to \infty, \qquad -\pi + \delta \leq \arg z \leq 2\pi - \delta. \tag{11.5-14}$$

In a completely analogous manner one obtains from (11.5-13)

$$H_\nu^{(2)}(z) \approx \sqrt{\frac{2}{\pi z}} \exp\left[-i\left(z - \frac{\nu\pi}{2} - \frac{\pi}{4}\right)\right]{}_2F_0\left(\frac{1}{2} + \nu, \frac{1}{2} - \nu; -\frac{1}{2iz}\right),$$

$$z \to \infty, \qquad -2\pi + \delta \leqslant \arg z \leqslant \pi - \delta. \tag{11.5-15}$$

As suggested in §11.4, these asymptotic series agree up to constant factors with the formal solutions (9.12-18) of Bessel's equation.

In view of the relations

$$J_\nu = \tfrac{1}{2}\{H_\nu^{(1)} + H_\nu^{(2)}\}, \qquad Y_\nu = \frac{1}{2i}\{H_\nu^{(1)} - H_\nu^{(2)}\},$$

it is now an easy matter to obtain asymptotic expansions for J_ν and Y_ν valid as $z \to \infty$, $z \in \hat{S}_{\pi-\delta}$. In view of the exponential factors, only $H_\nu^{(2)}$ is relevant if $0 < \arg z \leqslant \pi - \delta$, and only $H_\nu^{(1)}$ if $-\pi + \delta \leqslant \arg z < 0$.

However, the asymptotic behavior for $z \to \infty$, $z = x > 0$ is of special interest. Here both $H_\nu^{(1)}$ and $H_\nu^{(2)}$ have the same order of magnitude. It is customary to write the asymptotic formulas in a manner that renders this fact more explicit. Let

$$P_\nu(z) := \frac{1}{2}\sqrt{\frac{z}{2\pi}}\left\{\exp\left[-i\left(z - \frac{\nu\pi}{2} - \frac{\pi}{4}\right)\right]H_\nu^{(1)}(z) + \exp\left[i\left(z - \frac{\nu\pi}{2} - \frac{\pi}{4}\right)\right]H_\nu^{(2)}(z)\right\},$$

$$\tag{11.5-16}$$

$$Q_\nu(z) := \frac{i}{2}\sqrt{\frac{z}{2\pi}}\left\{\exp\left[-i\left(z - \frac{\nu\pi}{2} - \frac{\pi}{4}\right)\right]H_\nu^{(1)}(z) - \exp\left[i\left(z - \frac{\nu\pi}{2} - \frac{\pi}{4}\right)\right]H_\nu^{(2)}(z)\right\}.$$

It follows from (11.5-14) and (11.5-15) that

$$P_\nu(z) \approx \sum_{m=0}^{\infty} (-1)^m \frac{(1/2-\nu)_{2m}(1/2+\nu)_{2m}}{(2m)!(2z)^{2m}},$$

$$Q_\nu(z) \approx \sum_{m=0}^{\infty} (-1)^m \frac{(1/2-\nu)_{2m+1}(1/2+\nu)_{2m+1}}{(2m+1)!(2z)^{2m+1}}, \tag{11.5-17}$$

$$z \to \infty, \qquad z \in \hat{S}_{\pi-\delta}.$$

It is merely a matter of some algebra to verify that

$$J_\nu(z) = \sqrt{\frac{2}{\pi z}}\left\{P_\nu(z)\cos\left(z - \frac{\nu\pi}{2} - \frac{\pi}{4}\right) - Q_\nu(z)\sin\left(z - \frac{\nu\pi}{2} - \frac{\pi}{4}\right)\right\},$$

$$\tag{11.5-18}$$

$$Y_\nu(z) = \sqrt{\frac{2}{\pi z}}\left\{Q_\nu(z)\cos\left(z - \frac{\nu\pi}{2} - \frac{\pi}{4}\right) + P_\nu(z)\sin\left(z - \frac{\nu\pi}{2} - \frac{\pi}{4}\right)\right\}.$$

The expansions (11.5-17) can now be used to calculate numerical values of the Bessel functions for sufficiently large positive values of z, and also to determine the approximate location of the zeros. For many purposes it is sufficient to consider the leading terms of the asymptotic expansions only. For $x \to \infty$, $x > 0$ we have

$$J_\nu(x) = \sqrt{\frac{2}{\pi x}} \left\{ \cos\left(x - \frac{\nu\pi}{2} - \frac{\pi}{4}\right) + O(x^{-1}) \right\},$$

$$Y_\nu(x) = \sqrt{\frac{2}{\pi x}} \left\{ \sin\left(x - \frac{\nu\pi}{2} - \frac{\pi}{4}\right) + O(x^{-1}) \right\}. \tag{11.5-19}$$

From an orthodox numerical point of view, all the foregoing expansions are useless unless strict numerical bounds for the remainder terms can be given. Although the construction of such bounds is implicitly contained in the proof of Theorem 11.5, this construction is probably difficult to carry through in many concrete cases. We return to this question in §12.12-13, where expressions in terms of convergent continued fractions are given for the remainders in many of these expansions.

In the three applications of Theorem 11.5 given thus far the asymptotic series for ϕ was convergent for sufficiently small values of $|t|$. In the following example this series is divergent for all $t \neq 0$.

EXAMPLE 4   **Products of Hankel Functions.**

An integral equivalent to

$$H_\nu^{(1)}(z)H_\nu^{(2)}(z) = \frac{i \exp(\nu\pi i/2)}{\pi} \int_0^\infty e^{-z^2\tau} \tau^{-1} e^{1/2\tau} H_\nu^{(1)}\left(\frac{i}{2\tau}\right) d\tau \tag{11.5-20}$$

is established in Watson's treatise ([1944], p. 439). The integral is of the form (11.5-1) where

$$\phi(t) := \frac{i \exp(i\nu\pi/2)}{\pi} t^{-1} e^{1/2t} H_\nu^{(1)}\left(\frac{i}{2t}\right).$$

By (11.5-14),

$$\phi(t) \approx \frac{2}{\pi^{3/2}} t^{-1/2} {}_2F_0(\tfrac{1}{2} - \nu, \tfrac{1}{2} + \nu; -t), \qquad t \to 0, \qquad t \in \hat{S}_{3\pi/2 - \delta},$$

hence it follows from Theorem 11.5 that

$$H_\nu^{(1)}(z)H_\nu^{(2)}(z) \approx \frac{2}{\pi} \sum_{n=0}^\infty (-1)^n \frac{(1/2-\nu)_n(1/2+\nu)_n(1/2)_n}{n! z^{2n+1}}$$

$$= \frac{2}{\pi z} {}_3F_0\left(\frac{1}{2} - \nu, \frac{1}{2} + \nu, \frac{1}{2}; -\frac{1}{z^2}\right) \tag{11.5-21}$$

for $z \to \infty$, $z^2 \in \hat{S}_{2\pi - \delta}$, that is, $z \in \hat{S}_{\pi - \delta}$. The same result could have been obtained by direct multiplication of the asymptotic series for $H_\nu^{(1)}$ and $H_\nu^{(2)}$, using Dixon's formula (see §1.6, problem 12). It should be noted that the divergence of the series for ϕ

causes the series for $\mathscr{L}\phi$ to diverge more strongly than in the examples previously considered.

PROBLEMS

1. Justify the asymptotic expansion of the function (11.1-4) given in §11.1 by Watson's lemma.

2. The **Fresnel integrals** $C(u)$ and $S(u)$ are for real u defined by

$$C(u) := \int_0^u \cos(t^2)\, dt, \qquad S(u) := \int_0^u \sin(t^2)\, dt.$$

Show that

$$C(u) + iS(u) \approx \frac{(1+i)\sqrt{\pi}}{2\sqrt{2}} - i\frac{e^{iu^2}}{2u} \sum_{n=0}^{\infty} \frac{(1/2)_n}{(iu^2)^n}, \qquad u \to \infty, \qquad u > 0,$$

and describe the curve $z := C(u) + iS(u)$, $-\infty < u < \infty$ (*Cornu spiral*).

3. Deduce from (10.7-15) the asymptotic expansion

$$W_{\kappa,\mu}(z) \approx e^{-z/2} z^{\kappa} \,_2F_0\left(\mu - \kappa + \frac{1}{2}, -\mu - \kappa + \frac{1}{2}; -\frac{1}{z}\right),$$

$$z \to \infty, \qquad z \in \hat{S}_{3\pi/2-\delta}.$$

(11.5-22)

4. The function

$$\psi(z) := \frac{\Gamma'(z)}{\Gamma(z)}$$

can be shown to have the integral representation

$$\psi(z) = \int_0^\infty \left(\frac{e^{-\tau}}{\tau} - \frac{e^{-z\tau}}{1-e^{-\tau}}\right) d\tau, \qquad \operatorname{Re} z > 0.$$

Establish that

$$\psi(z) \approx \operatorname{Log} z + \frac{B_1}{z} - \frac{B_2}{2z^2} + \frac{B_4}{4z^4} - \frac{B_6}{6z^6} - \cdots, \qquad z \to \infty, \qquad z \in \hat{S}_{\pi-\delta},$$

where $B_1, B_2, \ldots$, are the Bernoulli numbers.

5. Let the function erfc_n be defined as in Problem 3, § 11.3, and obtain its asymptotic expansion from the integral representation

$$\operatorname{erfc}_n(z) = \frac{2}{\sqrt{\pi}\, n!} \int_z^\infty (t-z)^n e^{-t^2}\, dt$$

obtained from a well-known formula for repeated integration.

6. Equation (11.5-19) suggests that the Bessel function $J_\nu(x)$ has, for large values of n, zeros close to the points

$$x_n := (n + \tfrac{1}{2}\nu - \tfrac{1}{4})\pi, \qquad n = 1, 2, \ldots.$$

Show how to construct a formal series

$$X = x_n + \sum_{k=1}^{\infty} \frac{a_k}{x_n^{2k-1}}$$

such that, with the definitions of Example 3,

$$P_\nu(X) \cos\left(X - \frac{\nu\pi}{2} - \frac{\pi}{4}\right) - Q_\nu(X) \sin\left(X - \frac{\nu\pi}{2} - \frac{\pi}{4}\right) \approx 0, \qquad x_n \to \infty.$$

In particular establish the leading terms

$$a_1 = -\frac{4\nu^2 - 1}{8}, \qquad a_2 = -\frac{(4\nu^2 - 1)(28\nu^2 - 31)}{384}.$$

[That the series X indeed is an asymptotic expansion for the large zeros is shown by Watson [1944], p. 506.]

7. For $x > 0$ and $n = 0, 1, 2, \ldots$, let

$$E_n(x) := \int_1^\infty s^{-n} e^{-xs}\, ds.$$

Show that for each fixed n,

$$E_n(x) \approx e^{-x} \sum_{k=0}^{\infty} (-1)^k \frac{(n)_k}{x^{k+1}}, \qquad x \to \infty.$$

8. The **Sievert integral** is for $x > 0$ and $0 \le \theta \le \pi/2$ defined by

$$S(x, \theta) := \int_0^\theta e^{-x/\cos\phi}\, d\phi.$$

Show that

$$S\left(x, \frac{\pi}{2}\right) = e^{-x} \int_0^\infty e^{-x\tau} \phi(\tau)\, d\tau,$$

where $\phi(\tau) := (1+\tau)^{-1}(2\tau + \tau^2)^{-1/2}$, and deduce that

$$S\left(x, \frac{\pi}{2}\right) \approx \sqrt{\frac{\pi}{2x}}\, e^{-x} \{1 - \tfrac{5}{8} x^{-1} + \tfrac{129}{128} x^{-2} - \cdots\}, \qquad x \to \infty.$$

Show further that for any fixed $\theta > 0$ this also is the asymptotic expansion of $S(x, \theta)$ for $x \to \infty$.

9. As a special case of (10.7-15), obtain the representation

$$W_{\kappa,\mu}(z) = \frac{z^{\mu+1/2}\, e^{-z/2}}{\Gamma(\mu - \kappa + 1/2)} \int_0^\infty e^{-z\tau} \tau^{\mu-\kappa-1/2}(1+\tau)^{\mu+\kappa-1/2}\, d\tau, \qquad (11.5\text{-}23)$$

and obtain (11.5-22) in a different way. Also, obtain an estimate for the remainder by applying Taylor's formula (with remainder term) to $(1+\tau)^{\mu+\kappa-1/2}$.

§11.6. EXTENSION OF THE LEMMA

In many applications the function whose asymptotic behavior for $z \to \infty$ is to be investigated appears in the form

$$f(z) = \int_{\alpha}^{\beta} e^{-z\psi(\sigma)} \, d\sigma$$

or, more generally,

$$f(z) = \int_{\alpha}^{\beta} e^{-z\psi(\sigma)} \theta(\sigma) \, d\sigma. \tag{11.6-1}$$

Here ψ is a real function that increases monotonically on (α, β), $\psi(\sigma) \to \infty$ for $\sigma \to \beta$, and θ is a possibly complex valued function whose growth is suitably restricted so as to make the integral convergent at least for Re z sufficiently large.

Without loss of generality, we may assume that $\psi(\alpha) = 0$, for if this is not the case, we can divide the integral by $e^{-z\psi(\alpha)}$. Furthermore, by a mere shift of the independent variable the lower limit of integration can be assumed to be zero.

By substituting a new variable of integration, integrals such as the foregoing are easily reduced to a form in which the Watson–Doetsch lemma is applicable. We assume that ψ' exists and is positive on $(0, \beta)$. Then by letting $\psi(\sigma) := \tau$ in (11.6-1) we get

$$f(z) = \int_{0}^{\infty} e^{-z\tau} \theta(\psi^{[-1]}(\tau)) \psi^{[-1]\prime} \, d\tau, \tag{11.6-2}$$

where $\psi^{[-1]}$ denotes the inverse function of ψ. If the required conditions are satisfied, Theorem 11.5 thus becomes applicable with

$$\phi := \theta \circ \psi^{[-1]} \cdot \psi^{[-1]\prime},$$

which by virtue of $\psi^{[-1]\prime} = \psi'^{-1} \circ \psi^{[-1]}$ and by the associative law of composition is the same as

$$\phi = (\theta \psi'^{-1}) \circ \psi^{[-1]}. \tag{11.6-3}$$

The Lagrange–Bürmann formula affords an easy way to determine the coefficients of the required asymptotic power series for ϕ as $\tau \to 0$ if asymptotic power series for ψ and θ are known. We assume

$$\psi(\sigma) \approx \Psi = a_k \sigma^k + a_{k+1} \sigma^{k+1} + \cdots, \qquad \sigma \to 0, \qquad \sigma > 0,$$

where k is chosen so that $a_k \neq 0$. Because ψ grows monotonically, we have $a_k > 0$. Furthermore, we let

$$\theta(\sigma) \approx \Theta = b_0 + b_1 \sigma + b_2 \sigma^2 + \cdots, \qquad \sigma \to 0, \qquad \sigma > 0.$$

We then have

THEOREM 11.6a

If, in addition to the foregoing hypotheses, ψ' possesses an asymptotic power series as $\sigma \to 0$, $\sigma > 0$, then

$$(\theta\psi'^{-1}) \circ \psi^{[-1]}(\tau) \approx \frac{1}{k}\tau^{1/k-1} \sum_{m=0}^{\infty} \mathrm{res}(\Theta\Psi^{-(m+1)/k})\tau^{m/k}, \qquad \tau \to 0, \qquad \tau > 0.$$

$$(11.6\text{-}4)$$

It is understood that $\Psi^{-(1+m)/k} := (\Psi^{1/k})^{-1-m}$, where

$$\Psi^{1/k} := a_k^{1/k}\sigma\left(1 + \frac{a_{k+1}}{a_k}\sigma + \cdots\right)^{1/k}, \qquad (11.6\text{-}5)$$

$a_k^{1/k} > 0$. If Z is a formal Laurent series in x, say, res Z denotes, as usual, the coefficient of x^{-1}.

Proof of (11.6-4). Let ϕ be defined by (11.6-3). The assertion of the theorem is equivalent to the assertion that the function

$$\chi(v) := kv^{k-1}\phi(v^k)$$

admits the asymptotic expansion

$$\chi(v) \approx \sum_{m=0}^{\infty} \mathrm{res}(\Theta\Psi^{-(m+1)/k})v^m, \qquad v \to 0, \qquad v > 0.$$

By the distributive law of composition and by the formula for the derivative of the inverse function,

$$\phi = (\theta \circ \psi^{[-1]})(\psi^{-1} \circ \psi^{[-1]}) = (\theta \circ \psi^{[-1]})\psi^{[-1]\prime}.$$

Thus if π denotes the kth power function, $\pi(u) := u^k$ $(u \geq 0)$.

$$\chi = (\phi \circ \pi)\pi' = (\theta \circ \psi^{[-1]} \circ \pi)(\psi^{[-1]\prime} \circ \pi)\pi'.$$

Because the range of ψ is $[0, \infty)$, the function

$$\omega := \psi^{1/k} = \pi^{[-1]} \circ \psi$$

is well defined, and

$$\omega^{[-1]} = \psi^{[-1]} \circ \pi$$

yields

$$\chi = (\theta \circ \omega^{[-1]})\omega'. \qquad (11.6\text{-}6)$$

Let $\Omega := \Psi^{1/k}$ as defined in (11.6-5). By Problem 1, §11.2, and by Theorem 11.3e,

$$\omega(\sigma) \approx \Omega, \qquad \omega'(\sigma) \approx \Omega', \qquad \sigma \to 0, \qquad \sigma > 0$$

and by Problem 2,

$$\omega^{[-1]}(v) \approx \Omega^{[-1]}, \qquad v \to 0, \qquad v > 0.$$

By Theorem 11.2b (or its analog, in which v is substituted for z^{-1})

$$\chi(v) = (\theta\omega'^{-1}) \circ \omega^{[-1]}(v) \approx (\Theta\Omega'^{-1}) \circ \Omega^{[-1]}.$$

According to one version of the Lagrange–Bürmann theorem (Corollary 1.9c),

$$(\Theta\Omega'^{-1}) \circ \Omega^{[-1]} = \sum_{m=0}^{\infty} \mathrm{res}(\Theta\Omega^{-m-1})v^m,$$

establishing (11.6-4). ∎

Theorem 11.5 now yields immediately:

THEOREM 11.6b

Let the functions ψ and θ be such that $\phi := (\theta\psi'^{-1}) \circ \psi^{[-1]}$ satisfies the conditions (i) *of §11.5. Then the function f defined by* (11.6-1) *admits for every $\delta > 0$ the asymptotic power series*

$$f(z) \approx \frac{1}{k} \sum_{m=0}^{\infty} \mathrm{res}(\Theta\Psi^{-(m+1)/k}) \frac{\Gamma((m+1)/k)}{z^{(m+1)/k}}, \qquad z \to \infty, \qquad z \in \hat{S}_{\pi/2-\delta}.$$

$$(11.6\text{-}7)$$

We illustrate this generalized version of the Watson–Doetsch lemma by several examples.

EXAMPLE 1

Let

$$f(z) := \int_0^{\infty} e^{-z\sigma e^{\sigma}} \sigma e^{\sigma} \, d\sigma.$$

Here we have

$$\psi(\sigma) = \theta(\sigma) = \sigma e^{\sigma};$$

hence

$$\Psi = \Theta = \sigma e^{\sigma} = \sigma + \frac{\sigma^2}{1!} + \frac{\sigma^3}{2!} + \cdots.$$

Because $\theta \psi'^{-1}(\sigma) = \sigma/(1+\sigma)$, the function ϕ is bounded and certainly satisfies the conditions (i). Because $k = 1$,

$$\text{res}(\Theta \Psi^{-(1+m)/k}) = \text{res}(\Theta^{-m}) = \text{res}(\sigma^{-m} e^{-m\sigma}) = (-1)^{m-1} \frac{m^{m-1}}{(m-1)!},$$

$m = 1, 2, \ldots$; for $m = 0$ the residue is zero. Formula (11.6-7) thus yields

$$f(z) \approx \frac{1}{z} \sum_{m=1}^{\infty} (-1)^{m-1} \frac{m^{m-1}}{(m-1)!} m! z^{-m}$$

$$= \frac{1^1}{z^2} - \frac{2^2}{z^3} + \frac{3^3}{z^4} - \frac{4^4}{z^5} + \cdots, \qquad z \to \infty, \qquad z \in \hat{S}_{\pi/2-\delta}.$$

EXAMPLE 2   **Stirling's Formula.**

Theorem 11.6b affords still another approach to the asymptotic expansion of the Γ function. We start from Euler's integral,

$$\Gamma(z) = \int_0^{\infty} e^{-\tau} \tau^{z-1} \, d\tau, \qquad \text{Re } z > 0.$$

By the basic functional relation we also have

$$\Gamma(z) = z^{-1} \Gamma(z+1) = z^{-1} \int_0^{\infty} e^{-\tau} \tau^z \, d\tau.$$

Here we make the substitution $\tau = z\sigma$. Because the integrand vanishes exponentially at ∞, we may after a rotation of the path of integration again integrate along the real axis. This yields

$$\Gamma(z) = \int_0^{\infty} e^{-z\sigma} (z\sigma)^z \, d\sigma,$$

(principal values), and

$$f(z) := z^{-z} e^z \Gamma(z) = \int_0^{\infty} (\sigma e^{1-\sigma})^z \, d\sigma.$$

Clearly, this is of the form

$$f(z) = \int_0^{\infty} e^{-z\psi(\sigma)} \, d\sigma,$$

where

$$\psi(\sigma) := \sigma - 1 - \text{Log } \sigma.$$

The function ψ does not satisfy the hypotheses of Theorem 11.6b, because it does not increase monotonically. However, because

$$\lim_{\sigma \to 0+} \psi(\sigma) = \infty, \qquad \lim_{\sigma \to \infty} \psi(\sigma) = \infty, \qquad \psi(1) = 0, \quad \text{and} \quad \psi'(\sigma) = 1 - \frac{1}{\sigma},$$

we find that ψ decreases monotonically from ∞ to 0 on $(0, 1)$ and increases monotonically from 0 to ∞ on $(1, \infty)$. Thus we may write

$$f(z) = \int_0^1 \exp[-z\psi_1(\sigma)] \, d\sigma + \int_0^\infty \exp[-z\psi_2(\sigma)] \, d\sigma,$$

where

$$\psi_1(\sigma) := \psi(1-\sigma) = -\sigma - \text{Log}(1-\sigma),$$

$$\psi_2(\sigma) := \psi(1+\sigma) = \sigma - \text{Log}(1+\sigma).$$

Clearly, these functions satisfy the required conditions. We have

$$\psi_1(\sigma) \approx \Psi_1 = \frac{\sigma^2}{2} + \frac{\sigma^3}{3} + \frac{\sigma^4}{4} + \cdots,$$

$$\psi_2(\sigma) \approx \Psi_2 = \frac{\sigma^2}{2} - \frac{\sigma^3}{3} + \frac{\sigma^4}{4} - \cdots,$$

$\sigma \to 0$, $\sigma > 0$; hence $k = 2$; furthermore, $\theta = 1 \approx 1$. Because $\Psi_2 = \Psi_1 \circ (-X)$,

$$\text{res}(\Psi_2^{-(1+m)/2}) = (-1)^m \, \text{res}(\Psi_1^{-(1+m)/2}).$$

The odd terms in the asymptotic expansion thus cancel, and putting $m = 2q$, $\Gamma(\tfrac{1}{2}+q) = \sqrt{\pi}(\tfrac{1}{2})_q$, we find

$$\Gamma(z) \approx z^z \, e^{-z} \sqrt{\pi} \sum_{q=0}^\infty \text{res}(\Psi^{-1/2-q})(\tfrac{1}{2})_q z^{-1/2-q}, \qquad z \to \infty, \qquad z \in \hat{S}_{\pi-\delta}$$

$$(11.6\text{-}8)$$

where

$$\Psi := \frac{s^2}{2} + \frac{s^3}{3} + \frac{s^4}{4} + \cdots.$$

This, at last, is a "closed formula" for the asymptotic expansion of Γ.

The residues are computed easily by means of the J. C. P. Miller formula (Theorem 1.6c), and we find

$$\Gamma(z) \approx \sqrt{2\pi} \, z^{z-1/2} \, e^{-z} \{1 + \tfrac{1}{12}z^{-1} + \tfrac{1}{288}z^{-2} + \cdots\},$$

in agreement with (11.3-13).

EXAMPLE **3** **The Legendre polynomials $P_n(x)$ for $x > 1$.**

To examine the asymptotic behavior of these polynomials for $n \to \infty$, when $x > 1$ is fixed, we start from the integral representation (see Chapter 18 or Szegö [1959], p. 88)

$$P_n(x) = \frac{1}{\pi} \int_0^\pi (x + x' \cos \alpha)^n \, d\alpha, \qquad n = 0, 1, 2, \ldots, \qquad (11.6\text{-}9)$$

where $x' := \sqrt{x^2-1}$ for brevity. To apply Theorem 11.6b, we put

$$\xi := \frac{x'}{x+x}$$

and can write

$$P_n(x) = \frac{(x+x')^n}{\pi} \int_0^\pi [1-\xi(1-\cos\alpha)]^n \, d\alpha.$$

The integral is of the form (11.6-1), with z replaced by the discrete variable n, $\theta = 1$, and

$$\psi(\alpha) := -\text{Log}[1-\xi(1-\cos\alpha)], \qquad 0 \le \alpha \le \pi.$$

From

$$\psi(\alpha) = -\text{Log}\left[1-\xi\left(\frac{\alpha^2}{2}-\frac{\alpha^4}{24}+\frac{\alpha^6}{720}-\cdots\right)\right]$$

we find by expanding the logarithm and rearranging

$$\psi(\alpha) \approx \Psi = p_2\alpha^2 + p_4\alpha^4 + \cdots, \qquad \alpha \to 0,$$

where p_{2k} is a polynomial of degree k in ξ,

$$p_2(\xi) = \tfrac{1}{2}\xi,$$

$$p_4(\xi) = \tfrac{1}{8}\xi^2 - \tfrac{1}{24}\xi,$$

$$p_6(\xi) = \tfrac{1}{24}\xi^3 - \tfrac{1}{48}\xi^2 + \tfrac{1}{720}\xi, \ldots.$$

Evidently, $k = 2$. Because Ψ contains only even powers, the powers appearing in $\Psi^{-(1+m)/2}$ have the same parity as $1+m$, and the residue is therefore different from zero only for even values of m, $m = 2q$. By virtue of $\Gamma(\tfrac{1}{2}+q) = \sqrt{\pi}(\tfrac{1}{2})_q$ there follows

$$P_n(x) \approx \frac{(x+x')^n}{2\sqrt{\pi n}} \sum_{q=0}^\infty \text{res}(\Psi^{-q-1/2})(\tfrac{1}{2})_q n^{-q}, \qquad n \to \infty. \qquad (11.6\text{-}10)$$

Straightforward computation yields

$$\text{res}(\Psi^{-1/2}) = \sqrt{\frac{2}{\xi}}, \quad \text{res}(\Psi^{-3/2}) = \sqrt{\frac{2}{\xi}}\frac{1-3\xi}{4\xi};$$

thus the leading terms of the expansion are given by

$$P_n(x) \approx \frac{(x+x')^{n+1/2}}{\sqrt{2\pi n x'}}\left\{1 + \frac{x-2x'}{8x'n} + O(n^{-2})\right\}, \qquad n \to \infty.$$

This result was derived for x real, $x > 1$. It can be shown, however, that (11.6-10) also holds when x is replaced by any complex number outside the real interval $[-1, 1]$; see (11.9-14). For real values of x such that $-1 \le x \le 1$, the asymptotic behavior of $P_n(x)$ for n large is best studied by the method of Darboux; see §11.10.

PROBLEMS

1. Let $\operatorname{Re}(b-a)>0$. Use Euler's beta integral in the form

$$\frac{\Gamma(z+a)}{\Gamma(z+b)}=\frac{1}{\Gamma(b-a)}\int_0^1 \tau^{z+a-1}(1-\tau)^{b-a-1}\,d\tau$$

to obtain an asymptotic series

$$\frac{\Gamma(z+a)}{\Gamma(z+b)}\approx z^{a-b}\left\{1+\frac{(1-a-b)(b-a)}{2z}+\cdots\right\},\qquad z\to\infty,\qquad z\in S_{\pi-\delta}.$$

2. Let $E_n(x)$ be defined as in Problem 7, §11.5. Show that there exist polynomials $b_m(x)$ of degree m, $m=0, 1, 2, \ldots$, such that for fixed $x>0$

$$E_n(x)\approx e^{-x}\sum_{m=0}^\infty \frac{b_m(x)}{n^{m+1}},\qquad n\to\infty,$$

where, in particular,

$$b_0(x)=1,\qquad b_1(x)=1-x,\qquad b_2(x)=\frac{1-3x+x^2}{2}.$$

3. The coefficients in Adams' formula of numerical integration are given by

$$\gamma_n := \int_0^1 \frac{(\sigma)_n}{n!}\,d\sigma,\qquad n=0, 1, 2, \ldots.$$

Show that

$$\gamma_n = \frac{1}{\operatorname{Log} n}+O((\operatorname{Log} n)^{-2}),\qquad n\to\infty.$$

(Sidney Spital.)

4. By considering leading terms of appropriate asymptotic expansions, show that

$$\lim_{x\to\infty}\frac{1}{\Gamma(\lambda x)}\int_x^\infty e^{-\tau}\tau^{\lambda x-1}\,d\tau=\begin{cases}0, & 0<\lambda<1,\\ \frac12, & \lambda=1,\\ 1, & \lambda>1.\end{cases}$$

(John Whittlesey.)

5. Determine the asymptotic behavior of the modified Bessel function $I_n(z) := i^{-n}J_n(iz)$, $n=0, 1, 2, \ldots$, as $z\to\infty$, $z\in \hat{S}_{\pi/2-\delta}$ from the integral representation

$$I_n(z)=\frac{1}{2\pi}\int_0^{2\pi} e^{z\cos\tau}\cos n\tau\,d\tau.$$

6. Let θ_n be defined by

$$1+\frac{n}{1!}+\frac{n^2}{2!}+\cdots+\frac{n^{n-1}}{(n-1)!}+\frac{n^n}{n!}\theta_n=\frac12 e^n,\qquad n=1, 2, \ldots.$$

Show that

$$\theta_n = 1 + \frac{n}{2}\left\{\int_0^1 e^{-n\psi_1(\sigma)}\, d\sigma - \int_0^\infty e^{-n\psi_2(\sigma)}\, d\sigma\right\},$$

where ψ_1 and ψ_2 are defined as in Example 2, and deduce that

$$\theta_n \approx 1 + \tfrac{1}{2}n \sum_{m=0}^\infty \mathrm{res}(\Psi^{-m-1})m!n^{-m}$$

$$= \tfrac{1}{3} + \tfrac{4}{135}n^{-1} + \cdots, \qquad n \to \infty.$$

(Ramanujan.)

7. In one form of chess match, $2n$ games are played. Wins count 1 point each, draws $\tfrac{1}{2}$, losses are worth 0. To win the match, the defender needs to score at least n, whereas the challenger must achieve at least $n + \tfrac{1}{2}$. We suppose that the two players are of equal strength and that the probability of a draw is a constant δ. The probability of the defender's keeping his title then is $\tfrac{1}{2}(1 + \alpha_n)$, where

$$\alpha_n(\delta) := \sum_{k=0}^n \frac{(2n)!}{(k!)^2(2n-2k)!}\, \delta^{2n-2k}\left(\frac{1-\delta}{2}\right)^{2k}.$$

(a) Study the asymptotic behavior of $\pi_n(\delta)$ as $n \to \infty$, paying particular attention to small values of δ.

(b) Show that π_n is not a monotonic function of δ, that the unique point δ_n where $\pi_n(\delta)$ assumes its minimum satisfies

$$\delta_n = \frac{\mathrm{Log}\, 8n}{4n} - \left(\frac{\mathrm{Log}\, 8n}{4n}\right)^2 + O(n^{-2}), \qquad n \to \infty,$$

and that

$$\frac{\alpha_n(\delta_n)}{\alpha_n(0)} \to \frac{1}{2}, \qquad n \to \infty.$$

(c) Show that the unique point $\delta_n^* > 0$ such that $\pi_n(\delta_n^*) = \pi_n(0)$ satisfies

$$\delta_n^* = \frac{3}{4} - \frac{3}{32n} + O(n^{-2}), \qquad n \to \infty.$$

[Use the integral representation

$$\alpha_n(\delta) = \frac{1}{\pi}\int_0^\pi [\delta + (1-\delta)\cos\phi]^{2n}\, d\phi.]$$

8. Let n and m be integers, $0 < n \leqslant m$. The numbers

$$\delta(m, n) := \frac{1}{2}\left(\frac{n-1}{m} + \frac{n-1}{m}\frac{n-2}{m} + \frac{n-1}{m}\frac{n-2}{m}\frac{n-3}{m} + \cdots\right)$$

are of interest in the combinatorics of computer science, especially in the theory of hashing (see Knuth [1974]). Study the behavior of $\delta(m, \alpha m)$ as $m \to \infty$. In particular, show that for $0 < \alpha < 1$

$$\delta(m, \alpha m) = \frac{\alpha}{2(1-\alpha)} + O(m^{-1}),$$

and that

$$\delta(m, m) = \sqrt{\frac{\pi m}{8}} + O(1).$$

Also determine the asymptotic expansions of which the foregoing are the leading terms.
[Use

$$\delta(m, n+1) = \frac{1}{2} \int_0^\infty \left[\left(1 + \frac{\tau}{m} \right)^n - 1 \right] e^{-\tau} \, d\tau. \right]$$

§11.7. ASYMPTOTIC FORMULAS; LAPLACE'S METHOD

Let S be an unbounded set in the complex plane, let the functions f and g be defined on S, and let $g(z) \neq 0$ for sufficiently large values of $|z|$. If

$$\lim_{\substack{z \to \infty \\ z \in S}} \frac{f(z)}{g(z)} = 1,$$

then one also writes

$$f(z) \sim g(z), \qquad z \to \infty, \qquad z \in S \tag{11.7-1}$$

and calls (11.7-1) an **asymptotic formula** for f. The leading term of every asymptotic power series furnishes an example of an asymptotic formula. For instance, from (11.5-14) we can extract the asymptotic formula

$$H_\nu^{(1)}(z) \sim \sqrt{\frac{2}{\pi z}} \exp\left[i\left(z - \frac{\nu \pi}{2} - \frac{\pi}{4} \right) \right], \qquad z \to \infty, \qquad -iz \in \hat{S}_{3\pi/2 - \delta},$$

and from Stirling's expansion,

$$\Gamma(z) \sim \sqrt{2\pi} z^{z-1/2} e^{-z}, \qquad z \to \infty, \qquad z \in \hat{S}_{\pi - \delta}.$$

An asymptotic formula is a very much cruder tool for approximating a function for large values of the variable than an asymptotic power series. Yet if the mere qualitative behavior of a function is at issue, nothing more than an asymptotic formula may be required to settle the matter.

In some of the examples of the preceding section the algebraic machinery required to obtain complete asymptotic expansions tends to obscure a simple type of reasoning that is often sufficient to obtain the leading terms of the asymptotic series, that is, asymptotic formulas. The idea is due to Laplace; it first is described in intuitive terms.

Let $I := [\alpha, \beta]$ a finite or infinite interval, and let ψ and θ be two real-valued functions defined on I. We assume that $|\theta| \, e^{-x\psi}$ is integrable on I for x sufficiently large and wish to study the asymptotic behavior of the function

$$f(x) := \int_{\alpha}^{\beta} e^{-x\psi(\sigma)} \theta(\sigma) \, d\sigma \qquad (11.7\text{-}2)$$

for $x \to \infty$.

Contrary to the situation encountered at the beginning of §11.6, we now assume that the function ψ assumes its minimum value at an *interior* point γ of I, and that for every $\delta > 0$

$$\inf_{\substack{\sigma \in I \\ |\sigma - \gamma| \geq \delta}} \psi(\sigma) > \psi(\gamma).$$

For large values of x the function $e^{-x\psi(\sigma)}$ then has a sharp maximum at $\sigma = \gamma$. Only the vicinity of γ, then, will yield a significant contribution to the integral (11.7-2), and the asymptotic behavior of f will be determined by the behavior of ψ and θ in the vicinity of γ. Assuming that ψ is twice continuously differentiable in a neighborhood of γ, $\psi''(\gamma) > 0$, we approximate ψ near γ by its Taylor polynomial of degree 2,

$$\psi(\gamma) + \frac{(\sigma - \gamma)^2}{2} \psi''(\gamma).$$

If θ is continuous at γ and $\theta(\gamma) \neq 0$, the integral is then approximated by

$$\int_{\gamma - \delta}^{\gamma + \delta} \theta(\gamma) \exp\left\{ -x\left[\psi(\gamma) + \frac{(\sigma - \gamma)^2}{2} \psi''(\gamma) \right] \right\} d\sigma$$

$$= \theta(\gamma) \, e^{-x\psi(\gamma)} \int_{\gamma - \delta}^{\gamma + \delta} \exp\left[-\frac{x}{2}(\sigma - \gamma)^2 \psi''(\gamma) \right] d\sigma,$$

where δ is a suitably chosen small positive number. Because the behavior of ψ away from γ is immaterial, we subsequently replace the limits of integration by $-\infty$ and ∞. In view of the well-known result

$$\int_{-\infty}^{\infty} e^{-\kappa \sigma^2} \, d\sigma = \sqrt{\frac{\pi}{\kappa}}, \qquad \kappa > 0,$$

it is to be expected that

$$f(x) \sim \theta(\gamma) \, e^{-x\psi(\gamma)} \sqrt{\frac{2\pi}{x\psi''(\gamma)}}, \qquad x \to \infty.$$

EXAMPLE **1**

In the integral

$$x^{-x} e^x \Gamma(x) = \int_0^\infty e^{-x[\sigma - 1 - \text{Log}\,\sigma]} \, d\sigma$$

(see example **2** of §11.6) we have $\psi(\sigma) := \sigma - 1 - \text{Log }\sigma$, $\theta(\sigma) := 1$, $\gamma = 1$, $\psi(\gamma) = 0$, $\psi''(\gamma) = 1$, and the foregoing reasoning yields the correct result

$$x^{-x} e^x \Gamma(x) \sim \sqrt{\frac{2\pi}{x}}, \qquad x \to \infty.$$

We now justify the intuitive argument by establishing a multidimensional generalization of it.

Let S be a possibly unbounded region in euclidean n space $\mathbb{R}^n$ whose points we identify with the column vectors $\mathbf{s} = (s_1, s_2, \ldots, s_n)^T$. Let θ and ψ be real-valued functions defined on S, and let there exist a real number ξ such that $|\theta| e^{-x\psi}$ is integrable over S for $x \geq \xi$. The function

$$f(x) := \int_S \theta(\mathbf{s}) e^{-x\psi(\mathbf{s})} \, dv, \tag{11.7-3}$$

where $dv := ds_1 \cdots ds_n$, is then defined for $x \geq \xi$.

We assume that the function ψ assumes its infimum at an interior point of S, which we call $\mathbf{c}$, and that

$$\mu(\delta) := \inf_{\substack{\mathbf{s} \in S \\ \|\mathbf{s} - \mathbf{c}\| \geq \delta}} (\psi(\mathbf{s}) - \psi(\mathbf{c})) > 0 \tag{11.7-4}$$

for every $\delta > 0$. We also assume that the second partial derivatives of ψ exist and are continuous in a neighborhood of $\mathbf{c}$, and that the Hessean matrix of ψ at $\mathbf{c}$,

$$\mathbf{H} := \left(\frac{\partial^2 \psi}{\partial s_i \, \partial s_j}(\mathbf{c}) \right),$$

is positive definite. Finally, we assume

$$\theta(\mathbf{c}) \neq 0.$$

We then have:

THEOREM 11.7a

Under the above hypotheses, the function f defined by (11.7-3) satisfies

$$f(x) \sim \theta(\mathbf{c})(\det \mathbf{H})^{-1/2} e^{-x\psi(\mathbf{c})} \left(\frac{2\pi}{x} \right)^{n/2}, \qquad x \to \infty. \tag{11.7-5}$$

Proof. It evidently suffices to prove the assertion in the normalized situation where

$$\theta(\mathbf{c}) = 1, \qquad \psi(\mathbf{c}) = 0, \qquad \mathbf{c} = \mathbf{0},$$

to which the general case can be reduced by considering the functions $\theta(\mathbf{s})/\theta(\mathbf{c})$, $\psi(\mathbf{s}) - \psi(\mathbf{c})$ and the variable $\mathbf{s} - \mathbf{c}$. In the normalized situation, the formula

$$f(x) \sim (\det \mathbf{H})^{-1/2} \left(\frac{2\pi}{x}\right)^{n/2}, \qquad x \to \infty,$$

is to be proved.

For $\delta > 0$, let $S_\delta : \|\mathbf{s}\| < \delta$. Let δ be so small that $S_\delta \subset S$. We then clearly have

$$f(x) = I_1(\delta, x) + I_2(\delta, x),$$

where

$$I_1(\delta, x) := \int_{S_\delta} \theta(\mathbf{s})\, e^{-x\psi(\mathbf{s})}\, dv,$$

$$I_2(\delta, x) := \int_{S \setminus S_\delta} \theta(\mathbf{s})\, e^{-x\psi(\mathbf{s})}\, dv.$$

To estimate I_2, let

$$\kappa := \int_S |\theta(\mathbf{s})| e^{-\xi\psi(\mathbf{s})}\, dv,$$

which by hypothesis is finite. We then have

$$|I_2(\delta, x)| \leq \int_{S \setminus S_\delta} |\theta(\mathbf{s})| e^{-\xi\psi(\mathbf{s}) - (x-\xi)\psi(\mathbf{s})}\, dv$$

$$\leq e^{-(x-\xi)\mu(\delta)} \int_{S \setminus S_\delta} |\theta(\mathbf{s})| e^{-\xi\psi(\mathbf{s})}\, dv$$

$$\leq e^{-(x-\xi)\mu(\delta)} \kappa,$$

where, by hypothesis,

$$\mu(\delta) = \inf_{|\mathbf{s}| \geq \delta} \psi(\mathbf{s}) > 0.$$

Turning to I_1, we write for $\mathbf{s} \neq \mathbf{0}$

$$2\psi(\mathbf{s}) = \mathbf{s}^T \mathbf{H} \mathbf{s} + \eta(\mathbf{s})\mathbf{s}^T \mathbf{s}.$$

If for $\delta > 0$ we set

$$\eta(\delta) := \sup_{|\mathbf{s}| \leq \delta} |\eta(\mathbf{s})|,$$

it follows from the differentiability hypothesis on ψ that $\eta(\delta)$ tends to zero with δ. Similarly, if

$$\theta^+(\delta) := \sup_{\|\mathbf{s}\|\leqslant\delta} \theta(\mathbf{s}),$$

$$\theta^-(\mathbf{s}) := \inf_{\|\mathbf{s}\|\leqslant\delta} \theta(\mathbf{s}),$$

we have by the assumed continuity of θ

$$\lim_{\delta\to0} \theta^+(\delta) = \lim_{\delta\to0} \theta^-(\delta) = 1.$$

With these notations, I_1 is contained between the lower bound

$$\theta^-(\delta) \int_{S_\delta} \exp\left[-\frac{x}{2}(\mathbf{s}^T\mathbf{Hs} + \eta(\delta)\mathbf{s}^T\mathbf{s}) \right] dv$$

and the upper bound

$$\theta^+(\delta) \int_{S_\delta} \exp\left[-\frac{x}{2}(\mathbf{s}^T\mathbf{Hs} - \eta(\delta)\mathbf{s}^T\mathbf{s}) \right] dv.$$

We now require the following lemma, whose proof we postpone until after completing the proof of the theorem.

LEMMA 11.7b

Let the nth order matrix $\mathbf{A}$ be real, symmetric, and positive definite. Then

$$\int_{\mathbb{R}^n} \exp(-\mathbf{s}^T\mathbf{As}) \, dv = \frac{\pi^{n/2}}{(\det \mathbf{A})^{1/2}}. \tag{11.7-6}$$

Let δ be so small that the smallest eigenvalue of the matrix $\mathbf{H}-\eta(\delta)\mathbf{I}$ is already positive. It then follows from the lemma that I_1 is contained between the lower bound

$$\theta^-(\delta)\left\{ [\det(\mathbf{H}+\eta(\delta)\mathbf{I})]^{-1/2}\left(\frac{2\pi}{x}\right)^{n/2} - I_3(\delta, x) \right\},$$

where

$$I_3(\delta, x) := \int_{\mathbb{R}^n \setminus S_\delta} \exp\left[-\frac{x}{2}(\mathbf{s}^T\mathbf{Hs} + \eta(\delta)\mathbf{s}^T\mathbf{s}) \right] dv,$$

and the upper bound

$$\theta^+(\delta)[\det(\mathbf{H}-\eta(\delta)\mathbf{I})]^{-1/2}\left(\frac{2\pi}{x}\right)^{n/2}.$$

We set

$$g(x) := \sqrt{\det \mathbf{H}} \left(\frac{x}{2\pi}\right)^{n/2} f(x);$$

the assertion to be proved is equivalent to

$$g(x) \to 1, \qquad x \to \infty.$$

Our estimates for I_1 and I_2 so far show that for all sufficiently small values of δ, $g(x)$ lies between the lower bound

$$\theta^-(\delta)\left[\frac{\det \mathbf{H}}{\det(\mathbf{H} + \eta(\delta)\mathbf{I})}\right]^{1/2} - \sqrt{\det \mathbf{H}} \left(\frac{x}{2\pi}\right)^{n/2} I_3(\delta, x)$$

$$- \sqrt{\det \mathbf{H}} \left(\frac{x}{2\pi}\right)^{n/2} \kappa e^{-\mu(\delta)x}$$

and the upper bound

$$\theta^+(\delta)\left[\frac{\det \mathbf{H}}{\det(\mathbf{H} - \eta(\delta)\mathbf{I})}\right]^{1/2} + \sqrt{\det \mathbf{H}} \left(\frac{x}{2\pi}\right)^{n/2} \kappa e^{-\mu(\delta)x}.$$

Let $\epsilon > 0$ be given, and choose δ such that the numbers

$$\theta^\pm(\delta)\left[\frac{\det \mathbf{H}}{\det(\mathbf{H} \mp \eta(\delta)\mathbf{I})}\right]^{1/2}$$

lie between $1 \mp \epsilon/2$. It is clear that for this δ there exists x_0 such that $x > x_0$ implies

$$\sqrt{\det \mathbf{H}} \left(\frac{x}{2\pi}\right)^{n/2} \kappa e^{-\mu(\delta)x} < \frac{\epsilon}{4}.$$

It remains to be shown that for x sufficiently large also the remaining term

$$\sqrt{\det \mathbf{H}} \left(\frac{x}{2\pi}\right)^{n/2} I_3(\delta, x) \tag{11.7-7}$$

becomes $< \epsilon/4$. If α denotes the smallest eigenvalue of $\mathbf{H} - \eta(\delta)\mathbf{I}$, which by assumption is positive, we have

$$I_3(\delta, x) \le \int_{\mathbb{R}^n \setminus S_\delta} \exp\left(-\frac{x}{2}\alpha \mathbf{s}^T \mathbf{s}\right) dv.$$

We evaluate the last integral by introducing n-dimensional spherical coordinates. Let σ_n denote the surface area of the sphere $\|\mathbf{s}\| = 1$ (see Problem 2, §8.7). Then the integral equals

$$\sigma_n \int_\delta^\infty \exp\left(-\frac{x}{2}\alpha\rho^2\right)\rho^{n-1} d\rho.$$

Letting $\rho^2 = \delta^2 + \tau$, this becomes

$$\frac{1}{2}\sigma_n\delta^{n-2}\exp\left(-\frac{x}{2}\alpha\delta^2\right)\int_0^\infty \exp\left(-\frac{x}{2}\alpha\tau\right)\left(1+\frac{\tau}{\delta^2}\right)^{n-2} d\tau,$$

which by Watson's lemma is asymptotic to

$$\sigma_n\delta^{n-2}\exp\left(-\frac{x}{2}\alpha\delta^2\right)(x\alpha)^{-1}, \qquad x \to \infty.$$

It thus clearly follows that, for the fixed δ already chosen, the term (11.7-7) tends to zero for $x \to \infty$, as required. Thus for all sufficiently large values of x, $g(x)$ lies between the bounds $1 \pm \epsilon$, as required.

To complete the proof of Theorem 11.7a, it remains to establish the lemma. Let $\mathbf{A}^{1/2}$ denote the positive definite square root of $\mathbf{A}$. (If $\mathbf{A} = \mathbf{U}\Lambda\mathbf{U}^T$, where $\mathbf{U}$ is orthogonal and Λ is the diagonal matrix formed with the eigenvalues of $\mathbf{A}$, then $\mathbf{A}^{1/2} := \mathbf{U}\Lambda^{1/2}\mathbf{U}^T$, where $\Lambda^{1/2}$ is the diagonal matrix whose diagonal elements are the positive square roots of the diagonal elements of Λ.) Introducing the new variable of integration $\mathbf{t} := \mathbf{A}^{1/2}\mathbf{s}$, the integral becomes

$$\int_{\mathbb{R}^n} \exp(-\mathbf{s}^T\mathbf{A}\mathbf{s})\, ds_1 \cdots ds_n = \int_{\mathbb{R}^n} \exp(-\mathbf{t}^T\mathbf{t})\Delta(\mathbf{t})\, dt_1 \cdots dt_n,$$

where Δ is the Jacobian determinant of the transformation,

$$\Delta(\mathbf{t}) = \det\left(\frac{\partial s_i}{\partial t_j}\right) = \det(\mathbf{A}^{-1/2}) = (\det \mathbf{A})^{-1/2},$$

which is constant and can be pulled outside the integral. There remains

$$\int_{-\infty}^\infty \cdots \int_{-\infty}^\infty \exp(-t_1^2 - \cdots - t_n^2)\, dt_1 \cdots dt_n = \left\{\int_{-\infty}^\infty e^{-t^2}\, dt\right\}^n = \pi^{n/2},$$

which establishes the lemma and hence Theorem 11.7a. ∎

Some applications of Theorem 11.7a where $n = 1$ are given in the exercise section. Here we present an application where n is arbitrary, $n \geqslant 1$.

EXAMPLE 2

For $m = 1, 2, \ldots$ and $p = 1, 2, \ldots$, let

$$S(p, m) := \sum_{k=0}^m \binom{m}{k}^p, \tag{11.7-8}$$

the sum of the pth powers of the binomial coefficients. It is well known that

$$S(1, m) = 2^m,$$

$$S(2, m) = \binom{2m}{m};$$

however, no such closed formula is known for $p > 2$. Here we study the asymptotic behavior of $S(p, m)$ where p is fixed and $m \to \infty$.

Our first goal is to obtain an integral representation for $S(p, m)$. Let $n := p - 1 > 0$, and consider the polynomial in n indeterminates $z_1, \ldots, z_n$,

$$P(z_1, \ldots, z_n) := (1 + z_1)^m \cdots (1 + z_n)^m (1 + z_1 \cdots z_n)^m. \qquad (11.7\text{-}9)$$

We show that $S(n + 1, m)$ equals the coefficient of the monomial $(z_1 z_2 \cdots z_n)^m$ in the expansion of P. This monomial arises if the kth terms in the binomials $(1 + z_j)^m$ $(j = 1, \ldots, n)$ are multiplied by the $(m - k)$th term in the expansion of $(1 + z_1 \cdots z_n)^m$ $(k = 0, 1, \ldots, n)$. The coefficient therefore equals

$$\sum_{k=0}^{m} \binom{m}{k}^n \binom{m}{m-k} = \sum_{k=0}^{n} \binom{m}{k}^{n+1} = S(n + 1, m).$$

By Cauchy's formula for analytic functions of several variables (see Chapter 17),

$$S(n + 1, m) = \frac{1}{(2\pi i)^n} \int \cdots \int \frac{P(z_1, \ldots, z_n)}{(z_1 z_2 \cdots z_n)^{m+1}} \, dz_1 \cdots dz_n,$$

where the integrals are extended over the circles $|z_j| = 1$ $(j = 1, \ldots, n)$ in the positive sense. We parametrize the paths of integration by setting

$$z_j = e^{2i\phi_j},$$

where $-\pi/2 \leq \phi_j \leq \pi/2$, $j = 1, \ldots, n$. In view of

$$\frac{P(z_1, \ldots, z_n)}{(z_1 \cdots z_n)^m} = \{(z_1^{1/2} + z_1^{-1/2}) \cdots (z_n^{1/2} + z_n^{-1/2})[(z_1 \cdots z_n)^{1/2} + (z_1 \cdots z_n)^{-1/2}]\}^m,$$

$$\frac{dz_j}{z_j} = 2i\phi_j,$$

$j = 1, \ldots, n$, we obtain

$$S(n + 1, m) = \frac{2^{pm}}{\pi^n} \int \cdots \int_T [\cos \phi_1 \cdots \cos \phi_n \cos(\phi_1 + \cdots + \phi_n)]^m \, d\phi_1 \cdots d\phi_n,$$

$$(11.7\text{-}10)$$

where T denotes the hypercube $-\pi/2 \leq \phi_j \leq \pi/2$ in n space. The expression in brackets is positive within the set $S : |\phi_1 + \phi_2 + \cdots + \phi_n| < \pi/2$, achieving its maximum 1 only at $\phi_1 = \phi_2 = \cdots = \phi_n = 0$. Outside S the integrand is bounded by $(\cos \pi/2n)^m$. Thus asymptotically as $m \to \infty$, the dominant contribution arises from the integration over S. Here the integral is a special case of (11.7-3) where $\theta = 1$, $x = m$, and

$$\psi(\phi_1, \ldots, \phi_n) := -\text{Log}[\cos \phi_1 \cdots \cos \phi_n \cos(\phi_1 + \cdots + \phi_n)].$$

To apply (11.7-5), it remains to evaluate the determinant of the Hessean matrix at $\phi_1 = \cdots = \phi_n = 0$. We have

$$\frac{\partial \psi}{\partial \phi_i} = \tan \phi_i + \tan(\phi_1 + \cdots + \phi_n),$$

$$\frac{\partial^2 \psi}{\partial \phi_i \, \psi_j} = \delta_{ij} \frac{1}{(\cos \phi_i)^2} + \frac{1}{\cos(\phi_1 + \cdots + \phi_n)^2},$$

where δ_{ij} is the Kronecker symbol. The Hessean matrix at $\mathbf{0}$,

$$\mathbf{H} := \begin{pmatrix} 2 & 1 & \cdots & 1 \\ 1 & 2 & \cdots & 1 \\ & \cdot & \cdot & \\ 1 & 1 & \cdots & 2 \end{pmatrix},$$

is clearly positive definite, for if $\boldsymbol{\phi} := (\phi_1, \ldots, \phi_n)^T$, then

$$\boldsymbol{\phi}^T \mathbf{H} \boldsymbol{\phi} = \phi_1^2 + \cdots + \phi_n^2 + (\phi_1 + \cdots + \phi_n)^2 \geq 0,$$

with equality holding only for $\boldsymbol{\phi} = \mathbf{0}$. As a circulant matrix, $\mathbf{H}$ has the eigenvalues

$$\lambda_k = \sum_{j=1}^{n} a_j e^{2\pi i k (j-1)/n}, \qquad k = 1, 2, \ldots, n,$$

where $a_1 = 2, a_2 = \cdots = a_n = 1$ (see Marcus [1960], p. 9). It follows that $\lambda_1 = \cdots = \lambda_{n-1} = 1$, $\lambda_n = n + 1$; hence

$$\det \mathbf{H} = \lambda_1 \lambda_2 \cdots \lambda_n = n + 1,$$

and (11.7-5) yields

$$S(n+1, m) \sim \frac{2^{(n+1)m}}{\sqrt{n+1}} \left(\frac{2}{\pi m} \right)^{n/2}, \qquad m \to \infty. \qquad (11.7\text{-}11)$$

For $n = 0$ (for which the result was not proved), the asymptotic value is 2^m, which is exact. For $n = 1$ we obtain

$$S(2, m) \sim \frac{2^{2m}}{\sqrt{\pi m}},$$

in agreement with the closed formula. For $p > 2$, the asymptotic formula may be used to test conjectures about closed formulas for $S(p, m)$.

PROBLEMS

1. Using the formula

$$\int_{-\pi/2}^{\pi/2} (\cos x)^{2n} \, dx = \frac{(1/2)_n}{n!} \pi,$$

show that

$$\frac{(1/2)_n}{n!} \sim \frac{1}{\sqrt{n\pi}}, \qquad n \to \infty.$$

2. Determine the asymptotic behavior of the **Erlang loss coefficient** defined by

$$f_n(A) := \frac{\dfrac{A^n}{n!}}{1 + \dfrac{A}{1!} + \dfrac{A^2}{2!} + \cdots + \dfrac{A^n}{n!}}$$

as $n \to \infty$, $A \to \infty$, if $q := A/n$ is fixed.
[Consider the integral

$$\int_0^\infty \left[\left(1 + \frac{x}{q}\right) e^{-x}\right]^n dx;$$

cf. H. Störmer, *Arch. Elektr. Uebertragung* **17** (1963), 476–478.]
3. Let q be fixed, $q > -1$. Show that

$$\int_1^\infty \tau^q \left(\frac{xe}{\tau}\right)^\tau d\tau \sim \sqrt{2\pi} x^{q+1/2} e^x, \qquad x \to \infty, \qquad x > 0.$$

(Polya and Szegö).
4. For $m = 1, 2, \ldots$ and $p = 1, 2, \ldots$, let

$$T(p, m) := \sum_{k=0}^{m} (-1)^k \binom{m}{k}^p.$$

It is clear that $T(p, m) = 0$ when m is odd. Show that

$$(-1)^m T(p, 2m) \sim \frac{2}{\sqrt{\pi}} \left(\sqrt{2} \cos\frac{\pi}{2p}\right)^{p-1} (2\pi m)^{1-p/2} \left(2 \cos\frac{\pi}{2p}\right)^{2mp}, \qquad m \to \infty.$$

Compare the result with the closed formula

$$(-1)^m T(p, 2m) = \frac{(pm)!}{(m!)^p}, \qquad m = 1, 2, \ldots$$

known to hold for $p = 2, 3$ (see Problems 11 and 14, §1.6), and show that this formula cannot hold for any $p > 3$ (De Bruijn).

§11.8. THE METHOD OF STEEPEST DESCENT

Here we consider integrals of the form

$$f(z) := \int_\Lambda e^{zg(\zeta)} h(\zeta) \, d\zeta, \qquad (11.8\text{-}1)$$

where the functions g and h are analytic in a region S of the ζ plane and where Λ is an appropriate path in S. Again we wish to determine the asymptotic behavior of f as a function of the parameter z, which to begin with we assume real and positive.

The *method of steepest descent* (also called *saddle point method* for reasons to become apparent subsequently) is a pattern of thought that is frequently helpful in solving the foregoing problem. It makes essential use of the freedom in selecting the exact shape of Λ, which is at our disposal by virtue of Cauchy's theorem. We first describe the method in intuitive terms.

From our discussion of Laplace's method it should be clear that, for any choice of Λ, the main contribution to the integral will arise from those portions of Λ where $|e^{zg(\zeta)}| = e^{z \operatorname{Re} g(\zeta)}$ is largest. Since for $z > 0$, $e^{z \operatorname{Re} g(\zeta)}$ is a monotonic function of $\operatorname{Re} g(\zeta)$, these will also be the portions where $\operatorname{Re} g(\zeta)$ is largest. The contributions of these maxima will be most pronounced if the path is selected in such a way that $\operatorname{Re} g(\zeta)$ changes as rapidly as possible along it.

It is easy to determine the general equation of any path of most rapid change of $\operatorname{Re} g(\zeta)$. Let $\zeta = \xi + i\eta$, $\omega(\xi, \eta) := \operatorname{Re} g(\zeta)$. By calculus, the direction of strongest increase (or steepest ascent) of ω is given by the vector grad ω or, in complex notation, $\operatorname{grd} \omega := \partial\omega/\partial\xi + i(\partial\omega/\partial\eta)$. By the Cauchy–Riemann equations, $\operatorname{grd} \omega = \overline{g'(\zeta)}$. We call **steepest path** for $g(\zeta)$ any path that at *every* point where $g'(\zeta) \neq 0$ runs in the direction of $\overline{g'(\zeta)}$. Let $\zeta = \zeta(\tau)$, $\alpha \leq \tau \leq \beta$, be a piecewise regular steepest path. Then $\arg \zeta'(\tau) = -\arg g'(\zeta(\tau))$ for all $\tau \in [\alpha, \beta]$ except at isolated points. Thus

$$\frac{d}{d\tau} g(\zeta(\tau)) = g'(\zeta(\tau))\zeta'(\tau) \tag{11.8-2}$$

is real wherever defined, which implies that

$$\operatorname{Im} g(\zeta(\tau)) = \text{const.}, \qquad \alpha \leq \tau \leq \beta.$$

We thus have established that *along any steepest path the function g has constant imaginary part.*

Along a steepest path Λ, where do the local maxima of $\operatorname{Re} g(\zeta(\tau))$ occur? By the rules of calculus, such maxima can occur at the end-points of the interval $[\alpha, \beta]$, that is,

(i) at the endpoints of Λ.

Further maxima can occur where $(d/d\tau)g(\zeta(\tau)) = 0$, which by (11.8-2) means that $g'(\zeta(\tau))\zeta'(\tau) = 0$. Because the parametrization of a piecewise regular steepest path may be assumed to be such that ζ' exists and is different from zero at all points where $g'(\zeta) \neq 0$, this leaves only

(ii) the points where $g'(\zeta) = 0$.

By the maximum principle, unless at the same time $g(\zeta) = 0$, these are the points where the surface $\zeta \to |g(\zeta)|$ has a *saddle point*—hence the name **saddle-point method** for the method of steepest descent.

It is now clear that any steepest path Λ is a sum of arcs Λ_j with the following properties: Any two consecutive arcs Λ_{j-1} and Λ_j meet at a point ζ_j

where $g'(\zeta_j) = 0$, and the function $\mathrm{Re}\, g(\zeta)$ is monotone along any Λ_j. If Λ is a steepest path, the integral (11.8-1) can be represented as

$$f(z) = \sum_j f_j(z),$$

where

$$f_j(z) := \int_{\Lambda_j} e^{zg(\zeta)} h(\zeta)\, d\zeta. \tag{11.8-3}$$

The asymptotic evaluation of the functions f_j is easily accomplished by the Watson–Doetsch lemma. Without loss of generality we may assume that the integration is in the direction of decreasing values of $\mathrm{Re}\, g$, for if the orientation of Λ_j is otherwise it suffices to change the sign of the integral. If Λ_j begins at ζ_j, we parametrize Λ_j by setting

$$\tau := g(\zeta_j) - g(\zeta).$$

This parameter is legitimate, for since $\mathrm{Im}\, g(\zeta) = \mathrm{Im}\, g(\zeta_j)$ on Λ_j, τ is real, and because $\mathrm{Re}\, g(\zeta)$ is decreasing, τ is increasing. The function $\zeta = \zeta(\tau)$ decreasing Λ_j thus is well defined. If τ increases to β, we have

$$f_j(z) = e^{zg(\zeta_j)} \int_0^\beta e^{-z\tau} h(\zeta(\tau))\zeta'(\tau)\, d\tau, \tag{11.8-4}$$

and to apply the Watson–Doetsch lemma it only remains to determine the asymptotic series of $h(\zeta(\tau))\zeta'(\tau)$ as $\tau \to 0$, $\tau > 0$.

Let $\omega := \zeta - \zeta_j$, and put

$$g(\zeta_j) - g(\zeta) = a_k \omega^k + a_{k+1} \omega^{k+1} + \cdots =: G_j, \tag{11.8-5}$$

$$h(\zeta) = b_0 + b_1\omega + b_2\omega^2 + \cdots =: H_j, \tag{11.8-6}$$

where $|\omega|$ is sufficiently small, $k \geq 1$, and $a_k \neq 0$. (If ζ_j is a saddle point, we have $k \geq 2$, but some of the integrations may begin at a nonsaddle point.) Let $e^{i\phi_j}$ denote the direction under which Λ_j emanates from ζ_j. Because $a_k \omega^k \{1 + O(\omega)\}$ is positive along Λ_j, $-\phi_j$ is one of the arguments of $(a_k)^{1/k}$; its precise value, however, can be ascertained only by study of the geometric situation on hand. If ζ is restricted to Λ_j, then

$$\sigma := [g(\zeta_j) - g(\zeta)]^{1/k} = c_1\omega + c_2\omega^2 + \cdots = G_j^{1/k}, \tag{11.8-7}$$

where the series $G_j^{1/k}$ is uniquely determined by

$$c_1 e^{i\phi_j} > 0. \tag{11.8-8}$$

Letting $\zeta(\tau) =: \hat{\zeta}(\sigma)$, we get by the implicit function theorem (Theorem 2.4d)

$$h(\hat{\zeta}(\sigma))\hat{\zeta}'(\sigma) = \sum_{m=0}^{\infty} \mathrm{res}(H_j G_j^{-(m+1)/k})\sigma^m,$$

which in view of $\sigma = \tau^{1/k}$, $\hat{\zeta}(\tau^{1/k}) = \zeta(\tau)$ yields

$$h(\zeta(\tau))\zeta'(\tau) = \frac{1}{k} \sum_{m=0}^{\infty} \text{res}(H_j G_j^{-(m+1)/k})\tau^{(m+1)/k-1}.$$

Because this series is convergent for $\tau > 0$ sufficiently small, it is asymptotic, and the Watson–Doetsch lemma directly furnishes the desired asymptotic expansion

$$f_j(z) \approx \frac{1}{k} e^{zg(\zeta_j)} \sum_{m=0}^{\infty} \text{res}(H_j G_j^{-(m+1)/k}) \frac{\Gamma((m+1)/k)}{z^{(m+1)/k}}, \qquad (11.8\text{-}9)$$

$$z \to \infty, \qquad z \in \hat{S}_{\pi/2-\delta}.$$

In summary, the method of steepest descent is seen to consist of the following three steps:

(A) Determine the saddle points, that is, the points where $g'(\zeta) = 0$.

(B) Deform the original path of integration into a piecewise regular path along which $\text{Im } g(\zeta) = \text{const}$. This step may be facilitated by drawing, for each saddle point ζ_j, the curves $\text{Im } g(\zeta) = \text{Im } g(\zeta_j)$. The new path should be composed of such curves.

(C) Evaluate the integrals along the differentiable sub-arcs of Λ by the Watson–Doetsch lemma.

Step (A), unless the function g is unduly complicated, is easy. The step that is likely to give the most trouble is (B). It may be necessary to introduce auxiliary subarcs along which the condition $\text{Im } g(\zeta) = \text{const}$. is violated. If so, the contribution of these subarcs must be estimated separately and shown to be negligible compared to the other contributions, or they must be evaluated accurately. Due regard must also be paid to singularities, if any, of the functions g and h.

Step (C) is completely mechanized by formula (11.8-9) once the series G_j and H_j and the directions $\exp(i\phi_j)$ of the steepest paths at the saddle points are known.

We illustrate the method by several examples.

EXAMPLE **1**

A function equivalent to

$$\text{Ai}(z) := \frac{1}{\pi} \int_0^{\infty} \cos(\tfrac{1}{3}t^3 + zt)\, dt \qquad (z \text{ real})$$

occurs in a paper by Sir George B. Airy [1838], "On the intensity of light in the neighborhood of a caustic." The function $\text{Ai}(z)$ is therefore called **Airy's integral**. We discuss its behavior if $z \to \infty$, $z > 0$. Let $t =: z^{1/2}\zeta$, $x := z^{3/2}$, then we obtain $\text{Ai}(x) = (2\pi)^{-1}x^{1/2}f(x^{3/2})$, where

$$f(x) := \int_{-\infty}^{\infty} \exp[ix(\tfrac{1}{3}\zeta^3 + \zeta)]\, d\zeta. \qquad (11.8\text{-}10)$$

This is of the form (11.8-1) where $h(\zeta) := 1$,

$$g(\zeta) := i(\tfrac{1}{3}\zeta^3 + \zeta).$$

The integral converges very slowly—it is not even absolutely convergent—and thus at first sight seems to present a difficult case for asymptotic evaluation. However, it can be evaluated easily by the saddle-point method.

    Following the procedure outlined, we first determine the saddle points. These are given by $g'(\zeta) = i(\zeta^2 + 1) = 0$, and hence are

$$\zeta_1 := i, \qquad \zeta_2 := -i.$$

We next determine the curves $\operatorname{Im} g(\zeta) = \operatorname{Im} g(\zeta_k)$ $(k = 1, 2)$ to find a suitable path of integration. Setting $\zeta = \xi + i\eta$, we have

$$\operatorname{Im} g(\zeta) = \operatorname{Im} i\,(\tfrac{1}{3}\zeta^3 + \zeta) = \tfrac{1}{3}\xi(\xi^2 - 3\eta^2 + 3)$$

and hence $\operatorname{Im} g(\zeta_{1,2}) = 0$. Thus the desired curves are given by the equation $\xi(\xi^2 - 3\eta^2 + 3) = 0$ (see Fig. 11.8a). They consist of the η axis and of a hyperbola with the asymptotes $\xi = \pm\sqrt{3}\eta$. It is now necessary to find the directions in which $\operatorname{Re} g(\zeta)$ decreases on the curves. These directions are indicated by arrows in Fig. 11.8a; they are easily determined from $g(\zeta) \sim \tfrac{1}{3}i\zeta^3(\zeta \to \infty)$.

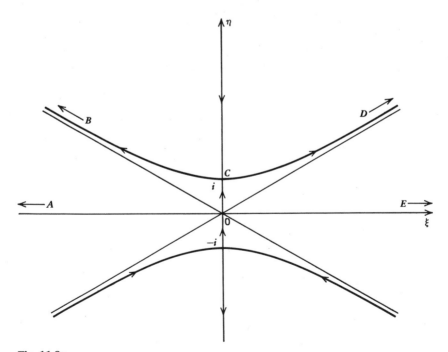

Fig. 11.8a.

It appears that the curve from B to D through C is a candidate for Λ, because $\operatorname{Re} g(\zeta)$ tends to $-\infty$ along both branches of the hyperbola. To show that it is equivalent to the original path from A to E, we must verify that the integrals from A to B and from D to E, taken, say, along circular arcs $|\zeta| = \rho$, tend to zero as $\rho \to \infty$. Writing $\zeta = \rho e^{i\alpha}$, we have

$$\left| \int_A^B \exp[ix(\tfrac{1}{3}\zeta^3 + \zeta)]\, d\zeta \right| = \left| \int_{-\pi}^{-\pi+\pi/6} \exp[ix(\tfrac{1}{3}\rho^3 e^{3i\alpha} + \rho e^{i\alpha})]i\rho\, e^{i\alpha}\, d\alpha \right|$$

$$\leqslant \rho \int_0^{\pi/6} \exp[-x(\tfrac{1}{3}\rho^3 \sin 3\alpha + \rho \sin \alpha)\, d\alpha.$$

Using $\sin \alpha \geqslant 0$ and $\sin 3\alpha \geqslant (6/\pi)\alpha$, we find that the last integral is bounded by

$$\int_0^{\pi/6} e^{-x\rho^3(2/\pi)\alpha}\, d\alpha \leqslant \frac{\pi}{2\rho^2}, \qquad x \geqslant 1,$$

proving the assertion. The integral from D to E can be treated similarly. This completes step (B).

Turning to step (C), we first consider the integral along the path Λ_1 from C to D. We have $h(\zeta) = 1$; hence $H = I$. Writing $\omega = \zeta - i$, we easily find

$$g(i) - g(\zeta) = \omega^2 - \frac{i}{3}\omega^3.$$

Hence $k = 2$, and because $\phi = 0$ for the path from C to D,

$$G_1^{1/2} = \omega\left(1 - \frac{i}{3}\omega\right)^{1/2}.$$

Thus by the binomial series,

$$\operatorname{res} G_1^{-(m+1/2)} = \operatorname{res}\left[\omega^{-m-1}\left(1 - \frac{i}{3}\omega\right)^{-(m+1/2)}\right]$$

$$= \left(\frac{i}{3}\right)^m \frac{(m+1/2)_m}{m!}, \qquad m = 0, 1, 2, \ldots.$$

For the path Λ_0 from B to C the result is the same, except that we now integrate in the ascending direction, and because $\phi = \pi$,

$$G_0^{1/2} = -\omega\left(1 - \frac{i}{3}\omega\right)^{1/2}.$$

The foregoing residue thus is to be multiplied by $(-1)^{m+1}$. Hence only the terms with m even appear in the final asymptotic formula, and the total integral thus has the asymptotic series

$$e^{-2x/3} \sum_{m=0}^{\infty} \left(\frac{i}{3}\right)^{2m} \frac{(m+1/2)_{2m}\Gamma(m+1/2)}{(2m)!} x^{-m-1/2}$$

After some simplification we thus find

$$f(x) \approx \sqrt{\pi} x^{-1/2} e^{-2x/3} \sum_{m=0}^{\infty} (-1)^m \frac{(1/2)_{3m}}{(2m)!} \frac{1}{(9x)^m}, \qquad x \to \infty, \qquad x > 0. \qquad (11.8\text{-}11)$$

EXAMPLE **2**

The function

$$f(x) := \int_0^\infty \exp[ix(\tfrac{1}{3}\zeta^3 + \zeta)] \qquad (11.8\text{-}12)$$

differs from the f of Example 1 only by the limits of integration. The function g, and hence the saddle points, are the same, and Fig. 11.8a still applies. Using the same estimate as before, we see that the original path of integration OE may be replaced by the path OCD. It passes through the saddle point C at $\zeta = i$, but because $\operatorname{Re} g(i) = -\tfrac{2}{3} < \operatorname{Re} g(0) = 0$, the contribution of the saddle point is exponentially small compared to the contribution of the vicinity of O. Estimating the contribution of the segment OC by means of (11.8-9), we find in a straightforward manner

$$f(x) \approx \frac{i}{x} \sum_{m=0}^\infty \frac{(3m)!}{m!(3x^2)^m}, \qquad x \to \infty, \qquad x > 0. \qquad (11.8\text{-}13)$$

EXAMPLE **3**   **Bessel function of large order.**

The results of §11.5 by means of the Watson–Doetsch lemma exhaustively describe the asymptotic behavior of the Bessel functions of large argument z when the order ν is fixed. They do not answer the question as to the asymptotic behavior when ν and z tend to infinity simultaneously. This problem is tackled here.

Because all Bessel functions can be expressed linearly in terms of the Hankel functions of the first and second kind, we first consider our problem for the Hankel functions. Our starting point is *Sommerfeld's integrals* (see §10.12). We denote by Λ_1 a path lying in the strip $-\pi < \operatorname{Re} \zeta < 0$ and leading from $-\pi + i\infty$ to $-i\infty$, and by Λ_2 a path in $0 < \operatorname{Re} \zeta < \pi$ leading from $-i\infty$ to $\pi + i\infty$. We then have for $\nu > 0$, $\operatorname{Re} z > 0$

$$H_\nu^{(k)}(z) = \frac{1}{\pi} \int_{\Lambda_k} e^{-iz \sin \zeta + i\nu\zeta} \, d\zeta, \qquad k = 1, 2.$$

Consequently, if $z = \nu q$, where q is a constant, $\operatorname{Re} q > 0$,

$$H_\nu^{(k)}(\nu q) = \frac{1}{\pi} \int_{\Lambda_k} e^{\nu g(\zeta)} \, d\zeta, \qquad k = 1, 2, \qquad (11.8\text{-}14)$$

where

$$g(\zeta) := i(\zeta - q \sin \zeta). \qquad (11.8\text{-}15)$$

From (10.12-20) we also have

$$J_\nu(\nu q) = \frac{1}{2\pi} \int_{\Lambda_1 + \Lambda_2} e^{\nu g(\zeta)} \, d\zeta, \qquad (11.8\text{-}16)$$

$$Y_\nu(\nu q) = \frac{1}{2\pi i} \int_{\Lambda_1 - \Lambda_2} e^{\nu g(\zeta)} \, d\zeta. \qquad (11.8\text{-}17)$$

These integrals are all of the form (11.8-1). They are now used to determine the asymptotic behavior of the functions they represent for $\nu \to \infty$, $\nu > 0$. For simplicity, it is assumed that q is real, $q > 0$.

The saddle points of the integrand are located where $g'(\zeta) = 0$ or

$$q \cos \zeta = 1. \qquad (11.8\text{-}18)$$

The location of the saddle points in the strip $-\pi < \operatorname{Re} \zeta < \pi$ depends on the size of q. If $q > 1$, there are two real saddle points; if $0 < q < 1$, there are two complex saddle points that are purely imaginary, and if $q = 1$, there is exactly one saddle point, at $\zeta = 0$. Accordingly, we distinguish three cases.

(i) $q > 1$. Let α be defined by $\cos \alpha = q^{-1}$, $0 < \alpha < \pi/2$. Then the two saddle points are $\zeta = \pm\alpha$. The lines of steepest descent are given by

$$\operatorname{Im} g(\zeta) = \operatorname{Im} g(\pm\alpha) = \pm(\alpha - \tan \alpha).$$

An elementary discussion shows that they proceed as shown in Fig. 11.8b, with arrows indicating the direction of decrease of $\operatorname{Re} g(\zeta)$. Clearly, the path ABC may be taken as Λ_1, and the path DEF as Λ_2.

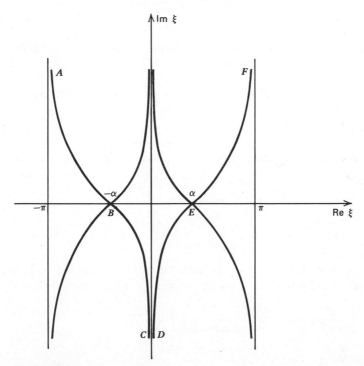

Fig. 11.8b.

We first consider the integral along Λ_1. The curve passes through the sole saddle point $\zeta = -\alpha$. Letting $\zeta = -\alpha + \omega$, we have

$$g(-\alpha) - g(\zeta) = i\{\tan\alpha(1 - \cos\omega) + (\sin\omega - \omega)\}.$$

The expansion of this function in powers of ω is given by

$$G := i\frac{\tan\alpha}{2}\omega^2\left\{1 - \frac{2\cot\alpha}{3!}\omega - \frac{2}{4!}\omega^2 + \frac{2\cot\alpha}{5!}\omega^3 + \cdots\right\}.$$

For the descent from B to C, the angle ϕ defined after (11.8-6) is $-\pi/4$; consequently

$$G^{1/2} := \exp\left(i\frac{\pi}{4}\right)\sqrt{\frac{\tan\alpha}{2}}\,\omega\left\{1 - \frac{2\cot\alpha}{3!}\omega - \cdots\right\}^{1/2}$$

is the correct choice of the root. It is convenient to write

$$H := H(t) := \omega\left\{1 - \frac{2t}{3!}\omega - \frac{2}{4!}\omega^2 + \frac{2t}{5!}\omega^3 + \cdots\right\}^{1/2},$$

where t is a parameter, and to define

$$A_n(t) := \mathrm{res}[H(t)]^{-n-1}. \tag{11.8-19}$$

It is clear that A_n is a polynomial of degree $\le n$ in t. In this notation, the residues required in (11.8-9) are

$$\mathrm{res}\,G^{-(m+1)/2} = e^{-i(m+1)\pi/4}\left(\frac{\tan\alpha}{2}\right)^{-(m+1)/2}A_m(\cot\alpha).$$

For the path from B to A the choice of the sign of $G^{1/2}$ is opposite. Combining the contributions of the two paths, we find that the terms corresponding to odd values of m cancel, and we obtain, after simplification,

$$H_\nu^{(1)}\left(\frac{\nu}{\cos\alpha}\right) \approx \exp\left(-i\frac{\pi}{4}\right)\left(\frac{2}{\pi\nu\tan\alpha}\right)^{1/2}e^{i\nu(\tan\alpha - \alpha)}$$

$$\cdot \sum_{m=0}^{\infty} A_{2m}(\cot\alpha)\left(\frac{1}{2}\right)_m\left(\frac{2}{i\nu\tan\alpha}\right)^m, \qquad \nu \to \infty, \qquad \nu > 0. \tag{11.8-20}$$

The term in the first line is the principal term in the expansion. The integral along Λ_2 can be treated in exactly the same manner; the result is

$$H_\nu^{(2)}\left(\frac{\nu}{\cos\alpha}\right) \approx \exp\left(i\frac{\pi}{4}\right)\left(\frac{2}{\pi\nu\tan\alpha}\right)^{1/2}e^{i\nu(\alpha - \tan\alpha)}$$

$$\cdot \sum_{m=0}^{\infty} A_{2m}(\cot\alpha)\left(\frac{1}{2}\right)_m\left(\frac{2}{-i\nu\tan\alpha}\right)^m, \qquad \nu \to \infty, \qquad \nu > 0. \tag{11.8-21}$$

The same result could have been deduced from the relation $H_\nu^{(2)}(q\nu) = \overline{H_\nu^{(1)}(q\nu)}$. The formulas for $J_\nu(q\nu)$ and $Y_\nu(q\nu)$ are easily deduced from the foregoing by means of (11.5-11). The leading terms are

$$J_\nu\left(\frac{\nu}{\cos\alpha}\right) \sim \sqrt{\frac{2}{\pi\nu\tan\alpha}}\cos\left[\nu(\tan\alpha - \alpha) - \frac{\pi}{4}\right],$$

$$\text{(11.8-22)}$$

$$Y_\nu\left(\frac{\nu}{\cos\alpha}\right) \sim \sqrt{\frac{2}{\pi\nu\tan\alpha}}\sin\left[\nu(\tan\alpha - \alpha) - \frac{\pi}{4}\right], \qquad \nu \to \infty, \qquad \nu > 0.$$

(ii) $0 < q < 1$. Let α now be defined by $\operatorname{Cos}\alpha = q^{-1}$, $\alpha > 0$.[1] The saddle points are now located at $\zeta = \pm i\alpha$. Because the values

$$g(\pm i\alpha) = \mp(\alpha - \operatorname{Tg}\alpha)$$

are real, the curves of steepest descent through the saddle points are now given by $\operatorname{Im} g(\zeta) = 0$. They consist of the imaginary axis $\operatorname{Re}\zeta = 0$ and of two curved branches through the points $\pm i\alpha$ having the vertical lines $\operatorname{Re}\zeta = \pm\pi$ as asymptotes (see Fig. 11.8c; arrows mark directions of descent). It is clear that the path $ACDE$ is equivalent to Λ_1, and the path $EDCB$ equivalent to Λ_2.

We first consider the integral along Λ_1. There are two saddle points on the path; however, because $g(-i\alpha) > g(i\alpha)$, only the lower point needs to be taken into consideration. Letting $\zeta = -i\alpha + \omega$ we have

$$g(-i\alpha) - g(\zeta) = \operatorname{Tg}\alpha(\cos\omega - 1) + i(\sin\omega - \omega);$$

hence the expansion in powers of ω is

$$G = -\frac{\operatorname{Tg}\alpha}{2}\omega^2\left\{1 + \frac{2i\operatorname{Ctg}\alpha}{3!}\omega - \frac{2}{4!}\omega^2 - \frac{2i\operatorname{Ctg}\alpha}{5!}\omega^3 + \cdots\right\}.$$

Because $\phi = -\pi/2$ for the descent from D to E, the correct choice of the square root is

$$G^{1/2} = i\sqrt{\frac{\operatorname{Tg}\alpha}{2}}\omega\left\{1 + \frac{2i\operatorname{Ctg}\alpha}{3!}\omega - \cdots\right\}^{1/2}.$$

Still using the abbreviation (11.8-19), we have

$$\operatorname{res} G^{-(m+1)/2} = i^{-m-1}\left(\frac{\operatorname{Tg}\alpha}{2}\right)^{-(m+1)/2}A_m(-i\operatorname{Ctg}\alpha),$$

[1] Following Central European practice, we denote the hyperbolic functions by Sin, Cos, Tg, Ctg rather than by the more cumbersome symbols sinh, cosh, tanh, cotanh.

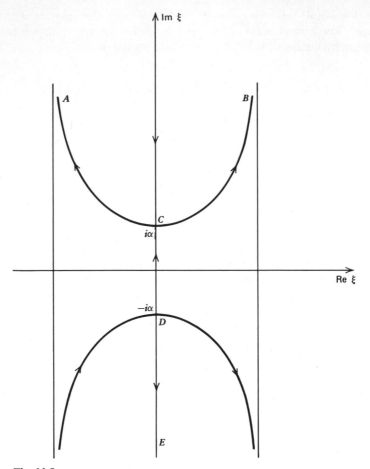

Fig. 11.8c.

and hence from (11.8-9), after the usual cancellation of odd terms and other simplification,

$$H_\nu^{(1)}\left(\frac{\nu}{\text{Cos }\alpha}\right) \approx -i\sqrt{\frac{2}{\pi\nu \text{ Tg }\alpha}}\, e^{\nu(\alpha-\text{Tg }\alpha)}$$

$$\cdot \sum_{m=0}^{\infty} A_{2m}(i\text{ Ctg }\alpha)\left(\frac{1}{2}\right)_m \left(\frac{-2}{\nu \text{ Tg }\alpha}\right)^m, \qquad \nu \to \infty, \qquad \nu > 0. \tag{11.8-23}$$

Because Λ_2 runs opposite to Λ_1 on the relevant part of the path of integration, the expression on the right of (11.8-23) also gives the asymptotic expansion of $-H_\nu^{(2)}(\nu/\text{Cos }\alpha)$, and hence of $iY_\nu(\nu/\text{Cos }\alpha)$.

If we tried to obtain the asymptotic expansion of $J_\nu(\nu/\text{Cos}\,\alpha)$ from these results, we would obtain the zero series, indicating that this function is exponentially small compared to the leading term in (11.8-23). The explanation lies in the fact that we now integrate along $\Lambda_1 + \Lambda_2$; hence the contributions of the segments CDE and EDC cancel each other, and the higher saddle point does not come into play. The appropriate path of integration is the curve ACB, with the sole saddle point at $\zeta = i\alpha$. Here we have, putting $\zeta = {}'i\alpha + \omega$,

$$g(i\alpha) - g(\zeta) = \text{Tg}\,\alpha\,(1 - \cos\omega) + i(\sin\omega - \omega),$$

and hence, expanding in powers of ω,

$$G = \frac{\text{Tg}\,\alpha}{2}\omega^2\left\{1 - \frac{2i\,\text{Ctg}\,\alpha}{\omega}\omega - \frac{2}{4!}\omega^2 + \frac{2i\,\text{Ctg}\,\alpha}{5!}\omega^3 + \cdots\right\}.$$

Because $\phi = 0$ for the descent from C to B, the appropriate root is

$$G^{1/2} = \sqrt{\frac{\text{Tg}\,\alpha}{2}}\,\omega\left\{1 - \frac{2i\,\text{Ctg}\,\alpha}{3!}\omega - \cdots\right\}^{1/2},$$

and

$$\text{res}\,G^{-(m+1)/2} = \left(\frac{\text{Tg}\,\alpha}{2}\right)^{-(m+1)/2}A_m(i\,\text{Ctg}\,\alpha).$$

There results

$$J_\nu\left(\frac{\nu}{\text{Cos}\,\alpha}\right) \approx \frac{1}{\sqrt{2\pi\nu\,\text{Tg}\,\alpha}}e^{\nu(\text{Tg}\,\alpha - \alpha)}$$

$$\cdot \sum_{m=0}^{\infty} A_{2m}(i\,\text{Ctg}\,\alpha)\left(\frac{1}{2}\right)_m\left(\frac{2}{\nu\,\text{Tg}\,\alpha}\right)^m, \qquad \nu \to \infty, \qquad \nu > 0.$$

(11.8-24)

For convenience a short table of the polynomials $A_{2m}(t)$ occurring in (11.8-20, 21, 23, 24) is appended.

m	$A_{2m}(t)$
0	1
1	$\dfrac{1}{8} + \dfrac{5}{24}t^2$
2	$\dfrac{3}{128} + \dfrac{77}{576}t^2 + \dfrac{385}{3456}t^4$

(iii) $q = 1$. This case is exceptional because there now is only one saddle point, at $\zeta = 0$. Because $g(0) = 0$, the curves of steepest descent are given by Im $g(\zeta) = 0$, and a simple discussion reveals that they run as shown in Fig. 11.8d, with directions of descent indicated by arrows. Evidently, ABC is

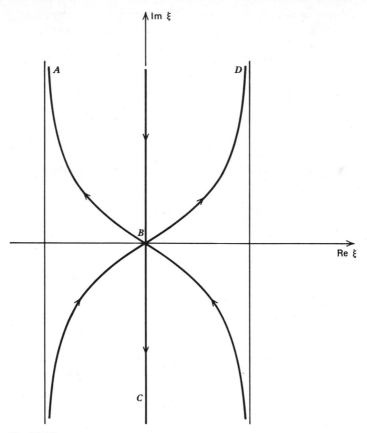

Fig. 11.8d.

equivalent to Λ_1, CBD to Λ_2, and ABD to $\Lambda_1 + \Lambda_2$. Because $g(0) = 0$,

$$g(0) - g(\zeta) = i(\sin \zeta - \zeta),$$

and the expansion in powers of $\omega := \zeta - 0$ is

$$G := -\frac{i\omega^3}{6}\left\{1 - \frac{6}{5!}\omega^2 + \frac{6}{7!}\omega^4 - \cdots\right\}. \qquad (11.8\text{-}25)$$

This is an expansion where $k = 3$ (in addition to g', g'' vanishes at the saddle point, which is therefore said to be of order two). To apply (11.8-9), we thus must form

$$G^{1/3} = \theta 6^{-1/3} \omega \left\{ 1 - \frac{6}{5!} \omega^2 + \cdots \right\}^{1/3},$$

where θ is a third root of $-i$. More precisely, letting

$$\epsilon := \exp\left(i\frac{\pi}{6}\right),$$

we have

for descent from B to	θ
D	ϵ^{-1}
A	ϵ^{-5}
C	ϵ^{-9}

It is convenient to define the rational numbers

$$B_n := \operatorname{res} \omega^{-n-1} \left\{ 1 - \frac{6}{5!} \omega^2 + \frac{6}{7!} \omega^4 - \cdots \right\}^{-(n+1)/3}, \qquad (11.8\text{-}26)$$

$$n = 0, 1, 2, \ldots,$$

where the series in { } is the same as in (11.8-25). Evidently, $B_n = 0$ when n is odd; some values for n even are as follows:

m	0	1	2	3
B_{2m}	1	$\dfrac{1}{20}$	$\dfrac{1}{280}$	$\dfrac{1}{3600}$

For the residues required in (11.8-9) we find

$$\operatorname{res} G^{-(m+1)/3} = \theta^{-m-1} 6^{(m-1)/3} B_m, \qquad m = 0, 1, 2, \ldots.$$

Appropriately combining the contributions of the various paths of descent and simplifying, we obtain

$$H_\nu^{(1)}(\nu) \approx \frac{2}{3} \exp\left(-i\frac{\pi}{3}\right)\left(\frac{6}{\nu}\right)^{1/3} \sum_{m=0}^{\infty} \frac{B_{2m}}{\Gamma\left(\dfrac{2-2m}{3}\right)} \left[\frac{\exp\left(i\frac{\pi}{3}\right) 6^{2/3}}{\nu^{2/3}} \right]^m,$$

$$(11.8\text{-}27)$$

$$H_\nu^{(2)}(\nu) \approx \frac{2}{3}\exp\left(i\frac{\pi}{3}\right)\left(\frac{6}{\nu}\right)^{1/3}\sum_{m=0}^{\infty}\frac{B_{2m}}{\Gamma\left(\frac{2-2m}{3}\right)}\left[\frac{\exp\left(-i\frac{\pi}{3}\right)6^{2/3}}{\nu^{2/3}}\right]^m,$$

(11.8-28)

$$J_\nu(\nu) \approx \frac{1}{3\pi}\left(\frac{6}{\nu}\right)^{1/3}\sum_{m=0}^{\infty}B_{2m}\sin\frac{(m+2)\pi}{3}\Gamma\left(\frac{2m+1}{3}\right)\left[\frac{6^{2/3}}{\nu^{2/3}}\right]^m,$$

(11.8-29)

$$\nu \to \infty, \qquad \nu > 0.$$

Considering leading terms only, we find, in particular,

$$H_\nu^{(1)}(\nu) = \overline{H_\nu^{(2)}(\nu)} \sim \exp\left(i\frac{\pi}{3}\right)\frac{2^{1/3}\Gamma(1/3)}{3^{1/6}\nu^{1/3}},$$

$$J_\nu(\nu) \sim \frac{\Gamma(1/3)}{2^{2/3}3^{1/6}\nu^{1/3}}$$

as $\nu \to \infty$, the error in each case being $O(\nu^{-5/3})$.

The formulas given in Example 3 do not describe the asymptotic behavior of the Bessel functions in the "transitional region," when z and ν tend to ∞ in such a manner that $z/\nu \to 1$ while not being exactly equal to 1. For the case in which $\nu = z(1-\sigma)$, $\sigma = O(z^{-2/3})$, it is possible to derive asymptotic expansions by similar methods (see Watson [1944], p. 245). The problem can also be approached from the point of view of differential equations with a large parameter. The formulas appropriate to the case $z \sim \nu$ then involve Airy functions; see Problem 6.

PROBLEMS

1. Establish the result of Example 2.
2. For $\alpha > 0$ and $x > 0$ let

$$f(x) := \int_0^1 e^{ix\zeta^\alpha}\,d\zeta.$$

Show that when α is fixed and $x \to \infty$,

$$f(x) \approx \exp\left(\frac{i\pi}{2\alpha}\right)\Gamma\left(\frac{1}{\alpha}+1\right)x^{-1/\alpha} + \frac{e^{ix}}{iax}\sum_{m=0}^{\infty}\frac{[1-(1/\sigma)]_m}{(ix)^m}.$$

3. Let Λ denote a curve as indicated in Fig. 11.8e. Obtain the asymptotic expansion of $1/\Gamma(z)$ for $z \to \infty$ from Hankel's integral

$$\frac{1}{\Gamma(z)} = \frac{1}{2\pi i}\int_\Lambda e^{\zeta}\zeta^z\,d\zeta.$$

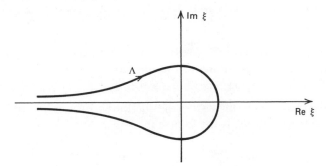

Fig. 11.8e.

4. For $s > 0$, $\beta > 0$ let

$$G(s, \beta) := \frac{s\, e^{-2\beta s^2}}{2\pi i} \int_\Lambda e^{4\beta s^2 \zeta} [1 - (1/\zeta)]^\beta \, d\zeta,$$

where the ζ plane is cut between $\zeta = 0$ and $\zeta = 1$ and where Λ is a path encircling the cut counterclockwise. Show that for $s = 1$ there holds the asymptotic expansion

$$G(1, \beta) \approx \frac{4}{3\pi} \sum_{m=1}^\infty b_m \cos\left(\beta\pi + \frac{\beta\pi m}{3}\right) \sin\frac{m\pi}{3} \Gamma\left(\frac{m}{3}\right)\left(\frac{3}{2\beta}\right)^{m/3},$$

$\beta \to \infty$, where $b_1 = 1$, $b_3 = -(3/5)$, $b_5 = 3/35$, $b_2 = b_4 = \cdots = 0$. Also show that for fixed θ, $0 < \theta < \pi/2$,

$$G(\cos\theta, \beta) = \frac{1}{2\sqrt{\pi\beta}\,\sin\theta} \sin\left[\beta(\pi - 2\theta + \sin 2\theta) - \frac{\pi}{4}\right] + O(\beta^{-1}),$$

5. A series of the form

$$\sum_{n=0}^\infty a_n J_n(nz),$$

where $\{a_n\}$ is a sequence of complex numbers, is called a **Kapteyn series**. Show that a Kapteyn series satisfying

$$\limsup_{n \to \infty} |a_n|^{1/n} = 1$$

converges and represents an analytic function in the set

$$S := \left\{ z : \left| \frac{z \exp\sqrt{1 - z^2}}{1 + \sqrt{1 - z^2}} \right| < 1 \right\}.$$

6. Obtain an approximate representation of $J_n(x)$ in terms of Airy's integral when n and x are large and nearly equal by approximating the function $\sin\tau$ in Bessel's integral,

$$J_n(x) = \frac{1}{\pi} \int_0^\pi \cos[n\tau - x\sin\tau]\, d\tau,$$

by its Taylor polynomial of degree four.

§11.9. GENERAL ASYMPTOTIC EXPANSIONS; ASYMPTOTIC FACTORIAL SERIES

In the first eight sections of the present chapter we considered formal series whose terms were proportional to the functions

$$g_n(z) := z^{-n}, \qquad n = 0, 1, 2, \dots. \tag{11.9-1}$$

An essential property of this system of functions, a property used for proving some of the basic results, is the fact that

$$\lim_{z \to \infty} \frac{g_{n+1}(z)}{g_n(z)} = 0, \qquad n = 0, 1, 2, \dots. \tag{11.9-2}$$

Although many applications of formal series built from these functions have been given, there are situations in which series built from other functions are more convenient. We adopt the following definition: Let S be an unbounded set in the complex plane and let $\{g_n\}$ be a sequence of functions defined on S. The sequence $\{g_n\}$ is called an **asymptotic sequence** in S, if each $g_n(z) \neq 0$ for $|z|$ sufficiently large, $z \in S$, and if (11.9-2) holds for $n = 0, 1, 2, \dots$.

It is clear that the notion of an asymptotic sequence can also be defined for the approach to a finite point, but there is no need for doing so.

The reader is asked to verify that each of the sequences $\{g_n\}$ given below satisfies the foregoing definition in the indicated domain.

EXAMPLE **1**

Let $0 = \lambda_0 < \lambda_1 < \lambda_2 < \cdots, 0 < \beta < \pi$. The functions

$$g_n(z) := z^{-\lambda_n}, \qquad n = 0, 1, 2, \dots$$

(principal values) form an asymptotic sequence in S_β. [The sequence (11.9-1) evidently is a special case.]

EXAMPLE **2**

The functions

$$g_n(z) := \frac{1}{(z)_n}, \qquad n = 0, 1, 2, \dots,$$

form an asymptotic sequence in every region S_β with $0 < \beta < \pi$, but not in the whole complex plane.

EXAMPLE **3**

Let $0 < \nu_0 < \nu_1 < \cdots$. The functions

$$g_n(z) := \nu_n^{-z}, \quad n = 0, 1, 2, \dots$$

(principal values) form an asymptotic sequence in every region S_β where $0 < \beta < \pi/2$, but not in $S_{\pi/2}$. More generally, let $\{\xi_n\}$ be a sequence of complex numbers lying on a ray through 0, $\arg \zeta_n = \alpha$, and satisfying $|\zeta_{n+1}| > |\zeta_n|$, $n = 0, 1, \ldots$. Then the functions

$$g_n(z) := \zeta_n^z = e^{-z[\text{Log}|\zeta_n| + i\alpha]}, \qquad n = 0, 1, \ldots,$$

form an asymptotic sequence in S_β where $\beta < \pi/2$.

We now proceed as in §11.1. Let S be an unbounded set, and let $\{g_n\}$ be an asymptotic sequence in S. If f is defined on S and if $\{a_n\}$ is a sequence of complex numbers, then the functions

$$F_n(z) := \sum_{k=0}^{n} a_k g_k(z), \qquad n = 0, 1, 2, \ldots, \qquad (11.9\text{-}3)$$

may or may not have one of the following properties.

PROPERTY (A)

For $n = 0, 1, 2, \ldots$,

$$f(z) - F_n(z) = O(g_{n+1}(z)), \qquad z \to \infty, \qquad z \in S. \qquad (11.9\text{-}4)$$

PROPERTY (B)

For $n = 0, 1, 2, \ldots$,

$$\lim_{\substack{z \to \infty \\ z \in S}} \frac{1}{g_n(z)} [f(z) - F_n(z)] = 0. \qquad (11.9\text{-}5)$$

PROPERTY (C)

For $n = 0, 1, 2, \ldots$,

$$\lim_{\substack{z \to \infty \\ z \in S}} \frac{1}{g_n(z)} [f(z) - F_{n-1}(z)] = a_n. \qquad (11.9\text{-}6)$$

(Here $F_{-1} := 0$.)

THEOREM 11.9a

The properties (A), (B), (C) *are equivalent.*

Proof. We shall show $(A) \Rightarrow (B) \Rightarrow (C) \Rightarrow (A)$. Written out in full, (11.9-4) means that for each n there exist numbers κ_n and ρ_n such that

$$|f(z) - F_n(z)| \leq \kappa_n |g_{n+1}(z)|, \qquad |z| > \rho_n, \qquad z \in S.$$

Because $g_n(z) \neq 0$, it follows that for the same z

$$\left| \frac{1}{g_n(z)} [f(z) - F_n(z)] \right| \leq \kappa_n \left| \frac{g_{n+1}(z)}{g_n(z)} \right|.$$

The expression on the right tends to zero for $z \to \infty$ by (11.9-2); hence (B) follows. Now for each n,

$$\frac{1}{g_n(z)} [f(z) - F_n(z)] = \frac{1}{g_n(z)} [f(z) - F_{n-1}(z)] - a_n.$$

Because this tends to zero, (C) follows. Using (C) with the index $n + 1$, we see by the same decomposition that

$$\lim_{\substack{z \to \infty \\ z \in S}} \frac{1}{g_{n+1}(z)} [f(z) - F_{n+1}(z)] = 0.$$

Hence there exists ρ_n such that

$$\left| \frac{1}{g_{n+1}(z)} [f(z) - F_{n+1}(z)] \right| < 1, \qquad |z| > \rho_n, \qquad z \in S$$

or, by the triangle inequality, for the same z,

$$\left| \frac{1}{g_{n+1}(z)} [f(z) - F_n(z)] \right| < 1 + |a_{n+1}|.$$

(A) now is an immediate consequence. ∎

Theorem 11.9a justifies the following:

DEFINITION

*Let $\{g_n\}$ be an asymptotic sequence in the unbounded set S, let f be defined on S and let $\{a_n\}$ be a sequence of complex numbers. If the sequence of functions $\{F_n\}$ defined by (11.9-3) enjoys any of the properties (A), (B), (C), then f is said to admit the **asymptotic expansion** $\sum a_n g_n$ as $z \to \infty$ in S. This is expressed by writing*

$$f(z) \approx \sum_{n=0}^{\infty} a_n g_n(z), \qquad z \to \infty, \qquad z \in S.$$

Because the numbers a_n must satisfy (11.9-6), a given function can possess at most one asymptotic expansion in terms of a given sequence $\{g_n\}$ and in a given set S. Naturally, it need not possess an asymptotic expansion at all.

In the remainder of this section, we deal in some generality with asymptotic expansions in terms of the asymptotic sequence

$$g_n(z) := \frac{1}{(z+b)_n}, \qquad n = 0, 1, 2, \ldots,$$

where b is a fixed complex number. We begin by describing a general method for obtaining such expansions. This method bears the same relationship to Euler's beta integral which the Watson–Doetsch lemma bears to Euler's gamma integral.

THEOREM 11.9b

Let $\text{Re } b > -1$, *and let* ϕ *be a continuous, complex-valued function defined on* $(0, 1]$ *satisfying*

$$\phi(\tau) \approx \Phi := \tau^b \{a_0 + a_1 \tau + a_2 \tau^2 + \cdots\}, \qquad \tau \to 0, \qquad \tau > 0. \qquad (11.9\text{-}7)$$

Then the integral

$$f(z) := \int_0^1 (1-\tau)^{z-1} \phi(\tau) \, d\tau \qquad (11.9\text{-}8) \text{ -}$$

defines a function that is analytic in $\text{Re } z > 0$ *and for every* δ, $0 \leq \delta < \pi/2$, *admits the asymptotic expansion*

$$f(z) \approx F := \frac{\Gamma(z)}{\Gamma(z+b+1)} \sum_{m=0}^{\infty} a_m \frac{\Gamma(b+m+1)}{(z+b+1)_m}, \qquad z \to \infty, \qquad z \in \hat{S}_\delta. \quad (11.9\text{-}9)$$

Proof. That f is analytic in $\text{Re } z > 0$ follows from Theorem 4.1 (integration with respect to a parameter). The asymptotic nature of the series (11.9-9) is established by considerations similar to those used in the proof of the Watson–Doetsch lemma. The series F can be obtained formally by substituting (11.9-7) into (11.9-8) and using the beta integral in the form

$$\int_0^1 \tau^{b+m}(1-\tau)^{z-1} \, d\tau = \frac{\Gamma(b+m+1)\Gamma(z)}{\Gamma(z+b+m+1)}.$$

Consequently, denoting by Φ_m and F_m the mth partial sums of the series Φ and F, respectively, we have

$$F_m(z) = \int_0^1 (1-\tau)^{z-1} \Phi_m(\tau) \, d\tau, \qquad m = 0, 1, 2, \ldots.$$

To show that F has (say) property (B) with respect to f, it thus suffices to show that for $m = 0, 1, 2, \ldots$,

$$\frac{\Gamma(z+b+1)}{\Gamma(z)}(z+b+1)_m \int_0^1 (1-\tau)^{z-1}[\phi(\tau) - \Phi_m(\tau)] \, d\tau \to 0$$

$$(11.9\text{-}10)$$

as $z \to \infty$, $z \in \hat{S}_\delta$.

Let m and $\epsilon > 0$ be given. By (11.9-7) there exists $\eta > 0$ such that

$$|\tau^{-b-m}[\phi(\tau) - \Phi_m(\tau)]| < \epsilon \qquad (11.9\text{-}11)$$

for $0 < \tau \leq \eta$. We set

$$\int_0^1 = \int_0^\eta + \int_\eta^1$$

and estimate the two integrals separately. Using (11.9-11), we find for the first integral, setting $b = \beta + i\beta'$, $z = x + iy$,

$$\left| \int_0^\eta (1-\tau)^{z-1}[\phi(\tau) - \Phi_m(\tau)]\, d\tau \right| \leq \epsilon \int_0^\eta (1-\tau)^{x-1}\tau^{\beta+m}\, d\tau$$

$$\leq \frac{\Gamma(x)\Gamma(\beta+m+1)}{\Gamma(x+\beta+m+1)}.$$

Thus the contribution of the first integral to the expression on the left of (11.9-10) is bounded by $\epsilon\Gamma(\beta+m+1)\kappa(z)$, where

$$\kappa(z) := \left| \frac{\Gamma(z+b+m+1)}{\Gamma(z)} \frac{\Gamma(x)}{\Gamma(x+\beta+m+1)} \right|.$$

By Stirling's formula, if $\psi := \operatorname{Arg} z$,

$$\kappa(z) \sim e^{-\beta'\psi - (\beta+m+1)\operatorname{Log}(\cos\psi)}, \qquad z \to \infty,$$

which remains bounded for $z \to \infty$ because $\psi \leq \delta < \pi/2$. It follows that the contribution of the first integral to (11.9-10) is bounded by $\epsilon\kappa$, where κ is independent of z and ϵ.

We now estimate the contribution of the second integral. Because $\phi - \Phi_m$ is continuous on $[\eta, 1]$, $|\phi - \Phi_m| < \mu$ there (say), and we find immediately

$$\left| \int_\eta^1 (1-\tau)^{z-1}[\phi(\tau) - \Phi_m(\tau)]d\tau \right| \leq \mu \int_\eta^1 (1-\tau)^{x-1}\, d\tau$$

$$= \mu x^{-1}(1-\eta)^x.$$

Once more, by Stirling's formula,

$$\left| \frac{\Gamma(z+b+m+1)}{\Gamma(z)} \right| \sim e^{-\beta'\psi}|z|^{\beta+m+1}.$$

We thus see that the whole expression (11.9-10) is for $|z|$ sufficiently large, $z \in \hat{S}_\delta$, bounded by

$$\epsilon\kappa + \kappa'|z|^{\beta+m}(1-\eta)^{|z|\cos\delta}.$$

Here the first term can be made arbitrarily small by choosing ϵ sufficiently small, and the second by making $|z| \to \infty$. This establishes (11.9-10), hence property (B) and thus the theorem. ■

As an application, we can determine the asymptotic behavior of the hypergeometric function in its dependence on the third (denominator) parameter.

THEOREM 11.9c

Let $0 \le \alpha < \pi/2$. *For complex numbers* a, b, c, z *satisfying* $z \in \hat{S}_\alpha$, $|\arg(1-c)| < \pi$, *let* $F(a, b, z, c)$ *denote the value of the hypergeometric function obtained by analytic continuation of the hypergeometric series along the ray from 0 to c. Then the asymptotic behavior of F for* $z \to \infty$, $z \in \hat{S}_\alpha$ *and fixed* a, b, c *is described by the hypergeometric series* (even though it may be divergent); *that is, there holds the asymptotic expansion in* $1/(z)_n$,

$$F(a, b; z; c) \approx \sum_{n=0}^{\infty} \frac{(a)_n (b)_n c^n}{n!} \frac{1}{(z)_n}, \qquad z \to \infty, \qquad z \in \hat{S}_\alpha. \qquad (11.9\text{-}12)$$

Proof. We first assume $\operatorname{Re} b > 0$. For $|z|$ sufficiently large we also have $\operatorname{Re} z > \operatorname{Re} b$ since $z \in \hat{S}_\alpha$. Under these circumstances there holds Riemann's integral representation (8.7-8),

$$F(a, b; z; c) = \frac{\Gamma(z)}{\Gamma(b)\Gamma(z-b)} \int_0^1 (1-c\tau)^{-a} \tau^{b-1} (1-\tau)^{z-b-1} \, d\tau.$$

Apart from the Γ factor, this is precisely of the form (11.9-8) where z is replaced by $z - b$ and

$$\phi(\tau) := \tau^{b-1} (1-c\tau)^{-a}.$$

For $\tau < |c|^{-1}$ we have

$$\phi(\tau) = \tau^{b-1} \sum_{n=0}^{\infty} \frac{(a)_n}{n!} c^n \tau^n,$$

which as a convergent power series certainly is asymptotic. Formula (11.9-9) now immediately yields the result (11.9-12).

If $\operatorname{Re} b < 0$, let k be an integer such that $\operatorname{Re} b + k > 0$. We use the decomposition

$$F(a, b; z; c) = \sum_{n=0}^{k-1} \frac{(a)_n (b)_n c^n}{n!} \frac{1}{(z)_n} + \frac{(a)_k (b)_k c^k}{k!(z)_k} F(a+k, b+k; z+k; c),$$

and on applying the foregoing result to the hypergeometric function on the right the desired formula again follows. ■

We utilize Theorem 11.9c to determine the asymptotic behavior as $n \to \infty$ of the Legendre polynomials $P_n(z)$ for a fixed complex value of z that is not real and outside the interval $(-1, 1)$. (For the excepted case the asymptotic behavior was discussed by means of Watson's Lemma in § 11.5.) We have

$$P_n(z) = F\left(-n, n+1; 1; \frac{1-z}{2}\right), \qquad n = 0, 1, 2, \ldots.$$

The hypergeometric function appearing here is not of the form to which Theorem 11.9c is directly applicable, because the large parameter appears in the numerator and not in the denominator. However, because of the large number of transformations of the hypergeometric series at our disposal, we may hope to find another hypergeometric representation of $P_n(z)$ in which n occurs in the denominator only.

It is convenient to consider separately the cases in which n is even and where n is odd. If n is even, $n = 2m$, then P_n is an even polynomial. By Goursat's formula (9.10-4) we then have

$$P_{2m}(z) = \frac{1}{2}\left\{ F\left(-2m, 2m+1; 1; \frac{1+z}{2}\right) + F\left(-2m, 2m+1; 1; \frac{1-z}{2}\right)\right\}$$

$$= (-1)^m \frac{(1/2)_m}{m!} F\left(-m, m+\frac{1}{2}; \frac{1}{2}; z^2\right).$$

By Euler's first transformation this becomes

$$(-1)^m \frac{(1/2)_m}{m!} w^{-2m-1} F\left(m+\frac{1}{2}, m+\frac{1}{2}; \frac{1}{2}; \frac{z^2}{z^2-1}\right),$$

where $w := (1+z)^{1/2}(1-z)^{1/2}$ with $|\arg(1+z)| < \pi$, $|\arg(1-z)| < \pi$. Applying Goursat now in the opposite direction, we obtain

$$(-1)^m \frac{(2m)!}{2^{2m}(3/2)_{2m}} w^{-2m-1}$$

$$\cdot \left\{ F\left(2m+1, 2m+1; 2m+\frac{3}{2}; \frac{1}{2}-\frac{iz}{2w}\right) + F\left(\cdots; \frac{1}{2}+\frac{iz}{2w}\right)\right\},$$

where the parameters are the same in both cases. We can get rid of the large numerator parameters by Euler's second transformation. This yields

$$(-1)^m \frac{\sqrt{2}}{\pi} \frac{(2m)!}{(3/2)_{2m}} w^{-1/2} \cdot \left\{ (w+iz)^{-2m-1/2} F\left(\frac{1}{2}, \frac{1}{2}; 2m+\frac{3}{2}; \frac{1}{2}-\frac{iz}{2w}\right)\right.$$

$$\left. + (w-iz)^{-2m-1/2} F\left(\frac{1}{2}, \frac{1}{2}; 2m+\frac{3}{2}; \frac{1}{2}+\frac{iz}{2w}\right)\right\}.$$

An analogous computation can be carried through for odd values of n using (9.10-5) twice. The two results may be combined into the single formula

$$P_n(z) = \frac{\sqrt{2}}{\pi} \frac{n!}{(3/2)_n} w^{-1/2} \cdot \left\{ e^{-i\pi/4} (z+iw)^{n+1/2} F\left(\frac{1}{2}, \frac{1}{2}; \frac{3}{2}+n; \frac{1}{2} - \frac{iz}{2w}\right) \right.$$
$$\left. + e^{i\pi/4} (z-iw)^{n+1/2} F\left(\frac{1}{2}, \frac{1}{2}; \frac{3}{2}+n; \frac{1}{2} + \frac{iz}{2w}\right) \right\}. \qquad (11.9\text{-}13)$$

This is a representation of the required type, in which the third parameter of the hypergeometric series is large and positive and all other quantities are fixed. To discuss the leading terms of the asymptotic expansion, we discuss two cases.

(i) If z is real, $-1 < z < 1$, let $z = \cos \theta$, where $0 < \theta < \pi$. Then $z \pm iw = e^{\pm i\theta}$, and the foregoing reads

$$P_n(\cos \theta) = \frac{2^{3/2}}{\pi} \frac{n!}{(3/2)_n} \frac{1}{\sqrt{\sin \theta}} \operatorname{Re}\left\{ \exp\left[i\left(n+\frac{1}{2}\right)\theta - i\frac{\pi}{4}\right] \right.$$
$$\left. \cdot F\left(\frac{1}{2}, \frac{1}{2}; \frac{3}{2}+n; \frac{1}{2} - \frac{\cot \theta}{2}\right) \right\}.$$

In view of

$$\frac{n!}{(3/2)_n} \sim \frac{1}{2} \sqrt{\frac{\pi}{n}}, \qquad n \to \infty,$$

the leading term yields the asymptotic formula

$$P_n(\cos \theta) \sim \left[\frac{2}{\pi n \sin \theta} \right]^{1/2} \left\{ \cos\left(\left(n+\frac{1}{2}\right)\theta - \frac{\pi}{4}\right) + O(n^{-1}) \right\}, \qquad n \to \infty. \qquad (11.9\text{-}14)$$

(ii) If z is not real, we first assume $\operatorname{Im} z > 0$. Then $z + iw = z + \sqrt{z^2 - 1}$, where $iw = \sqrt{z^2 - 1} = \sqrt{z+1}\sqrt{z-1}$, $0 < \arg(z \pm 1) < \pi$. Thus $\operatorname{Im} iw > 0$, $|z + iw| > |z - iw|$. Hence the second term in (11.9-13) is exponentially small compared to the first, and the asymptotic behavior is given by

$$P_n(z) \approx \frac{\sqrt{2}}{\pi} \frac{n!}{(3/2)_n} \frac{(z+\sqrt{z^2-1})^{n+1/2}}{(z^2-1)^{1/4}} F\left(\frac{1}{2}, \frac{1}{2}; \frac{3}{2}+n; \frac{\sqrt{z^2-1}+z}{2\sqrt{z^2-1}}\right), \qquad n \to \infty. \qquad (11.9\text{-}15)$$

If $\operatorname{Im} z < 0$, then $iw = -\sqrt{z^2-1}$, $\operatorname{Im} iw < 0$; hence $|z+iw| < |z-iw|$. Only the second term in (11.9-13) is now significant. However, in view of the reversal of sign the appearance of the asymptotic expansion is as in (11.9-15).

PROBLEMS

1. Let ν be a complex number, but not an integer. Show: The sequence $g_0(n)$, $g_1(n), \ldots$, where

$$g_k(n) := \frac{(\nu - k)_n}{n!},$$

is asymptotic.

2. For $k = 0, 1, 2, \ldots$, let

$$-(1-z)^k \, \text{Log}(1-z) = \sum_{n=0}^{\infty} \delta_k(n) z^n, \qquad |z| < 1.$$

Show that the sequence of functions $\delta_0(n), \delta_1(n), \delta_2(n), \ldots$ is asymptotic. [See Problem 5, §1.7, for a representation of $\delta_k(n)$.]

3. So that $f(z) \approx \sum_{n=0}^{\infty} a_n g_n(z)$, it is not necessary that one of the properties (A), (B), (C) holds for all integers n. It suffices that the property holds for infinitely many n.

4. Let $\{a_n\}$ be a sequence of complex numbers such that

$$0 < \rho := \limsup_{n \to \infty} |a_n|^{1/n} < 1.$$

Show that the series

$$\sum_{n=0}^{\infty} a_n P_n(z)$$

converges locally uniformly (and hence represents an analytic function) in the interior of the ellipse with foci at ± 1 whose sum of semiaxes equals ρ^{-1}, and that the series diverges outside that ellipse.

5. Find an asymptotic representation similar to (11.9-13) for the ultraspherical polynomials

$$C_n^{\nu}(z) := \frac{(2\nu)_n}{n!} F\left(-n, n + 2\nu; \nu + \frac{1}{2}; \frac{1-z}{2}\right).$$

6. If Q_{ν}^{μ} denotes the Legendre function of the second kind, as defined by (9.10-12), show that for z and μ fixed, $|\arg(z \pm 1)| < \pi$,

$$Q_{\nu}^{\mu}(z) \approx \sqrt{\frac{\pi}{2}} \frac{\Gamma(\mu + \nu + 1)}{\Gamma(\nu + 3/2)} \frac{(z - \sqrt{z^2 - 1})^{\nu + 1/2}}{(z^2 - 1)^{1/4}}$$

$$\cdot F\left(\mu + \frac{1}{2}, -\mu + \frac{1}{2}; \nu + \frac{3}{2}; \frac{-z + \sqrt{z^2 - 1}}{2\sqrt{z^2 - 1}}\right), \qquad \nu \to \infty, \qquad \nu \in \hat{S}_{\pi/2 - \epsilon}.$$

7. Let $\{g_n\}$, $\{h_n\}$ be two asymptotic sequences in the same region S. The sequence $\{g_n\}$ is called **coasymptotic** to the sequence $\{h_n\}$ if there exists an upper triangular matrix $\mathbf{T} = (t_{nm})$ where $t_{nn} \neq 0$, $n = 0, 1, 2, \ldots$, such that for $n = 0, 1, 2, \ldots$

$$g_n(z) \approx \sum_{m=n}^{\infty} t_{nm} h_m(z), \qquad z \to \infty, \qquad z \in S. \qquad (11.9\text{-}16)$$

Introducing the infinite column vectors

$$\mathbf{g} := \begin{pmatrix} g_0 \\ g_1 \\ \vdots \end{pmatrix}, \qquad \mathbf{h} := \begin{pmatrix} h_0 \\ h_1 \\ \vdots \end{pmatrix},$$

relation (11.9-16) is formally the same as $\mathbf{g} = \mathbf{Th}$. Prove:

(a) If $\{g_n\}$ is coasymptotic to $\{h_n\}$, then $\{h_n\}$ is coasymptotic to $\{g_n\}$, and

$$\mathbf{h} = \mathbf{T}^{-1} \mathbf{g}.$$

(b) Let

$$f(z) \approx \sum_{n=0}^{\infty} c_n g_n(z), \qquad z \to \infty, \qquad z \in S.$$

Then

$$f(z) \approx \sum_{m=0}^{\infty} d_m h_m(z), \qquad z \to \infty, \qquad z \in S,$$

where

$$d_m := \sum_{n=0}^{m} c_m t_{nm}, \qquad m = 0, 1, 2, \ldots,$$

which, on defining

$$\mathbf{c} := \begin{pmatrix} c_0 \\ c_1 \\ \vdots \end{pmatrix}, \qquad \mathbf{d} := \begin{pmatrix} d_0 \\ d_1 \\ \vdots \end{pmatrix}$$

may be written

$$\mathbf{d}^T = \mathbf{c}^T \mathbf{T}.$$

(c) Show that the asymptotic sequences

$$\frac{1}{z^n} \quad \text{and} \quad \frac{1}{(z)_n}, \qquad n = 0, 1, 2, \ldots,$$

are coasymptotic in every sector $\hat{S}_\alpha$ where $\alpha < \pi$.

8. Applying Theorem 11.9b with

$$\phi(\tau) := -[\text{Log}(1-\tau)]^{p-1},$$

show that for every $\alpha < \pi$ and for $p = 1, 2, \ldots,$

$$\frac{1}{z^p} \approx \sum_{m=p}^{\infty} \frac{S_{m-p}^{m-1}}{(z)_m}, \qquad z \to \infty, \qquad z \in \hat{S}_\alpha,$$

where the S_m^n are the **Stirling numbers** defined by

$$(z)_n = S_0^n z^n + S_1^n z^{n-1} + \cdots + S_{n-1}^n z,$$

$n = 0, 1, 2, \ldots$.

[To calculate derivatives of f, note that for every sufficiently differentiable function g,

$$\frac{d^n}{d\sigma^n} g(\text{Log } \sigma) = \sigma^{-n} \sum_{r=0}^{n-1} (-1)^r \gamma_r^n g^{(n-r)}(\text{Log } \sigma),$$

where the numbers γ_r^n are independent of g. Identify the γ_r^n in terms of Stirling numbers by considering the special case $g(\text{Log } \sigma) := \sigma^{-c}$.]

§11.10. GENERATING FUNCTIONS; SUBTRACTED SINGULARITIES

Here we are concerned with the asymptotic behavior as $n \to \infty$ of quantities p_n that are defined as elements of the sequence of Taylor coefficients at 0 of a given function $p(t)$:

$$p(t) = \sum_{n=0}^{\infty} p_n t^n, \qquad |t| < \rho. \tag{11.10-1}$$

We recall that p is referred to as the **generating function** of the sequence $\{p_n\}$.

We quote a few among the numerous examples of generating functions that already have been encountered.

EXAMPLE **1**

The *Fibonacci numbers* $f_0, f_1, \ldots,$ ordinarily defined by the recurrence relation $f_0 := 1, f_1 := 1, f_n := f_{n-1} + f_{n-2}$ can also be defined by the generating function

$$\frac{1}{1-t-t^2} = \sum_{n=0}^{\infty} f_n t^n. \tag{11.10-2}$$

EXAMPLE **2**

The *Bernoulli numbers* $B_0, B_1, \ldots$ may be defined by the generating function (see Problem 4, §1.2)

$$\frac{t}{e^t - 1} = \sum_{n=0}^{\infty} \frac{B_n}{n!} t^n \tag{11.10-3}$$

Strictly speaking, the function of Example 2 "generates" the sequence $\{B_n/n!\}$ and not the sequence $\{B_n\}$; however, it is customary to extend the notion of generating functions to relations of the form

$$p(t) = \sum_{n=0}^{\infty} a_n p_n t^n,$$

where $\{a_n\}$ is some known sequence.

In many cases, the function p, and hence the coefficients p_n, depend on one or several parameters. For instance, many known systems of polynomials can be defined in this way.

EXAMPLE 3

The *Legendre polynomials* $P_n(z)$ may be defined by the generating function (see Example 4, §1.9)

$$\frac{1}{\sqrt{1-2zt+t^2}} = \sum_{n=0}^{\infty} P_n(z)t^n. \qquad (11.10\text{-}4)$$

More generally, the ultraspherical polynomials $C_n^{\nu}(z)$ can for arbitrary ν be defined by

$$(1-2zt+t^2)^{-\nu} = \sum_{n=0}^{\infty} C_n^{\nu}(z)t^n, \qquad (11.10\text{-}5)$$

so that $P_n(z) = C_n^{1/2}(z)$.

EXAMPLE 4

The Hermite polynomials[2] $H_n(z)$ can be generated as follows:

$$e^{2zt-t^2} = \sum_{n=0}^{\infty} H_n(z)\frac{t^n}{n!}. \qquad (11.10\text{-}6)$$

EXAMPLE 5

The Laguerre polynomials $L_n^{(\alpha)}(z)$ are given by

$$\frac{1}{(1-t)^{\alpha+1}} \exp\left(-\frac{zt}{1-t}\right) = \sum_{n=0}^{\infty} L_n^{(\alpha)}(z); \qquad (11.10\text{-}7)$$

see problem 9, §10.5.

---

[2] Several notations for the Hermite polynomials are in common use. The polynomials here denoted by $H_n(z)$ are identical with the polynomials $He_n^*(z)$ of Magnus and Oberhettinger [1958], and with the polynomials $H_n(z)$ of Szegö [1959] and Hilbert-Courant [1930]. Polya and Szegö [1925] employ a different notation, as we have done in Example 1 of § 1.6.

If the sequence $\{p_n\}$ possesses the generating function p, then we immediately have by Cauchy's formula

$$p_n = \frac{1}{2\pi i} \int_\Lambda p(\zeta)\zeta^{-n-1}\, d\zeta, \qquad (11.10\text{-}8)$$

$n = 0, 1, 2, \ldots$, where the closed curve Λ winds around 0 once positively, and where p is analytic in a simply connected region containing Λ. By setting $h(\zeta) := p(\zeta)\zeta^{-1}$, $g(\zeta) := \log \zeta$, this integral can be cast into the form (11.8-1), and it thus seems plausible to investigate the behavior of p_n for $n \to \infty$ by the method of steepest descent. However, g has no saddle points, and Λ is closed. The methods of §11.8 are thus not directly applicable; for some exceptions, see De Bruijn [1961].

We mention in passing that the saddle-point method may be fruitfully applied to Cauchy's integral in the study of the asymptotic behavior of p_n if n and one of the parameters simultaneously become large.

EXAMPLE **6**

Replacing z by $-2nq$ in (11.10-7) and applying (11.10-8), we obtain

$$L_n^{(\alpha)}(-2nq) = \frac{1}{2\pi i} \int_\Lambda (1-\zeta)^{-\alpha-1}\exp\left(2nq\frac{\zeta}{1-\zeta}\right)\zeta^{-n-1}\, d\zeta,$$

which may be written as

$$L_n^{(\alpha)}(-2nq) = \frac{1}{2\pi i} \int_\Lambda \zeta^{-1}(1-\zeta)^{-\alpha-1}\exp\left[n\left(\frac{2q\zeta}{1-\zeta}-\log \zeta\right)\right]\, d\zeta. \quad (11.10\text{-}9)$$

EXAMPLE **7**

Letting $z := \sqrt{2nq}$ in (11.10-6), we obtain in a similar fashion

$$H_n(\sqrt{2nq}) = \frac{n!}{2\pi i} \int_\Lambda e^{2\sqrt{2nq}t - t^2}t^{-n-1}\, dt,$$

which on letting $t = \sqrt{n/2}\,\zeta$ turns into

$$H_n(\sqrt{2nq}) = \frac{n!}{2\pi i}\left(\frac{n}{2}\right)^{n/2}\int_\Lambda \zeta^{-1}\, e^{ng(\zeta)}\, d\zeta, \qquad (11.10\text{-}10)$$

where

$$g(\zeta) := 2\zeta q - \tfrac{1}{2}\zeta^2 - \log \zeta.$$

As an alternate method for discussing the asymptotic behavior of sequences defined by generating functions, we now discuss the **method of subtracted singularities**. In its simplest form the method can already be illustrated by Example 1. Decomposing the generating function of the Fibonacci numbers into partial fractions, we obtain

$$\frac{1}{1-t-t^2}=\frac{1/\sqrt{5}}{(\sqrt{5}-1)/2-t}+\frac{1/\sqrt{5}}{(\sqrt{5}+1)/2+t}.$$

The first partial fraction has a pole at $t=t_1:=(\sqrt{5}-1)/2$, the second at $t_2=(-\sqrt{5}-1)/2$. We note that $|t_2|>|t_1|$.

Ignoring that the complete partial fraction decomposition is known, we write the foregoing in the form

$$\frac{1}{1-t-t^2}=\frac{1/\sqrt{5}}{(\sqrt{5}-1)/2-t}+g(t),$$

where about g we merely need to know that it is analytic for $|t|\leq\rho$ where $\rho>|t_1|$. Now the first term can immediately be expanded in a power series:

$$\frac{1/\sqrt{5}}{(\sqrt{5}-1)/2-t}=\frac{\sqrt{5}+1}{2\sqrt{5}}\sum_{n=0}^{\infty}\left(\frac{2t}{\sqrt{5}-1}\right)^n.$$

As to the power series of g, we know by the Cauchy estimate that its coefficients are bounded by $\mu\rho^{-n}$, where μ is a constant. It thus follows that

$$f_n=\frac{\sqrt{5}+1}{2\sqrt{5}}\left(\frac{2}{\sqrt{5}-1}\right)^n+O(\rho^{-n}),$$

or

$$f_n\sim\frac{1}{\sqrt{5}}\left(\frac{\sqrt{5}+1}{2}\right)^{n+1},\qquad n\to\infty.$$

We next consider a more general case:

THEOREM 11.10a

Let the function p be meromorphic, with simple poles at the points t_m, $m=1,2,\ldots$, where $|t_{m+1}|>|t_m|$, $m=1,2,\ldots$, and let r_m be the residue at t_m. Then the coefficients p_n defined by

$$p(t)=\sum_{n=0}^{\infty}p_n t^n$$

possess the following asymptotic expansion in terms of the asymptotic sequence $\{t_m^{-n}\}$:

$$p_n\approx-\sum_{m=1}^{\infty}\frac{r_m}{t_m^{n+1}},\qquad n\to\infty. \tag{11.10-11}$$

Proof.   For any positive integer m, the function

$$p(t) - \frac{r_1}{t-t_1} - \frac{r_2}{t-t_2} - \cdots - \frac{r_m}{t-t_m}$$

is analytic in $|t| < |t_{m+1}|$. Its nth Taylor coefficient,

$$p_n + \frac{r_1}{t_1^{n+1}} + \cdots + \frac{r_m}{t_m^{n+1}},$$

thus is $O(\rho^{-n})$ for $n \to \infty$, where ρ is any number $< |t_{m+1}|$. There follows

$$\lim_{n \to \infty} t_m^n \left[p_n + \frac{r_1}{t_1^{n+1}} + \cdots + \frac{r_m}{t_m^{n+1}} \right] = 0$$

for $m = 1, 2, \ldots$. The formal series (11.10-11) thus satisfies property (B) of §11.9, which is equivalent to the statement of the theorem. ∎

EXAMPLE **8**

Let $\{p_n\}$ be the sequence of Taylor coefficients of $\Gamma(z)$ at $z = 1$,

$$\Gamma(1+t) = \sum_{n=0}^{\infty} p_n t^n.$$

It is known that $p_0 = 1$, $p_1 = -\gamma$ (the Euler constant); no simple formula for the general coefficient exists. However, an asymptotic expansion is easily found. The function $\Gamma(1+t)$ has simple poles at the points $t_m = -m$, $m = 1, 2, \ldots$, with residues $r_m = (-1)^{m-1}(m-1)!$; hence Theorem 11.10a yields

$$p_n \approx \sum_{k=1}^{\infty} (-1)^{n+k-1} \frac{(k-1)!}{k^{n+1}}, \qquad n \to \infty.$$

It is easy to see by means of the ratio test that the above series diverges for every n. However, as an asymptotic series it has a definite meaning.

   Theorem 11.10a can be extended to the case in which there are poles of order higher than 1, or where several poles have equal moduli. We leave these generalizations to the imagination of the reader and turn instead to the situation, also of frequent occurrence in practice, in which the generating function has singularities other than poles on the boundary of its disk of convergence. Because there is no partial fraction expansion in such cases, the simple device of subtracting singularities no longer works. However, asymptotic expansions can frequently be obtained by a method originally due to Darboux. It makes use of certain elementary properties of Fourier series.

   Before stating Darboux' result in a simple special case, we recall from §11.9 that, for any complex number ν that is not an integer, the sequence of functions defined on the positive integers $n = 1, 2, \ldots$ by

$$g_k(n) := \frac{(\nu - k)_n}{n!}, \qquad k = 0, 1, 2, \ldots \tag{11.10-12}$$

forms an asymptotic sequence as $n \to \infty$. In fact, it is easily seen from Stirling's formula that

$$g_k(n) \sim \frac{1}{\Gamma(\nu-k)} n^{\nu-k-1}, \qquad n \to \infty. \tag{11.10-13}$$

This fact permits us to state:

THEOREM 11.10b (Theorem of Darboux)

Let t_1 and ν be complex numbers, $t_1 \neq 0$, and let ν not be an integer. For $|\arg(1 - t t_1^{-1})| < \pi$, let

$$p(t) = \left(1 - \frac{t}{t_1}\right)^{-\nu} r(t), \tag{11.10-14}$$

where the power has its principal value, and where r is analytic in $|t| < \rho$ for some $\rho > |t_1|$. For $|t - t_1| < \rho - |t_1|$, let

$$r(t) = \sum_{k=0}^{\infty} b_k (t - t_1)^k.$$

Then the coefficients p_n generated by p admit the following asymptotic expansion in terms of the asymptotic sequence (11.10-12):

$$p_n \approx t_1^{-n} \sum_{k=0}^{\infty} (-1)^k b_k t_1^k \frac{(\nu-k)_n}{n!}, \qquad n \to \infty. \tag{11.10-15}$$

Proof. Let m be an integer, $m > \max(0, \operatorname{Re} \nu)$. We have

$$r(t) = b_0 + b_1(t - t_1) + \cdots + b_{m-1}(t - t_1)^{m-1} + (t - t_1)^m r_m(t),$$

where r_m is analytic in $|t| < \rho$. Using this representation of r in (11.10-14) and transposing terms, we get

$$p(t) - \sum_{k=0}^{m-1} (-1)^k b_k t_1^k \left(1 - \frac{t}{t_1}\right)^{k-\nu} = (-1)^m t_1^m \left(1 - \frac{t}{t_1}\right)^{m-\nu} r_m(t). \tag{11.10-16}$$

Using the binomial theorem in the form

$$\left(1 - \frac{t}{t_1}\right)^{-\mu} = \sum_{n=0}^{\infty} \frac{(\mu)_n}{n!} \left(\frac{t}{t_1}\right)^n,$$

the nth Taylor coefficient at $t = 0$ of the function on the left is readily found to be

$$p_n - \sum_{k=0}^{m-1} (-1)^k b_k t_1^{k-n} \frac{(\nu-k)_n}{n!}, \qquad n = 0, 1, 2, \ldots.$$

To estimate the nth Taylor coefficient of the function on the right, we use the following two facts on Fourier series:

(i) Let the function

$$f(t) := \sum_{n=0}^{\infty} a_n t^n$$

be analytic $|t| < \tau$ and continuous on $|t| \leq \tau$. Then a_n equals τ^{-n} times the nth Fourier coefficient of the function $f(\tau e^{i\phi})$. This follows much as in §4.5 by expressing a_n by Cauchy's integral, extended over the circle $|t| = \tau_1 < \tau$ and letting $\tau_1 \to \tau$.

(ii) If a periodic function is d times continuously differentiable, its nth Fourier coefficient is $O(n^{-d})$ as $n \to \infty$.

The function on the right of (11.10-16) is continuous for $|t| \leq |t_1|$, hence its Taylor coefficients are the Fourier coefficients of

$$g(\phi) := (-1)^m t_1^m (1 - \epsilon e^{i\phi})^{m-\nu} r_m(|t_i| e^{i\phi}),$$

where $\epsilon := |t_1| t_1^{-1}$. The function g evidently is $[m - \operatorname{Re} \nu]$ times continuously differentiable; hence its Fourier coefficients are $O(n^{-[m-\operatorname{Re}\nu]})$. We thus find

$$p_n = t_1^{-n} \sum_{k=0}^{m-1} b_k (-t_1)^k \frac{(\nu-k)_n}{n!} + O(t_1^{-n} n^{-[m-\operatorname{Re}\nu]}),$$

$n \to \infty$. This relation in itself is not yet quite sufficient to establish the assertion, because the O term is not small compared to the last member of the sum. However, by replacing m by $m+1$ and using (11.10-13), we readily find

$$p_n = t_1^{-n} \left\{ \sum_{k=0}^{m-1} b_k (-t_1)^k \frac{(\nu-k)_n}{n!} + O\left(\frac{(\nu-m)_n}{n!} \right) \right\},$$

$n \to \infty$, which is equivalent to property (B) of §11.9 and thus establishes the theorem. ■

An analogous result can, of course, be derived if p has any finite number of singularities of type (11.10-14) on the boundary of the disk of convergence of its Taylor series.

EXAMPLE **9**

We apply Theorem 11.10b to obtain an asymptotic formula for the ultraspherical polynomials $C_n^\nu(z)$ for $n \to \infty$ if ν and z are fixed complex numbers, ν not an integer, and z outside the real interval $[-1, 1]$. We have

$$1 - 2zt + t^2 = \left(1 - \frac{t}{t_1}\right)\left(1 - \frac{t}{t_2}\right),$$

where $t_{1,2} = z \pm \sqrt{z^2 - 1}$. If z is not in the real interval $[-1, 1]$, then the numbers t_1 and t_2 are not conjugate and have different moduli. We choose the indices such that $|t_1| < |t_2|$. The generating function of the ultraspherical polynomials,

$$(1 - 2zt + t^2)^{-\nu} = \left(1 - \frac{t}{t_1}\right)^{-\nu}\left(1 - \frac{t}{t_2}\right)^{-\nu},$$

is now precisely of the form (11.10-14), where

$$r(t) := \left(1 - \frac{t}{t_2}\right)^{-\nu}.$$

To obtain the required expansion around the point t_1 we write

$$r(t) = \left(1 - \frac{t - t_1 + t_1}{t_2}\right)^{-\nu} = \left(1 - \frac{t_1}{t_2}\right)^{-\nu}\left(1 - \frac{t - t_1}{t_2 - t_1}\right)^{-\nu}$$

$$= \left(1 - \frac{t_1}{t_2}\right)^{-\nu} \sum_{k=0}^{\infty} \frac{(\nu)_k}{k!}\left(\frac{t - t_1}{t_2 - t_1}\right)^k$$

and thus get

$$b_k = \left(1 - \frac{t_1}{t_2}\right)^{-\nu} \frac{(\nu)_k}{k!}(t_2 - t_1)^{-k},$$

where the first power again has its principal value. After some simplification we find from (11.10-15)

$$C_n^{\nu}(z) \approx \left(1 - \frac{t_1}{t_2}\right)^{-\nu} t_2^n \sum_{k=0}^{\infty} \frac{(\nu)_k}{k!} \frac{(\nu - k)_n}{n!}\left(\frac{t_1}{t_1 - t_2}\right)^k, \qquad (11.10\text{-}17)$$

$n \to \infty$. By using the identity

$$(\nu - k)_n = \frac{(\nu)_n (1 - \nu)_k}{(1 - \nu - n)_k},$$

this may be written in the form

$$C_n^{\nu}(z) \approx \left(1 - \frac{t_1}{t_2}\right)^{-\nu} \frac{(\nu)_n}{n!} F\left(\nu, 1 - \nu; 1 - \nu - n; \frac{t_1}{t_1 - t_2}\right), \qquad (11.10\text{-}18)$$

which expresses the asymptotic expansion as a hypergeometric series. The series (11.10-18) can also be obtained from the representation

$$C_n^{\nu}(z) = \frac{(\nu)_n}{n!} z^n F\left(-\frac{n}{2}, -\frac{n}{2} + \frac{1}{2}; 1 - \nu - n; z^{-2}\right) \qquad (11.10\text{-}19)$$

by means of a quadratic transformation. The series is not of the type considered in Theorem 11.9b, because the denominator parameter tends to $-\infty$ instead of $+\infty$. Hence the derivation from (11.10-19) does not divulge its asymptotic character as $n \to \infty$, which persists even if $|t_1(t_1 - t_2)^{-1}| > 1$ and the series is divergent.

PROBLEMS

1. Let q_n denote the probability that in n tosses of an ideal coin no run of three consecutive heads appears. Clearly, $q_0 = q_1 = q_2 = 1$, and in probability theory it is established that $q_n = \frac{1}{2}q_{n-1} + \frac{1}{4}q_{n-2} + \frac{1}{8}q_{n-3}$. Show that the q_n possess the generating function

$$\sum_{n=0}^{\infty} q_n t^n = \frac{2t^2 + 4t + 8}{8 - 4t - 2t^2 - t^3}$$

and deduce the asymptotic formula

$$q_n \sim \frac{1.2368398446}{(1.0873780254)^{n+1}}, \qquad n \to \infty$$

    (Feller [1957], p. 260.)

2. For $n = 1, 2, 3, \ldots$, let

$$a_n := 1 + \frac{1-n}{1!} + \frac{(2-n)^2}{2!} + \cdots + \frac{(-1)^{n-1}}{(n-1)!}.$$

    Find the generating function of the sequence $\{a_n\}$ and show that $a_n \to 0, n \to \infty$.

3. What is the asymptotic behavior of the numbers p_n defined by

$$\frac{\sqrt{z}}{\sin\sqrt{z}} = \sum_{n=0}^{\infty} p_n z^n?$$

4. Let $x > 1$. From the generating function of the Jacobi polynomials,

$$\sum_{n=0}^{\infty} P_n^{\alpha,\beta}(x)t^n = 2^{\alpha+\beta}R^{-1}(1-t+R)^{-\alpha}(1+t+R)^{-\beta},$$

    where $R := (1-2tx+t^2)^{1/2}$, deduce by Darboux' method that for $n \to \infty$,

$$P_n^{\alpha,\beta}(x) \sim (x-1)^{-\alpha/2}(x+1)^{-\beta/2}\{\sqrt{x-1}+\sqrt{x+1}\}^{\alpha+\beta}$$

$$\cdot (x^2-1)^{-1/4}(2\pi n)^{-1/2}[x+\sqrt{x^2-1}]^{n+1/2}.$$

5. Develop an analog of Darboux' Theorem for sequences with generating functions of the form

$$p(t) = \mathrm{Log}\left(1 - \frac{t}{t_1}\right)r(t),$$

    where r is analytic in $|t| \le |t_1|$.

§11.11. THE EULER–MACLAURIN SUMMATION FORMULA

The objective in this section is the derivation of a classical formula, due to Euler and Maclaurin, that concerns functions f defined on the positive real line and, roughly speaking, provides a link between $\sum_{k=0}^{n} f(n)$ and $\int_{0}^{n} f$. The formula as such does not have the form of an asymptotic expansion, but it is frequently useful for *obtaining* asymptotic expansions of sums of the above form as $n \to \infty$.

The starting point is a problem in numerical integration. Let f be a continuous complex-valued function on $[0, 1]$ that has continuous derivatives of all orders. How can the integral

$$\int_{0}^{1} f(x)\, dx \tag{11.11-1}$$

be expressed in terms of the values of f and of the derivatives of f at the endpoints of the interval of integration?

The problem can be solved in many different ways by the following general procedure. Let p_1 be any polynomial such that $p_1'(x) = 1$. Applying integration by parts to (11.11-1) we then get[3]

$$\int_{0}^{1} f(x)\, dx = [p_1(x)f(x)]_0^1 - \int_{0}^{1} p_1(x)f'(x)\, dx.$$

The procedure may be repeated. If p_2 is any polynomial such that $p_2'(x) = p_1(x)$, we have

$$\int_{0}^{1} p_1(x)f'(x)\, dx = [p_2(x)f'(x)]_0^1 - \int_{0}^{1} p_2(x)f''(x)\, dx.$$

Continuing in this manner, we get the following preliminary result: Let $\{p_k\}$ denote any sequence of polynomials such that

(i) $\qquad p_0(x) = 1,\ p_{k+1}'(x) = p_k(x), \qquad k = 0, 1, 2, \ldots.$

Then there holds for $m = 0, 1, 2, \ldots$ and for any infinitely differentiable function f:

$$\int_{0}^{1} f(x)\, dx = \sum_{k=0}^{m} (-1)^k [p_{k+1}(x)f^{(k)}(x)]_0^1$$

$$+ (-1)^{m+1} \int_{0}^{1} p_{m+1}(x)f^{(m+1)}(x)\, dx. \tag{11.11-2}$$

This formula is now simplified by subjecting the sequence $\{p_k\}$ to the additional condition that

(ii) $\qquad \int_{0}^{1} p_k(x)\, dx = 0, \qquad k = 1, 2, \ldots.$

---

[3] Here and in the following we use the notation $[g(x)]_a^b := g(b) - g(a)$.

By (i), each p_k may be determined from p_{k-1} up to an additive constant. Condition (ii) now fixes that constant. Thus there exists one and only one sequence of polynomials satisfying both conditions (i) and (ii).

In view of (i) it follows from (ii) that

$$p_k(1) = p_k(0), \qquad k = 2, 3, \ldots. \tag{11.11-3}$$

Even more can be said. For $k = 1$, (ii) in view of $p_1'(x) = 1$ clearly implies

$$p_1(x) = x - \tfrac{1}{2}. \tag{11.11-4}$$

Thus $p_1(x + \tfrac{1}{2})$ is an odd function. It follows that $p_2(x + \tfrac{1}{2})$ is even and [using (ii)] that $p_3(x + \tfrac{1}{2})$ again is odd. Generally, $p_k(x + \tfrac{1}{2})$ has the same parity as k. For odd $k \geqslant 3$ this means that $p_k(0) = -p_k(1)$, which is compatible with (11.11-3) only if $p_k(0) = p_k(1) = 0$. Thus (11.11-3) may be replaced by the more precise statement

$$p_{2k}(0) = p_{2k}(1),$$
$$p_{2k+1}(0) = p_{2k+1}(1) = 0, \qquad k = 1, 2, \ldots. \tag{11.11-5}$$

With this choice of the polynomials p_k, the integration formula (11.11-2) now reads as follows:

$$\int_0^1 f(x)\, dx = \tfrac{1}{2}[f(0) + f(1)]$$
$$- \sum_{k=1}^m p_{2k}(0)[f^{(2k-1)}(1) - f^{(2k-1)}(0)] \tag{11.11-6}$$
$$- \int_0^1 p_{2m+1}(x) f^{(2m+1)}(x)\, dx, \qquad m = 1, 2, \ldots.$$

The first term on the right is the result of evaluating the integral by one application of the trapezoidal formula. The remaining terms express the error of that formula in terms of the derivatives of odd orders of the integrand at the two endpoints, and in terms of an integral involving a derivative of arbitrarily high order.

Let us now identify the polynomials p_k and the numbers $p_k(0)$. We introduce the generating power series

$$F(x) := \sum_{k=0}^\infty p_k(x) t^k.$$

If F has a positive radius of convergence for all $x \in [0, 1]$ and thus represents a function $f(x, t)$, termwise differentiation with respect to x in view of (i) yields

$$f_x(x, t) = \sum_{k=0}^{\infty} p'_k(x)t^k = \sum_{k=1}^{\infty} p_{k-1}(x)t^k = t \sum_{k=0}^{\infty} p_k(x)t^k$$

$$= tf(x, t).$$

Solving the ordinary differential equation in x, there follows

$$f(x, t) = g(t) e^{xt} \tag{11.11-7}$$

where g is yet to be determined. The condition (ii), interpreted in terms of f, yields

$$\int_0^1 f(x, t) \, dx = \sum_{k=0}^{\infty} \int_0^1 p_k(x) \, dx \cdot t^k = 1$$

for all t. On the other hand, from (11.11-7), if $t \neq 0$,

$$\int_0^1 f(x, t) \, dx = g(t) \frac{e^t - 1}{t}.$$

Together with the previous result this yields

$$g(t) = \frac{t}{e^t - 1},$$

hence

$$f(x, t) = \frac{t e^{tx}}{e^t - 1}.$$

The singularity at $t = 0$ is obviously removable, and on setting $f(x, 0) := 1$ the function f becomes analytic at $t = 0$. Its Taylor series has radius of convergence 2π for all x, and its Taylor coefficients are seen to satisfy (i) and (ii). Because these properties determine the p_k uniquely, there follows

$$\sum_{k=0}^{\infty} p_k(x)t^k = \frac{t e^{tx}}{e^t - 1}, \qquad |t| < 2\pi.$$

For historical reasons it is customary to work with the polynomials

$$B_k(x) := k! p_k(x), \qquad k = 0, 1, 2, \ldots, \tag{11.11-8}$$

called **Bernoulli polynomials**, in place of the p_k. In view of (i) $B_k(x)$ is a polynomial of degree k and leading coefficient 1. In terms of the Bernoulli polynomials the generating relation reads

$$\frac{t e^{tx}}{e^t - 1} = \sum_{k=0}^{\infty} \frac{B_k(x)}{k!} t^k, \qquad |t| < 2\pi. \tag{11.11-9}$$

The generating series enables us to calculate the numbers $p_{2k}(0)$ required in (11.11-6). Setting $x = 0$ in (11.11-9) yields

$$\frac{t}{e^t - 1} = \sum_{k=0}^{\infty} \frac{B_k(0)}{k!} t^k.$$

By comparison with the generating function (2.5-12) of the Bernoulli *numbers*[4] there follows

$$B_k(0) = B_k, \qquad k = 0, 1, 2, \ldots. \tag{11.11-10}$$

We now can state (11.11-3) as follows:

$$\int_0^1 f(x) \, dx = \tfrac{1}{2}[f(0) + f(1)]$$

$$- \sum_{k=1}^{m} \frac{B_{2k}}{(2k)!}[f^{(2k-1)}(1) - f^{(2k-1)}(0)]$$

$$\tag{11.11-11}$$

$$- \frac{1}{(2m+1)!} \int_0^1 B_{2m+1}(x) f^{(2m+1)}(x) \, dx.$$

To obtain the desired summation formula, let f be defined and have continuous derivatives up to order $2m+1$ for all $x \geq 0$. By a shift of variable we then obtain from (11.11-11) for $l = 0, 1, 2, \ldots$

$$\int_l^{l+1} f(x) \, dx = \tfrac{1}{2}[f(l) + f(l+1)]$$

$$- \sum_{k=1}^{m} \frac{B_{2k}}{(2k)!}[f^{(2k-1)}(l+1) - f^{(2k-1)}(l)]$$

$$\tag{11.11-12}$$

$$- \frac{1}{(2m+1)!} \int_l^{l+1} B_{2m+1}(x - [x]) f^{(2m+1)}(x) \, dx.$$

It is convenient to define the **Bernoulli function** of order k by

$$B_k^*(x) := B_k(x - [x]), \qquad k = 0, 1, 2, \ldots. \tag{11.11-13}$$

The Bernoulli function B_k^* is not a polynomial; rather, it is the periodic continuation with period 1 of the restriction of the polynomial $B_k(x)$ to the interval $[0, 1]$.

---

[4] The Bernoulli *polynomials* and the Bernoulli *numbers* are traditionally denoted by the same symbol, B_n. To avoid confusion, an argument is always used to denote the polynomial.

Summing the expressions (11.11-12) from $l = 0$ to $l = n - 1$, where n is an arbitrary positive integer, we find that the integrals can be combined into integrals between the limits 0 and n, and that the values of the derivatives at intermediate points cancel. We thus find

THEOREM 11.11 **(Euler–Maclaurin summation formula)**

Let f be a complex valued function defined for $x \geq 0$ having $2m + 1$ continuous derivatives. Then the following identity holds for $n = 1, 2, \ldots$:

$$\tfrac{1}{2}f(0) + f(1) + f(2) + \cdots + f(n-1) + \tfrac{1}{2}f(n)$$

$$= \int_0^n f(x)\, dx + \sum_{k=1}^m \frac{B_{2k}}{(2k)!}[f^{(2k-1)}(n) - f^{(2k-1)}(0)] \quad (11.11\text{-}14)$$

$$+ \frac{1}{(2m+1)!} \int_0^n B^*_{2m+1}(x) f^{(2m+1)}(x)\, dx.$$

It is obvious that this formula may convey information concerning the behavior of $\sum_{l=0}^n f(l)$ as $n \to \infty$ (the factors $\tfrac{1}{2}$ are, of course, of trivial consequence), and several examples to this effect follow. The formula in itself, however, is not asymptotic. It will become so only if the functions $f^{(2k-1)}(n)$, $k = 1, 2, \ldots$, form an asymptotic sequence, and if the remainder integrals are suitably bounded.

EXAMPLE **1**

Let $f(x) := (1+x)^{-1}$. We have

$$f^{(2k-1)}(x) = -(2k-1)!(1+x)^{-2k},$$

$k = 1, 2, \ldots$, and therefore, replacing n by $n-1$ in (11.11-14) and transposing terms,

$$1 + \frac{1}{2} + \frac{1}{3} + \cdots + \frac{1}{n-1} + \frac{1}{2}\frac{1}{n} - \mathrm{Log}\, n$$

$$= \frac{1}{2} + \sum_{k=1}^m \frac{B_{2k}}{2k}[1 - n^{-2k}] - \int_1^n B^*_{2m+1}(x) x^{-2m-2}\, dx.$$

Here we let $n \to \infty$ while keeping m fixed. We know from (8.4-6) that the limit of the expression on the left equals Euler's constant γ. The integral on the right converges in view of the boundedness of $B^*_{2m+1}(x)$. We thus obtain

$$1 + \frac{1}{2} + \frac{1}{3} + \cdots + \frac{1}{n-1} + \frac{1}{2}\frac{1}{n} - \mathrm{Log}\, n$$

$$(11.11\text{-}15)$$

$$= \gamma - \sum_{k=1}^m \frac{B_{2k}}{2k} n^{-2k} + \int_n^\infty B^*_{2m+1}(x) x^{-2m-2}\, dx.$$

The integral is bounded by

$$\max_{0 \leq x \leq 1} |B^*_{2m+1}(x)| \frac{1}{2m+1} n^{-2m-1} = \text{const} \cdot n^{-2m-1},$$

thus the series obtained by letting $m \to \infty$ has property (A) of §11.1 where $z = n$. We thus get

$$1 + \frac{1}{2} + \frac{1}{3} + \cdots + \frac{1}{n-1} + \frac{1}{2n} - \text{Log } n \approx \gamma - \sum_{k=1}^{\infty} \frac{B_{2k}}{2k} n^{-2k}, \qquad n \to \infty. \qquad (11.11\text{-}16)$$

This series is useful for the numerical calculation of Euler's constant; see §11.12.

EXAMPLE 2 Once again: **Binet's function.**

Let $z > 0$, $f(x) := \text{Log}(z + x)$. We have

$$f^{(2k-1)}(x) = \frac{(2k-2)!}{(z+x)^{2k-1}}, \qquad k = 1, 2, \ldots.$$

In view of

$$\int_0^n f(x) \, dx = (z+n) \text{Log}(z+n) - z \text{Log } z - n,$$

the sum formula yields for $m = 0, 1, 2, \ldots$

$$\sum_{l=0}^{n} \text{Log}(z+l) = (z+n+\tfrac{1}{2}) \text{Log}(z+n) - (z-\tfrac{1}{2}) \text{Log } z - n$$

$$+ \sum_{k=1}^{m} \frac{B_{2k}}{2k(2k-1)} \left[\frac{1}{(z+n)^{2k-1}} - \frac{1}{z^{2k-1}} \right]$$

$$+ \frac{1}{2m+1} \int_0^n \frac{B^*_{2m+1}}{(z+x)^{2m+1}} \, dx.$$

Subtracting from this the same formula where $z = 1$, $m = 0$, and n is diminished by 1, and also subtracting $n \text{ Log } z$, we get

$$\text{Log} \frac{(z)_{n+1}}{n! n^z} = \left(z + n + \frac{1}{2} \right) \text{Log} \frac{z+n}{n} - \left(z - \frac{1}{2} \right) \text{Log } z - 1 - \int_0^{n-1} \frac{B^*_1(x)}{1+x} \, dx$$

$$+ \sum_{k=1}^{m} \frac{B_{2k}}{2k(2k-1)} \left[\frac{1}{(z+n)^{2k-1}} - \frac{1}{z^{2k-1}} \right] + \frac{1}{2m+1} \int_0^n \frac{B^*_{2m+1}(x)}{(z+x)^{2m+1}} \, dx.$$

Here we let $n \to \infty$, keeping z and m fixed. By (II), §8.4, the limit on the left exists and equals $-\text{Log } \Gamma(z)$. Thus the limit must exist on the right, which in particular implies the existence of

$$\beta := \int_0^{\infty} \frac{B^*_1(x)}{1+x} \, dx.$$

We thus find

$$\text{Log } \Gamma(z) = (z - \tfrac{1}{2})\text{Log } z - z + 1 + \beta$$

$$+ \sum_{k=1}^{m} \frac{B_{2k}}{2k(2k-1)} \frac{1}{z^{2k-1}} + \frac{1}{2m+1} \int_0^\infty \frac{B_{2m+1}^*(x)}{(z+x)^{2m+1}} \, dx.$$

Recalling (IV), §8.5, this may be recast as a formula for Binet's function $J(z)$. Because we know that $J(z) \to 0$ for $z \to \infty$, there follows $1 + \beta = \text{Log}\sqrt{2\pi}$ and thus

$$J(z) = \sum_{k=1}^{m} \frac{B_{2k}}{2k(2k-1)} \frac{1}{z^{2k-1}} + \frac{1}{2m+1} \int_0^\infty \frac{B_{2m+1}^*(x)}{(z+x)^{2m+1}} \, dx, \qquad (11.11\text{-}17)$$

which may be regarded as a quantitative version (with remainder term) of the asymptotic expansion of Binet's function.

EXAMPLE **3** **Numerical integration.**

Let g be $2m+1$ times continuously differentiable on $[0, 1]$. We apply the summation formula to

$$f(x) := g(hx),$$

where $h := 1/n$. Writing $x_l := lh$, $l = 0, 1, \ldots, n$, there results

$$h[\tfrac{1}{2} g(x_0) + g(x_1) + \cdots + g(x_{n-1}) + \tfrac{1}{2} g(x_n)]$$

$$= \int_0^1 g(x) \, dx + \sum_{k=1}^{m} \frac{B_{2k}}{(2k)!} [g^{(2k-1)}(1) - g^{(2k-1)}(0)] h^{2k} \qquad (11.11\text{-}18)$$

$$+ \frac{h^{2m+1}}{(2m+1)!} \int_0^1 B_{2m+1}^* \left(\frac{x}{h} \right) g^{(2m+1)}(x) \, dx.$$

The first term is the result of integrating $\int_0^1 g$ numerically, applying the trapezoidal rule to n equal subintervals. We call these numbers the **trapezoidal values** of the integral. The formula exhibits the error of these trapezoidal values. The integral on the right clearly is $O(h^{2m+1})$. If g has derivatives of all orders, the formula holds for $m = 1, 2, \ldots$ and thus expresses the fact that the trapezoidal values admit the asymptotic expansion

$$h[\tfrac{1}{2} g(x_0) + g(x_1) + \cdots + g(x_{n-1}) + \tfrac{1}{2} g(x_n)]$$

$$\approx \int_0^1 g(x) \, dx + \sum_{k=1}^{\infty} \frac{B_{2k}}{(2k)!} [g^{(2k-1)}(1) - g^{(2k-1)}(0)] h^{2k} \qquad (11.11\text{-}19)$$

as $h \to 0$, $h > 0$. Unless the derivatives of g are easily calculated, it is not a practical proposition to increase the accuracy of the trapezoidal values by evaluating some terms of this expansion. However, we see in §11.12 that the mere fact that the trapezoidal values *admit* an asymptotic expansion of known structure greatly facilitates the numerical evaluation of integrals.

PROBLEMS

1. Show that

$$B_{2k}^*(x) = (-1)^{k+1} \frac{2(2k)!}{(2\pi)^{2k}} \sum_{n=1}^{\infty} \frac{\cos(2\pi nx)}{n^{2k}}, \qquad k = 1, 2, \ldots,$$

$$B_{2k+1}^*(x) = (-1)^{k+1} \frac{2(2k+1)!}{(2\pi)^{2k+1}} \sum_{n=1}^{\infty} \frac{\sin(2\pi nx)}{n^{2k+1}}, \qquad k = 0, 1, 2, \ldots,$$

   (a) by verifying that the series on the right define functions that enjoy the properties (i), (ii);
   (b) by the calculus of residues (see §4.9).

2. Deduce from Problem 1 that

$$|B_{2k}^*(x)| \leqslant |B_{2k}|, \qquad k = 1, 2, \ldots.$$

3. Show that the remainder integral in the Euler–Maclaurin summation formula may be written

$$-\frac{1}{(2m)!} \int_0^n B_{2m}^*(x) f^{(2m)}\, dx$$

   and hence is bounded by

$$\frac{|B_{2m}|}{(2m)!} \int_0^n |f^{(2m)}(x)|\, dx.$$

4. By multiplying the Maclaurin expansions of e^{xt} and $t(e^t - 1)^{-1}$, establish the following representation of the Bernoulli polynomials in terms of the Bernoulli numbers:

$$B_n(x) = \sum_{m=0}^{n} \binom{n}{m} B_m x^{n-m}, \qquad n = 0, 1, 2, \ldots.$$

   Deduce the special formulas

$$B_0(x) = 1,$$
$$B_1(x) = x - \tfrac{1}{2},$$
$$B_2(x) = x^2 - x + \tfrac{1}{6},$$
$$B_3(x) = x^3 - \tfrac{3}{2}x^2 + \tfrac{1}{2}x,$$

   and show that the polynomials $B_n(x) - \tfrac{1}{2}nx^{n-1}$ have the same parity as n.

5. Find representations of the sums

$$s_p(n) = 1^p + 2^p + \cdots + n^p$$

   ($p = 1, 2, \ldots$) by applying the summation formula to $f(x) = x^p$.

6. Let

$$s_n := 2^1 3^{1/2} 4^{1/3} \ldots n^{1/(n-1)}, \qquad n = 2, 3, \ldots.$$

Show that for some constant C,

$$s_n \sim Cn^{(1/2)\mathrm{Log}\,n}, \qquad n \to \infty.$$

7. A good approximation to

$$1 + \frac{1}{3} + \frac{1}{5} + \cdots + \frac{1}{2n-1}$$

is $\frac{1}{2}\mathrm{Log}\,n + \mathrm{Log}\,2 + \frac{1}{2}\gamma$, $\gamma :=$ Euler's constant (A. C. Aitken; see Minc [1974]).

8. Let

$$s_n := 1^1 2^2 3^3 \cdots n^n, \qquad n = 1, 2, \ldots.$$

Show that there exists a constant C such that

$$s_n \sim C e^{-n^2/4} n^{n^2/2 + n/2 + 1/12}, \qquad n \to \infty.$$

(It can be shown that

$$C = (2\pi)^{1/12} \left[\prod_{k=1}^{\infty} k^{1/k^2} \right]^{1/2\pi^2},$$

see Hardy [1910], p. 52.)

9. Establish an integration formula analogous to (11.11-6),

$$\int_0^1 f(x)\,dx = \sum_{k=0}^{m-1} \frac{E_{2k+1}(0)}{(2k+1)!} [f^{(2k)}(0) + f^{(2k)}(1)]$$

$$+ \frac{1}{(2m+1)!} \int_0^1 E_{2m+1}(x) f^{(2m)}(x)\,dx,$$

where the functions $E_n(x)$, known as **Euler polynomials**, are defined by

$$\frac{e^{xt}}{e^t + 1} = \sum_{n=0}^{\infty} \frac{E_n(x)}{n!} t^n, \qquad |t| < \pi.$$

10. Show that the Euler polynomials considered in Problem 9 satisfy

$$E_{n-1}(x) = \frac{2}{n} \left[B_n(x) - 2^n B_n\left(\frac{x}{2}\right) \right], \qquad n = 1, 2, \ldots,$$

and hence that

$$E_{2k-1}(0) = -\frac{1}{2k} (2^{2k} - 1) B_{2k}, \qquad k = 1, 2, \ldots.$$

11. If $E_n^*(x) := E_n(x - [x])$, prove that

$$E_{2k-1}^*(x) = \frac{(-1)^k 4(2k-1)!}{\pi^{2k}} \sum_{n=0}^{\infty} \frac{\cos\{(2n+1)\pi x\}}{(2n+1)^{2k}},$$

$$E_{2k}^*(x) = \frac{(-1)^k 4(2k)!}{\pi^{2k+1}} \sum_{n=0}^{\infty} \frac{\sin\{(2n+1)\pi x\}}{(2n+1)^{2k+1}},$$

$k = 1, 2, \ldots$. Deduce the inequality

$$|E^*_{2k+1}(x)| \leqslant |E_{2k+1}(0)|, \qquad k = 1, 2, \ldots.$$

12. Let $\zeta(z)$ denote the Riemann zeta function. If $z \neq 1$, prove that for $n \to \infty$

(i) $\displaystyle \sum_{k=1}^{n} k^{-z} = \frac{n^{1-z}}{1-z} + \frac{1}{2} n^{-z} + \zeta(z) + O(n^{-1-z}),$

(ii) $\displaystyle \sum_{k=1}^{n} k^{-z} \operatorname{Log} k = \frac{n^{1-z}}{1-z} \operatorname{Log} n - \frac{n^{1-z}}{(1-z)^2} + \frac{1}{2} n^{-z} \operatorname{Log} n$

$$- \zeta'(z) + O(n^{-1-z} \operatorname{Log} n).$$

(De Bruijn [1961], p. 39.)

13. For $n = 2, 3, \ldots$, let

$$\lambda_n := \frac{2}{n} \sum_{l=1}^{n-1} \frac{1}{\sin \dfrac{l\pi}{n}}.$$

Show that

$$\frac{\pi}{4} \lambda_n - \operatorname{Log} \frac{2n}{\pi} \approx \gamma + \sum_{k=1}^{\infty} \frac{(2^{2k-1} - 1)}{k(2k)!} B^2_{2k} \left(-\frac{\pi^2}{n^2} \right)^k$$

$$= \gamma - \frac{1}{72} \frac{\pi^2}{n^2} + \frac{7}{43200} \frac{\pi^4}{n^4} - \cdots, \qquad n \to \infty.$$

(J. Waldvogel, private communication.)

14. Let r be a rational function with a zero of order $\geqslant 2$ at ∞ and no zeros at the positive integers. Show that the partial sums

$$s_n := \sum_{k=1}^{n} r(k), \qquad n = 1, 2, \ldots,$$

admit an asymptotic expansion in powers of $1/n$ as $n \to \infty$. [Partial fraction decomposition.]

15. Let $s_n := 1 + 1/2 + \cdots + 1/n$, the nth partial sum of the harmonic series. It is well known that $s_n \to \infty$ for $n \to \infty$, but for which n does s_n reach a given real number $\alpha \geqslant 1$? Show that $s_n > \alpha$ if

$$\operatorname{Log} n > \alpha - \gamma - \frac{1}{2n} + \frac{1}{8n^2},$$

where γ denotes Euler's constant, and $s_n < \alpha$ if

$$\operatorname{Log} n < \alpha - \gamma - \frac{1}{2n}.$$

(See Boas and Wrench [1971], where more refined results are given.)

16. Let g be analytic on the set E_ρ of all points having a distance $< \rho$ from the real segment $[0, 1]$. Show that the remainder in the asymptotic formula (11.11-18) for the trapezoidal values of $\int_0^1 g(x)\, dx$ is for every $\rho_1 < \rho$ bounded by

$$\text{const.}(2m)! \left(\frac{h}{2\pi\rho_1} \right)^{2m}, \qquad m = 1, 2, \ldots.$$

[Estimate $f^{(2m)}(x)$ by Cauchy's integral formula.]

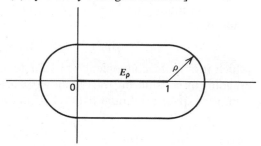

Fig. 11.11a.

§11.12. THE NUMERICAL EVALUATION OF LIMITS: ROMBERG'S ALGORITHM

In computational analysis one is frequently faced with the problem of determining the limit a_0 as $h \to 0$ of a quantity $q(h)$ that can be calculated only for positive values of h or even only for a set of discrete values of h having 0 as a point of accumulation. A typical example is the numerical evaluation of the integral

$$a_0 := \int_0^1 f(t)\, dt$$

as limit of the trapezoidal values

$$h\{\tfrac{1}{2}f(0) + f(h) + f(2h) + \cdots + f((n-1)h) + \tfrac{1}{2}f(nh)\}$$

$(nh = 1)$ for $h \to 0$. Here h can take only the discrete values $1, \frac{1}{2}, \frac{1}{3}, \ldots$. In this and similar problems, the amount of numerical work required to calculate $q(h)$ is roughly inversely proportional to h. One thus is interested in algorithms that permit an accurate determination of a_0 without necessitating the evaluation of $q(h)$ for exceedingly small values of h.

In the present section we describe such a procedure that is valid under the condition that the function $q(h)$ admits an asymptotic power series (not assumed convergent) as $h \to 0$:

$$q(h) \approx a_0 + a_1 h + a_2 h^2 + \cdots, \qquad h \to 0, \qquad h > 0. \qquad (11.12\text{-}1)$$

It is not required that the coefficients $a_1, a_2, \ldots$ be known. However, if explicit error estimates are desired, it is necessary to know *bounds* for the constants $\mu_1, \mu_2, \ldots$ in the formulas

$$|q(h) - [a_0 + a_1 h + \cdots + a_{n-1} h^{n-1}]| \leq \mu_n h^n, \qquad (11.12\text{-}2)$$

$h > 0, n = 1, 2, \ldots$. The existence of these μ_n is implied by property (A') of §11.1.

The idea of the algorithm is as follows. Let ρ be a fixed number, $0 < \rho < 1$. (In applications, ρ will frequently be $\frac{1}{2}$ or $\frac{1}{4}$.) We calculate $q(h)$ for $h = \rho^n$, $n = 0, 1, 2, \ldots$, and set

$$q_{n,0} := q(\rho^n).$$

(A situation in which it is more convenient to evaluate q for $h = h_0 \rho^n$ is reduced to the foregoing by considering the function $q(h_0 h)$ in place of $q(h)$, which has an asymptotic expansion similar to (11.12-2).) From (11.12-2) we have

$$q_{n,0} \approx a_0 + a_1 \rho^n + a_2 \rho^{2n} + \cdots, \qquad n \to \infty,$$

thus, in particular, $q_{n,0} - a_0 = O(\rho^n)$. From the sequence $\{q_{n,0}\}$ we now form a new sequence

$$q_{n,1} := \alpha_0 q_{n,0} + \beta_0 q_{n-1,0},$$

where we try to determine α_0 and β_0 such that $q_{n,1}$ converges to a_0 faster. Assuming $a_1 \neq 0$, this will in view of

$$\alpha_0 q_{n,0} + \beta_0 q_{n-1,0} \approx (\alpha_0 + \beta_0) a_0 + (\rho \alpha_0 + \beta_0) a_1 \rho^{n-1} + (\rho^2 \alpha_0 + \beta_0) a_2 \rho^{2n-2} + \cdots$$

be the case if and only if

$$\alpha_0 + \beta_0 = 1, \qquad \rho \alpha_0 + \beta_0 = 0,$$

that is, if

$$\alpha_0 = \frac{1}{1-\rho}, \qquad \beta_0 = -\frac{\rho}{1-\rho}.$$

With this choice of α_0 and β_0 we obtain

$$q_{n,1} = \frac{q_{n,0} - \rho q_{n-1,0}}{1-\rho},$$

and there holds the asymptotic expansion

$$q_{n,1} \approx a_0 + \sum_{k=2}^{\infty} \frac{1-\rho^{1-k}}{1-\rho} a_k \rho^{kn}, \qquad n \to \infty;$$

in particular, $q_{n,1} - a_0 = O(\rho^{2n})$. Obviously, this process can be repeated. By forming a linear combination

$$q_{n,2} = \alpha_1 q_{n,1} + \beta_1 q_{n-1,1}$$

we can eliminate also the term in ρ^{2n} from the asymptotic expansion. A similar calculation shows that the quantities

$$q_{n,2} := \frac{q_{n,1} - \rho^2 q_{n-1,1}}{1-\rho^2}$$

satisfy

$$q_{n,2} \approx a_0 + \sum_{k=3}^{\infty} \frac{(1-\rho^{1-k})(1-\rho^{2-k})}{(1-\rho)(1-\rho^2)} a_k \rho^{kn}, \qquad n \to \infty.$$

Continuing in this manner, we construct a triangular array of numbers $q_{n,m}$ by means of the formulas

$$q_{n,0} := q(\rho^n),$$

$$q_{n,m} := \frac{q_{n,m-1} - \rho^m q_{n-1,m-1}}{1-\rho^m}, \qquad n = 0, 1, 2, \ldots; \qquad m = 1, 2, \ldots, n. \quad (11.12\text{-}3)$$

The array

$$
\begin{array}{l}
q_{00} \\
\quad \searrow \\
q_{10} \to q_{11} \\
\quad \searrow \quad\ \searrow \\
q_{20} \to q_{21} \to q_{22} \\
\quad \searrow \quad\ \searrow \quad\ \searrow \\
q_{30} \to q_{31} \to q_{32} \to q_{33}
\end{array}
$$

$$\cdot \quad \cdot \quad \cdot \quad \cdot \quad \cdot \quad \cdot$$

is called the **Romberg array** (after its inventor) of the function $q(h)$, calculated with the ratio ρ. Arrows indicate the flow of computation. The entries in the first column are particular values of $q(h)$. The remaining entries of the array are calculated from two entries immediately to the left of it by trivial arithmetic operations. Each row of the array can be completed as soon as its leftmost element has been calculated.

By induction it is now easy to prove:

THEOREM 11.12a

For each fixed $m = 0, 1, 2, \ldots$, the following asymptotic expansion holds:

$$q_{n,m} \approx a_0 + \sum_{k=m+1}^{\infty} \prod_{j=1}^{m} \frac{1-\rho^{j-k}}{1-\rho^j} \frac{1-\rho^{j-k}}{1-\rho^j} a_k \rho^{kn}, \qquad n \to \infty.$$

$$(11.12\text{-}4)$$

In particular,

$$q_{n,m} = a_0 + \text{const.} \ a_{m+1} \rho^{(m+1)n} + O(\rho^{(m+2)n}),$$

$n \to \infty$, which immediately yields

COROLLARY 11.12b

If all $a_k \neq 0$ in (11.12-1), then each column of the Romberg array converges to a_0 faster than the preceding column.

Because the work required to calculate additional columns of the array is usually negligible compared to the work required to calculate the first column, the corollary is already very useful for numerical purposes. However, the convergence of each column is still linear; that is, the error tends to zero like γ^n, where γ is a fixed quantity for each column (although its value becomes smaller for each successive column). It is now shown that under certain conditions on the constants μ_k occurring in (11.12-2), the *diagonals* of the Romberg array (i.e., the sequences $\{q_{n+k,n}\}$, where k is fixed) converge to a_0 faster than linearly. In some instances, we are able to show that the error tends to zero like γ^{n^2}, where γ is a fixed number <1. In the terminology of numerical analysis, we thus obtain *quadratic convergence*.

We begin by noting that, by virtue of the relations (11.12-3), each entry $q_{n,m}$ of the Romberg array is a linear combination of the quantities $q_{n-k,0}$ ($k = 0, 1, \ldots, m$), with coefficients that depend only on k and m:

$$q_{n,m} = \sum_{k=0}^{m} c_{m,k} q_{n-k,0}, \qquad \begin{array}{l} n = 0, 1, 2, \ldots; \\ m = 0, 1, \ldots, n. \end{array} \qquad (11.12\text{-}5)$$

The second relation (11.12-3) yields

$$q_{n,m} = \sum_{k=0}^{m} c_{m,k} q_{n-k,0}$$

$$= \frac{1}{1-\rho^m} q_{n,m-1} - \frac{\rho^m}{1-\rho^m} q_{n-1,m-1}$$

$$= \sum_{k=0}^{m} \left[\frac{1}{1-\rho^m} c_{m-1,k} - \frac{\rho^m}{1-\rho^m} c_{m-1,k-1} \right] q_{n-k,0}$$

($c_{m-1,-1} := 0$). Because this relation must hold for arbitrary sequences $\{q_{n,0}\}$, we have

$$c_{m,k} = \frac{c_{m-1,k} - \rho^m c_{m-1,k-1}}{1-\rho^m}. \qquad (11.12\text{-}6)$$

For the *generating polynomials* of the $c_{m,k}$,

$$t_m(z) := \sum_{k=0}^{m} c_{m,k} z^k, \qquad m = 0, 1, \ldots,$$

we find after multiplying (11.12-6) by z^k and summing with respect to k,

$$t_m(z) = \frac{1 - z\rho^m}{1-\rho^m} t_{m-1}(z), \qquad m = 1, 2, \ldots,$$

which by virtue of $t_0(z) = 1$ immediately yields the explicit representation

$$t_m(z) = \frac{1-z\rho}{1-\rho} \frac{1-z\rho^2}{1-\rho^2} \cdots \frac{1-z\rho^m}{1-\rho^m}. \qquad (11.12\text{-}7)$$

Evidently,
$$t_m(\rho^{-k}) = 0, \qquad k = 1, 2, \ldots, m. \qquad (11.12\text{-}8)$$

The zeros of t_m are thus positive and because $t_m(0) > 0$ there follows
$$(-1)^k c_{m,k} > 0, \qquad k = 1, 2, \ldots, m. \qquad (11.12\text{-}9)$$

Let now the functions r_m ($m = 0, 1, 2, \ldots$) be defined by
$$q(h) = a_0 + a_1 h + a_2 h^2 + \cdots + a_{m-1} h^{m-1} + r_m(h). \quad (11.12\text{-}10)$$

By (11.12-2) we have for $h > 0$
$$|r_m(h)| \leqslant \mu_m h^m, \qquad m = 0, 1, 2, \ldots. \qquad (11.12\text{-}11)$$

Evaluating the numbers $q_{n-k,0} = q(\rho^{n-k})$ by (11.12-10), substituting in (11.12-5) and rearranging the summation we get

$$\begin{aligned}
q_{n,m} &= \sum_{k=0}^{m} c_{m,k} \left\{ \sum_{p=0}^{m-1} a_p \rho^{(n-k)p} + r_m(\rho^{n-k}) \right\} \\
&= \sum_{p=0}^{m-1} a_p \sum_{k=0}^{m} c_{m,k} \rho^{(n-k)p} + \sum_{k=0}^{m} c_{m,k} r_m(\rho^{n-k}) \\
&= \sum_{p=0}^{m-1} a_p \rho^{np} t_m(\rho^{-p}) + \sum_{k=0}^{m} c_{m,k} r_m(\rho^{n-k}).
\end{aligned}$$

By (11.12-8), all terms of the first sum except the first term vanish, and we are left with

$$q_{n,m} = a_0 + \sum_{k=0}^{m} c_{m,k} r_m(\rho^{n-k}). \qquad (11.12\text{-}12)$$

Using the estimate (11.12-11), this yields

$$|q_{n,m} - a_0| \leqslant \mu_m \sum_{k=0}^{m} |c_{m,k}| \rho^{(n-k)m},$$

which by virtue of (11.12-9) may be written

$$|q_{n,m} - a_0| \leqslant \mu_m \rho^{nm} t_m(-\rho^{-m}). \qquad (11.12\text{-}13)$$

The representation (11.12-7) yields

$$\begin{aligned}
t_m(-\rho^{-m}) &= \frac{(1 + \rho^{1-m})(1 + \rho^{2-m}) \cdots (1 + \rho^{m-m})}{(1 - \rho)(1 - \rho^2) \cdots (1 - \rho^m)} \\
&= \rho^{-[m(m-1)]/2} \frac{(\rho^{m-1} + 1)(\rho^{m-2} + 1) \cdots (1 + 1)}{(1 - \rho)(1 - \rho^2) \cdots (1 - \rho^m)} \\
&= \rho^{-[m(m-1)]/2} \frac{2}{1 + \rho^m} \frac{(1 + \rho)(1 + \rho^2) \cdots (1 + \rho^m)}{(1 - \rho)(1 - \rho^2) \cdots (1 - \rho^m)}.
\end{aligned}$$

Defining the convergent infinite product

$$t(\rho) := \prod_{m=1}^{\infty} \frac{1+\rho^m}{1-\rho^m}, \tag{11.12-14}$$

we thus have

$$t_m(-\rho^{-m}) \leqslant 2t(\rho)\rho^{-[m(m-1)]/2}, \qquad m = 1, 2, \ldots.$$

Using this in (11.12-13), we obtain

THEOREM 11.12c

Let the constants μ_m satisfy (11.12-2), and let $t(\rho)$ be defined by (11.12-14). Then the following estimate holds for the elements $q_{n,m}$ of the Romberg array defined by (11.12-3):

$$|q_{n,m} - a_0| \leqslant 2t(\rho)\mu_m\rho^{m[n-(m-1)/2]}, \tag{11.12-15}$$
$$n = 0, 1, 2, \ldots; \qquad m = 1, 2, \ldots, n.$$

By taking logarithms it is easily shown that

$$t(\rho) < \left(\frac{1+\rho}{1-\rho}\right)^{1/(1-\rho)};$$

more accurate values of $t(\rho)$ may be computed as shown in Problem 6. We thus find, for instance,

$$t(\tfrac{1}{2}) = 8.256\ldots,$$
$$t(\tfrac{1}{4}) = 1.969\ldots.$$

Thus if bounds for the constants μ_m are known, explicit error estimates for the Romberg scheme are easily obtained from (11.12-15). Moreover the following is a direct consequence:

COROLLARY 11.12d

Let $\mu_m^{1/m}\rho^{m/2} \to 0$, $m \to \infty$. Then each diagonal of the Romberg scheme converges to a_0 faster than linearly. That is, for any $\lambda > 0$ (no matter how large) and for $k = 0, 1, 2, \ldots$, there holds

$$\lim_{m\to\infty} \lambda^m|q_{m+k,m} - a_0| = 0.$$

Proof. Letting $n = m + k$ in (11.12-15) we get

$$\lambda^m|q_{m+k,m} - a_0| \leqslant 2t(\rho)\mu_m\lambda^m\rho^{m(m/2+k+1/2)}.$$

But

$$\mu_m \lambda^m \rho^{m(m/2+k+1/2)} = [(\mu_m^{1/m} \rho^{m/2})(\lambda \rho^{k+1/2})]^m,$$

and the conclusion follows from the fact that, by virtue of the hypothesis, the expression in brackets is <1 for sufficiently large values of m. ∎

The hypothesis made on the growth of μ_m is a very mild one. Stirling's formula shows that it is satisfied, for any $\rho < 1$, for such strongly diverging series where $\mu_m = O((pm)!)$ or $\mu_m = O((m!)^p)$, where p is an arbitrary fixed integer.

We now present several applications of the Romberg algorithm.

EXAMPLE 1 **Numerical integration over a finite interval**

Let f satisfy the conditions of Example 3, §11.11, and let $q(h)$ denote the trapezoidal value of $\int_0^1 f(x)\,dx$ calculated with step h. Then by the result of the example $q(h)$ admits an asymptotic expansion of the form (11.12-1) where h is to be replaced by h^2. Successive halving of the integration step thus corresponds to the value $\rho = \frac{1}{4}$ in the Romberg scheme. By Problem 16, §11.11, the constants μ_m are no larger than

$$\text{const.}(2m)!\,(2\pi\rho_1)^{-2m}.$$

Thus at most

$$\mu_m^{1/m} \rho^{m/2} \sim (\pi e \rho_1)^{-2} m^2 2^{-m},$$

and the hypothesis of the corollary is clearly fulfilled. The resulting method of integration is known as **Romberg integration** and is widely used in numerical practice. For algorithmic details the reader is referred to numerical analysis texts.

EXAMPLE 2 **Calculation of natural logarithms**

Let $x > 1$, and let it be desired to calculate $\text{Log } x$. By calculus,

$$\text{Log } x = \lim_{h \to 0} q(h),$$

where

$$q(h) := \frac{x^h - x^{-h}}{2h}, \qquad h > 0.$$

Here q admits the *convergent* expansion

$$q(h) = \sum_{n=0}^{\infty} \frac{(\text{Log } x)^{2n+1\cdot}}{(2n+1)!} h^{2n}.$$

Although the coefficients of this expansion depend on $\text{Log } x$, the quantity to be computed, the Romberg algorithm is applicable, because no knowledge of these coefficients is required. Because the series proceeds in powers of h^2, a successive halving of h is consistent with $\rho = \frac{1}{4}$. The elements in the first column of the array are

$$l_n := q_{n,0} = q(2^{-n}) = 2^n \operatorname{Sin}(2^{-n} \operatorname{Log} x). \qquad (11.12\text{-}16)$$

The evaluation of hyperbolic functions is avoided by using the recurrence relation

$$l_0 = \frac{x^2 - 1}{2x}, \qquad l_1 = \frac{x - 1}{\sqrt{x}},$$

$$\qquad\qquad\qquad\qquad (11.12\text{-}17)$$

$$l_{n+1} = l_n \sqrt{\frac{2l_n}{l_n + l_{n-1}}}, \qquad n = 1, 2, \dots,$$

which is advantageous also from the point of view of numerical stability.

By way of illustration, we show in Table 11.12a the scheme resulting for $x = 10$.

Table 11.12a. Romberg Scheme for Log 10

4.950000000				
2.846049894	2.144733192			
2.431876170	2.293818262	2.303757266		
2.334508891	2.302053132	2.302602123	2.302583788	
2.310541291	2.302552090	2.302585354	2.302585088	2.302585093

The constants μ_n are easily estimated as follows: If $h \leq 1$, then (11.12-2) holds with h^2 in place of h and

$$\mu_n := \sum_{m=n}^{\infty} \frac{(\operatorname{Log} x)^{2m+1}}{(2m+1)!} \leq \frac{(\operatorname{Log} x)^{2n+1}}{(2n+1)!} e^{\operatorname{Log} x} = x \frac{(\operatorname{Log} x)^{2n+1}}{(2n+1)!}.$$

Using $t(\frac{1}{4}) < 2$, Theorem 11.12c thus yields the following estimate for the diagonal elements, simplified by approximating the factorials by Stirling's formula:

$$|q_{m,m} - \operatorname{Log} x| \leq 4x \left[\frac{e \operatorname{Log} x}{2^{m/2}(2m+1)} \right]^{2m+1}.$$

For $m = 5$ (which requires the evaluation of four square roots) and $x \leq 10$ the error thus can be guaranteed to be less than 10^{-9}.

EXAMPLE **3** **Calculation of Arcsin x**

Let $0 \leq x \leq 1$, and let $\alpha := \operatorname{Arcsin} x$. Clearly,

$$\alpha = \lim_{h \to 0} q(h),$$

where

$$q(h) := \frac{\sin(h\alpha)}{h}.$$

In view of

$$q(h) = \sum_{m=0}^{\infty} \frac{\alpha^{2m+1}}{(2m+1)!} (-h^2)^m,$$

the limit α of $q(h)$ can be computed from the Romberg scheme with left column

$$s_n := q_{n,0} = q(2^{-n}) = 2^n \sin(2^{-n}\alpha),$$

where $\rho = 1/4$. The values s_n are obtained most conveniently from the recursive scheme

$$s_{-1}(x) = x\sqrt{1-x^2}, \qquad s_0 = x, \qquad (11.12\text{-}18a)$$

$$s_{n+1} = s_n\sqrt{\frac{2s_n}{s_n + s_{n-1}}}, \qquad n = 0, 1, 2, \ldots, \qquad (11.12\text{-}18b)$$

which differs from (11.12-17) only in the initial conditions. For the error of the diagonal elements we find as in example 3

$$|q_{m,m} - \alpha| \leqslant 4\left[\frac{e\alpha}{2^{m/2}(2m+1)}\right]^{2m+1}, \qquad m = 0, 1, 2, \ldots.$$

EXAMPLE 4 **Euler's constant**

For $n = 1, 2, \ldots$, let

$$\gamma(n) := 1 + \frac{1}{2} + \frac{1}{3} + \cdots + \frac{1}{n-1} + \frac{1}{2n} - \text{Log } n.$$

The definition of Euler's constant shows that $\lim_{n\to\infty} \gamma(n) = \gamma$. More precisely, by (11.11-16),

$$\gamma(n) \approx \gamma - \sum_{k=1}^{\infty} \frac{B_{2k}}{2k}\frac{1}{n^{2k}}, \qquad n \to \infty.$$

Thus the function

$$q(h) := \gamma(h^{-1/2}), \qquad h = 1, \frac{1}{4}, \frac{1}{9}, \frac{1}{16}, \ldots,$$

satisfies (11.12-1). To evaluate γ by the Romberg algorithm, we use the initial values

$$q_{n,0} := \gamma(2^n) = 1 + \frac{1}{2} + \frac{1}{3} + \cdots + \frac{1}{2^n - 1} + \frac{1}{2}\frac{1}{2^n} - n\,\text{Log } 2,$$

which means that we are taking $\rho = \frac{1}{4}$. Part of the resulting scheme is shown in Table 11.12b. The last entries in the last row are correct to all places.

Table 11.12b. Romberg Scheme for Euler's Constant

0.5000000000				
0.5568528194	0.5758037592			
0.5720389722	0.5771010231	0.5771875074		
0.5759156012	0.5772078108	0.5772149300	0.5772153653	
0.5768902710	0.5772151609	0.5772156509	0.5772156624	0.5772156635
0.5771342926	0.5772156332	0.5772156647	0.5772156649	0.5772156649

By Problem 3, §11.11, we have $\mu_n \leqslant |B_{2n}| n^{-1}$. Using an obvious bound for the Bernoulli numbers, Theorem 11.12c thus yields the following estimate for the errors of the diagonal entries of the array:

$$|q_{n,n} - \gamma| \leqslant 8\sqrt{\pi n} \left[\frac{n}{2^{(n+1)/2}\pi e} \right]^{2n}, \qquad n = 1, 2, \ldots.$$

For instance (neglecting rounding errors) no more than 187 steps of the algorithm are necessary to compute γ to 10,000 decimal places, without using explicit values of Bernoulli numbers.

PROBLEMS

1. Calculate $\pi/2$ by choosing $x = 1$ in Example **3**.
2. Let x be a given real number. Show how to compute

$$\beta := \operatorname{Arctan} x = \lim_{h \to 0} \frac{\tan(h \operatorname{Arctan} x)}{h}$$

by means of the Romberg algorithm, using the starting values

$$q_{n,0} := t_n := 2^n \tan(2^{-n} \operatorname{Arctan} x).$$

Also show that

$$t_0 = x, \qquad t_1 = \frac{2x}{1+\sqrt{1+x^2}}, \qquad t_{n+1} = \frac{2t_n}{1+\sqrt{2-(t_n/t_{n-1})}}, \qquad n = 1, 2, \ldots.$$

3. Let r be a rational function with a zero of order 2 at ∞ and with no poles at the positive integers. Using the fact that the partial sums of

$$\sum_{n=1}^{\infty} r(n)$$

admit an asymptotic expansion in powers of n^{-1} as $n \to \infty$, show how to speed up the convergence of the series by means of the Romberg algorithm.
4. Invent a speed-up algorithm for computing $a_0 := \lim q_n$, if q_n is known to admit the generalized asymptotic expansion

$$q_n \approx \sum_{k=0}^{\infty} a_k \gamma_k^n, \qquad n \to \infty,$$

where $1 = \gamma_0 > \gamma_1 > \cdots > 0$, and where the γ_k are known and the a_k are unknown.
5. In the notations of the preceding problem, speed up the convergence of q_n if both the a_k and the γ_k are unknown.
6. In the notation of §8.2,

$$t(\rho) = \frac{p(\rho)}{q(\rho)}.$$

Show that

$$t(\rho) = \frac{q(\rho^2)}{[q(\rho)]^2}$$

and use the known power series for the infinite product q to calculate accurate numerical values of $t(\frac{1}{2})$ and $t(\frac{1}{4})$.

SEMINAR ASSIGNMENTS

1. Study the numerical effectiveness of an asymptotic expansion such as (11.1-1) or (11.1-16) by plotting, in the complex plane, the curves where an accuracy of 10^{-p} $(p = 1, 2, \ldots)$ can be attained by breaking off the expansion at the most favorable term.

2.[5] Study the probability density $F_n(\tau)$ of a sum $\eta = \sum_{k=1}^{n} \xi_k$ of independent random variables, each having the probability density

$$F_1(\tau) := \begin{cases} 1, & 0 \leqslant \tau \leqslant 1, \\ 0, & \text{otherwise.} \end{cases}$$

Start from the recurrence relation

$$F_{n+1} = F_n * F_1, \qquad n = 1, 2, \ldots,$$

to obtain the correspondence

$$F_n(\tau) \circ\!\!-\!\!\bullet f_n(s) = \left[\frac{1 - e^{-s}}{s} \right]^n.$$

(a) Obtain a piecewise polynomial representation of F_n.
(b) Use the complex inversion formula to obtain

$$F(\tau) = \frac{1}{2\pi} \int_{-\infty}^{\infty} \left[\frac{\sin(\omega/2)}{\omega/2} \right]^n \cos \omega\tau \, d\omega$$

and study the asymptotic behavior of $F_n(\tau)$ as $n \to \infty$. In addition to the main saddle point at $\omega = 0$ the subsidiary saddle points at the solutions of $\omega/2 = \text{tg } \omega/2$ should be considered. Graph some of the results and compare with the explicit values obtained under (a).

NOTES

The following are some specialized texts dealing with asymptotic methods: Erdélyi [1956], De Bruijn [1961], Berg [1968], Sirovich [1971], Nayfeh [1973], Dingle [1973], Olver [1974]. The last work and Wasow [1965] are

---

[5] Suggested by Dr. J. Waldvogel.

oriented toward differential equations. A major portion of Doetsch [1955] is devoted to asymptotic methods.

§11.1. The enveloping property defined in Problem 8 is discussed by Polya and Szegö [1925], Vol. I, p. 26. In some older texts this is taken as the defining property of an asymptotic expansion.

§11.2. Problems 6 and 7 are suggested by, but not identical with, Shafer [1975].

§11.3. For Theorem 11.3b see Ritt [1916]. The remaining results are proved as in Erdélyi [1956].

§11.4. The proof of Theorem 11.4a is an amalgamation of ideas of Coddington and Levinson [1955], Chapter 4, and Wasow [1965], Chapter 4.

§11.5. For an even more general version of the Watson–Doetsch lemma (Theorem 11.5) see Doetsch [1955], p. 45. For the incomplete Γ function both the definition by (11.5-7) and the definition given in Problem 2, §10.5, are in use.

§11.6. See Buckholtz [1963] for an interesting generalization of Problem 6. For Problem 7 see Henrici and Hoffmann [1975]. Concerning the origin of Problem 8 see Knuth [1974].

§11.7. Theorem 11.7 is from De Bruijn [1961], and Example **2** is a simplified version of Problem 4, which is taken from the same source.

§11.8. For a generalization of the method of steepest descent see Chester, Friedman, and Ursell [1957]. Many further applications of the method occur in Szegö [1959] and in most books on special functions.

§11.9. General asymptotic series are discussed in Erdélyi [1956].

§11.10. For Problem 1 see Feller [1957], p. 260.

§11.11. For the history of the Euler–Maclaurin sum formula see Whittaker and Watson [1927], p. 127. Some interesting examples are in the appendix of Hardy [1910]; Problem 8 is taken from there. For Problem 6 see Wilkinson [1961], p. 284. Problem 14 was used for numerical calculations in the ETH dissertation of W. Gander (unpublished). Boas and Stutz [1971] compare several methods of estimating sums by integrals.

§11.12. Extrapolation to the limit was used informally by Richardson and Gaunt [1927]. A systematic algorithm was proposed by Romberg [1955]. A rigorous analysis of the Romberg process, as applied to integration, was given by Bauer, Rutishauser, and Stiefel [1962]. Rutishauser [1963] discusses applications to numerical differentiation, Filippi [1966] to the computation of elementary functions. The polynomials $t(\rho)$ are expressible in terms of the "Gauss binomial coefficients" defined by

$$\begin{bmatrix} n \\ r \end{bmatrix} := \frac{1-\gamma^n}{1-\gamma} \frac{1-\gamma^{n-1}}{1-\gamma^2} \cdots \frac{1-\gamma^{n-r+1}}{1-\gamma^r}$$

(Zurich lecture by G. Polya on January 29, 1970). For a variant of Romberg quadrature see Laurie [1975].

12
CONTINUED FRACTIONS

Like infinite series and infinite products, continued fractions can be considered as arising through the composition of certain types of Moebius transformations. Again like infinite series and products, continued fractions can be used to represent certain types of analytic functions. Contrary to representations by power series, continued fraction representations may converge in regions that contain isolated singularities of the function to be represented, and contrary to representations by infinite products, continued fraction representations in many cases converge very rapidly.

In addition to the representation of analytic functions, continued fractions also have interesting applications in stability theory, in asymptotics, and in number theory.

Wherever possible, the treatment of continued fractions given here is based on Moebius transformations. Indeed, the very definition of continued fractions makes use of that concept. In comparison to some earlier presentations of the theory, the explicit use of Moebius transformations, in addition to simplifying some concepts and some proofs, injects a geometric note into the theory, which is deemed desirable.

§12.1. DEFINITION AND BASIC PROPERTIES

A continued fraction is often defined as "an expression of the form"

$$
\cfrac{a_1}{b_1 + \cfrac{a_2}{b_2 + \cfrac{a_3}{b_3 + \cdots}}} ,
\tag{12.1-1}
$$

which it has become customary to write in a typographically more convenient form as follows:

$$
\frac{a_1}{|\,b_1\,} + \frac{a_2}{|\,b_2\,} + \frac{a_3}{|\,b_3\,} + \cdots .
\tag{12.1-2}
$$

Such an "expression," however, is at best a prescription to perform certain algebraic operations. As a formal definition, it stands on the same level as the definition of an infinite series as "an expression of the form"

$$a_1 + a_2 + a_3 + \cdots .$$

To obtain an operative definition of a continued fraction, we recall the definitions of an infinite series and of an infinite product given at the beginning of §8.1. To prepare for the analogy to continued fractions, the definition of an infinite series given there might also be phrased as follows: An infinite series is a pair of sequences $[\{a_n\}_1^\infty, \{s_n\}_1^\infty]$, where the a_n are complex, and where, denoting by t_k the Moebius translation (T)

$$t_k : u \to a_k + u, \qquad k = 1, 2, \ldots,$$

$$s_n := t_1 \circ \cdots \circ t_n(0), \qquad n = 1, 2, \ldots .$$

Similarly, an infinite product is a pair of sequences $[\{a_n\}_1^\infty \{p_n\}_1^\infty]$, where the a_n are again complex, and where, denoting by t_k the Moebius rotation (R)

$$t_k : u \to a_k u, \qquad k = 1, 2, \ldots,$$

$$p_n := t_1 \circ t_2 \circ \cdots \circ t_n(1), \qquad n = 1, 2, \ldots .$$

Infinite series and products thus both are built up from entire Moebius transformations that leave the point at infinity fixed.

In a similar vein, continued fractions are built up from Moebius transformations that involve a rotation (R), an inversion (I), and a translation (T)— so-called *RIT* transforms—and thus do not leave the point at infinity fixed.

DEFINITION

An (infinite) **continued fraction** (c.f.) *is a triple* $[\{a_n\}_1^\infty, \{b_n\}_1^\infty, \{w_n\}_1^\infty]$ *of sequences, where* $a_1, a_2, \ldots$ *and* $b_1, b_2, \ldots$ *are complex numbers*; $a_n \neq 0$, $n = 1, 2, \ldots$; *and where* w_n *is the element of the extended complex plane defined as follows: If* t_k *denotes the RIT transform*

$$t_k : u \to \frac{a_k}{b_k + u}, \qquad k = 1, 2, \ldots, \tag{12.1-3}$$

then

$$w_n := t_1 \circ t_2 \circ \cdots \circ t_n(0), \qquad n = 1, 2, \ldots . \tag{12.1-4}$$

Writing out the first few w_n will convince the reader that w_n equals the value of the finite compound fraction obtained from (12.1-1) by setting $a_{n+1} = 0$. The following terminology follows the standard works by Wall [1948] and Perron [1957]:

$$\left. \begin{array}{c} a_n \\ b_n \end{array} \right\} \text{ is called the } n\text{th } \textbf{partial} \left\{ \begin{array}{l} \textbf{numerator} \\ \textbf{denominator} \end{array} \right.$$

whereas w_n is called the nth **approximant** of the continued fraction $[\{a_n\}, \{b_n\}, \{w_n\}]$. The partial numerators and denominators are also called the **elements** of the continued fraction, and the transformations t_k its **constituents**.

Once this definition is understood, there is no reason why the continued fraction $[\{a_n\}, \{b_n\}, \{w_n\}]$ should not be denoted by such a well-established symbol as (12.1-2), or also by a symbol such as

$$\mathop{\Phi}_{n=1}^{\infty} \frac{a_n}{b_n},\qquad (12.1\text{-}5)$$

patterned after the familiar symbols Σ and Π.

If the sequence of approximants of a continued fraction C converges, which is to say that

$$w := \lim_{n\to\infty} w_n$$

exists and is finite, then C is said to be (properly) **convergent**, and w is called the **value** of C. If $\lim w_n = \infty$ (in the sense that, no matter how large μ, $|w_n| > \mu$ for all sufficiently large n), C is called **improperly convergent**. It is somewhat ambiguous but frequently convenient to use the symbols (12.1-2) and (12.1-5) to denote not only a continued fraction, but also its value, if defined. (A similarly imprecise usage is made of the symbols Σ and Π in the theory of infinite series and products.)

In addition to infinite continued fractions we also must consider **terminating continued fractions**. These are triples $[\{a_k\}, \{b_k\}, \{w_k\}]$ of finite sequences, with indices running from 1 to n for some integer $n \geqslant 1$, where the a_k and b_k are complex numbers, $a_k \neq 0$, and where w_k is defined by (12.1-4) for $k = 1, 2, \ldots, n$. In this case w_n is called the value of the continued fraction. It always exists as an element of the extended complex plane, and is, like the fraction itself, denoted by one of the symbols

$$\frac{a_1}{\mid b_1} + \frac{a_2}{\mid b_2} + \cdots + \frac{a_n}{\mid b_n}$$

or

$$\mathop{\Phi}_{k=1}^{n} \frac{a_k}{b_k}. \qquad (12.1\text{-}6)$$

Evaluation of continued fractions. There are two essentially different methods for computing the value of a continued fraction, called, respectively, the ascending and the descending method.

(a) *The ascending method* is suitable only for terminating continued fractions. It consists in computing the numbers

$$w_{m,n} := t_m \circ t_{m+1} \circ \cdots \circ t_n(0),$$

$m = 1, 2, \ldots, n$, by the backward recurrence

$$w_{n,n} = t_n(0) = \frac{a_n}{b_n},$$

$$w_{m,n} = t_m(w_{m+1,n}) = \frac{a_m}{b_m + w_{m+1,n}}, \qquad m = n-1, n-2, \ldots, 1.$$

Evidently, $w_{1,n} = w_n$, the value of (12.1-6). The evaluation of this continued fraction by the ascending method requires no more than $n - 1$ additions and n divisions.

EXAMPLE 1

To evaluate

$$\frac{2\,|}{|-1} + \frac{7\,|}{|\,6} - \frac{1\,|}{|\,5} + \frac{3\,|}{|\,4} - \frac{2\,|}{|\,3} - \frac{5\,|}{|\,2}.$$

We obtain

m	6	5	4	3	2	1
a_m	-5	-2	3	-1	7	2
b_m	2	3	4	5	6	-1
$w_{m,6}$	$-\frac{5}{2}$	4	∞	0	$\frac{7}{6}$	12

The value of the continued fraction thus is 12. The fact that an intermediate $w_{m,n}$ has the value ∞ does no harm, because Moebius transformations are defined on the extended complex plane.

(b) *The descending method* is suitable for the evaluation of both terminating and nonterminating (i.e., infinite) continued fractions. It is of interest far beyond the immediate purpose of evaluating a continued fraction numerically. By Theorem 5.2c, the functions

$$t_{1,m} := t_1 \circ t_2 \circ \cdots \circ t_m, \qquad m = 1, 2, \ldots$$

all are Moebius transformations. Moreover, denoting the matrix of the transformation t_k by

$$\mathbf{M}_k := \begin{pmatrix} 0 & a_k \\ 1 & b_k \end{pmatrix},$$

then we have

$$t_{1,m}(u) = \frac{r_m u + p_m}{s_m u + q_m},$$

where

$$\mathbf{M}_{1,m} := \begin{pmatrix} r_m & p_m \\ s_m & q_m \end{pmatrix} = \mathbf{M}_1 \mathbf{M}_2 \cdots \mathbf{M}_m.$$

Because $\mathbf{M}_{1,m} = \mathbf{M}_{1,m-1} \mathbf{M}_m$, there follows

$$\begin{pmatrix} r_m & p_m \\ s_m & q_m \end{pmatrix} = \begin{pmatrix} r_{m-1} & p_{m-1} \\ s_{m-1} & q_{m-1} \end{pmatrix} \begin{pmatrix} 0 & a_m \\ 1 & b_m \end{pmatrix},$$

$m = 2, 3, \ldots$, which on carrying out the matrix product yields

$$r_m = p_{m-1}, \qquad s_m = q_{m-1}, \tag{12.1-7}$$

and

$$p_m = a_m r_{m-1} + b_m p_{m-1},$$

$$q_m = a_m s_{m-1} + b_m q_{m-1},$$

$m = 2, 3, \ldots$. In view of (12.1-7) the latter relations may be written (in an obvious abbreviated form)

$$\begin{pmatrix} p_m \\ q_m \end{pmatrix} = a_m \begin{pmatrix} p_{m-2} \\ q_{m-2} \end{pmatrix} + b_m \begin{pmatrix} p_{m-1} \\ q_{m-1} \end{pmatrix}. \tag{12.1-8a}$$

Together with the initial conditions $p_0 = 0$, $p_1 = a_1$ and $q_0 = 1$, $q_1 = b_1$ following from (12.1-7) and from the fact that $\mathbf{M}_{1,1} = \mathbf{M}_1$, these relations permit the recursive computation of the sequences $\{p_m\}$ and $\{q_m\}$. Under the simpler initial conditions

$$\begin{matrix} p_{-1} = 1, & p_0 = 0 \\ q_{-1} = 0, & q_0 = 1 \end{matrix} \tag{12.1-8b}$$

the relations (12.1-8a) are seen to hold already from $m = 1$ onward. The numbers p_m and q_m are called the mth **numerator** and the mth **denominator** of the continued fraction (12.1-5). Both the numerators and the denominators satisfy the same recurrence relation; only the initial conditions are different. For reference we note

THEOREM 12.1a

Let t_k be defined by (12.1-3). Then for $n = 1, 2, \ldots$, the Moebius transformation

$$t_{1,n} := t_1 \circ t_2 \circ \cdots \circ t_n$$

is associated with the matrix

$$\mathbf{M}_{1,n} = \begin{pmatrix} p_{n-1} & p_n \\ q_{n-1} & q_n \end{pmatrix}, \tag{12.1-9}$$

where the p_n and the q_n are defined by the recurrence relations (12.1-8).

Because the nth approximant is given by

$$w_n = t_{1,n}(0),$$

we immediately obtain

COROLLARY 12.1b

For $n = 1, 2, \ldots$, the nth approximant of the continued fraction (12.1-5) *is*

$$w_n = \frac{p_n}{q_n}. \tag{12.1-10}$$

The evaluation of w_n by the descending method, that is, by the evaluation of the recurrence relations (12.1-8) and the quotient (12.1-10), normally requires one division, $2n - 3$ additions, and $4n - 6$ multiplications. The number of multiplications is cut in half if either all partial numerators or all partial denominators are equal to 1. This can always be achieved by an equivalence transformation (see below).

EXAMPLE **2**

To evaluate the c.f. of example **1** by the descending method, we form the following scheme:

m	-1	0	1	2	3	4	5	6
a_m			2	7	-1	3	-2	-5
b_m			-1	6	5	4	3	2
p_m	1	0	2	12	58	268	688	36
q_m	0	1	-1	1	6	27	69	3

We again see that the continued fraction has the value $w_6 = 36/3 = 12$. At the same time, the algorithm yields the values of all approximants,

$$w_1 = \frac{2}{-1} = -2, \qquad w_2 = \frac{2|}{|-1} + \frac{7|}{|6} = \frac{12}{1} = 12, \text{ etc.}$$

Equivalent Continued Fractions. Two continued fractions are called **equivalent** if they have the same sequence of approximants. The equivalence of two continued fractions C and C^* is denoted by the symbol

$$C \cong C^*.$$

It is clear that if one of two equivalent continued fractions is convergent, the other is likewise convergent, and has the same value. It will now be shown that any continued fraction is equivalent to a continued fraction whose partial numerators are 1, and any continued fraction whose partial denominators are different from zero is equivalent to a continued fraction with partial denominators 1.

Let $\{c_m\}_1^\infty$ be any sequence of nonzero complex numbers. The Moebius transformation

$$t_m : u \to \frac{a_m}{b_m + u}$$

evidently can be represented in the form

$$t_m(u) = \frac{c_m a_m}{c_m b_m + c_m u}$$

and this may be regarded as the composition of two transformations, $t_m = s_m \circ r_m$, where

$$s_m : u \to \frac{c_m a_m}{c_m b_m + u},$$

$$r_m : u \to c_m u.$$

Thus we have

$$t_1 \circ t_2 \circ \cdots \circ t_m = s_1 \circ r_1 \circ s_2 \circ r_2 \circ \cdots \circ s_m \circ r_m$$

$$= s_1 \circ (r_1 \circ s_2) \circ (r_2 \circ s_3) \circ \cdots \circ (r_{m-1} \circ s_m) \circ r_m.$$

Here the transforms

$$t_m^* := r_{m-1} \circ s_m : u \to \frac{c_{m-1} c_m a_m}{c_m b_m + u}$$

may be regarded as a single *RIT* transform. Because $r_m(0) = 0$ we have, if $c_0 := 1$,

$$w_m = t_1 \circ t_2 \circ \cdots \circ t_m(0) = t_1^* \circ t_2^* \circ \cdots \circ t_m^*(0).$$

Thus we obtain the important relation

$$\overset{\infty}{\underset{m=1}{\Phi}} \frac{a_m}{b_m} \cong \overset{\infty}{\underset{m=1}{\Phi}} \frac{c_{m-1} c_m a_m}{c_m b_m}, \tag{12.1-11}$$

valid for an arbitrary sequence of nonzero complex numbers $c_0 = 1$, c_1, $c_2, \ldots$.

It should be noted that although equivalent continued fractions have the same approximants, their numerators and denominators are, in general, not the same. If we denote by p_m, q_m the numerators and denominators of the fraction on the left of (12.1-11) and by p_m^*, q_m^* the corresponding quantities of the fraction on the right, then it follows from (12.1-8) that

$$p_m^* = c_1 c_2 \cdots c_m p_m,$$
$$q_m^* = c_1 c_2 \cdots c_m q_m,$$

(12.1-12)

$m = 1, 2, \ldots$.

To find for the c. f. on the left of (12.1-11) an equivalent c. f. with *partial denominators* 1, it evidently suffices to pick the c_m such that $c_m b_m = 1$, which is always possible if $b_m \neq 0$. We thus find

$$\mathop{\Phi}_{m=1}^{\infty} \frac{a_m}{b_m} \cong \mathop{\Phi}_{m=1}^{\infty} \frac{b_{m-1}^{-1} b_m^{-1} a_m}{1}$$

(12.1-13)

($b_0 := 1$). To find an equivalent c. f. with *partial numerators* 1, the c_m have to be chosen such that

$$c_1 a_1 = 1, \qquad c_{m-1} c_m a_m = 1, \qquad m = 2, 3, \ldots.$$

This implies

$$c_1 = \frac{1}{a_1}, \qquad c_m = \frac{1}{a_m c_{m-1}},$$

hence

$$c_{2m} = \frac{a_1 a_3 \cdots a_{2m-1}}{a_2 a_4 \cdots a_{2m}}, \qquad c_{2m+1} = \frac{a_2 a_4 \cdots a_{2m}}{a_1 a_3 \cdots a_{2m+1}}. \quad (12.1\text{-}14)$$

We thus have obtained

$$\mathop{\Phi}_{m=1}^{\infty} \frac{a_m}{b_m} \cong \mathop{\Phi}_{m=1}^{\infty} \frac{1}{b_m c_m},$$

(12.1-15)

where the c_m are defined by (12.1-14). The fraction on the right is called the **reduced form** of the fraction on the left.

Improper Continued Fractions. In some contexts it is advantageous to consider continued fractions of the form

$$b_0 + \frac{a_1 \vert}{\vert b_1} + \frac{a_2 \vert}{\vert b_2} + \cdots$$

(12.1-16)

where b_0 is some complex number. Such fractions might be called **improper** continued fractions. Their formal definition can be given along the same lines as that of the "proper" c. f. considered above, with the following differences: The sequences $\{b_n\}$ and $\{w_n\}$ now begin with the elements b_0 and w_0, and we have

$$w_n := t_0 \circ t_1 \circ \cdots \circ t_n(0),$$

where $t_1, t_2, \ldots,$ are defined as before, but where $t_0: u \to b_0 + u$. All the remaining definitions and terminology remain unchanged. We leave it to the reader to verify that

$$t_{0,n}(u) := t_0 \circ t_1 \circ \cdots \circ t_n(u) = \frac{p_{n-1}u + p_n}{q_{n-1}u + q_n},$$

$n = 0, 1, 2, \ldots,$ where the sequences $\{p_n\}$ and $\{q_n\}$ again satisfy the difference equation (12.1-8a), but now obey the less symmetric initial conditions

$$p_{-1} = 1, \qquad p_0 = b_0$$
$$q_{-1} = 0, \qquad q_0 = 1$$

$$(12.1\text{-}17)$$

in place of (12.1-8b). Thus the nth approximant $w_n = t_{0,n}(0)$ of the improper fraction (12.1-16) is again given by

$$w_n = \frac{p_n}{q_n}, \qquad n = 0, 1, 2, \ldots.$$

Even and Odd Part of a Continued Fraction. By the **even [odd] part** of a proper or improper continued fraction C is meant a continued fraction whose sequence of approximants consists of the approximants of even [odd] order of C. Thus if $\{w_n\}$ denotes the sequence of approximants of C, the sequence of approximants of an even part of C is $\{w_{2n}\}$, and that of an odd part $\{w_{2n+1}\}$. Thus if C converges, both its even and its odd part converge and have the same value; on the other hand, if both an even and an odd part of C converge, then C itself converges if and only if both parts have the same value.

In later applications we mainly require the even and odd parts of the continued fraction

$$C := \frac{a_1 \,|}{|\, 1} + \frac{a_2 \,|}{|\, 1} + \frac{a_3 \,|}{|\, 1} + \cdots . \qquad (12.1\text{-}18)$$

If

$$t_k(u) := \frac{a_k}{1 + u},$$

then

$$t_k \circ t_{k+1}(u) = \frac{a_k}{1 + a_{k+1}/(1+u)} = \frac{a_k(1+u)}{1 + a_{k+1} + u}$$

$$= a_k - \frac{a_k a_{k+1}}{1 + a_{k+1} + u} = s_k \circ t_{k+1}^*(u),$$

where

$$s_k(u) := a_k - u, \qquad t_{k+1}^*(u) := \frac{a_k a_{k+1}}{1 + a_{k+1} + u}.$$

Thus because $s_k(0) = t_k(0)$, if $n = 1, 2, \ldots$, the $2n$th approximant of C equals

$$w_{2n} = t_1 \circ t_2 \circ \cdots \circ t_{2n-1} \circ t_{2n}(0)$$

$$= t_1 \circ (s_2 \circ t_3^*) \circ (s_4 \circ t_5^*) \circ \cdots \circ (s_{2n-2} \circ t_{2n-1}^*) \circ s_{2n}(0)$$

$$= (t_1 \circ s_2) \circ (t_3^* \circ s_4) \circ \cdots \circ (t_{2n-1}^* \circ s_{2n})(0)$$

$$= t_1^e \circ t_2^e \circ \cdots \circ t_n^e(0),$$

where

$$t_1^e(u) := t_1 \circ s_2(u) = \frac{a_1}{1 + a_2 + u},$$

$$t_k^e(u) := t_{2k-1}^* \circ s_{2k}(u) = \frac{a_{2k-2} a_{2k-1}}{1 + a_{2k-1} + a_{2k} + u}.$$

Thus an even part of C is

$$C^e := \frac{a_1}{|1 + a_2|} - \frac{a_2 a_3}{|1 + a_3 + a_4|} - \frac{a_4 a_5}{|1 + a_5 + a_6|} - \cdots. \qquad (12.1\text{-}19)$$

A similar computation yields the odd part

$$C^o := a_1 - \frac{a_1 a_2}{|1 + a_2 + a_3|} - \frac{a_3 a_4}{|1 + a_4 + a_5|} - \cdots. \qquad (12.1\text{-}20)$$

The Difference between Consecutive Approximants. It was shown in Theorem 12.1a that the matrix

$$\mathbf{M}_{1,m} := \begin{pmatrix} p_{m-1} & p_m \\ q_{m-1} & q_m \end{pmatrix}$$

equals $\mathbf{M}_1 \mathbf{M}_2 \cdots \mathbf{M}_m$, where

$$\mathbf{M}_k := \begin{pmatrix} 0 & a_k \\ 1 & b_k \end{pmatrix}, \qquad k = 1, 2, \ldots.$$

From the fact that the determinant of a product of matrices equals the product of the determinants of the factors, we immediately obtain the important relation

$$p_{m-1} q_m - p_m q_{m-1} = (-1)^m a_1 a_2 \cdots a_m, \qquad (12.1\text{-}21)$$

$m = 1, 2, \ldots$. If $q_{m-1}q_m \neq 0$, we find on dividing that

$$w_m - w_{m-1} = \frac{p_m}{q_m} - \frac{p_{m-1}}{q_{m-1}} = (-1)^{m+1} \frac{a_1 a_2 \cdots a_m}{q_{m-1}q_m}. \qquad (12.1-22)$$

Obviously, $w_m = (w_m - w_{m-1}) + (w_{m-1} - w_{m-2}) + \cdots + (w_1 - w_0) + w_0$. Thus if no denominator vanishes, we find that the approximants of a proper continued fraction can by virtue of $w_0 = 0$ be written in the form

$$w_m = \frac{a_1}{q_0 q_1} - \frac{a_1 a_2}{q_1 q_2} + \cdots + (-1)^{m+1} \frac{a_1 a_2 \cdots a_m}{q_{m-1}q_m}.$$

A continued fraction C is called **equivalent to an infinite series** S if for $n = 1, 2, \ldots$, the nth approximant of C equals the nth partial sum of S. We again indicate the equivalence of C and S by the symbol $C \cong S$. This result permits us to state that if the denominators $q_1, q_2, \ldots$, of the continued fraction on the left are all different from zero, then

$$\overset{\infty}{\underset{m=1}{\Phi}} \frac{a_m}{b_m} \cong \sum_{m=1}^{\infty} (-1)^{m+1} \frac{a_1 a_2 \cdots a_m}{q_{m-1}q_m}. \qquad (12.1-23)$$

There follows, in particular, that the fraction on the left is convergent if and only if the series on the right is convergent.

The Convergence of a Continued Fraction with Positive Elements. We now discuss a general criterion for the convergence of continued fractions with positive elements. Because any such fraction is equivalent to a reduced fraction (partial numerators one), we begin by discussing such fractions. The simple main result is as follows.

THEOREM 12.1c

Let $\beta_m > 0$, $m = 1, 2, \ldots$. Then the continued fraction

$$C := \overset{\infty}{\underset{m=1}{\Phi}} \frac{1}{\beta_m}$$

is convergent if and only if the infinite series

$$S := \sum_{m=1}^{\infty} \beta_m$$

is divergent.

Proof. The denominators q_m of C satisfy the recurrence relation

$$q_0 = 1, \qquad q_1 = \beta_1, \qquad q_m = q_{m-2} + \beta_m q_{m-1}, \qquad (12.1-24)$$

$m = 2, 3, \ldots$. It is evident that $q_m > 0$, $m = 0, 1, 2, \ldots$; thus as a special case of (12.1-24),

$$\Phi_{m=1}^{\infty} \frac{1}{\beta_m} \cong \sum_{m=1}^{\infty} \frac{(-1)^{m+1}}{q_{m-1}q_m}.$$

The fraction C converges if and only if the alternating series on the right converges. From (12.1-24) there follows $q_{m+1} > q_{m-1}$, $m = 1, 2, \ldots$; hence

$$q_m q_{m+1} > q_{m-1}q_m,$$

$m = 1, 2, \ldots$; hence the absolute values of the terms of the alternating series on the right are monotonically decreasing. Thus by the Leibniz criterion the series is convergent if and only if

$$\lim_{m \to \infty} q_{m-1}q_m = \infty. \tag{12.1-25}$$

Now again by (12.1-24), because $q_m \geq \gamma := \min(1, \beta_1)$, $m = 1, 2, \ldots$,

$$q_{m-1}q_m = q_{m-2}q_{m-1} + \beta_m q_{m-1}^2$$
$$\geq q_{m-2}q_{m-1} + \beta_m \gamma^2$$
$$\geq (\beta_1 + \beta_2 + \cdots + \beta_m)\gamma^2.$$

Thus if S diverges, (12.1-25) holds, and C is convergent. On the other hand, once more from (12.1-24),

$$q_{m-1} + q_m = q_{m-2} + (1 + \beta_m)q_{m-1} \leq (1 + \beta_m)(q_{m-1} + q_{m-2});$$

hence by induction, because $q_0 + q_1 = 1 + \beta_1$,

$$q_{m-1} + q_m \leq (1 + \beta_1)(1 + \beta_2) \cdots (1 + \beta_m) < e^{\beta_1 + \beta_2 + \cdots + \beta_m}.$$

Hence if S converges, and if σ denotes the sum of S,

$$q_{m-1} + q_m \leq e^\sigma, \qquad m = 1, 2, \ldots;$$

thus by the inequality of the geometric and arithmetic mean,

$$q_{m-1}q_m \leq \tfrac{1}{4}(q_{m-1} + q_m)^2 \leq \tfrac{1}{4}e^{2\sigma}$$

for all $m = 1, 2, \ldots$. Consequently, (12.1-25) does not hold, and C fails to converge. ■

We leave it to the reader to deduce from Theorem 12.1c convergence criteria for continued fractions

$$\Phi_{m=1}^{\infty} \frac{\alpha_m}{\beta_m}$$

where $\alpha_m > 0$, $\beta_m > 0$, by using the equivalence relations (12.1-13) and (12.1-15).

PROBLEMS

1. Evaluate the continued fraction

$$\frac{1\,|}{|\,i} + \frac{1\,|}{|\,i} + \frac{1\,|}{|\,i} + \frac{1\,|}{|\,1}$$

by both the ascending and the descending method.

2. Let $\alpha_m > 0$, $\beta_m > 0$, $m = 1, 2, \ldots$, and let

$$\sum_{m=1}^{\infty} \sqrt{\frac{\beta_m \beta_{m+1}}{\alpha_{m+1}}} = \infty.$$

Show that the continued fraction

$$\mathop{\Phi}_{m=1}^{\infty} \frac{\alpha_m}{\beta_m}$$

is convergent.

3. Discuss the convergence of the continued fraction

$$\mathop{\Phi}_{m=1}^{\infty} \frac{a}{b},$$

where a and b are complex numbers, $a \neq 0$, by explicitly solving the difference equation for the numerators and the denominators. Check the result of Problem 2 when a and b are positive.

4. Show that the numerators p_n and the denominators q_n of the improper continued fraction

$$b_0 + \mathop{\Phi}_{m=1}^{\infty} \frac{a_m}{b_m}$$

can be represented by determinants, as follows:

$$p_n = \begin{vmatrix} b_0 & 1 & & & & & 0 \\ a_1 & b_1 & 1 & & & & \\ & a_2 & b_2 & 1 & & & \\ & & & \ddots & \ddots & \ddots & \\ & & & & a_{n-1} & b_{n-1} & 1 \\ 0 & & & & & a_n & b_n \end{vmatrix},$$

$$q_n = \begin{vmatrix} b_1 & 1 & & & & & 0 \\ a_2 & b_2 & 1 & & & & \\ & a_3 & b_3 & 1 & & & \\ & & & \ddots & \ddots & \ddots & \\ & & & & a_{n-1} & b_{n-1} & 1 \\ 0 & & & & & a_n & b_n \end{vmatrix}.$$

(Elements not shown are zero.) [Expand the determinants in terms of the elements of the last row and thus show that they satisfy the recurrence relations (12.1-8a).]

5. Prove that no two of any three consecutive approximants of an infinite continued fraction with nonzero elements can have the same value.

6. The numerators q_n of the infinite continued fraction

$$\frac{1}{|b_1|} + \frac{1}{|b_2|} + \frac{1}{|b_3|} + \cdots$$

are on occasion denoted by $[b_1, b_2, \ldots, b_n]$ and are called **Gauss brackets**. The empty bracket is given the value 1. Prove the following relations:

(a) $[b_1, b_2, \ldots, b_n] = [b_n, b_{n-1}, \ldots, b_1]$;

(b) $[b_1, b_2, \ldots, b_n] = b_1[b_2, b_3, \ldots, b_n] + [b_3, b_4, \ldots, b_n]$;

(c) $\begin{vmatrix} [b_1, \ldots, b_n] & [b_1, \ldots, b_{n-1}] \\ [b_2, \ldots, b_n] & [b_2, \ldots, b_{n-1}] \end{vmatrix} = (-1)^n, \qquad n \geq 2$

7. Let $c_i \neq 0$, $i = 0, 1, \ldots$. Prove Euler's relations

$$c_0 + c_1 + c_2 + \cdots \cong c_0 + \frac{c_1}{|1|} - \frac{c_2/c_1}{|1 + c_2/c_1|} - \frac{c_3/c_2}{|1 + c_3/c_2|} - \cdots$$

$$\cong c_0 + \frac{c_1}{|1|} - \frac{c_2}{|c_1 + c_2|} - \frac{c_1 c_3}{|c_2 + c_3|} - \frac{c_2 c_4}{|c_3 + c_4|} - \cdots .$$

8. Show that for arbitrary $z \neq 0$

$$e^z = 1 + \frac{z}{|1|} - \frac{1 \cdot z}{|2 + z|} - \frac{2 \cdot z}{|3 + z|} - \frac{3 \cdot z}{|4 + z|} - \cdots .$$

9. Prove:

$$\mathop{\Phi}_{m=1}^{\infty} \frac{m}{m} = \frac{1}{e - 1}.$$

10. Show that for $-1 \leq x \leq 1$

$$\text{Arctan } x = \frac{x}{|1|} + \frac{1 \cdot x^2}{|3 - x^2|} + \frac{9 \cdot x^2}{|5 - 3x^2|} + \frac{25 \cdot x^2}{|7 - 5x^2|} + \cdots$$

and, consequently,

$$\frac{\pi}{4} = \frac{1}{|1|} + \frac{1}{|2|} + \frac{9}{|2|} + \frac{25}{|2|} + \frac{49}{|2|} + \cdots .$$

11. Let $\alpha_m > 0$, $\beta_m > 0$, $m = 1, 2, \ldots$. Show that the even and the odd part of the fraction

$$\mathop{\Phi}_{m=1}^{\infty} \frac{\alpha_m}{\beta_m}$$

always converge.

12. Let $\beta > 0$. Show that the continued fraction

$$\mathop{\Phi}_{m=1}^{\infty} \frac{m^k}{\beta}$$

converges for $k \leq 2$ and diverges for $k > 2$.

13. Let the elements of the fraction $\Phi(a_m/b_m)$ be positive. Show that there exist constants $c_1, c_2, \ldots$, such that the denominators of the equivalent fraction (12.1-11) all are 1. [Solution:

$$c_n := \frac{1}{\mid b_n} + \frac{a_n}{\mid b_{n-1}} + \frac{a_{n-1}}{\mid b_{n-2}} + \cdots + \frac{a_2}{\mid b_1}.]$$

§12.2. CONTINUED FRACTIONS IN NUMBER THEORY

We expect the reader to be familiar with the **Euclidean algorithm**. Let $k = k_0$ and $l = k_1$ be two positive integers. The Euclidean algorithm consists in determining integers $k_2 > k_3 > \cdots \geq 0$ and $b_0, b_1, \ldots (b_0 \geq 0, b_i \geq 1, i = 1, 2, \ldots)$ such that

$$k_0 = b_0 k_1 + k_2,$$
$$k_1 = b_1 k_2 + k_3,$$
$$\cdots \qquad\qquad (12.2\text{-}1)$$
$$k_m = b_m k_{m+1} + k_{m+2},$$
$$\cdots$$

Because a decreasing sequence of nonnegative integers necessarily reaches the value zero in a finite number of steps, the Euclidean algorithm necessarily terminates; that is, there exists an index n such that $k_{n+2} = 0$, the last relation (12.2-1) having the form

$$k_n = b_n k_{n+1}. \qquad\qquad (12.2\text{-}2)$$

It follows from the elements of number theory that for $m = 0, 1, \ldots, n-1$ every divisor of k_m and k_{m+1} is also a divisor of k_{m+2}. Thus every divisor of k_0 and k_1 is a divisor of k_{n+1}. Conversely, every divisor of k_{n+1} is seen to be a divisor of k_0 and k_1. Thus k_{n+1} is the *greatest common divisor* of k_0 and k_1, and the Euclidean algorithm is commonly regarded as a convenient tool (as opposed to the decomposition in prime factors) to determine this greatest common divisor.

EXAMPLE 1

$k_0 = 63, k_1 = 24$ yields

$$63 = 2 \cdot 24 + 15$$
$$24 = 1 \cdot 15 + 9$$
$$15 = 1 \cdot 9 + 6$$
$$9 = 1 \cdot 6 + 3$$
$$6 = 2 \cdot 3 + 0$$

Thus the greatest common divisor of 63 and 24 is 3.

EXAMPLE **2**

$k_0 = 64$, $k_1 = 25$ yields

$$64 = 2 \cdot 25 + 14$$
$$25 = 1 \cdot 14 + 11$$
$$14 = 1 \cdot 11 + 3$$
$$11 = 3 \cdot 3 + 2$$
$$3 = 1 \cdot 2 + 1$$
$$2 = 2 \cdot 1 + 0$$

Thus the greatest common divisor of 64 and 25 is 1.

In a slightly different vein, we here look at the Euclidean algorithm as a tool to obtain a certain standardized representation for rational numbers. For $m = 0, 1, \ldots, n$ let

$$\rho_m := \frac{r_m}{r_{m+1}}.$$

Then the relations (12.2-1) may be written

$$\rho_0 = b_0 + \frac{1}{\rho_1},$$

$$\rho_1 = b_1 + \frac{1}{\rho_2},$$

$$\cdots \qquad\qquad\qquad (12.2\text{-}3)$$

$$\rho_m = b_m + \frac{1}{\rho_{m+1}},$$

$$\cdots$$

$$\rho_n = b_n.$$

Starting with any rational number $\rho := k/l$, the numbers ρ_i and b_i resulting in the foregoing manner through the Euclidean algorithm may also be characterized by the conditions $\rho_0 = \rho$,

$$b_m := [\rho_m], \qquad \frac{1}{\rho_{m+1}} = \rho_m - [\rho_m], \qquad m = 0, 1, \ldots \qquad (12.2\text{-}4)$$

Here, for any real number ξ the symbol $[\xi]$ denotes the greatest integer not exceeding ξ.

Let us now define the Moebius transformations

$$t_0(u) := b_0 + u, \qquad t_m(u) := \frac{1}{b_m + u}, \qquad m = 1, 2, \ldots, n.$$

Then the relations (12.2-3) show that

$$\frac{1}{\rho_n} = t_n(0), \qquad \frac{1}{\rho_m} = t_m\left(\frac{1}{\rho_{m+1}}\right), \qquad m = 1, 2, \ldots, \qquad n-1,$$

$$\rho_0 = t_0\left(\frac{1}{\rho_1}\right).$$

Thus evidently

$$\rho = t_0 \circ t_1 \circ t_2 \circ \cdots \circ t_n(0),$$

which we also write

$$\rho = b_0 + \frac{1}{\lfloor b_1} + \frac{1}{\lfloor b_2} + \cdots + \frac{1}{\lfloor b_n}. \tag{12.2-5}$$

Thus the rational number ρ has been represented as the value of an improper, terminating, reduced continued fraction whose partial denominators are integers, $b_i > 0$ ($i = 1, 2, \ldots, n-1$), $b_n > 1$. Such a continued fraction is called **simple**. The foregoing illustrations of the Euclidean algorithm yield the following examples of representations of rational numbers by terminating simple continued fractions:

$$\frac{63}{24} = 2 + \frac{1}{\lfloor 1} + \frac{1}{\lfloor 1} + \frac{1}{\lfloor 1} + \frac{1}{\lfloor 2},$$

$$\frac{64}{25} = 2 + \frac{1}{\lfloor 1} + \frac{1}{\lfloor 1} + \frac{1}{\lfloor 3} + \frac{1}{\lfloor 1} + \frac{1}{\lfloor 2}.$$

Because the value of any terminating simple continued fraction obviously is a rational number, we have

THEOREM 12.2a

A real number is rational if and only if it is the value of a terminating simple continued fraction.

It is easy to see that a rational number can be represented in one way only by a simple terminating continued fraction.

Diophantine Equations. Let k and l be two positive integers that are mutually prime, that is, whose greatest common divisor is 1. Here we consider the problem of finding all pairs of integers (x, y) solving the equation

$$kx - ly = 1, \tag{12.2-6}$$

called a **Diophantine equation**. It follows from simple facts in the theory of ideals that this equation always has ·a solution. The theory of continued fractions provides an algorithm for actually constructing the solution.

Let the number k/l be expanded in a terminating simple continued fraction:

$$\frac{k}{l} = b_0 + \frac{1}{\lfloor b_1} + \frac{1}{\lfloor b_2} + \cdots + \frac{1}{\lfloor b_n}.$$

Let p_k, q_k denote the numerators and denominators of this fraction. Then by (12.1-21), because all $a_i = 1$,

$$p_m q_{m-1} - p_{m-1} q_m = (-1)^{m-1}, \qquad m = 1, 2, \ldots, n. \qquad (12.2\text{-}7)$$

Naturally,

$$\frac{p_n}{q_n} = \frac{k}{l},$$

the value of the continued fraction. Because (12.2-7) for $m = n$ shows that p_n and q_n have no common divisors, there follows $p_n = k$, $q_n = l$.

To find a *special* solution of (12.2-6), we distinguish two cases according to the partity of n. If n is odd, let $x := q_{n-1}$, $y := p_{n-1}$. Then by (12.2-7)

$$kx - ly = p_n x - q_n y = (-1)^{n-1} = 1,$$

and the equation is solved. If n is even, let $x := l - q_{n-1}$, $y := k - p_{n-1}$. Then

$$kx - ly = p_n(q_n - q_{n-1}) - q_n(p_n - p_{n-1}) = -(-1)^{n-1} = 1,$$

and the equation is again solved.

The *general* solution is now easily found by observing that if (x_0, y_0) and (x_1, y_1) are distinct special solutions, then $k(x_1 - x_0) - l(y_1 - y_0) = 0$, hence

$$\frac{y_1 - y_0}{x_1 - x_0} = \frac{k}{l},$$

and thus, because k and l are mutually prime,

$$y_1 = y_0 + km, \qquad x_1 = x_0 + lm,$$

where m is some integer. Conversely, any such (x_1, y_1) is a solution, hence the general solution is given by

$$(x, y) = (x_0, y_0) + m(l, k),$$

where m is an arbitrary integer and (x_0, y_0) is any special solution, for instance the solution found above.

EXAMPLE **3**

$61x - 48y = 1$. From the expansion in a simple continued fraction

$$\frac{61}{48} = 1 + \frac{1}{\lfloor 3} + \frac{1}{\lfloor 1} + \frac{1}{\lfloor 2} + \frac{1}{\lfloor 4}$$

we find the numerators and denominators

b_m	1	3	1	2	4
p_m	1	4	5	14	61
q_m	1	3	4	11	48

Because $n = 4$ is even, a special solution is $(x_0, y_0) = (48 - 11, 61 - 14) = (37, 47)$, and the general solution is

$$(x, y) = (37, 47) + m(48, 61).$$

Representation of Irrational Numbers. Let the number ξ now be irrational. Then it still can be subjected to the algorithm (12.2-4); that is, we can still determine a sequence of integers b_i ($b_i > 0$ for $i > 0$) and of real numbers ξ_i ($\xi_i > 0$ for $i > 0$) by the conditions

$$\xi_0 = \xi,$$

$$b_m = [\xi_m], \qquad \xi_m = b_m + \frac{1}{\xi_{m+1}},$$

(12.2-8)

$m = 0, 1, 2, \ldots$. The process cannot terminate, for if it would, then ξ would be value of a terminating simple continued fraction, and thus rational, which would be contrary to our hypothesis. By analogy with the rational case, we expect that

$$\xi = b_0 + \frac{1}{\lceil b_1} + \frac{1}{\lceil b_2} + \frac{1}{\lceil b_3} + \cdots.$$

(12.2-9)

By Theorem 12.1c, the continued fraction on the right certainly converges. The problem is to show that its value is ξ. By the definitions of the numbers b_i, if the t_i denote the same Moebius transformations as before, we have for $m = 1, 2, \ldots$

$$\xi = t_0 \circ t_1 \circ t_2 \circ \cdots \circ t_m \left(\frac{1}{\xi_{m+1}} \right).$$

Thus by Theorem 12.1a, if the numerators and denominators of the fraction are denoted by p_m and q_m,

$$\xi = \frac{p_{m-1} + p_m \xi_{m+1}}{q_{m-1} + q_m \xi_{m+1}}, \qquad m = 1, 2, \ldots.$$

There follows

$$\xi - \frac{p_m}{q_m} = \frac{(p_{m-1} + p_m \xi_{m+1})q_m - p_m(q_{m-1} + q_m \xi_{m+1})}{q_m(q_{m-1} + q_m \xi_{m+1})}$$

$$= \frac{p_{m-1}q_m - p_m q_{m-1}}{q_m(q_{m-1} + q_m \xi_{m+1})},$$

and thus by (12.1-21), because all $q_i > 0$ and all $\xi_i > 1$,

$$\left| \xi - \frac{p_m}{q_m} \right| \leq \frac{1}{q_m(q_{m-1} + q_m)}.$$

Because $q_m \to \infty$ (see the proof of Theorem 12.1c), there follows

$$\xi = \lim_{m \to \infty} \frac{p_m}{q_m},$$

which is the same as (12.2-9).

We call a continued fraction such as the one appearing on the right of (12.2-5) a **nonterminating simple continued fraction**. We have just shown that every irrational real number can be represented as a nonterminating simple continued fraction. It is easy to see that this representation is unique. Let

$$\xi = b_0 + \frac{1}{\left| b_1 \right.} + \frac{1}{\left| b_2 \right.} + \cdots$$

$$= c_0 + \frac{1}{\left| c_1 \right.} + \frac{1}{\left| c_2 \right.} + \cdots$$

be two such representations. Because the value of a proper simple continued fraction always lies between 0 and 1, there follows immediately

$$b_0 = c_0 = [\xi].$$

Having proved that $b_i = c_i$ for $i = 0, 1, \ldots, m - 1$, we may similarly conclude from

$$b_m + \frac{1}{\left| b_{m+1} \right.} + \frac{1}{\left| b_{m+2} \right.} + \cdots = c_m + \frac{1}{\left| c_{m+1} \right.} + \cdots$$

that $b_m = c_m$.

Every nonterminating simple continued fraction converges by Theorem 12.1c. Its value must be irrational, by Theorem 12.2a. In summary, we thus have

THEOREM 12.2b

Every irrational number can be represented by a nonterminating simple continued fraction, and the value of every nonterminating simple continued fraction is irrational. The representation of a given irrational ξ is (12.2-9), where the b_i are determined by the algorithm (12.2-8).

EXAMPLE 4

For $\xi = \sqrt{6}$, the algorithm begins as follows:

$$b_0 = [\sqrt{6}] = 2, \qquad \xi_1 = \frac{1}{\sqrt{6}-2} = \frac{\sqrt{6}+2}{2}$$

$$b_1 = \left[\frac{\sqrt{6}+2}{2}\right] = 2, \qquad \xi_2 = \frac{2}{\sqrt{6}-2} = \sqrt{6}+2$$

$$b_2 = [\sqrt{6}+2] = 4, \qquad \xi_3 = \frac{1}{\sqrt{6}-2} = \xi_1.$$

Because $\xi_3 = \xi_1$, there follows $\xi_4 = \xi_2$; hence $b_1 = b_3 = b_5 = \cdots = 2$, $b_2 = b_4 = b_6 = \cdots = 4$. Hence the desired representation is

$$\sqrt{6} = 2 + \frac{1}{\mid 2} + \frac{1}{\mid 4} + \frac{1}{\mid 2} + \frac{1}{\mid 4} + \cdots,$$

the partial denominators periodically repeated.

EXAMPLE 5

In §12.6 it will be shown accidentally that

$$\frac{e-1}{e+1} = \frac{1}{\mid 2} + \frac{1}{\mid 6} + \frac{1}{\mid 10} + \frac{1}{\mid 14} + \frac{1}{\mid 18} + \cdots,$$

showing that e is not rational.

Ultimately Periodic Continued Fractions. A nonterminating simple continued fraction

$$b_0 + \frac{1}{\mid b_1} + \frac{1}{\mid b_2} + \frac{1}{\mid b_3} + \cdots \qquad (12.2\text{-}10)$$

is called **ultimately periodic** if there exist integers $k \geq 0$ and $m \geq 1$ such that

$$b_{n+m} = b_n \quad \text{for all } n \geq k. \qquad (12.2\text{-}11)$$

For instance, the continued fraction for $\sqrt{6}$ given above is ultimately periodic with $k = 1$ and $m = 2$. We now investigate the nature of the irrational numbers that are represented by ultimately periodic simple continued fractions.

Let ξ be the value of (12.2-10). This means that the sequence of approximants

$$w_n := t_0 \circ t_1 \circ \cdots \circ t_n(0)$$

converges to ξ. Thus a fortiori the sequence w_k, w_{k+m}, w_{k+2m}, ... converges to ξ. Putting

$$r := t_0 \circ t_1 \circ \cdots \circ t_{k-1},$$

$$s := t_k \circ t_{k+1} \circ \cdots \circ t_{k+m-1},$$

we have $w_{k+nm} = r \circ s^{[n]}(0)$, where $s^{[n]} = s \circ s \circ \cdots \circ s$ (n times), the n-fold iterate of s. Thus $r \circ s^{[n]}(0) \to \xi$ for $n \to \infty$, which is to say that $\eta_n := s^{[n]}(0) \to r^{[-1]}(\xi)$, properly or improperly. From the fact that $\eta_n = s(\eta_{n-1})$, it follows by letting $n \to \infty$ that $\eta = s(\eta)$, and thus that η is a fixed point of s (see §6.12).

The fixed points of s are easily determined. Let p_i^* and q_i^* denote the numerators and denominators of the continued fraction

$$C := \frac{1}{\lceil b_k} + \frac{1}{\lceil b_{k+1}} + \frac{1}{\lceil b_{k+2}} + \cdots .$$

Then the p_i^* and q_i^* are positive integers, and by Theorem 12.1a

$$s(u) = \frac{p_{m-1}^* u + p_m^*}{q_{m-1}^* u + q_m^*}.$$

Thus η is a solution of the quadratic equation

$$p_{m-1}^* \eta + p_m^* = \eta(q_{m-1}^* \eta + q_m^*),$$

or

$$q_{m-1}^* \eta^2 + (q_m^* - p_{m-1}^*)\eta - p_m^* = 0, \qquad (12.2\text{-}12)$$

which has integer coefficients. Both solutions of this equation are real and $\neq 0$, and precisely one solution is positive. Because all approximants of C are positive, η equals the positive solution of (12.2-12).

Once more, by Theorem 12.1a,

$$r(u) = t_0 \circ t_1 \circ \cdots \circ t_{k-1}(u) = \frac{p_{k-2}u + p_{k-1}}{q_{k-2}u + q_{k-1}},$$

where the p_i and q_i denote the numerators and denominators of the full fraction (12.2-10), and thus are again integers. The value of the full fraction thus equals

$$\xi = r(\eta) = \frac{p_{k-2}\eta + p_{k-1}}{q_{k-2}\eta + q_{k-1}}, \qquad (12.2\text{-}13)$$

where η denotes the positive solution of (12.2-12).

A solution of an algebraic equation of degree 2 whose coefficients are integers is called an **algebraic number of degree** 2 or a **quadratic irrationality**. Clearly, the number $r(\eta)$ defined by (12.2-13) shares with η the property of being a quadratic irrationality. We thus have proved one-half of the following celebrated theorem due to Lagrange:

THEOREM 12.2c

Every simple continued fraction that is ultimately periodic represents a quadratic irrationality. Conversely, the representation of any real quadratic irrationality by a simple continued fraction is ultimately periodic.

Proof. It only remains to prove the second assertion. Let the quadratic equation satisfied by ξ be

$$h\xi^2 + k\xi + l = 0, \tag{12.2-14}$$

where h, k, l are integers, $k^2 - 4hl > 0$, and let

$$\xi = b_0 + \cfrac{1}{\vert b_1} + \cfrac{1}{\vert b_2} + \cfrac{1}{\vert b_3} + \cdots \tag{12.2-15}$$

be the simple continued fraction representation of ξ. We wish to show that the fraction (12.2-15) is ultimately periodic.

For $n = 0, 1, 2, \ldots$, let

$$\xi_n := b_n + \cfrac{1}{\vert b_{n+1}} + \cfrac{1}{\vert b_{n+2}} + \cdots,$$

so that, by Theorem 12.1a,

$$\xi = t_0 \circ t_1 \circ \cdots \circ t_n\left(\frac{1}{\xi_{n+1}}\right) = \frac{p_{n-1} + p_n\xi_{n+1}}{q_{n-1} + q_n\xi_{n+1}},$$

p_n and q_n denoting, as usual, the (integral) numerators and denominators of (12.2-15). From (12.2-14) we have

$$h\left[\frac{p_{n-1} + p_n\xi_{n+1}}{q_{n-1} + q_n\xi_{n+1}}\right]^2 + k\frac{p_{n-1} + p_n\xi_{n+1}}{q_{n-1} + q_n\xi_{n+1}} + l = 0$$

or, after removing denominators,

$$h_n\xi_{n+1}^2 + k_n\xi_{n+1} + l_n = 0, \tag{12.2-16}$$

where

$$h_n := hp_n^2 + kp_nq_n + lq_n^2,$$
$$k_n := 2hp_{n-1}p_n + k(p_{n-1}q_n + p_nq_{n-1}) + 2lq_{n-1}q_n,$$
$$l_n := hp_{n-1}^2 + kp_{n-1}q_{n-1} + lq_{n-1}^2.$$

These numbers are integers for every n.

Equation (12.2-16) can be simplified by using the fact that for $n = 0, 1, 2, \ldots$,

$$\left| \xi - \frac{p_n}{q_n} \right| < \frac{1}{q_n q_{n+1}},$$

as follows from the proof of Theorem 12.1c, and hence

$$p_n = q_n \xi + \frac{\theta_n}{q_{n+1}}, \qquad n = 1, 2, \ldots,$$

where $|\theta_n| < 1$. Expressing all p_n by q_n, we thus get

$$h_n = h\left(q_n \xi - \frac{\theta_n}{q_{n+1}}\right)^2 + kq_n\left(q_n \xi - \frac{\theta_n}{q_{n+1}}\right) + lq_n^2$$

$$= q_n^2(h\xi^2 + k\xi + l) - \frac{q_n}{q_{n+1}}\theta_n(2h\xi + k) + h\frac{\theta_n^2}{q_{n+1}^2},$$

which by virtue of (12.2-14) simplifies. Using the fact that the q_n increase monotonically and $q_0 = 1$, we thus find for h_n the bound

$$|h_n| \leqslant |h|(2|\xi| + 1) + |k|.$$

In a similar manner one may establish that

$$|k_n| \leqslant 4|h|\,|\xi| + 2|k| + 2|\rho|,$$

$$|l_n| \leqslant |h|(2|\xi| + 1) + |k|.$$

Thus for each of the coefficients h_n, k_n, l_n of (12.2-16) there are only finitely many possibilities, which implies that eventually a triple of coefficients must repeat itself. In fact, there exists a triple (h^*, k^*, l^*) that occurs infinitely many times among the triples (h_n, k_n, l_n). In particular, there exist three distinct indices n_i such that $(h_{n_i}, k_{n_i}, l_{n_i}) = (h^*, k^*, l^*)$. At least two of the three solutions of the corresponding quadratics (12.2-16) must be identical. If, say, $\xi_{n_1+1} = \xi_{n_2+1}$, then by the algorithm (12.2-8) $\xi_{n_1+m} = \xi_{n_2+m}$ for all positive integers m, showing that the simple continued fraction representing ξ is ultimately periodic. ∎

Some examples of such ultimately periodic representations of quadratic irrationals, in addition to **4**, are

5
$$\frac{1+\sqrt{5}}{2} = 1 + \frac{1}{|1} + \frac{1}{|1} + \frac{1}{|1} + \cdots,$$

6
$$\frac{24-\sqrt{15}}{17} = 1 + \frac{1}{|5} + \frac{1}{|2} + \frac{1}{|3} + \frac{1}{|2} + \frac{1}{|3} + \cdots,$$

7
$$\sqrt{7} = 2 + \cfrac{1}{\mid 1} + \cfrac{1}{\mid 1} + \cfrac{1}{\mid 1} + \cfrac{1}{\mid 4} + \cfrac{1}{\mid 1} + \cfrac{1}{\mid 1} + \cdots .$$

Some of these fractions are **purely periodic**, that is, there exists an integer m such that

$$b_{n+m} = b_n \quad \text{for all } n \geq 0;$$

another example is

$$\sqrt{7} + 2 = 4 + \cfrac{1}{\mid 1} + \cfrac{1}{\mid 1} + \cfrac{1}{\mid 1} + \cfrac{1}{\mid 4} + \cfrac{1}{\mid 1} + \cfrac{1}{\mid 1} + \cdots .$$

Which algebraic property characterizes quadratic irrationals with purely periodic continued fraction representations?

Let

$$\xi = b_0 + \cfrac{1}{\mid b_1} + \cfrac{1}{\mid b_2} + \cdots + \cfrac{1}{\mid b_m} + \cfrac{1}{\mid b_0} + \cfrac{1}{\mid b_1} + \cdots \quad (12.2\text{-}17)$$

be such a quadratic irrational. By the above, ξ satisfies the equation

$$\xi = \frac{p_{m-1} + p_m \xi}{q_{m-1} + q_m \xi},$$

that is,

$$q_m \xi^2 + (q_{m-1} + p_m)\xi - p_{m-1} = 0. \quad (12.2\text{-}18)$$

Now consider the fraction

$$\eta := b_m + \cfrac{1}{\mid b_{m-1}} + \cfrac{1}{\mid b_{m-2}} + \cdots + \cfrac{1}{\mid b_0} + \cfrac{1}{\mid b_m} + \cfrac{1}{\mid b_{m-1}} + \cdots ,$$

with the period reversed. This likewise is a simple purely periodic continued fraction, and its value is a fixed point of the Moebius transformation

$$s := s_m \circ s_{m-1} \circ \cdots \circ s_0,$$

where

$$s_k : u \to b_k + \frac{1}{u}, \qquad k = 0, 1, \ldots, m.$$

Let us compute the matrix associated with s. If

$$\mathbf{M}_k := \begin{pmatrix} b_k & 1 \\ 1 & 0 \end{pmatrix}$$

is the matrix associated with s_k, then the matrices

$$\mathbf{M}_{k,1} := \begin{pmatrix} e_k & f_k \\ g_k & h_k \end{pmatrix} = \mathbf{M}_k \mathbf{M}_{k-1} \cdots \mathbf{M}_0$$

satisfy the recurrence relation

$$\mathbf{M}_{k,1} = \mathbf{M}_k \mathbf{M}_{k-1,1},$$

that is,

$$\begin{pmatrix} e_k & j_k \\ g_k & h_k \end{pmatrix} = \begin{pmatrix} b_k & 1 \\ 1 & 0 \end{pmatrix} \begin{pmatrix} e_{k-1} & f_{k-1} \\ g_{k-1} & h_{k-1} \end{pmatrix}$$

which on comparing elements yields

$$e_k = b_k e_{k-1} + g_{k-1}, \qquad f_k = b_k f_{k-1} + h_{k-1},$$

$$g_k = e_{k-1}, \qquad h_k = f_{k-1}.$$

Thus

$$\mathbf{M}_{k,1} = \begin{pmatrix} e_k & f_k \\ e_{k-1} & f_{k-1} \end{pmatrix},$$

where

$$e_k = b_k e_{k-1} + e_{k-2}, \qquad f_k = b_k f_{k-1} + f_{k-2},$$

$$e_0 = b_0, \qquad e_{-1} = 1, \qquad f_0 = 1, \qquad f_{-1} = 0.$$

It thus follows that the e_k and f_k satisfy exactly the same recurrence relations and initial conditions as the numerators and denominators p_k and q_k of the fraction for ξ. Hence $e_k = p_k$, $f_k = q_k$, $k = 0, 1, \ldots$, and

$$s(u) = \frac{p_m u + q_m}{p_{m-1} u + q_{m-1}}.$$

The fixed point relation $\eta = s(\eta)$ thus becomes

$$p_{m-1}\eta^2 + (q_{m-1} - p_m)\eta - q_m = 0. \qquad (12.2\text{-}19)$$

The last equation is related to (12.2-18), the equation satisfied by ξ, in the following obvious manner: If η is a solution of (12.2-19), then $-1/\eta$ is a solution of (12.2-18). We conclude that $\eta = -1/\xi'$, where ξ' is one of the solutions of (12.2-18). Now ξ' cannot be the solution represented by the fraction (12.2-17), because both ξ and $\eta > 0$. It follows that ξ' is the second solution of (12.2-18). This is known as the (algebraic) **conjugate** of the quadratic irrational ξ. Because $\eta > 1$, we have $-1 < \xi' < 0$.

We thus have obtained:

THEOREM 12.2d

Any purely periodic simple continued fraction represents a quadratic irration-ality $\xi > 1$ whose algebraic conjugate satisfies $-1 < \xi' < 0$. There holds

$\xi' = -1/\eta$, where η is the value of the continued fraction with the period reversed.

As an illustration, recall the fraction for $\xi = (1+\sqrt{5})/2$ considered in **5**. The equation satisfied by ξ is $\xi^2 - \xi - 1 = 0$. The algebraic conjugate is $\xi' = (1-\sqrt{5})/2$. It so happens that $\xi'\xi = -1$, and thus $\eta = \xi$, which obviously must be the case if the period is 1.

The algebraic conjugate of the number $2+\sqrt{7}$ considered after example **7** is $2-\sqrt{7}$. By virtue of Theorem 12.2d,

$$-\frac{1}{2-\sqrt{7}} = \frac{\sqrt{7}+2}{3} = 1 + \frac{1}{|1|} + \frac{1}{|1|} + \frac{1}{|4|} + \frac{1}{|1|} + \frac{1}{|1|} + \frac{1}{|1|} + \frac{1}{|4|} + \cdots.$$

Remarkably, the converse of Theorem 12.2d is also true. Thus *every* quadratic irrationality $\xi > 1$ whose algebraic conjugate ξ' satisfies $-1 < \xi' < 0$ is represented by a purely periodic simple continued fraction. The proof of this result, which was first given by Galois, is too long to be included here. Assuming the truth of Galois' theorem, we briefly determine the nature of the continued fraction representation of the numbers $\sqrt{l}$, where $l > 1$ is not a perfect square. The algebraic conjugate $\xi' = -\sqrt{l}$ does not satisfy the condition of Theorem 12.2d. However, if we let $b_0 := [\sqrt{l}]$, and if $\xi := \sqrt{l} + b_0$, then $\xi' = -\sqrt{l} + b_0$ satisfies $-1 < \xi' < 0$, by the definition of b_0, and thus the expansion of $\sqrt{l} + b_0$ is purely periodic. It thus follows that

$$\sqrt{l} = b_0 + \frac{1}{|b_1|} + \frac{1}{|b_2|} + \cdots + \frac{1}{|b_m|} + \frac{1}{|2b_0|} + \frac{1}{|b_1|} + \cdots, \quad (12.2\text{-}20)$$

the period consisting of the integers $b_1, b_2, \ldots, 2b_0$. It is seen that the expansion in example **7** is of this special form.

Pell's Equation. This is the name given to the equation

$$x^2 - ly^2 = 1, \quad (12.2\text{-}21)$$

where l is an integer, not a perfect square. The equation is to be solved in integers x and y. References to this equation are scattered throughout the history of number theory; among the mathematicians who contributed to the theory of Pell's equation we find Archimedes, Fermat, Euler (but no Pell). Here we wish to record an algorithm for solving Pell's equation, which is based on continued fractions. Let the expansion of $\sqrt{l}$ be given by (12.2-20), and let p_i, q_i denote the numerators and denominators of this fraction. Then by Theorem 12.1a,

$$\sqrt{l} = \frac{p_{m-1} + p_m \xi_{m+1}}{q_{m-1} + q_m \xi_{m+1}},$$

where

$$\xi_{m+1} := 2b_0 + \cfrac{1}{|b_1|} + \cfrac{1}{|b_2|} + \cdots = \sqrt{l} + b_0.$$

Hence

$$\sqrt{l}\{q_{m-1} + q_m(\sqrt{l} + b_0)\} = p_{m-1} + p_m(\sqrt{l} + b_0)$$

or

$$\sqrt{l}\{q_{m-1} + q_m b_0 - p_m\} = p_{m-1} + p_m b_0 - q_m l,$$

implying that both sides of this equation are zero. Hence $q_{m-1} = p_m - q_m b_0$, $p_{m-1} = q_m l - p_m b_0$, which on substitution into the basic relation (12.1-21) yields

$$p_m(p_m - q_m b_0) - q_m(q_m l - p_m b_0) = (-1)^{m-1}$$

or

$$p_m^2 - l q_m^2 = (-1)^{m-1}.$$

Hence $x := p_m$, $y := q_m$ is a solution of Pell's equation if m is odd. If m is even, then replacing m by $2m + 1$ in the foregoing computation yields

$$p_{2m+1}^2 - l q_{2m+1}^2 = 1,$$

hence $x := p_{2m+1}$, $y := q_{2m+1}$ is a solution in this case.

EXAMPLE **8**

To solve $x^2 - 7y^2 = 1$. From the expansion given in **7** we see that $m = 3$, and the approximant p_3/q_3 is required. We find

b_i	2	1	1	1
p_i	2	3	5	8
q_i	1	1	2	3

This yields $x = 8$, $y = 3$, and indeed $8^2 - 7 \cdot 3^2 = 1$.

PROBLEMS

The reader will probably want to test the algorithms of this section in some examples of his own invention. In addition, we recommend:

1. Rediscover Euler's simple continued fraction expansion for $e = 2.71828\ 18284\ldots$.
2. Find the simple continued fraction expansions for $\sqrt{8}, \sqrt{15}, \sqrt{24}, \sqrt{35}$.

3. Show that for $m = 2, 3, \ldots$

$$\sqrt{m^2-1} = m - 1 + \cfrac{1}{\vert 1} + \cfrac{1}{\vert 2(m-1)} + \cfrac{1}{\vert 1} + \cfrac{1}{\vert 2(m-1)} + \cdots.$$

4. Solve Pell's equation

$$x^2 - (m^2 - 1)y^2 = 1$$

for $m = 2, 3, \ldots$.

§12.3. CONVERGENCE OF CONTINUED FRACTIONS WITH COMPLEX ELEMENTS

In mathematics much can be learned from a careful study of special examples. To form an idea about the convergence of continued fractions with complex elements, we consider the special fraction

$$C := \cfrac{a}{\vert 1} + \cfrac{a}{\vert 1} + \cfrac{a}{\vert 1} + \cdots, \tag{12.3-1}$$

where a is any complex number. Here the recurrence relations (12.1-8) satisfied by the numerators p_n and the denominators q_n are

$$p_n = a p_{n-2} + p_{n-1},$$
$$q_n = a q_{n-2} + q_{n-1}. \tag{12.3-2}$$

The standard method for solving such difference equations with constant coefficients is to assume a solution in the form $p_n = (-s)^n$—the minus sign is inserted for later convenience—that on substituting into (12.3-2) yields

$$s^2 + s - a = 0. \tag{12.3-3}$$

If this equation has two distinct solutions s_1 and s_2—which happens if and only if $a \neq -\frac{1}{4}$—then the first equation (12.3-2) has the general solution

$$p_n = c_1(-s_1)^n + c_2(-s_2)^n.$$

Determining the constants c_1 and c_2 from the initial conditions (12.1-8b), we get

$$p_n = \frac{1}{s_2 - s_1}[s_2(-s_1)^{n+1} - s_1(-s_2)^{n+1}], \qquad n = -1, 0, 1, \cdots.$$

In a similar way one finds

$$q_n = \frac{1}{s_2 - s_1}[(-s_1)^{n+1} - (-s_2)^{n+1}], \qquad n = -1, 0, 1, \cdots.$$

Thus if $a \neq -\frac{1}{4}$, the nth approximant of C is

$$w_n = \frac{p_n}{q_n} = s_1 s_2 \frac{(-s_1)^n - (-s_2)^n}{(-s_2)^{n+1} - (-s_1)^{n+1}},$$

and the sequence $\{w_n\}$ is seen to converge if and only if $|s_1| \neq |s_2|$. If, say,

$$|s_2| > |s_1|, \qquad (12.3\text{-}4)$$

then by writing

$$w_n = \frac{s_1 - s_2 \left(\frac{s_1}{s_2}\right)^{n+1}}{1 - \left(\frac{s_1}{s_2}\right)^{n+1}},$$

we see that the value of C is

$$w := \lim_{n \to \infty} w_n = s_1.$$

To see what the condition (12.3-4) means in terms of a, we note that the two solutions of the characteristic equation (12.3-2) are

$$s_{1,2} = -\tfrac{1}{2} \pm \sqrt{\tfrac{1}{4} + a}.$$

They have distinct moduli if and only if the square root is not pure imaginary, which is the case if and only if a is not real and $\leq -\frac{1}{4}$. The solution of smaller modulus is then given by

$$s_1 = -\tfrac{1}{2} + \sqrt{\tfrac{1}{4} + a},$$

where the root symbol denotes the principal value of the square root.

It remains to discuss the case $a = -\frac{1}{4}$ where the characteristic equation (12.3-2) has the sole solution $s_1 = -\frac{1}{2}$. Then the general solution of the difference equation for p_n is

$$p_n = (c_1 + c_2 n)(\tfrac{1}{2})^n,$$

and the initial conditions (12.18b) yield

$$p_n = -n(\tfrac{1}{2})^{n+1}.$$

Similarly, we find

$$q_n = (n+1)(\tfrac{1}{2})^{n+1}.$$

Thus

$$w_n = \frac{p_n}{q_n} = -\frac{n}{2(n+1)}, \qquad (12.3\text{-}5)$$

and $w = \lim w_n$ again exists and equals $-\frac{1}{2}$. In summary, we have obtained:

THEOREM 12.3a

If a is complex, the continued fraction (12.3-1) converges if and only if a is not real and $< -\frac{1}{4}$. The value of the fraction, if convergent, is

$$w := -\tfrac{1}{2} + \sqrt{\tfrac{1}{4} + a} \text{ (principal value)}.$$

This method settles the question of convergence of C in a rather pedestrian way. By availing ourselves of the theory of fixed points (§6.12) and of circular arithmetic (§6.6), we now treat the same problem by a method that will lead to an important generalization.

For the fraction C, the *RIT* transformations (12.1-3) defining the approximants all are identical,

$$t_k = t : u \to \frac{a}{1+u}.$$

Hence the approximants w_n of C satisfy

$$w_0 = 0, \qquad w_{n+1} = t(w_n), \qquad n = 0, 1, 2, \ldots; \qquad (12.3\text{-}6)$$

they are obtained by iterating the function t. Thus if the sequence $\{w_n\}$ converges, its limit w satisfies $w = t(w)$, that is, w is a fixed point of t. Because $t(w) = a(1+w)^{-1}$, the fixed points of t are precisely the solutions s_1 and s_2 of (12.3-3).

To decide whether the sequence $\{w_n\}$ converges, and to which one of the two fixed points it converges if it converges, we apply Theorem 6.12a. According to this result, convergence will occur if there exists a simply connected region S with closure S' such that $t(S') \subset S$ and $w_0 = 0 \in S'$.

To determine a suitable S, assume $a \neq -\frac{1}{4}$, so that $s_1 \neq s_2$, and let the s_i be numbered such that

$$\theta := \frac{|s_1|}{|s_2|} \leq 1. \qquad (12.3\text{-}7)$$

For $0 \leq \mu < \infty$, let

$$D_\mu := \left\{ u : \frac{|u - s_1|}{|u - s_2|} \leq \mu \right\}. \qquad (12.3\text{-}8)$$

By Lemma 5.4a, D_μ is a circular region. Its boundary, the set of all u such that

$$\left| \frac{u - s_1}{u - s_2} \right| = \mu,$$

is a straight line (the bisector of s_1 and s_2) if $\mu = 1$, and a circle (one of the so-called circles of Apollonius associated with s_1 and s_2) if $\mu \neq 1$. It follows directly from the definition (12.3-8) that

$$D_\nu \subset D_\mu \text{ if and only if } \nu \leqslant \mu. \tag{12.3-9}$$

We next determine the image of D_μ under the map t. From the fact that the matrix associated with t,

$$\mathbf{T} = \begin{pmatrix} 0 & a \\ 1 & 1 \end{pmatrix},$$

has the Jordan decomposition

$$\mathbf{T} = \mathbf{V}^{-1}\mathbf{R}\mathbf{V},$$

where

$$\mathbf{V} := \begin{pmatrix} 1 & -s_1 \\ 1 & -s_2 \end{pmatrix}, \qquad \mathbf{R} := \begin{pmatrix} -s_1 & 0 \\ 0 & -s_2 \end{pmatrix},$$

and

$$\mathbf{V}^{-1} = \text{const.} \begin{pmatrix} s_2 & -s_1 \\ 1 & -1 \end{pmatrix},$$

there follows $t = v^{[-1]} \circ r \circ v$, where

$$v(u) = \frac{u - s_1}{u - s_2}, \qquad r(u) = \frac{s_1}{s_2}u, \qquad v^{[-1]}(u) = \frac{s_2 u - s_1}{u - 1}.$$

(We study the transformation t by means of an auxiliary map v sending the fixed points to 0 and ∞, respectively.) By the definition of D_μ, $v(D_\mu)$ is the disk of radius μ about the origin. The rotation r maps this disk onto a disk of radius $\theta\mu$, which under the map $v^{[-1]}$ yields $D_{\theta\mu}$. Thus there follows

$$t(D_\mu) = D_{\theta\mu} \tag{12.3-10}$$

for all nonnegative μ (see Fig. 12.3a).

The origin, $u = 0$, is on the boundary of the set D_μ where

$$\mu = \left| \frac{0 - s_1}{0 - s_2} \right| = \theta.$$

If $\theta = 1$, this set is mapped onto itself, and the iteration sequence (12.3-6) need not converge. If $\theta < 1$, then $t(D_\theta) = D_{\theta^2}$ is contained in the interior of D_θ. By Theorem 6.12a the sequence $\{w_n\}$ thus converges to s_1, the unique fixed point of t in D_θ. Under the hypothesis that $|s_1| < |s_2|$, that is, that a is not real and $\leqslant -\frac{1}{4}$, we thus have proved once again the convergence of the

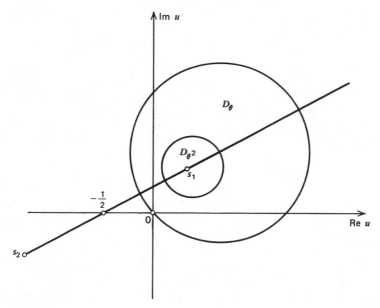

Fig. 12.3a.

continued fraction C and determined its value. The method may seem far-fetched in the present simple situation, but it is capable of generalization.

The main subject of the present section is the study of the convergence of continued fractions

$$C = \frac{a_1 \mid}{\mid b_1} + \frac{a_2 \mid}{\mid b_2} + \frac{a_3 \mid}{\mid b_3} + \cdots$$

with arbitrary complex elements. As befits an analytic theory, the emphasis is on *uniform convergence*. Let $w_n(C)$ and $w(C)$, respectively, denote the nth approximant and the value of a convergent continued fraction C. A family $\mathscr{F}$ of continued fractions is called **uniformly convergent**, if the limits

$$w(C) = \lim_{n \to \infty} w_n(C)$$

exist *uniformly* for all $C \in \mathscr{F}$, that is, if

$$\lim_{n \to \infty} \sup_{C \in \mathscr{F}} |w_n(C) - w(C)| = 0.$$

All convergence results given below are derived by means of a single basic principle. We consider families $\mathscr{F}$ of continued fractions C and denote by t_k the constituents of C; $t_{1,n} := t_1 \circ t_2 \circ \cdots \circ t_n$.

LEMMA 12.3b (Contraction principle for continued fractions)

Let D be a closed set in the extended complex plane enjoying the following two properties:

$$0 \in D; \tag{12.3-11a}$$

$$t_k(D) \subset D, \qquad k = 1, 2, \ldots, \text{for all } C \in \mathcal{F}. \tag{12.3-11b}$$

If the sets $t_{1,n}(D)$ are bounded for sufficiently large n, and if their diameters δ_n satisfy

$$\lim_{n \to \infty} \delta_n = 0 \tag{12.3-12}$$

uniformly for all $C \in \mathcal{F}$, then the family $\mathcal{F}$ converges uniformly, and the value of any $C \in \mathcal{F}$ lies in D.

Proof. We show that the approximants form a Cauchy sequence. Let $\epsilon > 0$ be given, and let m be chosen such that $\delta_m < \epsilon$ for all $C \in \mathcal{F}$. Let $D^{(m)} := t_{1,m}(D)$. In view of $0 \in D$, $w_m = t_{1,m}(0) \in t_{1,m}(D) = D^{(m)}$. If $n > m$, then by (12.3-11) we also have

$$w_n = t_1 \circ t_2 \circ \cdots \circ t_n(0) \in t_1 \circ t_2 \circ \cdots \circ t_n(D)$$

$$= t_{1,m} \circ t_{m+1} \circ \cdots \circ t_n(D) \subset t_{1,m}(D) = D^{(m)}$$

for all $C \in \mathcal{F}$. Thus both w_m and w_n belong to $D^{(m)}$, hence $|w_n - w_m| \leq \delta_m < \epsilon$ for all $n > m$, showing that the sequence $\{w_n\}$ is a Cauchy sequence. Because D is closed and $w_m \in D$ for all m, it follows that $w = \lim w_m$ lies in D. ∎

A closed set having the two properties (12.3-11) for a family $\mathcal{F}$ of continued fractions is called an **eigendomain** of $\mathcal{F}$. In all applications that follow, the eigendomains are circular regions, and the estimation of the diameters δ_n is facilitated by circular arithmetic (see §6.6).

As a first application of the contraction principle we prove the following classical result.

THEOREM 12.3c (Worpitzky's theorem)

The family $\mathcal{F}$ of continued fractions

$$C = \frac{a_1 \mid}{\mid 1} + \frac{a_2 \mid}{\mid 1} + \frac{a_3 \mid}{\mid 1} + \cdots$$

such that

$$|a_k| \leq \tfrac{1}{4}, \qquad k = 1, 2, \ldots, \tag{12.3-13}$$

converges uniformly. The value w of any $C \in \mathcal{F}$ satisfies $|w| \leqslant \frac{1}{2}$, and for the approximants w_n there holds

$$|w_n - w| \leqslant \frac{1}{2n+1}, \qquad n = 0, 1, \ldots. \qquad (12.3\text{-}14)$$

Proof. Here we have

$$t_k : u \to \frac{a_k}{1+u}, \qquad k = 1, 2, \ldots.$$

Let $D := [0; \frac{1}{2}]$. Evidently $0 \in D$. By the rules of circular arithmetic,

$$t_k(D) = \frac{a_k}{1+D} = a_k[1; \tfrac{1}{2}]^{-1} = [\tfrac{4}{3} a_k ; \tfrac{2}{3}|a_k|].$$

In view of $|a_k| \leqslant \frac{1}{4}$, $t_k(D) \subset D$. It follows that D is an eigendomain of the family $\mathcal{F}$. Let m be any positive integer, and let $D^{(0)} := D$, $D^{(k+1)} := t_{m-k}(D^{(k)})$, $k = 0, 1, \ldots, m-1$, so that $D^{(m)} = t_{1,m}(D)$. The $D^{(k)}$ all are circular regions contained in D, hence they are disks. We set

$$d_k := \operatorname{mid} D^{(k)}, \qquad \delta_k := |d_k|, \qquad \rho_k := \operatorname{rad} D^{(k)}.$$

Evidently, $\delta_0 = 0$, $\rho_0 = \frac{1}{2}$. We assert that

$$\delta_k \leqslant \frac{1}{2k+1}, \qquad \rho_k \leqslant \frac{1}{2(2k+1)},$$

$k = 0, 1, \ldots, m$. This is correct for $k = 0$, and if correct for some $k \geqslant 0$ then we have from

$$D^{(k+1)} = t_{m-k}(D^{(k)}) = \frac{a_{m-k}}{1 + D^{(k)}}$$

$$= \left[\frac{a_{m-k}(1 + \bar{d}_k)}{|1 + d_k|^2 - \rho_k^2} ; \frac{|a_{m-k}|\rho_k}{|1 + d_k|^2 - \rho_k^2} \right]$$

that

$$\delta_{k+1} \leqslant \frac{1}{4} \frac{1 - \delta_k}{(1 - \delta_k)^2 - \rho_k^2}, \qquad \rho_{k+1} \leqslant \frac{1}{4} \frac{\rho_k}{(1 - \delta_k)^2 - \rho_k^2}.$$

The expressions on the right are increasing functions of δ_k and ρ_k. Thus, using the hypothesis of induction,

$$\delta_{k+1} \leqslant \frac{k+1}{2k+3}, \qquad \rho_{k+1} \leqslant \frac{1}{2(2k+3)},$$

which is the same hypothesis with k increased by 1. Thus there follows

$$\rho_m = \operatorname{rad} D^{(m)} = \operatorname{rad} t_{1,m}(D) \leq \frac{1}{2(2m+1)},$$

implying that

$$|w_m - w| \leq 2\rho_m \leq \frac{1}{2m+1}, \qquad m = 0, 1, \ldots,$$

as asserted. That $|w| \leq \frac{1}{2}$ follows from the fact that $D = [0; \frac{1}{2}]$ is eigendomain of $\mathscr{F}$. ∎

If all $a_k = -\frac{1}{4}$, then $w = -\frac{1}{2}$ and by (12.3-5)

$$w_n - w = \frac{1}{2(n+1)}, \qquad n = 0, 1, \ldots,$$

showing that the truncation error estimate (12.3-14) is close to best possible. The estimate may in fact be sharpened to

$$|w_n - w| \leq \frac{1}{2(n+1)}, \qquad n = 0, 1, \ldots; \qquad (12.3\text{-}15)$$

see Problem 2. Thus the fraction (12.3-1) where $a = -\frac{1}{4}$ represents the case of slowest convergence in Worpitzky's theorem.

Our next application of the contraction principle is motivated by the following heuristic consideration. The example at the beginning of this section showed that the fraction

$$C = \frac{a_1}{\lceil 1} + \frac{a_2}{\lceil 1} + \frac{a_3}{\lceil 1} + \cdots \qquad (12.3\text{-}16)$$

converges if all $a_k = a$, where a is a complex number such that $a + \frac{1}{4}$ is in the cut plane. Is it not reasonable to expect that C still converges if the a_k do not differ too much from a fixed such a? The following theorem confirms this expectation quantitatively.

THEOREM 12.3d (Perron–Pringsheim theorem)

Let a be a complex number that is not real and $\leq -\frac{1}{4}$. Then there exists $\epsilon > 0$, which depends only on a, such that the family $\mathscr{F}$ of continued fractions (12.3-16) where

$$|a_k - a| \leq \epsilon, \qquad k = 1, 2, \ldots, \qquad (12.3\text{-}17)$$

converges uniformly.

Proof. If $|a| < \frac{1}{4}$, then by the Worpitzky theorem the assertion is true for $\epsilon = \frac{1}{4} - |a|$. We therefore may assume $|a| \geqslant \frac{1}{4}$. In fact, the following proof applies to any $a \neq 0$.

Let again $s_{1,2} = -\frac{1}{2} \pm \sqrt{\frac{1}{4} + a}$ denote the two fixed points of the transformation

$$t : u \rightarrow \frac{a}{1+u},$$

numbered such that

$$\theta := \left| \frac{s_1}{s_2} \right| < 1.$$

The assertion of the theorem will be proved for

$$\epsilon = \frac{(1-\theta)^3}{64} |a|. \tag{12.3-18}$$

Some preliminary material on the circular regions D_μ described by (12.3-8) is required. If $0 \leqslant \mu < 1$, then it follows from the proof of Lemma 5.4a that, in the notation of circular arithmetic, $D_\mu = [c_\mu; \rho_\mu]$ where

$$c_\mu = \frac{s_1 - \mu^2 s_2}{1 - \mu^2}, \qquad \rho_\mu = \frac{\mu |s_1 - s_2|}{1 - \mu^2}.$$

The center c_μ is located on the prolongation beyond s_1 of the straight line segment from s_2 to s_1. The circle bounding D_μ intersects that straight line segment in the point

$$u_\mu := \frac{s_1 + \mu s_2}{1 + \mu}.$$

The circles bounding two distinct disks D_μ and D_ν $(0 \leqslant \nu < \mu < 1)$ are nearest to each other at the points u_μ and u_ν. Thus the distance of the circles is

$$|u_\mu - u_\nu| = \frac{\mu - \nu}{(1+\mu)(1+\nu)} |s_1 - s_2|;$$

hence

$$|u_\mu - u_\nu| \geqslant \frac{\mu - \nu}{4} |s_1 - s_2|. \tag{12.3-19}$$

Let now again

$$t_k : u \rightarrow \frac{a_k}{1+u}.$$

The following lemma describes the effect of a transformation t_k on the disks D_μ.

LEMMA 12.3e

If $a \neq 0$, and if $|a_k - a| \leq \alpha|a|$ where

$$\alpha \leq \frac{(1-\theta)^2}{8\theta},\tag{12.3-20}$$

then there exists $\gamma < 1$ such that for any μ satisfying

$$\frac{8\theta}{(1-\theta)^2}\alpha \leq \mu \leq 1\tag{12.3-21}$$

there holds

$$t_k(D_\mu) \subset D_{\gamma\mu}.\tag{12.3-22}$$

Proof of the Lemma.   In view of (12.3-10),

$$t_k(D_\mu) \subset t(D_\mu) + \frac{a_k - a}{1 + D_\mu} = D_{\theta\mu} + \frac{a_k - a}{a}D_{\theta\mu},$$

thus $t_k(D_\mu)$ protrudes by at most

$$\delta := \alpha \sup_{u \in D_{\theta\mu}} |u|$$

beyond $D_{\theta\mu}$. Now $D_{\theta\mu} \subset D_\theta$, a disk whose boundary passes through 0. Thus

$$\delta \leq 2\alpha\rho_\theta = \frac{2\alpha\theta}{1-\theta^2}|s_2 - s_1|.$$

The disk $t_k(D_\mu)$ is contained in $D_{\gamma\mu}$ if $|u_{\theta\mu} - u_{\gamma\mu}| \geq \delta$, which by (12.3-19) holds if

$$\frac{2\alpha\theta}{1-\theta^2} \leq \frac{(\gamma - \theta)\mu}{4}.$$

If $\gamma := \frac{1}{2}(1+\theta)$, then this is easily seen to be the case for all μ satisfying (12.3-21). The condition (12.3-20) assures that this set is not empty.   ∎

We now turn to the proof of Theorem 12.3d. We have $0 \in D_1$. Thus if α satisfies (12.3-20) and $\epsilon := \alpha|a|$, Lemma 12.3e shows that D_1 is eigendomain for the family $\mathscr{F}$. By Lemma 12.3b it thus merely remains to be shown that

$$\lim_{n \to \infty} \mathrm{rad}\, t_{1,n}(D_1) = 0$$

uniformly for all $C \in \mathscr{F}$.

Let $m > 0$ be an integer, and let

$$D^{(0)} := D_1, \qquad D^{(k+1)} := t_{m-k}(D^{(k)}), \qquad d_k := \text{mid } D^{(k)},$$

$$\rho_k := \text{rad } D^{(k)},$$

$k = 0, 1, \ldots, m$. Lemma 12.3e shows that $D^{(1)} \subset D_\gamma$, $D^{(2)} \subset D_{\gamma^2}$, and generally

$$D^{(k)} \subset D_{\gamma^k} \text{ for } k = 1, 2, \ldots, l,$$

where l is the largest integer such that $\mu = \gamma^{l-1}$ satisfies (12.3-21), that is, the smallest integer such that

$$\gamma^l < \nu := \frac{8\theta}{(1-\theta)^2}\alpha. \tag{12.3-23}$$

Because $D^{(l)} \subset D_\nu \subset D_{\gamma^{l-1}}$, there follows $D^{(l+1)} = t_{m-1}(D^{(l)}) \subset t_{m-1}(D_{\gamma^{l-1}}) \subset D_\nu$ and generally

$$D^{(k)} \subset D_\nu = \left[\frac{s_1 - \nu^2 s_2}{1 - \nu^2}; \frac{\nu|s_1 - s_2|}{1 - \nu^2}\right], \qquad k > l.$$

Hence

$$|d_k - s_1| \leq \sigma, \qquad \rho_k \leq \sigma, \qquad k > l,$$

where

$$\sigma := \frac{\nu|s_1 - s_2|}{1 - \nu^2}. \tag{12.3-24}$$

By circular arithmetic,

$$\rho_{k+1} \leq \frac{|a_{m-k}|\rho_k}{|1 + d_k|^2 - \rho_k^2} = \frac{|d_{k+1}|}{|1 + d_k|}\rho_k.$$

Using the fact that $1 + s_1 = -s_2$, we have for $k > l$ $|1 + d_k| \geq |1 + s_1| - |d_k - s_1| \geq |s_2| - \sigma$, hence

$$\rho_{k+1} \leq \lambda\rho_k, \qquad k > l,$$

where

$$\lambda := \frac{|s_1| + \sigma}{|s_2| - \sigma}. \tag{12.3-25}$$

The desired relation $\rho_m \to 0$ $(m \to \infty)$ uniformly in $\mathscr{F}$ is assured if the α in (12.3-23) is chosen such that the σ in (12.3-24) becomes so small that $\lambda < 1$. This requires $2\sigma < |s_2| - |s_1|$, which in turn holds if

$$\frac{2\nu}{1 - \nu^2} < \frac{1 - \theta}{1 + \theta},$$

which in turn holds, for instance, if $\nu^2 \leq \frac{1}{2}$ and $8\nu < 1 - \theta$. Both these inequalities, as well as (12.3-20), are satisfied if

$$\alpha \leq \frac{(1-\theta)^3}{64},$$

completing the proof of Theorem 12.3d with the value of ϵ specified by (12.3-18). ■

One easy consequence of the theorem is as follows:

THEOREM 12.3f

Let a be a complex number that is not real and $\leq -\frac{1}{4}$, and let $\{a_m\}$ be any sequence of nonzero complex numbers that converges to a. Then the continued fraction (12.3-16) *converges, either properly or improperly.*

Proof.   Let $\epsilon = \epsilon(a)$ be defined as in the preceding theorem, and let m be such that

$$|a_n - a| \leq \epsilon(a) \quad \text{for } n \geq m.$$

Then, by Theorem 12.3d, the fraction

$$C^* := \frac{a_{m+1}}{\lvert\ 1} + \frac{a_{m+2}}{\lvert\ 1} + \cdots$$

converges. Denoting its approximants by $w_1^*, w_2^*, \ldots$ and its value by w^*, we thus have

$$\lim_{n \to \infty} w_n^* = w^*.$$

The approximants of C are

$$w_n = t_1 \circ t_2 \circ \cdots \circ t_m(w_{n-m}^*), \qquad n \geq m,$$

or, in view of Theorem 12.1a,

$$w_n = t_{1,m}(w_{n-m}^*)$$

where, in the usual notation for the numerators and the denominators of C,

$$t_{1,m} : u \to \frac{p_{m-1}u + p_m}{q_{m-1}u + q_m}.$$

Thus depending on whether $t_{1,m}(w^*)$ is finite or not, the sequence $\{w_n\}$ either converges or tends to ∞. ■

PROBLEMS

1. Prove Theorem 12.3a by means of Lemma 12.3b, without using Theorem 6.12a.

2. Using the hypotheses and the notation of the proof of Worpitzky's theorem, let

$$c_k := t_{m-k+1} \circ t_{m-k+2} \circ \cdots \circ t_m(0), \qquad \epsilon_k := |d_k - c_k|,$$

$k = 1, 2, \ldots, m$, so that $c_m = w_m$. Show that

$$\epsilon_k \leqslant \frac{k}{2(k+1)(2k+1)}, \qquad k = 1, 2, \ldots, m,$$

consequently for $n \geqslant m$

$$|w_n - w_m| \leqslant |w_n - d_m| + |d_m - c_m| \leqslant \rho_m + \epsilon_m \leqslant \frac{1}{2(m+1)}$$

and thus

$$|w - w_m| \leqslant \frac{1}{2m+2}, \qquad m = 0, 1, \ldots.$$

3. Show that the set $D := [0; 1]$ is eigendomain of the family $\mathcal{F}$ of continued fractions

$$C = \frac{a_1}{|b_1} + \frac{a_2}{|b_2} + \frac{a_3}{|b_3} + \cdots$$

defined by

$$|b_k| \geqslant 1 + |a_k|, \qquad k = 1, 2, \ldots. \tag{12.3-26}$$

4. Let $\{\epsilon_k\}$ be any sequence of positive numbers such that $\epsilon_0 = 1$ and

$$\sum_{k=0}^{\infty} \epsilon_k = \infty.$$

Show that the subset of the family $\mathcal{F}$ of Problem 3 consisting of those fractions C for which, in addition to (12.3-26),

$$|b_k| \geqslant 1 + \frac{\epsilon_k}{\epsilon_{k-1}}, \qquad k = 1, 2, \ldots, \tag{12.3-27}$$

converges uniformly.
[See Perron [1957], p. 63. Using the representation

$$t_{1,m}(u) = \frac{p_{m-1}u + p_m}{q_{m-1}u + q_m} = \frac{p_{m-1}}{q_{m-1}} + \frac{q_{m-1}p_m - p_{m-1}q_m}{q_{m-1}} \frac{1}{q_{m-1}u + q_m}$$

$$= w_{m-1} + (-1)^{m-1} \frac{\prod_{i=1}^m a_i}{q_{m-1}} \frac{1}{q_{m-1}u + q_m},$$

show that

$$\operatorname{rad} t_{1,m}(D) = \frac{\prod_{i=1}^m |a_i|}{|q_m|^2 - |q_{m-1}|^2}.$$

Equation (12.3-26) implies

$$|q_m| - |q_{m-1}| \geq |a_1 \cdots a_m|,$$

Equation (12.3-27) implies

$$|q_m| \geq 1 + \epsilon_1 + \cdots + \epsilon_m;$$

hence

$$2 \operatorname{rad} t_{1,m}(D) \leq \left(\sum_{i=0}^{m} \epsilon_i \right)^{-1}. \Bigg]$$

5.   Prove: The half plane

$$H := \{u : \operatorname{Re} u \geq -\tfrac{1}{2}\}$$

is eigendomain of the family $\mathscr{F}$ of continued fractions

$$C = \frac{a_1 \vert}{\vert 1} + \frac{a_2 \vert}{\vert 1} + \frac{a_3 \vert}{\vert 1} + \cdots$$

satisfying

$$\tfrac{1}{2} + \operatorname{Re} a_k \geq |a_k|, \qquad k = 1, 2, \ldots.$$

(Note. The celebrated **parabola theorem** due to Scott and Wall [1940] states that a fraction $C \in \mathscr{F}$ converges if at least one of the series

$$\sum \frac{a_2 a_4 \cdots a_{2k}}{a_3 a_5 \cdots a_{2k+1}}, \qquad \sum \frac{a_3 a_5 \cdots a_{2k-1}}{a_2 a_4 \cdots a_{2k}}$$

diverges. It does not seem possible to prove this by means of Lemma 12.3b. See also Perron [1957], p. 77; Thron [1958].)

6.   For $\alpha > 0$, let $\mathscr{F}_\alpha$ denote the family of continued fractions

$$C = -\frac{z\epsilon_1 \vert}{\vert \beta_1 + z} - \frac{z\epsilon_2 \vert}{\vert \beta_2 + z} - \frac{z\epsilon_3 \vert}{\vert \beta_3 + z} - \cdots$$

where $\epsilon_i > 0$, $\beta_i > 0$, $\beta_i - \epsilon_i \geq \alpha$, $i = 1, 2, \ldots$, and where the parameter z lies in the parabolic region

$$P_\alpha : \operatorname{Re} z + 2\alpha \geq |z|.$$

Show that all approximants of all $C \in \mathscr{F}_\alpha$ are finite.
[Decompose the constituents t_i of C to show that the nth approximant of C is

$$w_n = s_1 \circ s_2 \circ \cdots \circ s_n(z) - z,$$

where

$$s_i : u \to z\left(1 - \frac{\epsilon_i}{\beta_i + u}\right).$$

Letting $H : \operatorname{Re} u > -\alpha$, show that $s_i(H) \subset H$, $i = 1, 2, \ldots$. Because $\infty \notin H$ and $z \in H$, there follows $w_n + z \in H$.]

7. Let z, β_i, ϵ_i satisfy the conditions of Problem 6 for some $\alpha > 0$, and let the sequence of polynomials $\{p_k\}$ be defined by $p_{-1} := 0$, $p_0 := 1$,

$$p_k = (\beta_k + z)p_{k-1} - z\epsilon_k p_{k-2}, \qquad k = 1, 2, \ldots. \tag{12.3-28}$$

Show that no p_k vanishes in the set P_α.

8. Let F be a formal power series,

$$F = \sum_{k=0}^{\infty} \alpha_k z^k,$$

where all $\alpha_k > 0$ and

$$\alpha := \inf_{k \geqslant 0} \left(\frac{\alpha_k}{\alpha_{k+1}} - \frac{\alpha_{k-1}}{\alpha_k} \right) > 0$$

($\alpha_{-1} := 0$). Show that for $z \in P_\alpha$ all partial sums of F are different from zero.
[Let $s_k(z) := \sum_{m=0}^{k} \alpha_m z^m$. The polynomials $p_k := \alpha_k^{-1} s_k$ satisfy (12.3-28) where

$$\beta_k := \frac{\alpha_{k-1}}{\alpha_k}, \qquad \epsilon_k := \beta_{k-1}.]$$

9. No partial sum of the exponential series has a zero in the set P_1.
10. Determine values of α, β, γ such that the confluent hypergeometric series $_1F_1(\beta; \gamma; z)$ is different from zero in the set P_α.
    [Use the theorem of Hurwitz (Theorem 4.10f) in connection with the preceding problems.]
11. Let C be a continued fraction belonging to the family $\mathscr{F}_\alpha$ defined in Problem 6. If $\beta_k \to \infty (k \to \infty)$, C converges for all $z \in P_\alpha$, uniformly on compact subsets.

§12.4. *RITZ* FRACTIONS: FORMAL THEORY; PADÉ TABLE

It was shown at the beginning of §12.1 that infinite series and infinite products can be defined by forming compositions of Moebius transformations of the types (T) and (R), respectively. By letting the transformations t_m occurring in the definitions depend on a parameter z, basic results on the representation of analytic functions are obtained. For instance, if

$$t_m : u \to a_m + zu, \qquad m = 0, 1, 2, \ldots,$$

then the value of the composite Moebius transformation $t_0 \circ t_1 \circ \cdots \circ t_n$ at $u = 0$ is a polynomial in z,

$$t_0 \circ t_1 \circ \cdots \circ t_n(0) = a_0 + a_1 z + \cdots + a_n z^n.$$

[The backward evaluation of this expression, beginning with $t_n(0)$, yields the familiar Horner rule for the evaluation of a polynomial (see §6.1).] Letting $n \to \infty$ we are led to power series, and thus to the very foundations of analytic functions. Similarly by the composition of rotations of the special form

$$t_m : u \to (1 + a_m z)u$$

we can obtain the infinite product representations of the kind considered in §8.3.

In a similar spirit we now consider *RIT* transformations

$$t_m : u \to \frac{z a_m}{1 + u}, \qquad m = 1, 2, \ldots, \tag{12.4-1}$$

where $a_1, a_2, \ldots$, are arbitrary nonzero complex constants, and where z is a complex parameter. By composing such transformations we obtain continued fractions whose approximants depend on the parameter z. It is customary to omit the factor z from the transformation t_1. The resulting *RIT* fraction

$$C(z) = t_1 \circ t_2 \circ t_3 \circ \cdots \dot{:} (0)$$

$$= \frac{a_1}{\mid 1} + \frac{z a_2}{\mid 1} + \frac{z a_3}{\mid 1} + \cdots \tag{12.4-2}$$

$$= \frac{1}{z} \mathop{\Phi}_{m=1}^{\infty} \frac{z a_m}{1}$$

will be called a **RITZ fraction**.

The sequences of the numerators $\{p_n\}$ and of the denominators $\{q_n\}$ of C satisfy the initial conditions

$$p_0 = 0, \qquad p_1 = a_1$$
$$q_0 = 1, \qquad q_1 = 1 \tag{12.4-3a}$$

and the recurrence relations

$$p_n = z a_n p_{n-2} + p_{n-1},$$
$$q_n = z a_n q_{n-2} + q_{n-1}, \tag{12.4-3b}$$

where $n = 2, 3, \ldots$. These relations show that both p_n and q_n now are polynomials in z. An easy induction argument shows that p_n has degree

$$\left[\frac{n-1}{2} \right]$$

and q_n has degree

$$\left[\frac{n}{2} \right].$$

From the initial conditions and from the recurrence relations it is also clear that

$$q_n(0) = 1, \qquad n = 0, 1, 2, \dots. \tag{12.4-4}$$

We shall require:

THEOREM 12.4a

For $n = 0, 1, 2, \dots$, the polynomials p_n and q_n have no common zeros.

Proof. The relation (12.1-21) shows that for $n = 1, 2, \dots$, and for all z,

$$p_{n-1}(z)q_n(z) - p_n(z)q_{n-1}(z) = (-1)^n a_1 a_2 \cdots a_n z^{n-1}. \tag{12.4-5}$$

If $z \neq 0$ were a common zero of p_n and q_n, the expression on the left would have to vanish. This is impossible, because the expression on the right is $\neq 0$. It follows from (12.4-4) that $z = 0$ is not a common zero of p_n and q_n, because it is not a zero of q_n. ∎

Being the ratio of two polynomials, the nth approximant of $C(z)$ is a rational function of z. By virtue of (12.4-4) this rational function is analytic at $z = 0$ and thus can be expanded in a Taylor series,

$$\frac{p_n(z)}{q_n(z)} = c_0^{(n)} + c_1^{(n)} z + c_2^{(n)} z^2 + \cdots, \tag{12.4-6}$$

that converges for $|z|$ sufficiently small. At first sight it would appear that the coefficients $c_k^{(n)}$ depend on n as well as on k. This, however, is true only up to a point, as the following result shows.

THEOREM 12.4b

For all positive integers k, the coefficients $c_k^{(n)}$ have the same values for all $n > k$.

Proof. In (12.4-5) we replace n by $n+1$ and divide by $q_n(z)q_{n+1}(z)$. This yields

$$\frac{p_{n+1}(z)}{q_{n+1}(z)} - \frac{p_n(z)}{q_n(z)} = (-1)^n \frac{a_1 a_2 \cdots a_{n+1} z^n}{q_n(z)q_{n+1}(z)}. \tag{12.4-7}$$

Here we expand both sides in powers of z. Because $q_n(0)q_{n+1}(0) \neq 0$, the expansion on the right begins with the term in z^n. Thus the same must be true for the expansion of the expression on the left. From (12.4-6), the coefficient of z^k in the expansion on the left is $c_k^{(n+1)} - c_k^{(n)}$. We thus find

$$c_k^{(n+1)} - c_k^{(n)} = 0, \qquad n > k,$$

and hence by induction

$$c_k^{(n)} = c_k^{(k+1)}, \qquad n = k+1, k+2, \ldots, \qquad (12.4\text{-}8)$$

as was to be shown. ■

If the coefficients $c_k^{(n)}$ are arranged in a two-dimensional array (the index k running horizontally), then Theorem 12.4b states that in each column of the array all entries below the main diagonal are the same. We denote by c_k the ultimate values of the entries in the kth column, $c_k := c_k^{(k+1)}$. Let

$$P := c_0 + c_1 z + c_2 z^2 + \cdots$$

be the formal power series formed with these coefficients. Then the series P and the fraction C are said to **correspond** to each other. There is no assurance, of course, that the formal power series corresponding to a given *RITZ* fraction converges. In some of the most interesting applications the series P is in fact divergent for all $z \neq 0$. Theorem 12.4b merely implies that for $n = 0, 1, 2, \ldots$, the nth approximant w_n of C, if expanded in a series of powers of z, satisfies

$$P - w_n = O(z^n), \qquad (12.4\text{-}9)$$

where here and in the following $O(z^n)$ denotes a formal power series that begins (at the earliest) with the term in z^n.

The study of the analytical relationship between C and the corresponding series P is pursued in §12.5 and §12.9 et seq. At this point we wish to consider the following formal problem: Given a formal power series P, does there exist a *RITZ* fraction C corresponding to P? The answer, given in Theorem 12.4c, reveals a connection between C and the quotient-difference scheme associated with P (see §7.6).

THEOREM 12.4c

Given a formal power series

$$P = c_0 + c_1 z + c_2 z^2 + \cdots,$$

there exists at most one RITZ fraction corresponding to P. There exists precisely one such fraction if and only if the Hankel determinants

$$H_k^{(n)} := \begin{vmatrix} c_n & c_{n+1} & \cdots & c_{n+k-1} \\ c_{n+1} & c_{n+2} & \cdots & c_{n+k} \\ & & \cdot \quad \cdot \quad \cdot & \\ c_{n+k-1} & c_{n+k} & \cdots & c_{n+2k-2} \end{vmatrix}$$

satisfy

$$H_k^{(n)} \neq 0 \quad \text{for } n = 0, 1 \text{ and } k = 1, 2, \ldots. \qquad (12.4\text{-}10)$$

If $q_k^{(n)}$, $e_k^{(n)}$ denote the elements of the quotient-difference scheme associated with P, the RITZ fraction corresponding to P is given by

$$C = \frac{c_0}{|1} - \frac{q_1^{(0)}z}{|1} - \frac{e_1^{(0)}z}{|1} - \frac{q_2^{(0)}z}{|1} - \frac{e_2^{(0)}z}{|1} - \cdots . \qquad (12.4\text{-}11)$$

EXAMPLE **1**

From example **1**, §7.6, we conclude that the *RITZ* fraction corresponding to the series

$$P := 0! - 1! z + 2! z^2 - 3! z^3 + \cdots$$

is

$$C = \frac{1}{|1} + \frac{z}{|1} + \frac{z}{|1} + \frac{2z}{|1} + \frac{2z}{|1} + \frac{3z}{|1} + \cdots .$$

Quite generally, Theorem 12.4c enables us to construct the *RITZ* fraction corresponding to a given series P easily by means of the *qd* algorithm.

Proof of Theorem 12.4c. We first show that condition (12.4-10) is *sufficient* for the existence of a corresponding *RITZ* fraction. Let us call a rational function $r = p/q$ of **type** (k, l) if $\deg p \leq k$ and $\deg q \leq l$. We then have:

LEMMA 12.4d

If the series P satisfies (12.4-10), then for $n = 0, 1, 2, \ldots$, there exists precisely one rational function $w_n = p_n/q_n$ of type

$$\left(\left[\frac{n-1}{2} \right], \left[\frac{n}{2} \right] \right)$$

such that

$$P - w_n = O(z^n). \qquad (12.4\text{-}12)$$

Proof. (a) Let n be even, $n = 2m$. Because w_n must be analytic at 0, the required polynomials p_n and q_n may be assumed in the form

$$q_n(z) = 1 + d_1 z + d_2 z^2 + \cdots + d_m z^m,$$
$$p_n(z) = e_0 + e_1 z + e_2 z^2 + \cdots + e_{m-1} z^{m-1}.$$

Condition (12.4-12) is equivalent to

$$P q_n - p_n = O(z^{2m})$$

or

$$(c_0 + c_1 z + \cdots + c_{2m} z^{2m} + \cdots)(1 + d_1 z + \cdots + d_m z^m)$$
$$- (e_0 + e_1 z + \cdots + e_{m-1} z^{m-1}) = O(z^{2m}). \qquad (12.4\text{-}13a)$$

Comparing the coefficients of z^m, $z^{m+1}, \ldots, z^{2m-1}$ yields for the d_i the system of linear equations

$$c_i d_m + c_{i+1} d_{m-1} + \cdots + c_{i+m-1} d_1 + c_{i+m} = 0, \qquad (12.4\text{-}14a)$$

$i = 0, 1, \ldots, m-1$, whose determinant $H_m^{(0)} \neq 0$ by hypothesis. Thus the d_i are uniquely determined. Comparing coefficients of $z^0, z^1, \ldots, z^{m-1}$ then uniquely determines the e_i.

(b) If n is odd, $n = 2m+1$, we must assume

$$q_n(z) = 1 + d_1 z + d_2 z^2 + \cdots + d_m z^m,$$

$$p_n(z) = e_0 + e_1 z + e_2 z^2 + \cdots + e_m z^m.$$

The required identity now is

$$(c_0 + c_1 z + \cdots + c_{2m+1} z^{2m+1} + \cdots)(1 + d_1 z + \cdots + d_m z^m)$$

$$- (e_0 + e_1 z + \cdots + e_m z^m) = O(z^{2m+1}). \qquad (12.4\text{-}13b)$$

We now first compare the coefficients of z^{m+1}, $z^{m+2}, \ldots, z^{2m}$ to get the system

$$c_i d_m + c_{i+1} d_{m-1} + \cdots + c_{i+m-1} d_1 + c_{i+m} = 0, \qquad (12.4\text{-}14b)$$

where $i = 1, 2, \ldots, m$. The determinant now is $H_m^{(1)} \neq 0$, thus again a unique solution exists, which in turn determines the e_i by comparing the coefficients of $z^0, z^1, \ldots, z^m$ in (12.4-13b). ∎

For the following we require the leading coefficients of the O terms on the right of the relation (12.4-12). We write

$$P q_n - p_n = s_n z^n + \cdots, \qquad n = 1, 2, \ldots. \qquad (12.4\text{-}15)$$

For $n = 2m$ we get from (12.4-13a)

$$c_m d_m + c_{m+1} d_{m-1} + \cdots + c_{2m} = s_{2m}.$$

Together with (12.4-14a) this may be considered a system of $m+1$ equations for the $m+1$ unknowns $d_1, d_2, \cdots, d_m$; s_{2m}. The determinant is $-H_m^{(0)}$ and thus $\neq 0$ by hypothesis. Solving for s_{2m} by Cramer's rule yields

$$s_{2m} = \frac{H_{m+1}^{(0)}}{H_m^{(0)}}, \qquad m = 0, 1, \ldots. \qquad (12.4\text{-}16a)$$

For $n = 2m+1$ we get from (12.4-13b)

$$c_{m+1} d_m + c_{m+2} d_{m-1} + \cdots + c_{2m+1} = s_{2m+1}.$$

Together with (12.4-14b) this forms a system of $m+1$ linear equations for the unknowns $d_1, d_2, \ldots, d_m$; s_{2m+1}. The determinant now is $-H_m^{(1)} \neq 0$.

Solving as before by Cramer's rule now yields

$$s_{2m+1} = \frac{H_{m+1}^{(1)}}{H_m^{(1)}}, \qquad m = 0, 1, 2, \ldots . \qquad (12.4\text{-}16b)$$

We note that under the hypothesis (12.4-10) all $s_n \neq 0$, $n = 0, 1, \ldots$. Thus we may define

$$a_1 := s_0 = c_0,$$

$$\qquad\qquad\qquad\qquad (12.4\text{-}17)$$

$$a_{n+1} := -\frac{s_n}{s_{n-1}}, \qquad n = 1, 2, \ldots ,$$

which implies

$$s_n = a_1 a_2 \cdots a_{n+1}(-1)^n, \qquad n = 0, 1, \ldots . \qquad (12.4\text{-}18)$$

We now assert:

LEMMA 12.4e

The rational functions $w_n := p_n/q_n$ are precisely the approximants of the continued fraction

$$C := \frac{a_1}{|\ 1\ } + \frac{a_2 z}{|\ 1\ } + \frac{a_3 z}{|\ 1\ } + \frac{a_4 z}{|\ 1\ } + \cdots .$$

Proof. We show that the polynomials p_n and q_n constructed in the proof of Lemma 12.4d satisfy the appropriate recurrence relations (12.4-3). Clearly, the correct initial conditions (12.4-3a) are satisfied. By the relations (12.4-15) and (12.4-18) we have for $n = 0, 1, 2, \ldots$

$$Pq_n - p_n = (-1)^n a_1 a_2 \cdots a_{n+1} z^n + \cdots ,$$

thus for $n > 1$

$$P(q_n - q_{n-1} - za_n q_{n-2}) - (p_n - p_{n-1} - za_n p_{n-2})$$
$$= (-1)^n a_1 a_2 \cdots a_n z^{n-1}\{a_{n+1} z - 1 + 1\} + O(z^{n+1})$$
$$= (-1)^n a_1 a_2 \cdots a_{n+1} z^n + O(z^{n+1}).$$

The expression $q_n - q_{n-1} - za_n q_{n-2}$ is a polynomial of degree $[n/2]$ with constant term zero. It thus may be assumed in the form $d_1 z + \cdots + d_m z^m$ where $m := [n/2]$. Similarly $p_n - p_{n-1} - za_n p_{n-2}$ is a polynomial of degree $[(n-1)/2]$ and thus may be written $e_0 + e_1 z + \cdots + e_{m'} z^{m'}$ where $m' := [(n-1)/2]$. In the relation

$$P(q_n - q_{n-1} - za_n q_{n-2}) - (p_n - p_{n-1} - za_n p_{n-2}) = O(z^n) \quad (12.4\text{-}19)$$

we now compare the coefficients of $z^{m'+1}, z^{m'+2}, \ldots, z^{m'+m}$. This yields for the d_i the m homogeneous equations

$$c_i d_m + c_{i+1} d_{m-1} + \cdots + c_{i+m-1} d_1 = 0,$$

$i = m' - m + 1, \ldots, m'$. Because the determinant $H_m^{(m'-m+1)} \neq 0$ by hypothesis, it follows that all $d_i = 0$. Comparing coefficients of z^0, $z^1, \ldots, z^{m'}$ shows that also all $e_i = 0$. Thus the two expressions in parentheses in (12.4-19) vanish identically for all $n > 1$, which is to say that the recurrence relations (12.4-3b) hold. ∎

From (12.4-17) and (12.4-16) there now follows immediately

$$a_{2m+1} = -\frac{s_{2m}}{s_{2m-1}} = -\frac{H_{m+1}^{(0)}}{H_m^{(0)}} \frac{H_{m-1}^{(1)}}{H_m^{(1)}} = -e_m^{(0)},$$

$$a_{2m+2} = -\frac{s_{2m+1}}{s_{2m}} = -\frac{H_{m+1}^{(1)}}{H_m^{(1)}} \frac{H_m^{(0)}}{H_{m+1}^{(0)}} = -q_m^{(0)},$$

by the basic relations (7.6-5) and (7.6-6). Thus the *RITZ* fraction (12.4-11) certainly is a fraction corresponding to P. To show that it is the *only* fraction corresponding to P under the hypothesis (12.4-10), let

$$C^* = \frac{a_1^*}{|1} + \frac{a_2^* z}{|1} + \frac{a_3^* z}{|1} + \cdots$$

be any fraction corresponding to P. Then its approximants w_n^* by the proof of Theorem 12.4b satisfy

$$P - w_n^* = (-1)^n a_1^* a_2^* \cdots a_{n+1}^* z^n + O(z^{n+1}) \qquad (12.4\text{-}20)$$

for $n = 0, 1, 2, \ldots$. Because w_n^* is of type

$$\left(\left[\frac{n-1}{2} \right], \left[\frac{n}{2} \right] \right),$$

there follows $w_n^* = p_n/q_n$, where p_n and q_n are the unique polynomials constructed in the proof of Lemma 12.4d. There furthermore follows

$$(-1)^n a_1^* a_2^* \cdots a_{n+1}^* = s_n = (-1)^n a_1 a_2 \cdots a_{n+1}, \qquad (12.4\text{-}21)$$

$n = 0, 1, 2, \ldots$; hence, because $a_1^* = a_1 = c_0$ and all $a_n \neq 0$, $a_n^* = a_n$ for all $n \geq 1$.

We finally show that the condition (12.4-10) is *necessary* for the existence of a corresponding *RITZ* fraction. Thus let the condition be violated, and in the sequence

$$H_1^{(0)}, H_1^{(1)}, H_2^{(0)}, H_2^{(1)}, H_3^{(0)}, H_3^{(1)}, \ldots$$

let the kth element be the first that is zero. We assume that k is even, $k = 2m$;

the proof for k odd is similar. Thus we have

$$H_1^{(0)} \neq 0, H_1^{(1)} \neq 0, \ldots, H_m^{(0)} \neq 0, H_m^{(1)} = 0, \tag{12.4-22}$$

and the existence of a unique rational function w_n of type

$$\left(\left[\frac{n-1}{2}\right], \left[\frac{n}{2}\right]\right)$$

satisfying (12.4-12) can be proved for $n = 0, 1, \ldots, 2m$, as in the proof of Lemma 12.4d. Assume now that a corresponding *RITZ* fraction

$$C^* = \frac{a_1^*}{|\,1} + \frac{a_2^* z}{|\,1} + \frac{a_3^* z}{|\,1} + \cdots$$

exists, which implies that

$$a_i^* \neq 0, \qquad i = 1, 2, \ldots. \tag{12.4-23}$$

Then the approximants w_i^* of C^* satisfy (12.4-20), and by the modified Lemma 12.4d $w_i^* = p_i / q_i$ follows for $i = 0, 1, \ldots, 2m$. However, in view of (12.4-22) the coefficient s_{2m-1} in

$$Pq_{2m-1} - p_{2m-1} = s_{2m-1} z^{2m-1} + \cdots$$

now equals zero. There follows

$$-a_1^* a_2^* \cdots a_{2m}^* = s_{2m-1} = 0,$$

contradicting (12.4-23). ∎

The following is a trivial extension of Theorem 12.4c that is frequently used.

COROLLARY 12.4f

Let the series P be normal, that is, let $H_k^{(n)} \neq 0$ for all $n = 0, 1, 2, \ldots$ and all $k = 1, 2, \ldots$. Then for $n = 0, 1, 2, \ldots$ the formal power series

$$P_n := c_n + c_{n+1} z + c_{n+2} z^2 + \cdots$$

possesses the corresponding RITZ fraction

$$C_n := \frac{c_n}{|\,1} - \frac{q_1^{(n)} z}{|\,1} - \frac{e_1^{(n)} z}{|\,1} - \frac{q_2^{(n)} z}{|\,1} - \frac{e_2^{(n)} z}{|\,1} - \cdots.$$

This simply follows from Theorem 12.4c by considering the series $P := c_0^* + c_1^* z + \cdots$ where $c_k^* = c_{n+k}$. ∎

The Padé Table. Let the series P be normal, as defined in Corollary 12.4f. Then the method used in the proof of Lemma 12.4d more generally yields the following:

THEOREM 12.4g

If P is normal, then for every pair (k, l) of nonnegative integers there exists precisely one rational function $w_{k,l}$ of type (k, l) such that

$$P - w_{k,l} = O(z^{k+l}). \tag{12.4-24}$$

Thus with every normal series P we may associate the two-dimensional array of rational functions $w_{k,l}$. The basic formal properties of this array were established by Frobenius [1881]. Today the array is commonly known as the **Padé table** associated with P (Padé [1892]). Among the more recent contributors to the formal theory of the Padé table we mention Wynn [1966]. See also the excellent survey article by Gragg [1972]. A considerable number of applications of aspects of the Padé table to problems in computational physics are discussed by Baker and Gammel [1970]. From the point of view of the Padé table, the study of *RITZ* fractions merely amounts to a study of the two diagonals $w_{n-1,n}$ and $w_{n,n}$ of the Padé table. In fact, many of the results about the Padé table that appear in the literature are merely results about *RITZ* fractions in disguise. Because the geometric point of view introduced by the Moebius transformations t_m is not applicable to the Padé table, we do not pursue this subject here.

Expansions in z^{-1}. On occasion it is convenient to denote the parameter in the transformation t_m by z^{-1} in place of z. This yields a continued fraction

$$C = \frac{a_1}{\vert 1} + \frac{a_2 z^{-1}}{\vert 1} + \frac{a_3 z^{-1}}{\vert 1} + \cdots$$

$$= z \mathop{\Phi}_{m=1}^{\infty} \frac{a_m z^{-1}}{1} \tag{12.4-25}$$

which corresponds to the formal series

$$P = c_0 + c_1 z^{-1} + c_2 z^{-2} + \cdots$$

in the following sense: If the nth approximant of C is expanded in powers of z^{-1}, the expansion agrees with P through the term $c_{n-1} z^{-n+1}$.

The nth numerator and the nth denominator of C now are polynomials (of degree $[(n-1)/2]$ and $[n/2]$, respectively) in z^{-1}. However, if we subject the fraction C to an equivalence transformation of type (12.1-11), where

$c_1 = c_3 = c_5 = \cdots = z$, $c_0 = c_2 = c_4 = \cdots = 1$, we obtain the equivalent fraction

$$C := \frac{a_1 z}{|\ z\ } + \frac{a_2}{|\ 1\ } + \frac{a_3}{|\ z\ } + \frac{a_4}{|\ 1\ } + \cdots, \qquad (12.4\text{-}26)$$

whose approximants are unchanged, but whose numerators and denominators by (12.1-12) now are

$$z^{[(n+1)/2]} p_n\left(\frac{1}{z}\right) \quad \text{and} \quad z^{[(n+1)/2]} q_n\left(\frac{1}{z}\right),$$

respectively, and thus are polynomials (of degree $[(n+1)/2]$) in z. It is customary to drop the factor z in the fraction (12.4-26). This yields the continued fraction (to be called *RITZ*$^{-1}$ **fraction**)

$$C := \frac{a_1}{|\ z\ } + \frac{a_2}{|\ 1\ } + \frac{a_3}{|\ z\ } + \frac{a_4}{|\ 1\ } + \cdots. \qquad (12.4\text{-}27)$$

It corresponds to the formal series

$$P := c_0 z^{-1} + c_1 z^{-2} + c_2 z^{-3} + \cdots$$

in the following sense: For $n = 1, 2, \ldots$, the expansion of the nth approximant in powers of z^{-1} agrees with P through the term $c_{n-1} z^{-n}$. We now denote the numerators and denominators of the *RITZ*$^{-1}$ fraction (12.4-27) by $\hat{p}_n$ and $\hat{q}_n$, respectively. They are polynomials of the respective degrees $[(n-1)/1]$ and $[(n+1)/2]$ that satisfy the recurrence relations

$$\hat{p}_0 = 0, \qquad \hat{p}_1 = a_1,$$
$$\hat{q}_0 = 1, \qquad \hat{q}_1 = z, \qquad\qquad (12.4\text{-}28a)$$

and

$$\hat{p}_n = a_n \hat{p}_{n-2} + b_n \hat{p}_{n-1},$$
$$\hat{q}_n = a_n \hat{q}_{n-2} + b_n \hat{q}_{n-1}, \qquad\qquad (12.4\text{-}28b)$$

where $b_n = z$ for n odd and $b_n = 1$ for n even.

It is of interest to consider the even and the odd parts of the *RITZ*$^{-1}$ fraction (12.4-27). Because the formulas of §12.1 apply only to the case in which the partial denominators are 1, we first carry through the computation for the fraction (12.4-25). Its even and odd parts are by (12.1-19) and (12.1-20)

$$\frac{a_1}{|\ 1 + a_2 z^{-1}\ } - \frac{a_2 a_3 z^{-2}}{|\ 1 + a_3 z^{-1} + a_4 z^{-1}\ } - \frac{a_4 a_5 z^{-2}}{|\ 1 + a_5 z^{-1} + a_6 z^{-1}\ } - \cdots.$$

and

$$a_1 - \cfrac{a_1 a_2 z^{-2}}{\Big|\,1 + a_2 z^{-1} + a_3 z^{-1}} - \cfrac{a_3 a_4 z^{-2}}{\Big|\,1 + a_4 z^{-1} + a_5 z^{-1}} - \cdots .$$

Treating these fractions to an equivalence transformation (12.1-11) where $c_m = z$, $m = 1, 2, \ldots$, and dividing by z, we obtain the even and the odd parts of the $RITZ^{-1}$ fraction (12.4-27):

$$C^e := \cfrac{a_1}{\Big|\,z + a_2} - \cfrac{a_2 a_3}{\Big|\,z + a_3 + a_4} - \cfrac{a_4 a_5}{\Big|\,z + a_5 + a_6} - \cdots \quad \text{(even part)},$$

$$C^o := \frac{a_1}{z} \left\{ 1 - \cfrac{a_2 a_3}{\Big|\,z + a_2 + a_3} - \cfrac{a_4 a_5}{\Big|\,z + a_4 + a_5} - \cdots \right\} \quad \text{(odd part)}.$$

The foregoing relations are of interest if viewed in connection with the quotient-difference scheme associated with a normal formal power series

$$P = c_0 + c_1 z + c_2 z^2 + \cdots .$$

By Corollary 12.4f, the even part of the $RITZ^{-1}$ fraction corresponding to

$$P_n := c_n z^{-1} + c_{n+1} z^{-2} + \cdots$$

is given by

$$E_n(z) := \cfrac{c_n}{\Big|\,z - q_1^{(n)}} - \cfrac{q_1^{(n)} e_1^{(n)}}{\Big|\,z - e_1^{(n)} - q_2^{(n)}} - \cfrac{q_2^{(n)} e_2^{(n)}}{\Big|\,z - e_2^{(n)} - q_3^{(n)}} - \cdots , \quad (12.4\text{-}29)$$

and the odd part by

$$O_n(z) := \frac{c_n}{z} \left\{ 1 + \cfrac{q_1^{(n)}}{\Big|\,z - q_1^{(n)} - e_1^{(n)}} - \cfrac{e_1^{(n)} q_2^{(n)}}{\Big|\,z - q_2^{(n)} - e_2^{(n)}} - \cdots \right\} \quad (12.4\text{-}30)$$

Thus for $m = 1, 2, \ldots$, the mth approximant of the fraction $O_n(z)$, if expanded in powers of z^{-1}, agrees with P_n through the term $c_{n+2m} z^{-2m-1}$. This implies that the mth approximant of

$$z O_n(z) - c_n = \cfrac{c_{n+1}}{\Big|\,z - q_1^{(n)} - e_1^{(n)}} - \cfrac{e_1^{(n)} q_2^{(n)}}{\Big|\,z - q_2^{(n)} - e_2^{(n)}} - \cdots$$

agrees with

$$z P_n - c_n = c_{n+1} z^{-1} + c_{n+2} z^{-2} + \cdots = P_{n+1}$$

through the term $c_{n+2m} z^{-2m}$. The same is true of the fraction

$$E_{n+1}(z) = \cfrac{c_{n+1}}{\Big|\,z - q_1^{(n+1)}} - \cfrac{q_1^{(n+1)} e_1^{(n+1)}}{\Big|\,z - e_1^{(n+1)} - q_2^{(n+1)}} - \cdots .$$

Making the plausible assumption that the even part (12.4-29) of the $RITZ^{-1}$ fraction corresponding to a given series P is uniquely determined—we have not proved this—we are thus led to conjecture that

$$q_m^{(n)} + e_m^{(n)} = e_{m-1}^{(n+1)} + q_m^{(n+1)},$$

$$e_m^{(n)} q_{m+1}^{(n)} = q_m^{(n+1)} e_m^{(n+1)}$$

$$(12.4\text{-}31)$$

for $n = 0, 1, \ldots$ and $m = 1, 2, \ldots$. These relations are indeed true; they are identical with the so-called **rhombus rules** that were proved by an entirely different method in §7.6. It is likely, however, that the rhombus rules were discovered by Rutishauser in the foregoing manner.

Another connection with the qd scheme is as follows. The mth denominator of the fraction $E_n(z)$ is the polynomial $\hat{q}_{2m}(z)$ defined by (12.4-28). The basic formulas (12.1-8) thus yield

$$\hat{q}_{2m+2}(z) = (z - e_m^{(n)} - q_{m+1}^{(n)})\hat{q}_{2m}(z) - e_m^{(n)} q_m^{(n)} \hat{q}_{2m-2}(z),$$

$m = 1, 2, \ldots$. Exactly the same recurrence relations are also satisfied by the Hadamard polynomials $p_m^{(n)}(z)$ associated with the qd scheme (see §7.7). Because

$$\hat{q}_0(z) = 1 = p_0^{(n)}(z),$$

$$\hat{q}_2(z) = z - q_1^{(n)} = p_1^{(n)}(z),$$

there follows

$$\hat{q}_{2m}(z) = p_m^{(n)}(z), \qquad m = 0, 1, 2, \ldots. \qquad (12.4\text{-}32)$$

If the identity (12.1-23) is applied to the fraction $E_n(z)$, we finally obtain the important formula

$$E_n(z) \cong \sum_{m=1}^{\infty} \frac{s_{m-1}^{(n)}}{p_{m-1}^{(n)}(z) p_m^{(n)}(z)}, \qquad (12.4\text{-}33)$$

where

$$s_{m-1}^{(n)} := c_n q_1^{(n)} e_1^{(n)} \cdots q_{m-1}^{(n)} e_{m-1}^{(n)},$$

or by (7.7-14)

$$s_{m-1}^{(n)} = \frac{H_m^{(n)}}{H_{m-1}^{(n)}}. \qquad (12.4\text{-}34)$$

We recall that the equivalence symbol used in (12.4-33) means that for $m = 1, 2, \ldots$, the mth approximant of the fraction on the left equals the mth partial sum of the series on the right. Convergence neither of the series nor of the fraction is implied.

PROBLEMS

1. Show that the continued fraction

$$\frac{1}{|1|} + \frac{z}{|1|} + \frac{z}{|1|} + \frac{z}{|1|} + \cdots$$

corresponds to the power series

$$P := \sum_{n=0}^{\infty} \frac{(-4)^n (1/2)_n}{(n+1)!} z^n.$$

2. Show that to the power series

$$1 + (1-a)z + (1-a)(1-aq)z^2 + (1-a)(1-aq)(1-aq^2)z^3 + \cdots$$

there corresponds the continued fraction

$$\frac{1}{|1|} - \frac{(1-a)z}{|1|} - \frac{a(1-q)z}{|1|} - \frac{q(1-aq)z}{|1|} - \frac{aq(1-q^2)z}{|1|} \cdots.$$

3. Show that the formal power series

$$P := \sum_{n=0}^{\infty} t^{n^2} z^n$$

gives rise to a qd scheme where

$$q_k^{(n)} = t^{2n+4k-3}, \qquad e_k^{(n)} = t^{2n+2k-1}(t^{2k} - 1)$$

and that it therefore corresponds to Eisenstein's continued fraction

$$\frac{1}{|1|} - \frac{tz}{|1|} - \frac{t(t^2-1)z}{|1|} - \frac{t^5 z}{|1|} - \frac{t^3(t^4-1)z}{|1|} - \cdots.$$

4. Show that the qd scheme associated with the series

$$P := 1 + \sum_{n=1}^{\infty} \left\{ \prod_{k=0}^{n-1} \frac{1-t^{\alpha+k}}{1-t^{\gamma+k}} \right\} z^n$$

is given by

$$q_k^{(n)} = -t^{k-1} \frac{(1-t^{\alpha+n+k-1})(1-t^{\gamma+n+k-2})}{(1-t^{\gamma+n+2k-3})(1-t^{\gamma+n+2k-2})},$$

$$e_k^{(n)} = -t^{\alpha+n+k-1} \frac{(1-t^k)(1-t^{\gamma-\alpha+k-1})}{(1-t^{\gamma+n+2k-2})(1-t^{\gamma+n+2k-1})}.$$

5. Let the formal power series

$$P := \frac{c_0}{z} + \frac{c_1}{z^2} + \frac{c_2}{z^3} + \cdots$$

possess a corresponding $RITZ^{-1}$ fraction. Show that the equivalent reduced fraction is given by

$$C = \frac{1}{\mid b_1 z} - \frac{1}{\mid b_2} - \frac{1}{\mid b_3 z} - \frac{1}{\mid b_4} - \cdots ,$$

where

$$b_{2m} = \frac{[H_m^{(0)}]^2}{H_{m-1}^{(1)} H_m^{(1)}}, \qquad b_{2m+1} = \frac{[H_m^{(1)}]^2}{H_m^{(0)} H_{m+1}^{(0)}}.$$

6. A continued fraction of the form

$$\frac{a_1}{\mid z + b_1} - \frac{a_2}{\mid z + b_2} - \frac{a_3}{\mid z + b_3} - \cdots$$

is called a **J fraction** (*J* for Jacobi). Show that with every *J* fraction we can associate a formal power series

$$P = \frac{c_0}{z} + \frac{c_1}{z^2} + \frac{c_2}{z^3} + \cdots$$

with the following property: For $m = 1, 2, 3, \ldots$ the expansion of the *m*th approximant of the fraction in powers of z^{-1} agrees with *P* through the term involving z^{-2m}.

7. Prove that given a formal series $P = c_0 z^{-1} + c_1 z^{-2} + \cdots$, there exists at most one *J* fraction to which it is associated. There exists precisely one such fraction if and only if $H_m^{(0)} \neq 0$, $m = 1, 2, \ldots$. In the affirmative case the associated *J* fraction is equivalent to the even part of the $RITZ^{-1}$ fraction corresponding to *P*.

§12.5. THE CONVERGENCE OF *RITZ* FRACTIONS: EXAMPLES

In §12.4 we have dealt exclusively with the *formal* aspects of the correspondence between *RITZ* fractions and power series. At no point did we assume the convergence either of the fraction *C* or of the corresponding series *P*. The reader now undoubtedly expects to learn about the mutual implications of convergence of either *C* or *P*. The conjecture would seem hopeful, for instance, that *C* (as a function of *z*) converges in some region containing the origin if and only if *P* has a positive radius of convergence.

Disappointingly, an example constructed by Perron, ([1957], II, p. 158) proves the "if" part of this conjecture to be false. Thus we either must *assume* the convergence of *C*, or we can assert the convergence of *C* under additional hypotheses on the elements of *C* (in addition, possibly, to assumptions on *P*). All results given subsequently are of this kind. Fortunately, they are still strong enough to permit us to deal very completely with a number of concrete special cases. In §12.9 et seq. we see that *C* may converge in large regions (not containing 0) even if P has radius of convergence 0. A connection between C and P then exists via the theory of asymptotic expansions.

Throughout this section we assume that

$$P(z) = c_0 + c_1 z + c_2 z^2 + \cdots \qquad (12.5\text{-}1)$$

is a power series satisfying the determinant condition (12.4-10), and that

$$C(z) = \frac{a_1}{|1|} + \frac{a_2 z}{|1|} + \frac{a_3 z}{|1|} + \frac{a_4 z}{|1|} + \cdots \qquad (12.5\text{-}2)$$

is its corresponding *RITZ* fraction.

THEOREM 12.5a

Let S be a region containing 0, and let $C(z)$ converge locally uniformly on S. Then if $w(z)$ denotes the value of C at z, the function $f : z \to w(z)$ is analytic on S; moreover, the corresponding series P is the Taylor series of f at 0.

Proof. The hypothesis implies that for every compact subset $T \subset S$ the approximants $w_n(z)$ of $C(z)$ are defined for all $z \in T$ and for all sufficiently large n, and that $w_n(z) \to w(z)$ uniformly for $z \in T$. Because the approximants are rational, they are analytic wherever defined; hence by the general form of the Weierstrass double series theorem (Theorem 3.4b) the limit function f is analytic. By the same theorem, if as in (12.4-6) $c_k^{(n)}$ denotes the kth Taylor coefficient of $w_n(z)$ at 0, then for each k the limit $\lim_{n \to \infty} c_k^{(n)}$ exists and equals the kth Taylor coefficient of f at 0. But, by the very definition of the corresponding series, $\lim_{n \to \infty} c_k^{(n)} = c_k$ [in fact, $c_k^{(n)} = c_k$ for $n > k$], hence the Taylor series of f at 0 equals P. ∎

The following theorem asserts the convergence of $C(z)$ under special hypotheses on the a_i.

THEOREM 12.5b

Let $|a_i| \le \mu$, $i = 1, 2, \ldots$. Then $C(z)$ converges uniformly for $|z| \le 1/4\mu$.

Proof. If $|z| \le 1/4\mu$ and $|a_i| \le \mu$, then $|a_i z| \le \frac{1}{4}$. The uniform convergence (uniform with respect to z, for fixed a_i) follows from Worpitzki's theorem 12.3c. ∎

By the preceding theorem, the function defined by $C(z)$ is analytic in the disk $|z| < 1/4\mu$, and the corresponding series P thus has radius of convergence $\ge 1/4\mu$.

THEOREM 12.5c

*Let the sequence $\{a_i\}$ tend to zero. Then $C(z)$ converges, properly or impro-
perly, for all z, and the function f represented by it is analytic at 0 and
meromorphic in the whole complex plane.*

Proof. Let $\mu > 0$ be arbitrary. Let m be such that $|a_n| < \mu$ for all $n > m$.
Denote by $w_1^*(z)$, $w_2^*(z)$, ... the approximants of the continued fraction

$$\frac{a_{m+1}z}{\vert\quad 1} + \frac{a_{m+2}z}{\vert\quad 1} + \cdots .$$

By the preceding theorem, $w_j^*(z) \to w^*(z)$, an analytic function, uniformly
for all z in $D_\mu : |z| \leqslant 1/4\mu$. The nth approximant of $C(z)$ for $n > m$ is

$$w_n(z) = t_1 \circ t_2 \circ \cdots \circ t_m (w_{n-m}^*(z))$$

or, by virtue of Theorem 12.1a,

$$w_n(z) = \frac{p_{m-1}(z)w_{n-m}^*(z) + p_m(z)}{q_{m-1}(z)w_{n-m}^*(z) + q_m(z)}. \tag{12.5-3}$$

The limits

$$p(z) := \lim_{n\to\infty} [p_{m-1}(z)w_{n-m}^*(z) + p_m(z)] = p_{m-1}(z)w^*(z) + p_m(z),$$

$$q(z) := \lim_{n\to\infty} [q_{m-1}(z)w_{n-m}^*(z) + q_m(z)] = q_{m-1}(z)w^*(z) + q_m(z)$$

exist and are analytic in D_μ. Because $q(0) = 1$, q is not identically zero. Thus
for all $z \in D_\mu$ such that $q(z) \neq 0$,

$$\lim_{n\to\infty} w_n(z) = \frac{p(z)}{q(z)}$$

exists. The limit function is analytic at 0 and, being the ratio of two analytic
functions in D_μ, it is meromorphic in D_μ. Because this holds for every $\mu > 0$,
the assertion of Theorem 12.5c follows. ∎

THEOREM 12.5d

Let $a \neq 0$, and let S denote the region

$$\arg\left(z + \frac{1}{4a}\right) \neq \arg\left(-\frac{1}{4a}\right). \tag{12.5-4}$$

*If $\{a_i\}$ is any sequence of nonzero complex numbers such that $\lim_{i\to\infty} a_i = a$,
then $C(z)$ converges, properly or improperly, for all $z \in S$, and represents a
function f meromorphic in S whose Taylor series at 0 is P.*

Proof.   Let $z_0 \in S$. Then az_0 is not real and $\leq -\frac{1}{4}$. Thus by Perron's theorem 12.3d, there exists $\epsilon_0 = \epsilon_0(z_0)$, $0 < \epsilon_0 \leq 1$, such that the fractions

$$C(b_1, b_2, \ldots) := \frac{b_1}{\lceil 1} + \frac{b_2}{\lceil 1} + \cdots$$

converge, uniformly in the b_i, for all b_i such that

$$|b_i - az_0| \leq \epsilon_0(z_0).\tag{12.5-5}$$

Let T be a compact subset of S. Then the open disks

$$D(z_0) := \left\{ z : |z - z_0| < \frac{\epsilon_0}{1 + |a| + |z_0|} \right\}$$

form an open covering of T. By the Heine–Borel lemma (Theorem 3.4a) we can extract a finite subcovering. Let $\delta > 0$ be the minimum of the radii of the covering disks, and let m be such that in the given fraction (12.5-2)

$$|a_n - a| < \delta\tag{12.5-6}$$

for all $n > m$. We assert that *the continued fraction*

$$C^*(z) := \frac{a_{m+1}z}{\lceil 1} + \frac{a_{m+2}z}{\lceil 1} + \cdots$$

converges uniformly for $z \in T$.

Let $z \in T$. Then

$$|z - z_0| < \frac{\epsilon_0(z_0)}{1 + |a| + |z_0|}\tag{12.5-7}$$

for a suitable center z_0 of a covering disk; hence if $n > m$

$$|a_n z - az_0| \leq |a_n - a| |z_0| + |z - z_0| |a| + |a_n - a| |z - z_0|$$

$$\leq \left\{ \frac{|z_0|}{1 + |a| + |z_0|} + \frac{|a|}{1 + |a| + |z_0|} + \frac{1}{(1 + |a| + |z_0|)^2} \right\} \epsilon_0(z_0)$$

$$\leq \epsilon_0(z_0).$$

Thus by the Perron theorem $C^*(z)$ converges uniformly in $D(z_0)$, and, because finitely many such disks cover T, uniformly in T.

The remainder of the proof is similar to the proof of Theorem 12.5c. Let $w^*(z)$ be the analytic function defined by C^*, and let p_n and q_n denote the numerators and the denominators of C. Then if $\{w_k^*(z)\}$ is the sequence of approximants of C^*, the relation (12.5-3) again holds. Because the limits of the numerator and of the denominator exist,

$$C(z) = \frac{p_{m-1}(z)w^*(z) + p_m(z)}{q_{m-1}(z)w^*(z) + q_m(z)},$$

a meromorphic function analytic at 0. ■

These results possess interesting illustrations. A first group of examples may be drawn from the series

$$P(z) := {}_2F_1(a, 1; c; bz) = \sum_{n=0}^{\infty} \frac{(a)_n}{(c)_n}(bz)^n,$$

where a, b, c are complex, $b \neq 0$, $c \neq 0, -1, -2, \ldots$. The qd scheme associated with this series can be constructed explicitly. The initial conditions (7.6-3) readily yield

$$q_1^{(n)} = \frac{a+n}{b+n}b, \qquad n = 0, 1, 2, \ldots. \tag{12.5-8}$$

The continuation rules (7.6-4) then furnish, as may be verified by a somewhat tedious calculation,

$$q_k^{(n)} = \frac{(a+n+k-1)(c+n+k-2)}{(c+n+2k-3)(c+n+2k-2)}b, \qquad k = 2, 3, \ldots,$$

$$\tag{12.5-9 a}$$

$$e_k^{(n)} = \frac{k(c-a+k-1)}{(c+n+2k-2)(c+n+2k-1)} \cdot b, \qquad k = 1, 2, \ldots,$$

$$\tag{12.5-9b}$$

$n = 0, 1, 2, \ldots$. Thus the corresponding continued fraction where $a_{2k} = -q_k^{(0)}$, $a_{2k+1} = -e_k^{(0)}$ can be written down explicitly. We see that

$$\lim_{k \to \infty} a_k = -\frac{b}{4}$$

exists. We thus may apply Theorem 12.5d. The region S here is the complex plane cut along the ray $\arg(z - 1/b) = \arg(1/b)$. We know from §9.9 that the function defined by ${}_2F_1(a, 1; c; bz)$ for $|z| < 1$ can be continued to a function f that is analytic in all of S. It follows that $C(z)$ converges (properly) and equals $f(z)$ for all $z \in S$.

We consider some illustrative special cases.

EXAMPLE **1**

For $|z| < 1$,

$$\text{Log}(1+z) = z \, {}_2F_1(1, 1; 2; -z).$$

For $a = 1$, $c = 2$, $b = -1$ we find from (12.5-9)

$$a_{2k} = -q_k^{(0)} = \frac{k}{2(2k-1)}, \qquad a_{2k+1} = -e_k^{(0)} = \frac{k}{2(2k+1)}.$$

Hence for all z that are not real and $\leqslant -1$,

$$\text{Log}(1+z) = \frac{z}{|1} + \frac{z/2}{|1} + \frac{z/6}{|1} + \frac{2z/6}{|1} + \frac{2z/10}{|1} + \frac{3z/10}{|1} + \cdots,$$

which by an equivalence transformation may be thrown into the form

$$\text{Log}(1+z) = \frac{z}{|1} + \frac{1^2 z}{|2} + \frac{1^2 z}{|3} + \frac{2^2 z}{|4} + \frac{2^2 z}{|5} + \cdots \qquad (12.5\text{-}10)$$

To check the numerical performance of this fraction, we list in Table 12.5 some numerators, denominators, and approximants resulting for $z = 1$.

Table 12.5. Approximants of (12.5-10) for $z = 1$

a_n	0	1	1	1	4	4	9	9
b_n	0	1	2	3	4	5	6	7
p_n	0	1	2	7	36	208	1572	12876
q_n	1	1	3	10	52	300	2268	18576
w_n	0	1.00000	0.66667	0.70000	0.69231	0.69333	0.69312	0.69315

Because the elements of the continued fraction are positive, we may deduce from the proof of Theorem 12.1c that the approximants alternately furnish upper and lower bounds for the value of the fraction. (Truncation error estimates for complex z are given in §12.11.) We thus see that already the seventh approximant yields an accuracy that would require as many as 50,000 terms of the series

$$\text{Log } 2 = \text{Log}(1+1) = 1 - \tfrac{1}{2} + \tfrac{1}{3} - \tfrac{1}{4} + \cdots.$$

EXAMPLE **2**

For $|z| < 1$,

$$\text{Arctan } z = z - \tfrac{1}{3}z^3 + \tfrac{1}{5}z^5 - \tfrac{1}{7}z^7 + \cdots$$

$$= {}_2F_1(\tfrac{1}{2}, 1; \tfrac{3}{2}; -z^2).$$

Here the partial numerators of the corresponding continued fraction are

$$a_{2k} = \frac{(2k-1)^2}{(4k-3)(4k-1)}, \qquad a_{2k+1} = \frac{(2k)^2}{(4k-1)(4k+1)}.$$

The fraction converges for all z such that z^2 is not real and $\leqslant -1$, that is, for all z not of the form $z = iy$ where y is real and $y^2 \geqslant 1$. After an equivalence transformation we get

$$\text{Arctan } z = \frac{z}{|1} + \frac{1z^2}{|3} + \frac{4z^2}{|5} + \frac{9z^2}{|7} + \frac{16z^2}{|9} + \cdots. \qquad (12.5\text{-}11)$$

EXAMPLE **3**

In a similar manner one finds

$$\text{Log}\frac{1+z}{1-z} = \frac{2z}{|1} - \frac{1z^2}{|3} - \frac{4z^2}{|5} - \frac{9z^2}{|7} - \frac{16z^2}{|9} - \cdots, \quad (12.5\text{-}12)$$

which holds for all z that are not real and ≤ -1 or ≥ 1.

EXAMPLE **4**

The general binomial series is

$$(1+z)^{-a} = {}_2F_1(a, 1; 1; -z).$$

Here (12.5-9) yields $a_2 = a$,

$$a_{2k+1} = \frac{k-a}{2(2k-1)}, \qquad a_{2k+2} = \frac{k+a}{2(2k+1)},$$

$k = 1, 2, \ldots$. After an equivalence transformation to remove denominators, we thus find

$$(1+z)^{-a} = \frac{1}{|1} + \frac{az}{|1} + \frac{(1-a)z}{|2} + \frac{(1+a)z}{|3} + \frac{(2-a)z}{|2} + \frac{(2+a)z}{|5}$$
$$+ \frac{(3-a)z}{|2} + \frac{(3+a)z}{|7} + \cdots,$$

which converges for all z not real and ≤ -1. For $a = \frac{1}{2}$, $z = 1$ we can recover the fraction for $\sqrt{2}$ given in §12.2.

Another group of examples may be deduced from the series

$$P(z) := {}_1F_1(1; c; bz) = \sum_{n=0}^{\infty} \frac{(bz)^n}{(c)_n},$$

where b and c are again complex, $c \neq 0, -1, -2, \ldots$. Again the qd scheme is known explicitly here; we find

$$q_1^{(n)} = \frac{b}{c+n},$$

$$e_k^{(n)} = -\frac{kb}{(c+n+2k-2)(c+n+2k-1)}, \qquad k = 1, 2, \ldots,$$

$$\hspace{10cm} (12.5\text{-}13)$$

$$q_k^{(n)} = \frac{(c+n+k-2)b}{(c+n+2k-3)(c+n+2k-2)}, \qquad k = 2, 3, \ldots.$$

Thus again the corresponding continued fraction $C(z)$ can be written down explicitly with

$$a_2 = -\frac{b}{c}, \qquad a_{2k+1} = \frac{kb}{(c+2k-2)(c+2k-1)}, \qquad a_{2k+2} = -\frac{(c+k-1)b}{(c+2k-1)(c+2k)},$$

$k = 1, 2, \ldots.$ It is evident that

$$\lim_{K \to \infty} a_k = 0.$$

Thus Theorem 12.5c is applicable. Because $_1F_1(1; c; bz)$ defines an entire function, equality between function and corresponding series holds for all complex z.

EXAMPLE 5

The most familiar special case is

$$e^z = {}_1F_1(1; 1; z).$$

There follows

$$e^z = \frac{1|}{|1} - \frac{z|}{|1} + \frac{z/2|}{|1} - \frac{z/6|}{|1} + \frac{z/6|}{|1} - \frac{z/10|}{|1} + \frac{z/10|}{|1} - \cdots$$

for all complex z. After an equivalence transformation this becomes

$$e^z = \frac{1|}{|1} - \frac{z|}{|1} + \frac{z|}{|2} - \frac{z|}{|3} + \frac{z|}{|2} - \frac{z|}{|5} + \frac{z|}{|2} - \frac{z|}{|7} + \cdots. \qquad (12.5\text{-}14)$$

In conclusion we present applications of some of the foregoing expansions to two problems in numerical analysis.

EXAMPLE 6 **Difference Operators of Maximum Order.**

A multistep method for the numerical solution of ordinary differential equations is defined by two polynomials without common zeros,

$$\rho(\zeta) = \alpha_k \zeta^k + \alpha_{k-1} \zeta^{k-1} + \cdots + \alpha_0,$$

$$\sigma(\zeta) = \beta_k \zeta^k + \beta_{k-1} \zeta^{k-1} + \cdots + \beta_0$$

($\alpha_k \neq 0$), which for maximum accuracy should be chosen such that the order p of the resulting method is as large as possible. Here p is the largest integer such that for all sufficiently differentiable functions $y(\tau)$, as $h \to 0$,

$$\alpha_k y(\tau + kh) + \alpha_{k-1} y(\tau + (k-1)h) + \cdots + \alpha_0 y(\tau)$$

$$- h\{\beta_k y'(\tau + kh) + \cdots + \beta_0 y'(\tau)\} = O(h^{p+1}). \qquad (12.5\text{-}15)$$

The relation (12.5-15) must hold, in particular, for $y(\tau) = e^\tau$, which after cancelling e^τ yields

$$\rho(e^h) - h\sigma(e^h) = O(h^{p+1}).$$

On letting $z := e^h - 1$, we see that this holds if and only if

$$\rho_1(z) - \sigma_1(z) \operatorname{Log}(1+z) = O(z^{p+1}) \tag{12.5-16}$$

as $z \to 0$, where $\rho_1(z) := \rho(1+z)$, $\sigma_1(z) := \sigma(1+z)$. If $p > 0$, then $\rho_1(0) = 0$. Because ρ_1 and σ_1 have no common zeros, $\sigma_1(0) \neq 0$. Thus (12.5-16) is equivalent to

$$\operatorname{Log}(1+z) - \frac{\rho_1(z)}{\sigma_1(z)} = O(z^{p+1}),$$

where p should be as large as possible. It follows from Lemma 12.4d that to achieve order p, $z^{-1}\rho_1(z)$ and $\sigma_1(z)$ should be chosen as the numerator and the denominator of the pth approximant of the continued fraction corresponding to

$$\frac{1}{z}\operatorname{Log}(1+z) = 1 - \tfrac{1}{2}z + \tfrac{1}{3}z^2 - \cdots,$$

given by (12.5-10). In view of the factor z this produces polynomials ρ_1 and σ_1 of the respective degrees $[(p+1)/2]$ and $[p/2]$. Thus if k is given, the order $p = 2k$ can be achieved. Unfortunately, the operators thus obtained cannot normally be used, because they are subject to numerical instability (see Henrici [1962]).

EXAMPLE **7 Rational Approximations to $e^{\tau \mathbf{A}}$.**

The linear initial value problem for time-independent systems in both ordinary and partial differential equations has the form

$$\frac{d\mathbf{x}}{d\tau} = \mathbf{A}\mathbf{x}, \qquad \mathbf{x}(0) = \mathbf{s}. \tag{12.5-17}$$

Here $\mathbf{x}$ is either a vector, of dimension m say, and accordingly $\mathbf{A}$ is an $m \times m$ matrix, or $\mathbf{x}$ is a function defined on a certain region of the space variables, and $\mathbf{A}$ is a linear operator acting in the space of these functions. In either case the solution of (12.5-17) may be written symbolically as

$$\mathbf{x}(\tau) = e^{\tau \mathbf{A}}\mathbf{s}. \tag{12.5-18}$$

To compute $e^{\tau \mathbf{A}}$ by means of the Jordan canonical form as described in §2.6 is possible only if $\mathbf{A}$ is a matrix, and may not always be feasible even then. On the other hand, computation by power series may not be practical because it may be too expensive to compute high powers of $\mathbf{A}$. One thus is led to approximate $e^{\tau \mathbf{A}}$ by a rational expression,

$$e^{\tau \mathbf{A}} \sim \frac{p_n(\tau \mathbf{A})}{q_n(\tau \mathbf{A})},$$

where p_n and q_n are chosen such as to produce maximum agreement with the exponential series. By Lemma 12.4d, choosing for p_n and q_n the numerators and denominators of the continued fraction corresponding to e^z produces agreement up to $O(z^n)$, and this is the only choice featuring this property. Table 12.5 gives the polynomials p_n and q_n.

Table 12.5.   Approximants of *RITZ* Fraction Corresponding to e^z

n	$p_n(z)$	$q_n(z)$
1	1	1
2	1	$1-z$
3	$1+\frac{1}{2}z$	$1-\frac{1}{2}z$
4	$1+\frac{1}{3}z$	$1-\frac{2}{3}z+\frac{1}{6}z^2$
5	$1+\frac{1}{2}z+\frac{1}{12}z^2$	$1-\frac{1}{2}z+\frac{1}{12}z^2$
6	$1+\frac{2}{5}z+\frac{1}{20}z^2$	$1-\frac{3}{5}z+\frac{3}{20}z^2-\frac{1}{60}z^3$
7	$1+\frac{1}{2}z+\frac{1}{10}z^2+\frac{1}{120}z^3$	$1-\frac{1}{2}z+\frac{1}{10}z^2-\frac{1}{120}z^3$

To compute the value of the approximate solution at τ,

$$\mathbf{y} := \frac{p_n(\tau\mathbf{A})}{q_n(\tau\mathbf{A})}\mathbf{s},$$

it is not necessary to compute the inverse of the operator $q_n(\tau\mathbf{A})$, because $\mathbf{y}$ may be found as the solution of the linear equation

$$q_n(\tau\mathbf{A})\mathbf{y}=p_n(\tau\mathbf{A})\mathbf{s}. \tag{12.5-19}$$

In certain applications to partial differential equations $\mathbf{A}$ is a sparse matrix of very high order that is not stored explicitly. To solve (12.5-19) by elimination may then be impractical, and an iterative method may be preferred. In this case it would be wasteful to compute the vectors $\mathbf{y}_n := q_n(\tau\mathbf{A})\mathbf{y}$ and $\mathbf{s}_n := p_n(\tau\mathbf{A})\mathbf{s}$ by first forming the polynomials $q_n(\tau\mathbf{A})$ and $p_n(\tau\mathbf{A})$ using matrix–matrix multiplications. It is more economical to use the recurrence relation

$$p_k(\tau\mathbf{A})=p_{k-1}(\tau\mathbf{A})+a_k\tau\mathbf{A}p_{k-2}(\tau\mathbf{A})$$

$(k=2,3,\ldots)$ which on multiplying from the right by $\mathbf{s}$ and taking into account the initial conditions (12.4-3a) yields

$$\mathbf{s}_0=\mathbf{0}, \qquad \mathbf{s}_1=\mathbf{s}, \qquad \mathbf{s}_k=\mathbf{s}_{k-1}+a_k\tau\mathbf{A}\mathbf{s}_{k-2},$$

$k=2,3,\ldots$. Similarly for the $\mathbf{y}_n$ we find

$$\mathbf{y}_0=\mathbf{y}, \qquad \mathbf{y}_1=\mathbf{y}, \qquad \mathbf{y}_k=\mathbf{y}_{k-1}+a_k\tau\mathbf{A}\mathbf{y}_{k-2},$$

$k=2,3,\ldots$. This method has the further advantage that it is not necessary to specify n in advance.

PROBLEMS

1. If

$$e^z = 1 + \frac{z}{1!} + \frac{z^2}{2!} + \cdots + \frac{z^{n-1}}{(n-1)!} + \frac{z^n}{n!} r_n(z),$$

show that

$$r_n(z) = \frac{1}{\vert 1} - \frac{z}{\vert n+1} + \frac{z}{\vert n+2} - \frac{(n+1)z}{\vert n+3} + \frac{2z}{\vert n+4} - \frac{(n+2)z}{\vert n+5} + \cdots$$

for all complex z. Obtain (12.5-14) as a special case.

2. Discuss the convergence of Eisenstein's fraction given in Problem 3, §12.4, and show, for instance, that for $|t| < 1$

$$\sum_{n=0}^{\infty} t^{n^2} = \frac{1}{\vert 1} - \frac{t}{\vert 1} - \frac{t^3 - t}{\vert 1} - \frac{t^5}{\vert 1} - \frac{t^7 - t^3}{\vert 1} - \cdots.$$

3. Show that for $n = 1, 2, \ldots$

$$\int_0^z \frac{1}{1 + t^n} \, dt = \frac{z}{\vert 1} + \frac{1^2 z^n}{\vert n+1} + \frac{n^2 z^n}{\vert 2n+1} + \frac{(n+1)^2 z^n}{\vert 3n+1}$$

$$+ \frac{(2n)^2 z^n}{\vert 4n+1} + \frac{(2n+1)^2 z^n}{\vert 5n+1} + \frac{(3n)^2 z^n}{\vert 6n+1} + \frac{(3n+1)^2 z^n}{\vert 7n+1} + \cdots$$

[Expand the integral in powers of z.]

§12.6.   THE DIVISION ALGORITHM. RATIONAL *RITZ* FRACTIONS

It has been shown in §12.4 how to construct the *RITZ* fraction corresponding to a normal formal power series in a purely arithmetic manner by means of the *qd* algorithm. Here we show that the *RITZ* fraction corresponding to a *convergent* formal power series P can also be obtained directly from the *function* represented by P. This construction is based on the following fact.

THEOREM 12.6a

Let $a_i \neq 0$, $i = 1, 2, \ldots$, and let the RITZ fraction

$$C(z) := \frac{a_1}{\vert 1} + \frac{a_2 z}{\vert 1} + \frac{a_3 z}{\vert 1} + \cdots$$

be uniformly convergent in a neighborhood N of 0. For $z \neq 0$, let

$$t_k : u \to \frac{a_k z}{1 + u}, \qquad k = 1, 2, \ldots,$$

so that

$$t_k^{[-1]} : v \to \frac{za_k}{v} - 1, \qquad k = 1, 2, \ldots.$$

If $f(z)$ denotes the value of $C(z)$ for $z \in N$, define for $z \in N$, $z \neq 0$

$$f_0(z) := zf(z), \qquad f_k(z) := t_k^{[-1]}(f_{k-1}(z)), \qquad k = 1, 2, \ldots \qquad (12.6\text{-}1)$$

Then for $k = 0, 1, 2, \ldots$

$$a_{k+1} = \lim_{z \to 0} \frac{1}{z} f_k(z). \qquad (12.6\text{-}2)$$

Note: We have changed the earlier definition of t_1 (see §12.4) to avoid an exceptional situation for $k = 1$.

The idea of Theorem 12.6a may be simply stated thus: Letting $z \to 0$ in a convergent *RITZ* fraction evidently yields its first partial numerator. From

$$f_0(z) = t_1 \circ t_2 \circ \cdots \circ t_n \circ \cdots (0)$$

there follows

$$f_k(0) = t_k^{[-1]} \circ t_{k-1}^{[-1]} \circ \cdots \circ t_1^{[-1]}(f_0(z))$$

$$= t_{k+1} \circ t_{k+2} \circ t_{k+n} \circ \cdots (0)$$

$$= \frac{za_{k+1}}{\mid 1} + \frac{za_{k+2}}{\mid 1} + \cdots,$$

whence (12.6-2) follows immediately. The justification of these steps is somewhat more elaborate than expected because the transformations t_k as well as f_0 depend on z.

Proof of Theorem 12.6a.   Let

$$C_k(z) := \frac{a_{k+1}}{\mid 1} + \frac{a_{k+2}z}{\mid 1} + \frac{a_{k+3}z}{\mid 1} + \cdots$$

and suppose that $C_k(z)$ converges uniformly in a neighborhood N_k of 0, and that for $z \in N_k$, $z \neq 0$

$$C_k(z) = \frac{1}{z} f_k(z). \qquad (12.6\text{-}3)$$

By hypothesis, these assumptions are true for $k = 0$. Assuming that they are true for some integer $k \geqslant 0$, we shall show that they are true for $k + 1$.

Let $\{w_n^*(z)\}$ denote the sequence of approximants of $C_k(z)$,

$$w_n^*(z) = \frac{1}{z} t_{k+1} \circ t_{k+2} \circ \cdots \circ t_{k+n}(0), \qquad (12.6\text{-}4)$$

$n = 1, 2, \ldots$. The uniform limit of this sequence by Theorem 12.5a is an analytic function whose value at 0 is a_{k+1}. Hence by (12.6-3), (12.6-2) follows for the index k. Furthermore, because $a_{k+1} \neq 0$, there exist $\rho > 0$ and n_0 such that

$$|w_n^*(z) - a_{k+1}| < \frac{|a_{k+1}|}{2}$$

for $0 < |z| \leq \rho$ and $n > n_0$. It follows that the sequence of functions

$$t_{k+1}^{[-1]}(zw_n^*(z)) = \frac{a_{k+1}}{w_n^*(z)} - 1 \tag{12.6-5}$$

converges, uniformly for $0 < |z| \leq \rho$, to

$$t_{k+1}^{[-1]}(f_k(z)) = f_{k+1}(z).$$

The functions (12.6-5) all have removable singularities at $z = 0$ with value 0. Hence as $n \to \infty$,

$$\frac{1}{z} t_{k+1}^{[-1]}(zw_n^*(z)) \to \frac{1}{z} f_{k+1}(z)$$

likewise holds uniformly for $0 < |z| \leq \rho$. (This follows from the fact that the absolute value of the function

$$\frac{1}{z} t_{k+1}^{[-1]}(zw_n^*(z)) - \frac{1}{z} f_{k+1}(z),$$

which has a removable singularity with value 0 at 0, takes its maximum on $|z| = \rho$.) However, by (12.6-4) the functions

$$\frac{1}{z} t_{k+1}^{[-1]}(zw_n^*(z)) = \frac{1}{z} t_{k+2} \circ t_{k+3} \circ \cdots \circ t_{k+n}(0)$$

are the approximants of $C_{k+1}(z)$. We thus have verified the induction hypothesis with k increased by 1, and in doing so have proved (12.6-2). ■

Theorem 12.6a suggests the following **division algorithm** for computing the *RITZ* fraction corresponding to a series P representing a function f analytic at 0. Let

$$f_0(z) := zf(z), \tag{12.6-6 a}$$

$$a_1 := \lim_{z \to 0} \frac{1}{z} f_0(z) \tag{12.6-6b}$$

(which of course means $a_1 := f(0)$). If f_{k-1} and a_k have been determined and if $a_k \neq 0$, put

$$f_k(z) := t_k^{[-1]}(f_{k-1}(z)) = \frac{a_k z}{f_{k-1}(z)} - 1, \qquad (12.6\text{-}7a)$$

$$a_{k+1} := \lim_{z \to 0} \frac{1}{z} f_k(z), \qquad (12.6\text{-}7b)$$

and let $k + 1 \to k$. The division algorithm is said to *terminate after k steps* if $a_{k+1} = 0$.

THEOREM 12.6b

Let f be analytic at 0, *and let the division algorithm described above never terminate. Then the RITZ fraction*

$$C(z) = \frac{a_1}{|1} + \frac{a_2 z}{|1} + \frac{a_3 z}{|1} + \cdots \qquad (12.6\text{-}8)$$

corresponds to the power series representing f at 0. (Convergence of C is not implied.)

Proof. Denoting by p_n and q_n the numerators and denominators of C and by

$$w_n(z) = \frac{p_n(z)}{q_n(z)}, \qquad n = 0, 1, 2, \ldots$$

its approximants, it is to be shown that for $n = 0, 1, 2, \ldots$

$$f(z) - w_n(z) = (-1)^n a_1 a_2 \cdots a_{n+1} z^n + O(z^{n+1}). \qquad (12.6\text{-}9)$$

Evidently, this is true for $n = 0$. For an induction proof we assume the truth of (12.6-9) with n replaced by an integer $n - 1 \geqslant 0$. Clearly by (12.6-7a)

$$f_n(z) = t_n^{[-1]} \circ t_{n-1}^{[-1]} \circ \cdots \circ t_1^{[-1]}(f_0(z)).$$

By the fundamental Theorem 12.1a, taking into account that $t_1(z)$ contains a factor z,

$$t_1 \circ t_2 \circ \cdots \circ t_n(u) = z \frac{p_{n-1}(z) u + p_n(z)}{q_{n-1}(z) u + q_n(z)};$$

hence

$$t_n^{[-1]} \circ t_{n-1}^{[-1]} \circ \cdots \circ t_1^{[-1]}(v) = \frac{q_n(z) v - z p_n(z)}{-q_{n-1}(z) v + z p_{n-1}(z)}.$$

In view of (12.6-6a) there follows

$$f_n(z) = -\frac{q_n(z)f(z) - p_n(z)}{q_{n-1}(z)f(z) - p_{n-1}(z)}.$$

Because the polynomials $q_m(0) = 1$, $m = 0, 1, \ldots$, we may write for $z \neq 0, |z|$ sufficiently small,

$$f_n(z) = -\frac{q_n(z)}{q_{n-1}(z)} \frac{f(z) - w_n(z)}{f(z) - w_{n-1}(z)}. \qquad (12.6\text{-}10)$$

In view of the induction hypothesis,

$$f(z) - w_{n-1}(z) = (-1)^{n-1} a_1 a_2 \cdots a_n z^{n-1} + O(z^n).$$

This shows that f_n is analytic for $z \neq 0$, $|z|$ sufficiently small. By (12.6-7b), the singularity is removable, and

$$f_n(z) = a_{n+1}z + O(z^2).$$

Because $q_n(z)/q_{n-1}(z) = 1 + O(z)$, (12.6-10) now implies that (12.6-9) holds, and the induction step is complete. ■

Classical examples for the division algorithm arise in the theory of hypergeometric series. Perhaps the simplest of these is the following. For $s \neq 0, -1, -2, \ldots$ let

$$F_s(z) := {}_0F_1(s; z) = \sum_{n=0}^{\infty} \frac{z^n}{(s)_n n!}. \qquad (12.6\text{-}11)$$

By comparing coefficients in the power series expansion, the entire function F_s is easily seen to satisfy the recurrence relation

$$F_s(z) = F_{s+1}(z) + \frac{z}{s(s+1)} F_{s+2}(z).$$

On dividing by $F_{s+1}(z)$ there follows

$$\frac{z}{s(s+1)} \frac{F_{s+2}(z)}{F_{s+1}(z)} = \frac{1}{\dfrac{F_{s+1}(z)}{F_s(z)}} - 1. \qquad (12.6\text{-}12)$$

If for some fixed $s \neq 0, -1, -2, \ldots$ and for $k = 0, 1, 2, \ldots$ we define

$$f_k(z) := \frac{z}{(s+k)(s+k+1)} \frac{F_{s+k+2}(z)}{F_{s+k+1}(z)}, \qquad (12.6\text{-}13)$$

$$a_{k+1} := \frac{1}{(s+k)(s+k+1)}, \qquad (12.6\text{-}14)$$

then replacing s by $s+k$ in (12.6-12) yields

$$f_k(z) = \frac{a_k z}{f_{k-1}(z)} - 1, \qquad k = 1, 2, \ldots,$$

and from (12.6-13) there follows

$$\lim_{z \to 0} \frac{1}{z} f_k(z) = a_{k+1}.$$

Thus applying the division algorithm to the function

$$f(z) := \frac{1}{z} f_0(z) = \frac{1}{s(s+1)} \frac{{}_0F_1(s+2; z)}{{}_0F_1(s+1; z)} \qquad (12.6\text{-}15)$$

yields the constants a_k defined by (12.6-14). It follows that the power series representing f at 0 has the corresponding *RITZ* fraction

$$\frac{\dfrac{1}{s(s+1)}}{\mid 1} + \frac{\dfrac{z}{(s+1)(s+2)}}{\mid 1} + \frac{\dfrac{z}{(s+2)(s+3)}}{\mid 1} + \cdots .$$

Because the a_k tend to zero, Theorem 12.5c is applicable, and the continued fraction converges, to $f(z)$, for all complex z. The convergence is proper for all z at which $f(z)$ is finite. After an equivalence transformation we thus obtain

$$\frac{1}{s+1} \frac{{}_0F_1(s+2; z)}{{}_0F_1(s+1; z)} = \frac{1}{\mid s+1} + \frac{z}{\mid s+2} + \frac{z}{\mid s+3} + \frac{z}{\mid s+4} + \cdots \qquad (12.6\text{-}16)$$

$(s \neq -1, -2, \ldots)$, the convergence being proper for all z such that ${}_0F_1(s+1; z) \neq 0$.

The following special cases are worth mentioning.

EXAMPLE **1**

In view of

$$\sin z = z {}_0F_1\left(\frac{3}{2}; -\frac{z^2}{4}\right), \qquad \cos z = {}_0F_1\left(\frac{1}{2}; -\frac{z^2}{4}\right)$$

we get from (12.6-16) for $s = -\frac{1}{2}$ after a further equivalence transformation

$$\tan z = \frac{z}{\mid 1} - \frac{z^2}{\mid 3} - \frac{z^2}{\mid 5} - \frac{z^2}{\mid 7} - \cdots \qquad (12.6\text{-}17)$$

or, replacing z by iz,

$$\frac{e^z - e^{-z}}{e^z + e^{-z}} = \frac{z}{\mid 1} + \frac{z^2}{\mid 3} + \frac{z^2}{\mid 5} + \frac{z^2}{\mid 7} + \cdots . \qquad (12.6\text{-}18)$$

EXAMPLE **2**

For arbitrary ν the Bessel function of order ν is defined (see §9.7) by

$$J_\nu(z):=\frac{(z/2)^\nu}{\Gamma(\nu+1)}{}_0F_1\left(\nu+1;\,-\frac{z^2}{4}\right).$$

Thus

$$\frac{J_{\nu+1}(z)}{J_\nu(z)}=\frac{z}{2\nu+2}\frac{{}_0F_1(\nu+2;\,-z^2/4)}{{}_0F_1(\nu+1;\,-z^2/4)},$$

and (12.6-16) yields

$$\frac{J_{\nu+1}(z)}{J_\nu(z)}=\frac{z}{\lceil\,2\nu+2\,}-\frac{z^2}{\lceil\,2\nu+4\,}-\frac{z^2}{\lceil\,2\nu+6\,}-\frac{z^2}{\lceil\,2\nu+8\,}-\,\cdots. \qquad (12.6\text{-}19)$$

Setting $\nu=-\tfrac{1}{2}$ in view of (9.7-26) once again yields (12.6-17).

Rational RITZ fractions. A terminating *RITZ* fraction

$$C(z)=\frac{a_1}{\lceil\,1\,}+\frac{a_2z}{\lceil\,1\,}+\,\cdots\,+\frac{a_nz}{\lceil\,1\,} \qquad (12.6\text{-}20)$$

where $a_i\neq0$, $i=1,2,\ldots,n$, will be called an n-**terminating *RITZ* fraction.** It is clear that every n-terminating *RITZ* fraction represents a rational function r; in fact,

$$r(z)=\frac{p_n(z)}{q_n(z)},$$

where p_n and q_n denote the nth numerator and the nth denominator of C. We call a rational function $r=p/q$ of **precise type** (m,n) if the polynomials p and q have no factors in common, and if the degrees of p and of q are precisely m and n, respectively. It then follows from Theorem 12.4a and from the material immediately preceding it that the n-terminating *RITZ* fraction (12.6-20) is of precise type $([(n-1)/2],[n/2])$.

The question naturally arises, to what extent rational functions of type $([(n-1)/2],[n/2])$ can be represented by an n-terminating *RITZ* fraction. The proof of Theorem 12.6a shows that if

$$r(z)=\frac{a_1}{\lceil\,1\,}+\frac{a_2z}{\lceil\,1\,}+\,\cdots\,+\frac{a_nz}{\lceil\,1\,},$$

where $a_i\neq0$, $i=1,2,\ldots,n$, then

$$a_{k+1}=\lim_{z\to0}\frac{1}{z}r_k(z),\qquad k=0,1,\ldots,n-1, \qquad (12.6\text{-}21)$$

where the functions r_k are defined recursively by

$$r_0(z) := zr(z),$$

$$(12.6\text{-}22)$$

$$r_k(z) := \frac{a_k}{r_{k-1}(z)} - 1, \qquad k = 1, 2, \ldots, n.$$

This means that the division algorithm (12.6-7) can be carried out for the function r for at least n steps. Trivially, there exist rational functions for which the division algorithm breaks down earlier; for instance, $a_1 = 0$ if $r(0) = 0$. However, the following holds:

THEOREM 12.6c

Let $n > 0$, and let r be a rational function of precise type $([(n-1)/2], [n/2])$. Then r can be represented by an n-terminating RITZ fraction if and only if the limits (12.6-21) (with the r_k defined by (12.6-22)) all exist and are different from zero. In the affirmative case,

$$r(z) = \frac{a_1}{|1} + \frac{a_2 z}{|1} + \cdots + \frac{a_n z}{|1}. \qquad (12.6\text{-}23)$$

Proof. It remains to establish the sufficiency of the existence of the nonzero limits (12.6-21). We assert: For $k = 0, 1, \ldots, n$,

$$r_k(z) \text{ is of precise type } \left(\left[\frac{n-k+1}{2}\right], \left[\frac{n-k}{2}\right]\right). \qquad (12.6\text{-}24)$$

Because $r_0(z) = zr(z)$ and 0 is not a pole of r, this is true for $k = 0$. Assume the truth of (12.6-24) for some $k \geq 0$. Then

$$\frac{a_{k+1}}{r_k(z)} \text{ is of precise type } \left(\left[\frac{n-k}{2}\right], \left[\frac{n-k+1}{2}\right]\right).$$

In view of the existence of (12.6-21), $r_k(0) = 0$, thus

$$\frac{a_{k+1}z}{r_k(z)} \text{ is of precise type } \left(\left[\frac{n-k}{2}\right], \left[\frac{n-k-1}{2}\right]\right),$$

and the same holds for

$$r_{k+1}(z) = \frac{a_{k+1}z}{r_k(z)} - 1,$$

establishing (12.6-24) with k increased by 1. Thus the assertion holds up to $k = n$. It follows that r_{n-1} is of precise type $(1, 0)$, that is, $r_{n-1}(z) = a_n z$, and $r_n(z) = 0$. Defining the transforms t_k as in Theorem 12.6a, we have

$$r_k(z) = t_k^{[-1]}(r_{k-1}(z)),$$

and thus evidently

$$r_0(z) = t_1 \circ t_2 \circ \cdots \circ t_n(0),$$

which on interpreting the composition as a continued fraction and dividing by z yields (12.6-23). ∎

It remains to discuss the details of the division algorithm defined by (12.6-6) and (12.6-7). We assume that f in a neighborhood of 0 is defined by

$$f(z) = \frac{c_0 + c_1 z + c_2 z^2 + \cdots}{d_0 + d_1 z + d_2 z^2 + \cdots}, \qquad d_0 \neq 0.$$

Evidently,

$$f_0(z) = \frac{\sum_{n=0}^{\infty} c_n z^{n+1}}{\sum_{n=0}^{\infty} d_n z^n}$$

and

$$a_1 = \frac{c_0}{d_0}.$$

We now set

$$f_k(z) = \frac{\sum_{n=0}^{\infty} c_n^{(k)} z^{n+1}}{\sum_{n=0}^{\infty} d_n^{(k)} z^n}, \qquad k = 0, 1, 2, \ldots$$

Then (12.6-7b) yields

$$a_{k+1} = \frac{c_0^{(k)}}{d_0^{(k)}}, \tag{12.6-25}$$

and from (12.6-7a) we get

$$\frac{\sum c_n^{(k+1)} z^{n+1}}{\sum d_n^{(k+1)} z^n} = \frac{a_{k+1} \sum d_n^{(k)} z^n}{\sum c_n^{(k)} z^n} - 1,$$

which shows that $d_n^{(k+1)} = c_n^{(k)}$ and thus

$$c_n^{(k+1)} = a_{k+1} c_{n+1}^{(k-1)} - c_{n+1}^{(k)}, \qquad n = 0, 1, 2, \ldots. \tag{12.6-26}$$

The relation (12.6-25) now becomes

$$a_{k+1} = \frac{c_0^{(k)}}{c_0^{(k-1)}}, \qquad k = 0, 1, 2, \ldots. \tag{12.6-27}$$

The algorithm of determining the a_{k+1} in this manner is a variant of the so-called **Routh algorithm**. The computation starts from the first two rows of coefficients

$$c_n^{(-1)} = d_n, \qquad c_n^{(0)} = c_n, \qquad n = 0, 1, 2, \ldots, \qquad (12.6\text{-}28)$$

and then proceeds as indicated in Table 12.6a.

Table 12.6a   The Routh Algorithm (Arrows indicate flow of computation)

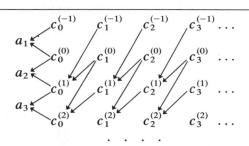

EXAMPLE 3

The rational function

$$r(z) = \frac{1 + z + z^2}{1 + 2z + 3z^2 + 4z^3}$$

is of precise type ($[(n-1)/2]$, $[n/2]$) for $n = 6$. We seek its representation by a 6-terminating *RITZ* fraction. The Routh algorithm yields the following scheme:

		1	2	3	4
1					
		1	1	1	
1					
		1	2	4	
−1					
		−1	−3		
−1					
		1	−4		
7					
		7			
−4					
		−28			

Thus the desired representation is

$$r(z) = \frac{1}{\vert\ 1\ } + \frac{z}{\vert\ 1\ } - \frac{z}{\vert\ 1\ } - \frac{z}{\vert\ 1\ } + \frac{7z}{\vert\ 1\ } - \frac{4z}{\vert\ 1\ },$$

as can be verified directly.

The coefficients $c_n^{(k)}$, and thus the elements a_k of the corresponding *RITZ* fraction, can be expressed in terms of determinants involving only the coefficients c_n and d_n, much like the elements of the qd scheme can be expressed in terms of Hankel determinants. It readily follows from (12.6-26) and (12.6-28) that for $n = 0, 1, 2, \ldots$

$$c_n^{(1)} = \frac{1}{d_0} \begin{vmatrix} c_0 & c_{n+1} \\ d_0 & d_{n+1} \end{vmatrix}, \qquad c_n^{(2)} = \frac{1}{d_0 c_0} \begin{vmatrix} c_0 & c_1 & c_{n+2} \\ d_0 & d_1 & d_{n+2} \\ 0 & c_0 & c_{n+1} \end{vmatrix},$$

$$c_n^{(3)} = \frac{1}{d_0} \frac{\begin{vmatrix} c_0 & c_1 & c_2 & c_{n+3} \\ d_0 & d_1 & d_2 & d_{n+3} \\ 0 & c_0 & c_1 & c_{n+2} \\ 0 & d_0 & d_1 & d_{n+2} \end{vmatrix}}{\begin{vmatrix} c_0 & c_1 \\ d_0 & d_1 \end{vmatrix}}.$$

To formulate a general result, let $\mathbf{h}_k^{(n)}$ denote the k-dimensional column vector whose elements are the first k elements of the sequence $c_n, d_n, c_{n-1}, d_{n-1}, \ldots$, coefficients with negative indices being zero. Let

$$D_0^{(n)} := 0, \quad D_k^{(n)} := \det(\mathbf{h}_k^{(0)}, \mathbf{h}_k^{(1)}, \ldots, \mathbf{h}_k^{(k-2)}, \mathbf{h}_k^{(k-1+n)}),$$

$$\text{(12.6-29)}$$

where $k = 1, 2, \ldots$; $n = 0, 1, 2, \ldots$, and set

$$D_k := D_k^{(0)}, \qquad k = 0, 1, 2, \ldots. \tag{12.6-30}$$

Then, if f possesses a corresponding *RITZ* fraction, $D_k \neq 0$ for $k = 0, 1, 2, \ldots$, and

$$c_n^{(k)} = \frac{1}{d_0} \frac{D_{k+1}^{(n)}}{D_{k-1}}, \qquad k = 1, 2, \ldots, \tag{12.6-31}$$

thus

$$c_0^{(k)} = \frac{1}{d_0} \frac{D_{k+1}}{D_{k-1}}, \qquad k = 1, 2, \ldots. \tag{12.6-32}$$

With the ad hoc definition $D_{-1} := d_0^{-1}$, $D_{-2} := d_0^{-2}$ (12.6-32) remains true even for $k = -1$ and $k = 0$. Then generally

$$a_k = \frac{c_0^{(k-1)}}{c_0^{(k-2)}} = \frac{D_k}{D_{k-2}} \frac{D_{k-3}}{D_{k-1}}, \qquad k = 1, 2, \ldots. \tag{12.6-33}$$

If f is a rational function representable by an n-terminating *RITZ* fraction, this statement holds for $k = 1, 2, \ldots, n$.

The *proof* of (12.6-31) is based on the fact that the formula evidently holds for $k = 1$ and $k = 2$. The verification of the recurrence relation (12.6-26) then requires the determinant identity

$$D_{k+1}D_k^{(n+1)} - D_{k+1}^{(n+1)}D_k = D_{k-1}D_{k+2}^{(n)}, \tag{12.6-34}$$

which can be established by a method similar to that used in the proof of Jacobi's identity (Theorem 7.5a).

PROBLEMS

1. Show that

$$\frac{1 + 2z + 2z^2}{1 + 3z + 6z^2 + 6z^3} = \frac{1}{|1|} + \frac{z}{|1|} - \frac{2z}{|1|} + \frac{2z}{|1|} - \frac{z}{|1|} + \frac{3z}{|1|}.$$

2. For $n = 0, 1, 2, \ldots$, let

$$p_n(z) := 1 + \frac{1}{1!}z + \frac{1}{2!}z^2 + \cdots + \frac{1}{n!}z^n,$$

the nth partial sum of the exponential series. Show that

$$\frac{p_n(z)}{p_{n+1}(z)} = \frac{n+1}{|z|} + \frac{1}{|1|} - \frac{n}{|z|} + \frac{2}{|1|} - \frac{n-1}{|z|} + \frac{3}{|1|} - \frac{n-2}{|z|} + \cdots.$$

3. Let α, β be arbitrary complex numbers. Show that the formal power series

$$_2F_0(\alpha, \beta; -z)/{}_2F_0(\alpha, \beta - 1; -z)$$

corresponds to the *RITZ* fraction

$$\frac{1}{|1|} + \frac{\alpha z}{|1|} + \frac{\beta z}{|1|} + \frac{(\alpha+1)z}{|1|} + \frac{(\beta+1)z}{|1|} + \cdots.$$

4. Show: The nth approximant of the $RITZ^{-1}$ fraction arising from (12.6-16),

$$\frac{1}{\beta z}\frac{{}_0F_1(\beta+1; z^{-2})}{{}_0F_1(\beta; z^{-2})} = \frac{1}{|\beta z|} + \frac{1}{|(\beta+1)z|} + \frac{1}{|(\beta+2)z|} + \cdots \tag{12.6-35}$$

is $q_{n-1}(\beta+1, z)/q_n(\beta, z)$, where

$$q_n(\beta, z) := \sum_{k=0}^{[n/2]} \binom{n-k}{k}(\beta+k)_{n-2k}z^{n-2k}$$

$$= (\beta)_n z^n {}_2F_3\left[\begin{array}{cc} -n/2, & -n/2+1/2; \quad 4z^{-2} \\ -n, & -n-\beta+1, \quad \beta \end{array}\right]$$

(The polynomials $q_n(\beta, z)$ are related to the Lommel polynomials; see Watson [1944], p. 297.)

5. Establish the recurrence relation

$$F(\alpha, \beta; \gamma; z) = F(\alpha, \beta+1; \gamma+1; z) - \frac{\alpha(\gamma-\beta)}{\gamma(\gamma+1)}zF(\alpha+1, \beta+1; \gamma+2; z)$$

$(F := {}_2F_1)$, and deduce the continued fraction of Gauss,

$$\frac{F(\alpha, \beta+1; \gamma+1; z)}{F(\alpha, \beta; \gamma; z)} = \frac{1}{|1|} + \frac{a_2 z}{|1|} + \frac{a_3 z}{|1|} + \frac{a_4 z}{|1|} + \cdots,$$

where

$$a_{2k} := -\frac{(\alpha+k-1)(\gamma-\beta+k-1)}{(\gamma+2k-2)(\gamma+2k-1)}, \qquad a_{2k+1} := -\frac{(\beta+k)(\gamma-\alpha+k)}{(\gamma+2k-1)(\gamma+2k)}.$$

Show that the fraction converges for all z that are not real and ≥ 1. Deduce (12.5-9) as a special case.

6. For $|z| < 1$, if principal values are taken,

$$(1+z)^\mu + (1-z)^\mu = 2F\left(\frac{1-\mu}{2}, -\frac{\mu}{2}; \frac{1}{2}; z^2\right),$$

$$(1+z)^\mu - (1-z)^\mu = 2\mu z F\left(\frac{1-\mu}{2}, \frac{2-\mu}{2}; \frac{3}{2}; z^2\right).$$

Use this fact to show that for all z for which both principal values are defined,

$$\frac{(1+z)^\mu - (1-z)^\mu}{(1+z)^\mu + (1-z)^\mu} = \frac{\mu z}{|1|} + \frac{(\mu^2-1)z^2}{|3|}$$

$$+ \frac{(\mu^2-4)z^2}{|5|} + \frac{(\mu^2-9)z^2}{|7|} + \cdots.$$

7. Let q not be a root of unity, and let

$$P(z) := 1 + \sum_{n=1}^{\infty} \frac{q^{n^2}}{(1-q)(1-q^2)\cdots(1-q^n)} z^n.$$

Show that

$$\frac{P(qz)}{P(z)} \sim \frac{1}{|1|} + \frac{qz}{|1|} + \frac{q^2 z}{|1|} + \frac{q^3 z}{|1|} + \cdots.$$

§12.7. *SITZ* FRACTIONS: APPROXIMANTS, STABLE POLYNOMIALS

The theory of *RITZ* fractions described in the preceding three sections, although leading to a number of interesting special results and expansions, is incomplete in one important aspect: In all special cases that have been discussed, the convergence of the *RITZ* fraction corresponding to a given power series had to be verified a posteriori. We did not identify any important class of functions for which the convergence (let alone the region of convergence) of the fraction was assured a priori.

This unsatisfactory situation changes to the better if we restrict our attention to a special class of *RITZ* fractions called Stieltjes or *SITZ* fractions. A **SITZ fraction** is a *RITZ* fraction

$$C(z) = \frac{\alpha_1}{|1|} + \frac{\alpha_2 z}{|1|} + \frac{\alpha_3 z}{|1|} + \cdots \qquad (12.7\text{-}1)$$

where all $\alpha_i > 0$, $i = 1, 2, \ldots$. [The rotations (R) in the RIT transformations occurring in a $RITZ$ fraction are replaced by stretchings (S).] The theory of $SITZ$ fractions was first expounded by Stieltjes in a celebrated posthumous memoir which appeared in 1894, the year of his death. Stieltjes' ultimate motivation for discussing $SITZ$ fractions was to provide a solution for the so-called moment problem (see §12.14). The treatment of $SITZ$ fractions given here is oriented toward the representation of functions and the "summing" of divergent asymptotic expansions. After discussing the convergence of $SITZ$ fractions (§12.8) we identify a class of functions (the so-called Stieltjes transforms) that can occur as limits of convergent $SITZ$ fractions (§12.9). Stieltjes transforms possess simple asymptotic expansions as $z \to \infty$. It is shown in §12.10 that every such asymptotic series possesses a corresponding $SITZ^{-1}$ fraction (the parameter z being replaced by z^{-1}), which under weak additional conditions converges to the given Stieltjes transform (§12.11). Applications to special functions and to Laplace transforms are discussed in the two subsequent paragraphs.

The present section is devoted to a discussion of *terminating SITZ* fractions and to some of their applications. Let

$$C(z) := \frac{\alpha_1 |}{| 1} + \frac{\alpha_2 z |}{| 1} + \cdots + \frac{\alpha_n z |}{| 1} \tag{12.7-2}$$

be an n-terminating $SITZ$ fraction. As an n-terminating $RITZ$ fraction, $C(z)$ is a rational function of precise type $([(n-1)/2], [n/2])$. We thus have

  (i) $C(z)$ is analytic at infinity.

A further obvious property is

  (ii) $C(x)$ is positive and finite for $x \geq 0$.

To state a third property, let U and L denote the half planes Im $z > 0$ and Im $z < 0$, respectively. We then have

  (iii) $C(U) \subset L$, $C(L) \subset U$.

To prove (iii), let

$$t_1 : u \to \frac{\alpha_1}{1+u}, \qquad t_k : u \to \frac{\alpha_k z}{1+u}, \qquad k = 2, \ldots, n.$$

We recall that for any $z \neq 0$

$$C(z) = t_1 \circ t_2 \circ \cdots \circ t_n (0).$$

Let $z \in U$. Then $\phi := \arg z$ can be chosen in the interval $(0, \pi)$. Because all $\alpha_i > 0$, there follows

$$\arg t_n(0) = \arg \alpha_n z = \phi,$$

hence $\arg(1 + t_n(0)) = \arg(1 + \alpha_n z) \in (0, \phi)$, $\arg(1 + t_n(0))^{-1} \in (-\phi, 0)$, thus again

$$\arg t_{n-1} \circ t_n(0) = \arg \alpha_{n-1} z (1 + \alpha_n z)^{-1} \in (0, \phi).$$

It follows inductively that
$$\arg(t_2 \circ t_3 \circ \cdots \circ t_n(0)) \in (0, \phi)$$
and hence, since t_1 lacks the factor z,
$$\arg(t_1 \circ t_2 \circ \cdots \circ t_n(0)) \in (-\phi, 0),$$
proving the first assertion (iii). The second assertion is proved in a similar manner; in view of (ii) it also follows from the symmetry principle (Theorem 5.11b).

A rational function r having the properties (i) r is analytic at infinity, (ii) $r(x) > 0$ for $x \geq 0$, (iii) $r(U) \subset L$, $r(L) \subset U$ is called a **positive symmetric rational function**. We have just shown that terminating *SITZ* fractions are positive symmetric rational functions.

Let us now compute the partial fraction decomposition of a positive symmetric rational function r. The poles of r cannot lie at ∞ because of (i). They cannot lie at 0 or on the positive real axis because of (ii). They cannot lie in U or L because of (iii). They thus must lie on the negative real axis. Let $\xi < 0$ be a pole of r, and let $n \geq 1$ be its order. Then for $|z - \xi|$ sufficiently small,

$$\frac{1}{r(z)} = c_n (z - \xi)^n \{1 + \text{power series in } (z - \xi)\},$$

where $c_n \neq 0$. Because $r(z)$ is real for z real, c_n must be real. We wish to show that $n = 1$. Let us look at the preimage of the real axis under the map $z \to 1/r(z)$. By the general inverse function theorem (Theorem 2.4b), the preimage consists of $2n$ arcs emanating from 0 under the angles π/n. If $n > 1$, at least one of these arcs initially lies in U. This is a contradiction, since by (iii) the image of U under $z \to 1/r(z)$ lies in U. There follows

$$\frac{1}{r(z)} = c_1 (z - \xi)\{1 + O(z - \xi)\}.$$

In view of $1/r(U) \subset U$, $c_1 > 0$. Thus all poles of r are real, negative, and simple, and have positive residues. It thus follows that the partial fraction expansion of a positive symmetric rational function has the form

$$r(z) = \alpha_0 + \sum_{k=1}^{m} \frac{\alpha_k}{z + \xi_k}, \qquad (12.7\text{-}3)$$

where $\alpha_0 \geq 0$, $\alpha_k > 0$ $(k = 1, 2, \ldots, m)$, and $0 < \xi_1 < \xi_2 < \cdots < \xi_m$.

A rational function whose partial fraction expansion has the special form (12.7-3) is called a **rational Stieltjes transform**. We have just shown that positive symmetric rational functions, and thus terminating *SITZ* fractions, are rational Stieltjes transforms. We now close the circle by showing that *every rational Stieltjes transform can be expressed as a terminating SITZ fraction*.

Our proof is based on subjecting the function (12.7-3) to the division algorithm (12.6-7). The function $r(z) = z^{-1} r_0(z)$ by hypothesis is a rational Stieltjes transform. We now show: *If $z^{-1} r_k(z)$ is a rational Stieltjes transform, then $a_{k+1} > 0$, and $z^{-1} r_{k+1}(z)$ is a rational Stieltjes transform.*

The function $z^{-1} r_k(z)$ is either of type (m, m) or of type $(m-1, m)$ for a suitable integer m. Accordingly, we distinguish two cases.

(a) Let the function $z^{-1} r_k(z)$ be of type (m, m). Then it has the partial fraction decomposition

$$z^{-1} r_k(z) = \alpha_0 + \sum_{i=1}^{m} \frac{\alpha_i}{z + \xi_i},$$

where $\alpha_i > 0$, $i = 0, 1, \ldots, m$; $\xi_i > 0$, $i = 1, \ldots, m$. (These α_i and ξ_i are not necessarily the same as in (12.7-3).) Evidently,

$$a_{k+1} = \lim_{z \to 0} z^{-1} r_k(z) = \alpha_0 + \sum_{i=1}^{m} \frac{\alpha_i}{\xi_i} > 0.$$

By (12.6-7a),

$$z^{-1} r_{k+1}(z) = \frac{1}{z} \left\{ \frac{a_{k+1}}{z^{-1} r_k(z)} - 1 \right\}. \tag{12.7-4}$$

By the proof of Theorem 12.6c, the function $z^{-1} r_{k+1}(z)$ is of type $(m-1, m)$. Let us find its partial fraction expansion. Its poles are the zeros of $z^{-1} r_k(z)$. [$z = 0$ is not a pole, in view of the definition of a_{k+1}.] The real function $x^{-1} r_k(x)$ jumps from $-\infty$ to ∞ at each pole $-\xi_i$, decreases monotonically in between, and tends to the limit $\alpha_0 > 0$ for $x \to \pm \infty$. The m zeros $-\eta_1, \ldots, -\eta_m$ of $z^{-1} r_k(z)$ thus are all real; moreover they satisfy $-\infty < -\eta_m < -\xi_m < -\eta_{m-1} < -\xi_{m-1} < \cdots < -\xi_2 < -\eta_1 < -\xi_1 < 0$. At each of these zeros $z^{-1} r_k(z)$ changes from positive to negative values. Thus in view of the factor z^{-1} in (12.7-4), which is negative at these zeros, the function $x^{-1} r_{k+1}(x)$ jumps from $-\infty$ to ∞. Since $z^{-1} r_{k+1}(z) \to 0$ for $z \to \infty$, its partial fraction expansion thus has the form

$$z^{-1} r_{k+1}(z) = \sum_{i=1}^{m} \frac{\beta_i}{z + \eta_i},$$

where all $\beta_i > 0$, showing that $z^{-1} r_{k+1}(z)$ is a rational Stieltjes transform.

(b) If $z^{-1} r_k(z)$ is of type $(m-1, m)$, the proof is essentially the same. The function $z^{-1} r_k(z)$ now has $m-1$ finite zeros, which all are real and satisfy $-\xi_m < -\eta_{m-1} < \cdots < -\xi_2 < -\eta_1 < -\eta_1$. Because $z^{-1} r_k(z)$ now at ∞ has a zero of order 1 with a positive Taylor coefficient, $z^{-1} r_{k+1}(z)$ has a positive limit as $z \to \infty$. Thus the partial fraction decomposition now is

$$z^{-1} r_{k+1}(z) = \beta_0 + \sum_{i=1}^{m-1} \frac{\beta_i}{z + \eta_i},$$

where all $\beta_i > 0$, showing again that $z^{-1} r_{k+1}(z)$ is a rational Stieltjes transform.

It follows from the foregoing that the division algorithm (12.6-7) can be carried through for $2n$ or $2n - 1$ steps, depending on the type of $r(z)$, and that all resulting $a_k > 0$. Thus $r(z)$ can be represented as a terminating *SITZ* fraction. Altogether we have obtained:

THEOREM 12.7a

For each integer $n > 0$, the following three classes of rational functions are identical:
   (a) *the functions represented by n-terminating SITZ fractions;*
   (b) *the positive symmetric functions of type $([(n-1)/2], [n/2])$;*
   (c) *the rational Stieltjes transforms of the same type.*

Stable Polynomials. An interesting application of positive symmetric rational functions can be made in connection with the following problem, considered already in §6.7: Given a polynomial p of degree n with real coefficients,

$$p(z) = \alpha_0 + \alpha_1 z + \cdots + \alpha_n z^n \tag{12.7-5}$$

($\alpha_n \neq 0$), how can we determine whether the polynomial p is **stable**, that is, whether the zeros $z_1, z_2, \cdots, z_n$ of p satisfy Re $z_i < 0$, $i = 1, 2, \ldots, n$?

We have already seen in §6.7 that for a real polynomial to be stable it is necessary, but not sufficient, that all coefficients of p have the same sign.

To obtain a necessary and sufficient condition, we here consider the **Hurwitz alternant** of p. Thus we call the rational function

$$r(z) := \frac{\alpha_1 + \alpha_3 z + \alpha_5 z^2 + \cdots}{\alpha_0 + \alpha_2 z + \alpha_4 z^2 + \cdots} \tag{12.7-6}$$

$$= \sum_{k=0}^{[(n-1)/2]} \alpha_{2k+1} z^k \bigg/ \sum_{k=0}^{[n/2]} \alpha_{2k} z^k.$$

THEOREM 12.7b

The polynomial p is stable if and only if its Hurwitz alternant is a positive symmetric rational function.
Proof. (a) Let p be stable. If $z \neq 0$ and $\sqrt{z}$ denotes a fixed value of the square root, then the Hurwitz alternant of p is also given by

$$r(z) = \frac{1}{\sqrt{z}} \frac{p(\sqrt{z}) - p(-\sqrt{z})}{p(\sqrt{z}) + p(-\sqrt{z})}. \tag{12.7-7}$$

Let now z not be real and negative, and let $\mathrm{Re}\sqrt{z} > 0$. The factor representation of p then shows

$$|p(\sqrt{z})| > |p(-\sqrt{z})|;$$

hence

$$r(z) = \frac{1}{\sqrt{z}} \frac{1 - [p(-\sqrt{z})/p(\sqrt{z})]}{1 + [p(-\sqrt{z})/p(\sqrt{z})]} \neq 0.$$

Because $z = 0$ is not a pole of r, the poles of r thus can lie only on the negative real axis. Let $z = -\xi^2$ where $\xi > 0$ be a pole of r. This requires

$$p(i\xi) + p(-i\xi) = 0. \tag{12.7-8}$$

If the pole had multiplicity > 1, this would require that also

$$p'(i\xi) - p'(-i\xi) = 0.$$

Combining this with (12.7-8), we would find

$$\frac{p'(i\xi)}{p(i\xi)} + \frac{p'(-i\xi)}{p(-i\xi)} = 2\,\mathrm{Re}\,\frac{p'(i\xi)}{p(i\xi)} = 0$$

which contradicts that in view of p being stable,

$$\mathrm{Re}\,\frac{p'(i\xi)}{p(i\xi)} = \mathrm{Re}\,\sum_{k=1}^{n} \frac{1}{i\xi - z_k} > 0.$$

Thus the poles of r are simple, and the residue at the pole $z = -\xi^2$ is found to be

$$\mathrm{res}\,r(-\xi^2) = \left[\frac{1}{2}\,\mathrm{Re}\,\frac{p'(i\xi)}{p(i\xi)}\right]^{-1} > 0.$$

Because r is analytic at ∞ and obviously $r(x) > 0$ for $x \geq 0$, r is a rational Stieltjes transform.

    (b) Let r, the Hurwitz alternant of the real polynomial p, be a rational Stieltjes transform. Denoting by s and t the numerator and the denominator of r, we evidently have

$$p(z) = t(z^2) + zs(z^2). \tag{12.7-9}$$

It is to be shown that p is stable, that is, that all zeros of p have negative real parts. Now evidently $z = 0$ is not a zero of p, since 0 is not a pole of r. Let $z \neq 0$ be a zero of p. Then from (12.7-9)

$$r(z^2) = \frac{s(z^2)}{t(z^2)} = -\frac{1}{z};$$

on the other hand, by virtue of the hypotheses on r,

$$r(z^2) = \alpha_0 + \sum_{i=1}^{m} \frac{\alpha_i}{z^2 + \xi_i}$$

where $\alpha_0 \geq 0$, $\alpha_i > 0$, $\xi_i > 0$, $i = 1, \ldots, m$. We thus have

$$-\frac{1}{z} = \alpha_0 + \sum_{i=1}^{m} \frac{\alpha_i}{z^2 + \xi_i}. \tag{12.7-10}$$

If $z > 0$, (12.7-10) is evidently contradictory. If $0 < \arg z < \pi/2$, then Im $z^2 > 0$, and the imaginary part of the expression on the right is negative while $\text{Im}(-1/z) > 0$. Similarly $-\pi/2 < \arg z < 0$ is impossible. If, finally, z is pure imaginary, then the expression on the left of (12.7-10) is pure imaginary, and that on the right is real. There follows Re $z < 0$, as desired. ∎

The Theorems 12.7a and 12.7b now immediately suggest the following method for testing whether a given real polynomial is stable: Form the Hurwitz alternant of p and subject it to the division algorithm. The polynomial is stable if and only if the algorithm can be carried to its conclusion and all partial numerators a_k of the resulting *RITZ* fraction are > 0.

EXAMPLE 1

Is the polynomial

$$p(z) := 1 + 3z + 6z^2 + 12z^3 + 11z^4 + 11z^5 + 6z^6$$

stable? The Hurwitz alternant is

$$r(z) = \frac{3 + 12z + 11z^2}{1 + 6z + 11z^2 + 6z^3}.$$

The following Routh scheme results:

		1	6	11	6
3					
		3	12	11	
2					
		6	22	18	
$\frac{1}{3}$					
		2	4		
$\frac{5}{3}$					
		$\frac{10}{3}$	6		
$\frac{1}{5}$					
		$\frac{2}{3}$			

All elements in the first column are positive; the polynomial is stable.

EXAMPLE 2

The polynomial

$$p(z) := 1 + 4\tau z + 6\tau z^2 + 4\tau z^3 + z^4$$

is surely stable for $\tau = 1$, unstable for $\tau = 0$. For precisely which real values of τ is p stable? The Hurwitz alternant is

$$r(z) = \frac{4\tau + 4\tau z}{1 + 6\tau z + z^2}.$$

There results the Routh scheme

$$
\begin{array}{c|ccc}
 & 1 & 6\tau & 1 \\
4\tau & & & \\
 & 4\tau & 4\tau & \\
6\tau - 1 & & & \\
 & 24\tau^2 - 4\tau & & \\
\dfrac{6\tau - 2}{6\tau - 1} & & & \\
 & 24\tau^2 - 8\tau & & \\
\dfrac{1}{6\tau - 1} & & & \\
 & 4\tau\dfrac{6\tau - 2}{6\tau - 1} & &
\end{array}
$$

All $a_k > 0$ if and only if $4\tau > 0$ and $6\tau - 1 > 0$ and $6\tau - 2 > 0$, which is the case if and only if $3\tau > 1$. Thus the given polynomial is stable for all $\tau > \frac{1}{3}$.

By virtue of the formulas (12.6-33) the condition that a polynomial be stable can also be expressed directly in terms of determinants involving the coefficients of the polynomial. The determinants D_k defined by (12.6-30) here are

$$D_1 = \alpha_1, \quad D_2 = \begin{vmatrix} \alpha_1 & \alpha_3 \\ \alpha_0 & \alpha_2 \end{vmatrix}, \quad D_3 = \begin{vmatrix} \alpha_1 & \alpha_3 & \alpha_5 \\ \alpha_0 & \alpha_2 & \alpha_4 \\ 0 & \alpha_1 & \alpha_3 \end{vmatrix}$$

and generally

$$D_k = \det(\mathbf{h}_k^{(1)}, \mathbf{h}_k^{(3)}, \ldots, \mathbf{h}_k^{(2k-1)}), \tag{12.7-11}$$

where $\mathbf{h}_k^{(m)}$ denotes the k-dimensional column vector whose components are the first k elements of the sequence $\alpha_m, \alpha_{m-2}, \alpha_{m-4}, \ldots$ ($\alpha_m := 0$ for $m > n$ and for $m < 0$). We also recall the ad hoc definitions

$$D_0 := 1, \qquad D_{-1} := \alpha_0^{-1}, \qquad D_{-2} := \alpha_0^{-2}.$$

The Hurwitz alternant then possesses a corresponding n-terminating *RITZ* fraction if $D_k \neq 0$, $k = 1, 2, \ldots, n$. The *RITZ* fraction is a *SITZ* fraction if

$$a_k = \frac{D_k}{D_{k-1}} \frac{D_{k-3}}{D_{k-2}} > 0, \qquad k = 1, 2, \ldots, n.$$

In terms of the determinants D_k this result reads as follows:

THEOREM 12.7c (Hurwitz criterion for stability)

The real polynomial

$$p(z) = \alpha_0 + \alpha_1 z + \cdots + \alpha_n z^n$$

$(\alpha_n \neq 0)$ *is stable if and only if $\alpha_0 \neq 0$ and the determinants* (12.7-11) *satisfy*

$$D_{2k} > 0, k = 1, \ldots, \left[\frac{n}{2}\right]; \text{ sign } D_{2k+1} = \text{sign } \alpha_0, k = 0, \ldots, \left[\frac{n-1}{2}\right].$$

From the numerical point of view it should be noted that it is easier to test the stability by the division algorithm than by computing the determinants D_k.

PROBLEMS

1. If μ and δ are real, the polynomial

    $$p(z) := z^3 + \mu z^2 + \delta z + 1$$

    is stable if and only if $\delta > 0$ and $\delta \mu > 1$.
2. The real polynomial

    $$p(z) := z^4 + \alpha z^3 + \beta z^2 + \gamma z + 1$$

    is stable if and only if $\gamma > 0$ and $\beta \gamma - \alpha > 0$ and $\alpha > \gamma^2/(\beta \gamma - \alpha)$.
3. Show that the nth partial sum of the exponential series,

    $$p_n(x) := 1 + \frac{x}{1!} + \frac{x^2}{2!} + \cdots + \frac{x^n}{n!}$$

    is stable for $n = 1, 2, 3, 4$, but not for any $n \geq 5$.
4. *Alternate criterion for stability.* If $p(z) := \alpha_n z^n + \alpha_{n-1} z^{n-1} + \cdots + \alpha_0$ is a real polynomial, $\alpha_n \neq 0$, let its **stability test function** be defined by

    $$s(z) := \frac{\alpha_{n-1} z^{n-1} + \alpha_{n-3} z^{n-3} + \cdots}{\alpha_n z^n + \alpha_{n-2} z^{n-2} + \cdots}.$$

Prove that p is stable if and only if $s(z)$ has the $RITZ^{-1}$ fraction representation

$$s(z) = \frac{1}{\lfloor b_1 z} + \frac{1}{\lfloor b_2 z} + \cdots + \frac{1}{\lfloor b_n z},$$

where all $b_i > 0$. Establish the connection with the stability criterion of §6.7. [p is stable if and only if the reciprocal polynomial $p^*(z) := z^n p(z^{-1})$ is stable. If r^* denotes the Hurwitz alternant of p^*, then

$$s(z) = \{z r^*(z^2)\}^{(-1)^n}.]$$

5. *Wall's criterion for stability.* With p as in the preceding problem, let

$$q(z) := \alpha_{n-1} z^{n-1} + \alpha_{n-3} z^{n-3} + \cdots .$$

Prove that p is stable if and only if

$$\frac{q(z)}{p(z)} = \frac{1}{\lfloor 1 + c_1 z} + \frac{1}{\lfloor c_2 z} + \cdots + \frac{1}{\lfloor c_n z},$$

where all $c_i > 0$. (Wall [1945].)
[Note that the stability test function is

$$s(z) = \frac{q(z)}{p(z) - q(z)}.]$$

6. *Stability of the Bessel polynomials.* The polynomials

$$p_n(z) := \sum_{k=0}^{n} \frac{(n+k)!}{(n-k)! k!} \left(\frac{z}{2}\right)^k, \qquad n = 0, 1, 2, \ldots,$$

are called **Bessel polynomials**. Show that the stability test function of p_n is

$$s_n(z) = \frac{1}{\lfloor z} + \frac{1}{\lfloor 3z} + \frac{1}{\lfloor 5z} + \cdots + \frac{1}{\lfloor (2n-1)z};$$

hence that p_n is stable for all n.
[The Bessel polynomials satisfy the recurrence relation

$$p_{n+1}(z) = (2n+1)z p_n(z) + p_{n-1}(z).$$

The same recurrence relation is satisfied by the numerators

$$q_n(z) := \tfrac{1}{2}\{p_n(z) - (-1)^n p_n(-z)\}$$

of s_n, hence also by the denominators $p_n - q_n$. Taking into account the values for $n = 0$ and 1, $q_n/(p_n - q_n)$ thus is the nth approximant of

$$\frac{1}{\lfloor z} + \frac{1}{\lfloor 3z} + \frac{1}{\lfloor 5z} + \cdots .]$$

7. Let $q_n(\beta, z)$ be defined as in Problem 4, §12.6. Show that for all $\alpha > 0, \beta > 0$ and for $n = 1, 2, \ldots$ the polynomial

$$p(z) := q_n(\beta, z) + \alpha q_{n-1}(\beta + 1, z)$$

is stable.
[The stability test function of p is just the nth approximant of (12.6-35).]

§12.8. *S* FRACTIONS: GENERALIZED VALUE FUNCTIONS, CONVERGENCE

The theory of *SITZ* fractions can be developed more smoothly if in place of *SITZ* fractions we consider $SITZ^{-1}$ fractions. These are obtained from *SITZ* fractions as $RITZ^{-1}$ fractions were obtained from *RITZ* fractions (see §12.4): The parameter z is replaced by z^{-1}, and the fraction is divided by z. Thus a $\boldsymbol{SITZ^{-1}}$ **fraction** is given by

$$C(z) = \frac{\alpha_1}{|z} + \frac{\alpha_2}{|1} + \frac{\alpha_3}{|z} + \frac{\alpha_4}{|1} + \cdots, \qquad (12.8\text{-}1)$$

where all $\alpha_i > 0$. $SITZ^{-1}$ fractions are also called **S fractions**.

It should be recalled from §12.4 that the *numerators* and *denominators* of an S fraction, which we now again denote by $p_n(z)$ and $q_n(z)$, are polynomials of the degrees $[(n-1)/2]$ and $[(n+1)/2]$, respectively; they satisfy the recurrence relations

$$p_{-1}(z) = 1, \qquad p_0(z) = 0, \qquad p_n(z) = \alpha_n p_{n-2}(z) + \epsilon_n(z) p_{n-1}(z),$$

$$\qquad (12.8\text{-}2)$$

$$q_{-1}(z) = 0, \qquad q_0(z) = 1, \qquad q_n(z) = \alpha_n q_{n-2}(z) + \epsilon_n(z) q_{n-1}(z),$$

$n = 1, 2, \ldots$, where $\epsilon_n(z) := 1$ for n even and $\epsilon_n(z) := z$ for n odd. The *approximants*

$$w_n(z) = \frac{p_n(z)}{q_n(z)} \qquad (12.8\text{-}3)$$

are rational functions that vanish at infinity. They *correspond* to a formal power series in z^{-1},

$$P = \frac{\gamma_0}{z} + \frac{\gamma_1}{z^2} + \frac{\gamma_2}{z^3} + \cdots \qquad (12.8\text{-}4)$$

in the sense that for $n = 1, 2, \ldots$ the Laurent expansion of $w_n(z)$ at $z = \infty$ agrees with P through the term $\gamma_{n-1} z^{-n}$. By Theorem 12.7a it follows after the necessary change of variables that the *partial fraction expansion* of $w_n(z)$ has the form

$$w_n(z) = \sum_{i=1}^{m} \frac{\rho_i}{z + \xi_i}, \qquad (12.8\text{-}5)$$

where $m := [(n+1)/2]$, $\rho_i > 0$, $i = 1, \ldots, m$, and $0 \leqslant \xi_1 < \xi_2 < \cdots < \xi_m$. It may be seen from (12.8-5) that the expansion at ∞ of $w_n(z)$ in powers of z^{-1} begins with $\sum \rho_i z^{-1}$. On the other hand, because C corresponds to P, the series begins with $\gamma_0 z^{-1}$. There follows

$$\rho_1 + \rho_2 + \cdots + \rho_m = \gamma_0; \qquad (12.8\text{-}6)$$

we conclude, in particular, that the sum on the left is independent of n.

Let S denote the complex plane cut along the negative real axis, $S := \{z : |\arg z| < \pi\}$, and let T denote a compact subset of S. Let $\delta > 0$ denote the distance of T from the complement of S (see Fig. 12.8). Then if $\xi \geqslant 0$ and $z \in T$, $|z + \xi| \geqslant \delta$, and thus for any n

$$|w_n(z)| \leqslant \sum_{i=1}^{m} \frac{\rho_i}{|z + \xi_i|} \leqslant \frac{1}{\delta} \sum_{i=1}^{m} \rho_i = \frac{\gamma_0}{\delta}, \qquad (12.8\text{-}7)$$

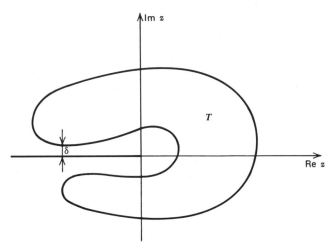

Fig. 12.8.

by virtue of (12.8-6). Thus for every compact subset $T \subset S$ there exists a constant μ (here $\mu = \gamma_0 \delta^{-1}$) such that all approximants on T are bounded by one and the same constant μ. [It is trivial that every individual w_n is bounded by some constant μ_n. The interest lies in the fact that all w_n are bounded by the same μ.] One briefly says that the family of functions $\{w_n\}$ is **uniformly bounded on every compact subset** of S. [The word uniform refers to the functions w_n, not to the compact subsets. There may be a different bound for each choice of T.]

The fact that the approximants of a Stieltjes fraction C are uniformly bounded on every compact subset of the cut plane S has the important implication that a subsequence of these approximants will always converge on S; indeed, the convergence is locally uniform, that is, uniform on every compact subset of S. We now study this implication in a somewhat more general context. This is required again in Chapter 15.

Let S be a region in the complex plane, and let $\mathcal{F}$ be a family of functions f that are analytic in S. The family $\mathcal{F}$ is called **normal** if every sequence of functions $f \in \mathcal{F}$ contains a subsequence that converges uniformly on every compact subset $T \subset S$.

THEOREM 12.8a (Montel's theorem)

Let $\mathcal{F}$ be a family of analytic functions defined on a region S and uniformly bounded on every compact subset of S. Then $\mathcal{F}$ is normal.

Proof. As a preparation we show that the functions $f \in \mathcal{F}$ are **equicontinuous** on every compact subset $T \subset S$, which means that, given any compact $T \subset S$ and any $\epsilon > 0$, there exists $\delta = \delta(\epsilon, T)$ such that

$$|f(z') - f(z'')| < \epsilon \tag{12.8-8}$$

for all $z', z'' \in T$ such that $|z' - z''| < \delta$ *and for all* $f \in \mathcal{F}$. [That (12.8-8) holds for any individual $f \in \mathcal{F}$ trivially follows from the fact that a function that is continuous on a compact set automatically is uniformly continuous. The emphasis in the definition of equicontinuity lies in the statement that one and the same $\delta > 0$ will do for all $f \in \mathcal{F}$.]

Let $z_0 \in S$, and let $D := D(z_0, \rho)$, the closed disk of radius ρ about z_0, be contained in S. Then by Cauchy's formula, if Γ is the circumference (positively oriented) of D, and if z' and z'' are both interior to Γ,

$$f(z') = \frac{1}{2\pi i} \int_\Gamma \frac{f(t)}{t - z'} \, dt, \qquad f(z'') = \frac{1}{2\pi i} \int_\Gamma \frac{f(t)}{t - z''} \, dt,$$

thus by subtraction

$$f(z') - f(z'') = \frac{z' - z''}{2\pi i} \int_\Gamma \frac{f(t)}{(t - z')(t - z'')} \, dt. \tag{12.8-9}$$

Because the family $\mathcal{F}$ is uniformly bounded on compact subsets of S, there exists μ (depending only on z_0 and ρ, but not on f) such that

$$|f(z)| \leq \mu \text{ for all } z \in D(z_0, \rho) \text{ and all } f \in \mathcal{F}.$$

Thus if $z', z'' \in D(z_0, \frac{1}{2}\rho)$ we find by estimating the integral in (12.8-9) that

$$|f(z') - f(z'')| \leq \frac{4\mu}{\rho} |z' - z''| \quad \text{for all } f \in \mathcal{F}. \tag{12.8-10}$$

Thus for the disk $D(z_0, \frac{1}{2}\rho)$, (12.8-8) is proved with $\delta = \rho\epsilon/4\mu$.

Let now T be any compact subset of S. Each point $z_0 \in T$ is center of a disk $D(z_0, \rho)$ contained in S. Thus the open disks $N(z_0, \frac{1}{4}\rho)$ form an open covering of T. By the Heine–Borel lemma (Theorem 3.4a) we can select a finite subcovering, consisting, say, of the disks $N(z_j, \frac{1}{4}\rho_j)$, $j = 1, 2, \ldots, k$. Let ρ denote the smallest ρ_j and μ the maximum of $|f(z)|$ in the union of all $D(z_j, \rho_j)$. Then (12.8-10) holds for all disks $D(z_j, \frac{1}{2}\rho_j)$. Let now ϵ be given, and let

$$\delta := \min\left(\frac{1}{4}\rho, \frac{\epsilon\rho}{4\mu}\right).$$

Then if z', $z'' \in T$ and $|z' - z''| < \delta$, because $|z'' - z_j| < \frac{1}{4}\rho$ for some j it follows that $|z' - z_j| < \frac{1}{2}\rho$ for the same j. Hence (12.8-10) is applicable, yielding (12.8-8) and establishing equicontinuity of $\mathscr{F}$ on T.

To prove normality, let $\{f_n\}$ be a sequence of functions of $\mathscr{F}$. To extract a subsequence that converges on every compact subset of S we use a nonconstructive logical device known as **Cantor's diagonal process**. Let $\{z_k\}_1^\infty$ be a sequence of points that is everywhere dense in S, for instance the points with rational coordinates. The point z_i forms a compact subset of S. Therefore the numbers $\{f_n(z_1)\}$ form a bounded set, and thus there exists an increasing sequence of integers $\{n_1(i)\}$ ($i = 1, 2, \ldots$) such that the sequence $\{f_{n_1(i)}\}$ converges at z_1. Because the numbers $\{f_{n_1(i)}(z_2)\}$ again form a bounded set, a subsequence $\{n_2(i)\}$ of $\{n_1(i)\}$ exists such that the sequence $\{f_{n_2(i)}\}$, in addition to converging at z_1, converges also at z_2. Continuing in this manner we can construct sequences $\{n_3(i)\}$, $\{n_4(i)\}, \ldots$ such that $\{f_{n_k}(i)\}$ converges at z_1, $z_2, \ldots, z_k$. The "diagonal sequence" $\{n_i(i)\}$ then is a subsequence of all sequences thus constructed. Thus the sequence $\{f_{n_i(i)}\}$ converges at all points z_k. For convenience we write $n_i := n_i(i)$.

Let now T be a compact subset of S. To show that the sequence $\{f_{n_i}\}$ converges uniformly on T, let $\epsilon > 0$ be given. Because the functions f_{n_i} are equicontinuous on T, there exists $\delta > 0$ such that z', $z'' \in T$, $|z' - z''| < \delta$ implies $|f_{n_i}(z') - f_{n_i}(z'')| < \epsilon/3$ for all i. The neighborhoods $N(z_k, \delta)$ cover T. We extract a finite subcovering. There exists n_0 such that $i > n_0$, $j > n_0$ implies

$$\left| f_{n_i}(z_k) - f_{n_j}(z_k) \right| < \frac{\epsilon}{3}$$

for all centers z_k of the finite covering. Each $z \in T$ lies within a distance δ from one of these z_k. Hence using equicontinuity, if $i, j > n_0$,

$$|f_{n_i}(z) - f_{n_j}(z)| \leq |f_{n_i}(z) - f_{n_i}(z_k)| + |f_{n_i}(z_k) - f_{n_j}(z_k)| + |f_{n_j}(z_k) - f_{n_j}(z)|$$

$$\leq \frac{\epsilon}{3} + \frac{\epsilon}{3} + \frac{\epsilon}{3} = \epsilon,$$

establishing uniform convergence on T. ∎

As mentioned earlier, the approximants of a Stieltjes fraction satisfy the hypotheses of Theorem 12.8a where S denotes the cut plane. Thus as a corollary of Montel's theorem we have

THEOREM 12.8b

The sequence of approximants of any S fraction contains a subsequence that converges uniformly on every compact subset of the cut plane.

By the fundamental theorem on sequences of analytic functions (Theorem 3.4b) the limit function of any such convergent subsequence of approximants is analytic in *S*. We shall call **generalized value function** of an *S* fraction *C* any function *f* that is the limit of a subsequence of the sequence of approximants of *C*. If *C* converges for $z \in S$, then all subsequences converge to the limit of the sequence of all approximants; thus the only generalized value function of *C* is the function defined by the values of *C*. This function is also called the **value function** of *C*.

The next result links the generalized value functions of an *S* fraction to the formal power series corresponding to *C*.

THEOREM 12.8c

Let C be an S fraction, let

$$P = \frac{\gamma_0}{z} + \frac{\gamma_1}{z^2} + \frac{\gamma_2}{z^3} + \cdots$$

be the series in powers of z^{-1} corresponding to C, and let f be any generalized value function of C. Then the asymptotic expansion

$$f(z) \approx P(z) \tag{12.8-11}$$

holds for $z \to \infty$, $z > 0$; indeed, P envelops $f(x)$ for $x > 0$.

Proof. The enveloping property (see Problem 8, §11.1) requires that the functions $\theta_m(z)$ defined for $m = 0, 1, 2, \ldots$ and for all $z \in S$ by

$$f(z) - \left\{ \frac{\gamma_0}{z} + \frac{\gamma_1}{z^2} + \cdots + \frac{\gamma_{m-1}}{z^m} \right\} = \theta_m(z) \frac{\gamma_m}{z^{m+1}} \tag{12.8-12}$$

satisfy $0 \leq \theta_m(x) \leq 1$ for all $x > 0$. This property obviously implies (12.8-11).

If $\{w_n\}$ is the sequence of approximants of *C*, then by the definition of the corresponding series, if $n \geq m$,

$$w_n(z) = \frac{\gamma_0}{z} + \frac{\gamma_1}{z^2} + \cdots + \frac{\gamma_{m-1}}{z^m} + O(z^{-m-1}) \tag{12.8-13}$$

as $z \to \infty$. On the other hand, by applying the identity

$$\frac{1}{z + \xi} = \frac{1}{z} + \frac{-\xi}{z^2} + \cdots + \frac{(-\xi)^{m-1}}{z^m} + \frac{(-\xi)^m}{z^m(z + \xi)}$$

in the partial fraction expansion (12.8-5), we get ($k := [(n+1)/2]$)

$$w_n(z) = \sum_{i=1}^{k} \rho_{ni} \left\{ \frac{1}{z} - \frac{\xi_{ni}}{z^2} + \cdots + \frac{(-\xi_{ni})^{m-1}}{z^m} + \frac{(-\xi_{ni})^m}{z^m(z + \xi_{ni})} \right\}$$

$$= \sum_{l=1}^{m} \frac{(-1)^{l-1}}{z^l} \sum_{i=1}^{k} \rho_{ni} \xi_{ni}^{l-1} + \frac{(-1)^m}{z^m} \sum_{i=1}^{k} \frac{\rho_{ni} \xi_{ni}^m}{z + \xi_{ni}}.$$

Comparing the coefficients of $z^{-1}, z^{-2}, \ldots, z^{-m}$ with those in (12.8-13), we obtain in view of the uniqueness of the Laurent series

$$\sum_{i=1}^{k} \rho_{ni}\xi_{ni}^{l} = (-1)^{l}\gamma_{l}, \qquad l = 0, 1, \ldots, m-1. \qquad (12.8\text{-}14)$$

The representation for w_n may now be written

$$w_n(z) = \frac{\gamma_0}{z} + \frac{\gamma_1}{z^2} + \cdots + \frac{\gamma_{m-1}}{z^m} + \theta_{m,n}(z)\frac{\gamma_m}{z^{m+1}},$$

where

$$\theta_{m,n}(z) := \frac{\sum_{i=1}^{k}[\rho_{ni}\xi_{ni}^{m}/(1+\xi_{ni}/z)]}{\sum_{i=1}^{k}\rho_{ni}\xi_{ni}^{m}}.$$

Clearly, $x > 0$ implies $0 \le \theta_{m,n}(x) \le 1$. By letting $n \to \infty$ through the sequence $\{n_i\}$ defining the generalized value function f, we obtain (12.8-12) where $0 \le \theta_m(x) \le 1$. ∎

If C converges, Theorem 12.8c simply states that the value function of C admits the asymptotic expansion P as $z \to \infty$ through positive values. If C diverges, it may have many different generalized value functions. Theorem 12.8c then implies the remarkable fact that all these functions have the *same* asymptotic expansion as $z \to \infty$ through positive values.

Under what condition does C converge for all $z \in S$, that is, when is a generalized value function a value function? The key to the answer is supplied by a general theorem on locally bounded families of analytic functions.

THEOREM 12.8d (Vitali's theorem)

Let $\mathscr{F}$ be a family of analytic functions defined in a region S and uniformly bounded on every compact subset of S. Let $\{f_n\}$ be a sequence of functions of $\mathscr{F}$ that converges on a set $Q \subset S$ having a point of accumulation $q \in S$. Then $\{f_n\}$ converges in all of S, uniformly on every compact subset $T \subset S$.

Proof. The family $\mathscr{F}$ being normal by Theorem 12.8a, there exists an increasing sequence of integers $\{n_i\}$ such that the sequence $\{f_{n_i}\}$ converges to some f analytic on S; the convergence is uniform on every compact subset of $T \subset S$. Thus in particular

$$\lim_{i \to \infty} f_{n_i}(z) = f(z), \qquad z \in Q.$$

But on Q the whole sequence converges; because a subsequence converges to f, the whole sequence must do likewise:

$$\lim_{n \to \infty} f_n(z) = f(z), \qquad z \in Q. \qquad (12.8\text{-}15)$$

Let T be a compact subset of S. By enlarging T, if necessary, we may assume that T contains infinitely many points of Q. We set

$$\sigma_n := \sup_{z \in T} |f_n(z) - f(z)|,$$

$n = 1, 2, \ldots$, and wish to show that $\lim \sigma_n = 0$. In any case, if

$$\sigma := \limsup_{n \to \infty} \sigma_n$$

there exists an increasing sequence of integers $\{m_j\}$ such that

$$\sigma = \lim_{j \to \infty} \sigma_{m_j}.$$

The corresponding sequence $\{f_{m_j}\}$, again by the normality of $\mathscr{F}$, contains a subsequence $\{f_{k_j}\}$ which converges, uniformly on compact subsets of S, to some function f^* analytic on S. In particular,

$$\lim_{j \to \infty} f_{k_j}(z) = f^*(z), \qquad z \in T \cap Q.$$

In conjunction with (12.8-15) this implies (because a convergent sequence and any of its subsequences have identical limits)

$$f^*(z) = f(z), \qquad z \in T \cap Q.$$

By the fundamental lemma on analytic continuation (Theorem 3.2d) we conclude

$$f^*(z) = f(z), \qquad z \in S,$$

hence

$$\sigma_{m_j} = \sup_{z \in T} |f_{m_j}(z) - f^*(z)| \to 0,$$

implying that $\sigma = \lim_{j \to \infty} \sigma_{m_j} = 0$, as was to be shown. ∎

It is now an easy step to obtain

THEOREM 12.8e

Let the S fraction (12.8-1) *be convergent for $z = 1$. It then converges in the whole cut plane S, uniformly on every compact subset.*

Proof. By the equivalence relation (12.1-15),

$$C(z) = \cfrac{1}{\vert\ z\beta_1} + \cfrac{1}{\vert\ \beta_2} + \cfrac{1}{\vert\ z\beta_3} + \cfrac{1}{\vert\ \beta_4} + \cdots$$

where

$$\beta_{2m} = \frac{\alpha_1 \alpha_2 \cdots \alpha_{2m-1}}{\alpha_2 \alpha_4 \cdots \alpha_{2m}}, \qquad \beta_{2m+1} = \frac{\alpha_2 \alpha_4 \cdots \alpha_{2m}}{\alpha_1 \alpha_3 \cdots \alpha_{2m+1}}.$$

If $C(1)$ converges, then by Theorem 12.1c the series

$$\sum \beta_m$$

diverges. Hence for each $\xi > 0$ at least one of the series

$$\sum \beta_{2m}, \qquad \xi \sum \beta_{2m+1}$$

diverges, and by the same theorem we conclude that $C(\xi)$ converges for every $\xi > 0$. Hence the sequence of approximants of $C(z)$ converges on a set having a point of accumulation in S. Because the approximants form a normal family, the assertion of Theorem 12.8e now follows from Theorem 12.8d. ∎

A simple sufficient condition for convergence of $C(1)$ can be obtained by observing that for $m = 1, 2, \ldots$, using the inequality of the geometric and arithmetic mean,

$$\beta_m + \beta_{m+1} \geqslant 2\sqrt{\beta_m \beta_{m+1}} = \frac{2}{\sqrt{\alpha_{m+1}}}.$$

Hence if the series $\sum \alpha_m^{-1/2}$ diverges, then the series $\sum \beta_m$ is certainly divergent, and we find

COROLLARY 12.8f

In the S fraction (12.8-1) let $\sum \alpha_m^{-1/2} = \infty$. Then the fraction converges for all z in the cut plane S, uniformly on every compact subset.

EXAMPLE

Let $\alpha_m > 0$, $\lim \alpha_m = 0$. Then the condition of the corollary is satisfied, and $C(z)$ converges for all z in the cut plane, in agreement with what we found in §12.5.

§12.9. *S* FRACTIONS: THE REPRESENTATION OF THEIR GENERALIZED VALUE FUNCTIONS BY STIELTJES TRANSFORMS

Having disposed of the question of convergence of an S fraction, we now obtain an analytic representation of its generalized value functions (and thus, if the fraction converges, of its value function). This formula enables us, among other things, to describe in a simple manner the class of all functions that can be represented by convergent S fractions, and to estimate the truncation error in these representations. A generalization, due to Stieltjes, of the classical concept of the Riemann integral is required.

I. The Stieltjes Integral

Let $[\alpha, \beta]$ be a closed finite interval, let ψ be a real function defined on $[\alpha, \beta]$, and let f be a complex valued (not necessarily analytic) function defined on $[\alpha, \beta]$. A finite sequence of real numbers τ_i ($i = 0, 1, \ldots, m$) where

$$\alpha = \tau_0 < \tau_1 < \tau_2 \cdots < \tau_m = \beta$$

is called a **subdivision** of $[\alpha, \beta]$ and may, as in §4.2, be denoted by a single letter such as Δ. The **norm** of the subdivision Δ is

$$\|\Delta\| := \max_{1 \leq k \leq m} |\tau_k - \tau_{k-1}|. \tag{12.9-1}$$

A **set of pivotal points** Θ consistent with the subdivision Δ is a set of numbers τ_k' satisfying $\tau_{k-1} \leq \tau_k' \leq \tau_k$, $k = 1, 2, \ldots, m$. Given any subdivision Δ and any consistent set of pivotal points Θ we form the sum

$$S(\Delta, \Theta) := \sum_{k=1}^{m} f(\tau_k')[\psi(\tau_k) - \psi(\tau_{k-1})] \tag{12.9-2}$$

If there exists a complex number S such that, given any $\epsilon > 0$, a number $\delta = \delta(\epsilon) > 0$ exists such that, for all subdivisions Δ with $\|\Delta\| < \delta$ and all consistent choices of Θ,

$$|S(\Delta, \Theta) - S| < \epsilon, \tag{12.9-3}$$

then S is called the **Stieltjes integral** of f with respect to ψ from α to β and is customarily denoted by

$$\int_\alpha^\beta f(\tau) \, d\psi(\tau) \qquad \text{or simply} \qquad \int_\alpha^\beta f \, d\psi.$$

Some special cases of the Stieltjes integral are already familiar.

EXAMPLE 1

If $\psi(\tau) = \tau$ or, more generally, $\psi(\tau) = \tau + \gamma$ for some constant γ, the Stieltjes integral is identical with the Riemann integral between the limits α and β.

EXAMPLE 2

If ψ has a continuous derivative on $[\alpha, \beta]$, then by the mean value theorem

$$\psi(\tau_k) - \psi(\tau_{k-1}) = \psi'(\tau_k^*)(\tau_k - \tau_{k-1}),$$

where τ_k^* is a suitable point in (τ_{k-1}, τ_k). If f is continuous on $[\alpha, \beta]$, we thus have by the definition of the Riemann integral

$$\int_\alpha^\beta f(\tau) \, d\psi(\tau) = \int_\alpha^\beta f(\tau) \psi'(\tau) \, d\tau.$$

EXAMPLE 3

A further noteworthy case of the Stieltjes integral is encountered if ψ is a step function with jumps at the finitely many points $\xi_1, \xi_2, \ldots, \xi_n$, given analytically by

$$\psi(\tau) = \begin{cases} 0, \, \alpha \leqslant \tau \leqslant \xi_1, \\ \pi_1, \xi_1 < \tau \leqslant \xi_2, \\ \pi_1 + \pi_2, \xi_2 < \tau \leqslant \xi_3, \\ \cdot \quad \cdot \quad \cdot \quad \cdot \\ \pi_1 + \pi_2 + \cdots + \pi_n, \xi_n < \tau \leqslant \beta, \end{cases}$$

where $\pi_1, \pi_2, \ldots, \pi_n$ are arbitrary real numbers. Only the intervals $[\tau_{k-1}, \tau_k)$ containing a jump point then can make a nonzero contribution to the sum $S(\Delta, \theta)$. For $\|\Delta\| < \min(\xi_k - \xi_{k-1})$ the sum thus reduces to

$$\sum_{k=1}^n f(\xi_k^*) \pi_k,$$

where $|\xi_k^* - \xi_k| \leqslant \|\Delta\|$. Assuming f to be continuous, we have $f(\xi_k^*) \to f(\xi_k)$ as $\|\Delta\| \to 0$, and thus

$$\int_\alpha^\beta f(\tau) \, d\psi(\tau) = \sum_{k=1}^n f(\xi_k) \pi_k.$$

We now deal with a more general situation where the existence of the Stieltjes integral can be asserted.

THEOREM 12.9a

If f is continuous and ψ is nondecreasing on the closed finite interval $[\alpha, \beta]$, then

$$\int_\alpha^\beta f \, d\psi$$

exists.

Proof. By considering separately the integrals of the real and of the imaginary part of f, it suffices to prove the assertion for *real* functions f. We denote by

$$\omega(\delta) := \sup_{|\tau^* - \tau| \leqslant \delta} |f(\tau^*) - f(\tau)|$$

the **modulus of continuity** of f. It is well known that by virtue of f being continuous on a closed and bounded interval,

$$\lim_{\delta \to 0} \omega(\delta) = 0. \qquad (12.9\text{-}4)$$

If Δ is the subdivision of $[\alpha, \beta]$ defined by the points $\tau_0, \tau_1, \ldots, \tau_m$, we let

$$\mu_{\Delta k} := \inf_{\tau_{k-1} \leqslant \tau \leqslant \tau_k} f(\tau), \quad \sigma_{\Delta k} := \sup_{\tau_{k-1} \leqslant \tau \leqslant \tau_k} f(\tau)$$

and define

$$\mu_\Delta := \sum_{k=1}^{m} \mu_{\Delta k} [\psi(\tau_k) - \psi(\tau_{k-1})],$$

$$\sigma_\Delta := \sum_{k=1}^{m} \sigma_{\Delta k} [\psi(\tau_k) - \psi(\tau_{k-1})].$$

Let

$$\mu := \sup_\Delta \mu_\Delta, \qquad \sigma := \inf_\Delta \sigma_\Delta,$$

where the sup and the inf are taken with respect to all subdivisions of $[\alpha, \beta]$. If Δ_1 and Δ_2 are any two subdivisions, and if Δ_3 is the union of Δ_1 and Δ_2 (i.e., the subdivision defined by the points of division of both subdivisions Δ_1 and Δ_2), then clearly

$$\mu_{\Delta_1} \leqslant \mu_{\Delta_3} \leqslant \sigma_{\Delta_3} \leqslant \sigma_{\Delta_2};$$

hence in particular $\mu_{\Delta_1} \leqslant \sigma_{\Delta_2}$ and consequently

$$\mu \leqslant \sigma. \qquad (12.9\text{-}5)$$

On the other hand, if Δ is any subdivision,

$$\sigma_\Delta - \mu_\Delta = \sum_{k=1}^{m} (\sigma_{\Delta k} - \mu_{\Delta k})[\psi(\tau_k) - \psi(\tau_{k-1})]$$

$$\leqslant \omega(\|\Delta\|) \sum_{k=1}^{m} [\psi(\tau_k) - \psi(\tau_{k-1})],$$

thus

$$\sigma_\Delta - \mu_\Delta \leqslant \omega(\|\Delta\|)[\psi(\beta) - \psi(\alpha)]. \qquad (12.9\text{-}6)$$

It follows from (12.9-4) that $\sigma_\Delta - \mu_\Delta$ tends to zero as $\|\Delta\| \to 0$ and hence by (12.9-5) that $\sigma = \mu$. We now show that

$$\int_\alpha^\beta f(\tau) \, d\psi(\tau) = \sigma.$$

Let $\epsilon > 0$ be given and let δ_0 be such that

$$\omega(\delta)[\psi(\beta) - \psi(\alpha)] < \epsilon$$

for all $\delta < \delta_0$. Then, if Δ is any subdivision such that $\|\Delta\| < \delta_0$, we have for any consistent system of pivotal points Θ

$$\mu_\Delta \leq S(\Delta, \Theta) \leq \sigma_\Delta,$$

implying

$$\mu_\Delta - \sigma \leq S(\Delta, \Theta) - \sigma \leq \sigma_\Delta - \sigma,$$

implying a fortiori

$$\mu_\Delta - \sigma_\Delta \leq S(\Delta, \Theta) - \sigma = \sigma_\Delta - \mu_\Delta;$$

thus by (12.9-6)

$$|S(\Delta, \Theta) - \sigma| \leq \sigma_\Delta - \mu_\Delta \leq \omega(\|\Delta\|)[\psi(\beta) - \psi(\alpha)] < \epsilon,$$

satisfying (12.9-3). ■

We note without proof the following properties of the Stieltjes integral, which are entirely analogous to those of the Riemann integral:

$$\int_\alpha^\beta d\psi := \int_\alpha^\beta 1 \, d\psi = \psi(\beta) - \psi(\alpha) \qquad (12.9\text{-}7)$$

$$\int_\alpha^\gamma f \, d\psi + \int_\gamma^\beta f \, d\psi = \int_\alpha^\beta f \, d\psi \quad (\alpha < \gamma < \beta) \qquad (12.9\text{-}8)$$

For any complex number c,

$$\int_\alpha^\beta cf \, d\psi = c \int_\alpha^\beta f \, d\psi. \qquad (12.9\text{-}9)$$

For real $f_1(\tau) \leq f_2(\tau)$ $(\alpha \leq \tau \leq \beta)$ we have, if ψ is nondecreasing,

$$\int_\alpha^\beta f_1 \, d\psi \leq \int_\alpha^\beta f_2 \, d\psi. \qquad (12.9\text{-}10)$$

This implies, even if f is complex valued,

$$\left| \int_\alpha^\beta f \, d\psi \right| \leq \int_\alpha^\beta |f| \, d\psi. \qquad (12.9\text{-}11)$$

Improper Stieltjes integrals are defined as in the case of Riemann integration. For instance, if f is continuous and ψ is nondecreasing on $[\alpha, \infty)$, we set

$$\int_\alpha^\infty f \, d\psi := \lim_{\beta \to \infty} \int_\alpha^\beta f \, d\psi,$$

provided that the limit exists. A simple criterion for the existence of improper Stieltjes integrals is as follows:

THEOREM 12.9b

On the interval $[\alpha, \infty)$, *let the complex function f be continuous and bounded, and let the real function ψ be nondecreasing and bounded. Then*

$$\int_\alpha^\infty f \, d\psi$$

exists.

Proof. For every $\beta \geqslant \alpha$ the integral

$$g(\beta) := \int_\alpha^\beta f \, d\psi$$

exists by Theorem 12.9a. We show that the function $g(\tau)$ satisfies the Cauchy condition as $\tau \to \infty$. Let $|f(\tau)| \leqslant \gamma$ for $\tau \geqslant \alpha$, and let $\psi(\infty) := \lim_{\tau \to \infty} \psi(\tau)$. If $\epsilon > 0$ is given, there exists τ_0 such that

$$\gamma[\psi(\infty) - \psi(\tau)] < \epsilon$$

for all $\tau > \tau_0$. Hence, if $\tau_0 < \tau_1 < \tau_2$, then by (12.9-11), (12.9-9), and (12.9-7),

$$|g(\tau_2) - g(\tau_1)| = \left| \int_{\tau_1}^{\tau_2} f \, d\psi \right|$$

$$\leqslant \gamma \int_{\tau_1}^{\tau_2} d\psi = \gamma[\psi(\tau_2) - \psi(\tau_1)] < \epsilon,$$

as requested. ∎

The next result concerns Stieltjes integrals formed with *sequences* of functions $\{\psi_n\}$.

THEOREM 12.9c (Helly's theorem)

Let $[\alpha, \beta]$ be a finite interval, let $\{\psi_n\}$ be a converging sequence of uniformly bounded, nondecreasing functions on $[\alpha, \beta]$, and let f be continuous on $[\alpha, \beta]$. Then if $\psi := \lim \psi_n$,

$$\int_\alpha^\beta f \, d\psi_n \to \int_\alpha^\beta f \, d\psi \quad (n \to \infty).$$

Proof. It is not assumed that the convergence of the sequence $\{\psi_n\}$ is uniform. Nevertheless the hypothesis implies that the limit function ψ is nondecreasing and bounded. Hence the integral $\int_\alpha^\beta f\, d\psi$ exists; moreover,

$$\sigma := \sup_n \, [\psi_n(\beta) - \psi_n(\alpha)] < \infty.$$

Let $\epsilon > 0$ be given, and let, as before, $\omega(\delta)$ denote the modulus of continuity of f. We choose a subdivision Δ of $[\alpha, \beta]$ such that

$$\omega(\|\Delta\|) < \frac{\epsilon}{4\sigma}. \tag{12.9-12}$$

If the points of subdivision are $\alpha = \tau_0 < \tau_1 < \cdots < \tau_m = \beta$, we have

$$d_n := \int_\alpha^\beta f\, d\psi_n - \int_\alpha^\beta f\, d\psi$$

$$= \sum_{k=1}^m \left\{ \int_{\tau_{k-1}}^{\tau_k} f(\tau)\, d\psi_n(\tau) - \int_{\tau_{k-1}}^{\tau_k} f(\tau)\, d\psi(\tau) \right\};$$

hence

$$d_n = \sum_{k=1}^m \int_{\tau_{k-1}}^{\tau_k} [f(\tau) - f(\tau_k)]\, d\psi_n(\tau)$$

$$+ \sum_{k=1}^m \int_{\tau_{k-1}}^{\tau_k} [f(\tau_k) - f(\tau)]\, d\psi(\tau)$$

$$+ \sum_{k=1}^m \left\{ f(\tau_k) \int_{\tau_{k-1}}^{\tau_k} d\psi_n(\tau) - f(\tau_k) \int_{\tau_{k-1}}^{\tau_k} d\psi(\tau) \right\}.$$

The terms in brackets can be estimated by the modulus of continuity, and the expression on the last line can be evaluated as

$$\sum_{k=1}^m f(\tau_k) \{ [\psi_n(\tau_k) - \psi(\tau_k)] - [\psi_n(\tau_{k-1}) - \psi(\tau_{k-1})] \}.$$

Hence if $\gamma := \sup_{\alpha \le \tau \le \beta} |f(\tau)|$,

$$|d_n| \le 2\omega(\|\Delta\|)\sigma + 2\gamma \sum_{k=0}^m |\psi_n(\tau_k) - \psi(\tau_k)|.$$

Now let n_0 be such that for all $n > n_0$ and for $k = 0, 1, \ldots, m$,

$$|\psi_n(\tau_k) - \psi(\tau_k)| < \frac{\epsilon}{4(m+1)\gamma}.$$

(Because there are only finitely many τ_k, n_0 exists even though the convergence of the sequence need not be uniform.) Then by (12.9-12)

$$|d_n| < \epsilon \quad \text{for all } n > n_0,$$

proving the assertion of Theorem 12.9c. ∎

It is easy to see by means of counterexamples that the conclusion of Theorem 12.9c need not hold if the interval of integration is infinite.

The next result provides a method to realize the hypothesis of Theorem 12.9c even if the sequence $\{\psi_n\}$ is originally divergent.

THEOREM 12.9d (Helly's selection principle)

Let $\{\psi_n\}$ be a sequence of nondecreasing real functions on the (finite or infinite) *interval I, and let there be a constant μ such that*

$$|\psi_n(\tau)| \leq \mu \tag{12.9-13}$$

for all $\tau \in I$ and all $n = 0, 1, 2, \ldots$. Then there exist an increasing sequence of indices $\{n_i\}$ and a nondecreasing function χ such that

$$\lim_{i \to \infty} \psi_{n_i}(\tau) = \chi(\tau) \tag{12.9-14}$$

for all $\tau \in I$.

Proof. Let $\tau_1, \tau_2, \tau_3, \ldots$ be the rational points of I, numbered in some fashion. By virtue of (12.9-13) the sequence $\{\psi_i(\tau_1)\}$ has a point of accumulation χ_1, and there exists a sequence $\{n_{1i}\}$ such that

$$\lim_{i \to \infty} \psi_{n_{1i}}(\tau_1) = \chi_1.$$

The sequence $\{\psi_{n_{1i}}(\tau_2)\}$ having a point of accumulation τ_2, we can extract from the sequence $\{n_{1i}\}$ a subsequence $\{n_{2i}\}$ such that

$$\lim_{i \to \infty} \psi_{n_{2i}}(\tau_2) = \chi_2.$$

Because $\{n_{2i}\}$ is a subsequence of $\{n_{1i}\}$, there still holds

$$\lim_{i \to \infty} \psi_{n_{2i}}(\tau_1) = \chi_1.$$

Continuing in the same manner, we can consecutively find numbers χ_k and extract subsequences $\{n_{ki}\}$ ($k = 1, 2, \ldots$) such that

$$\lim_{i \to \infty} \psi_{n_{ki}}(\tau_j) = \chi_j$$

holds for $j = 1, 2, \ldots, k$. Now let $n_k := n_{kk}$. The sequence $\{n_k\}$ ultimately being a subsequence of every sequence $\{n_{ki}\}$, there holds

$$\lim_{k \to \infty} \psi_{n_k}(\tau_j) = \chi_j \tag{12.9-15}$$

for all $j = 1, 2, 3, \ldots$.

We next show that $\tau_i < \tau_j$ implies $\chi_i \leq \chi_j$. Indeed, let $\epsilon > 0$ be arbitrary. Then by (12.9-15) both

$$\chi_i < \psi_{n_k}(\tau_i) + \epsilon$$

and

$$\chi_j > \psi_{n_k}(\tau_j) - \epsilon$$

hold for all sufficiently large k, thus also

$$\chi_j - \chi_i > \psi_{n_k}(\tau_j) - \psi_{n_k}(\tau_i) - 2\epsilon \geq -2\epsilon,$$

because the functions ψ_n are nondecreasing. Because ϵ was arbitrary, $\chi_j - \chi_i \geq 0$ follows.

On I, we now define a function χ by

$$\chi(\tau) := \sup_{\tau_i \leq \tau} \chi_i.$$

Clearly, χ is nondecreasing, and $|\chi(\tau)| \leq \mu$ for all $\tau \in I$. Moreover, $\chi(\tau_i) = \chi_i$ for all i. Thus by (12.9-15)

$$\lim_{k \to \infty} \psi_{n_k}(\tau) = \chi(\tau) \qquad (12.9\text{-}16)$$

holds at all rational points τ. We next show that this in fact holds at all points τ where χ is continuous. Let τ be such a point, and let $\epsilon > 0$ be arbitrary. The set of rational points being dense, there exist two such points, τ_i and τ_j say, such that

$$\tau_i < \tau < \tau_j$$

and

$$\chi(\tau) - \epsilon < \chi(\tau_i), \qquad \chi(\tau_j) > \chi(\tau) + \epsilon.$$

By virtue of (12.9-15) we have for all sufficiently large k

$$\chi(\tau_i) - \epsilon < \psi_{n_k}(\tau_i) \leq \psi_{n_k}(\tau) \leq \psi_{n_k}(\tau_j) < \chi(\tau_j) + \epsilon.$$

The last two relations imply

$$\chi(\tau) - 2\epsilon \leq \psi_{n_k}(\tau) \leq \chi(\tau) + 2\epsilon$$

for all sufficiently large k, which is equivalent to (12.9-16).

It remains to consider the convergence of the sequence $\{\psi_{n_k}\}$ at irrational points that are not points of continuity of χ. Because χ is nondecreasing and bounded, its points of discontinuity are at most denumerable in number (see Natanson [1961], p. 229). Let $\xi_1, \xi_2, \ldots$ denote the irrational points of discontinuity of χ. Letting $\{n_k\} := \{n_{0k}\}$ we can extract subsequences $\{n_{jk}\}$

$(j = 1, 2, \ldots)$ such that the limits

$$\gamma_i := \lim_{k \to \infty} \psi_{n_{jk}}(\xi_i)$$

exist for $i = 1, 2, \ldots, j$, although they are not necessarily equal to $\chi(\xi_i)$. Again the diagonal sequence $\{\psi_{n_{kk}}\}$ converges at all points ξ_i. If we amend the definition of the function χ by setting

$$\chi(\xi_i) := \gamma_i,$$

$i = 1, 2, \ldots$, then the diagonal sequence converges to χ at all points $\tau \in I$. Because the functions $\psi_{n_{kk}}$ are nondecreasing, their limit function χ likewise is nondecreasing. This completes the proof of Theorem 12.9d. ∎

II. Representation of Generalized Value Functions

Let

$$C(z) = \frac{\alpha_1 \,|}{|\, z} + \frac{\alpha_2 \,|}{|\, 1} + \frac{\alpha_3 \,|}{|\, z} + \frac{\alpha_4 \,|}{|\, 1} + \cdots$$

be an S fraction. It was shown in §12.8 that the approximants $w_n(z)$ of $C(z)$ can be represented in the form

$$w_n(z) = \sum_{i=1}^{m} \frac{\rho_{ni}}{z + \xi_{ni}}, \qquad (12.9\text{-}17)$$

$n = 1, 2, \ldots$, where $m := [(n+1)/2]$, $\rho_{ni} > 0$, $i = 1, 2, \ldots, m$, and $0 \le \xi_{n1} < \xi_{n2} < \cdots < \xi_{nm}$. Let us define the step function

$$\psi_n(\tau) = \begin{cases} 0, & \tau \le \xi_1 \\ \rho_{n1}, & \xi_1 < \tau \le \xi_2 \\ \rho_{n1} + \rho_{n2}, & \xi_2 < \tau \le \xi_3 \\ \quad \cdots \cdots \\ \rho_{n1} + \rho_{n2} + \cdots + \rho_{nm}, & \xi_m < \tau; \end{cases} \qquad (12.9\text{-}18)$$

then by example **2**, if z is not real and negative,

$$w_n(z) = \int_0^\infty \frac{1}{z + \tau} \, d\psi_n(\tau).$$

Thus certainly for every terminating S fraction C there exists a nondecreasing function ψ on $[0, \infty)$ such that the value function of C is given by the formula

$$f(z) = \int_0^\infty \frac{1}{z + \tau} \, d\psi(\tau). \qquad (12.9\text{-}19)$$

It is next shown that a similar representation holds for the generalized value functions of any nonterminating S fraction, and thus in particular for the value function of any convergent S fraction.

THEOREM 12.9e

Let C be an S fraction, and let f be any generalized value function of C. Then there exists a nondecreasing, bounded function ψ defined on $[0, \infty)$ such that for all $z \in S$

$$f(z) = \int_0^\infty \frac{1}{z+\tau} \, d\psi(\tau). \tag{12.9-20}$$

Proof. Let $\{w_n\}$ be the sequence of approximants of C, and let $\{n_i\}$ be a sequence of indices such that

$$\lim_{i \to \infty} w_{n_i}(z) = f(z)$$

locally uniformly for $z \in S$. For each i the representation

$$w_{n_i}(z) = \int_0^\infty \frac{1}{z+\tau} \, d\psi_{n_i}(\tau)$$

holds, where ψ_n is defined by (12.9-18). The functions ψ_{n_i} are nondecreasing, and they are uniformly bounded, because by (12.8-6) their values lie between 0 and

$$\rho_{n_i 1} + \rho_{n_i 2} + \cdots + \rho_{n_i m} = \gamma_0,$$

which is independent of i. Thus by Theorem 12.9d there exist a subsequence $\{n_{1i}\}$ of the sequence $\{n_i\}$ and a nondecreasing function ψ such that

$$\lim_{i \to \infty} \psi_{n_{1i}}(\tau) = \psi(\tau) \tag{12.9-21}$$

for all $\tau \in [0, \infty)$. One now is tempted to finish the proof very quickly by appealing to Theorem 12.9c; however, this theorem does not hold for infinite intervals of integration.

Because $\int_0^\infty d\psi(\tau) = \gamma_0 < \infty$, it is clear that the integral

$$\int_0^\infty \frac{1}{z+\tau} \, d\psi(\tau)$$

exists for every $z \in S$. To prove that (12.9-20) holds with ψ defined by (12.9-21), we first show that the sequence $\{w_{n_{1i}}\}$ converges to f, uniformly on every compact subset of S. Let T be such a subset, and let $\epsilon > 0$ be given. It is to be shown that

$$\left| \int_0^\infty \frac{1}{z+\tau} d\psi_{n_{1i}}(\tau) - \int_0^\infty \frac{1}{z+\tau} d\psi(\tau) \right| < \epsilon \qquad (12.9\text{-}22)$$

for all $z \in T$ and all sufficiently large i.

Let $\tau_0 > 0$ be a point of continuity of ψ such that

$$|z + \tau| \geqslant \frac{4\gamma_0}{\epsilon}$$

for all $z \in T$ and all $\tau \geqslant \tau_0$. (The existence of τ_0 follows from the boundedness of T.) Then by (12.9-11) there holds for all $z \in T$ and all i

$$\left| \int_{\tau_0}^\infty \frac{1}{z+\tau} d\psi_{n_{1i}}(\tau) - \int_{\tau_0}^\infty \frac{1}{z+\tau} d\psi(\tau) \right|$$

$$\leqslant \int_{\tau_0}^\infty \frac{1}{|z+\tau|} d\psi_{n_{1i}}(\tau) + \int_{\tau_0}^\infty \frac{1}{|z+\tau|} d\psi(\tau)$$

$$\leqslant \frac{\epsilon}{4\gamma_0} \int_{\tau_0}^\infty d\psi_{n_{1i}}(\tau) + \frac{\epsilon}{4\gamma_0} \int_{\tau_0}^\infty d\psi(\tau)$$

$$\leqslant \frac{\epsilon}{2}.$$

To prove (12.9-22) it remains to be shown that

$$\left| \int_0^{\tau_0} \frac{1}{z+\tau} d\psi_{n_{1i}}(\tau) - \int_0^{\tau_0} \frac{1}{z+\tau} d\psi(\tau) \right| < \frac{\epsilon}{2}$$

for all $z \in T$ and all sufficiently large i. Because the interval of integration now is finite, and because $(z+\tau)^{-1}$ as a function of τ has a uniform modulus of continuity for $\tau \in [0, \tau_0]$ and $z \in T$, this follows from Theorem 12.9c.

It has now been shown that $w_{n_{1i}} \to f$, uniformly on every compact subset of S. However, because by hypothesis the sequence $\{w_{n_i}\}$ itself converges, $w_{n_i} \to f$. The generalized value function defined by the subsequence $\{w_{n_i}\}$ thus has the representation (12.9-20), as was to be shown. ∎

If ψ is any nondecreasing real function, we call the **point of increase** of ψ any ξ such that ψ is not constant in any interval $[\xi - \epsilon, \xi + \epsilon]$ where $\epsilon > 0$. If ξ is an isolated point of increase, then there exists $\epsilon > 0$ such that ψ is constant in each of the intervals $(\xi - \epsilon, \xi)$ and $(\xi, \xi + \epsilon)$. Thus if an increasing function has only finitely many points of increase, it is a step function with only a finite number of steps, such as the function defined by (12.9-18). We conclude that the function ψ occurring in the representation (12.9-20) cannot have a mere finite number of points of increase, for otherwise f would be rational, and therefore could not be represented by a nonterminating S fraction.

COROLLARY 12.9f

The function ψ occurring in the representation (12.9-20) *has infinitely many points of increase.*

III. Stieltjes Transforms

Let Ψ denote the class of all real, nondecreasing, bounded functions defined on $[0, \infty)$ having infinitely many points of increase. It was shown in the proof of Theorem 12.9e that for every $\psi \in \Psi$ and all z in the cut plane S the integral

$$f(z) := \int_0^\infty \frac{d\psi(\tau)}{z + \tau} \qquad (12.9\text{-}23)$$

exists. The function f thus defined is called the **Stieltjes transform** of ψ; it is denoted by

$$f = \mathcal{S}\psi.$$

Examples of Stieltjes transforms abound in classical analysis.

EXAMPLE **4**

Let $\psi(\tau) := \tau, 0 \leqslant \tau \leqslant 1; \psi(\tau) := 1, \tau \geqslant 1$. Then $f = \mathcal{S}\psi$ is the function

$$f(z) = \int_0^1 \frac{1}{z + \tau} \, d\tau = \text{Log}\left(1 + \frac{1}{z}\right).$$

EXAMPLE **5**

For $\psi(\tau) := \sqrt{\tau}, 0 \leqslant \tau \leqslant 1; \psi(\tau) := 1, \tau \geqslant 1$ we get by Example **2**

$$f(z) = \frac{1}{2} \int_0^1 \frac{\tau^{-1/2}}{z + \tau} \, d\tau = \frac{1}{\sqrt{z}} \text{Arctan} \frac{1}{\sqrt{z}}.$$

EXAMPLE **6**

Exponential integral is the name given to the function E defined for $z \in S$ by

$$E(z) := \int_z^\infty \frac{e^{-t}}{t} \, dt,$$

the path of integration being a straight line parallel to the real axis. Setting $t = z + \tau$ yields

$$e^z E(z) = \int_0^\infty \frac{e^{-\tau}}{z + \tau} \, d\tau,$$

the Stieltjes transform of $\psi(\tau) := 1 - e^{-\tau}$.

A number of additional examples of Stieltjes transforms are discussed in §12.12.

THEOREM 12.9g

The Stieltjes transform of any function $\psi \in \Psi$ assumes positive values on the positive real axis and is analytic in the cut plane S.

Proof. Only the second statement requires proof. Let $\psi \in \Psi$. We first show that for every $\beta > 0$ the function

$$f(\beta; z) := \int_0^\beta \frac{1}{z+\tau} d\psi(\tau)$$

is analytic in S. By the definition of the Stieltjes integral, $f(\beta; z)$ is the limit as $n \to \infty$ of the functions

$$f_n(\beta; z) := \sum_{k=1}^n \frac{\psi(\tau_k) - \psi(\tau_{k-1})}{z+\tau_k},$$

where $\tau_k := k/n$, $k = 0, 1, \ldots, n$. Each f_n is rational, with poles only on the negative real axis, and thus is clearly analytic in S. To show that f is analytic, it suffices by Theorem 3.4b that $f_n \to f$ uniformly on every compact subset $T \subset S$. Indeed we have

$$f(\beta; z) - f_n(\beta; z) = \sum_{k=1}^n \int_{\tau_{k-1}}^{\tau_k} \left(\frac{1}{z+\tau} - \frac{1}{z+\tau_k} \right) d\psi(\tau)$$

$$= \sum_{k=1}^n \int_{\tau_{k-1}}^{\tau_k} \frac{\tau_k - \tau}{(z+\tau)(z+\tau_k)} d\psi(\tau);$$

hence by (12.9-11), if $\delta > 0$ denotes the distance of T from the negative real axis,

$$|f(\beta; z) - f_n(\beta; z)| \leq \frac{\beta}{n\delta^2} \int_0^\beta d\psi(\tau).$$

The expression on the right does not depend on z and tends to zero for $n \to \infty$, proving uniform convergence. We next show that $f(\beta; z) \to f(z)$ for $\beta \to \infty$ uniformly in z on every compact subset $T \subset S$. This simply follows from

$$|f(z) - f(\beta; z)| = \left| \int_\beta^\infty \frac{1}{z+\tau} d\psi(\tau) \right| \leq \frac{1}{\delta} \int_\beta^\infty d\psi(\tau),$$

where δ has the same meaning as before. As a uniform limit of analytic functions, f is itself analytic. ∎

Theorem 12.9e shows that every generalized value function is a Stieltjes transform. From Theorem 12.8c we remember that every generalized value function of an S fraction admits an asymptotic power series as $z \to \infty$, $z > 0$. It is now shown that this latter property imposes a severe restriction on those $\psi \in \Psi$ whose Stieltjes transforms can be generalized value functions.

THEOREM 12.9h

Let $\psi \in \Psi$, let $f := \mathcal{S}\psi$, and let P be a formal power series,

$$P = \frac{\gamma_0}{z} + \frac{\gamma_1}{z^2} + \frac{\gamma_2}{z^3} + \cdots,$$

such that

$$f(z) \approx P(z), \qquad z \to \infty, \qquad z > 0. \tag{12.9-24}$$

Then the integrals

$$\mu_k := \int_0^\infty \tau^k \, d\psi(\tau), \qquad k = 0, 1, 2, \ldots, \tag{12.9-25}$$

all exist; moreover,

$$\mu_k = (-1)^k \gamma_k, \qquad k = 0, 1, 2, \ldots. \tag{12.9-26}$$

The integrals μ_k are called the **moments** of the function ψ. If $d\psi(\tau)$ is interpreted as a mass distribution on the positive real line, the numbers μ_1 and μ_2 are identical with the first and the second moment of the mass distribution, as considered in mechanics.

Proof of Theorem 12.9h.   We prove the following assertion (A_m) by induction with respect to m: The moments μ_k exist for $k = 0, 1, \ldots, m$ and are equal to $(-1)^k \gamma_k$ for $k = 0, 1, \ldots, m-1$. Clearly, (A_0) is true, because the boundedness of ψ implies that μ_0 exists. Assuming the truth of (A_m) for some integer $m \geq 0$, we thus have to show that

$$(\alpha) \ \mu_m = (-1)^m \gamma_m; \qquad (\beta) \ \mu_{m+1} \text{ exists.}$$

Using

$$\frac{1}{z+\tau} = \frac{1}{z} + \frac{-\tau}{z^2} + \frac{(-\tau)^2}{z^3} + \cdots + \frac{(-\tau)^{m-1}}{z^m} + \frac{(-\tau)^m}{z^m(z+\tau)}$$

in (12.9-23), we obtain by the induction hypothesis

$$f(z) = \frac{\gamma_0}{z} + \frac{\gamma_1}{z^2} + \cdots + \frac{\gamma_{m-1}}{z^m} + \frac{1}{z^m} \int_0^\infty \frac{(-\tau)^m}{z+\tau} \, d\psi(\tau).$$

On the other hand, (12.9-24) by the definition of an asymptotic expansion implies that for $z > 0$

$$f(z) = \frac{\gamma_0}{z} + \frac{\gamma_1}{z^2} + \cdots + \frac{\gamma_{m-1}}{z^m} + \frac{\gamma_m}{z^{m+1}} + \frac{\kappa_{m+1}(z)}{z^{m+2}},$$

where $\kappa_{m+1}(z)$ is bounded for $z \to \infty$. Comparing the two expressions for $f(z)$ yields

$$\int_0^\infty \frac{z(-\tau)^m}{z + \tau}\, d\psi(\tau) = \gamma_m + \frac{\kappa_{m+1}(z)}{z}$$

or, because μ_m exists,

$$\int_0^\infty (-\tau)^m\, d\psi(\tau) + \int_0^\infty \frac{(-\tau)^{m+1}}{z + \tau}\, d\psi(\tau) = \gamma_m + O(z^{-1}). \quad (12.9\text{-}27)$$

Here we let $z \to \infty$. The limit on the right exists and equals γ_m; we shall have proved (α) if it is shown that

$$\lim_{z \to \infty} \int_0^\infty \frac{(-\tau)^{m+1}}{z + \tau}\, d\psi(\tau) = 0. \quad (12.9\text{-}28)$$

Because μ_m exists, given any $\epsilon > 0$ there exists σ such that

$$\left| \int_\sigma^\infty (-\tau)^m\, d\psi(\tau) \right| < \frac{\epsilon}{2}.$$

Then for every $z > 0$

$$\left| \int_\sigma^\infty \frac{(-\tau)^{m+1}}{z + \tau}\, d\psi(\tau) \right| < \frac{\epsilon}{2}.$$

Letting

$$\kappa := \left| \int_0^\sigma (-\tau)^{m+1}\, d\psi(\tau) \right|,$$

we have for $z > 2\kappa\epsilon^{-1}$

$$\left| \int_0^\sigma \frac{(-\tau)^{m+1}}{z + \tau}\, d\psi(\tau) \right| < \frac{1}{z} \left| \int_0^\sigma (-\tau)^{m+1}\, d\psi(\tau) \right| \leq \frac{\epsilon}{2};$$

hence

$$\left| \int_0^\infty \frac{(-\tau)^{m+1}}{z + \tau}\, d\psi(\tau) \right| < \epsilon,$$

establishing (12.9-28) and hence (α).

To establish (β) we note that (12.9-27) now implies

$$\int_0^\infty \frac{z(-\tau)^{m+1}}{z+\tau}\, d\psi(\tau) = \kappa_{m+1}(z).$$

Thus if $\sigma > 0$ is arbitrary and $z > \sigma$,

$$\int_0^\sigma \tau^{m+1}\, d\psi(\tau) \leqslant 2 \int_0^\sigma \frac{z\tau^{m+1}}{z+\tau}\, d\psi(\tau) \leqslant 2\kappa_{m+1}(\sigma),$$

where $\kappa_{m+1}(\sigma) := \sup_{z>\sigma}|\kappa_{m+1}(z)|$. The integral

$$\int_0^\sigma \tau^{m+1}\, d\psi(\tau)$$

thus is bounded by a quantity that has a finite limit as $\sigma \to \infty$, showing that the improper integral μ_{m+1} exists.

On the basis that (A_m) is true, we thus have established the truth of (A_{m+1}). Because (A_0) is true, it follows that (A_m) is true for all $m \geqslant 0$. ■

If $0 \leqslant \alpha < \pi$, we continue to denote by $\hat{S}_\alpha$ the set $\{z : |\arg z| \leqslant \alpha\}$, a closed wedge of opening 2α symmetric about the real axis.

COROLLARY 12.9i

Under the hypotheses of Theorem 12.9h, the asymptotic expansion (12.9-24) holds not only as $z \to \infty$ on the positive real line, but also, for every $\alpha \in (0, \pi)$, if $z \in \hat{S}_\alpha$.

Proof. Using once more the formula for the terminating geometric series, we have for $m = 0, 1, \ldots$ and for $z \in S$, using (12.9-26),

$$f(z) = \int_0^\infty \frac{1}{z+\tau}\, d\psi(\tau)$$

$$= \frac{\gamma_0}{z} + \frac{\gamma_1}{z^2} + \cdots + \frac{\gamma_{m-1}}{z^m} + \frac{1}{z^{m+1}}r_m(z),$$

where

$$r_m(z) := z \int_0^\infty \frac{(-\tau)^m}{z+\tau}\, d\psi(\tau).$$

If $z \in \hat{S}_\alpha$, then $|z+\tau| \geqslant |z| \sin[\max(\alpha, \pi/2)]$; hence

$$|r_m(z)| \leqslant \frac{\mu_m}{\sin[\max(\alpha, \pi/2)]},$$

proving the desired result. ■

EXAMPLE 7

If, as in Example **6**, $d\psi(\tau) = e^{-\tau} d\tau$, then

$$\mu_k = \int_0^\infty \tau^k e^{-\tau} d\tau = k!, \qquad k = 0, 1, 2, \ldots.$$

Thus for $z \to \infty$, $z \in \hat{S}_\alpha$,

$$E(z) = e^{-z} \int_0^\infty \frac{e^{-\tau}}{z + \tau} d\tau \approx e^{-z} \left\{ \frac{0!}{z} - \frac{1!}{z^2} + \frac{2!}{z^3} - \cdots \right\},$$

as found already in §11.1.

The gist of what has been accomplished in the present section may be stated briefly as follows: Let *GVF* denote the set of all functions that are generalized value functions of a nonterminating Stieltjes fraction. Let Ψ^* denote the subset of those $\psi \in \Psi$ for which all moments (12.9-25) exist, and by *ST* the set of all Stieltjes transforms of functions $\psi \in \Psi^*$. Then the Theorems 12.8c, 12.9e and 12.9h imply that

$$GVF \subset ST. \tag{12.9-29}$$

The question arises whether this inclusion is proper; it is seen in §12.11 that it is. Moreover, one may ask for a characterization of the set *VF* of value functions of *convergent* Stieltjes fractions. No simple characterization seems to exist. However, we identify (by means of the Carleman condition; see subsection III of §12.11) a large subclass $ST^* \subset ST$ of functions that belong to *VF* and hence can be represented by convergent *S* fractions.

PROBLEMS

1. Let $\alpha > 0$. By solving the difference equations for the numerators and the denominators, determine the approximants of the *S* fraction

$$C(z) = \frac{1}{|z|} + \frac{\alpha/4}{|1|} + \frac{\alpha/4}{|z|} + \frac{\alpha/4}{|1|} + \cdots,$$

showing that

$$\frac{p_n(z)}{q_n(z)} = \frac{2}{\sqrt{z}} \frac{(\sqrt{z} + \sqrt{z+\alpha})^n - (\sqrt{z} - \sqrt{z+\alpha})^n}{(\sqrt{z} + \sqrt{z+\alpha})^{n+1} - (\sqrt{z} - \sqrt{z+\alpha})^{n+1}}. \tag{12.9-30}$$

2. Determine the partial fraction decomposition of (12.9-30), and by letting $n \to \infty$ and using the definition of the Stieltjes integral show that the value function of *C* is

$$f(z) = \frac{2}{\alpha\pi} \int_0^\alpha \frac{\sqrt{(\alpha - \tau)/\tau}}{z + \tau} d\tau = \frac{2}{z + \sqrt{z^2 + \alpha z}}, \qquad z \in S.$$

§12.10. POSITIVE SYMMETRIC FUNCTIONS AND THEIR REPRESENTATION AS STIELTJES TRANSFORMS

In this section we characterize the class ST in a more intrinsic, less computational manner.

Every function $f \in ST$, in addition to being analytic (but not rational) in the set $S := \{z : |\arg z| < \pi\}$, possesses the following three properties:

(i) $f(x) > 0$ for $x > 0$;

(ii) if U and L, respectively, denote the upper and the lower half plane, then
$$f(L) \subset U$$
[and consequently by (i) and the reflection principle, $f(U) \subset L$];

(iii) f admits an asymptotic expansion in negative powers of z (beginning with z^{-1}) as $z \to \infty$, $z \in \hat{S}_{\pi-\epsilon}$, for every $\epsilon > 0$.

A function enjoying these properties is called a (nonrational) **positive symmetric function**, and the set of all such functions is denoted by PSA. It is our goal to show that $PSA = ST$. Some preliminary work is required.

I. Representations of Functions with Values in a Half Plane

We denote by $\mathcal{H}$ the class of all functions g that are analytic in the unit disk $D : |w| < 1$, and satisfy Re $g(w) > 0$ for all $w \in D$. We then have

THEOREM 12.10a (Herglotz representation theorem)

For every $g \in \mathcal{H}$ there exists a nondecreasing, bounded real function σ on $[-\pi, \pi]$ such that
$$g(w) = i \operatorname{Im} g(0) + \int_{-\pi}^{\pi} \frac{e^{i\theta} + w}{e^{i\theta} - w} \, d\sigma(\theta) \qquad (12.10\text{-}1)$$
for all $w \in D$.

The Herglotz representation theorem is remarkable for the fact that it provides a Cauchy-like representation of a function analytic in a disk, assuming solely that its values lie in the right half plane but making no hypotheses on the boundary behavior of the function.

Proof. We require a formula that under strong regularity assumptions represents the values of an analytic function g in a disk in terms of the values of the real part of g on the boundary of the disk.

LEMMA 12.10b (Schwarz formula)

Let g be analytic in a region containing the disk $|w| \leq \rho$, and let $h := $ Re g. Then for all w such that $|w| < \rho$.
$$g(w) = -i\beta + \frac{1}{2\pi} \int_{-\pi}^{\pi} \frac{\rho e^{i\theta} + w}{\rho e^{i\theta} - w} h(\rho e^{i\theta}) \, d\theta, \qquad (12.10\text{-}2)$$
where $\beta := -\operatorname{Im} g(0)$.

The Schwarz formula is best understood in the larger context of potential theory (see Chapter 14), where it is proved under weaker hypotheses. For our present needs the following *verification* suffices. The hypothesis implies that the Taylor series of g at 0,

$$g(w) = \sum_{n=0}^{\infty} a_n w^n, \qquad (12.10\text{-}3)$$

converges uniformly for $|w| \leq \rho$. There follows

$$h(\rho e^{i\theta}) = \operatorname{Re} a_0 + \frac{1}{2} \sum_{n=1}^{\infty} (a_n \rho^n e^{in\theta} + \bar{a}_n \rho^n e^{-in\theta}).$$

For $|w| < \rho$,

$$\frac{\rho e^{i\theta} + w}{\rho e^{i\theta} - w} = 1 + 2 \sum_{n=1}^{\infty} \rho^{-n} e^{-in\theta} w^n.$$

Substituting both expansions into (12.10-2), multiplying, integrating term by term, and adding $-i\beta$, we get (12.10-3). ∎

Under the hypotheses of Theorem 12.10a, the Schwarz formula is applicable for $\rho < 1$. Letting

$$\sigma_\rho(\theta) := \frac{1}{2\pi} \int_{-\pi}^{\theta} h(\rho e^{i\theta}) \, d\theta,$$

it may be written as

$$g(w) = -i\beta + \int_{-\pi}^{\pi} \frac{\rho e^{i\theta} + w}{\rho e^{i\theta} - w} \, d\sigma_\rho(\theta), \qquad (12.10\text{-}4)$$

where the function σ_ρ is nondecreasing by virtue of the positivity of h and bounded by

$$\frac{1}{2\pi} \int_{-\pi}^{\pi} h(\rho e^{i\theta}) \, d\theta = \operatorname{Re} g(0),$$

independently of ρ. Let $\{\rho_n\}$ be any sequence of real numbers tending to 1 from below. Applying Helly's selection principle (Theorem 12.9d) to the sequence $\{\sigma_{\rho_n}\}$, we can extract a subsequence converging to a nondecreasing function σ. We denote the subsequence again by $\{\rho_n\}$. Replacing ρ by ρ_n in (12.10-4), using the fact that

$$\frac{\rho_n e^{i\theta} + w}{\rho_n e^{i\theta} - w} \to \frac{e^{i\theta} + w}{e^{i\theta} - w}$$

uniformly in θ for fixed $|w| < 1$, and applying Helly's theorem 12.9c, we obtain the desired representation (12.10-1). ∎

We next transform the Herglotz formula into a representation for functions analytic in U and having their values in U. This may be accomplished by means of the Moebius transformation

$$t : z \to w := \frac{i-z}{i+z}$$

mapping U onto the unit disk. If f is analytic and $f(U) \subset U$, then the function

$$g(w) := -if(t^{[-1]}(w)) = -if\left(i\frac{1-w}{1+w}\right)$$

is in $\mathcal{H}$ and therefore has a representation of the form (12.10-1). It follows that for $z \in U$,

$$f(z) = ig\left(\frac{i-z}{i+z}\right)$$

$$= \beta + i \int_{-\pi}^{\pi} \frac{(i+z)\,e^{i\theta} + i - z}{(i+z)\,e^{i\theta} - i + z}\, d\sigma(\theta)$$

$$= \beta + \int_{-\pi}^{\pi} \frac{1 + z\,\tan(\theta/2)}{\tan(\theta/2) - z}\, d\sigma(\theta).$$

Here we substitute $\tau := \tan(\theta/2)$, which by changing the range of integration to $(-\infty, \infty)$ makes the integral improper. A possible contribution from the points $\theta = \pm\pi$, because not necessarily do $\sigma(-\pi+) = \sigma(-\pi)$ and $\sigma(\pi-) = \sigma(\pi)$, has to be accounted for separately, yielding a contribution αz where $\alpha \geq 0$. Setting $\chi(\tau) := \sigma(2\,\mathrm{Arctan}\,\tau)$ we thus get the **Nevanlinna representation formula** for any function f analytic in U and having its values in U:

$$f(z) = \alpha z + \beta + \int_{-\infty}^{\infty} \frac{1 + \tau z}{\tau - z}\, d\chi(\tau). \qquad (12.10\text{-}5)$$

Here α, β are real, $\alpha \geq 0$, and χ is nondecreasing and bounded.

Now let f be an analytic function from L to U. By the above, $f(-z)$ has a representation of the form (12.10-5). Thus there exists on $(-\infty, \infty)$ a nondecreasing and bounded function χ such that

$$f(z) = -\alpha z + \beta + \int_{-\infty}^{\infty} \frac{1 - \tau z}{\tau + z}\, d\chi(\tau) \qquad (12.10\text{-}6)$$

holds for all $z \in L$.

Of the properties characterizing PSA, only (ii) has been used thus far. We now simplify further by using (iii) in the very weak form that

$$\lim_{z \to \infty} zf(z)$$

exists, where z approaches ∞ through pure imaginary values. This implies the existence of $\gamma > 0$ such that

$$|yf(-iy)| \leq \gamma \tag{12.10-7}$$

for all $y \geq 1$, say. From (12.10-6) there follows,

$$f(-iy) = i\alpha y + \beta + \int_{-\infty}^{\infty} \frac{1 + i\tau y}{\tau - iy} \, d\chi(\tau),$$

and therefore

$$\text{Re}\{yf(-iy)\} = \beta y + y \int_{-\infty}^{\infty} \frac{\tau(1 - y^2)}{\tau^2 + y^2} \, d\chi(\tau),$$

$$\text{Im}\{yf(-iy)\} = \alpha y^2 + y^2 \int_{-\infty}^{\infty} \frac{1 + \tau^2}{\tau^2 + y^2} \, d\chi(\tau).$$

By (12.10-7), $\text{Im}\{yf(-iy)\} \leq \gamma \int_{-\infty}^{\infty} (1 + \tau^2) \, d\chi(\tau) \leq \gamma$, which implies $\alpha = 0$ and

$$y^2 \int_{-\infty}^{\infty} \frac{1 + \tau^2}{y^2 + \tau^2} \, d\chi(\tau) \leq \gamma \tag{12.10-8}$$

for all $y \geq 1$. This in turn implies

$$\int_{-\infty}^{\infty} (1 + \tau^2) \, d\chi(\tau) \leq \gamma, \tag{12.10-9}$$

for if this inequality were untrue, then we would have

$$\int_{-\mu}^{\mu} (1 + \tau^2) \, d\chi(\tau) = \gamma + \epsilon > \gamma$$

for some $\mu > 0$, and therefore

$$\int_{-\mu}^{\mu} \frac{1 + \tau^2}{1 + y^{-2}\tau^2} \, d\chi(\tau) \geq \gamma + \frac{\epsilon}{2}$$

for sufficiently large y, contradicting (12.10-8). Again by (12.10-7), $\text{Re}\{yf(-iy)\} \leq \gamma$ for $y \geq 1$. Therefore

$$\int_{-\infty}^{\infty} \frac{\tau(1 - y^2)}{\tau^2 + y^2} \, d\chi(\tau) + \beta = O(y^{-1}), \qquad y \to \infty,$$

which by a similar estimation shows that

$$\beta = \int_{-\infty}^{\infty} \tau \, d\chi(\tau).$$

Thus (12.10-6) turns into

$$f(z) = \int_{-\infty}^{\infty} \left(\tau + \frac{1-\tau z}{\tau + z} \right) d\chi(\tau) = \int_{-\infty}^{\infty} \frac{1+\tau^2}{\tau + z} d\chi(\tau)$$

Defining a new bounded, nondecreasing function ψ by

$$\psi(\tau) := \int_{-\infty}^{\tau} (1+\tau^2) \, d\chi(\tau),$$

this appears as

$$f(z) = \int_{-\infty}^{\infty} \frac{1}{\tau + z} \, d\psi(\tau). \tag{12.10-10}$$

We thus have proved:

THEOREM 12.10c (Hamburger's representation)

Let f be an analytic function from L to U, and let $yf(-iy) = O(1)$ as $y \to \infty$. Then there exists a bounded, nondecreasing function ψ on $(-\infty, \infty)$ such that the representation (12.10-10) holds for all $z \in L$.

The integral (12.10-10) defines a function, f_1 say, also for $z \in U$. This function is related to f by the formula

$$f_1(z) = \overline{f(\bar{z})}, \qquad z \in U.$$

However, unless ψ happens to be constant on some piece of the real axis, the functions f and f_1 are not necessarily analytic continuations of one and the same function.

II. The Stieltjes–Perron Inversion Formula

It is essential to know to what extent the function ψ in Hamburger's representation (12.10-10) is determined by f. It is clear for two reasons that ψ is not uniquely determined. First, the very definition of the Stieltjes integral already shows that two functions ψ that merely differ by a constant yield the same function f. Second, ψ may have discontinuities, even infinitely many. If τ is a point of discontinuity, the value of ψ at τ has no influence on an integral such as (12.10-10). What matters are only the one-sided limits $\psi(\tau+)$ and $\psi(\tau-)$, which both exist, because ψ is nondecreasing.

The following theorem shows that apart from the above provisos ψ is indeed uniquely determined by f.

THEOREM 12.10d (Stieltjes–Perron inversion formula)

Let ψ be a bounded, nondecreasing real function defined on $(-\infty, \infty)$, and let f be defined by

$$f(z) := \int_{-\infty}^{\infty} \frac{1}{z + \tau} \, d\psi(\tau), \qquad z \in L. \qquad (12.10\text{-}11)$$

Then for arbitrary real σ and τ

$$\tfrac{1}{2}\{\psi(\tau+) + \psi(\tau-)\} - \tfrac{1}{2}\{\psi(\sigma+) + \psi(\sigma-)\}$$
$$= \frac{1}{\pi} \lim_{\eta \to 0+} \operatorname{Im} \int_{-\tau}^{-\sigma} f(\lambda - i\eta) \, d\lambda. \qquad (12.10\text{-}12)$$

The proof requires

LEMMA 12.10e

If ψ satisfies the conditions of the previous theorem, then

$$\lim_{\eta \to 0+} \int_{0}^{\infty} \operatorname{Arctan} \frac{\xi}{\eta} \, d\psi(\xi) = \frac{\pi}{2} [\psi(\infty) - \psi(0+)]. \qquad (12.10\text{-}13)$$

Proof. Assuming $0 < \eta < 1$, we subdivide the interval of integration as follows:

$$\int_{0}^{\infty} = \int_{0}^{\eta^2} + \int_{\eta^2}^{\sqrt{\eta}} + \int_{\sqrt{\eta}}^{\infty}.$$

Because $0 \leqslant \operatorname{Arctan} \xi/\eta \leqslant \xi/\eta$, we have

$$0 \leqslant \int_{0}^{\eta^2} \operatorname{Arctan} \frac{\xi}{\eta} \, d\psi(\xi) \leqslant \eta[\psi(\eta^2) - \psi(0)];$$

hence [although possibly $\psi(0+) \neq \psi(0)$]

$$\lim_{\eta \to 0+} \int_{0}^{\eta^2} \operatorname{Arctan} \frac{\xi}{\eta} \, d\psi(\xi) = 0.$$

It thus suffices to show that

$$\lim_{\eta \to 0+} \int_{\eta^2}^{\infty} \operatorname{Arctan} \frac{\xi}{\eta} \, d\psi(\tau) = \frac{\pi}{2} [\psi(\infty) - \psi(0+)]$$

or, because

$$\psi(\infty) - \psi(0+) = \lim_{\eta \to 0+} \int_{\eta^2}^{\infty} d\psi(\xi),$$

that

$$\lim_{\eta \to 0+} \int_{\eta^2}^{\infty} \left[\frac{\pi}{2} - \text{Arctan} \frac{\xi}{\eta} \right] d\psi(\xi) = 0. \qquad (12.10\text{-}14)$$

Because $\lim \psi(\sqrt{\eta}) = \lim \psi(\eta^2) = \psi(0+)$,

$$0 \le \int_{\eta^2}^{\sqrt{\eta}} \left[\frac{\pi}{2} - \text{Arctan} \frac{\xi}{\eta} \right] d\psi(\xi) \le \frac{\pi}{2} \int_{\eta^2}^{\sqrt{\eta}} d\psi(\xi) = \frac{\pi}{2} [\psi(\sqrt{\eta}) - \psi(\eta^2)] \to 0.$$

On the other hand,

$$0 \le \int_{\sqrt{\eta}}^{\infty} \left[\frac{\pi}{2} - \text{Arctan} \frac{\xi}{\eta} \right] d\psi(\xi) \le \left[\frac{\pi}{2} - \text{Arctan} \frac{1}{\sqrt{\eta}} \right] \int_{\sqrt{\eta}}^{\infty} d\psi(\xi) \to 0,$$

because

$$\lim_{\eta \to 0+} \text{Arctan} \frac{1}{\sqrt{\eta}} = \frac{\pi}{2} \quad \text{and} \quad \int_{\sqrt{\eta}}^{\infty} d\psi(\xi) \le \pi(\infty) - \psi(0) < \infty.$$

This proves (12.10-14), hence the lemma. ∎

To prove Theorem 12.10d, we assume $\sigma < \tau$ without loss of generality and have, by the definition of f,

$$\int_{-\tau}^{-\sigma} f(\lambda - i\eta) \, d\lambda = \int_{\sigma}^{\tau} f(-\lambda - i\eta) \, d\lambda = \int_{\sigma}^{\tau} \left[\int_{-\infty}^{\infty} \frac{1}{\xi - \lambda - i\eta} \, d\psi(\xi) \right] d\lambda.$$

In the last integral it is permissible to change the order of integrations, because the inner integral is uniformly convergent (for fixed $\eta > 0$) with respect to λ. We then get, expressing the imaginary part of the logarithms by the principal value of the Arctan function,

$$\text{Im} \int_{-\tau}^{-\sigma} f(\lambda - i\eta) \, d\lambda$$

$$= \text{Im} \int_{-\infty}^{\infty} [\text{Log}(\xi - \sigma - i\eta) - \text{Log}(\xi - \tau - i\eta)] \, d\psi(\xi)$$

$$= \int_{-\infty}^{\infty} \text{Arctan} \frac{\xi - \sigma}{\eta} \, d\psi(\xi) - \int_{-\infty}^{\infty} \text{Arctan} \frac{\xi - \tau}{\eta} \, d\psi(\xi).$$

In view of

$$\int_{-\infty}^{\infty} \text{Arctan} \frac{\xi - \sigma}{\eta} \, d\psi(\xi) = \int_{-\infty}^{\sigma} + \int_{\sigma}^{\infty}$$

$$= \int_{0}^{\infty} \text{Arctan} \frac{\xi}{\eta} \, d\psi(\xi + \sigma) + \int_{0}^{\infty} \text{Arctan} \frac{\xi}{\eta} \, d\psi(\sigma - \xi)$$

we find by two applications of Lemma 12.10e

$$\lim_{\eta \to 0+} \int_{-\infty}^{\infty} \text{Arctan} \frac{\xi - \sigma}{\eta} d\psi(\xi) = \frac{\pi}{2} [\psi(\infty) + \psi(-\infty) - \{\psi(\sigma+) + \psi(\sigma-)\}].$$

A similar result holds when σ is replaced by τ. Subtracting the two formulas we obtain (12.10-12). ■

Let ψ_1 and ψ_2 be two nondecreasing bounded functions such that, for all $z \in L$,

$$f(z) = \int_{-\infty}^{\infty} \frac{1}{\tau + z} d\psi_1(\tau) = \int_{-\infty}^{\infty} \frac{1}{\tau + z} d\psi_2(\tau). \qquad (12.10\text{-}15)$$

Theorem 12.10d then shows that for all real σ and τ,

$$\tfrac{1}{2}[\psi_1(\tau+) + \psi_1(\tau-)] - \tfrac{1}{2}[\psi_1(\sigma+) + \psi_1(\sigma-)]$$
$$= \tfrac{1}{2}[\psi_2(\tau+) + \psi_2(\tau-)] - \tfrac{1}{2}[\psi_2(\sigma+) + \psi_2(\sigma-)].$$

Here we let $\sigma \to -\infty$. Because the functions ψ_k are monotonic and bounded, the limits of $\psi_k(\sigma)$, $\psi_k(\sigma+)$, $\psi_k(\sigma-)$ exist and are equal for each k. Calling the common limit $\psi_k(-\infty)$, we now normalize the ψ_k so that $\psi_k(-\infty) = 0$, $k = 1, 2$. Then the foregoing relation yields

$$\tfrac{1}{2}[\psi_1(\tau+) + \psi_1(\tau-)] = \tfrac{1}{2}[\psi_2(\tau+) + \psi_2(\tau-)]$$

for all real τ. Thus at every τ where both ψ_1 and ψ_2 are continuous,

$$\psi_1(\tau) = \psi_2(\tau).$$

Because both ψ_1 and ψ_2 can be discontinuous only on a denumerable set, the points where both ψ_1 and ψ_2 are continuous are everywhere dense. It follows that for *all* real τ

$$\psi_1(\tau-) = \psi_2(\tau-).$$

Thus if the functions ψ_k are further standardized such that $\psi_k(\tau-) = \psi_k(\tau)$ at all points of discontinuity, then $\psi_1(\tau) = \psi_2(\tau)$ for all τ. We call a nondecreasing, bounded function ψ **normalized** if $\psi(-\infty) = 0$ and $\psi(\tau) = \psi(\tau-)$ for all τ. In this terminology we have obtained the following uniqueness result:

THEOREM 12.10f.

Let ψ_1 and ψ_2 be two normalized, nondecreasing, bounded functions such that (12.10-15) holds for all $z \in L$. Then $\psi_1 = \psi_2$.

We refer to this result briefly by saying that in Hamburger's representation (12.10-10) the function ψ is **essentially uniquely** determined by f.

III. Representation of *PSA* Functions as Stieltjes Transforms

Now let f be a function of the class *PSA*. It will then satisfy the hypothesis of Theorem 12.10c, and therefore possess a representation of the form

$$f(z) = \int_{-\infty}^{\infty} \frac{1}{\tau + z} \, d\psi(\tau),$$

where ψ is nondecreasing and bounded. Of the three properties characterizing *PSA*, this uses only (ii) and part of (iii). We now use (i) : $f(x) > 0$ (and thus real) for $x > 0$. Let $0 < \sigma < \tau$. By the Stieltjes–Perron inversion formula we then have

$$\tfrac{1}{2}[\psi(-\sigma-) + \psi(-\sigma+)] - \tfrac{1}{2}[\psi(-\tau-) + \psi(-\tau+)] = \frac{1}{\pi} \lim_{.\eta \to 0+} \operatorname{Im} \int_{\sigma}^{\tau} f(\lambda - i\eta) \, d\lambda.$$

Being analytic in the cut plane S, f is uniformly continuous on compact subsets of S. Therefore the last limit is zero. It follows that ψ is constant on the negative real line. If $\psi(0) = \psi(0-)$, f therefore has the representation

$$f(z) = \int_0^{\infty} \frac{1}{\tau + z} \, d\psi(\tau). \tag{12.10-16}$$

Using the full property (iii), we conclude by Theorem 12.9h that all moments of ψ exist. Thus f belongs to the class *ST*.

THEOREM 12.10g

The classes ST and PSA are identical. That is, every nonrational analytic function in S possessing the properties (i), (ii), (iii) *stated at the beginning of this section is the Stieltjes transform of a bounded, nondecreasing function ψ with infinitely many points of increase, all of whose moments exist.*

PROBLEMS

1. Let $f(z) := i\gamma$ where $\gamma > 0$. Show that the Nevanlinna representation of this constant function is

   $$f(z) = \int_{-\infty}^{\infty} \frac{1 + \tau z}{\tau - z} \, d\chi(\tau),$$

   where

   $$\chi(\tau) := \frac{\gamma}{\pi} \operatorname{Arctan} \tau.$$

2. The totality of functions f analytic in U such that $f(U) \subset U$ and the values of f on the imaginary axis are pure imaginary is identical with the totality of all functions

   $$f(z) = \alpha z + z \int_0^{\infty} \frac{1}{\tau^2 - z^2} \, d\psi(\tau),$$

where $\alpha \geq 0$ and ψ is nondecreasing and bounded. In this representation, $\alpha > 0$ if and only if $f(iy)$ is unbounded as $y \to \infty$.

3. Let $\mathcal{A}$ denote the class of all functions f such that
    (a) f is analytic in the left half plane, and its values lie in the unit disk;
    (b) $f(z)$ is real for z real;
    (c) $f(z) \approx \beta_0 + \beta_1 z + \beta_2 z^2 + \cdots$, $z \to 0-$, $z < 0$.
    Show that for every $f \in \mathcal{A}$ there exist $\alpha \geq 0$ and a bounded, nondecreasing function ψ such that f is represented in the form

$$f(z) = 1 + \frac{2z}{\alpha - z + z^2 g(z)}, \tag{12.10-17}$$

where

$$g(z) := \int_0^\infty \frac{1}{1 + \tau^2 z^2} \, d\psi(\tau). \tag{12.10-18}$$

Also show that $\alpha > 0$ if and only if $\beta_0 = 1$.

4. For $f(z) := e^z$, show that the representation (12.10-17) yields

$$e^z = 1 + \frac{2z}{2 - z + z^2 g(z)},$$

where

$$g(z) := 2 \sum_{n=1}^\infty \frac{1}{z^2 + 4\pi^2 n^2}.$$

That is, (12.10-18) holds with

$$\psi(\tau) := \frac{1}{12} - \frac{1}{2\pi^2} \sum_k \frac{1}{k^2},$$

the sum comprising all integers such that

$$0 < k < \frac{1}{2\pi\sqrt{\tau}}.$$

5. Prove that an analytic function that maps the right half plane R into itself and is real on the real line and bounded near the origin can be represented as

$$g(z) = z \int_0^\infty \frac{1}{1 + \tau^2 z^2} \, d\psi(\tau),$$

where ψ is nondecreasing and bounded. It possesses an asymptotic expansion

$$g(z) \approx \gamma_0 z + \gamma_1 z^3 + \gamma_2 z^5 + \ldots, \quad z \to 0, \quad z > 0 \tag{12.10-19}$$

if and only if all moments

$$\mu_k := \int_0^\infty \tau^{2k} \, d\psi(\tau), \quad k = 0, 1, 2, \ldots,$$

exist, in which case $\gamma_k = (-1)^k \mu_k$.

6. *A theorem of Dahlquist.* For the numerical integration of ordinary differential equations one requires real rational functions r such that $r(\infty) \neq 0$ and

$$r(w) = \text{Log } w + O((w-1)^{p+1}), \qquad w \to 1, \qquad (12.10\text{-}20)$$

where p, the order of the method defined by r, is as large as possible (see Henrici [1962], Chap. 5). The method is called A stable if r maps the set $|w| > 1$ into the right half plane. Show that the order p of an A stable method is at most 2, and that $p = 2$ is achieved for

$$r(w) = 2\frac{w-1}{w+1}.$$

[The function

$$r_1(z) := r\left(\frac{1+z}{1-z}\right)$$

satisfies the hypotheses of Problem 5. The condition (12.10-20) means that $r_1(z)$ should agree as closely as possible with $\text{Log}(1+z)/(1-z) = 2(z + \frac{1}{3}z^3 + \frac{1}{5}z^5 + \cdots)$, but this is inconsistent with (12.10-19) in view of $\gamma_1 < 0$ if $p > 2$.]

§12.11. EXISTENCE AND CONVERGENCE OF THE S FRACTION CORRESPONDING TO A STIELTJES TRANSFORM

After it has been established that $GVF \subset ST$ and $PSA = ST$, it is natural also to ask whether $GVF = ST$. We also wish to characterize the set VF of those functions in GVF that are value functions of *convergent* Stieltjes fractions.

It will emerge that the inclusion $GVF \subset ST$ is proper; that is, not every Stieltjes transform is a generalized value function of a Stieltjes fraction. Moreover, it does not seem possible to characterize the class VF by some property that is independent of the notion of a continued fraction. However, we are able to define a large subclass ST^* of ST, containing all examples of practical interest, that is contained in VF.

I. Existence of the Corresponding S Fraction

Let $f \in ST$, $f = \mathscr{S}\psi$. We know that f then admits an asymptotic expansion,

$$f(z) \approx P\left(\frac{1}{z}\right) = \frac{\mu_0}{z} - \frac{\mu_1}{z^2} + \frac{\mu_2}{z^3} - \cdots, \qquad (12.11\text{-}1)$$

as $z \to \infty$, $z \in \hat{S}_\alpha$ for every $\alpha \in [0, \pi)$, where μ_k is the kth moment of ψ.

Now let, in addition, $f \in GVF$; that is, let f be a generalized value function of a nonterminating Stieltjes fraction C. This fraction corresponds to the power series that represents f asymptotically as $z \to \infty$, $z \in \hat{S}_\alpha$. By the uniqueness of asymptotic power series, C corresponds to P. Thus if an $f \in ST$ also is a generalized value function of a Stieltjes fraction, that fraction can only be the unique fraction corresponding to P. The following theorem affirms that this fraction exists for any $f \in ST$.

THEOREM 12.11a

Let $f = \mathscr{S}\psi \in ST$, and let $\mu_0, \mu_1, \ldots$ be the moments of ψ. Then the $RITZ^{-1}$ fraction corresponding to the series (12.11-1) exists and is an S fraction.

Proof. To show that the corresponding $RITZ^{-1}$ fraction exists it suffices, by Theorem 12.4c, that the Hankel determinants

$$H_k^{(n)} := \begin{vmatrix} \gamma_n & \gamma_{n+1} & \cdots & \gamma_{n+k-1} \\ \gamma_{n+1} & \gamma_{n+2} & \cdots & \gamma_{n+k} \\ & & \cdots & \\ \gamma_{n+k-1} & \gamma_{n+k} & \cdots & \gamma_{n+2k-2} \end{vmatrix},$$

where $\gamma_m := (-1)^m \mu_m$, satisfy

$$H_k^{(n)} \neq 0, \qquad n = 0, 1; \qquad k = 1, 2, \ldots. \tag{12.11-2}$$

To show that the $RITZ^{-1}$ fraction is a $SITZ^{-1}$ fraction, we must establish that $\gamma_0 > 0$, and that in the qd scheme associated with the series (12.11-1)

$$q_k^{(0)} < 0, \qquad e_k^{(0)} < 0, \qquad k = 1, 2, \ldots.$$

By the formulas (7.6-5) and (7.6-6) these inequalities are in turn implied by

$$\operatorname{sign} H_k^{(n)} = \begin{cases} 1, & n = 0, \\ (-1)^k, & n = 1. \end{cases} \tag{12.11-3}$$

For $k = 1, 2, \ldots$, let $\boldsymbol{\xi} := (\xi_0, \xi_1, \ldots, \xi_{k-1})$, where the ξ_i are real, and consider the quadratic form

$$Q_k^{(n)}(\boldsymbol{\xi}) = \sum_{i,j=0}^{k-1} \mu_{n+i+j} \xi_i \xi_j.$$

By expanding the square and using the definition of the μ_i, we see that

$$Q_k^{(n)}(\boldsymbol{\xi}) = \int_0^\infty \tau^n [\xi_0 + \xi_1 \tau + \cdots + \xi_{k-1} \tau^{k-1}]^2 \, d\psi(\tau).$$

If $\boldsymbol{\xi} \neq \boldsymbol{0}$, the integrand vanishes at most at $k-1$ points and is positive elsewhere. Because ψ has infinitely many points of increase, there follows

$$Q_k^{(n)}(\boldsymbol{\xi}) > 0.$$

Thus $Q_k^{(n)}$ is positive definite, which by a well-known result from linear algebra implies that the determinant of the coefficients of $Q_k^{(n)}$,

$$M_k^{(n)} = \begin{vmatrix} \mu_n & \mu_{n+1} & \cdots & \mu_{n+k-1} \\ \mu_{n+1} & \mu_{n+2} & \cdots & \mu_{n+k} \\ & & \cdots & \\ \mu_{n+k-1} & \mu_{n+k} & \cdots & \mu_{n+2k-2} \end{vmatrix} > 0.$$

Expressing the μ_m by the γ_m and changing first the signs of the 2nd, 4th, ... column and then that of the 2nd, 4th, ... row, we find

$$0 < M_k^{(n)} = \begin{cases} H_k^{(n)}, & n \text{ even} \\ (-1)^k H_k^{(n)}, & n \text{ odd}, \end{cases}$$

which proves both (12.11-2) and (12.11-3). ∎

II. The Approximants of the Corresponding *S* Fraction

Let $\psi \in \Psi^*$, let $\mu_0, \mu_1, \ldots$ denote the moments of ψ, and let $p_n(z), q_n(z)$ now denote the numerators and the denominators of the S fraction

$$C = \frac{\alpha_1}{\mid z} + \frac{\alpha_2}{\mid 1} + \frac{\alpha_3}{\mid z} + \frac{\alpha_4}{\mid 1} + \cdots \qquad (12.11\text{-}4)$$

corresponding to the series

$$P = \frac{\mu_0}{z} - \frac{\mu_1}{z^2} + \frac{\mu_2}{z^3} - \cdots. \qquad (12.11\text{-}5)$$

The existence of C is assured by Theorem 12.11a. We now wish to know whether C converges and, if so, to what. To this end we require

THEOREM 12.11b

For $n = 0, 1, 2, \ldots$ and for all $z \in S$,

$$\int_0^\infty \frac{q_n(z) - q_n(-\tau)}{z + \tau} d\psi(\tau) = p_n(z); \qquad (12.11\text{-}6)$$

furthermore, if $m := [(n+1)/2]$ is the degree of q_n,

$$\int_0^\infty \tau^k q_n(-\tau) d\psi(\tau) = \begin{cases} 0, & k = 0, 1, \ldots, n-m-1; \\ (-1)^m \alpha_1 \alpha_2 \cdots \alpha_{n+1}, & k = n-m. \end{cases} \qquad (12.11\text{-}7)$$

Proof. The integrand in (12.11-6), considered as a function of z and defined as $q_n'(-\tau)$ for $z = -\tau$, is a polynomial in z, of degree $m-1$, with coefficients that are polynomials in τ. The integral, too, is therefore a polynomial in z. We prove (12.11-6) by studying the asymptotic behavior of the difference of the polynomials on the left and on the right as $z \to \infty$.

By the proof of Lemma 12.4e (remembering that here we are dealing with a $RITZ^{-1}$ fraction), if $\gamma_m = (-1)^m \mu_m$,

$$\frac{p_n(z)}{q_n(z)} = \frac{\gamma_0}{z} + \frac{\gamma_1}{z^2} + \frac{\gamma_2}{z^3} + \cdots + \frac{\gamma_{n-1}}{z^n} + \frac{\gamma_n'}{z^{n+1}} + \cdots,$$

where

$$\gamma_n - \gamma_n' = (-1)^n \alpha_1 \alpha_2 \cdots \alpha_{n+1}.$$

Subtracting from this the asymptotic relation

$$\int_0^\infty \frac{1}{z + \tau} \, d\psi(\tau) \approx \frac{\gamma_0}{z} + \frac{\gamma_1}{z^2} + \cdots + \frac{\gamma_{n-1}}{z^n} + \frac{\gamma_n}{z^{n+1}} + \cdots,$$

we see that

$$\int_0^\infty \frac{1}{z + \tau} \, d\psi(\tau) - \frac{p_n(z)}{q_n(z)} = (-1)^n \frac{\alpha_1 \alpha_2 \cdots \alpha_{n+1}}{z^{n+1}} + O(z^{-n-2}).$$

Multiplication by $q_n(z) = z^m + \cdots$ yields

$$\int_0^\infty \frac{q_n(z)}{z + \tau} \, d\psi(\tau) - p_n(z) = (-1)^n \frac{\alpha_1 \alpha_2 \cdots \alpha_{n+1}}{z^{n+1-m}} + O(z^{-n+m-2})$$

From this we subtract

$$\int_0^\infty \frac{q_n(-\tau)}{z + \tau} \, d\psi(\tau),$$

which is at most $O(z^{-1})$ as $z \to \infty$. In view of $n + 1 - m \geq 1$ this yields

$$\int_0^\infty \frac{q_n(z) - q_n(-\tau)}{z + \tau} \, d\psi(\tau) - p_n(z) = O(z^{-1}).$$

The expression on the left being a polynomial, it can only be the zero polynomial, which proves (12.11-6).

We now know that

$$\int_0^\infty \frac{q_n(-\tau)}{z + \tau} \, d\psi(\tau) := \int_0^\infty \frac{q_n(z)}{z + \tau} \, d\psi(\tau) - p_n(z)$$

$$:= (-1)^n \frac{\alpha_1 \alpha_2 \cdots \alpha_{n+1}}{z^{n+1-m}} + O(z^{-n+m-2}). \quad (12.11\text{-}8)$$

The asymptotic expansion of the function on the left can be computed as in the proof of Corollary 12.9i. Writing the integrand in the form

$$\left[\frac{1}{z}+\frac{-\tau}{z^2}+\cdots+\frac{(-\tau)^{k-1}}{z^k}+\frac{(-\tau)^k}{z^k(z+\tau)}\right]q_n(-\tau)$$

and integrating term by term, we find

$$\int_0^\infty \frac{q_n(-\tau)}{z+\tau}\, d\psi(\tau) \approx \sum_{k=0}^\infty \frac{(-1)^k}{z^{k+1}}\int_0^\infty \tau^k q_n(-\tau)\, d\psi(\tau).$$

Comparing with (12.11-8) we find the desired relations (12.11-7). ∎

As an easy application of Theorem 12.11b we get:

COROLLARY 12.11c

The polynomials $q_{2k}(-\tau)$, $k = 0, 1, \ldots$, form an orthogonal set of polynomials with respect to the weight function $d\psi(\tau)$.

Proof. We are to show that

$$\int_0^\infty q_{2k}(-\tau)q_{2n}(-\tau)\, d\psi(\tau)=0 \tag{12.11-9}$$

for $k \neq n$. This follows readily by expanding the polynomial of lower degree in powers of τ and integrating term by term, using the first relation (12.11-7). Because the leading term of $q_{2n}(-\tau)$ is $(-\tau)^n$, using the second relation (12.11-7) similarly yields for $n = 0, 1, \ldots$

$$\int_0^\infty [q_{2n}(-\tau)]^2\, d\psi(\tau)=\alpha_1\alpha_2\cdots\alpha_{n+1}. ∎ \tag{12.11-10}$$

We recall from (12.4-32) that the polynomials $q_{2k}(\tau)$ are identical with the Hadamard polynomials $p_k^{(0)}(\tau)$ associated with the qd scheme of the series P. It is well known that there exists (up to constant factors) at most one set of orthogonal polynomials for a given weight function. Corollary 12.11c thus enables us in certain cases to identify the polynomials $p_k^{(0)}$ with known sets of orthogonal polynomials.

Concerning the convergence of C, we now can state

THEOREM 12.11d

Let $\psi \in \Psi^*$, $f := \mathcal{S}\psi$, and let C be the S fraction corresponding to the series (12.11-1). Then for $x > 0$ the approximants of C satisfy

$$\frac{p_{2m}(x)}{q_{2m}(x)}<f(x)<\frac{p_{2m+1}(x)}{q_{2m+1}(x)}, \qquad m = 0, 1, 2, \ldots. \tag{12.11-11}$$

Proof. To prove the left inequality, we consider the identity

$$[q_{2m}(x)]^2 \int_0^\infty \frac{1}{x+\tau} d\psi(\tau) - q_{2m}(x) \int_0^\infty \frac{q_{2m}(x) - q_{2m}(-\tau)}{x+\tau} d\psi(\tau)$$

$$-\int_0^\infty q_{2m}(-\tau) \frac{q_{2m}(x) - q_{2m}(-\tau)}{x+\tau} d\psi(\tau) = \int_0^\infty \frac{q_{2m}(-\tau)^2}{x+\tau} d\psi(\tau).$$

By Theorem 12.11b, the second integral on the left equals $p_{2m}(x)$. The third integral vanishes, because

$$\frac{q_{2m}(x) - q_{2m}(-\tau)}{x+\tau}$$

is a polynomial in τ of degree $m-1$. Hence after dividing by $[q_{2m}(x)]^2$ there remains

$$\int_0^\infty \frac{1}{x+\tau} d\psi(\tau) - \frac{p_{2m}(x)}{q_{2m}(x)} = \int_0^\infty \left[\frac{q_{2m}(-\tau)}{q_{2m}(x)}\right]^2 \frac{1}{x+\tau} d\psi(\tau) > 0.$$

To prove the right inequality in (12.11-11), we consider

$$q_{2m+1}(x) \int_0^\infty \frac{q_{2m+1}(x) - q_{2m+1}(-\tau)}{x+\tau} d\psi(\tau) - [q_{2m+1}(x)]^2 \int_0^\infty \frac{1}{x+\tau} d\psi(\tau)$$

$$+ x \int_0^\infty q_{2m+1}(-\tau) \frac{x^{-1} q_{2m+1}(x) + \tau^{-1} q_{2m+1}(-\tau)}{x+\tau} d\psi(\tau)$$

$$= x \int_0^\infty \tau^{-1}[q_{2m+1}(-\tau)]^2 \frac{1}{x+\tau} d\psi(\tau),$$

where we keep in mind that $q_{2m+1}(0) = 0$. [This follows from the recurrence relations (12.8-2).] The first integral equals $p_{2m+1}(x)$, and the third integral is again zero, because the fraction is a polynomial in τ of degree $m-1$. After dividing by $[q_{2m+1}(x)]^2$ there remains

$$\frac{p_{2m+1}(x)}{q_{2m+1}(x)} - \int_0^\infty \frac{1}{x+\tau} d\psi(\tau) = x \int_0^\infty \tau^{-1} \left[\frac{q_{2m+1}(-\tau)}{q_{2m+1}(x)}\right]^2 \frac{1}{x+\tau} d\psi(\tau),$$

establishing (12.11-11). ∎

COROLLARY 12.11e

If C converges, then its value function is f.

Proof. If C converges, then by (12.11-11) the value function equals $f(x)$ for $x > 0$. Because both the value function and f are analytic in the cut plane S, the value function equals $f(z)$ for all $z \in S$. ∎

Now let C be divergent. By the proof of Theorem 12.1c the approximants of C for $z = x > 0$ behave like the partial sums of an alternating series with terms whose absolute values decrease monotonically. Thus C has precisely two value functions: the limits of the approximants of even and of odd order. Each of these value functions is of the form $\mathscr{S}\psi_j$, where the functions $\psi_j \in \Psi^*$ both have the same moment sequence $\{\mu_k\}$ formed with the coefficients of the power series corresponding to C. If ψ_1 and ψ_2 were the only functions in Ψ^* with this moment sequence, then it could be asserted that $GVF = ST$. However, it is established in the theory of the moment problem that for every diverging C there exist infinitely many essentially different functions $\psi \in \Psi^*$ whose moment sequence is $\{\mu_k\}$. Each of these ψ gives rise to a function $f = \mathscr{S}\psi \in ST$, but only two of these are value functions. This shows that the class ST is larger than the class GVF.

III. A Priori Bounds for the Truncation Error

We have just learned that if a function $\psi \in \Psi^*$ gives rise to a continued fraction C that converges, then $f := \mathscr{S}\psi$ is represented by C throughout the cut plane S. Thus to decide whether a given function $f \in ST$ can be represented by a convergent continued fraction, we must be able to predict the convergence of the continued fraction. We subsequently establish, by elementary methods, some inequalities that estimate the difference of any two consecutive approximants in terms of quantities that depend only on the moment sequence $\{\mu_k\}$. If these bounds tend to zero for $z = x > 0$, the convergence of the continued fraction follows by Theorem 12.11d. Using the results of subsection IV, the bounds can also be used to estimate the truncation error of the fraction for complex $z \in S$.

THEOREM 12.11f

Let

$$C(z) = \frac{\alpha_1}{|\ z\ } + \frac{\alpha_2}{|\ 1\ } + \frac{\alpha_3}{|\ z\ } + \frac{\alpha_4}{|\ 1\ } + \cdots \qquad (12.11\text{-}12)$$

be an S fraction with the corresponding power series

$$P = \frac{\mu_0}{z} - \frac{\mu_1}{z^2} + \frac{\mu_2}{z^3} - \cdots . \qquad (12.11\text{-}13)$$

Let $\phi := \arg z \in (-\pi, \pi)$, $\sigma := \operatorname{Re}\sqrt{z} > 0$, and

$$\kappa := \frac{\alpha_1}{|z| \cos(\phi/2)} \sqrt{\frac{1 + \sqrt{5}}{2}} .$$

Then the approximants w_1, w_2, ... of C satisfy the following inequalities for $n = 2, 3, \ldots$:

(A) $$|w_n - w_{n-1}| \leq \kappa \prod_{k=2}^{n} (1 + 2\sigma\alpha_k^{-1/2})^{-1/2},$$

(B) $$|w_{n+1} - w_n| \leq \kappa \left[1 + 2\sigma \left(\frac{\mu_0}{\mu_n} \right)^{1/2n} \right]^{-n/2},$$

(C) $$|w_{n+1} - w_n| \leq \kappa \left[1 + 2\frac{\sigma}{e} \sum_{k=1}^{n} \left(\frac{\mu_0}{\mu_k} \right)^{1/2k} \right]^{-1/2}.$$

Before beginning with the proof, we state three classical inequalities that are required. The common hypothesis is that n is a positive integer and that $\gamma_k \geq 0$ for $k = 1, 2, \ldots, n$.

(I) **Bernoulli's inequality:**

$$\prod_{k=1}^{n} (1 + \gamma_k) \geq 1 + \sum_{k=1}^{n} \gamma_k.$$

This simply follows by expanding the product on the left.

(II) **Minkowski's inequality:**

$$\prod_{k=1}^{n} (1 + \gamma_k) \geq (1 + \sqrt[n]{\gamma_1 \gamma_2 \cdots \gamma_n})^n.$$

For a proof see Hardy, Polya, and Littlewood [1934], Theorem 64.

(III) **Carleman's inequality:**

$$\sum_{k=1}^{n} (\gamma_1 \gamma_2 \cdots \gamma_k)^{1/k} \leq e \sum_{k=1}^{n} \gamma_k$$

$(e := \lim(1 + 1/n)^n)$. Strict inequality holds unless all $\gamma_k = 0$.

Proof. The inequality between the geometric and the arithmetic mean yields for arbitrary $\beta_k > 0$

$$\sum_{k=1}^{n} (\gamma_1 \gamma_2 \cdots \gamma_k)^{1/k} = \sum_{k=1}^{n} \left[\frac{\gamma_1 \beta_1 \gamma_2 \beta_2 \cdots \gamma_k \beta_k}{\beta_1 \beta_2 \cdots \beta_k} \right]^{1/k}$$

$$\leq \sum_{k=1}^{n} \frac{1}{(\beta_1 \beta_2 \cdots \beta_k)^{1/k}} \frac{\gamma_1 \beta_1 + \cdots + \gamma_k \beta_k}{k}$$

$$= \sum_{k=1}^{n} \gamma_k \beta_k \sum_{m=k}^{n} \frac{1}{m(\beta_1 \beta_2 \cdots \beta_m)^{1/m}}.$$

We now let

$$\beta_k := \frac{(k+1)^k}{k^{k-1}}, \qquad k = 1, 2, \ldots, n,$$

to obtain

$$\beta_1\beta_2\cdots\beta_m = \frac{2}{1}\cdot\frac{3^2}{2^1}\cdots\frac{(m+1)^m}{m^{m-1}} = (m+1)^m,$$

$$(\beta_1\beta_2\cdots\beta_m)^{1/m} = m+1,$$

and therefore

$$\sum_{m=k}^{n}\frac{1}{m(\beta_1\beta_2\cdots\beta_m)^{1/m}} = \sum_{m=k}^{n}\frac{1}{m(m+1)} = \sum_{m=k}^{n}\left(\frac{1}{m}-\frac{1}{m+1}\right) < \frac{1}{k}.$$

Substituting, we get

$$\sum_{k=1}^{n}(\gamma_1\gamma_2\cdots\gamma_k)^{1/k} \le \sum_{k=1}^{n}\left(\frac{k+1}{k}\right)^k\gamma_k,$$

which in view of

$$\left(\frac{k+1}{k}\right)^k = \left(1+\frac{1}{k}\right)^k < e$$

yields (III). ■

We now proceed to the proof of Theorem 12.11f. Let $s := \sigma+i\tau := \sqrt{z}$, $\sigma > 0$. It is convenient to work with the equivalent fraction

$$C^*(z) = \frac{s^{-1}}{|\beta_1 s|} + \frac{1}{|\beta_2 s|} + \frac{1}{|\beta_3 s|} + \cdots$$

in place of C, where

$$\beta_{2m} = \frac{\alpha_1\alpha_3\cdots\alpha_{2m-1}}{\alpha_2\alpha_4\cdots\alpha_{2m}}, \qquad \beta_{2m+1} = \frac{\alpha_2\alpha_4\cdots\alpha_{2m}}{\alpha_1\alpha_3\cdots\alpha_{2m+1}}, \quad (12.11\text{-}14)$$

$m = 0, 1, 2, \ldots$. If the sequence of denominators of C^* is denoted by $\{r_n\}$, then by (12.1-23)

$$C^*(z) \cong \sum_{m=1}^{\infty}(-1)^{m-1}\frac{1}{sr_{m-1}r_m}. \qquad (12.11\text{-}15)$$

The recurrence relations satisfied by the r_m are $r_0 = 1$, $r_1 = \beta_1 s$,

$$r_m = r_{m-2}+\beta_m sr_{m-1}, \qquad m = 2, 3, \ldots. \qquad (12.11\text{-}16)$$

Letting $v_m := r_m\overline{r_{m-1}} = \xi_m+i\eta_m$, we find on multiplying by $\overline{r_{m-1}}$ that

$$v_m = v_{m-1}+\beta_m s|r_{m-1}|^2$$

or, separating real and imaginary parts,

$$\xi_m = \xi_{m-1} + \beta_m \sigma |r_{m-1}|^2, \qquad \eta_m = -\eta_{m-1} + \beta_m \tau |r_{m-1}|^2,$$

$m = 1, 2, \ldots$. Because $\xi_0 = \eta_0 = 0$, there follows

$$\xi_n = \sigma \sum_{m=1}^{n} \beta_m |r_{m-1}|^2 > 0, \qquad (12.11\text{-}17a)$$

showing that $\xi_n > \xi_{n-1} > 0$. Incidentally, we also get

$$\eta_n = \tau \sum_{m=1}^{n} (-1)^{n+m} \beta_m |r_{m-1}|^2. \qquad (12.11\text{-}17b)$$

From (12.11-15) we have

$$|w_n - w_{n-1}| = \frac{1}{|s|} \frac{1}{|v_n|} \leq \frac{1}{|s|} \frac{1}{\xi_n}. \qquad (12.11\text{-}18)$$

We thus want good *lower* bounds for ξ_n. From (12.11-17a),

$$\xi_n - \xi_{n-2} = \sigma[\beta_{n-1}|r_{n-2}|^2 + \beta_n|r_{n-1}|^2].$$

By the inequality of the arithmetic and geometric mean,

$$\beta_{n-1}|r_{n-2}|^2 + \beta_n|r_{n-1}|^2 \geq 2\sqrt{\beta_{n-1}\beta_n} |r_{n-1}r_{n-2}|$$
$$= 2\sqrt{\beta_{n-1}\beta_2}\, \xi_{n-1}.$$

In view of $\beta_{n-1}\beta_n = \alpha_n^{-1}$,

$$\xi_n \geq \xi_{n-2} + 2\sigma\alpha_n^{-1/2}\xi_{n-1}$$

or, because the sequence $\{\xi_n\}$ is increasing,

$$\xi_n\xi_{n-1} \geq (1 + 2\sigma\alpha_n^{-1/2})\xi_{n-1}\xi_{n-2};$$

hence

$$\xi_n\xi_{n-1} \geq \xi_2\xi_1 \prod_{k=3}^{n} (1 + 2\sigma\alpha_k^{-1/2})$$

or, again using $\xi_n > \xi_{n-1}$,

$$\xi_n \geq \sqrt{\xi_2\xi_1} \prod_{k=3}^{n} (1 + 2\sigma\alpha_k^{-1/2})^{1/2}, \qquad n = 3, 4, \ldots . \quad (12.11\text{-}19)$$

From (12.11-16) we have $r_0 = 1$, $r_1 = \beta_1 s$, $r_2 = r_0 + \beta_2 s r_1 = 1 + \beta_1\beta_2 s^2$. There follows

$$\xi_1 = \operatorname{Re} r_1 = \beta_1\sigma,$$

$$\xi_2 = \operatorname{Re} r_2\bar{r}_1 = \operatorname{Re} \beta_1\bar{s}(1 + \beta_1\beta_2 s^2) = \beta_1\sigma(1 + \beta_1\beta_2|z|),$$

and we find

$$\frac{\xi_2\xi_1}{1+2\sigma\alpha_2^{-1/2}} \geq \frac{\beta_1^2\sigma^2(1+\beta_1\beta_2\sigma^2)}{1+2\sqrt{\beta_1\beta_2}\,\sigma} \geq \beta_1^2\sigma^2\frac{\sqrt{5}-1}{2},$$

because the minimum of $f(x) := (1+x^2)/(1+2x)$ for $x \geq 0$ is $(\sqrt{5}-1)/2$, as is easily verified by calculus. We thus have from (12.11-19)

$$\xi_n \geq \beta_1\sigma\sqrt{\frac{\sqrt{5}-1}{2}} \prod_{k=2}^{n} (1+2\sigma\alpha_k^{-1/2})^{1/2}; \qquad (12.11\text{-}20)$$

hence by (12.11-18), since $\beta_1^{-1} = \alpha_1$,

$$|w_n - w_{n-1}| \leq \frac{1}{|s|\sigma}\sqrt{\frac{\sqrt{5}+1}{2}} \prod_{k=2}^{n} (1+2\sigma\alpha_k^{-1/2})^{-1/2},$$

proving (A).

To prove (B), we note that by Minkowski's inequality (II)

$$\prod_{k=2}^{n+1} (1+2\sigma\alpha_k^{-1/2}) \geq [1+2\sigma(\alpha_2\alpha_3 \cdots \alpha_{n+1})^{-1/2n}]^{n/2}. \quad (12.11\text{-}21)$$

By (12.4-18),

$$\alpha_1\alpha_2 \cdots \alpha_{n+1} = (-1)^n s_n,$$

where

$$s_{2m} = H_{m+1}^{(0)}/H_m^{(0)}, \qquad s_{2m+1} = H_{m+1}^{(1)}/H_m^{(1)}.$$

Thus in terms of the moment determinants introduced in the proof of Theorem 12.11a,

$$\alpha_1\alpha_2 \cdots \alpha_{n+1} = \begin{cases} M_{m+1}^{(0)}/M_m^{(0)}, & n = 2m, \\ M_{m+1}^{(1)}/M_m^{(1)}, & n = 2m+1. \end{cases} \quad (12.11\text{-}22)$$

We assert that for $m = 0, 1, 2, \ldots$

$$M_{m+1}^{(0)}/M_m^{(0)} \leq \mu_{2m}, \qquad M_{m+1}^{(1)}/M_m^{(1)} \leq \mu_{2m+1}. \quad (12.11\text{-}23)$$

To prove the first inequality, consider the quadratic form with the matrix

$$\begin{vmatrix} \mu_0 & \mu_1 & \cdots & \mu_{m-1} & \mu_m \\ \mu_1 & \mu_2 & \cdots & \mu_m & \mu_{m+1} \\ & & \cdots & & \\ \mu_{m-1} & \mu_m & \cdots & \mu_{2m-2} & \mu_{2m-1} \\ \mu_m & \mu_{m+1} & \cdots & \mu_{2m-1} & 0 \end{vmatrix}.$$

This form is not positive definite, one diagonal element being zero. The principal upper minors of order k (i.e., the determinants formed with the elements of the first k rows and columns) are positive for $k = 1, \ldots, m$. Hence the determinant of the full matrix cannot be positive. By expanding in terms of the elements of the last column, we find

$$M_{m+1}^{(0)} - \mu_{2m} M_m^{(0)} \leqslant 0,$$

proving the first relation (12.11-23). The second relation is proved similarly.

It now follows from (12.11-22) that

$$\alpha_1 \alpha_2 \cdots \alpha_{n+1} \leqslant \mu_n, \tag{12.11-24}$$

$n = 1, 2, \ldots$. In view of $\alpha_1 = \mu_0$ (12.11-21) now yields

$$\prod_{k=2}^{n+1} (1 + 2\sigma \alpha_k^{-1/2})^{1/2} \geqslant \left[1 + 2\sigma \left(\frac{\mu_0}{\mu_n} \right)^{1/2n} \right]^{n/2},$$

proving (B).

For the proof of (C) we again start from (12.11-20). Using Bernoulli's inequality (I) we get

$$\prod_{k=2}^{n+1} (1 + 2\sigma \alpha_k^{-1/2}) \geqslant 1 + 2\sigma \sum_{k=2}^{n+1} \alpha_k^{-1/2}.$$

By Carleman's inequality (III),

$$\sum_{k=2}^{n+1} \alpha_k^{-1/2} \geqslant \frac{1}{e} \sum_{k=2}^{n+1} (\alpha_2 \alpha_3 \cdots \alpha_{k+1})^{-1/2k} \geqslant \frac{1}{e} \sum_{k=1}^{n} \left(\frac{\mu_0}{\mu_k} \right)^{-1/2k},$$

where we have used once more (12.11-24). Inequality (C) now follows by (12.11-18), completing the proof. ∎

Some applications of Theorem 12.11f to special continued fractions are made in subsequent sections. Here we note some qualitative consequences. As a consequence of (A) we immediately recover Corollary 12.8f. Similarly, as a consequence of (B) we find

COROLLARY 12.11g

Let the power series P have radius of convergence $\rho > 0$. Then C at $z \in S$ converges at least like a geometric series with ratio $(1 + 2\sigma\sqrt{\rho})^{-1/2}$, where $\sigma := \mathrm{Re}\sqrt{z} > 0$.

Proof. By the Cauchy–Hadamard formula (Theorem 2.2a) we have

$$\rho = \liminf_{n \to \infty} \mu_n^{-1/n};$$

hence for every $\epsilon > 0$, if n is sufficiently large,

$$\left(\frac{\mu_0}{\mu_n}\right)^{1/2n} \geqslant \sqrt{\rho} - \epsilon. \quad \blacksquare$$

As a consequence of (C) we note

COROLLARY 12.11h (Carleman's condition)

A sufficient condition for the convergence of C is

$$\sum_{k=1}^{\infty} \mu_k^{-1/2k} = \infty. \tag{12.11-25}$$

It should be noted that (12.11-25) is a far weaker requirement than the convergence of P. For example, if $\mu_k = k!$, then by Stirling's formula

$$\mu_k^{-1/2k} \sim \left[\sqrt{2\pi k}\left(\frac{k}{e}\right)^k\right]^{-1/2k} \sim \left(\frac{e}{k}\right)^{1/2},$$

and Carleman's condition is clearly satisfied. In view of the divergence of the harmonic series the same is true even if $\mu_k = (k!)^2$ or if $\mu_k = (2k)!$.

We denote by ST^* the class of all $f = \mathcal{S}\psi \in ST$ such that the moment sequence of ψ satisfies the Carleman condition (12.11-25). Corollary 12.11h establishes that $ST^* \subset VF$; thus, every function in ST^* can be represented by a convergent Stieltjes fraction. The bounds (B) and (C) of Theorem 12.11f enable us to estimate the truncation errors in these representations, at least for real $z > 0$, without actually computing the continued fraction.

IV. A Posteriori Bounds

We now see what we can say about the truncation error of a Stieltjes fraction $C(z)$ if a finite number of the approximants w_i of C are already known. Specifically, the following problem is considered: Let $z \in S$, and let the complex numbers $w_j = w_j(z)$ be known for $j = 1, 2, \ldots, n$. What can be said about the possible values at z of the value function (or of a generalized value function) of C?

The answer to the above question is trivial if $z = x > 0$. Then the proof of Theorem 12.1c shows that C is equivalent to an alternating series with terms whose absolute values decrease monotonically. Thus any generalized value of C always lies in the interval spanned by two consecutive approximants.

Thus in the following we may assume that z is not real. A crude statement about the location of $f(z)$ can be made without knowing any approximants at all. Because every generalized value function belongs to the class PSA, it follows that

$$f(z) \in L \qquad \text{if } z \in U,$$
$$f(z) \in U \qquad \text{if } z \in L,$$

where U and L denote the upper and the lower half plane, respectively.

A more precise statement about the location of $f(z)$ is already possible if only the number $w_1(z)$ is known. By Theorem 12.9e there exists a nondecreasing function ψ such that

$$f(z) = \int_0^\infty \frac{1}{z + \tau} \, d\psi(\tau).$$

By the definition of the Stieltjes integral, this means that $f(z)$ lies in the closure of the set of points

$$v := \sum_{i=1}^m \frac{\omega_i}{z + \tau_i}, \qquad (12.11\text{-}26)$$

where m is an arbitrary positive integer, $0 \leqslant \tau_1 < \tau_2 < \cdots < \tau_m$, $\omega_i > 0$, $i = 1, 2, \ldots, m$, and

$$\omega_1 + \omega_2 + \cdots + \omega_m = \int_0^\infty d\psi(\tau) = \mu_0 = \alpha_1.$$

On writing

$$v = \frac{\sum\limits_{i=1}^m [\alpha_1/(z + \tau_i)]\omega_i}{\sum\limits_{i=1}^m \omega_i}$$

we see that v lies at the center of mass of a system of m masses ω_i attached to the points

$$t_i := \frac{\alpha_1}{z + \tau_i}, \qquad i = 1, 2, \ldots, m.$$

All points t_i lie on the curve

$$\Gamma_1 : t = t(\tau) := \frac{\alpha_1}{z + \tau}, \qquad 0 \leqslant \tau < \infty.$$

By the theory of Moebius transformations, Γ_1 is readily identified as part of the circle touching the real axis at the origin and passing through the point

$$t(0) = \frac{\alpha_1}{z} = w_1(z).$$

Because the circle also passes through

$$t(\alpha_2) = \frac{\alpha_1}{z + \alpha_2} = w_2(z),$$

Γ_1 may also be described as the circular arc from $0 = w_0(z)$ to $w_1(z)$ passing through $w_2(z)$. By virtue of being the center of mass of a system of masses attached to points of Γ_1, the point v, and with it the point $f(z)$, lies in the convex hull of the arc Γ_1, that is, in the set W_0 bounded by Γ_1 and by the straight line segment joining the terminal points $w_0 = 0$ and w_1 of Γ_1 (see Fig. 12.11a).

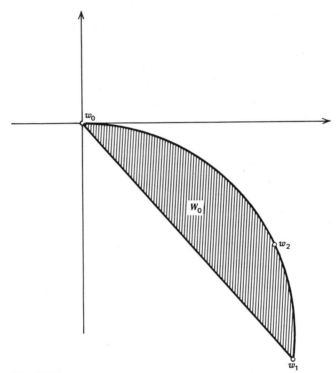

Fig. 12.11a.

To describe the set W_0 in a way that is useful in answering our question if an arbitrary number of approximants are known, we recall the concept of a **circular region** which was introduced in §6.7. Let a, b, c be any three distinct points of the extended complex plane, and let Λ denote the unique generalized circle (i.e., circle or straight line) passing through a, b, c. We denote by $[a, b, c]$ the closed set in the extended complex plane that is bounded by Λ and lies to the left of it in the orientation induced by the points a, b, c. Analytically, $[a, b, c]$ is defined as the circular region that is mapped onto the closed upper half plane by the Moebius transformation sending a, b, c to $0, 1, \infty$; that is,

$$[a, b, c] := \left\{ u : \text{Im} \left(\frac{u-a}{u-c} \frac{b-c}{b-a} \right) \geq 0 \right\},$$

subject to the obvious conventions if one of the points a, b, c is ∞.

EXAMPLE 1

Let $\text{Im } z > 0$, and let $w_{-1} = \infty$, $w_0 = 0$, $w_1 = \alpha_1/z$ be the initial approximants of C. Then $[w_{-1}, w_0, w_1]$ is a half plane bounded by a straight line through 0 and w_1. It is the half plane containing the point w_2, because for $u = w_2$

$$\frac{u - w_{-1}}{u - w_1} \frac{w_0 - w_1}{w_0 - w_{-1}} = \frac{w_0 - w_1}{w_2 - w_1} = \frac{z + \alpha_2}{\alpha_2},$$

hence

$$\text{Im} \frac{w_2 - w_{-1}}{w_2 - w_1} \frac{w_0 - w_1}{w_0 - w_{-1}} = \text{Im} \frac{z + \alpha_2}{\alpha_2} > 0.$$

EXAMPLE 2

Let again $\text{Im } z > 0$. Then $[w_0, w_1, w_2]$ is a circular region bounded by a generalized circle through w_0, w_1, w_2. This region does not contain ∞, because for $u = \infty$

$$\frac{u - w_0}{u - w_2} \frac{w_1 - w_2}{w_1 - w_0} = \frac{w_1 - w_2}{w_1 - w_0} = \frac{\alpha_2}{z + \alpha_2},$$

hence

$$\text{Im} \frac{u - w_0}{u - w_2} \frac{w_1 - w_2}{w_1 - w_0} < 0.$$

It follows that W is the closed disk bounded by the circle through w_0, w_1, w_2.

In the light of these examples the foregoing result concerning the location of $f(z)$ can be stated as follows:

THEOREM 12.11i

Let f be any generalized value function of the S fraction

$$C = \frac{\alpha_1}{\mid z} + \frac{\alpha_2}{\mid 1} + \frac{\alpha_3}{\mid z} + \frac{\alpha_4}{\mid 1} + \cdots$$

with the approximants $\{w_n\}$. If z is not real, then $w := f(z) \in W_0$ where

$$W_0 := [w_{-1}, w_0, w_1] \cap [w_0, w_1, w_2] \qquad \text{if } \operatorname{Im} z > 0,$$

$$W_0 := [w_1, w_0, w_{-1}] \cap [w_2, w_1, w_0] \qquad \text{if } \operatorname{Im} z < 0.$$

To locate the generalized values $w = f(z)$ for nonreal z when an arbitrary number of approximants are known, we let $s := z^{1/2}$, where $\operatorname{Re} s > 0$, and write C in the equivalent form

$$C = \frac{1}{s} \frac{\alpha_1 |}{| s} + \frac{\alpha_2 |}{| s} + \frac{\alpha_3 |}{| s} + \frac{\alpha_4 |}{| s} + \cdots$$

which does not require a distinction between approximants of an even and of an odd order. If

$$t_i : u \to \frac{\alpha_i}{s + u},$$

the approximants of C are

$$w_n = \frac{1}{s} t_1 \circ t_2 \circ \cdots \circ t_n (0), \qquad n = 1, 2, \dots.$$

Let $n > 0$ be arbitrary, and let

$$C^* := \frac{1}{s} \frac{\alpha_{n+1} |}{| s} + \frac{\alpha_{n+2} |}{| s} + \cdots.$$

This again is an S fraction. If w is a generalized value of C at the point z, then

$$w^* := \frac{1}{s} t_n^{[-1]} \circ t_{n-1}^{[-1]} \circ \cdots \circ t_1^{[-1]}(sw)$$

is a generalized value at z of C^*, and thus if $\operatorname{Im} z > 0$ by Theorem 12.11i is contained in the set

$$W_0^* := [w_{-1}^*, w_0^*, w_1^*] \cap [w_0^*, w_1^*, w_2^*],$$

where $\{w_i^*\}$ is the sequence of approximants of C^*. There follows

$$sw \in t_{1,n}(sW_0^*), \qquad (12.11\text{-}27)$$

where $t_{1,n} := t_1 \circ t_2 \circ \cdots \circ t_n$. Clearly,

$$sW_0^* = [sw_{-1}^*, sw_0^*, sw_1^*] \cap [sw_0^*, sw_1^*, sw_2^*].$$

We now use the fact that for any Moebius transformation t and three distinct points a, b, c of the extended complex plane

$$t([a, b, c]) = [t(a), t(b), t(c)].$$

Moreover, for any one-to-one mapping f and any subsets A, B of its domain of definition, $f(A \cap B) = f(A) \cap f(B)$. Thus we get

$$t_{1,n}(sW_0^*) = [t_{1,n}(sw_{-1}^*), \, t_{1,n}(sw_0^*), \, t_{1,n}(sw_1^*)]$$
$$\cap [t_{1,n}(sw_0^*), \, t_{1,n}(sw_1^*), \, t_{1,n}(sw_2^*)].$$

Now

$$t_{1,n}(sw_{-1}^*) = t_1 \circ t_2 \circ \cdots \circ t_{n-1}(0) = sw_{n-1},$$

$$t_{1,n}(sw_0^*) = t_1 \circ t_2 \circ \cdots \circ t_n(0) = sw_n,$$

$$t_{1,n}(sw_1^*) = t_1 \circ t_2 \circ \cdots \circ t_{n+1}(0) = sw_{n+1},$$

$$t_{1,n}(sw_2^*) = t_1 \circ t_2 \circ \cdots \circ t_{n+2}(0) = sw_{n+2}.$$

Thus if Im $z > 0$ (12.11-27) shows that $w \in W_n$ where

$$W_n := [w_{n-1}, \, w_n, \, w_{n+1}] \cap [w_n, \, w_{n+1}, \, w_{n+2}]. \qquad (12.11\text{-}28a)$$

If Im $z < 0$ an analogous argument shows that $w \in W_n$ where

$$W_n := [w_{n+1}, \, w_n, \, w_{n-1}] \cap [w_{n+2}, \, w_{n+1}, \, w_n]. \qquad (12.11\text{-}28b)$$

We assert that for $n = 1, 2, \ldots$ the circular regions $[w_{n-1}, w_n, w_{n+1}]$ do not contain the point ∞ if Im $z > 0$ and hence are disks, which follows as in example 2 if it is shown that

$$\operatorname{Im} \frac{w_n - w_{n+1}}{w_n - w_{n-1}} < 0. \qquad (12.11\text{-}29)$$

Reverting briefly to the notation of subsection III above, we have

$$\operatorname{Im} \frac{w_n - w_{n+1}}{w_n - w_{n-1}} = \operatorname{Im} \frac{r_{n-1}}{r_{n+1}} = \operatorname{Im} \frac{\overline{v_n}}{v_{n+1}}.$$

The sign of the last expression is opposite to the sign of $\xi_n \eta_{n+1} + \xi_{n+1} \eta_n$, which by the formulas (12.11-17) is seen to be positive. This proves (12.11-29). Similarly it is shown that the regions $[w_{n+1}, w_n, w_{n-1}]$ are disks for Im $z < 0$. Thus for $n \geq 1$, each W_n is the intersection of two disks, hence convex. The boundaries of $[w_{n-1}, w_n, w_{n+1}]$ and $[w_n, w_{n+1}, w_{n+2}]$ have the points w_n and w_{n+1} in common. Thus W_n is bounded by two circular arcs running from w_n to w_{n+1}. These arcs meet under the angle $|\arg z|$, because this is true of W_0^*, and W_n is the image of W_0^* under a Moebius transformation. It follows that the diameter of W_n is at most

$$|w_{n+1} - w_n|, \qquad \text{if } 0 < |\arg z| \leq \frac{\pi}{2},$$

$$|\tan(\tfrac{1}{2} \arg z)| \, |w_{n+1} - w_n|, \qquad \text{if } \frac{\pi}{2} < |\arg z| < \pi.$$

If C converges, then $|w_{n+1} - w_n| \to 0$; the diameters become arbitrarily small; and the inclusion regions W_n arbitrarily precise. In summary, we have obtained:

THEOREM 12.11j

Under the hypotheses of Theorem 12.11i, $w = f(z) \in W_n$ *for* $n = 0, 1, 2, \ldots,$ *where the sets W_n are defined by* (12.11-28). *Each W_n is a convex set bounded by two circular arcs from w_n to w_{n+1} that intersect under the angle* $|\arg z|$. *If C is convergent, the diameter of W_n tends to zero for* $n \to \infty$.

It can be shown that for each n the statement of inclusion of Theorem 12.11j is the best possible in the sense that every point of W_n can be value at z of an S-fraction the first $n+1$ approximants of which are w_1, $w_2, \ldots, w_{n+1}$.

Theorem 12.11j permits precise statements on the values of a convergent S fraction after only a finite number of its approximants have been computed. Another way to look at the sets W_n is as follows. Let Γ_n denote the circular arc from w_{n-1} to w_n, passing through w_{n+1}. The set W_n then is the moon-shaped region bounded by Γ_n and by that portion of Γ_{n-1} which lies between w_{n-1} and w_n. The arcs $\Gamma_1, \Gamma_2, \Gamma_3, \ldots$ all strung together form a piecewise smooth spiral, with corners at the approximants $w_0, w_1, w_2, \ldots$. Each Γ_n terminates somewhere on Γ_{n-1}. Any two consecutive pieces of the spiral form the boundary of a set W_n which contains every generalized value of the fraction. If the continued fraction converges, successive segments of the spiral wind more and more closely around the value of the fraction.

EXAMPLE **3**

The S fraction

$$C(z) := \frac{1}{\lceil z} + \frac{1/2}{\lceil 1} + \frac{1}{\lceil z} + \frac{3/2}{\lceil 1} + \cdots \qquad (12.11\text{-}30)$$

is convergent for all $z \in S$, and its values will in §12.13 be shown to be related to the error integral by the formula

$$C(z) = \frac{2e^z}{s} \int_s^\infty e^{-u^2}\, du \qquad (s^2 := z, \operatorname{Re} s > 0)$$

The first six approximants for $z = i$ are as follows:

n	1	2	3	4	5	6
w_n	$-i$	$\dfrac{2-4i}{5}$	$\dfrac{2-10i}{13}$	$\dfrac{38-124i}{145}$	$\dfrac{118-404i}{521}$	$\dfrac{1982-7278i}{8749}$

In Fig. 12.11*b* we show the resulting arcs Γ_n and inclusion sets W_n.

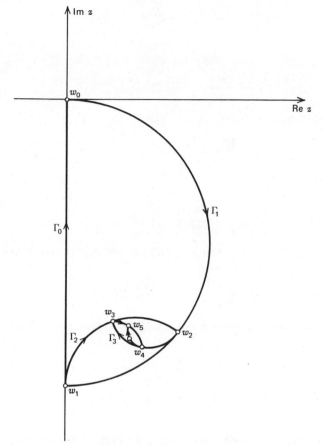

Fig. 12.11b.

PROBLEMS

1. Prove Minkowski's inequality (II) by induction with respect to n.
   [Apply the induction hypothesis to the first $n-1$ and the last $n-1$ factors in the product

   $$(1+\alpha_1)\cdots(1+\alpha_{n-1})(1+\alpha_n)(1+(\alpha_1\cdots\alpha_n)^{1/n})^{n-2}.]$$

2. Deduce from (12.11-17) that if z is not real and negative, $r_n \neq 0$ for $n = 0, 1, 2, \ldots$, and hence obtain a new proof that the poles of the approximants of a Stieltjes fraction lie on the negative real axis.

3. For $\xi \geq 0$, let $\omega(\xi) := \sqrt{1+\xi^2} + \xi$. Using the notation of the main text, obtain the following improvement of statement (A) of Theorem 12.11f: If the sequence

$\{\alpha_n\}$ is *de*creasing, then

$$\xi_n \geqslant \frac{4\beta_1\sigma}{3\sqrt{3}} \prod_{i=2}^{n} \omega(\sigma\alpha_i^{-1/2});$$

consequently,

$$|w_n - w_{n-1}| \leqslant \frac{\alpha_1}{|z|\cos(\phi/2)} \frac{3\sqrt{3}}{4} \prod_{k=2}^{n} [\omega(\sigma\alpha_k^{-1/2})]^{-1}.$$

4. (Continuation) If the sequence $\{\alpha_n\}$ is *increasing*, then

$$\xi_n \geqslant \frac{4\beta_1\sigma}{3\sqrt{3}} \prod_{i=3}^{n+1} \omega(\sigma\alpha_i^{-1/2});$$

consequently,

$$|w_n - w_{n-1}| \leqslant \frac{1}{|z|\cos(\phi/2)} \frac{3\sqrt{3}}{4} \prod_{k=3}^{n+1} [\omega(\sigma\alpha_k^{-1/2})]^{-1}.$$

5. Let f be analytic in the right half plane R, $f(R) \in R$, and let $f(z)$ be real for real z and possess an asymptotic expansion

$$f(z) \approx \frac{\mu_0}{z} - \frac{\mu_1}{z^3} + \frac{\mu_2}{z^5} - \cdots, \qquad z \to \infty, \qquad z > 0,$$

where the μ_k satisfy the Carleman condition. Show that there exist $\alpha_i > 0$ such that for all $z \in R$,

$$f(z) = \frac{\alpha_1|}{|z|} + \frac{\alpha_2|}{|z|} + \frac{\alpha_3|}{|z|} + \frac{\alpha_4|}{|z|} + \cdots.$$

Indicate how to construct the α_i by means of the *qd* algorithm.

§12.12 *S* FRACTIONS: EXPANSIONS OF STIELTJES TRANSFORMS

In this section we discuss some special Stieltjes transforms $f = \mathcal{S}\psi$ that can be expanded in an S fraction by the results described in §12.11. In place of expanding f, we also may expand a remainder in the asymptotic expansion of f.

Let $\psi \in \Psi^*$, and let μ_0, $\mu_1, \ldots$ denote the moments of ψ. Then for $m = 0, 1, \ldots$ the identity

$$f(z) = \int_0^{\infty} \frac{1}{z + \tau} \, d\psi(\tau)$$

$$= \frac{\mu_0}{z} - \frac{\mu_1}{z^2} + \cdots + \frac{(-1)^{m-1}\mu_{m-1}}{z^m} + \frac{(-1)^m}{z^m} r_m(z)$$

holds, where

$$r_m(z) := \int_0^\infty \frac{\tau^m}{z+\tau}\, d\psi(\tau).$$

The function r_m itself is a Stieltjes transform, $r_m = \mathscr{S}\psi^*$, where

$$\psi^*(\tau) := \int_0^\tau \sigma^m\, d\psi(\sigma).$$

Evidently, $\psi^* \in \Psi^*$. The moments $\mu_0^*, \mu_1^*, \ldots$ of ψ^* are $\mu_i^* = \mu_{m+1}$, $i = 0, 1, \ldots$. If the μ_i satisfy the Carleman condition (12.11-25), then so do the μ_i^*, thus the S fraction corresponding to the power series

$$P^* := \frac{\mu_0^*}{z} - \frac{\mu_1^*}{z^2} + \frac{\mu_2^*}{z^3} - \cdots.$$

converges to $r_m(z)$ for all $z \in S$. Its elements, by Corollary 12.4f, are the negatives of the elements in the mth diagonal of the qd scheme associated with the series

$$P = \frac{\mu_0}{z} - \frac{\mu_1}{z^2} + \frac{\mu_2}{z^3} - \cdots. \tag{12.12-1}$$

In summary, we thus have obtained:

THEOREM 12.12

Let $\psi \in \Psi^*$, and let the moments $\mu_0, \mu_1, \ldots$ of ψ satisfy the Carleman condition (12.11-25). If $f := \mathscr{S}\psi$, and if $q_n^{(m)}, e_n^{(m)}$ are the elements of the qd scheme associated with the formal series (12.12-1), then for $z \in S$ and for $m = 0, 1, 2, \ldots$ the following identity holds:

$$f(z) = \frac{\mu_0}{z} - \frac{\mu_1}{z^2} + \cdots + (-1)^{m-1}\frac{\mu_{m-1}}{z^m}$$

$$+ (-1)^m \frac{\mu_m}{z^{m+1}} \left\{ \frac{1}{1}\bigg|\ -\ \frac{q_1^{(m)}}{z}\bigg|\ -\ \frac{e_1^{(m)}}{1}\bigg|\ -\ \frac{q_2^{(m)}}{z}\bigg|\ -\ \frac{e_2^{(m)}}{1}\bigg|\ -\cdots \right\} \tag{12.12-2}$$

Ordinarily, to evaluate $f(z)$ from its asymptotic series one would use the series up to its smallest term and then stop, committing an unimprovable error. The foregoing identity exhibits the **converging factor**, by which the last-used term of the asymptotic series must be multiplied in order to get the exact value of $f(z)$.

Some illustrative examples follow.

I. Jacobi Distribution

Let $0 < \alpha < \gamma$ and

$$\psi(\tau) := \begin{cases} \dfrac{\Gamma(\gamma)}{\Gamma(\gamma - \alpha)\Gamma(\alpha)} \displaystyle\int_0^\tau \sigma^{\alpha-1}(1-\sigma)^{\gamma-\alpha-1}\, d\sigma, & 0 \leq \tau \leq 1, \\[2mm] \psi(1), & \tau > 1. \end{cases}$$

Evidently $\psi \in \Psi$; the moments (which automatically exist, because ψ is constant for $\tau \geq 1$) by Euler's beta integral (Theorem 8.7a) are

$$\mu_n = \frac{\Gamma(\gamma)}{\Gamma(\gamma - \alpha)\Gamma(\alpha)} \int_0^1 \tau^{\alpha+n-1}(1-\tau)^{\gamma-\alpha-1}\, d\tau = \frac{(\alpha)_n}{(\gamma)_n}, \qquad n = 0, 1, \ldots.$$

The corresponding series P thus in this case is

$$P(z) = \frac{1}{z}\,{}_2F_1\!\left(\alpha, 1; \gamma; -\frac{1}{z}\right)$$

and has radius of convergence 1. The associated qd scheme was constructed in §12.5 [see eq. (12.5-9)]. Thus, save for replacing z by z^{-1}, the resulting continued fraction expansion for $f = \mathscr{S}\psi$ formally agrees with the *RITZ* fraction for ${}_2F_1(\alpha, 1; \gamma; z)$ obtained in §12.5, which, in fact, was shown to hold even for complex α and γ. However, if $0 < \alpha < \gamma$ the Theorems 12.11f and 12.11j are now available for estimating the truncation error.

New light is shed on these expansions by identifying the denominator polynomials in terms of known functions. By Corollary 12.11c the polynomials $q_{2n}(-\tau) = p_n^{(m)}(-\tau)$ are identical with the orthogonal polynomials associated with $\tau^m\, d\psi(\tau)$, normalized such that the leading coefficient is $(-1)^n$. Now the **Jacobi polynomials**

$$P_n^{(\mu,\nu)}(x) := \frac{(\mu+1)_n}{n!}\,{}_2F_1\!\left(-n, n+\mu+\nu+1; \mu+1; \frac{1-x}{2}\right) \qquad (12.12\text{-}3)$$

are known to form an orthogonal set with respect to

$$d\chi(x) := (1-x)^\mu (1+x)^\nu\, dx, \qquad -1 \leq x \leq 1$$

(see Szegö [1959], Chapter 6). Upon setting $x = 1 - 2\tau$, $\tau = (1-x)/2$ we see that the polynomials $P_n^{(\mu,\nu)}(1-2\tau)$ are orthogonal on $[0, 1]$ with respect to $d\chi(\tau) := \tau^\mu(1-\tau)^\nu\, d\tau$. This agrees with $\tau^m\, d\psi(\tau)$ if $\mu = \alpha + m - 1$, $\nu = \gamma - \alpha - 1$. Thus $p_n^{(m)}(-\tau)$ is proportional to $P_n^{(\alpha+m-1,\gamma-\alpha-1)}(1-2\tau)$. The leading coefficient in the latter polynomial is

$$(-1)^n \frac{(n+m+\gamma-1)_n}{n!}.$$

We thus have

$$p_n^{(m)}(-\tau) = \frac{n!}{(n+m+\gamma-1)_n} P_n^{(\alpha+m-1,\gamma-\alpha-1)}(1-2\tau). \quad (12.12\text{-}4)$$

Interesting formulas are obtained by using the identity (12.4-33). From (12.5-9) we find

$$q_1^{(m)} e_1^{(m)} \cdots q_n^{(m)} e_n^{(m)} = \frac{n!(\alpha+m)_n(\gamma-\alpha)_n}{(\gamma+n+m-1)_n(\gamma+m)_{2n}}.$$

Expressing the even part of the continued fraction in (12.12-2) by (12.4-33) and taking into account (12.12-4), we thus get

$$\frac{\Gamma(\gamma)}{\Gamma(\alpha)\Gamma(\gamma-\alpha)} \int_0^1 \frac{\tau^{\alpha-1}(1-\tau)^{\gamma-\alpha-1}}{z+\tau} d\tau$$

$$= \frac{1}{z} - \frac{\alpha}{\gamma}\frac{1}{z^2} + \frac{(\alpha)_2}{(\gamma)_2}\frac{1}{z^3} - \cdots + (-1)^{m+1}\frac{(\alpha)_{m-1}}{(\gamma)_{m-1}}\frac{1}{z^m}$$

$$+ \frac{(-1)^m}{z^m} \sum_{n=0}^\infty \frac{(\alpha)_{m+n}(\gamma-\alpha)_n(\gamma+m+2n)}{(n+1)!(\gamma)_{m+n}}$$

$$\cdot \frac{1}{P_n^{(\alpha+m-1,\gamma-\alpha-1)}(1+2z)P_{n+1}^{(\cdots)}(1+2z)} \qquad (12.12\text{-}5)$$

The series (being equivalent to a convergent continued fraction) converges for all $z \in S$ and for $m = 0, 1, 2, \ldots$. Some special cases are as follows:

EXAMPLE **1**

For $\alpha = 1$, $\gamma = 2$ the integral is a logarithm, and we get

$$\mathrm{Log}\left(1+\frac{1}{z}\right) = \frac{1}{z} - \frac{1}{2z^2} + \frac{1}{3z^3} - \cdots + \frac{(-1)^{m-1}}{mz^m}$$

$$+ \frac{(-1)^m}{z^m} \sum_{n=0}^\infty \frac{2+m+2n}{(m+n+1)(n+1)} \frac{1}{P_n^{(m,0)}(1+2z)P_{n+1}^{(m,0)}(1+2z)}. \qquad (12.12\text{-}6)$$

For $m = 0$ the Jacobi polynomials reduce to Legendre polynomials, and we obtain

$$\mathrm{Log}\left(1+\frac{1}{z}\right) = \sum_{n=0}^\infty \frac{2}{(n+1)P_n(1+2z)P_{n+1}(1+2z)} \qquad (12.12\text{-}7)$$

By Corollary 12.11g the speed of convergence of this series is at least geometric. A more precise result can be obtained by using the asymptotic formulas for P_n derived by Darboux' method in §11.10. Those results imply that for fixed z outside the interval $[-1, 0]$, as $n \to \infty$,

$$P_n(1+2z) = \frac{[z^{1/2}+(1+z)^{1/2}]^{2n}}{2(\pi n)^{1/2}(z+z^2)^{1/4}}\left\{1+O\left(\frac{1}{n}\right)\right\}.$$

Thus the series (12.12-7) converges roughly like a geometric series with ratio $q := [(1+z)^{1/2}+z^{1/2}]^{-4}$. For $z = \frac{1}{2}$ (a value for which the ordinary power series for $\mathrm{Log}(1+1/z)$ diverges) we find $q = 0.0718$.

EXAMPLE 2

For $\alpha = \frac{1}{2}$, $\gamma = \frac{3}{2}$ we find

$$\mathscr{S}\psi(z^2) = \frac{\Gamma(1/2)}{\Gamma(1/2)\Gamma(1)} \int_0^1 \frac{\tau^{-1/2}}{z^2+\tau} d\tau = \frac{1}{z}\mathrm{Arctan}\frac{1}{z};$$

hence from (12.12-5)

$$\mathrm{Arctan}\frac{1}{z} = \frac{1}{z} - \frac{1}{3z^3} + \frac{1}{5z^5} - \cdots + \frac{(-1)^{m-1}}{(2m-1)z^{2m-1}}$$

$$+ \frac{(-1)^m}{z^{2m-1}} \sum_{n=0}^{\infty} \frac{2m+4n+3}{(2m+2n+1)(2n+2)} \frac{1}{P_n^{(m-1/2,0)}(1+2z^2)P_{n+1}^{(m-1/2,0)}(1+2z^2)}.$$

A simplification arises for $m = 0$. By a result in the theory of Legendre functions (see Szegö [1959], p. 59)

$$P_n^{(-1/2,0)}(1+2z^2) = (-1)^n P_{2n}(iz)$$

(Because P_{2n} is an even polynomial with real coefficients, the values of $P_{2n}(iz)$ are real for z real.) Replacing z by z^{-1} we thus get

$$\mathrm{Arctan}\, z = - \sum_{n=0}^{\infty} \frac{4n+3}{(2n+1)(2n+3)} \frac{1}{P_{2n}(i/z)P_{2n+2}(i/z)}. \tag{12.12-8}$$

It follows from §12.5 that this expansion holds for all z that are not of the form $z = ix$, x real, $|x| \geq 1$. The convergence for real z is comparable to that of a geometric series with ratio

$$q := \left[\frac{z}{z+\sqrt{z^2+1}}\right]^4.$$

II. Hypergeometric Distribution

Let $0 < \alpha < \gamma, 0 < \beta < 1$. We assert that the function defined by $\psi(0) = 0$,

$$d\psi(\tau) = \frac{\Gamma(\gamma)}{\Gamma(\alpha)\Gamma(\beta)\Gamma(\gamma-\alpha-\beta+1)} \tau^{\alpha-1}(1-\tau)^{\gamma-\beta-\alpha}$$

$$\cdot {}_2F_1(1-\beta, \gamma-\beta; \gamma-\alpha-\beta+1; 1-\tau)d\tau \quad 0 \leq \tau \leq 1,$$

$$d\psi(\tau) = 0, \quad \tau > 1,$$

$$\left.\begin{array}{c}\\\\\\\\\end{array}\right\} \tag{12.12-9}$$

belongs to the class Ψ^*. Indeed, the Γ factors are positive, and Riemann's representation (8.7-6),

$$_2F_1(1-\beta, \gamma-\beta; \gamma-\alpha.-\beta+1; 1-\tau)$$

$$= \frac{\Gamma(\gamma-\alpha-\beta+1)}{\Gamma(1-\beta)\Gamma(\gamma-\alpha)} \int_0^1 \sigma^{-\beta}(1-\sigma)^{\gamma-\alpha-1}[1-(1-\tau)\sigma]^{\gamma-\beta}\, d\sigma,$$

which holds for $1-\beta>0$, $\gamma-\alpha>0$ shows that the hypergeometric function is positive. Integrability of $d\psi(\tau)$ at $\tau=1$ is assured in view of $\gamma-\alpha>0$. From (9.9-26) there follows

$$_2F_1(1-\beta, \gamma-\beta; \gamma-\alpha-\beta+1; 1-\tau)$$

$$= \frac{\Gamma(\gamma-\alpha-\beta+1)\Gamma(\beta-\alpha)}{\Gamma(\gamma-\alpha)\Gamma(1-\alpha)}{}_2F_1(1-\beta, \gamma-\beta; \alpha-\beta+1; \tau)$$

$$+ \frac{\Gamma(\gamma-\alpha-\beta+1)\Gamma(\alpha-\beta)}{\Gamma(\gamma-\beta)\Gamma(1-\alpha)}\tau^{\beta-\alpha}{}_2F_1(\gamma-\alpha, 1-\alpha; \beta-\alpha+1; \tau),$$

showing that at $\tau=0$ $d\psi(\tau)$ behaves like $\tau^{\beta-1}\, d\tau$. For the moments μ_n we find

$$\frac{\Gamma(\alpha)\Gamma(\beta)\Gamma(\gamma-\beta-\alpha+1)}{\Gamma(\gamma)}\mu_n$$

$$= \int_0^1 \tau^{\alpha-1+n}(1-\tau)^{\gamma-\beta-\alpha}{}_2F_1(1-\beta, \gamma-\beta; \gamma-\alpha-\beta+1; 1-\tau)\, d\tau.$$

Integrating the hypergeometric series term by term and summing the resulting series by Gauss' formula (8.6-6), the integral becomes

$$\sum_{k=0}^{\infty} \frac{(1-\beta)_k(\gamma-\beta)_k}{(\gamma-\alpha-\beta+1)_k k!} \frac{\Gamma(\alpha+n)\Gamma(\gamma-\beta-\alpha+1+k)}{\Gamma(\gamma-\beta+k+n+1)}$$

$$= \frac{\Gamma(\alpha+n)\Gamma(\gamma-\beta-\alpha+1)}{\Gamma(\gamma-\beta+n+1)}{}_2F_1(1-\beta, \gamma-\beta; \gamma-\beta+n+1; 1)$$

$$= \frac{\Gamma(\alpha+n)\Gamma(\beta+n)\Gamma(\gamma-\beta-\alpha+1)}{\Gamma(\gamma+n)n!}.$$

Thus there follows

$$\mu_n = \frac{(\alpha)_n(\beta)_n}{(\gamma)_n n!}, \qquad n=0, 1, 2, \ldots. \tag{12.12-10}$$

Under the conditions on α, β, γ stated initially there follows

$$\frac{1}{z}{}_2F_1\left(\alpha, \beta; \gamma; -\frac{1}{z}\right) = \int_0^1 \frac{1}{z+\tau}\, d\psi(\tau). \tag{12.12-11}$$

It does not seem possible, except in the case $\beta = 1$ already noted, to give explicit formulas for the qd scheme associated with this series. However, the scheme can always be constructed numerically, and the continued fraction derived from it will converge, for all $z \in S$, to the foregoing series or its analytic continuation into S.

III. Laguerre Distribution

We next let

$$d\psi(\tau) := \frac{1}{\Gamma(\alpha)} \tau^{\alpha-1} e^{-\tau} d\tau, \qquad \tau \geq 0, \qquad (12.12\text{-}12)$$

where $\alpha > 0$. Clearly, $\psi \in \Psi^*$. For the moments we find

$$\mu_n = \frac{1}{\Gamma(\alpha)} \int_0^\infty \tau^{\alpha-1+n} e^{-\tau} d\tau = \frac{\Gamma(\alpha+n)}{\Gamma(\alpha)} = (\alpha)_n, \qquad n = 0, 1, \ldots.$$

The Carleman condition (12.11-25) is easily seen to be satisfied by Stirling's formula. It follows, first of all, that

$$f(z) := \frac{1}{\Gamma(\alpha)} \int_0^\infty \frac{\tau^{\alpha-1} e^{-\tau}}{z+\tau} d\tau \approx \frac{1}{z} - \frac{\alpha}{z^2} + \frac{(\alpha)_2}{z^3} - \cdots$$

as $z \to \infty$, $z \in S_\delta$ for any $\delta < \pi$. The qd scheme associated with the formal series is easily constructed, with the result

$$q_k^{(m)} = 1 - \alpha - m - k, \qquad e_k^{(m)} = -k. \qquad (12.12\text{-}13)$$

With these values of $q_k^{(m)}$ and $e_k^{(m)}$ we thus have

$$\frac{1}{\Gamma(\alpha)} \int_0^\infty \frac{\tau^{\alpha-1} e^{-\tau}}{z+\tau} d\tau$$

$$= \frac{1}{z} - \frac{\alpha}{z^2} + \frac{(\alpha)_2}{z^3} - \cdots + (-1)^{m-1} \frac{(\alpha)_{m-1}}{z^m} \qquad (12.12\text{-}14)$$

$$+ (-1)^m \frac{(\alpha)_m}{z^{m+1}} \left\{ \frac{1}{1} \left| - \frac{q_1^{(m)}}{z} \right| - \frac{e_1^{(m)}}{1} \left| - \frac{q_2^{(m)}}{z} \right| - \cdots \right\},$$

which holds for $m = 0, 1, 2, \ldots$ and for all $z \in S$.

EXAMPLE **3**

For $\alpha = 1$ there results the convergent representation for the exponential integral

$$e^z E(z) = \int_0^\infty \frac{e^{-\tau}}{z+\tau} d\tau = \frac{1}{z} - \frac{1}{z^2} + \frac{2!}{z^3} - \cdots + (-1)^{m-1} \frac{(m-1)!}{z^m}$$

$$+ (-1)^m \frac{m!}{z^{m+1}} \left\{ \frac{1}{1} \left| + \frac{m+1}{z} \right| + \frac{1}{1} \left| + \frac{m+2}{z} \right| + \frac{2}{1} \left| + \cdots \right\}.$$

Another way of writing (12.12-14) is by using the identity (12.4-33). To this end we must identify the polynomials $p_n^{(m)}(-\tau)$. By Corollary (12.11c) these are the orthogonal polynomials for the weight function $\tau^{\alpha+m-1}e^{-\tau}$ $(0 \leqslant \tau < \infty)$, normalized to have leading coefficient $(-1)^n$. The Laguerre polynomials (see Szegö [1959], p. 99)

$$L_n^{(\alpha+m-1)}(\tau):=\frac{(\alpha+m)_n}{n!}\,_1F_1(-n;\alpha+m;\tau)$$

are orthogonal with respect to the same weight function. Their leading coefficient being $(-1)^n(n!)^{-1}$, there follows

$$p_n^{(m)}(\tau)=n!L_n^{(\alpha+m-1)}(-\tau).$$

In view of

$$q_1^{(m)}e_1^{(m)}\cdots q_n^{(m)}e_n^{(m)}=n!(\alpha+m)_n,$$

(12.4-33) yields

$$\frac{1}{\Gamma(\alpha)}\int_0^\infty \frac{\tau^{\alpha-1}e^{-\tau}}{z+\tau}d\tau=\frac{1}{z}-\frac{\alpha}{z^2}+\cdots+(-1)^{m+1}\frac{(\alpha)_{m-1}}{z^m}$$

$$+(-1)^m\frac{(\alpha)_m}{z^m}\sum_{n=0}^\infty \frac{(\alpha+m)_n}{(n+1)!}\frac{1}{L_n^{(\alpha-1+m)}(-z)L_{n+1}^{(\alpha-1+m)}(-z)}. \tag{12.12-15}$$

EXAMPLE **4**

For $\alpha=1$, $m=0$ the following expansion for the exponential integral is obtained:

$$e^z E(z)=\int_0^\infty \frac{e^{-\tau}}{z+\tau}d\tau=\sum_{n=0}^\infty \frac{1}{(n+1)L_n(-z)L_{n+1}(-z)}. \tag{12.12-16}$$

Because the elements of the continued fraction are known explicitly, the truncation error of the fraction in (12.12-14) [or of the series in (12.12-15)] can be estimated by means of the bound (A) of Theorem 12.11f. However, the bound (B) is somewhat easier to apply. For instance if $\alpha=1$, $m=0$, then in view of $\mu_n=n!$ we get, using the notation of Theorem 12.11f,

$$|w_{n+1}-w_n|\leqslant\frac{1}{|z|\cos(\phi/2)}\left(\frac{1+\sqrt{5}}{2}\right)^{1/2}\{1+\sigma(n!)^{-1/2n}\}^{-n/2}.$$

By Stirling's formula $(n!)^{-1/2n}\sim(e/n)^{1/2}$; hence

$$\{1+\sigma(n!)^{-1/2n}\}^{-n/2}\sim e^{-(\sigma/2)\sqrt{en}} \qquad (n\to\infty).$$

Thus the estimate predicts a subgeometric rate of convergence for the continued fraction. A more detailed analysis involving the asymptotic behavior of the Laguerre polynomials shows that the truncation error after the nth approximant is indeed asymptotic to $\exp(-\text{const.}\sqrt{n})$. Thus to replace the error by its square, four times as many approximants are required.

IV. The Gamma Function

It was shown in §8.5 that the logarithm of the Γ function has for $z^2 \in S$ the representation

$$\log \Gamma(z) = (z - \tfrac{1}{2}) \operatorname{Log} z - z + \tfrac{1}{2} \operatorname{Log} 2\pi + J(z),$$

where $J(z)$ denotes Binet's function,

$$J(z) := \frac{1}{\pi} \int_0^\infty \frac{z}{z^2 + \tau^2} \operatorname{Log} \frac{1}{1 - e^{-2\pi\tau}} \, d\tau.$$

Thus if

$$f(z) := \frac{1}{\sqrt{z}} J(\sqrt{z}) \qquad \text{(principal values)}$$

then $f = \mathcal{S}\psi$ where

$$\psi(\tau) := \frac{1}{2\pi} \int_0^\tau \frac{1}{\sqrt{\sigma}} \operatorname{Log} \frac{1}{1 - e^{-2\pi\sqrt{\sigma}}} \, d\sigma. \qquad (12.12\text{-}17)$$

Clearly, $\psi \in \Psi^*$. The moments of ψ were calculated in §11.1 with the result (see also §11.2)

$$\mu_n = \frac{(-1)^n}{(2n+1)(2n+2)} B_{2n+2}, \qquad n = 0, 1, 2, \ldots, \qquad (12.12\text{-}18)$$

where B_m is the mth Bernoulli number. In view of

$$B_{2n} \sim (-1)^n \frac{2(2n)!}{(2\pi)^{2n}} \qquad (n \to \infty)$$

the Carleman condition is easily seen to be satisfied, and the formal series $\sum (-1)^n \mu_n z^{-n-1}$ may be converted into an S fraction converging for $z \in S$. The coefficients of the qd scheme cannot be expressed by simple formulas, nor are the polynomials $p_n^{(m)}$ related to any known set of orthogonal polynomials. Numerical computation yields the values in Table 12.12.

Table 12.12.   *qd* Scheme of Asymptotic Series for $f(z)$

n	B_{2n+2}	μ_n	$q_1^{(n)}$	$e_1^{(n)}$	$q_2^{(n)}$	$e_2^{(n)}$
0	$\dfrac{1}{6}$	$\dfrac{1}{12}$				
			$-\dfrac{1}{30}$			
1	$-\dfrac{1}{30}$	$\dfrac{1}{12\cdot 30}$		$-\dfrac{53}{210}$		
			$-\dfrac{2}{7}$		$-\dfrac{195}{371}$	
2	$\dfrac{1}{42}$	$\dfrac{1}{30\cdot 42}$		$-\dfrac{13}{28}$		$-\dfrac{22999}{22737}$
			$-\dfrac{3}{4}$		$-\dfrac{1841}{1716}$	
3	$-\dfrac{1}{30}$	$\dfrac{1}{56\cdot 30}$		$-\dfrac{263}{396}$		
			$-\dfrac{140}{99}$			
4	$\dfrac{5}{66}$	$\dfrac{5}{90\cdot 66}$				

It follows that for Re $z > 0$

$$J(z) = zf(z^2) = \cfrac{1}{|\,12z\,}} + \cfrac{2}{|\,5z\,}} + \cfrac{53}{|\,42z\,}} + \cfrac{1170}{|\,53z\,}} + \cdots \qquad (12.12\text{-}19)$$

Considering approximants, we find that for $x > 0$

$$J(x) < \frac{1}{12x}, \; J(x) > \frac{5x}{60x^2+2}, \; J(x) < \frac{210x+53x}{2520x^3+720x} \qquad (12.12\text{-}20)$$

The speed of convergence of (12.12-19) may be assessed by Theorem 12.11f. The estimate (A) cannot be used because the α_k are not known, and the estimate (B) does not tend to zero for $n \to \infty$. However, an application of Stirling's formula shows that

$$\left(\frac{\mu_0}{\mu_n}\right)^{1/2n} \geq \exp\left(-\frac{1}{4e}\right)\frac{\pi e}{n}, \qquad n = 1, 2, \ldots .$$

Thus

$$\frac{1}{e}\sum_{k=1}^{n}\left(\frac{\mu_0}{\mu_k}\right)^{1/2k} \geq \pi\,\exp\left(-\frac{1}{4e}\right)\sum_{k=1}^{n}\frac{1}{k} \geq \pi\,\exp\left(-\frac{1}{4e}\right)\text{Log } n,$$

and (C) for $z = x > 0$ yields the bound

$$|w_{n+1} - w_n| \leq \frac{1}{x}\sqrt{\frac{1+\sqrt{5}}{2}}[1 + 2x\pi e^{-1/4e}\,\text{Log } n]^{-1/2}.$$

This tends to zero as $n \to \infty$, although very slowly. Numerical experiments have shown that the convergence of the fraction is, in fact, very slow. It is interesting to note that the first few approximants (up to about $n = 7$) yield relatively good approximations. This follows from (B) by virtue of the fact that the first few Bernoulli numbers are abnormally small.

PROBLEMS

1. Show that

$$f(z) := \int_0^\infty \frac{e^{-\sqrt{\tau}}}{z + \tau} \, d\tau$$

is value function of a convergent S fraction and compute the first few approximants.

2. Let

$$f(z) := \int_0^\infty \frac{e^{-\tau}}{z + \tau} \, d\tau.$$

Draw the inclusion regions W_n described in Theorem 12.11j for the Stieltjes fraction representation of $f(1 + i)$, and obtain a numerical value that is in error by at most 10^{-2}.

3. Let $\psi \in \Psi^*$, with moments $\mu_0, \mu_1, \ldots$; let $f := \mathscr{S}\psi$; and let $s_0 := 0$, $s_1 := \mu_0/z$, $s_2, \ldots$ denote the partial sums of the asymptotic expansion

$$f(z) \approx \frac{\mu_0}{z} - \frac{\mu_1}{z^2} + \cdots.$$

Show that for $m = 1, 2, \ldots$ and for all $z \in S$, $f(z)$ is contained in the set $W^{(m)}$ bounded by the straight line segment from s_{m-1} to s_m and by the circular arc from s_{m-1} to s_m emanating from s_{m-1} in the direction $z(s_m - s_{m-1})$ (see Fig. 12.12a).

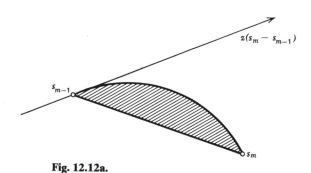

Fig. 12.12a.

[Consequence of Theorem 12.11i.]

§12.13 S FRACTIONS: EXPANSIONS OF ITERATED LAPLACE TRANSFORMS

We recall the *original space* Ω defined in §10.1 (roughly, the space of functions $\phi : \mathbb{R} \rightarrow \mathbb{C}$ that are zero for $\tau < 0$, piecewise continuous for $\tau > 0$ and absolutely integrable at $\tau = 0$). Let $\phi \in \Omega$, and let $f := \mathcal{L}\phi$, the Laplace transform of ϕ:

$$f(z) := \int_0^\infty e^{-sz} \phi(s)\, ds. \qquad (12.13\text{-}1)$$

Assume now that ϕ itself is a Laplace transform, $\phi = \mathcal{L}\omega$ for some $\omega \in \Omega$,

$$\phi(s) = \int_0^\infty e^{-\tau s} \omega(\tau)\, d\tau. \qquad (12.13\text{-}2)$$

Suppose this integral converges absolutely for $\text{Re } s \geq 0$. Then ϕ is analytic in $\text{Re } s > 0$ and bounded for $\text{Re } s \geq 0$. By Theorem 10.9f, $f = \mathcal{L}\phi$ is analytic in the cut plane S. In fact, by substituting (12.13-2) in (12.13-1) and changing the order of integrations in the double integral

$$f(z) = \int_0^\infty e^{-sz} \left\{ \int_0^\infty e^{-\tau s} \omega(\tau)\, d\tau \right\} ds$$

by appealing to Fubini's theorem (see Widder [1946], p. 335) we obtain

$$f(z) = \int_0^\infty \left\{ \int_0^\infty e^{-(z+\tau)s}\, ds \right\} \omega(\tau)\, d\tau = \int_0^\infty \frac{\omega(\tau)}{z+\tau}\, d\tau$$

(see also §10.11). Suppose now that ω is real and nonnegative, but not equivalent to the zero function in Ω. Then (because $\int_0^\infty \omega(\tau)\, d\tau < \infty$) the function

$$\psi(\tau) := \int_0^\infty \omega(\sigma)\, d\sigma$$

is a bounded, nondecreasing function with infinitely many points of increase, and f is the Stieltjes transform of ψ. Suppose further ω to be such that $\phi = \mathcal{L}\omega$ possesses an asymptotic expansion

$$\phi(s) \approx a_0 + a_1 s + a_2 s^2 + \ldots, \qquad s \rightarrow 0, \qquad s > 0. \qquad (12.13\text{-}3)$$

Then by the Watson–Doetsch lemma (Theorem 11.5) $f = \mathcal{L}\phi$ possesses the asymptotic expansion

$$f(z) \approx P(z) := \sum_{n=0}^\infty \frac{n!\, a_n}{z^{n+1}}, \qquad z \rightarrow \infty, \qquad z \in \hat{S}_{\pi/2-\epsilon} \qquad (12.13\text{-}4)$$

for every $\epsilon > 0$. In particular, this expansion holds if $z \to \infty$ through positive values. By Theorem 12.9h we conclude that the moments μ_n of ψ exist,

$$\mu_n = (-1)^n n! a_n, \qquad n = 0, 1, \ldots. \tag{12.13-5}$$

There follows $\psi \in \Psi^*$, and if the a_n are such that the μ_n satisfy the Carleman condition, then f (or any of the remainders of its asymptotic expansion) is represented by the S fraction corresponding to P (or to the appropriate remainder of P). By Stirling's formula, the Carleman condition is satisfied if and only if

$$\sum_{n=1}^{\infty} \frac{1}{\sqrt{n}} |a_n|^{-1/2n} = \infty. \tag{12.13-6}$$

In summary, we have

THEOREM 12.13a

Let ω be a real, nonnegative function in the class Ω, and let $\phi := \mathcal{L}\omega$ possess the asymptotic expansion (12.13-3) as $s \to 0$. Then $\psi := \mathcal{L}\omega \in \Psi^*$, $f := \mathcal{L}\phi = \mathcal{L}\psi$, and if (12.13-6) holds the conclusion of Theorem 12.12 applies to f.

The hypothesis that ϕ possesses the asymptotic expansion (12.13-3) is trivially satisfied if α_ω, the abscissa of simple convergence of $\mathcal{L}\omega$, is negative. By Theorem 10.1d, ϕ then is analytic for Re $s > \alpha_\omega$, and the series (12.13-3) is simply the Taylor series of ϕ at 0.

Some applications of Theorem 12.13a follow.

I.   Incomplete Gamma Function

Here we consider the incomplete gamma function defined, for arbitrary complex a and for $z \in S$, by

$$\Gamma(a, z) := \int_z^{\infty} t^{a-1} e^{-t} dt, \tag{12.13-7}$$

where the path of integration runs parallel to the real axis, and where t^{a-1} has its principal value. In §11.5 this was cast in the form

$$\Gamma(a, z) = z^a e^{-z} \int_0^{\infty} e^{-zs} (1+s)^{a-1} ds,$$

which permitted to obtain the asymptotic expansion (11.5-8) by the Watson–Doetsch lemma. For Re $a < 1$,

$$\phi(s) := (1+s)^{a-1}$$

is the Laplace transform of

$$\omega(\tau):=\frac{1}{\Gamma(1-a)}\,e^{-\tau}\tau^{-a},$$

which for real $a < 1$ satisfies the hypotheses of Theorem 12.12. For such a and for all $z \in S$ we thus have

$$\Gamma(a, z) = \frac{z^a e^{-z}}{\Gamma(1-a)} \int_0^\infty \frac{\tau^{-a} e^{-\tau}}{z + \tau}\,d\tau$$

and hence, as a special case of (12.12-14),

$$\Gamma(a, z) = z^a e^{-z} \left\{ \frac{1}{\mid z} + \frac{1-a}{\mid 1} + \frac{1}{\mid z} + \frac{2-a}{\mid 1} + \frac{2}{\mid z} + \frac{3-a}{\mid 1} + \frac{3}{\mid z} + \cdots \right\}.$$

$$(12.13-8)$$

Using the identity (12.4-33) there results

$$\Gamma(a, z) = z^a e^{-z} \sum_{n=0}^\infty \frac{(1-a)_n}{(n+1)!} \frac{1}{L_n^{(-a)}(-z) L_{n+1}^{(-a)}(-z)}, \qquad (12.13-9)$$

where $L_n^{(-a)}$ denotes the Laguerre polynomial. By analytic continuation, this expansion holds for all $z \in S$ not only if $a < 1$, but also for any complex a such that $-z$ is not a zero of some polynomial $L_n^{(-a)}$.

From the known asymptotic behavior of the Laguerre polynomials (see Szegö [1959], Theorem 8.22.5) it can be deduced that the nth term of the foregoing series (including the factor $z^a e^{-z}$) is asymptotic to

$$4\pi\sqrt{\frac{z}{n}}\,e^{-4\sqrt{nz}}.$$

The convergence of the series thus is sublinear. More precisely, the number n of terms required to obtain k correct decimal digits for $z = x > 0$ asymptotically equals

$$n \sim 0.33 \frac{k^2}{x}.$$

Numerical experiments have shown that the series is indeed well suited for the numerical computation of $\Gamma(a, x)$ except for very small values of $x > 0$. For purposes of computation the quantities

$$l_n := L_n^{(-a)}(-x)$$

are computed by means of the recurrence relation

$$l_{-1}=0, \qquad l_0=1,$$

$$l_n = \frac{1}{n}[(2n-a-1+x)l_{n-1}-(n-a-1)l_{n-2}]$$

$$= \frac{1}{n}[(n-a-1)(l_{n-1}-l_{n-2})+(n+x)l_{n-1}],$$

$n=1,2,\ldots$, which is a corollary of the recurrence relation satisfied by the Laguerre polynomials.

For the **exponential integral of order** k,

$$E_k(z):=\int_1^\infty \frac{e^{-zt}}{t^k}\,dt = z^{k-1}\Gamma(1-k,z),$$

$k=1,2,\ldots$, we obtain as a special case of (12.13-9)

$$E_k(z)=e^{-z}\sum_{n=0}^\infty \frac{(k)_n}{(n+1)!}\frac{1}{L_n^{(k-1)}(-z)L_{n+1}^{(k-1)}(-z)}, \qquad (12.13\text{-}10)$$

and for the error function complement

$$\operatorname{erfc}(z) = \frac{2}{\sqrt{\pi}}\int_z^\infty e^{-t^2}\,dt = \frac{1}{\sqrt{\pi}}\Gamma\left(\frac{1}{2},z^2\right)$$

we obtain similarly

$$\operatorname{erfc}(z) = \frac{z\,e^{-z^2}}{\sqrt{\pi}}\sum_{n=0}^\infty \frac{(1/2)_n}{(n+1)!}\frac{1}{L_n^{(-1/2)}(-z^2)L_{n+1}^{(-1/2)}(-z^2)}.$$

$$(12.13\text{-}11)$$

II. Confluent Hypergeometric Functions

Here the results of §12.11 are applied to the Whittaker function $W_{\kappa,\mu}(z)$ defined by (10.5-14), which covers all instances of the confluent hypergeometric function. For reasons of formal symmetry it is convenient in the present context to work with a function $\Phi_{\alpha,\beta}(z)$ which is elementarily related to $W_{\kappa,\mu}(z)$ by

$$\Phi_{\alpha,\beta}(z):=z^{-\kappa-1}\,e^{z/2}\,W_{\kappa,\mu}(z), \qquad (12.13\text{-}12)$$

where $2\kappa=-\alpha-\beta$, $2\mu=\alpha-\beta$. Conversely,

$$W_{\kappa,\mu}(z)=z^{\kappa+1}\,e^{-z/2}\Phi_{\mu-\kappa,-\mu-\kappa}(z) \qquad (12.13\text{-}13)$$

From (10.5-14) there follows the representation of $\Phi_{\alpha,\beta}$ by confluent hypergeometric series,

$$\Phi_{\alpha,\beta}(z) = \frac{\Gamma(\alpha-\beta)}{\Gamma(\alpha+1/2)} z^{\beta-1/2} {}_1F_1(\beta+\tfrac{1}{2}; \beta-\alpha+1; z)$$

$$(12.13\text{-}14)$$

$$+\frac{\Gamma(\beta-\alpha)}{\Gamma(\beta+1/2)} z^{\alpha-1/2} {}_1F_1(\alpha+\tfrac{1}{2}; \alpha-\beta+1; z),$$

making evident that

$$\Phi_{\alpha,\beta}(z) = \Phi_{\beta,\alpha}(z). \qquad (12.13\text{-}15)$$

By a special case of (10.7-15),

$$\Phi_{\alpha,\beta}(z) = \int_0^\infty e^{-sz} {}_2F_1(\alpha+\tfrac{1}{2}, \beta+\tfrac{1}{2}; 1; -s)\, ds \qquad (12.13\text{-}16)$$

holds for arbitrary α and β. By the Watson–Doetsch lemma there immediately follows the asymptotic expansion

$$\Phi_{\alpha,\beta}(z) \approx \sum_{n=0}^\infty (-1)^n \frac{(\alpha+1/2)_n (\beta+1/2)_n}{n! z^{n+1}} \qquad (12.13\text{-}17)$$

as $z \to \infty$, $z \in \hat{S}_{\pi/2-\epsilon}$. Using Euler's second transformation (9.9-23), we have

$${}_2F_1(\alpha+\tfrac{1}{2}, \beta+\tfrac{1}{2}; 1; -s) = (1+s)^{-\alpha-1/2} {}_2F_1\left(\alpha+\tfrac{1}{2}, \tfrac{1}{2}-\beta; 1; \frac{s}{s+1}\right),$$

showing that

$$\Phi_{\alpha,\beta}(z) = \int_0^\infty e^{-sz} (1+s)^{-\alpha-1/2} {}_2F_1\left(\alpha+\tfrac{1}{2}, \tfrac{1}{2}-\beta; 1; \frac{s}{s+1}\right) ds.$$

If $\alpha \geqslant -\tfrac{1}{2}, \beta \leqslant \tfrac{1}{2}$, the integrand is on the whole interval of integration represented by a convergent series whose terms are nonnegative. There follows

$$\Phi_{\alpha,\beta}(z) \geqslant 0 \quad \text{for} \quad z > 0, \alpha \geqslant -\tfrac{1}{2}, \beta \leqslant \tfrac{1}{2} \qquad (12.13\text{-}18)$$

(or $\alpha \leqslant \tfrac{1}{2}, \beta \geqslant -\tfrac{1}{2}$).

To apply Theorem 12.13a, we must show that the function ${}_2F_1$ is a Laplace transform, and that its original function is nonnegative. It turns out that ${}_2F_1$ is the Laplace transform of a function elementarily related to $\Phi_{-\alpha,-\beta}$. The following result is a special case of (10.5-15):

LEMMA 12.13b

If $\alpha > -\tfrac{1}{2}, \beta > -\tfrac{1}{2}, \operatorname{Re} s > -1$, then

$${}_2F_1(\tfrac{1}{2}+\alpha, \tfrac{1}{2}+\beta; 1; -s)$$

$$(12.13\text{-}19)$$

$$= \frac{1}{\Gamma(\alpha+1/2)\Gamma(\beta+1/2)} \int_0^\infty e^{-s\tau} e^{-\tau} \tau^{\alpha+\beta} \Phi_{-\alpha,-\beta}(\tau)\, d\tau.$$

By (12.13-18),

$$\omega(\tau) := e^{-\tau}\tau^{\alpha+\beta}\Phi_{-\alpha,-\beta}(\tau) \geq 0$$

if $\alpha \leq \frac{1}{2}$ and $\beta \leq -\frac{1}{2}$ or vice versa. The coefficients in the asymptotic series (12.13-17) are easily seen to satisfy the Carleman condition. Thus there follows

THEOREM 12.13c

Let $\alpha > -\frac{1}{2}$, $-\frac{1}{2} < \beta \leq \frac{1}{2}$ (*or vice versa*). *Then the function* $\Phi_{\alpha,\beta}(z)$ *is for all* $z \in S$ *represented by the S fraction corresponding to the series*

$$\frac{1}{z}{}_2F_0\left(\alpha + \frac{1}{2}, \beta + \frac{1}{2}; \; -\frac{1}{z}\right).$$

Explicit formulas for the *qd* scheme associated with this series $_2F_0$ are known only in the case $\alpha = \frac{1}{2}$ (or $\beta = \frac{1}{2}$). The scheme then is identical with the scheme for the Laguerre distribution. In general, the scheme must be generated numerically. To avoid problems of numerical instability, it is advisable to perform the necessary operations in rational arithmetic. An alternate possibility is to use the "incremental version" of the *qd* algorithm described by Gargantini and Henrici [1967].

An important special case of the function $\Phi_{\alpha,\beta}$ is obtained for $\beta = -\alpha$ in view of the relation

$$K_\nu(z) = \sqrt{2\pi z}\, e^{-z}\Phi_{\nu,-\nu}(2z). \tag{12.13-20}$$

Here $K_\nu(z)$ denotes the modified Bessel function,

$$K_\nu(z) = \frac{\pi}{2}i\, e^{\nu\pi i/2}H_\nu^{(1)}(iz),$$

where $H_\nu^{(1)}$ is the Hankel function. Table 12.13 presents an excerpt of the *qd* scheme associated with

$$\frac{1}{2z}{}_2F_0\left(\frac{1}{2}, \frac{1}{2}; \; -\frac{1}{2z}\right)$$

that is required to compute $K_0(z)$. A more extensive table, thus far unpublished, giving $q_k^{(n)}$ and $e_k^{(n)}$ for $k = 1, 2, \ldots, 26$ and $n = 1, 2, \ldots, 51 - k$ as both rational fractions and 80-digit decimal fractions was computed by Professor D. Cantor at UCLA. The numerator and the denominator of the fraction representing $q_{26}^{(0)} = -12.60535\ldots$ are integers with 1069 and 1068 decimal digits, respectively.

Table 12.13. qd Scheme Associated with $K_0(z)$; Entries Are $-q_k^{(n)}$, $-e_k^{(n)}$

q_1	e_1	q_2	e_2	q_3	e_3	q_4	e_4
$\frac{1}{8}$							
	$\frac{7}{16}$						
$\frac{9}{16}$		$\frac{69}{112}$					
	$\frac{23}{48}$		$\frac{2389}{2576}$				
$\frac{25}{45}$		$\frac{1175}{1104}$		$\frac{978215}{879152}$			
	$\frac{47}{96}$		$\frac{83847}{86480}$		$\frac{1519787307}{1068322576}$		
$\frac{49}{32}$		$\frac{11613}{7520}$		$\frac{164538619}{105088240}$		$\frac{51811311459}{317730045392}$	
	$\frac{79}{160}$		$\frac{145997}{148520}$		$\frac{66877367215}{457012657136}$		$\frac{20704360208551223}{10785526293606384}$
$\frac{81}{40}$		$\frac{3213}{1580}$		$\frac{660840633}{322945364}$		$\frac{32064911601493923}{15513841086389072}$	
	$\frac{119}{240}$		$\frac{520757}{526456}$		$\frac{50352189434685}{34060973958592}$		$\frac{37915193824338958585743}{1935078690874687753104}$
$\frac{121}{48}$		$\frac{101035}{39984}$		$\frac{70389656095}{27762597184}$		$\frac{10824875647938760199}{4248507157620375744}$	
	$\frac{167}{336}$		$\frac{4418223}{4451552}$		$\frac{7291897359213}{49084171835968}$		$\frac{616973275195929451466471}{312445928026957895996724}$
$\frac{169}{56}$		$\frac{113061}{37408}$		$\frac{23835215313}{7870327904}$		$\frac{21925695140964260529}{7219547503040384056}$	
	$\frac{223}{448}$		$\frac{1185183}{1191712}$		$\frac{13868413008257}{9309160515216}$		
$\frac{225}{64}$		$\frac{50225}{14272}$		$\frac{4966969945}{1409577648}$			
	$\frac{287}{576}$		$\frac{2549343}{2560040}$				
$\frac{289}{72}$		$\frac{103751}{25830}$					
	$\frac{359}{720}$						
$\frac{361}{80}$							

Table 12.13. qd Table Associated with $K_0(z)$ (decimal version); Entries are $-q_k^{(n)}$, $-e_k^{(n)}$

0.12500000						
0.43750000						
0.56250000	0.61607143					
0.47916667	0.92740683					
1.04166667	1.06431159	1.11268017				
0.48958333	0.96955365	1.42259215				
1.53125000	1.54428191	1.56571867	1.61083699			
0.49375000	0.98301239	1.46358609	1.91960742			
2.02500000	2.03354430	2.04629237	2.06685832	2.10966228		
0.49583333	0.98917478	1.47829564	1.95936186	2.41751258		
2.52083333	2.52688575	2.53541323	2.54792454	2.56781301	2.60884240	
0.49702381	0.99251295	1.48559048	1.97465616	2.45614043	2.91593105	
3.01785714	3.02237489	3.02849075	3.03699022	3.04929729	3.06863303	3.10823546
0.49776786	0.99452133	1.48975979	1.98265823	2.47171493	2.95356278	3.41487817
3.51562500	3.51912836	3.52372922	3.52988866	3.53835399	3.55048087	3.56935085
0.49826389	0.99582155	1.49236387	1.98741301	2.48018616	2.96926028	
4.01388889	4.01668602	4.02027154	4.02493780	4.03112713	4.03955499	
0.49861111	0.99671048	1.49409603	1.99046915	2.48537147		
4.51250000	4.51478539	4.51765710	4.52131092	4.52602945		
0.49886364	0.99734445	1.49530443	1.99254645			
5.01136364	5.01326620	5.01561709	5.01855293			
0.49905303	0.99781214	1.49617970				
5.51041667	5.51202531	5.51398464				
0.49919872	0.99816683					
6.00961538	6.01099342					
0.49931319						
6.50892857						

PROBLEMS

1. Work out the estimate (A) of Theorem 12.11f for the S fraction (12.13-8).

2. For $x > 0$ we have

$$J_0(x) + i Y_0(x) = -\frac{2i}{\pi} K_0(-ix)$$

$$\approx \frac{2\sqrt{2x}}{\sqrt{\pi}} e^{-3i\pi/4} e^{ix} \left\{ \frac{1}{-2ix} - \frac{[1/2]^2}{1!} \frac{1}{(-2ix)^2} \right.$$
$$\left. + \frac{[(1/2)_2]^2}{2!} \frac{1}{(-2ix)^3} - \cdots \right\}$$

(see §11.5). Let $s_n(x)$, $n = 0, 1, 2, \ldots$, be the result of taking into account n terms of the asymptotic series. Use Problem 3, §12.12, to show that the exact value of the function is contained in each of the circular semidisks

$$\frac{1}{2}(s_n + s_{n+1}) - \frac{w}{2}(s_{n+1} - s_n),$$

where $|w| \leq 1$, $0 \leq \arg w \leq \pi$. Draw the points s_n and the resulting semidisks for $x = 2$, until the smallest value of $s_{n+1} - s_n$ is reached, and obtain a more accurate value by observing that the exact value lies in the intersection of all semidisks. (Exact value: $J_0(2) + i Y_0(2) = 0.223891 + i0.510376$.)

§12.14. MOMENT PROBLEMS

Let $\{\mu_n\}_0^\infty$ be a sequence of real numbers. The **moment problem** for the interval $[0, \infty)$ consists in finding a function $\psi \in \Psi^*$ such that the μ_n are the moments of ψ; that is,

$$\mu_n = \int_0^\infty \tau^n \, d\psi(\tau), \qquad n = 0, 1, 2, \ldots. \qquad (12.14\text{-}1)$$

Any such function ψ called a solution solution of the moment problem for the sequence $\{\mu_n\}$ with respect to the interval $[0, \infty)$.

The following questions pose themselves:

(i) Does a solution exist?

(ii) Is the solution unique?

(iii) How can the solution be constructed?

We already know a necessary condition for the existence of a solution. Let, for $k = 1, 2, \ldots$ and $n = 0, 1, \ldots$,

$$M_k^{(n)} := \begin{vmatrix} \mu_n & \mu_{n+1} & \cdots & \mu_{n+k-1} \\ \mu_{n+1} & \mu_{n+2} & \cdots & \mu_{n+k} \\ & & \ddots & \\ \mu_{n+k-1} & \mu_{n+k} & \cdots & \mu_{n+2k-2} \end{vmatrix}.$$

The proof of Theorem ,12.11a shows that if a solution of the moment problem exists, then

$$M_k^{(n)} > 0 \quad \text{for} \quad n = 0, 1 \quad \text{and} \quad k = 1, 2, \dots. \tag{12.14-2}$$

Thus (12.14-2) is *necessary* for the existence of the solution of the moment problem. The following theorem shows that the condition is also *sufficient*.

THEOREM 12.14a

Let the sequence $\{\mu_n\}$ be such that (12.14-2) holds. Then the moment problem (12.14-1) possesses at least one solution ψ.

Proof. By the proof of Theorem 12.11a the formal power series

$$P(z) := \frac{\mu_0}{z} - \frac{\mu_1}{z^2} + \frac{\mu_2}{z^3} - \cdots \tag{12.14-3}$$

possesses a corresponding S fraction C. Let f be any generalized value function of C. Then by Theorem 12.8c,

$$f(z) \approx P(z), \qquad z \to \infty, \qquad z > 0.$$

By Theorem 12.9e there exists $\psi \in \Psi$ such that

$$f(z) = \int_0^\infty \frac{1}{z + \tau}\, d\psi(\tau), \qquad z \in S.$$

By Theorem 12.9h, all moments of ψ exist, and (12.14-1) holds. Because C does not terminate, none of its value functions can be rational; hence ψ has infinitely points of increase, and $\psi \in \Psi^*$. Thus ψ is a solution of the moment problem. ■

The question of uniqueness of the solution is equally easily answered by the theory of continued fractions. From the definition of the Stieltjes integral it is clear that the solution can be unique at best if it is *normalized* (as defined in §12.10) by the conditions $\psi(0) = 0$, $\psi(\tau) = \psi(\tau-)$ for all $\tau > 0$.

THEOREM 12.14b

Let the sequence $\{\mu_n\}$ satisfy (12.14-2). Then the moment problem (12.14-1) possesses a unique normalized solution if and only if the S fraction C corresponding to the series (12.14-3) converges for $z = 1$.

Proof. (a) Let C converge for $z = 1$, and let ψ be any normalized solution of the moment problem. Then by Theorem 12.8e the fraction C (which depends only on the given moment sequence) converges for all $z \in S$; by Corollary 12.11e its value function is $f := \mathcal{S}\psi$. By Theorem 12.10f, there exists only one normalized ψ such that $f = \mathcal{S}\psi$. Thus ψ is uniquely determined.

(b) Let C diverge for $z = 1$. Then the approximants of even and of odd order have different limits for $z > 0$. Thus C has at least two distinct value functions f_1 and f_2, which have representations $f_i = \mathcal{S}\psi_i$, $i = 1, 2$, in terms of distinct normalized functions $\psi_i \in \Psi^*$. By the proof of Theorem 12.14a, both ψ_i are solutions of the moment problem. ■

By the Carleman condition,

$$\sum_{n=1}^{\infty} \mu_n^{-1/2n} = \infty$$

is a sufficient condition for the solution of the moment problem to be unique.

To construct a solution of the moment problem one has to set up the fraction C corresponding to (12.14-3) and from the partial fraction decompositions of its approximants find ψ, as in the proof of Theorem 12.9e.

The moment problem is treated in the mathematical literature not only for the interval $[0, \infty)$, but also for bounded intervals $[0, \beta]$ where $\beta > 0$. Obviously the problem then consists in finding a nondecreasing function ψ on $[0, \beta]$ such that

$$\mu_n = \int_0^\beta \tau^n \, d\psi(\tau), \qquad n = 0, 1, 2, \ldots. \tag{12.14-4}$$

This problem is of interest in functional analysis. Let $C[0, \beta]$ denote the space of real continuous functions on $[0, \beta]$. By a theorem of F. Riesz, if ϕ is any positive linear functional on $C[0, \beta]$ there exists a nondecreasing function ψ on $[0, \beta]$ such that the value of ϕ at any $f \in C[0, \beta]$ is represented by the formula

$$\phi f = \int_0^\beta f(\tau) \, d\psi(\tau). \tag{12.14-5}$$

Conversely, every nondecreasing function ψ defines a positive linear functional ϕ by virtue of (12.14-5). Evidently, the moments μ_n are the values of the functional ϕ at the power functions $f(\tau) = \tau^n$. Thus the three questions posed earlier may in the present context be phrased as follows:

(i) Does there exist a positive linear functional on the space $C[0, \beta]$ that takes prescribed values at the functions $f(\tau) = \tau^n$?

(ii) Is a positive linear functional on $C[0, \beta]$ uniquely determined by its values at the functions $f(\tau) = \tau^n$, $n = 0, 1, \ldots$?

(iii) How can the foregoing positive linear functional be constructed if it exists?

Any solution ψ of the moment problem for the interval $[0, \beta]$ can be extended to a solution for the interval $[0, \infty)$ by letting $\psi(\tau) := \psi(\beta)$ for $\tau > \beta$. Thus the condition (12.14-2) is a fortiori necessary for the existence of a solution of the moment problem relative to $[0, \beta]$. The condition cannot be sufficient, however, because a sequence $\{\mu_n\}$ satisfying it will not, in general, give rise to a solution ψ that is constant on some interval $[\beta, \infty)$. To find a further necessary condition, assume a solution ψ exists. The generalized value function f of the continued fraction constructed in the proof of Theorem 12.14a then has the form

$$f(z) = \int_0^\beta \frac{1}{z+\tau} \, d\psi(\tau).$$

As was done many times before, we expand $(z+\tau)^{-1}$ in powers of z^{-1}. This converges for $|z| > \tau$ and thus, because $\tau \in [0, \beta]$, converges uniformly for $|z| \geq \beta_0 > \beta$. Integrating term by term then yields

$$f(z) = P(z) := \sum_{n=0}^{\infty} (-1)^n \frac{\mu_n}{z^{n+1}}. \tag{12.14-6}$$

It follows that the series P has a radius of convergence $\geq \beta^{-1}$. By the Cauchy–Hadamard formula (Theorem 2.2a)

$$\limsup_{n \to \infty} \mu_n^{1/n} \leq \beta. \tag{12.14-7}$$

Thus (12.14-7) is a further necessary condition for the solubility of the moment problem for the interval $[0, \beta]$. We now show that together with (12.14-2) this condition is also sufficient; furthermore, it automatically guarantees the uniqueness of the solution.

THEOREM 12.14c

Let the sequence $\{\mu_n\}$ be such that both conditions (12.14-2) and (12.14-7) are satisfied. Then there exists a unique normalized solution of the moment problem relative to the interval $[0, \beta]$.

Proof. The continued fraction C corresponding to the series P defined by (12.14-6) again exists and is an S fraction. By virtue of Corollary 12.11g, (12.11-7) implies the convergence of C for $z \in S$. Its value function can be represented as

$$f(z) = \int_0^\infty \frac{1}{z+\tau} \, d\psi(\tau),$$

where ψ is essentially uniquely determined, $\psi \in \Psi^*$, and has the moment sequence $\{\mu_n\}$. By the proof of Theorem 12.9e, ψ is limit of a subsequence of the step functions (12.9-18). Using the notation of §12.9(II), we now shall show that for every $\epsilon > 0$ the sum

$$\sigma_n := \sum_{\xi_{ni} \geqslant \beta + \epsilon} \rho_{ni} \tag{12.14-8}$$

of the jumps of ψ_n at points $\xi_{ni} \geqslant \beta + \epsilon$ tends to zero for $n \to \infty$. This, then, will prove that ψ has no points of increase for $\tau > \beta$, and thus, in fact, is a solution of the moment problem for the interval $[0, \beta]$.

To prove that $\sigma_n \to 0$, assume the contrary. Then there exist $\epsilon > 0$, $\eta > 0$ and an unbounded set of integers N such that

$$\sigma_n \geqslant \eta \quad \text{for all } n \in N. \tag{12.14-9}$$

We now use (12.8-14) for $m = n$, $l = n - 1$:

$$\sum_{i=1}^{k} \rho_{ni} \xi_{ni}^{n-1} = \mu_{n-1} \left(k := \left[\frac{n+1}{2} \right] \right).$$

By (12.14-7), $\mu_{n-1} \leqslant (\beta + \epsilon/2)^{n-1}$ for all sufficiently large n. Thus for all sufficiently large $n \in N$

$$\left(\beta + \frac{\epsilon}{2} \right)^{n-1} \geqslant \sum_{i=1}^{k} \rho_{ni} \xi_{ni}^{n-1} \geqslant \sum_{\xi_{ni} \geqslant \beta + \epsilon} \rho_{ni} \xi_{ni}^{n-1}$$

$$\geqslant (\beta + \epsilon)^{n-1} \sum_{\xi_{ni} \geqslant \beta + \epsilon} \rho_{ni} \geqslant (\beta + \epsilon)^{n-1} \eta.$$

This is contradictory for $n \to \infty$, establishing $\sigma_n \to 0$.

To prove uniqueness, we use the fact that C converges. By Theorem 12.14b, this proves uniqueness of the solution for the interval $[0, \infty)$, and thus a fortiori for $[0, \beta]$. ■

SEMINAR ASSIGNMENTS

1. Write a rigorous computer program for determining the nonperiodic terms and the period in the continued fraction expansion of a given quadratic irrationality.

2. Study the effectiveness of the continued fraction expansion of $\text{Log}(1 + 1/z)$, and of other continued fractions. Plot, as a function of the complex variable z, the number of approximants required to achieve full machine accuracy.

3. Using symbolic manipulation, construct the quotient-difference table for the function $K_\nu(z)$. (The entries are even, rational functions of ν. The fraction terminates for $\nu = n + \frac{1}{2}$, $n = 0, 1, 2, \ldots$.)

4. Compare the numerical effectiveness of various methods for computing the complex error integral,

$$\operatorname{erf}(z) := \frac{2}{\sqrt{\pi}} \int_0^z e^{-t^2} \, dt.$$

5. In statistics, the quantity

$$r(x) := e^{x^2/2} \int_x^\infty e^{-u^2/2} \, du$$

is known as **Mill's ratio**. Using continued fractions, obtain upper and lower bounds for Mill's ratio. Compare with **Birnbaum's inequality**,

$$r(x) \geqslant \tfrac{1}{2}(\sqrt{x^2+4} - x).$$

6. Find the solution of specific moment problems by carrying out the construction indicated in §12.14.

NOTES

Perron [1957] (2nd ed., 1929) is the classical reference for algebraic as well as analytic aspects of continued fractions. Wall [1948] is an original treatise dealing with the analytic theory.

§12.1. The definition of continued fractions given here, originally proposed by Henrici and Pfluger [1966], is widely used today. The numerical evaluation of continued fractions is dealt with by Blanch [1964] and by Jones and Thron [1974].

§12.2. The role of continued fractions in number theory is discussed by Hardy and Wright [1954] and, concisely and elegantly, by Davenport [1952]. Khintchin's constant (the geometric mean of the partial numerators in the expansion of "almost all" real numbers) was calculated by Khintchin [1934]; see also Lévy [1936], Stark [1971].

§12.3. Our Lemma 12.3b is related to work of Thron [1974]. For the parabola theorem see Scott and Wall [1940], Thron [1958], Lange [1966]. For Problems 6 through 9 see Saff and Varga [1975]. The zeros of partial sums of the exponential series are computed numerically by Iverson [1953] and Dejon and Nickel [1969], who give illustrative graphs.

§12.4. The connection between the qd algorithm and $RITZ$ fractions is due to Rutishauser [1954]. Gragg [1974] gives an excellent survey of the formal properties of $RITZ$ fractions. The Padé table and its relation to various algorithms in numerical analysis is discussed by Gragg [1972]. For a treatment oriented toward applications in physics see Baker [1975].

§12.5. Padé approximants for the exponential series are discussed by Fair and Luke [1970], Ehle [1973], and Saff and Varga [1975]. Jones and Thron [1975] discuss the convergence of more general Padé approximants.

§12.6. The division algorithm can be traced back at least to Kausler [1803]. A. A. Markov in his many papers on continued fractions made extensive use of it [1948]. Maehly [1960] is an early proponent of the practical use of continued fractions for purposes of numerical approximation. For some recent industrial applications see Patry [1972].

§12.7. Wall [1945] discusses the stability of polynomials by means of continued fractions. Positive symmetric functions as defined here are related to, but not identical with, the positive rational functions considered in circuit theory. These were used by Levinson and Redheffer ([1970], Chap. 5, Sect. 5) to give yet another, very elegant treatment of the stability problem. Various notations are in use for the Bessel polynomials discussed in Problem 6; see Krall and Frink [1949], Grosswald [1951], Al-Salam [1957], Luke [1969], and Barnes [1973].

§12.8, §12.9. Most of the material in these sections has its roots in the great posthumous memoir of Stieltjes [1894], in which the notion of the Stieltjes integral was introduced precisely to formulate Theorem 12.9e.

§12.10. A more detailed presentation is given in Chap. 3 of Akhiezer [1965], for which the paper by Nevanlinna [1922] is fundamental. For Problem 6 see Dahlquist [1963].

§12.11. For the Theorems 12.11f (here slightly improved) and 12.11j see Henrici and Pfluger [1966]. Corollary 12.11h goes back to Carleman [1923]. For other a priori estimates for the speed of convergence of continued fractions see Sweezy and Thron [1967] and Jones and Snell [1969]. A posteriori estimates for different types of continued fractions were given by Gragg [1968, 1970]. See also Merkes [1966], Baker [1975].

§12.12. The expansions in terms of reciprocals of orthogonal polynomials given here are mostly new; for (12.12-7), however, see Hobson [1931].

§12.13. Converging factors of asymptotic expansions were expressed in terms of continued fractions by Henrici [1963] and used for numerical purposes by Gargantini and Henrici [1967]. Conte and Fried [1961] computed a table of the complex error integral by means of continued fractions; see also Patry and Keller [1964]. Sophisticated summations of asymptotic series by continued fractions (in the disguise of Padé approximants) are discussed by Simon [1970].

§12.14. For comprehensive accounts of various moment problems (including the "indeterminate" case, in which the solution is not unique), see Akhiezer [1965] and Krein and Nydelmann [1973].

BIBLIOGRAPHY

Sections and chapters to which a reference is relevant are numbered at the end of each reference. G = general reference, S = seminar assignment.

Abramowitz, M., and I. A. Stegun [1965]. *Handbook of mathematical functions*. Dover, New York. G.

Ahlfors, L. [1966]. *Complex analysis*, 2nd ed. McGraw-Hill, New York, G. §§8.1, 8.4, 8.5, 9.9, 12.8.

Airy, G. B. [1838]. On the intensity of light in the neigborhood of a caustic. *Trans. Camb. Phil. Soc.* **6**, 379–402. §11.8.

Akhiezer, N. I. [1965]. *The classical moment problem*. Translated from the Russian by N. Kemmer. Oliver & Boyd, Edinburgh. §§12.10, 12.14.

Al-Salam, W. A. [1957]. The Bessel polynomials. *Duke Math. J.* **24**, 529–545. §12.7.

Bailey, W. N. [1935]. *Generalized hypergeometric series*. Cambridge Tracts in Mathematics and Mathematical Physics No. 32. University Press, Cambridge. §§8.6, 8.8.

Baker, G. A. Jr. [1975]. *Essentials of Padé approximants*. Academic, New York. §§12.4, 12.5, 12.9.

———, and J. L. Gammel (Eds.) [1970]: *The Padé approximants in theoretical physics*. Academic Press, New York. §12.4.

Barnes, C. W. [1973]. Remarks on the Bessel polynomials. *Amer. Math. Monthly* **80**, 1034–1040. §12.7.

Bauer, F. L., H. Rutishauser, and E. Stiefel [1963]. New aspects in numerical quadrature. *Proc. Symp. Appl. Math.* **15**, 199–218. Amer. Math. Soc., Providence. §11.12.

Bellmann, R., and K. L. Cooke [1963]. *Differential–difference equations*. Academic, New York. §10.4.

———, R. Kalaba, and J. Lockett [1966]. *Numerical inversion of the Laplace transform*. American Elsevier, New York. §10.7.

Berg, L. [1967]. *Introduction to the operational calculus*. North-Holland, Amsterdam. Chapter 10.

———, [1968]. *Asymptotische Darstellungen und Entwicklungen*. VEB Deutscher Verlag der Wissenschaften, Berlin. Chapter 11.

Birkhoff, G. D. [1909]. Singular points of ordinary differential equations. *Trans. Amer. Math. Soc.* **10**, 436–470. §§9.4, 9.5.

Blanch, G. [1964]. Numerical evaluation of continued fractions. *SIAM Rev.* **6**, 383–421. §12.1.

Boas, R. P. [1946]. Poisson summation formula in L^2. *J. London Math. Soc.* **21**, 102–105. §10.6.

642

—— [1964]. Periodic entire functions. *Amer. Math. Monthly* **71**, 782. §10.9.

—— [1972]. Summation formulas and band-limited signals. *Tôhoku Math. J.* **24**, 121–125. §10.6.

——, and H. Pollard [1973]. Continuous analogues of series. *Amer. Math. Monthly* **80**, 18–25. §10.6.

——, and C. Stutz [1971]. Estimating sums with integrals. *Amer. J. Physics* **39**, 745–753. §§10.6, 11.11.

——, and J. W. Wrench [1971]. Partial sums of the harmonic series. *Amer. Math. Monthly* **78**, 864–870. §11.11.

Bochner, S. (1955). *Harmonic analysis and the theory of probability.* U. of California Press, Berkeley, §10.6.

Buchholz, H. [1953]. *Die konfluente hypergeometrische Funktion.* Erg. d. angew. Math. vol. 2. Springer, Berlin. §§9.7, 10.5, 11.5, 12.13.

—— [1957]. *Elektrische und magnetische Potentialfelder.* Springer, Berlin. §§8.8, 10.12.

Buck, R. C. [1965]. *Advanced calculus.* MacGraw-Hill, New York. §§8.1, 12.1.

Buckholtz, J. D. [1963]. Concerning an approximation of Copson. *Proc. Amer. Math. Soc.* **14**, 564–568. §11.5.

Burkhill, J. D. [1951]. *The Lebesgue integral.* Cambridge Tracts in Mathematics and Mathematical Physics No. 40. University Press, Cambridge. §8.4.

Carleman, T. [1923]. Sur les fonctions quasi-analytiques. *Conférences faites au cinquième congrès des mathématiciens scandinaves*, pp. 181–186. Helsingfors. §12.11.

Carlslaw, H. S., and J. C. Jaeger [1947]. *Operational methods in applied mathematics.* University Press, Oxford. §10.3.

Chandrasekharan, K. [1968]. *Introduction to analytic number theory.* Springer, Berlin. §§10.7, 10.8.

Chester, C., B. Friedman, and F. Ursell [1957]. An extension of the method of steepest descent. *Proc. Camb. Phil. Soc.* **53**, 599–611. §11.8.

Coddington, E. A., and N. Levinson [1955]. *Theory of ordinary differential equations.* McGraw-Hill, New York. Chapter 9.

Conte, S. D., and B. D. Fried [1961]. *The plasma dispersion function.* Academic, New York. §12.13.

Copson, E. T. [1960]. *An introduction to the theory of functions of a complex variable.* Clarendon Press, Oxford. G, Chapter 8, §§8.8, 11.5, 11.6, 11.7, 11.8.

Courant, R., and D. Hilbert [1930]. *Methoden der mathematischen Physik*, Bd. I, 2. Aufl. Springer, Berlin. §10.6.

Dahlquist, G. [1963]. A special stability problem for linear multistep methods. *BIT* **3**, 27–43. §12.10.

Davenport, H. [1952]. *The higher arithmetic.* Hutchinson's University Library, London. §12.2.

Davis, H. T. [1962]. *Introduction to nonlinear differential and integral equations.* Dover, New York. 9S.

De Bruijn, N. G. [1961]. *Asymptotic methods in analysis.* North-Holland, Amsterdam. Chapter 11.

Dejon, B., and K. Nickel [1969]. A never failing, fast converging rootfinding algorithm. *Constructive aspects of the fundamental theorem of algebra*, B. Dejon and P. Henrici, Eds. Wiley-Interscience, London. §12.3.

Dingle, R. B. [1973]. *Asymptotic expansions. Their derivation and interpretation.* Academic, London. Chapter 11.

Doetsch, G. [1950, 1955, 1956]. *Handbuch der Laplace-Transformation,* 3 vols. Birkhäuser, Basel. Chapters 10, 11.

────── [1958]. *Einführung in Theorie und Anwendung der Laplace-Transformation.* Birkhäuser, Basel. Chapter 10.

────── [1967]. *Anleitung zum praktischen Gebrauch der Laplace- und der Z-Transformation,* 3. Auflage. Oldenburg, München. §10.10.

────── [1970]. *Einführung in Theorie und Anwendung der Laplace-Transformation,* 2. Aufl. Birkhäuser, Basel. Chapter 10.

Dubner, H., and J. Abate [1968]. Numerical inversion of Laplace transforms and the finite Fourier transform. *J. Assoc. Comput. Mach.* **15**, 115–123. §10.7.

Effertz, F. H., and F. Kolberg [1963]. *Einführung in die Dynamik selbsttätiger Regelungssysteme.* VDI-Verlag, Düsseldorf. §10.3.

Ehle, B. L. [1973]. A-stable methods and Padé approximations to the exponential. *SIAM J. Math. Anal.* **4**, 671–680. §12.5.

Erdelyi, A., et al. [1953, 1953, 1955]. *Higher transcendental functions,* 3 vols. McGraw-Hill, New York. G, §§8.6, 9.7, 9.9, 9.10.

────── [1954]. *Tables of integral transforms,* 2 vols. McGraw-Hill, New York, Chapter 10.

────── [1956]. *Asymptotic expansions.* Dover, New York. Chapter 11.

────── [1962]. *Operational calculus and generalized functions.* Holt, Rinehart and Winston, New York. Chapter 10.

Fair, W., and Y. L. Luke [1970]. Padé approximation to the operator exponential. *Num. Math.* **14**, 379–382. §12.5.

Feller, W. [1957]. *An introduction to probability theory and its applications,* vol. I, 2nd ed. Wiley, New York. §11.10.

Filippi, S. [1966]. Die Berechnung einiger elementarer Funktionen mittels des Rombergalgorithmus. *Computing* **1**, 127–132. §11.12.

Frobenius, G. [1873]. Ueber die Integration der linearen Differentialgleichungen durch Reihen. *J. für Math.* **76**, 214–235. §9.6.

────── [1881]. Ueber Relationen zwischen den Näherungsbrüchen von Potenzreihen. *J. für Math.* **90**, 1–17. §12.4.

Fuchs, L. [1866, 1868]. Zur Theorie der linearen Differential-gleichungen mit veränderlichen Koeffizienten. *J. für Math.* **66**, 121–160; **68**, 345–385. §9.8.

Gantmacher, F. R. [1959]. *The theory of matrices,* 2 vols. Chelsea, New York. §§9.3, 9.4.

Gargantini, I., and P. Henrici [1967]. A continued fraction algorithm for the computation of higher transcendental functions in the complex plane. *Math. Comp.* **21**, 18–29. §12.13.

Gautschi, W. [1974]. A harmonic mean inequality for the Gamma function. *SIAM J. Math. Anal.* **5**, 278–281; Some mean value inequalities for the Gamma function. *ibid.* **5**, 282–292. §8.4.

────── [1976]. Computational methods in special functions—A survey. In: *Special functions, Proceedings of an advanced seminar,* R. Askey, Ed. Academic, New York. G.

Gragg, W. B. [1968]. Truncation error bounds for *g*-fractions. *Numer. Math.* **11**, 370–379. §12.9.

────── [1970]. Truncation error bounds for *π*-fractions. *Bull. Amer. Math. Soc.* **76**, 1091–1094. §12.9.

—— [1972]. The Padé table and its relation to certain algorithms of numerical analysis. *SIAM Rev.* **14**, 1–62. §12.4.

—— [1974]. Matrix interpretations and applications of the continued fraction algorithm. *Rocky Mountain J. Math.* **4**, 213–225. §12.4.

Grosswald, E. [1951]. On some algebraic properties of the Bessel polynomials. *Trans. Amer. Math. Soc.* **71**, 197–210. §12.7.

Gutknecht, M. [1973]. *Ein Abstiegsverfahren für gleichmässige Approximation, mit Anwendungen.* Diss. ETH No. 5006. aku-Fotodruck, Zürich. §10.10.

Hardy, G. H. [1904]. On differentiation and integration of divergent series. *Trans. Cambridge Phil. Soc.* **19**, 297–321. §10.5.

—— [1910]. *Orders of Infinity.* University Press, Cambridge. §11.11.

——, J. E. Littlewood, and G. Polya [1934]. *Inequalities.* University Press, Cambridge. §12.11.

——, and E. M. Wright [1954]. *An introduction to the theory of numbers,* 3rd ed. Clarendon, Oxford. §§8.2, 10.8, 12.2.

Henrici, P. [1962]. *Discrete variable methods in ordinary differential equations.* Wiley, New York. §9.2, 12.10.

—— [1963]. Some applications of the quotient-difference algorithm. *Proc. Symp. Appl. Math.* **15**: *Experimental arithmetic, high-speed computing, and mathematics,* pp. 159–183. Amer. Math. Soc., Providence. §§12.12, 12.13.

—— [1970]. Upper bounds for the abscissa of stability of a stable polynomial. *SIAM J. Num. Anal.* **7**, 538–544. §§10.7, 12.7.

—— [1975]. Einige Anwendungen der Kreisscheibenarithmetik in der Kettenbruchtheorie. *Interval Mathematics, Lecture Notes in Computer Science* **29**, 19–30. §12.3.

——, and C. Hoffmann [1975]. Chess champ's chances: An exercise in asymptotics. *SIAM Rev.* **17**, 559–564. §11.6.

——, and P. Pfluger [1966]. Truncation error estimates for Stieltjes fractions. *Numer. Math.* **9**, 120–138. §§12.9, 12.11.

Hobson, E. W. [1931]. *The theory of spherical and ellipsoidal harmonics.* University Press, Cambridge. Reprint: Chelsea, New York. §9.10.

Hochstadt, H. [1971]. *The functions of mathematical physics.* Wiley-Interscience, New York. G, §8.8, Chapter 9.

Ince, E. L. [1926]. *Ordinary differential equations.* Reprint: Dover, New York (1956). Chapter 9.

Iverson, K. E. [1953]. The zeros of the partial sums of e^z. *Math. Tables Aids Comp.* **7**, 165–168. §12.3.

Jones, W. B., and R. I. Snell [1969]. Truncation error bounds for continued fractions. *SIAM J. Num. Anal.* **6**, 210–221. §12.11.

——, and W. J. Thron [1974]. Numerical stability in evaluating continued fractions. *Math. Comp.* **28**, 795–810. §12.1.

—— [1975]. On convergence of Padé approximants. *SIAM J. Math. Anal.* **6**, 9–16. §12.5.

Jury, E. I. [1964]. *Theory and application of the z-transform method.* Wiley, New York. §10.10.

Kaiser, J. F. [1966]. Digital filters. In: *System analysis by digital computers,* F. K. Kuo and J. F. Kaiser, Eds. Wiley, New York. §10.10.

Kamke, E. [1947]. *Differentialgleichungen reeller Funktionen.* Akad. Verlagsgesellschaft, Leipzig. Reprint: Chelsea, New York. §9.2.

Kausler, C. J. [1803]. *Die Lehre von den continirlichen Brüchen.* Löflund, Stuttgart. §12.6.

Khintchine, A. [1934]. Metrische Kettenbruchprobleme. *Compositio Math.* **1**, 361–382. §12.2.

Knuth, D. E. [1974]. Computer Science and its relation to mathematics. *Amer. Math. Monthly* **81**, 323–342. §11.6.

Krabbe, G. [1970]. *Operational calculus.* Springer, Berlin. Chapter 10.

Krall, H. L., and O. Frink [1949]. A new class of orthogonal polynomials, the Bessel polynomials. *Trans. Amer. Math. Soc.* **65**, 100–115. §12.7.

Krein, M. G., and A. A. Nydelman [1973]. *The Markov moment problem and extremal problems* (in Russian). Moscov. §12.14.

Lawrentjew, M. A., and B. V. Schabat [1967]. *Methoden der komplexen Funktionentheorie.* VEB Deutscher Verlag der Wissenschaften, Berlin. §§8.4, 8.7, Chapters 10, 11.

Lange, L. J. [1966]. On a family of twin convergence regions for continued fractions. *Ill. J. Math.* **10**, 97–108. §12.3.

Laurie, D. P. [1975]. Propagation of initial error in Romberg-like quadrature. *BIT* **15**, 277–282. §11.12.

Lebedev, N. N. [1965]. *Special functions and their applications.* Translated by R. A. Silverman. Prentice-Hall, Englewood Cliffs. Chapters 8, 9.

Levinson, N., and R. M. Redheffer [1970] *Complex variables.* Holden-Day, San Francisco. §12.7.

Lévy, P. [1936]. Sur le développement en fraction continue d'un nombre choisi au hasard. *Compositio Math.* **3**, 286–303. §12.2.

Luke, Y. L. [1969]. *The special functions and their approximations,* 2 vols. Academic, New York. Chapters 9, 11, §12.7.

MacRobert, T. M. [1967]. *Spherical Harmonics,* 3rd ed. Pergamon Press, Oxford. §9.10.

Maehly, H. J. [1960]. Rational approximations for transcendental functions. *Information Processing,* pp. 57–62. Unesco, Paris. §12.6.

Magnus, W., and F. Oberhettinger [1948]. *Formeln und Sätze für die speziellen Funktionen der mathematischen Physik,* 2. Aufl. Springer, Berlin. G.

——, and R. P. Soni [1966]. *Formulas and theorems for the special functions of mathematical physics,* 3rd ed. Springer, Berlin. G, §10.2.

Marcus, M. [1960]. *Basic theorems in matrix theory.* Appl. Math. Ser. **57**, National Bureau of Standards, Washington. §11.7.

Markov, A. A. [1948]. *Selected papers.* Moscow. Chapter 12.

Merkes, E. P. [1966]. On truncation errors for continued fraction computations. *SIAM J. Num. Anal.* **3**, 486–496. §12.9.

Mikusinski, J. G. [1953]. *Rachunek operatorow.* Warsaw. English edition: *Operational calculus.* Pergamon Press, London (1959). Chapter 10.

Minc, H. [1974]. Six letters from Alexander C. Aitken. *Linear Multilinear Algebra* **2**, 1–12. §11.11.

Moore, D. H. [1971]. *Heaviside operational calculus, an elementary foundation.* American Elsevier, New York. Chapter 10.

Natanson, I. P. [1961]. *Theorie der Funktionen einer reellen Veränderlichen.* Akademie-Verlag, Berlin. §§8.4, 10.8, 11.1, 12.9.

Nayfeh, A. H. [1973]. *Perturbation methods.* Wiley, New York. Chapter 11.

Nevanlinna, R. [1922]. Asymptotische Entwickelungen beschränkter Funktionen und das Stieltjessche Momentenproblem. *Ann. Acad. Sci. Fenn.* A **18**, No. 5. §12.10.

Oberhettinger, F. [1974]. *Table of Mellin transforms.* Springer, Berlin. §10.11.

———, and L. Badii [1973]. *Tables of Laplace transforms.* Springer, Berlin. Chapter 10, §§10.2, 10.5.

——— [1957]. *Tabellen zur Fourier-Transformation.* Springer, Berlin. §10.6.

Olver, F. W. J. [1974]. *Introduction to asymptotics and special functions.* Academic Press, New York. G, Chapters 8, 9, 11.

Padé, H. [1892]. Mémoire sur les développements en fractions continues de la fonction exponentielle pouvant servir d'introduction à la théorie des fractions continues algébriques. *Ann. Sci. Ecole Norm. Sup.* **16**, 395–426. §12.4.

Patry, J. [1972]. Utilisation des séries de puissances ou des fractions continues pour le calcul numérique des fonctions analytiques. *EIR-Bericht Nr.* 221, Eidgen. Inst. Reaktor-forschung, Würenlingen. §12.6.

———, and J. Keller [1964]. Zur Berechnung des Fehlerintegrals. *Numer. Math.* **6**, 89–97. §12.13.

Perron, O. [1957]. *Die Lehre von den Kettenbrüchen,* 3rd ed. (in 2 vols.). Teubner, Stuttgart. Chapter 12.

Piessens, R. [1975]. A bibliography on numerical inversion of Laplace transform and applications. *J. Comp. Appl. Math.* **1**, 115–128. §10.7.

Polya, G. [1929]. Untersuchungen über Lücken und Singularitäten von Potenzreihen. *Math. Z.* **29**, 549–640. §§10.9, 10.10.

———, and G. Szegö [1925]. *Aufgaben und Lehrsätze der Analysis,* 2 vols. Springer, Berlin. G, §§8.2, 11.7.

Poole, E. G. C. [1936]. *Introduction to the theory of linear differential equations.* University Press, Oxford. Chapter 9.

Richardson, L. F., and J. A. Gaunt [1927]. The deferred approach to the limit. *Trans. Roy. Soc. London* **226A**, 299–361. §11.12.

Riemann, B. [1857]. Beiträge zur Theorie der durch die Gauss'sche Reihe $F(\alpha, \beta, \gamma, x)$ dargestellten Funktionen. In: *Collected works of B. Riemann,* 2nd ed. 1902. Reprint: Dover, New York (1953). §§8.7, 9.9, 9.10.

Ritt, J. F. [1916]. On the derivatives of a function at a point. *Ann. Math.* **18**, 18–23. §11.3.

Robin, L. [1957–1959]. *Fonctions sphériques de Legendre et fonctions sphéroidales,* 3 vols. Dunod, Paris. §9.10.

Romberg, W. [1955]. Vereinfachte Numerische Integration. *Det. Kong. Norske Videnskabers Selskab Forhandlinger* **28**, Nr. 7.

Rutishauser, H. [1954]. Der Quotienten-Differenzen-Algorithmus. *Z. Angew. Math. Physik* **5**, 233–251. §12.4.

——— [1963]. Ausdehnung des Rombergschen Prinzips. *Numer. Math.* **5**, 48–54. §11.12.

Saff, E. B., and R. S. Varga [1975]. Zero-free parabolic regions for sequences of polynomials. *SIAM J. Math. Anal.* **7**, 344–357. §12.3.

Sauer, R., and I. Szabo (Eds.) [1967]. *Mathematische Hilfsmittel des Ingenieurs,* Bd. I. Springer, Berlin Chapter 10.

Saxer, W. [1958]. *Versicherungsmathematik,* 2. Teil. Springer, Berlin. §10.4.

Shafer, R. E. [1975]. Problem 6019. *Amer. Math. Monthly* **82**, 307. §11.2.

Schäfke, F. W. [1963]. *Einführung in die Theorie der speziellen Funktionen der mathematischen Physik.* Springer, Berlin. G, Chapters 8, 9.

Schwartz, L. [1950]. *Théorie des distributions.* Hermann, Paris. Chapter 10.

Scott, W. T., and H. S. Wall [1940]. A convergence theorem for continued fractions. *Trans. Amer. Math. Soc.* **47**, 155–172. §12.3.

Simon, B. [1970]. Coupling constant analyticity for the anharmonic oscillator. *Ann. Physics* **58**, 76–136. §12.13.

Sirovich, L. [1971]. *Techniques of asymptotic analysis*. Appl. Math. Sciences vol. 2, Springer, Berlin. Chapter 11.

Sommerfeld, A. [1947]. *Partielle Differentialgleichungen der Physik*. Geest und Portig, Leipzig. §10.12.

Stark, H. M. [1971]. An explanation of some exotic continued fractions found by Brillhart. *Computers in number theory*, A. O. L. Atkin and B. J. Birch (Eds.), Academic, London. §12.2.

Stehfest, H. [1970]. Algorithm 368. Numerical inversion of Laplace transform. *Comm. ACM* **13**, 47–49. §10.7.

Stieltjes, T. J. [1894]. Recherches sur les fractions continues. *Ann. Toulouse* **8**, 1–122; **9**, 1–47. §§12.7, 12.8, 12.9, 12.10.

Sweezy, W. B., and W. J. Thron [1967]. Estimates of the speed of convergence of certain continued fractions. *SIAM J. Num. Anal.* **4**, 254–270. §12.11.

Szegö, G. [1959]. *Orthogonal polynomials, revised edition*. Amer. Math. Soc. Coll. Publ. Vol. 23. Amer. Math. Soc., New York. G, §§11.6, 11.10, 12.12, 12.13.

Talbot, A. [1959]. The evaluation of integrals of products of linear systems responses, I. II. *Quart. J. Mech. Appl. Math.* **12**, 488–503; 504–520. §10.7.

———— [1965]. Some theorems on positive functions. *IEEE Transactions on circuit theory*, **CT 12**, 607–608. §12.7.

Thomé, L. W. [1872]. Zur Theorie der linearen Differentialgleichungen. *J. Math.* **74**.

Thron, W. J. [1958]. On parabolic convergence regions for continued fractions. *Math. Z.* **69**, 173–182. §12.3.

———— [1974]. A survey of recent convergence results for continued fractions. *Rocky Mountain J. Math.* **4**, 273–282. §12.3.

Titchmarsh, E. C. [1939]. *The theory of functions*. University Press, Oxford. §§8.1, 10.8.

———— [1937]. *Introduction to the theory of Fourier integrals*. Clarendon Press, Oxford. §10.6.

Todd, J. [1962]. *Survey of numerical analysis*. McGraw-Hill, New York. §11.6.

———— [1975]. The lemniscate constants. *Comm. ACM* **18**, 14–19. §8.7, 8S.

von Koch, N. H. F. [1892]. Sur les déterminants infinis et les équations différentielles linéaires. *Acta Math.* **16**, 217–295. §9.11.

Wall, H. S. [1945]. Polynomials whose zeros have negative real parts. *Amer. Math. Monthly* **52**, 308–322. §12.7.

———— [1948]. *Analytic theory of continued fractions*. Van Nostrand, Toronto. Chapter 12.

Walter, W. [1972]. *Gewöhnliche Differentialgleichungen*. Heidelberger Taschenbücher **110**, Springer, Berlin. Chapter 9, 9S.

Wasow, W. [1965]. *Asymptotic expansions for ordinary differential equations*. Wiley, New York. §§11.3, 11.4.

Watson, G. N. [1944]. *A treatise on the theory of Bessel functions*, 2nd ed. University Press, Cambridge. G, §§9.7, 9.12, 10.4, 10.5, 10.9, 10.12, 11.5, 11.8, 12.13.

Whittaker, E. T. [1928]. Oliver Heaviside, an historical foreword. *Bull. Calcutta Math. Soc.* **20**, 199–220. §10.0.

————, and G. N. Watson [1927]. *A course of modern analysis*, 4th ed. University Press, Cambridge. Chapter 8, §§9.7, 9.9, 9.10, 11.1, 11.5.

Widder, D. V. [1941]. *The Laplace transform*. University Press, Princeton. Chapter 10.

Wilkinson, J. H. [1961]. Error analysis of direct methods of matrix inversion. *J. Assoc. Comp. Mach.* **8**, 281–330. §11.11.

Wynn, P. [1964]. On some recent developments in the theory and application of continued fractions. *SIAM J. Num. Anal.* **1**, 177–197. §12.4.

————[1966]. Upon systems of recursions which obtain among the quotients of the Padé table. *Numer. Math.* **8**, 264–269. §12.4.

APPENDIX

Appendix: Some additional problems on vol. I.

§1.6.

18. Let $R = r_0 + r_1 x + r_2 x^2 + \cdots$ be any formal power series, and let

$$R_n(z) := r_0 + r_1 z + \cdots + r_n z^n, \qquad n = 0, 1, 2, \ldots,$$

denote its nth partial sum, evaluated at $z \in \mathbb{C}$. Show that the generating series of the numbers $R_n(z)$ is

$$\sum_{n=0}^{\infty} R_n(z) x^n = \frac{1}{1-x} R \circ (zX).$$

§1.7.

10. If $P := \frac{1}{2}(E_1 - E_{-1})$, $Q := P^{[-1]}$, show that

$$Q = x - \frac{1}{2} \cdot \frac{x^3}{3} + \frac{1 \cdot 3}{2 \cdot 4} \cdot \frac{x^5}{2} - \frac{1 \cdot 3 \cdot 5}{2 \cdot 4 \cdot 6} \cdot \frac{x^7}{7} + \cdots.$$

§1.9.

9. Let P be an almost unit, $R := P^{[-1]}$ and $R_n(z)$ the nth partial sum of R, evaluated at $z \neq 0$. Show that

(a) $\displaystyle \sum_{n=1}^{\infty} R_n(z) x^n = \frac{1}{1-x} \sum_{n=1}^{\infty} \operatorname{res}(XP'P^{-n-1}) z^n x^n,$

(b) $\displaystyle R_n(z) = \operatorname{res}\left\{ XP' z^n P^{-n-1} \left(1 - \frac{P}{z}\right)^{-1} \right\}, \qquad n = 1, 2, \ldots.$

(R. Brent.)

10. In the theory of waiting lines one considers formal power series

$$A = a_0 + a_1 x + a_2 x^2 + \cdots$$
$$B = b_1 x + b_2 x^2 + \cdots,$$

such that $B = X \cdot (A \circ B)$. Show that the series A and B uniquely determine

650

each other, and that

(a) $\quad b_n = \dfrac{1}{n} a_{n-1}^{(n)}, \qquad n = 1, 2, \ldots;$

(b) $\quad a_0 = b_1, a_1 = \dfrac{b_2}{b_1}, a_n = -\dfrac{1}{n-1} b_1^{(-n+1)}, n = 2, 3, \ldots.$

[See A. G. Konheim & B. Meister, *J. ACM* **21**, 470–490.]

§2.5.

16. Find the expansion in powers of x of $\mathrm{Log}(1 + ax + bx^2)$ for arbitrary complex a and b. [Factor the quadratic.]

17. Show that

$$\mathrm{Log}\frac{(8+4x+x^2)^2}{64+x^4} = \sum_{n=0}^{\infty} \frac{\epsilon_n}{(2n+1)2^{3n-1}} x^{2n+1},$$

where

$$\epsilon_n := \begin{cases} +1, \text{ if } n \equiv 0 \text{ or } 3 \pmod 4 \\ -1, \text{ if } n \equiv 1 \text{ or } 2 \pmod 4. \end{cases}$$

Convince yourself that the function on the left is odd.

§4.5.

12. Obtain the Fourier expansion

$$\mathrm{Arctan}(\alpha \cos \phi) = \sum_{n=0}^{\infty} \frac{(-1)^n}{2n+1} \left[\frac{\alpha}{1+\sqrt{1+\alpha^2}}\right]^{2n+1} \cos\{(2n+1)\phi\},$$

valid for α real, by means of the algorithm implied in Theorem 4.5.

INDEX

Principal references (referring to definitions) are printed in **bold face**.